THE GREEK ALPHABET

A	α	Alpha		N	ν	Nu
B	β	Beta		Ξ	ξ	Xi
Γ	γ	Gamma		O	o	Omicron
Δ	δ	Delta		Π	π	Pi
E	ε	Epsilon		P	ρ	Rho
Z	ζ	Zeta		Σ	σ	Sigma
H	η	Eta		T	τ	Tau
Θ	θ	Theta		Υ	υ	Upsilon
I	ι	Iota		Φ	φ	Phi
K	κ	Kappa		X	χ	Chi
Λ	λ	Lambda		Ψ	ψ	Psi
M	μ	Mu		Ω	ω	Omega

CELL BIOLOGY
STRUCTURE, BIOCHEMISTRY, AND FUNCTION

CELL BIOLOGY
STRUCTURE, BIOCHEMISTRY, AND FUNCTION

Second Edition

Phillip Sheeler

Professor of Biology
California State University Northridge

Donald E. Bianchi

Dean, School of Science and Mathematics
Professor of Biology
California State University Northridge

John Wiley & Sons, Inc.

New York   Chichester   Brisbane   Toronto   Singapore

Library of Congress Cataloging in Publication Data:

Sheeler, Phillip.
    Cell biology.

    Includes bibliographies and index.
    1. Cytology.   I. Bianchi, Donald E., 1933-
II. Title.
QH581.2.S53 1983   574.87   82-7037
ISBN 0-471-09308-4   AACR2

Printed in the United States of America

10 9 8 7 6 5 4 3

To my parents Barnett and Deborah,
to my wife Annette,
and to my children Wendy, Donna,
Lindsey, Paul and Carly.

P. S.

To my parents Ernest and Florence,
to my wife Georgia,
and to Dave, Bill, Diana and Jon.

D. E. B.

PREFACE

This book was written for sophomore and junior level courses in cell biology, molecular biology, and cellular physiology. We consider in detail the fine structure of eucaryotic and procaryotic cells (and viruses), the chemical composition and organization of cells, cell metabolism, and bioenergetics and, for each major cell organelle or structural component, its particular molecular and supermolecular organization and its functions. Special attention has also been paid to a description of the major research tools used by cell biologists to increase our knowledge of cell structure, biochemistry, and function.

In preparing this textbook, we have drawn on more than thirty-five years of collective university teaching experience in courses in cell biology, cell physiology, and molecular biology. We are hopeful that our early training in different areas of cell biology and our continuous maintenance of active research programs in these areas have helped us to present a balanced approach.

We have necessarily made certain assumptions with regard to the background of students reading this book. For example, we have assumed that students have had courses in introductory biology and introductory chemistry. Most portions of the book dealing with physical or biochemical principles or concepts are preceded by a discussion of the requisite fundamentals.

This edition of the book differs from the first in several ways. Despite the small number of years that have passed since the first edition appeared, there have been major advances in the fields of biochemistry, cell biology, molecular genetics, and genetic engineering. Therefore, the text has been expanded to incorporate discussions of the latest developments that bear directly on cell biology. Indeed, we have added an entire chapter dealing with molecular genetics. In addition to the new material, the contents of the first edition have been fully updated. Concepts that are no longer tenable in the light of recent findings have been discarded, while newly developing ideas are presented. The clear, informative diagrams of the first edition were one of the book's major strengths; most of these have been conserved in the second edition and a large number of new diagrams added. Once again we have searched the literature to obtain the best and clearest photomicrographs depicting cellular organelles and other cell structures, and this edition is replete with highly informative photomicrographs that have been obtained using the most recently evolved laboratory techniques.

Some illustrations are presented in stereoscopic form and, although individual members of each "stereo pair" may be viewed without an optical aid, the perspective effect requires the use of a stereo viewer. Plastic-frame viewers with adjustable lenses are available in most college and university campus bookstores or where graphic supplies are sold.

Many individuals provided invaluable help, suggestions, and guidance during the preparation, writing, and production of this book. We are indebted to them. Individual chapters were reviewed and critiqued by Betty D. Allamong (Ball State University), Allyn Bregman (SUNY, College at New Paltz), Richard Bowner (Idaho State University), Dennis E. Buetow (University of Illinois at Urbana-Champaign), Joseph Y. Cassim (Ohio State University), Janice E. Chambers (Mississippi State University), George B. Cline (University of Alabama, Birmingham), Ray Holton (University of Tennessee), Bonnie Lamvermeyer (Denison University), Harvard Lyman (State University of New York at Stony Brook), Janet L. Morgan (University of Kentucky), Eldon H. Newcomb (University of Wisconsin), Sally Nyquist (Bucknell University), Daniel G. Oldfield (DePaul University), David Rooney (Saint Louis University), Peter Snustad (University of Minnesota), Philip Stukus (Denison University), and Eugene Vigil (University of Maryland). We are also indebted to Robert Wolcott (Linfield College) and Les Ballou (St. Jude Children's Research Hospital, Memphis, Tennessee) for the assistance that they provided in specific sections.

In this edition, as in the first edition, we are especially grateful to those investigators who provided us with unique electron photomicrographs from their collections. Our special thanks to Keith Porter (University of Colorado), Keiichi Tanaka (Tottori University School of Medicine, Yonago, Tottori-Ken, Japan), John Possingham (CSIRO, Adelaide, Australia), and Richard Chao, Daisy A. Kuhn and Edward G. Pollock (California State University, Northridge). The spectacular scanning electron photomicrograph used for the cover of the book was kindly provided by K. Tanaka and T. Naguro.

The entire manuscript for this edition was prepared using word processing equipment, which greatly facilitated our need to edit continuously sections that deal with topics in which progress is so rapid. We are especially grateful to Lindsey Deborah Myers for entering the text onto diskettes and for providing us with her word processing expertise.

The staff at Wiley has always been most helpful and we are once again indebted to our editor Frederick C. Corey for his help and encouragement and to John Balbalis our illustrator for his extraordinary artistic talents and for his personal interest in and enthusiasm for this project. We should also like to thank Eugene Patti and Vivi Danser our copy editors and Linda Indig our production supervisor.

We apologize for any errors that we may have allowed into print and thank our readers in advance for bringing them to our attention.

Phillip Sheeler

Don Bianchi

Northridge, 1982

CONTENTS

CELL BIOLOGY
STRUCTURE, BIOCHEMISTRY, AND FUNCTION

Part 1
CELLS AND
CELL GROWTH

Chapter 1
THE CELL: AN INTRODUCTION

Development of the Cell Doctrine

Briefly summarized, the **cell doctrine** states that the cell is the fundamental unit of both *structure* and *function* in all living things; that all forms of life (animal, plant, and microbial) are composed of cells and their secretions; and that cells arise only from *preexisting* cells, each cell having a life of its own in addition to its integrated role in multicellular organisms.

This statement seems both elementary and obvious to any student with some background in the biological sciences. Nevertheless, it took several centuries for this concept to be developed and accepted. The very existence of cells was not even suspected until the seventeenth century, because most cells are too small to be discerned with the naked eye, and because instruments for significantly magnifying small objects did not exist. However, with the introduction of the first crude light **microscopes,** investigators began to examine small organisms, tissues cut from animals and plants, and the "animalcules" in pond water. The invention of the microscope and its gradual improvement went hand in hand with the development of the cell doctrine. It finally became apparent that a fundamental similarity existed in the structural organization of all the living things studied.

What follows is a brief description of but a few of the historical highlights that culminated in the cell doctrine. Although a great many individuals made contributions of varying significance to the development of this concept, the works of a certain small number of people stand out as milestones.

In 1558, the works of *Conrad Gesner* (Swiss, 1516–1565) on the structure of a group of protists called *foraminifera* were published. His sketches included so much detail that they could have been made only if Gesner had used some form of magnifying lens. This appears to be the earliest recorded use of a magnifying instrument in biology.

Francis and *Zacharias Janssens,* who manufactured eyeglasses in Holland, are generally credited with the construction of the first compound microscopes in 1590. Their microscopes had magnifying powers between $10\times$ and $30\times$ and were used primarily to examine small whole organisms such as fleas and other insects. The first microscopes were in fact referred to as "flea-glasses."

Although noted principally for his contributions in the fields of astronomy and physics, *Galileo Galilei* (Italian, 1564–1642) produced several important biological works. His own microscopes were constructed at about the same time as those of the Janssens (around 1610) and were employed for several extensive studies on the arrangements of the *facets* in the compound eyes of insects.

Among the earliest descriptions of the microanatomy of

3

tissues were those of *Marcello Malpighi* (Italian, 1628–1694), one of the first great animal and plant anatomists. He was the first to describe the existence of the capillaries, thereby completing the work on the circulation of the blood started by the great English physiologist William Harvey. Malpighi was among the first to use a microscope to examine and describe thin slices of animal tissues from such organs as the brain, liver, kidney, spleen, lungs, and tongue. His published works also include descriptions of the development of the chick embryo. In his later years, Malpighi turned to investigations of plant tissues and suggested that they were composed of structural units which he called "utricles."

Antony van Leeuwenhoek (Dutch, 1632–1723) was one of the most distinguished of all the early microscopists. Although it was only an avocation, Leeuwenhoek became an expert lens grinder and built numerous microscopes, some with magnifications approaching 300×. Leeuwenhoek was the first to describe microscopic organisms in rainwater collected from tubes inserted into the soil during rainfall. His sketches included numerous bacteria (bacilli, cocci, spirilla, etc.), protozoa, rotifers, and hydra. Leeuwenhoek was the first to describe sperm cells (of humans, dogs, rabbits, frogs, fish, and insects) and observe the movement of blood cells in the web capillaries of the frog's foot and the rabbit's ear. He described the blood cells of mammals, birds, amphibians, and fish, noting that those of fish and amphibians were oval in shape and contained a central body (i.e., the nucleus), while those of humans and other mammals were round. Leeuwenhoek's observations were recorded in a series of reports that he sent to the Royal Society of London.

Many of Leeuwenhoek's observations were confirmed in experiments conducted by *Robert Hooke* (English, 1635–1703), an architect and scientist employed by the Royal Society. Hooke popularized the use of microscopes among contemporary biologists in England and built several compound microscopes of his own. On one occasion, Hooke examined a thin slice cut from a piece of dried cork. In his description, Hooke wrote that he found the sections to be "all perforated and porous, much like a honeycombe" and referred to the boxlike structures as "cells." Thus, it is Hooke who introduced the term **cell** to biology. What he observed, of course, were not cork cells but rather the empty spaces left behind after the living portion of the cells had disintegrated.

Nehemiah Grew (English, 1641–1712), together with Marcello Malpighi, is recognized as one of the founders of plant anatomy. His publications included accounts of the microscopic examination of sections through the flowers, roots, and stems of plants and clearly indicate that he recognized the cellular nature of plant tissue. Grew was also the first to recognize that flowers are the sexual organs of plants.

In 1824 *Rene Dutrochet* (French, 1776–1847) wrote that all animal and plant tissues were "aggregates of globular cells," and in 1831 *Robert Brown* (English, 1773–1858) noted that the cells of plant epidermis, pollen grains, and stigmas contained certain "constant structures," which he called **nuclei,** thereby introducing this term to biology. Brown is also credited with the first description of the physical phenomenon now referred to as "Brownian motion." *Johannes E. Purkinje* (Czech, 1787–1869) coined the term *protoplasm* to describe the contents of cells.

Mathias J. Schleiden (German, 1804–1881) and *Theodor Schwann* (German, 1810–1882) are often credited, albeit incorrectly, with the first formal statement of a general cell theory. Their contributions to the development of the cell doctrine reside in the generalizations that they made based principally on the works of their predecessors. Schleiden and Schwann were particularly influential among their contemporaries and did, therefore, gain popular acceptance for the developing cell doctrine. Schleiden, a botanist, extended the studies begun by Robert Brown on the structure and function of the cell nucleus (which Schleiden called a "cytoblast") and was the first to describe nucleoli. Schleiden's writings clearly indicate his appreciation of the individual nature of cells. In 1838, he wrote that each cell leads a double life—one independent, pertaining to its own development, and another as an integral part of a plant. Schwann studied both plant and animal tissues. His work with connective tissues such as bone and cartilage led him to modify the evolving cell theory to include the notion that living things are composed of both cells *and the products of cells*. Schwann is also credited with introduction of the term **metabolism** to describe the activities of cells.

Rudolf Virchow (German, 1821–1902), a pathologist, recognized the cellular basis of disease. His writings, often in Latin, also reveal his appreciation of the cellular basis of life's continuity, as summarized in his now famous expression *omnis cellula e cellula*, "all cells arise from [preexisting] cells."

In the last part of the 1800s and certainly by the turn of the century, the light microscope approached its limit in terms of resolving power, and nearly all major cellular structures had at least been described.

In this century, especially during the past 25 years, we have witnessed unprecedented growth of our knowledge of the cell, its structural organization and diversity, its chemical organization, and the various functions of its component parts. This understanding is founded on the contributions of many thousands of scientists working in laboratories all over the world. Indeed, this textbook is a distillation of those contributions. To identify and credit all such contributors is, of course, beyond the scope of our concerns here, although many scientists and their particular contributions are identified in subsequent chapters of the book. Probably no symbol of recognition of the contributions made by scientists in this century has captured the imagination of the public (or for that matter scientists themselves) as has the Nobel Prize, an award recognizing specific contributions in diverse fields of human endeavor. Many such awards in the fields of *chemistry, physiology,* and *medicine* have been made for contributions that bear directly on cell biology (see Table 1–1), and you will encounter discussions of a number of these in subsequent reading.

Microscopy

Fundamentals of Light Microscopy and Transmission Electron Microscopy

Until the 1940s, most of our knowledge concerning the structure and organization of cells was obtained from studies conducted using light microscopy, and major structures and *organelles,* including the cell wall, nuclei, chromosomes, chloroplasts, mitochondria, vacuoles, centrioles, flagella, and cilia, had been described.

The smallest distance, *d,* between two points resolvable as separate points when viewed through lenses is given by the relationship

$$d = \frac{0.6\lambda}{n \sin \alpha} \qquad (1\text{--}1)$$

In this equation, λ is the **wavelength** of the light (radiation) employed to illuminate the specimen; *n* is the **refractive index** of the air or liquid between the specimen and the lens; and α is the **aperture angle.** The product $n \sin \alpha$ is

called the lens **numerical aperture,** and for a good microscope lens, its value is about 1.4.

Equation 1–1 also shows that the *resolving power* of a microscope varies with the wavelength of the source of illumination. The human eye cannot directly detect light having a wavelength of less than about 400 nm (see Table 1–2 for metric measurements). Therefore, in the case of the light microscope, the maximum resolving power is about 0.6(400/1.4), or about 0.17 μm. That is, points less than about 0.2 μm apart cannot be distinguished as separate points by light microscopy (in practice, the limit is closer to 0.5 μm). From a practical standpoint, this means that even when glass optics of the finest quality are used, it is possible to observe cells at magnifications no greater than about 2000$\times$. Resolution is improved when sources emitting rays that have shorter wavelengths are employed. For example, the resolving power of the ultraviolet light microscope (which requires quartz optics because glass does not transmit ultraviolet light) is approximately double that of the light microscope.

Much greater resolution has been obtained with the *electron microscope,* developed in the 1930s, and with which magnifications of several hundred thousands are a practical possibility. The wavelength of radiations used with the electron microscope is typically about 0.005 nm (0.05 Å). Although resolution of the order of an Angstrom or less is theoretically possible, the practical limit is about 5 Å. This is many thousand times greater than that attainable using microscopes with glass optics. The basic features of a *transmission* electron microscope (often simply abbreviated TEM) are shown in Figure 1–1, and a comparison between the component parts of the TEM and the light microscope is depicted diagrammatically in Figure 1–2. In recent years, the *scanning* electron microscope (SEM) has become an increasingly important tool of the cell biologist. The SEM employs quite different principles than the TEM and will be considered separately later.

In both the light and electron microscopes, the source of radiation is an electrically heated tungsten filament. In the light microscope, the light emitted from the glowing filament is focused by a **condenser** onto the specimen to be observed. In the transmission electron microscope, the condenser focuses **electrons** emitted by the excited tungsten atoms into a beam and electrodes accelerate the beam toward the specimen. While the condenser of a light microscope consists of one or a few glass lenses, the condenser

Table 1–1
Nobel Prizes for Cell Biology–Related Contributions

Year	Division	Name	Contribution
1902	Chemistry	Emil Fischer	Pioneering studies in the field of biochemistry, especially studies of proteins
1906	Physiology & Medicine	Camillo Golgi S. Ramon y Cajal	Studies of the organization of the nervous system; contributions to histology, especially the structure of nerve cells
1908	Physiology & Medicine	Elie Metchnikoff Paul Ehrlich	Phagocytosis of bacteria during infection; staining procedures for bacteria; studies on immunity
1910	Physics	J. van der Waals	Physical properties of liquids
1915	Chemistry	Richard Wilstätter	Studies of chlorophyll and other plant pigments
1922	Physiology & Medicine	Archibald V. Hill Otto Meyerhof	Metabolism of muscle tissue; relationship between muscle metabolism and lactic acid
1926	Chemistry	Theodor Svedberg	Properties of colloids, especially proteins; development of analytical ultracentrifugation
1930	Physiology & Medicine	Karl Landsteiner	Discovery of human blood groups; studies of cellular agglutinins or antigens
1931	Physiology & Medicine	Otto Warburg	Studies on the nature and mode of action of respiratory enzymes; studies of oxidation and reduction in metabolism
1932	Chemistry	Irving Langmuir	Studies of surface chemistry, especially surface properties of liquids
1933	Physiology & Medicine	Thomas Hunt Morgan	Discoveries concerning the role of chromosomes in the transmission of heredity
1935	Physiology & Medicine	Hans Spemann	Discoveries concerning the role of the organizer during egg development
1936	Physiology & Medicine	Henry Dale Otto Loewi	Studies on the chemical transmission of nerve impulses
1937	Physiology & Medicine	Albert von Szent-Gyorgyi	Studies on biological oxidation and the involvement of vitamin C
1938	Chemistry	Richard Kuhn	Studies on the chemistry of carotenoids and vitamins
1943	Chemistry	George de Hevesy	Development of the radioisotopic tracer technique for following the progress of chemical reactions
1946	Physiology & Medicine	Hermann Joseph Muller	Studies of genetic mutations produced by X-rays
1947	Physiology & Medicine	Carl F. Cori Gerty T. Cori	Studies of the metabolism of glycogen
1948	Chemistry	Arne Tiselius	Studies on the chemistry of proteins; development of electrophoresis
1952	Chemistry	Archer Morten Richard Synge	Development of chromatographic procedures for the separation of biological substances
1953	Chemistry	Hans A. Krebs	Elucidation of the tricarboxylic acid cycle (also called Krebs cycle)
		Fritz A. Lipman	Studies on coenzyme A.
1954	Chemistry	Linus Pauling	Studies on the nature of chemical bonds, especially the peptide bond of proteins
1958	Physiology & Medicine	George W. Beadle Edward L. Tatum Joshua Lederberg	Studies of the organization and actions of genes in bacteria; the ''one gene–one enzyme'' concept
	Chemistry	Frederick Sanger	Analysis of protein structure, especially insulin

Table 1–1   continued

Year	Division	Name	Contribution
1959	Physiology & Medicine	Severo Ochoa Arthur Kornberg	Studies on the synthesis of RNA and DNA
1961	Chemistry	Melvin Calvin	Studies on the assimilation of CO_2 by plants, photosynthesis, the ''Calvin cycle''
	Physiology & Medicine	Georg von Bekesy	Discoveries concerning the physiology of hearing
1962	Physiology & Medicine	James D. Watson Francis H. C. Crick Maurice Wilkins	Studies of the structure of genes, the double-helix model for DNA
	Chemistry	Max F. Perutz John C. Kendrew	Studies of the structure of globular proteins, especially myoglobin and hemoglobin
1963	Physiology & Medicine	John Eccles Alan Hodgkin Andrew Huxley	Role of sodium and potassium ions in the conduction of nerve impulses along the nerve cell membrane
1964	Physiology & Medicine	Konrad Bloch Feodor Lynen	Studies of the metabolism of cholesterol and fatty acids
1965	Physiology & Medicine	Francois Jacob Andre Lwoff Jacques Monod	Discovery of genes that regulate the actions of other genes; the ''Operon'' model
1968	Physiology & Medicine	Robert W. Holley H. C. Khorana Marshall W. Nirenberg	Studies deciphering the genetic code
1969	Physiology & Medicine	Max Delbrück Alfred D. Hershey Salvador E. Luria	Studies of virus mediated diseases
1970	Chemistry	Luis F. Leloir	Studies of the role of sugar nucleotides in the synthesis of carbohydrates
	Physiology & Medicine	Julius Axelrod Ulf von Euler Bernard Katz	Studies of the mechanisms of storage and release of neurohumors in nerve impulse transmission
1971	Physiology & Medicine	E. A. Sutherland	Mechanisms of action of hormones; role of cyclic AMP
1972	Physiology & Medicine	Maurice Edelman Rodney R. Porter	Chemical studies of immunoglobulins
	Chemistry	Christian B. Anfinsen	Studies of ribonuclease
		Stanford Moore William H. Stein	Chemical studies of the amino acid composition of proteins
1974	Physiology & Medicine	Albert Claude Christian de Duve George Palade	Isolation and characterization of subcellular organelles and other particles
1975	Physiology & Medicine	Howard Temin Renato Dulbecco David Baltimore	Studies of the interaction between tumor viruses and cells; discovery of reverse transcription
1978	Physiology & Medicine	Werner Abner Hamilton O. Smith Daniel Nathens	Studies on genetic engineering; discovery of restriction enzymes; mapping of DNA sequences
	Chemistry	Peter Mitchell	Studies of bioenergetics
1980	Chemistry	Paul Berg	Studies on gene splicing
		Frederick Sanger	Determination of nucleotide sequences of genes

Table 1–2
Metric Measurements of Size

1 meter (m)	= 39.4 inches (in.)
1 meter (m)	= 100 centimeters (cm)
1 centimeter (cm)	= 10 millimeters (mm)
1 millimeter (mm)	= 1000 micrometers (μm) or microns (μ)
1 micrometer (μm)	= 1000 nanometers (nm) or millimicrons (mμ)
1 nanometer (nm)	= 10 Angstroms (Å)

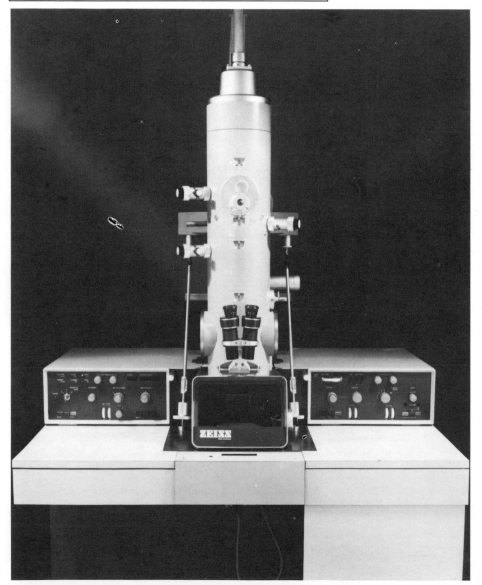

Figure 1–1
A transmission electron microscope. (Courtesy Carl Zeiss, Inc.)

of the electron microscope consists of several large, circular electromagnets. Indeed, all "lenses" of the electron microscope are electromagnets. In both microscopes, the radiation passes through the specimen and is then refocused by the *objective* lenses. The last lens of the light microscope is the **ocular,** through which the image may be viewed with the eye. The image of the electron microscope is viewed after its magnetic projection onto a zinc sulfide screen. The molecules of the screen are excited by the impinging electrons and emit visible light during their return to the ground state. Alternatively, the image may be captured on photographic film housed in a special camera mounted below the movable zinc sulfide screen.

The lenses of the light microscope have a fixed *focal length* and are focused by moving them nearer to or farther from the specimen. In the electron microscope, focusing is accomplished by manipulating the amount of current flowing through the windings of the series of electromagnetic lenses. This alters the electromagnetic field through which the electron beam must pass. The column through which the electron beam passes must be evacuated of air. If the

vacuum is inadequate, the electrons will be scattered by collisions with residual gas molecules. Consequently, the specimen, the filament, the electromagnets, and the zinc sulfide screen are all mounted within a sealed compartment connected to a vacuum pump. In order to avoid excessive scattering or absorption of electrons by the specimen itself, the material to be examined must be cut into extremely thin sections.

Two special forms of light microscopy warrant further description because of their widespread use and special application: phase-contrast microscopy and fluorescence microscopy.

Phase-Contrast Microscopy.   Although most regions of an unstained cell are transparent, they may have different densities and therefore different refractive indexes. Consequently, light rays travel through these regions at different velocities and may be refracted or bent to different extents. The phases of light rays that pass directly through an object and those that pass across its edges (i.e., at the interface where the refractive index changes) will necessarily be

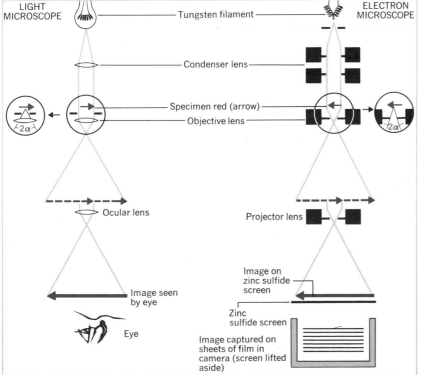

Figure 1–2
A comparison of the basic features of the light microscope and TEM.

altered. The change increases the contrast between the object in focus and its surroundings. In the phase-contrast microscope, the phases of light rays entering the object are shifted by an annular diaphragm below the condenser. The phases of the rays passing through and around the object are shifted again by a phase plate in the objective lens. The result is a striking increase in the contrast of the object as certain regions appear much brighter (owing to additive effects of rays brought into phase), while other regions appear much darker (owing to the canceling effects of rays shifted further out of phase). The effect can be seen in Figure 1–39. Since phase-contrast microscopy produces added contrast in the material being studied without the need to employ stains, the technique is especially useful when examining living material.

Fluorescence Microscopy. Certain chemical substances emit visible light when they are illuminated with ultraviolet light. The effect is termed **fluorescence** and is put to use in the fluorescence microscope in which ultraviolet light rays are focused on the specimen. Some cellular components possess a natural fluorescence and appear in various colors. Other, nonfluorescing structures can be made to fluoresce by staining them with fluorescent dyes (**fluorochromes**). One of the most popular contemporary uses of fluorescence microscopy involves the preparation of **antibodies** that will bind to specific cellular proteins (see Chap-

ter 4). The antibodies are first complexed with *fluorescein* (a fluorescent dye), and the *fluorescein-labeled* antibody is then applied to the cells. Cell structures containing the specific proteins capable of binding the fluorescein-labeled antibody are caused to fluoresce when examined with the fluorescence microscope, dramatically revealing their detail (Fig. 1–3).

Image intensifiers and microspectrofluorometers can also be coupled to fluorescence microscopes to provide images or detect the presence of very small quantities of labeled molecules. Such adaptations provide sensitivities to as few as 10^5–10^6 molecules. In recent years, there have also been amazing advances in computer enhancement of the microscopic images.

Preparation of Materials for Microscopy

The preparation of biological material for examination with either the light microscope or the transmission electron microscope involves a series of physical and chemical manipulations that include (1) fixation, (2) embedding, (3) sectioning, and (4) mounting.

Fixation. One notable advantage of the light microscope is the ability to observe whole, *living* cells. It is also possible to employ ''vital stains,'' which improve contrast

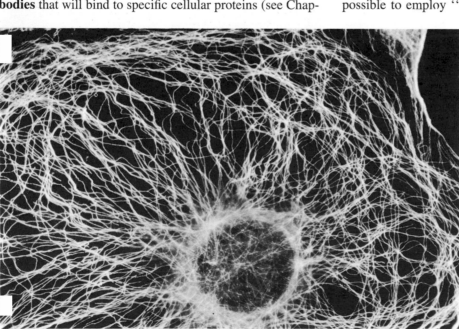

Figure 1–3
The network of microtubules present in this cell is clearly revealed using the combined techniques of fluorescent antibody labeling and fluorescence microscopy. (Photomicrograph courtesy of Drs. M. Osborn and K. Weber; copyright © 1977 by MIT Press, *Cell 12,* 561.)

but do not interfere with normal cell activity. More frequently, however, the cells are first killed and fixed. The fixation step is intended to preserve the structure of the material by preventing the growth of bacteria in the sample and by precluding postmortem changes. Formaldehyde and osmium tetroxide (OsO_4) are examples of fixatives most often employed for light microscopy. OsO_4 has a very high electron density, and since this gives contrast to the resulting image, OsO_4 has also found widespread use as a fixative in electron microscopy. Other popular fixatives include potassium permanganate and glutaraldehyde. After fixation for the required length of time, the samples are dehydrated by successive exposures to increasing concentrations of alcohol or acetone.

Embedding. Cells or tissues to be examined by light microscopy are usually embedded in warm liquid paraffin wax. The wax, which both surrounds the tissue and infiltrates it, hardens upon cooling, thereby supporting the tissue externally and internally. The resulting solid paraffin block is then trimmed to the appropriate shape before being sectioned. The ultrathin sections required for electron microscopy necessitate the use of harder embedding and infiltrating materials, such as methacrylate, Epon, or Vestopal. These initially are in liquid form and are poured into small molds containing pieces of the fixed tissue; upon heating, the liquid undergoes polymerization to form a hard plastic (Fig. 1–4).

Sectioning. The trimmed blocks containing the embedded samples are sectioned using a **microtome** (Fig. 1–5). In this instrument, the block is sequentially swept over the blade of a knife that cuts the block into a series of thin sections that adhere to one another end to end and thereby form a *ribbon*. Between each stroke, the block is advanced a short distance toward the knife. For light microscopy, the microtome knives are usually constructed of polished steel and can provide sections several micrometers thick. The sections for electron microscopy must be much thinner (typically 100 to 500 Å) and require more elaborate microtomes (called ultramicrotomes), such as the one shown in Figure 1–5, which either mechanically or thermally advance the plastic block much shorter distances with each stroke. Moreover, either diamond knives or knives prepared by fracturing plate glass are used in place of polished steel. Figure 1–6 illustrates the preparation of a ribbon of sections during ultramicrotomy.

Mounting. Sections prepared for light microscopy are mounted on glass slides and may be stained with dyes of various colors that specifically attach to different molecular constituents of the cells. Sections to be examined with the electron microscope are generally not stained (no colors are seen with the electron microscope), although contrast may be improved by "poststaining" with electron-dense materials such as uranyl acetate, uranyl nitrate, and lead citrate. The sections are mounted on copper "grids" (small disks perforated with numerous openings) that have been coated with a thin (sometimes monomolecular) film of carbon (Fig. 1–7). The grid supports the film, which in turn supports the thin section. Thus, the beam of electrons must pass through the spaces of the grid, the supporting film, and the section before striking the fluorescent screen. A comparison of photomicrographs obtained with the light microscope and the TEM is given in Figure 1–8.

Specialized Applications of Transmission Electron Microscopy

Shadow Casting. In shadow casting, the sample (usually containing small particles such as viruses or macromole-

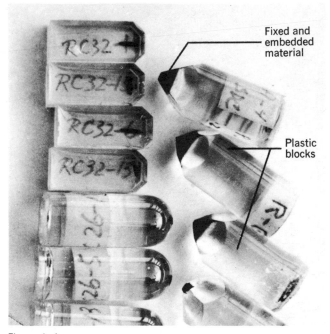

Figure 1–4
Plastic blocks of various shapes containing fixed and embedded tissue. (Photo courtesy of R. Chao.)

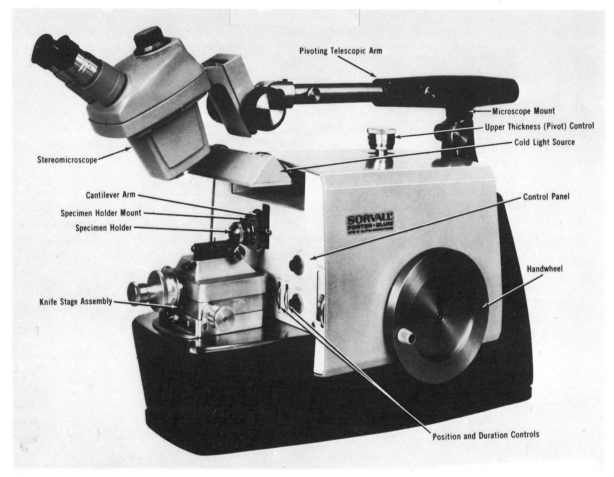

Figure 1–5
An ultramicrotome. (Photo courtesy of DuPont Instruments.)

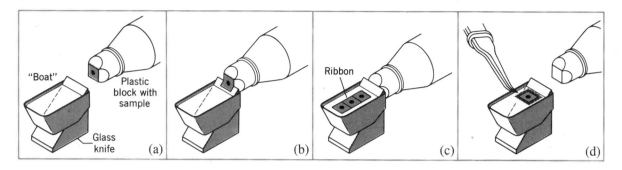

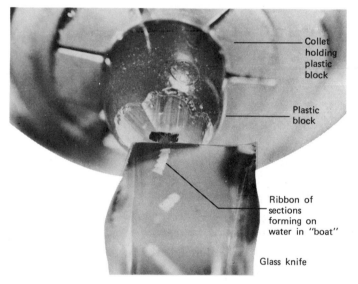

Figure 1–6
Sectioning of embedded tissue to form ribbon of thin sections floating on water in "boat" attached to glass microtome knife.

Figure 1–7
Various grids used for mounting sections for electron microscopy. (Photo courtesy of R. Chao.)

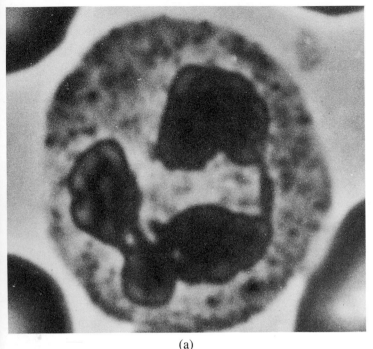

(a)

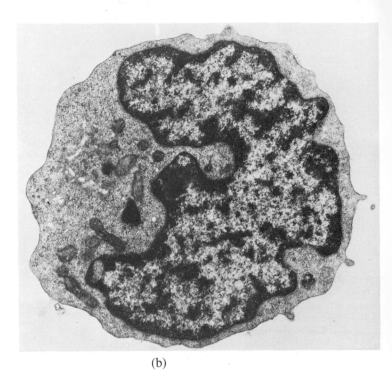

(b)

Figure 1–8

Comparison of photomicrographs of a white blood cell (leucocyte) obtained by (a) light and (b) transmission electron microscopy. (a, courtesy of M. Doolittle; b, courtesy of R. Chao.)

cules) is spread on a coated grid, which is then placed in an evacuated chamber. A chromium or platinum wire is heated until the metal is vaporized, and the vapor is deposited onto the sample *at a precise angle*. The metal piles up in front of the sample particles but leaves clear areas behind them. If the resulting electron photomicrographs are printed in reverse, the areas containing the electron-dense metal that had piled up against the particles appear light, while the electron-transparent areas behind the particles appear as dark shadows. Because the vaporized metal atoms tend to be projected in a straight line, the shadows are cast at precise angles, and in this manner the general shape and profile of a particle may be discerned (Fig. 1–9).

Negative Staining.   In the negative-staining procedure, the sample (again small particles such as viruses or macro-

molecules) is surrounded by an electron-dense material, such as phosphotungstic acid, that permeates the open superficial interstices of the sample. When the excess stain is carefully washed away, the sample particles appear as light (i.e., electron-transparent) areas surrounded by a dark background (Fig. 1–10).

Freeze-Fracturing.   Freeze-fracturing is a technique in which the tissue is first *fractured* (i.e., cracked) along planes of natural weakness that run through each cell. These planes generally occur between the two layers of lipid molecules that comprise part of the limiting membrane around the cells' various vesicular organelles (see Chapter 15). Figure 1–11 depicts the basic differences between sectioning and fracturing.

The tissue to be freeze-fractured is first impregnated with

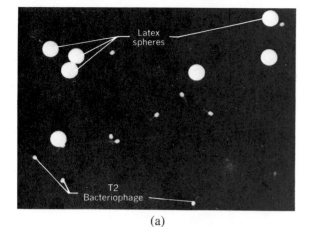

(a)

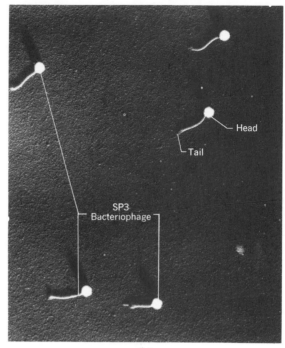

(b)

Figure 1–9
Shadow casting. (*a*) Metal shadowed mixture of
latex spheres and viruses (T2 bacteriophage). The
latex spheres are of known size (0.3 μm) and concentration, thereby permitting ready determination
of virus size and concentration. (*b*) Shadowed SP3
viruses. From the shadowing angle and shadow
contours, the size and shapes of the viral parts
may be determined. (Courtesy of Dr. F. A. Eiserling.)

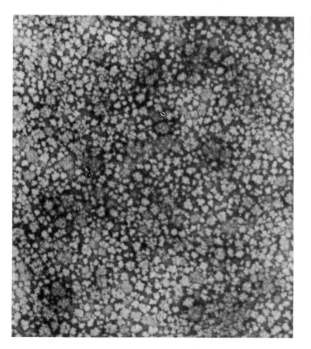

Figure 1–10
**Glycogen particles from liver tissue visualized by negative staining
(see also Chapter 5). Magnification 51,000×.**

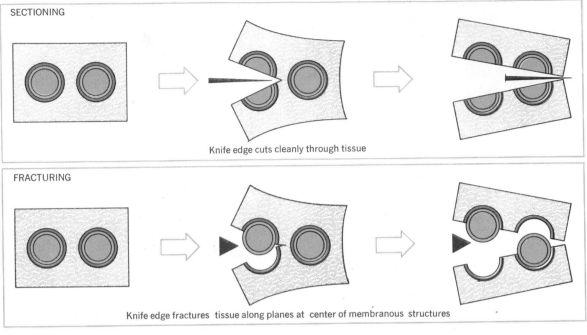

Figure 1–11
A comparison of sectioning and fracturing tissue.

glycerol and then frozen at −130°C in liquid Freon. The frozen tissue is transferred to an evacuated chamber containing a microtome and steel knife (also maintained at about −100°C using liquid nitrogen). The microtome knife is used to produce a fracture plane through the tissue (Fig. 1–12a and b). When the plane of the fracture intersects the membrane of a vesicular structure (e.g., nucleus, mitochondrion, vacuole), the membrane is split along its center,

producing two "half-membranes." These are called the E (for "exterior") half and P (for "protoplasmic") half. The E half formerly faced the cell's external phase (see below), and the P half faced the internal phase (i.e., the protoplasm). One surface of each half-membrane is the original membrane surface (the E and P surfaces of Fig. 1–12c and d), while the other surface is the newly exposed fracture face (the EF and PF surfaces of Fig. 1–12c and d). The

Figure 1–12
Stages in the freeze-fracturing procedure (see text for explanation).

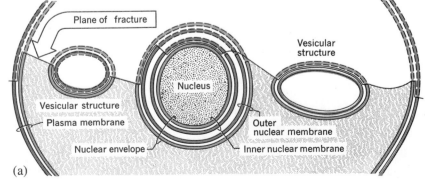

(a)

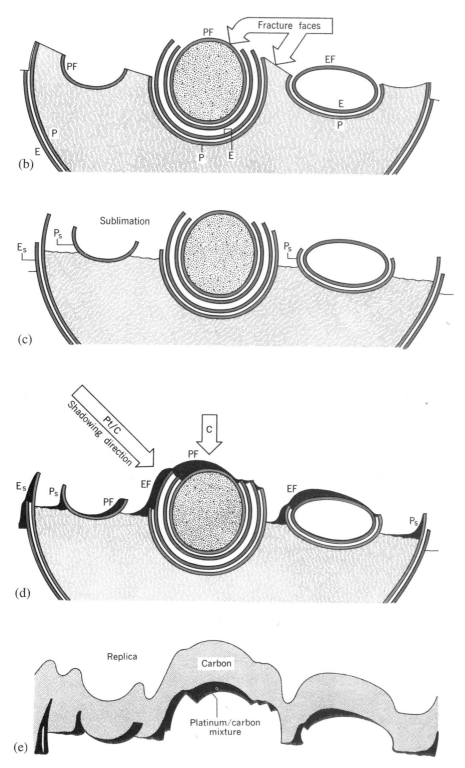

vacuum of the chamber is then used to *sublimate* water on the cut surface to a depth of several hundred Angstroms. New membrane faces exposed by sublimation are termed E_s and P_s (Fig. 1–12c). An electron-dense combination of metal (usually carbon and platinum) is then deposited on the cut surface at an angle and piles up in front of and behind projections from the surface, as well as in pits and depressions (Fig. 1–12d). Additional carbon is added to form an electron-transparent backing.

The shadowed and coated tissue is removed from the chamber, and the tissue itself is either floated off or dissolved away, thereby leaving only the carbon-platinum ''replica'' (Fig. 1–12e). The replica is trimmed to the proper size, placed on a grid, and examined with the trans-

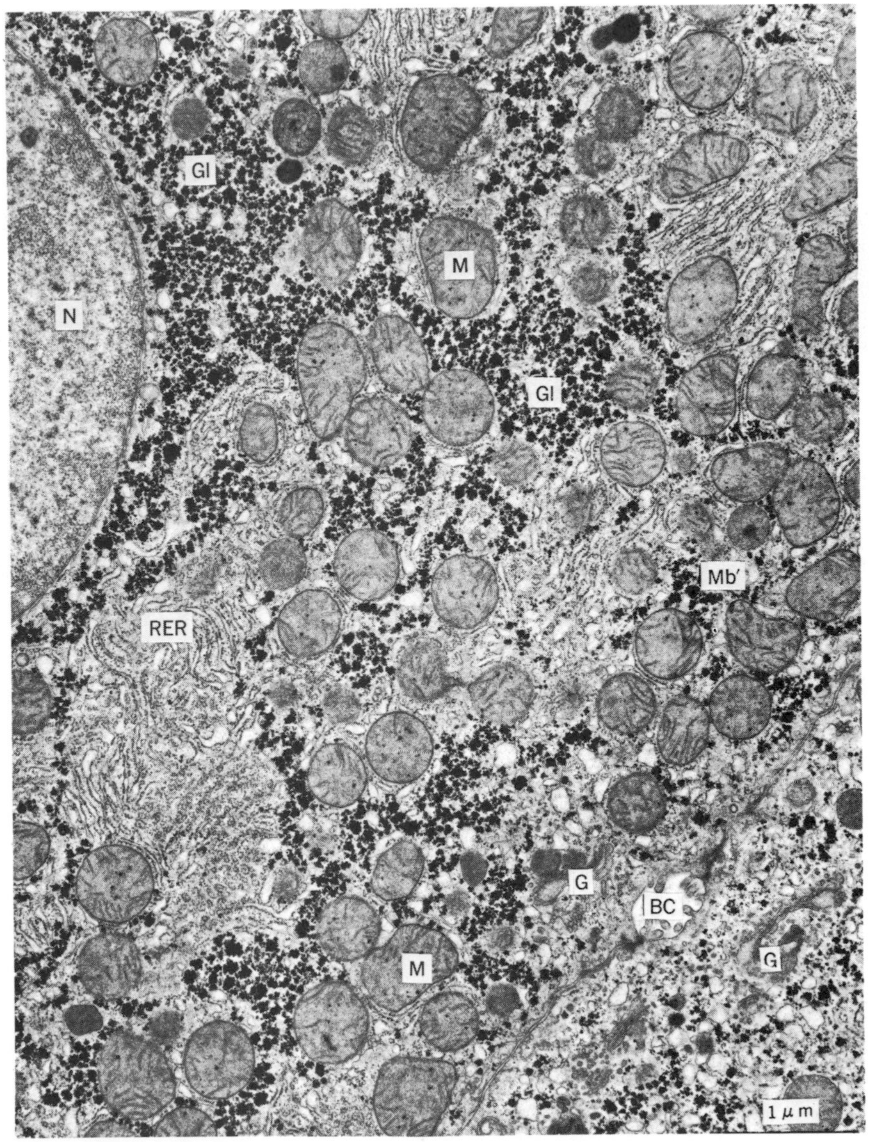

(a)

Figure 1–13

A comparison of electron photomicrographs of similar regions of liver cell obtained by sectioning (a) and freeze-fracturing (b). (N, nucleus; P, pores; RER, rough endoplasmic reticulum; SER, smooth endoplasmic reticulum; Go, Golgi body; M, mitochondria; Mb, microbody; BC, bile canaliculus; Gl, glycogen). The direction in which metal was applied to the fractured surface (i.e., the shadowing direction) in b is shown by the circled arrow. (Courtesy of Dr. L. Orci.)

mission electron microscope. Note that the replica is actually a templatelike impression of the distribution of particles in the original specimen. The electron beam readily passes through portions of the replica containing the carbon but is absorbed by the areas containing the platinum. The resulting images, which have a three-dimensional impact, are considerably different from those obtained with sectioned material (Fig. 1–13).

(b)

The Scanning Electron Microscope

Scanning electron microscopy has become an increasingly popular technique since its introduction as a biological tool in the 1960s. With this technique, the *surface topography* of a specimen may be examined in considerable detail. At the present time, resolution is of the order of 30 Å. The organization of the scanning electron microscope (SEM) is shown in Figure 1–14 and is similar in many respects to the TEM. However, instead of the electron beam passing through (i.e., being "transmitted" by) the specimen, the interaction of the electrons of the beam (called "primary" electrons) with the surface of the specimen causes the emission of "secondary" electrons from the surface. The beam rapidly *scans* back and forth over the surface of the specimen, thereby producing bursts of secondary electrons. Greater numbers of secondary electrons are produced when the beam strikes projections from the specimen surface than when the beam enters a pit or depression in the surface. Hence, the number of secondary electrons produced at each point on the specimen surface, as well as the direction in which scattering occurs, depends upon the surface topography. Therefore, there are quantitative and qualitative differences in the secondary electron bursts produced by the scanning electron beam. These ultimately give rise to an image in the following way.

Secondary electrons ejected at each point on the specimen surface are accelerated toward a postively charged scintillator located to one side of the specimen. Light scintillations created upon impact of these electrons with the scintillator are conducted by a light guide to the photocathode of a photomultiplier. Electrical pulses produced in the photomultiplier tube are then amplified, and the resulting signal is relayed to a cathode-ray tube. The result is an image much like that of a television picture, consisting of light and dark spots. The scanning of the specimen surface by the primary electron beam is synchronized with the projection of a beam on the television screen in such a way that each portion of the specimen is reproduced in a corresponding region of the television image.

Samples to be examined are usually coated first with a metal (typically a gold-palladium alloy), forming a layer 10 nm or more thick, and then are affixed to a supporting disk that is placed in the beam path. The metal coating efficiently reflects the primary electrons of the beam and also produces large numbers of secondary electrons. It is

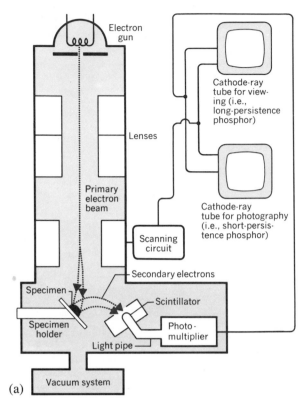

Figure 1–14

A scanning electron microscope (SEM). (Photo courtesy of International Scientific Instruments.)

to be noted that the thickness of the metal coating directly influences the maximum resolution attainable; for example, if the metal coating is 10 nm thick, then two particles that are less than 20 nm apart cannot be resolved because their metallic coatings are contiguous. Figure 1–15 contains examples of photomicrographs obtained with the SEM.

Since the specimen being examined with the SEM can be rotated, it is possible to obtain views from different angles. This provides additional information about the size, shape, and organization of the material being studied. For example, Figure 1–16 contains two scanning electron micrographs of the same cluster of chains of the bacterium *Simonsiella* taken from different angles.

Stereo Microscopy (Stereoscopy)

True three-dimensional (i.e., stereoscopic) images of the

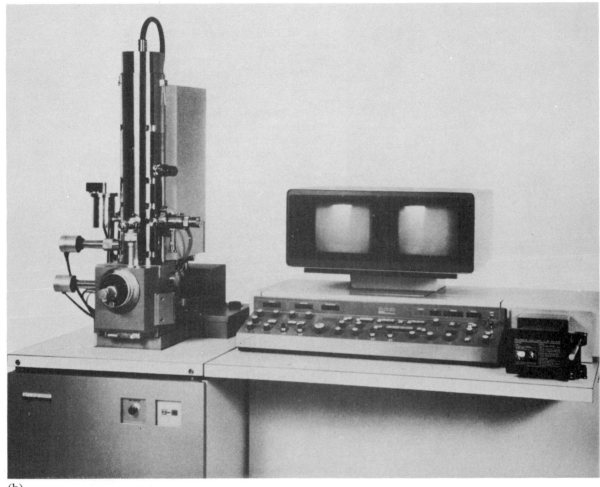

(b)

specimen being studied can be obtained if one photomicrograph is taken as though the specimen were being viewed with the left eye *only,* and a second is obtained representing the right eye view. (The two views are obtained by a minor tilting of the sample in the horizontal plane.) When the two photomicrographs are placed side by side and the *stereo pair* is viewed through the appropriate pair of lenses (called "stereo viewers"; see the Preface to the book), a striking three-dimensional impression is seen, revealing details and geometric relationships that cannot be discerned from a single photomicrograph. Stereo views of the surface topography of tissues and cells are readily obtained with specimens prepared for SEM study; illustrations are presented in Figure 1–17.

The internal organization of cells is revealed in three dimensions by *high-voltage transmission electron stereoscopy.* In this procedure, cells are placed or cultured on a conventional grid and then fixed and dehydrated. The grid and cells are sandwiched between layers of carbon and are examined in a TEM in which the accelerating voltage is great enough to cause electrons to penetrate the entire thickness of the cell (about 1 million volts). The cells are photographed at various tilt angles to produce the stereo pairs needed for the three-dimensional image (see Fig. 1–18).

The viewing of stereomicrographs may present some difficulties, especially for the novice. Generally, fewer problems are encountered with SEM stereoscopic views (i.e., Fig. 1–17), since the objects in the photomicrographs

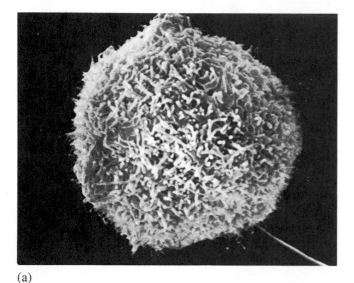

(a)

Figure 1–15
Examples of photomicrographs obtained using a scanning electron microscope. (*a*) Surface view of a teratoma tumor cell (note the numerous *microvilli*) that project from the cell surface; (*b*) mitochondria of heart muscle cell; in this view portions of the outer membrane of some of the organelles have been fractured away, revealing their internal membranous organization; (*c*) fractured pancreatic acinar cell. (*a* courtesy of Drs. S. B. Oppenheimer, E. G. Pollock, and R. Brenneman; *b* courtesy of Dr. Y. Masunaga, *J. Yonago Med. Assn. 30,* 519 (1979); *c* courtesy of Dr. T. Naguro, *Yonago Acta Medica 24* 23 (1980).)

(b)

(c)

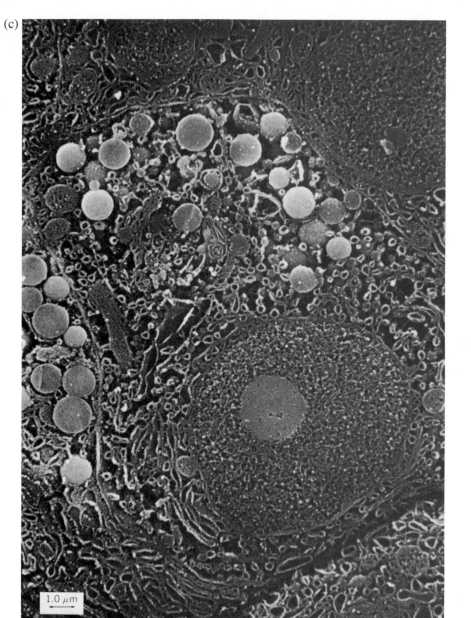

1.0 μm

are opaque (i.e., certain objects are clearly in front of others). Transmission stereomicrographs are more difficult to assimilate and interpret because most of the objects are translucent. However, no other procedure provides direct images of the three-dimensional morphology of the cell's interior. A single stereo pair can reveal the entire population of mitochondria, lysosomes, or other organelles (see below) distributed through the cell.

Cell Structure: A Preview

Free-living cells and the cells of multicellular organisms are subdivided into two major classes—**eucaryotes** (i.e., ''true nucleus'') and **procaryotes** (i.e., ''before nucleus''). In eucaryotes, the constituents of the cell nucleus are separated from the rest of the cell by a membranous envelope, whereas in procaryotes these materials are not separated.

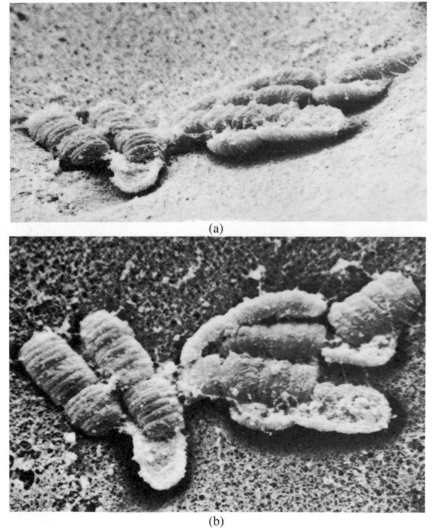

(a)

(b)

Figure 1–16
Scanning electron micrographs of chains of bacterial cells (*Simonsiella*) viewed from two (i.e., *a* and *b*) different angles showing the various perspectives attainable. Magnification 4000×. (Courtesy of Drs. D. A. Kuhn, J. Pangborn, and J. R. Woods.)

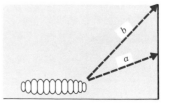

Figure 1–17
Stereoscopic electron photomicrographs obtained using SEM. To obtain the three-dimensional effect, these stereo pairs must be viewed through special stereo viewers (see the Preface). (*a*) Multiple lenses (*ommatidia*) of an insect eye (the spinelike structures are fine hairs); (*b*) a cluster of tertoma tumor cells; (*c*) a lymphocyte. (Photos courtesy of R. Chao.)

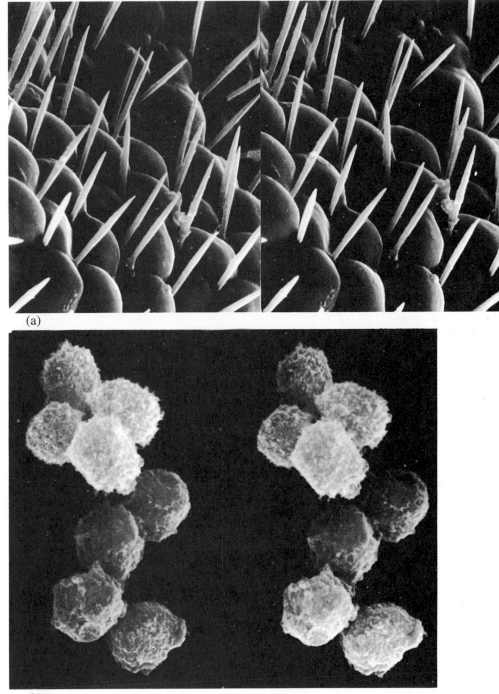

(a)

(b)

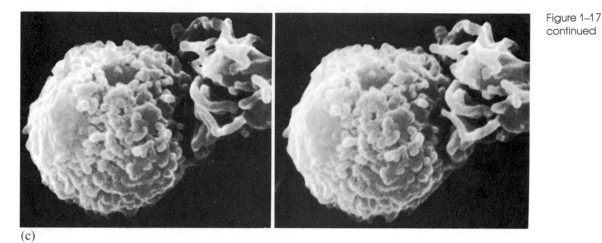

Figure 1–17
continued

(c)

Although the presence or absence of a true nucleus is the most obvious distinction between eucaryotic and procaryotic cells, it will soon become clear that these two groups of cells also differ in many other important respects. Es-

sentially all animal and plant cells are eucaryotic, whereas procaryotic cells include bacteria, blue-green algae (or cyanobacteria), and the so-called pleuropneumonia-like organisms (PPLO) or mycoplasmas.

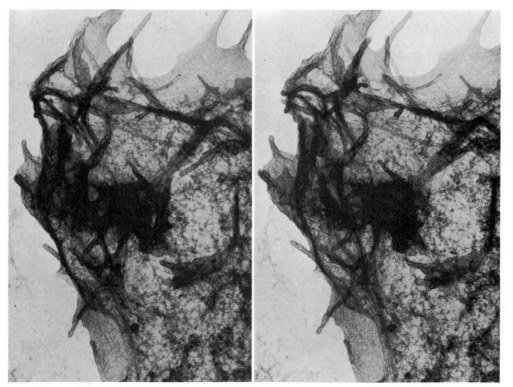

Figure 1–18
High-voltage stereo transmission electron photomicrograph. In this view, the ruffles and microvilli that cover the surface of the cell are clearly seen. A fine, irregular lattice of microtrabeculae (filamentous structures) can be seen to occupy the space between the upper and lower cell surfaces as well as the surface projections. Magnification, 11,000×. (Photomicrograph courtesy of Drs. J. J. Wolosewick and K. R. Porter.)

Eucaryotic Cells:
The Composite Animal Cell

Animal cells vary considerably in size, shape, organelle composition, and physiological roles. Consequently, there is no "typical" cell that can serve as an example of *all* animal cells. There are, however, a number of cell structures common to the majority of animal cells that are similar or identical in organization. These structures are depicted in the *composite* animal cell diagrammed in Figure 1–19 and described briefly in the following sections. They are dealt with in greater detail in later chapters individually

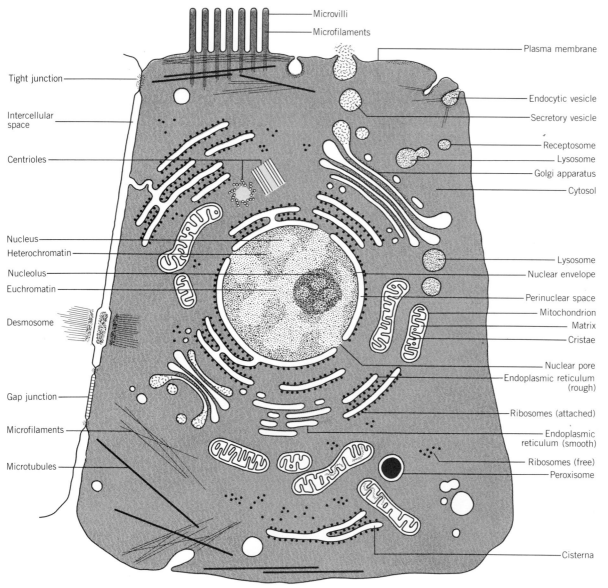

Figure 1–19
A *composite* animal cell.

devoted to the structure and functions of cell organelles. Figure 1–20 is an electron photomicrograph of an animal cell containing many of the structures to be discussed.

The Plasma Membrane. The contents of the cell (**cytoplasm** and cytoplasmic organelles) are separated from the external surroundings by a limiting membrane, the **plasma membrane** (also called **cell membrane** or **plasmalemma**), which is composed of protein, lipid, and carbohydrate. This structure regulates the passage of materials between the cell and its surroundings and in some tissues is involved in intercellular communication (e.g., nerve tissue). These subjects are treated in Chapters 15 and 24. In some tissue cells, portions of the plasma membrane are modified to form a large number of fingerlike projections called **microvilli** (Fig. 1–21a) because of their resemblance to the much

larger villi of the small intestine. The microvilli greatly increase the surface area of the cell and provide for the increased passage of materials across the plasma membrane. When large numbers of cells are in close contact with one another (as, for example, in a tissue), it is not unusual to observe special junctions between opposing plasma membranes. These take the form of **tight junctions, desmosomes,** and **gap junctions.** Some of these are illustrated in Figure 1–21 and are considered in greater depth in Chapter 15.

The plasma membrane should not be thought of as a homogeneous structure that has the same composition over its entire surface. Instead, the composition and organization vary in different regions of the membrane. Some areas of the plasma membrane of a liver cell, for example, face the plasma membranes of neighboring liver cells; other areas

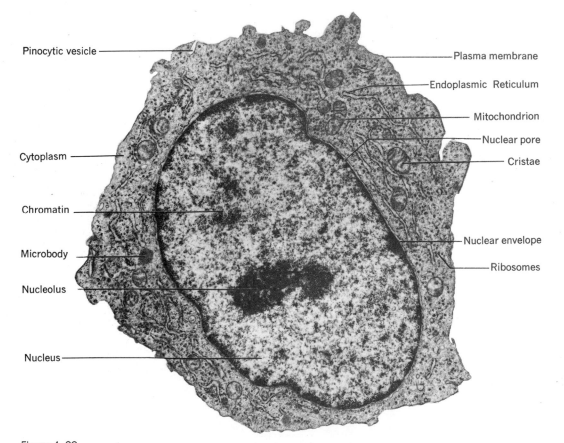

Figure 1–20
Electron photomicrograph of whole animal cell. Magnification, 13,000×.
(Courtesy of R. Chao.)

(a)

Figure 1–21
Specializations of the plasma membrane. The plasma membrane at the apical surface of secretory and absorptive tissues forms numerous tiny projections called *microvilli*. Seen in (*a*) is an SEM photomicrograph of the surface of apocrine tissue (i.e., sweat gland cells). Below their apical surfaces, the plasma membranes of adjacent cells form a variety of different junctions such as *desmosomes* (*b*) and *gap junctions* (*c*). (*a* courtesy of Dr. W. H. Wilborn; *b* and *c* courtesy of Dr. N. B. Gilula; from Gilula, N. B., *Cell Communications* (R. P. Cox, Ed.), John Wiley & Sons, 1974; reprinted with permission of The Rockefeller University Press.)

(b)

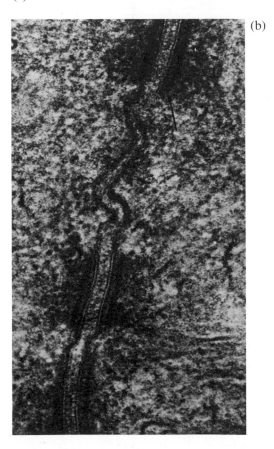

(c)

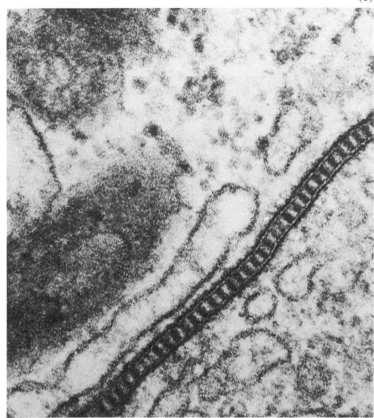

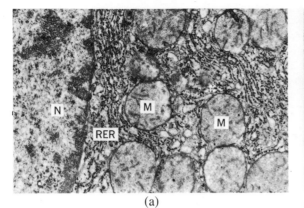

(a)

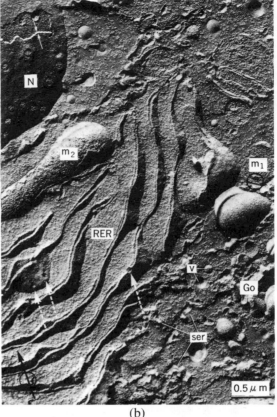

(b)

Figure 1–22

The endoplasmic reticulum. (a) and (b), Thin section and freeze-fracture through comparable regions of the cell. (c) High-magnification view. *N*, nucleus; *M*, mitochondria; *R*, ribosomes; *C*, cisternae; *H*, hyaloplasm (cytosol); *RER*, rough endoplasmic reticulum; *SER*, smooth endoplasmic reticulum; *Go*, Golgi body; *V*, vesicle. In (b) the dashed arrows note perforations in ER membranes. Also, one of the mitochondria (m₂) has been fractured in such a way that faces of both the outer and inner mitochondrial membranes may be seen. The space separating the two nuclear membranes is indicated by the white arrows. The direction of shadowing is indicated by the circled arrow. (d) Scanning photomicrograph of rough ER; (e) diagram showing the continuity that exists between adjacent membranes and cisternae. (b, courtesy of Dr. L. Orci; c, courtesy of R. Chao; d, courtesy of Dr. K. Tanaka; copyright © 1980 by Academic Press, *Intern. Rev. Cytol. 68*, 102.)

(c)

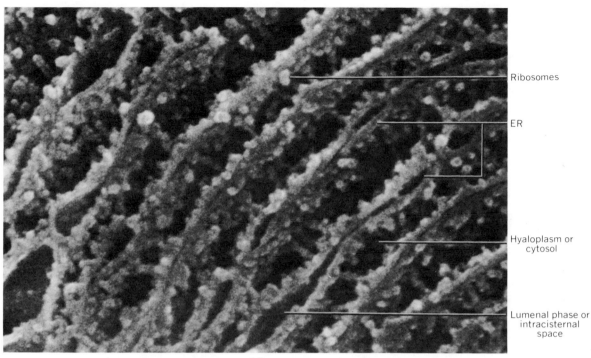

Ribosomes

ER

Hyaloplasm or cytosol

Lumenal phase or intracisternal space

(d)

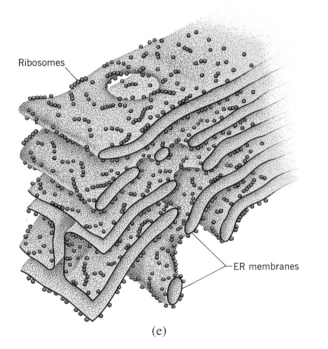

Ribosomes

ER membranes

(e)

face the bile channels (*bile canaliculi*) into which substances are secreted by the liver cell. Still other portions of the plasma membrane face the epithelial lining of *capillaries,* from which substances are absorbed. Each of these regions of the plasma membrane is differently composed and differently organized and, in fact, is continually undergoing change and reorganization.

The Endoplasmic Reticulum and Ribosomes. Within the cytoplasm of most animal cells is an extensive network of branching and anastomosing membrane-limited channels or cisternae collectively called the **endoplasmic reticulum** (Fig. 1–22). The membranes of the endoplasmic reticulum (usually abbreviated ER) divide the cytoplasm into two phases: the **lumenal** or **intracisternal** phase and the **hyaloplasmic** phase or **cytosol.** The lumenal phase consists of the material enclosed within the cisternae of the endoplasmic reticulum, while the cytosol surrounds the ER membranes. In the typical cell, the surface area that is represented by ER membranes is more than 10 times that of the plasma membrane.

In the cytosol are large numbers of small particles called

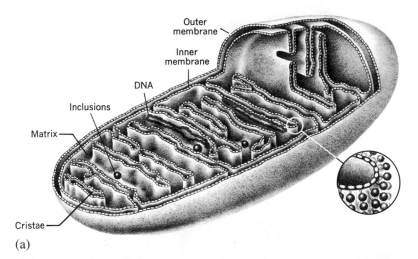

Outer membrane
Inner membrane
DNA
Inclusions
Matrix
Cristae

(a)

Figure 1–23
The *mitochondrion*. (Photomicrograph courtesy of Dr. K. Tanaka; copyright © 1980 by Academic Press, *Intern Rev. Cytol. 68,* 104.)

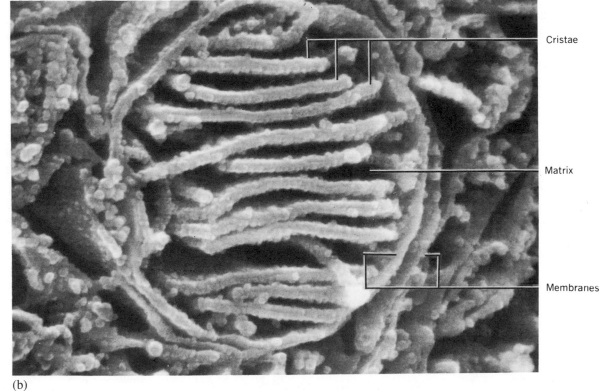

Cristae

Matrix

Membranes

(b)

ribosomes. These particles are either distributed along the hyaloplasmic surface of the endoplasmic reticulum ("attached" ribosomes) or free in the hyaloplasm ("free" ribosomes). There is some evidence that the free ribosomes are interconnected by fine filaments of the **microtrabecular lattice.** Endoplasmic reticulum with associated ribosomes is called **rough ER** (*RER*), whereas **smooth ER** (*SER*) is devoid of attached ribosomes. Ribosomes carry out the synthesis of the cell's proteins, a subject covered at some length in Chapter 22 (which also deals with the molecular architecture of ribosomes). Certain portions of the endoplasmic reticulum may be continuous with the plasma membrane and the nuclear envelope.

Mitochondria. Within the cytoplasm are numerous vesicular organelles called **mitochondria** (Figs. 1–13*a* and *b,* and 1–23). Each mitochondrion is bordered by two membranes. The *outer* membrane is smooth, but the *inner* membrane displays numerous infoldings called **cristae** that greatly increase the surface area of the inner membrane. The space between neighboring cristae is called the mitochondrial **matrix** and often contains inclusions. Mitochondria are engaged in numerous metabolic functions in the cell, including the energy-producing phases of carbohydrate and fat metabolism (called respiration), ATP synthesis, and porphyrin synthesis. The mitochondrion and its functions are considered in greater detail in Chapter 16.

The Golgi Apparatus. The **Golgi apparatus** (also called Golgi *body* or Golgi *complex*) consists of a set of smooth cisternae which often are stacked together in parallel rows; in this state, the organelle is sometimes referred to as a **dictyosome.** The Golgi apparatus is usually surrounded by vesicles of various sizes, some of which are discharged from the margins of the main body of the organelle (Fig. 1–24). As related in Chapter 18, a variety of functions are ascribed to the Golgi apparatus, including *secretory* activity (especially the secretion of enzymes), the sorting and glycosylation of proteins synthesized by the rough endoplasmic reticulum, and the proliferation of additional membranous material for intracellular organelles and the plasma membrane.

Lysosomes. Many cells contain vesicular structures that are generally smaller than mitochondria and are called **ly-sosomes** (Fig. 1–25). Lysosomes are bounded at their surface by a single membrane and contain quantities of various hydrolytic enzymes capable of digesting protein, nucleic acid, polysaccharide, and other materials. Under normal conditions, the activity of these enzymes is confined to the interior of the organelles and is therefore isolated from the cytoplasm. However, if the lysosomal membrane is ruptured, the released enzymes can quickly degrade the cell. Lysosomes are believed to be responsible for the intracellular digestion of particles that are ingested by the cell during *endocytosis,* the intracellular scavenging of worn and therefore poorly functioning organelles, and a number of other cell functions. Lysosomes and another class of organelles known as *microbodies* are considered in depth in Chapter 19.

Peroxisomes and Glyoxysomes. Many cells contain small numbers of **peroxisomes** and/or **glyoxysomes.** These small organelles, which are bounded by a single membrane, contain a number of enzymes whose functions are related to the metabolism of hydrogen peroxide and glyoxylic acid. The structure and functions of peroxisomes, glyoxysomes, and other microbodies are considered futher in Chapter 19.

The Nucleus. The **nucleus** is a relatively large and readily distinguished structure frequently but not always located near the center of the cell. The contents of the nucleus are separated from the cytoplasm by *two* membranes that together form the **nuclear envelope.** At various positions, the outer membrane of the envelope fuses with the inner membrane to form *pores* (Fig. 1–26). Nuclear pores provide continuity between the *cytosol* and the contents of the nucleus (usually referred to as the **nucleoplasm**). Often, the nuclear pores are plugged by a granular material (Fig. 1–27). The outer nuclear membrane may have ribosomes attached to its hyaloplasmic side (Fig. 1–28) and may also form continuities with the membranes of the endoplasmic reticulum. Since the latter may be continuous with the plasma membrane, the *perinuclear space* (i.e., the space between the inner and outer membranes of the nuclear envelope) corresponds to the lumenal phase of the cell and may be considered external to the cell (see Fig. 1–29).

The nuclear envelope and the pores that penetrate it are dramatically revealed in freeze-fracture preparations (Fig. 1–30). The cytosol-contacting half of the outer nuclear

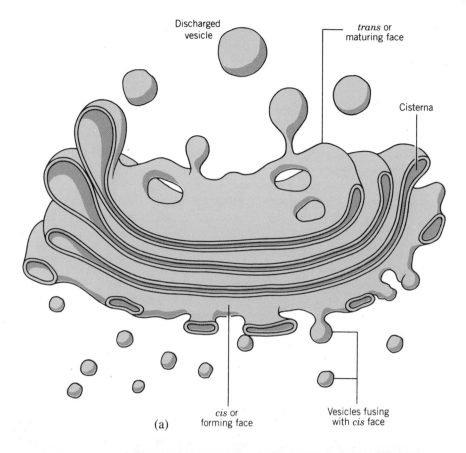

Discharged vesicle

trans or maturing face

Cisterna

cis or forming face

Vesicles fusing with *cis* face

(a)

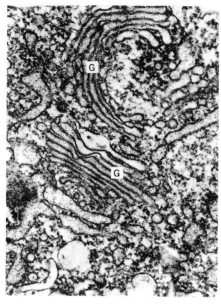

(b)

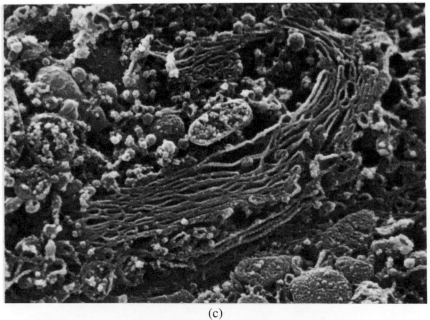

(c)

Figure 1–24

The *Golgi apparatus.* (*a*) Diagram depicting the three-dimensional relationship between the cisternae and peripheral vesicles. (*b*) Typical appearance of Golgi bodies (*G*) in thin sections. (*c*) SEM view of Golgi body in a freeze-fracture cell. (*b*, courtesy of Dr. E. G. Pollock; *c*, courtesy of Dr. Y. Kinose; *J. Yonago Med. Assn. 30,* 527 (1979).)

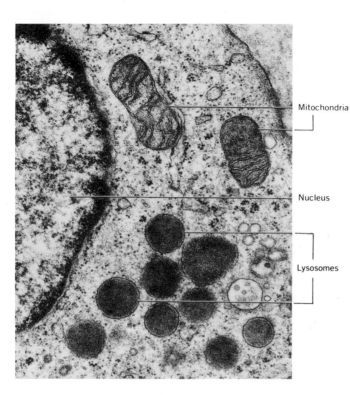

Figure 1–25
Lysosomes. In cells that have been especially treated to reveal these organelles, the lysosomes appear as electron-dense bodies somewhat smaller than mitochondria. (Courtesy of R. Chao.)

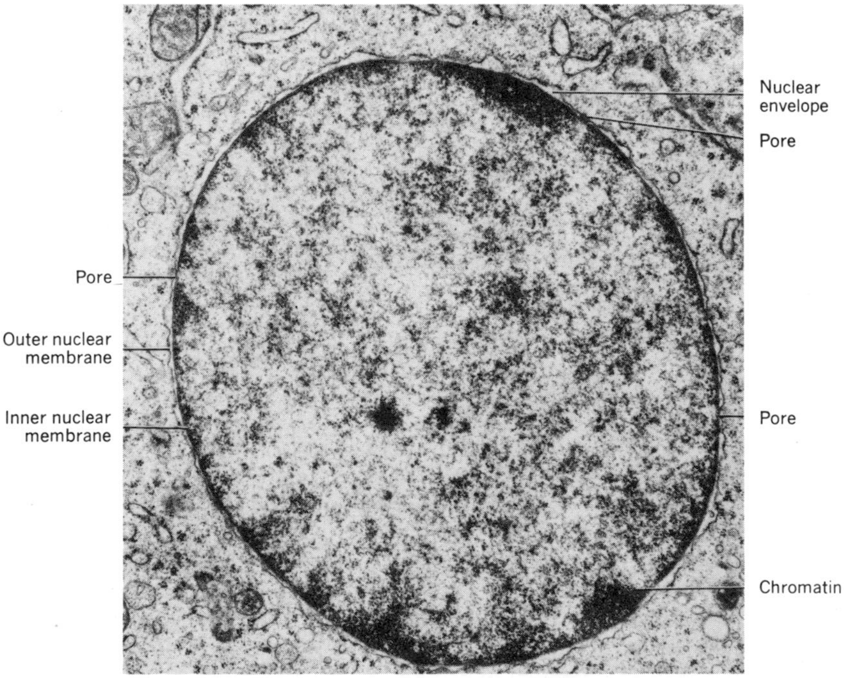

Mitochondria

Nucleus

Lysosomes

Figure 1–26
The cell *nucleus.* (Courtesy of R. Chao.)

Nuclear envelope

Pore

Pore

Outer nuclear membrane

Inner nuclear membrane

Pore

Chromatin

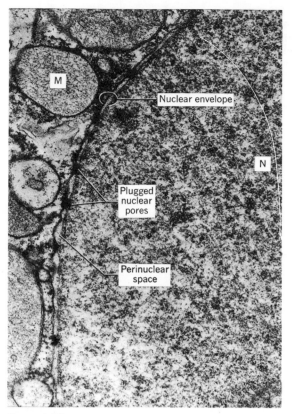

Figure 1–27
Plugged pores of the nuclear envelope. *N*, nucleus; *M*, mitochondrion. Magnification, 30,000×. (Photomicrograph courtesy of Dr. E. G. Pollock.)

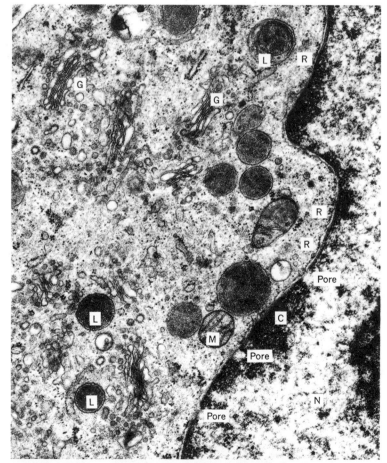

Figure 1–28
In this electron photomicrograph, one can clearly discern ribosomes (*R*) attached to the outer membrane of the nuclear envelope. Also seen are nuclear pores, mitochondria (*M*), lysosomes (*L*), Golgi bodies (*G*), and nuclear chromatin (*C*). Magnification, 20,000×. (Courtesy of R. Chao.)

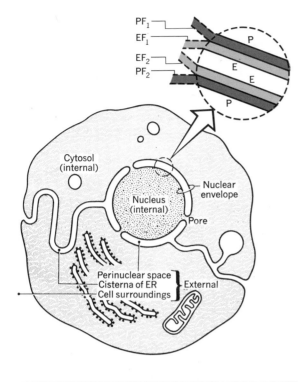

Figure 1–29
"Internal" and "external" phases of the cell. A region of the nuclear envelope is enlarged to illustrate the relationships among various faces of the inner and outer nuclear membranes. This illustration should be compared with Figure 1–12.

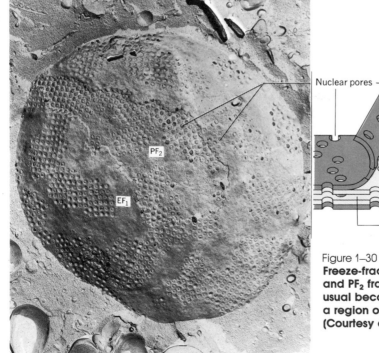

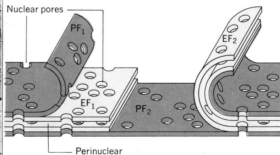

Figure 1–30
Freeze-fracture through the nuclear envelope exposing the EF_1 and PF_2 fracture faces. This freeze-fracture view is especially unusual because the fracture plane has passed through so large a region of the nuclear envelope. Magnification, 13,000×. (Courtesy of Dr. E. G. Pollock.)

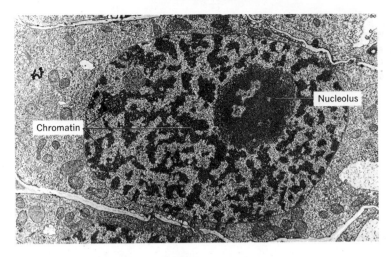

Figure 1–31
Dense, granular *nucleolus* in the nucleus of a plant cell. Magnification, 6000×. (Photomicrograph courtesy of Dr. E. G. Pollock.)

(a)

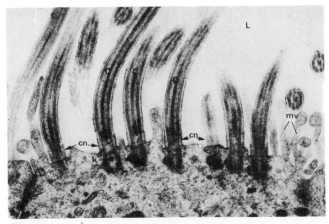

(b)

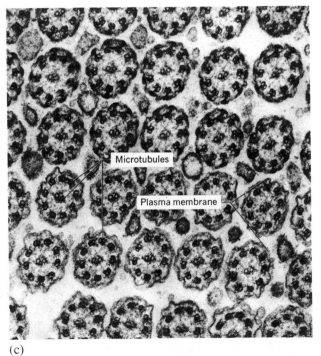

(c)

Figure 1–32
Cilia. The cilia have been fractured (*a*) and sectioned (*b*) in a plane parallel to the long axis of the organelles. *C*, cilium; *mv*, microvilli; *L*, lumen; *k*, kinetosome or basal body; *cn*, ciliary necklace. In (*c*), the cilia are seen in cross-section (note the arrangements of the microtubules). (*a* and *b* courtesy of Dr. E. Boisvieux-Ulrich copyright © 1977 *Biol. Cellulaire 30*, 245; *c* courtesy of R. Chao.)

membrane has been fractured away, exposing the inner half of that membrane (i.e., face EF_1 of Fig. 1–29). Also fractured away are pieces of the inner nuclear membrane (the half-membrane that faced the perinuclear space), leaving only the half-membrane that faced the nucleoplasm (i.e., PF_2 of Fig. 1–29). The nuclear pores penetrate both membranes and in Figure 1–30 appear to be nonrandomly distributed in the nuclear envelope.

The nucleus contains the genetic machinery of the cell (chromosomal DNA, histones, etc.), which is discussed in Chapters 20 and 21. Using either the light microscope or the electron microscope, the nucleus often reveals one or more dense, granular structures called **nucleoli** (Fig. 1–31). Nucleoli are not bounded by a membrane and are formed in part from localized concentrations of ribosomal precursors (see Chapters 20 and 22).

Flagella and Cilia. Many free-living cells (such as protozoa and other microorganisms) possess *locomotor* organelles that project from the cell surface; these are either **flagella** or **cilia** (Fig. 1–32). The tissue cells of multicellular animals may also contain cilia, but they are employed here to move a substrate across the cell surface (such as mucus in the respiratory tract or the egg cell during its passage through the oviduct) and not for cell locomotion. The organelles are called cilia when they are short but present in large numbers and are called flagella when long but few in number. Each cilium or flagellum is covered by an extension of the plasma membrane. Internally, these organelles contain a specific array of microtubules that run from the **basal body** or **kinetosome** toward the tip of the structure. This array consists of two central microtubules and nine pairs of peripheral (outer) microtubules (Fig. 1–32c). Just below the basal body are rows of membrane particles referred to as the *ciliary necklace*.

Other structural elements commonly found in animal cells include **cytoplasmic filaments** and **microtubules** that participate in intracellular movement and communication. These may be scattered through the cytoplasm, but many are located just under the plasma membrane and may be anchored to it. **Basal bodies** are found at the base of locomotor organelles and may give rise to the microtubules of these structures. In animal cells, pairs of **centrioles** are observed near the cell nucleus and may be involved in the mechanics of cell division. Cilia, flagella, and cytoplasmic filamentous and microtubular structures are treated in detail in Chapter 23.

Eucaryotic Cells: The Composite Plant Cell

All the organelles described in the preceding section as regular constituents of animal cells are also found in similar form in many plant cells. Several other organelles are unique to plant tissues and include the carbohydrate-rich *cell wall, plasmodesmata, chloroplasts,* and large *vacuoles*. A composite plant cell is depicted in Figure 1–33.

The Cell Wall. The **cell wall** is a thick polysaccharide-containing structure immediately surrounding the plasma membrane (Fig. 1–33). In multicellular plants, the plasma membranes of neighboring cells are separated by these walls, and adjacent plant cells have their walls fused together by a layer called the *middle lamella*. The cell wall serves both a protective and a supportive function for the plant. The degree to which the cell wall may be involved in the regulation of the exchange of materials between the plant cell and its surroundings is difficult to assess but is most likely restricted to macromolecules of considerable size. As in animal cells, most of the regulation of exchanges between the cytoplasm and the extracellular surroundings of plant cells is a function of the plasma membrane.

Plasmodesmata. At intervals the plant cell wall may be interrupted by cytoplasmic bridges between one cell and its neighbor (Fig. 1–34). These bridges are called **plasmodesmata** and represent regions in which channel-like extensions of the plasma membranes of neighboring cells merge. The channels serve in intercellular circulation of materials.

Chloroplasts. The ability to use light as a source of energy for sugar synthesis from water and carbon dioxide is a special feature of certain plant cells. This process, termed **photosynthesis,** is carried out in organelles called **chloroplasts** (Fig. 1–35). These organelles are commonly ellipsoidal structures bounded by an outer membrane but also containing a number of internal membranes. Internally, the chloroplast consists of a series of membranes arranged in parallel sheets called **lamellae** and supported in a homogeneous matrix called the **stroma.** The membranes are arranged as thin sacs (called **thylakoids**) that contain chlorophyll and may be stacked on top of one another, forming structures called **grana.** Lamellar membranes connecting the grana are called **stroma lamellae** (Fig. 1–35). A more detailed description of the structure of chloroplasts and the metabolic pathways of photosynthesis is found in Chapter 17.

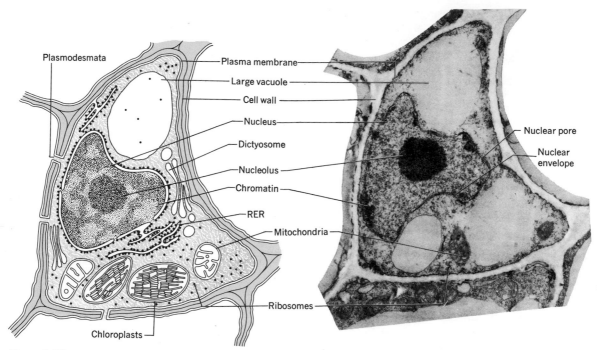

Plasmodesmata

Plasma membrane

Large vacuole

Cell wall

Nucleus

Dictyosome

Nucleolus

Chromatin

RER

Mitochondria

Ribosomes

Chloroplasts

Nuclear pore

Nuclear envelope

Figure 1–33

The *composite* plant cell. (Photomicrograph courtesy of R. Chao.)

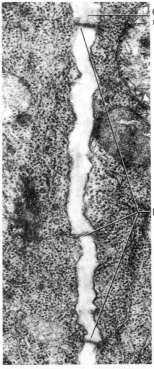

Cell wall (separating neighboring plant cells)

Plasmodesmata

Figure 1–34

***Plasmodesmata*. (Photomicrograph courtesy of N. Herzog.)**

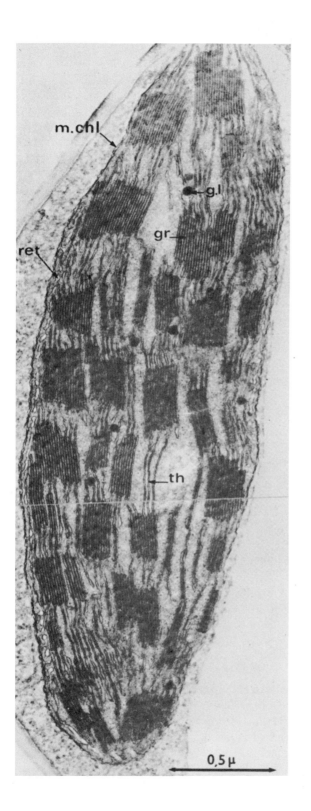

Vacuoles. Although **vacuoles** are present in both animal and plant cells, they are particularly large and abundant in plant cells (Fig. 1–33), often occupying a major portion of the cell volume and forcing the remaining intracellular structures into a thin peripheral layer. These vacuoles are bounded by a single membrane and are formed by the coalescence of smaller vacuoles during the plant's growth and development. Vacuoles serve to expand the plant cell without diluting its cytoplasm and also function as sites for the storage of water and cell products or metabolic intermediates.

Procaryotic Cells: Bacteria

The bacteria are structurally distinct from eucaryotic microorganisms such as protozoa and contain a number of unique cellular organelles. The typical bacterial cell is about the size of a mitochondrion of an animal or plant cell, and in view of this small size, it is to be expected that the organelles of bacteria would be correspondingly smaller. The composite structure of a bacterium is depicted in Figure 1–36.

The Bacterial Cell Wall. The bacterial cell is enclosed within a wall that differs chemically from the cell wall of plants in that it contains protein and lipid as well as polysaccharide. Its content of a particular "mucopeptide" (a protein–carbohydrate complex) has been the basis of the histochemical classification of bacteria, being high in the so-called gram-positive bacteria (such as *Bacillus subtilis*) and low in the gram-negative bacteria (such as *Escherichia coli* and the *Simonsiella*). In some bacteria, the cell wall is surrounded by an additional structure called a **capsule.** The cell wall and capsule confer shape and form to the bacterium and also act as a physical barrier to the cell membrane. This is important because osmotic forces usually result in a positive hydrostatic pressure inside the bacterium; in the absence of a cell wall and capsule, mechanically fragile bacteria would rupture.

Figure 1–35
Chloroplast of a corn cell; *m.chl*, chloroplast membrane; *g.l*, lipid globule; *gr*, grana; *ret*, marginal reticulum; *th*, thylakoid. (Courtesy of Dr. A. M. Lhoste-Bisch.)

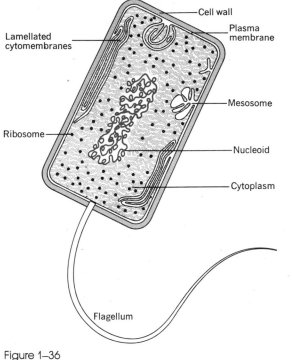

Figure 1–36
The *composite* bacterium.

Plasma Membrane Intrusions. Infoldings of the plasma membrane of gram-positive bacteria give rise to structures called **mesosomes** (or **chondrioids**) (Fig. 1–37*a*). Mesosomes may play a role in the division of the cell. Intrusions of the plasma membrane also form the photosynthetic organelles (**chromatophores**) of the photosynthetic bacteria.

Lamellate Cytomembranes. In some bacteria, there is a lamellar arrangement of membranes within the cytoplasm (Fig. 1–37*b*). However, there are no structures comparable to the endoplasmic reticulum of animal and plant cells. Bacteria contain large numbers of ribosomes. Most of the ribosomes are free in the bacterial cytoplasm, but some may be attached to the interior surface of the plasma membrane. Although bacterial ribosomes, like the ribosomes of eucaryotic cells, are the sites of protein synthesis, considerable differences exist between the organelles of these two groups (see Chapter 22). Lamellate membranes are particularly abundant in the *autotrophic* bacteria, which support their growth through photosynthesis or similar processes.

Nucleoids. In bacteria the nuclear material is not separated from the cytoplasm by membranes, as it is in eucaryotic cells. However, the nuclear material is usually confined to a specific region of the cell, referred to as a **nucleoid.** During bacterial cell division, the nuclear contents are distributed to the daughter cells without formation of observable chromosomes as occurs in eucaryotes. Nucleoli are not present in the nucleoid.

Bacterial Flagella. Many bacteria contain one or more **flagella** employed for cellular locomotion. These organelles arise from a small basal granule in the cytoplasm and penetrate the plasma membrane and cell wall. Bacterial flagella are smaller than those of animal and plant cells and are simpler in organization, containing a single filament of globular proteins (called *flagellin*) surrounded by a sheath. The multiflagellated (*peritrichous*) bacterium *Proteus mirabilis* is shown in Figure 1–38.

Some bacteria, such as *E. coli, P. mirabilis,* and *B. subtilis,* occur as separate, individual cells. However, in a number of groups, the daughter cells remain attached to each other following division, so that chains (e.g., *streptococci*) or filaments are formed. An example of a filamentous genus is *Simonsiella,* which colonizes the mucosal epithelial surface of the mouth (see Fig. 1–16). The individual cells of some filamentous bacteria reveal a dorsal–ventral differentiation; that is, the ventral surface (which in *Simonsiella* attaches to and glides along the epithelium) is structured differently than the dorsal surface (which faces away from the epithelium). This is apparent not only in the scanning electron micrographs of whole filaments shown in Figure 1–16 but also in transmission electron micrographs of thin sections through the filaments (Fig. 1–39), which reveal an internal differentiation. Individual cells of a filament exhibit features common to single-cell (i.e., nonfilamentous) forms like *E. coli* and *B. subtilis.*

Procaryotic Cells: Cyanobacteria (or Blue-green Algae)

The **cyanobacteria** (also called **blue-green algae**) are photosynthetic procaryotes and occur as individual cells, as small clusters or colonies of cells, or as long, filamentous chains (Figs. 1–40 and 1–41). Cyanobacteria lack locomotor organelles, and a gelatinous sheath replaces the capsule typical of bacteria. The photosynthetic apparatus con-

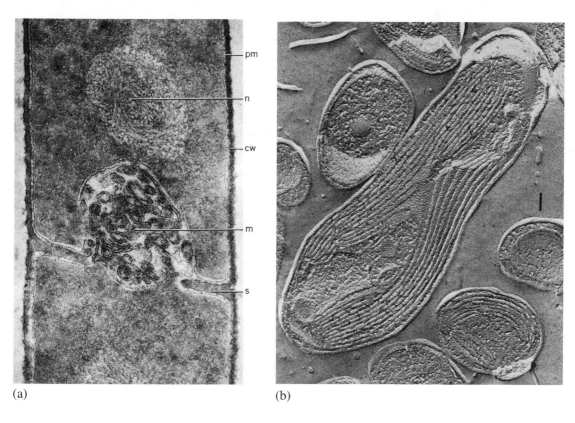

(a)

(b)

Figure 1–37
TEM photomicrographs of bacteria. (*a*) Thin section through *Bacillus subtilis*. *cw*, cell wall; *pm*, plasma membrane; *n*, nucleoid; *m*, mesosome. Note the formation of the transverse septum, *s*, as the cell divides. Magnification 27,000 ×. (*b*) Cross–fracture through *Rhodoseudomonas palustris* revealing the lamellar arrangement of photosynthetic membranes; bar = 1 μm (*a* courtesy of Dr. F. A. Eiserling; *b* courtesy of Drs. A. R. Varga and L. A. Staehelin.)

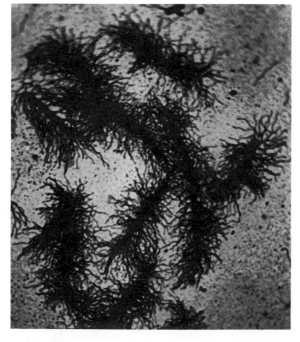

Figure 1–38
Some bacteria such as *Proteus mirabilis* shown here are multiflagellated (*peritrichous*). (Courtesy of Drs. D. A. Kuhn and G. Patane.)

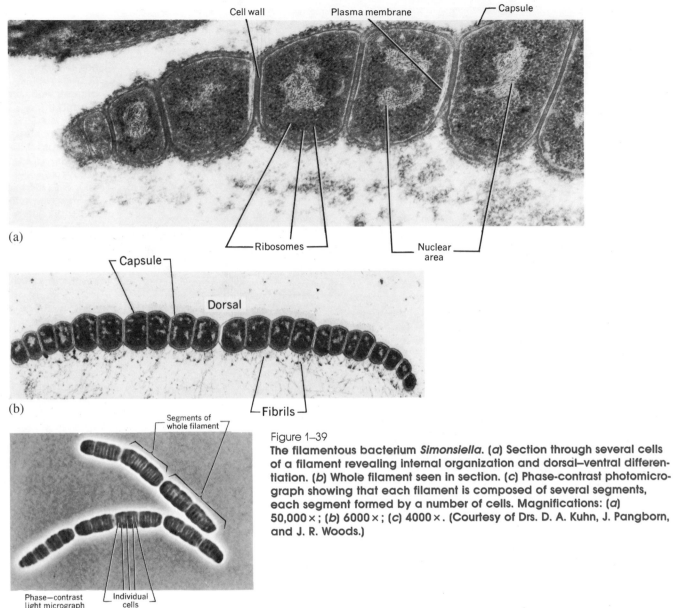

Figure 1-39
The filamentous bacterium *Simonsiella*. (*a*) Section through several cells of a filament revealing internal organization and dorsal–ventral differentiation. (*b*) Whole filament seen in section. (*c*) Phase-contrast photomicrograph showing that each filament is composed of several segments, each segment formed by a number of cells. Magnifications: (*a*) 50,000×; (*b*) 6000×; (*c*) 4000×. (Courtesy of Drs. D. A. Kuhn, J. Pangborn, and J. R. Woods.)

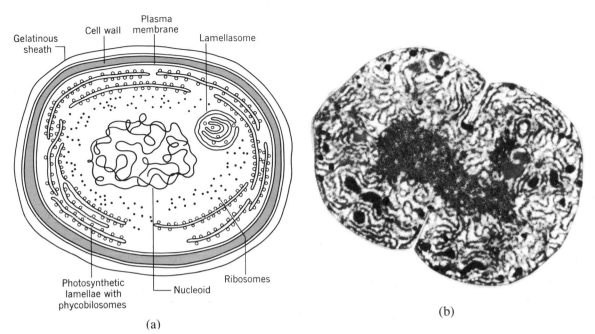

Gelatinous sheath
Cell wall
Plasma membrane
Lamellasome
Photosynthetic lamellae with phycobilosomes
Nucleoid
Ribosomes

(a)

(b)

Figure 1–40
(a) Generalized blue-green alga cell (cyanobacterium). (b) Electron photomicrograph of dividing Anabaena. Note the formation of the septum. Magnification, 15,000×. (Courtesy of Drs. H. W. Beams and R. G. Kessel.)

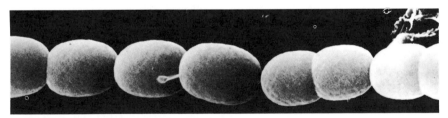

Figure 1–41
SEM photomicrograph of a filament of the blue-green alga Anabaena (compare with the thin section of Fig. 1–40b). Magnification, 6000×. (Courtesy of Drs. F. A. Eiserling and S. Eipert.)

sists of lamellae lined with pigment granules sometimes referred to as **phycobilosomes.**

Procaryotic Cells: Mycoplasmas (PPLO)

The **PPLO** (i.e., pleuropneumonia-like organism) or **mycoplasmas,** which cause a number of diseases in humans and other animals, are the smallest (i.e., about 0.1 μm in diameter) of all free-living cells. They are smaller even than some of the larger viruses. The PPLO is bounded at its surface by a membrane composed of proteins and lipid, but internally the cell's composition is more or less diffuse. The only microscopically discernible features within the cell are its genetic complement, which consists of a double-helical strand of circular DNA, and a number of ribosomes (Fig. 1–42). Mycoplasmas appear to contain the bare minimum of structural organization required for a viable, free-living cell and may represent a form intermediate between viruses and bacteria. The relative sizes of typical eucaryotic cells, bacteria, PPLOs, and viruses are compared in Figure 1–43, which dramatizes the differences that exist.

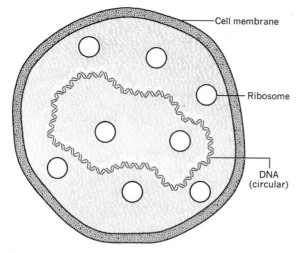

Figure 1–42
Structure of a PPLO or mycoplasma.

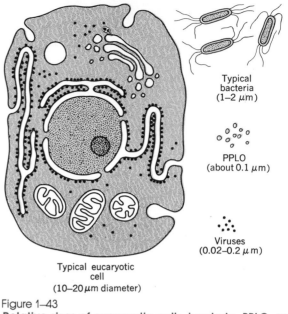

Typical eucaryotic
cell
(10–20 μm diameter)

Figure 1–43
Relative sizes of eucaryotic cells, bacteria, PPLO, and viruses.

Viruses

So far, the descriptions in this chapter have been restricted to various kinds of cells. In this section, we are concerned with the organization and activities of viruses, but it should be emphasized at the outset that *viruses are not cells* and that it is a moot question whether viruses constitute living systems. Viruses are described here because of their intimate association with cells and because of their contributions to our understanding of certain cellular phenomena. It will become apparent as we deal in later chapters with the structure and interactions of nucleic acids and proteins and with gene expression that much of our present-day understanding of these subjects is based on studies initiated with viruses.

Structure of Viruses

Although all viruses or **virions** are extremely small, they are diverse in size and in organization. Generally, viruses range in diameter (or length) from about 20 nm to about 200 nm. Thus the largest viruses are actually larger than the smallest cells. However, even the smallest of cells (bacteria, PPLOs, etc.) are subject to infection by viruses. Among those viruses that attack animal cells, the most notorious are the viruses that cause diseases in humans. Smallpox, chicken pox, herpes simplex, rabies, poliomy-

elitis, mumps, measles, influenza, hepatitis, and the "common cold" are all produced by viruses. Even certain leukemias and other cancers are of viral origin.

Most virions are either rod-shaped or quasi-spherical and contain a nucleic acid *core* surrounded by a specific geometric array of protein molecules that form a coat or *capsid* (Fig. 1–44). The proteins of the capsid are arranged to form either a *helical* pattern (when the virus is rodlike) or an *isometric* pattern (when the virus is globular). In the latter state, the virus appears much like a *polyhedron*. The viruses causing chicken pox, mononucleosis, herpes blisters, and colds are examples of virions having polyhedral capsids. Helical capsids are more common among viruses that infect plant cells and bacteria. The **tobacco mosaic virus** (TMV), which infects the leaves of the tobacco plant, is among the most extensively studied viruses and exhibits the helical capsid pattern.

In many animal viruses and in some plant viruses, a lipoprotein *envelope* surrounds the capsid (e.g., influenza virus, herpesvirus, and smallpox virus). Among the largest and most complex virions are those that attack bacteria (i.e., the **bacteriophages**). Most extensively studied among these are the T2, T4, and T6 (i.e., the "T-even") bacte-

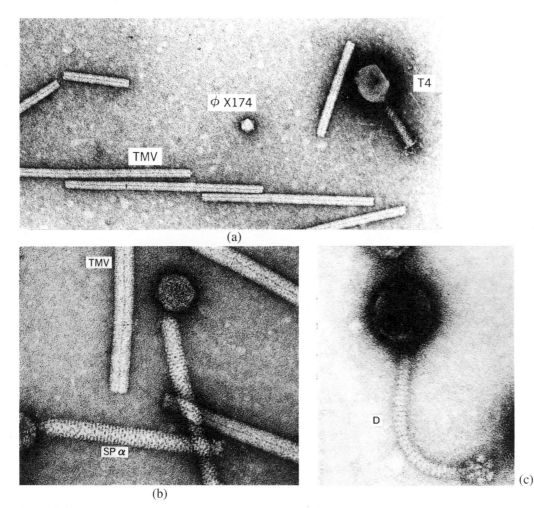

Figure 1–44
Viruses. (a) Negatively stained preparation of a mixture of tobacco mosaic virus (TMV) and the T4 and φX174 bacteriophages. **(b)** The SPα bacterio-phage of *Bacillus subtilis.* **(c)** Type *D* bacteriophage of *B. thuringiensis;* note the hexagonal array of the base plate (compare with Fig. 1–45). Magnifica-tions: **(a)** 98,000×; **(b)** 190,000×; **(c)** 297,000×. (*a* and *b*, courtesy of Dr. F. A. Eiserling; *c*, courtesy of Dr. H. W. Ackermann; copyright © 1978 *Canadian Jour. Microbiol.* **24,** 821.)

riophages (Fig. 1–44). These bacteriophages have a tail-like structure emerging from the capsid (Fig. 1–45). The tail is enclosed in a sheath of proteins arranged in a helical pattern, while the head of the virus is polyhedral. The end of the tail frequently reveals specialized structures (Fig. 1–45) involved in attachment to the surface of the host cell (see below).

Proliferative Cycle of a Virus

In the free or isolated state, viruses exhibit no metabolism and are incapable of proliferation. Proliferation of viruses requires a *host* cell and in its simplest and most direct form takes the following pattern. One or more viruses attach to specific sites on the surface of the host cell. Following

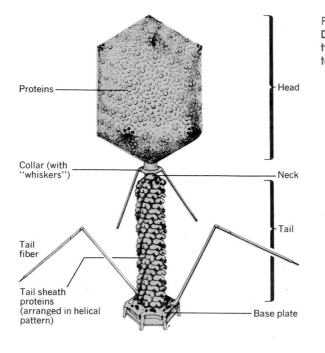

Proteins

Head

Collar (with "whiskers")

Neck

Tail

Tail fiber

Tail sheath proteins (arranged in helical pattern)

Base plate

Figure 1–45
Diagram of the T4 bacteriophage (Compare with Fig. 1–44). Only two of the six tail fibers and collar whiskers are depicted. (Courtesy of Dr. F. A. Eiserling.)

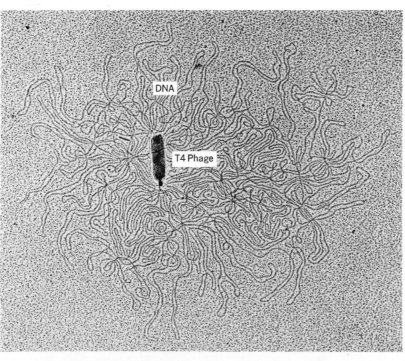

DNA

T4 Phage

Figure 1–46
DNA molecule released from a "giant" T4 bacteriophage particle. The single DNA molecule is more than 150 μm long. Magnification, 20,000×. (Courtesy of Dr. F. A. Eiserling.)

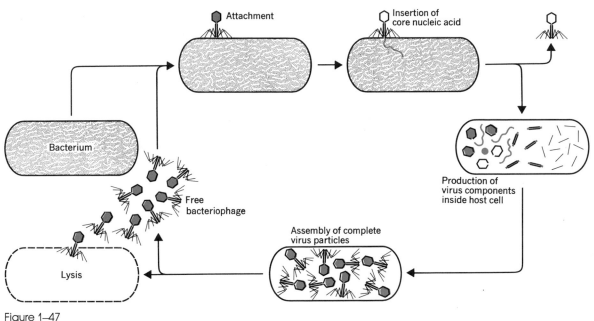

Figure 1–47
Proliferative cycle of the bacteriophage.

attachment or *adsorption* the viral nucleic acid is inserted through the plasma membrane into the host's cytoplasm. Release of the core nucleic acid from a virus can be achieved experimentally; Figure 1–46 dramatically reveals the uncoiled DNA molecule released by a T4 bacteriophage. Once inside the host cell, the viral nucleic acid redirects the metabolism of the host so that new viral proteins and new viral nucleic acids are formed. These viral components combine in the host to form large numbers of new virions that egress from the cell by disruption of its plasma membrane (i.e., cell lysis). Some viruses exit the host by budding off, enclosed in a small piece of the host cell's membrane; this does not lyse the host cell. The cycle of infection then repeats itself. The proliferative cycle of a virus is best understood for the bacteriophages and is depicted diagrammatically in Figure 1–47. The electron photomicrographs of Figure 1–48 show stages of the process.

On some occasions and only for certain viruses, the injected nucleic acid does not cause proliferation and release of new virions. Instead, the injected nucleic acid is incorporated into the host's genetic material, and the host cell continues to function in its normal manner. However, duplication of the host's genetic material prior to cell di-

vision is accompanied by duplication of the incorporated viral nucleic acid. Several generations of cells may be produced, each containing a copy of the viral nucleic acid. Viruses exhibiting this phenomenon are called *temperate* viruses, because they do not cause the death of the *immediate* host. Viruses that engage only in the cycle described earlier and that kill the host cell are called *virulent* viruses. The dormant viral nucleic acid within the host is referred to as a **provirus,** and the infected cell is said to be *lysogenic,* because sooner or later, in one of the generations of host cells, the provirus nucleic acid *will* begin to direct the replication of new virions, and this in turn will lead to cell lysis and release of new infective virus particles. Virus replication is considered further in Chapters 20 and 21.

Classification of Viruses and the Nature of Viral Nucleic Acids

The classification of viruses poses certain problems, and several different approaches have been used. One method is to classify the virus according to the type of host cell. Hence, there are *animal viruses, plant viruses,* and *bacterial viruses* (i.e., bacteriophages). This method is not al-

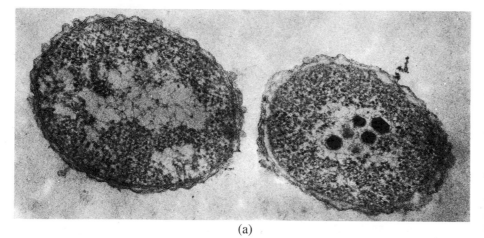

(a)

Figure 1—48
Stages in the proliferative cycle of a virus revealed by electron microscopy. (*a*) Two *E. coli* cells infected by T4 phage. In one cell, the thin section reveals viral DNA condensing within the head membranes of several T4 particles. (*b*) In this negatively stained preparation, more than 100 T4 particles are being released from the lysing *E. coli* cell. Along the upper right edge of the cell, several empty (contracted) phages, which began the infection by injecting their DNA into the host cell, can be seen. Magnifications: (*a*) 68,000×; (*b*) 26,000×. (*a,* courtesy of Dr. F. A. Eiserling; *b,* courtesy of Drs. L. Canedo and F. A. Eiserling.)

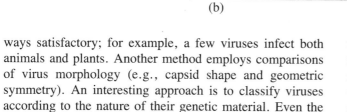

(b)

ways satisfactory; for example, a few viruses infect both animals and plants. Another method employs comparisons of virus morphology (e.g., capsid shape and geometric symmetry). An interesting approach is to classify viruses according to the nature of their genetic material. Even the beginning student of biology becomes quickly aware of the functional relationships among DNA, RNA, and proteins in cells. Genetic information stored in molecules of DNA is transcribed or copied into a corresponding RNA molecule (called a messenger), and the messenger then directs the synthesis of a specific protein. The protein then acts as an enzyme or structural component of the cell. *In some vi-*

ruses, however, the genetic information comprising the virion core is RNA and not DNA. Among the RNA viruses are TMV and the virions causing polio, mumps, measles, influenza, and colds. While RNAs of cells are single-stranded molecules (Chapter 7), viral RNAs can be single-stranded (e.g., TMV RNA) or double-stranded (e.g., the reoviruses). Like eucaryotic and procaryotic cells, the genetic information of many viruses is encoded in a DNA core. The DNA viruses include the T-even bacteriophages and those that cause chicken pox, herpes simplex, infectious mononucleosis, and shingles. However, the DNA may be *double-stranded linear* molecules (e.g., T5 and T7 bacteriophages), *double-stranded circular* molecules (polyoma and SV 40 viruses), *single-stranded linear* molecules (the parvoviruses), or *single-stranded circular* molecules (φX174 bacteriophage). The chemistry of these types of nucleic acids is considered in Chapter 7. Whatever form the nucleic acid takes, it includes the genetic information for the synthesis of the variety of proteins that either become components of new viruses or are involved in the redirection of the host cell's metabolism.

It was noted at the beginning of this discussion that while viruses themselves are not cells, the study of viruses has yielded a wealth of information about cells. Research with certain viruses has provided crucial and at times astounding information about the chemistry of and interactions among nucleic acids and proteins—subjects that are dealt with at length in Chapters 7, 20, 21, and 22—and these viruses should be specifically noted.

TMV.   Studies involving the tobacco mosaic virus (Fig. 1–44) began nearly a century ago and represent the starting point in the field of *virology*. In the 1950s, TMV was at the focal point of research that verified that nucleic acids and not proteins compose the genetic apparatus and that genetic information can be encoded in RNA as well as in DNA.

φX174.   Studies with the φX174 bacteriophage that infects *E. coli* revealed that the information for viral proliferation can be encoded in a *single* strand of DNA and does not require a double strand (i.e., double helix). Moreover, the structure of φX174 DNA has now been completely analyzed; that is, the entire base sequence is known. A most astounding finding yielded by these studies is that the coding sequences for several of the virus's proteins are included within sequences for other proteins. That is, certain genes *overlap.* Similar findings are being reported for other viruses, including simian virus 40 (SV 40), that possess double-stranded DNA and have both temperate and virulent phases.

Retroviruses.   A number of RNA and DNA viruses produce tumors in animals. Among these are the **retroviruses** which have RNA as their core nucleic acid (e.g., Rous sarcoma virus, avian leukemia virus). Unlike other RNA viruses, replication within the host cells requires that the inserted RNA be used for the preliminary synthesis of DNA, following which transcription and translation take the conventional pattern. The retroviruses have demonstrated that transcription can take place in the *reverse direction,* that is, from RNA to DNA.

Viroids

The smallest known agents of disease are the **viroids.** Viroids are much smaller than viruses and are also considerably simpler, for they consist of no more than a single strand of RNA. The RNA is not enclosed in any structure and except during infection is not associated with any other chemical substances. The typical viroid is an RNA molecule about 50 nm in length. At the present time, viroids are only suspected of being the agents of certain diseases in animals but are known to produce a number of plant diseases.

Summary

In the 300 years that followed the introduction of microscopy to biological science, the concept evolved that the cell is the fundamental unit of structure and function in all living things. This notion is referred to as the **cell doctrine.** Microscopy remains one of the cell biologist's most important and powerful research tools as new variations of light and electron microscopy have appeared, notably **freeze-fracture transmission electron microscopy, scanning electron microscopy,** and **stereoscopic electron microscopy.** With these tools, the detailed structure and organization of nearly all subcellular organelles have been revealed.

Free-living cells and the cells of multicellular organisms are divided into two classes: **eucaryotes** (nearly all animal and plant cells) and **procaryotes** (bacteria, cyanobacteria, mycoplasmas, etc.). Eucaryotic cells are characterized by a number of discrete organelles—especially the **nucleus, mitochondria, Golgi bodies, lysosomes, peroxisomes, rough** and **smooth endoplasmic reticulum,** and in plant cells, **chloroplasts** and **cell walls.** Like eucaryotic cells, many procaryotes possess a limiting (**plasma**) **membrane** and **ribosomes,** but lack the true nucleus and other discrete organelles characteristic of eucaryotes.

Although **viruses** are not cells, they are intimately associated with cells, and their study has made invaluable contributions to our understanding of cell function. In the isolated state, viruses are incapable of metabolizing or proliferating. Proliferation requires preliminary infection of a host cell, which takes the form of insertion of the viral nucleic acid (DNA or RNA). Following this, the metabolism of the host is redirected to make the components of the virus. Assembly of new virus particles within the host is ultimately followed by their egress to begin a new cycle. The **viroids**—naked pieces of RNA—are the smallest known independent agents of disease.

Most of the remaining chapters in the book are devoted to a detailed description of the biochemistry, structure, and physiological functions of cells and their organelles.

References and Suggested Reading

Articles and Reviews

Albersheim, P., The wall of growing plant cells. *Sci. Am. 232*(4), 80 (April 1975).

Allison, A., Lysosomes and disease. *Sci Am. 217*(5), 62 (Nov. 1967).

Brachet, J., The living cell. *Sci. Am. 205*(3), 50 (Sept. 1961).

Butler, P. J. G., and Klug, A., The assembly of a virus. *Sci Am 239*(5), 62 (Nov. 1978).

Campbell, A. M., How viruses insert their DNA into the DNA of the host cell. *Sci. Am. 235*(6), 102 (Dec. 1976).

da Silva, P. P., and Kachar, B., On tight junction structure. *Cell 28,* 441 (1982).

deDuve, C., The lysosome. *Sci. Am. 208*(5), 64 (May 1973).

Diener, T. O., Viroids. *Sci. Am. 244*(1), 66 (Jan. 1981).

Fiddes, J. C., The nucleotide sequence of viral DNA. *Sci. Am. 237*(6), 54 (Dec. 1977).

Fox, C. F., The structure of cell membranes. *Sci. Am. 226*(2), 30 (Feb. 1972).

Morowitz, H. J., and Tourtellotte, M. E., The smallest living cells. *Sci. Am. 206*(3), 30 (Mar. 1962).

Neutra, M., and Leblond, C. P., The Golgi apparatus. *Sci. Am. 202*(2), 100 (Feb. 1969).

Nomura, M., Ribosomes. *Sci. Am. 221*(4), 28 (Oct. 1969).

Orci, L., Matter, A., and Rouiller, C., A comparative study of freeze-etch replicas and thin sections of rat liver. *J. Ultr. Research 35,* 1 (1971).

Osborn, M., and Weber, K., The display of microtubules in transformed cells. *Cell 12,* 561 (1977).

Pangborn, J., Kuhn, D. A., and Woods, J. R., Dorsal–ventral differentiation in *Simonsiella* and other aspects of its morphology and ultrastructure. *Arch. Microbiol. 113,* 197 (1977).

Porter, K. R., and Tucker, J. B., The ground substance of the living cell. *Sci. Am. 244*(3), 56 (Mar. 1981).

Racker, E., The membrane of the mitochondrion. *Sci. Am. 218*(2), 32 (Feb. 1968).

Ravazzola, M., and Orci, L., Intercellular junctions in the rat parathyroid gland: A freeze-fracture study. *Rev. Cellulaire 28,* 137 (1977).

Satir, P., Cilia. *Sci. Am. 204*(2), 108 (Feb., 1961).

Satir, B., The final steps in secretion. *Sci. Am. 233*(4), 28 (Oct, 1975).

Schopf, J. W., The evolution of the earliest cells. *Sci. Am. 239*(3), 110 (Sept. 1978).

Sharon, N., The bacterial cell wall. *Sci. Am. 220*(5), 92 (May 1969).

Simons, K., Garoff, H., and Helenius, A., How an animal virus gets into and out of its host. *Sci. Am. 246*(2), 58 (Feb. 1982).

Staehelin, L. A., and Hull, B. E., Junctions between living cells. *Sci. Am. 238*(5), 140 (May 1978).

Stolinski, C., Freeze-fracture replication in biological research: development, current practice and future prospects. *Micron 8,* 87 (1977).

Tanaka, K., Scanning electron microscopy of intracellular organelles. *Intl. Revs. Cytol. 68,* 97 (1980).

Taylor, D. L., and Wang, Y., Fluorescent labelled molecules as probes of the structure and function of living cells. *Nature 284,* 405 (1979).

Woese, C. R., Archaeobacteria. *Sci. Am. 244*(6), 98 (June 1981).

Books, Monographs, and Symposia

Bodenheimer, F. S., *The History of Biology,* Wm. Dawson and Sons, London, 1958.

Echlin, P., The blue-green algae, in *Cellular and Organismal Biology* (D. Kennedy, Ed.), W. H. Freeman, San Francisco, 1974.

Fawcett, D. W., *Anatomy of Fine Structure,* Saunders, Philadelphia, 1966.

Florkin M., and Stotz, E. H. (Eds), *Comprehensive Biochemistry,* Vol. 3, *A History of Biochemistry,* Part IV, *Early Studies on Biosynthesis,* Elsevier, Amsterdam, 1977.

Gardner, E. L., *History of Biology,* Burgess, Minneapolis, 1965.

Haggis, G. H., *The Electron Microscope in Molecular Biology,* Wiley, New York, 1968.

Hayat, M. A., *Principles and Techniques of Electron Microscopy,* Vol. I., Van Nostrand Reinhold, New York, 1970.

Jensen, W. A., and Park, R. B., *Cell Ultrastructure,* Wadsworth, Belmont, Calif., 1967.

Luria, S. E., Darnell, J. E., Baltimore, D., and Campbell, A., *General Virology* (3rd ed.), Wiley, New York, 1978.

Murray, R. G. E., The organelles of bacteria, in *The General Physiology of Cell Specialization* (D. Mazia and A. Tyler, Eds.), McGraw-Hill, New York, 1963.

Nordenskiold, E., *The History of Biology,* Tudor, New York, 1928.

Orci, L., and Perrelet, A., *Freeze-Etch Histology,* Springer-Verlag, New York, 1977.

Singer, C. J., *A History of Biology,* Abelard-Schuman, London, 1959.

Swift, J. A., *Electron Microscopes,* Kegan-Paul, London, 1970.

Chapter 2
CELL GROWTH AND PROLIFERATION

It has been estimated that the various tissues and organs of the adult human contain more than 100,000,000,000,000 cells. All of these cells are derived initially from a single cell through growth and division. In many tissues (e.g., epithelium, blood-forming tissues, liver tissue), the cells continue to grow and divide for most of the person's life; however, in some tissues, such as muscle and nerve, cell division ceases some time after birth and subsequent tissue growth results from individual cell growth without division. The ongoing cell growth and division that characterizes most of the body's tissues accounts for the growth of the organism as a whole and the replacement of dying cells. With the exception of muscle and nerve, replacement of old tissue cells with new ones results in a complete turnover of the body's cellular composition every few years. A portion of the body's total mass is composed of noncellular materials that are secreted by cells; for example, most of the mass of bone and cartilage is represented by secreted calcium salts and proteins.

Other animals and plants also grow by the processes of cellular division followed by individual cell growth. When the term *growth* is applied to microbial organisms such as bacteria, it can refer either to an increase in the size of the individual organism (i.e., cell) or to an increase in the numbers of cells present (as for example in a culture).

If microorganisms such as bacteria or protozoa are placed in an appropriate nutrient medium, they may grow and divide until the medium is teeming with these cells. Similarly, cells isolated from certain tissues of higher animals and plants can be made to grow and divide in an artificial nutrient medium forming what is called a **tissue culture.** Much of our present understanding of the kinetics and mechanics of cell growth and division has been obtained from studies using such cultures of microorganisms and tissue cells, and cell-culturing techniques have become increasingly important research tools of the cell biologist.

Mathematical models can be derived that describe the rate at which a population of cells grows—that is, increases in cell numbers. Consider as an example the following hypothetical situation: Suppose that we begin with a culture medium containing a single cell that grows for some time period and then divides to yield two daughter cells; in turn, these two cells grow and after an identical period of time, they divide to yield four cells, and so on. In such a situation, the numbers of cells present in the population would increase exponentially in the following manner: 1, 2, 4, 8, 16, 32, 64 . . . (Fig. 2–1a). In other words, the population of cells would double in number with each generation. Consequently, after any specific length of time (say, from time t_1 to t_x) the number of cells in the population would be given by the equation

$$N_x = (N_1) (2^g) \qquad (2-1)$$

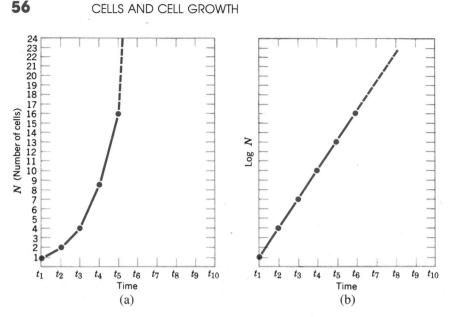

Figure 2–1
(a) A geometric expansion of cell numbers beginning with one, plotted in linear terms as a function of time. (b) The same geometric expansion plotted in semilog terms. Note that the latter plot yields a straight line.

where N_1 is the original number of cells present (at t_1), N_x is the number of cells present at t_x, and g is the number of generations that have occurred during the time interval $t_x - t_1$. Figures 2–1a and 2–1b depict graphically the exponential nature of the increase in population size during unrestrained growth.

The situation described above is ideal and does not occur in nature (except perhaps for some limited amount of time following the fertilization of an egg cell), but it can be approached artificially under conditions of **synchronous growth;** this will be considered later in the chapter. If a typical expanding population of cells is examined at any instant in time, some cells would be observed to be dividing, others would just have completed division, still others would be preparing to divide, and so on. Divisions of all cells present in the population would not occur during precisely the same time interval.

The Population Growth Cycle

Exponential Growth

If a large number of cells is cultured together in what is called a "batch" or "random" culture, the individual cells will be found in a variety of stages of their **growth-division cycle** or **cell cycle.** At any moment, the rate at which the number of cells in the culture increases is proportional to the number of cells present at that time. This, of course, presumes a steady state in which the needed cell nutrients are always available in adequate supply and in which cellular waste products excreted into the cells' environment do not interfere with the maintenance of normal growth and division. The growth of such a random culture is described by the following differential equation:

$$dN/dt = kN \qquad (2–2)$$

where N is the number of cells present at time t, dN/dt is the change in cell number with time, and k is a growth constant that is specific for the population. This equation may be solved by integration and yields the following expression

$$2.3 \log_{10}(N_2/N_1) = k(t_2 - t_1) \qquad (2–3)$$

If we let N_1 equal the number of cells present in the population at time t_1 and N_2 equal the number of cells present at time t_2, then equation 2–2 is solved by integration as follows.

$$\int_{N=N_1}^{N=N_2} dN/N = k \int_{t=t_1}^{t=t_2} dt$$

Thus, $\ln N_2 - \ln N_1 = k (t_2 - t_1)$

or

$$\ln (N_2/N_1) = k (t_2 - t_1)$$

By converting to the more familiar base 10, the last equation takes the form of equation 2–3.

Equation 2–3 indicates that the growth of the population (that is, the rate at which the numbers of cells in the population increases) is *exponential*.

Generation Time

Although the number of cells in a population increases exponentially with time, different types of cell populations (i.e., different species of microorganisms or cells from different tissues) grow at different rates. Even populations of the same types of cells may grow at different exponential rates if the temperature, nutrients, or other growth conditions vary. Differences in growth rates are reflected in the value of the constant, k, in equation 2–3. A convenient value that describes the specific rate of growth of a population of cells under a specified set of conditions is the **generation time.** The generation time is defined as the time required for the number of cells in the population to exactly double during exponential growth. Both the growth constant and the generation time are specific to a particular cell culture. Although the growth rate (i.e., kN) increases as the population of cells becomes larger, the generation time (and, of course, the value of k) remains constant. An equation for the generation time may be derived as follows.

After a time interval equal to the generation time has elapsed, the ratio N_2/N_1 is equal to 2; therefore, from equation 2–3,

$$2.3 \log 2 = kT \qquad (2\text{–}4)$$

where T is the generation time, $t_2 - t_1$. Hence,

$$0.693 = kT \qquad (2\text{–}5)$$

and

$$T = 0.693/k \qquad (2\text{–}6)$$

The actual value of k or T may easily be determined when experimental data are used to make a semilogarithmic plot of cell number versus time (Fig. 2–2).

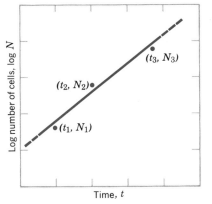

Figure 2–2
A hypothetical exponential increase in cell numbers. The generation time may be determined by the substitution of values for *t* and *N* into equation 2–3.

Sample Problem. Suppose that at time t_1, the number of cells in a population (i.e., N_1) is 62,400 and at time t_2, 18.5 hours later, there are 473,000 cells. What is the generation time for this population of cells?
From equation 2–3,

$$2.3 \log (473,000/62,400) = k (18.5)$$
$$2.3 \log 7.58 = 18.5 \, k$$
$$k = 0.1094$$

The dimensions of k in this instance are hr^{-1}; that is, the number of cells in the population increases by 10.94% per hour. Now from equation 2–6,

$$T = 0.693/0.1094 \; hr^{-1} = 6.33 \; hr$$

Therefore, during exponential growth, the number of cells in the population doubles every 6.33 hours.

The Lag Phase of Growth

Typically, when cells are placed in a nutrient medium that favors their growth and proliferation, exponential growth of the population does not begin immediately. Instead, there is a short interval of time in which there is little or no increase in the number of cells in the population. This time interval preceding exponential growth is known as the **lag phase.** The length of the lag phase is quite variable, even when comparing different cultures of the same type of cell.

A number of factors are believed to influence the length of the lag phase of the growth cycle.

Experiments with bacteria and other microorganisms have shown that variations in the concentrations of certain constituents of the growth medium, such as carbon dioxide and certain cations, such as H^+ (i.e., pH) markedly influence the length of the lag phase. Therefore, the chemical composition of the nutrient medium influences the time interval that precedes the onset of exponential population growth.

The cells that are used to initiate the growth of a new culture are obtained from a previous culture at some particular stage of the growth cycle. That portion of a culture used to start a new culture is called the **inoculum.** The stage of the parent culture used to provide the inoculum influences the length of the ensuing lag phase. For example, the lag phase of cultures of the bacterium *Aerobacter aerogenes* is longer when the inoculum is drawn from a previous culture in early exponential growth and shorter when drawn from a culture in late exponential growth. For the protozoan *Paramecium caudatum,* little or no lag period is observed when the inoculum consists of cells from a culture that had been growing exponentially; however, when the inoculum consists of cells from the **stationary phase** (see below), a clearly defined lag period is observed.

Generally, the greater the number of cells in the inoculum, the shorter the lag period. It has been suggested that this is due to the more rapid accumulation in the nutrient medium of a secretory substance that is required during exponential growth and that reaches a critical concentration earlier when there is a large number of initial cells.

The Stationary Phase

If cells are growing in a medium the size and contents of which are initially fixed, then it is clear that growth cannot continue indefinitely. One or more nutrients in the medium will eventually become depleted, and potentially harmful metabolic waste products excreted into the medium by the cells will accumulate in high concentration. Consequently, the cell population eventually reaches some limiting size, and following this the number of cells in the population no longer increases and may even decrease. It should be noted that the attainment of a maximum population size does not necessarily mean that cells are no longer growing and dividing, but rather that any additional cells produced by division are balanced by the death and disruption of other cells. The interval of time in which the number of cells in the population ceases to increase and remains fairly constant is known as the *quiescent* or **stationary phase.** Cells in healthy cultures that enter a stationary phase are in a "maintenance-only" state.

The Declining (or "Death") Phase

Stationary cultures eventually enter a **declining** or **death phase** in which the number of cells lost by death and disruption is greater than that being produced by cell divisions. Although the onset of a declining phase will be triggered by the exhaustion of nutrients in the medium or the accumulation of cellular excretory products, declining phases are ultimately observed even in cultures in which the growth medium is replaced with fresh medium. Such observations have lead to the notion that "programmed" into each type of cell is the capacity for a *limited* and *specific number* of divisions. For example, cultures of fibroblasts from human embryos will undergo about 50 divisions, after which the cells slowly stop dividing and eventually die. If the fibroblasts are frozen and stored in liquid nitrogen for months or even years after having undergone 20 generations of growth and later these cells are thawed and cultured again, they will resume division and growth for another 30 generations. In other words, the sum of the population doublings before and after freezing is still 50. It is as though the cells had a built-in clock or calculator that kept track of the number of divisions. Similar, finite life spans have been observed in cultures of liver cells, nerve cells, skin cells, and smooth muscle cells.

It is widely believed that aging of the body's tissues and the death of the individual that eventually ensues may be related to the cellular senescence that characterizes the declining phases of cell cultures. The underlying causes of these events are a subject of considerable controversy and are presently under intense investigation. The various phases of the growth cycle of a typical cell culture are summarized in Figure 2–3.

Contact Inhibition

A number of cells, especially those from mammalian tissue, are more easily cultured if they are allowed to grow on some sort of surface (usually glass). The increase in cell

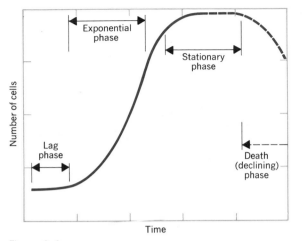

Figure 2–3
Various phases of the population growth curve of a typical random population of cells.

numbers generally halts when the population density is high enough for the cells to attain a state of **confluence,** that is, when they physically come in contact with one another along their edges. This phenomenon is called **contact inhibition** or **density-dependent regulation.** Since the cells will not grow on top of one another, they form a *monolayer.*

If, after contact inhibition has terminated the growth of the population, a strip of cells is removed from the monolayer, cells at the edges of the newly vacated space may divide until the exposed substratum is once again covered. Cells that are dislodged from the substratum generally fail to grow in suspension and soon die.

Growth Characteristics of Transformed Cells

Sometimes eucaryotic cells exhibit unusual and alternative growth characteristics both in culture and in vivo. Such cells are usually referred to as ''transformed'' cells. Although cellular transformation can take place spontaneously within a culture, one of the most common forms of transformation results from the infection of the cells with viruses through either deliberate or accidental contamination of the culture. For example, when cultures of human tissue cells are exposed to the SV 40 virus, the cells are transformed into a cell line that manifests a number of unusual characteristics including (1) the cells fail to exhibit contact inhibition, with the result that after the cells reach confluence they grow on top of one another, forming several layers, (2) the cells grow to an indefinite population size, (3) growth continues for many more generations than in the corresponding untransformed cells (i.e., the cells appear to ignore or to have lost the built-in counting mechanism keeping track of the number of cell doublings), and (4) the cells grow irregularly in size and in shape and often reveal an abnormal chromosome content. Because transformed cells may undergo an indefinite number of doublings, they are in effect immortal. One of the most famous (and useful) examples is the ''HeLa'' cell line that was originally derived from a cervical tumor in 1952 and has been continuously subcultured since that time. (HeLa is an acronym for Helen Lane, the woman from whose uterus the cervical cells were removed.) The relationship that exists between transformed cells and cancer should be apparent; indeed, when transformed cells are injected into laboratory animals, the cells form tumors, whereas normal cells do not. In tissue culture, transformed cells do not form conventional junctional complexes with neighboring cells. Hence, it is easy to understand why, when cellular transformation occurs in vivo, the transformed cells *metastasize,* that is, spread to other body tissues where they may form new tumors, growing and crowding out the tissue's normal cells.

The Quantitation of Cells

Fundamental to any quantitative study of the kinetics of cell population growth and to the accurate measurement of changes in the cell's physical or chemical parameters are methods for determining either the mass or the number of cells present in the population. A variety of techniques are available to do this; among those that are easily used are estimations of cell mass based on the total fresh (i.e., wet) weight or dry weight of the sample and the estimation of cell numbers from measurements of the relative turbidity of the cell suspension. However, these methods lack the level of accuracy generally considered necessary for quantitative studies. More frequently, the precise number of cells present in the sample is determined by the direct optical examination of an aliquot of cell suspension drawn from the culture, by electronic ''gating,'' or by ''flow cytometry.'' Although the last two procedures are far more accurate than weight determinations or turbidity methods, they require that the cells occur separately and, therefore, cannot be applied directly to tissues.

Optical Enumeration of Cells

The direct microscopic enumeration of cells requires a glass counting chamber (often referred to as a **hemacytometer** because of its historical and widespread use for counting various kinds of *blood cells*). This device consists of a glass microscope slide, the surface of which is etched with a number of squares of known dimensions. A drop of cell suspension is placed between the counting surface and an overlying cover slip; since the distance between the under-surface of the cover slip and the surface of the counting area is known, the total number of cells present in a known quantity of sample can be readily determined. By extrapolation, the number of cells in the entire culture (or in any portion of it) can easily be calculated. The enumeration of cells by direct optical examination is tedious but extremely useful, especially if the suspensions contain different types of cells (as in the case of blood) that may be distinguished visually. Hemacytometric methods are still used in many laboratories.

Electronic Enumeration of Cells

Electrical Gating. Rapid procedures for enumerating cells employ either *electrical gating* or *flow cytometry*. Among the most popular of the instruments that apply the gating approach is the ''Coulter Counter'' (after the inventor, Wallace Coulter), the main features of which are shown in Figure 2–4. A glass probe containing a small aperture near its base is submerged in a sample of cell suspension containing electrolyte (usually a physiological salt solution); the aperture tube is also filled with the same electrolyte. Current is caused to flow from a platinum electrode inside the tube through the small aperture to a second electrode immersed in the cell suspension. At the same time, the suspension is drawn through the aperture and into the probe by a small pump. The passage of a cell through the aperture temporarily interferes with the flow of current through the aperture, since the cell displaces a volume of the electrolyte and generally is not as conductive as the electrolyte. The resulting sudden drop in voltage is recorded by the instrument as a pulse (which is displayed on an associated oscilloscope) and is counted by a *scaler*. A mercury manometer connected to the probe (and therefore also to the pump) operates two switches through two electrodes that are inserted into the manometer along its length. With the pump

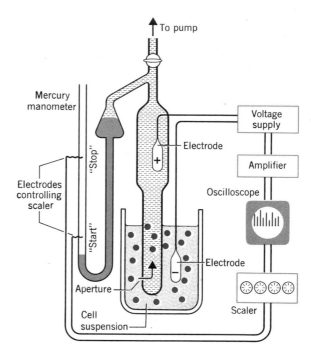

Figure 2–4

Basic components of an electronic cell-gating instrument such as the Coulter Counter (see text for details).

on and a vacuum applied to the aperture tube, not only is cell suspension drawn through the aperture but the mercury in the left arm of the manometer is also drawn downward below the first (i.e., the ''start'') electrode. When the vacuum is halted (by closing the stopcock) the mercury automatically rises in the left arm of the manometer, turning on the scaler as it passes the ''start'' electrode and turning off the scaler as it passes the ''stop'' electrode. With the stopcock closed, the movement of mercury upward in the left arm of the manometer draws cell suspension through the aperture, so that the cells present in a volume of suspension precisely equal to the volume of mercury between the ''start'' and ''stop'' electrodes are enumerated. Since this volume is known (i.e., the manometer is accurately calibrated), the number of cells present per unit volume of suspension is readily determined. With the electronic cell counter it is possible to count thousands of cells in just a few seconds. However, if the suspension contains different types of cells, those that are of similar size cannot be distinguished (and separately counted) by this procedure. In such a case, it may be possible to use chemical methods to eliminate certain types of cells. For example, during a clinical measurement of the white cell count in a blood sample, it is customary to eliminate the red blood cells by adding a few drops of a specific lysing agent to the cell suspension.

The magnitude of each pulse produced by the electronic counter is directly proportional to the size (i.e., volume) of the cell passing through the aperture. Two pulse-height threshold controls (upper and lower) may be separately adjusted on the instrument so that only those pulses whose peak height falls above the lower threshold and below the upper threshold will be counted. Consequently, in a cell suspension in which cell sizes vary considerably, only those cells within a selected size range will be counted. By simultaneously increasing both the lower and upper threshold values by the same increment, it is possible to obtain a pulse-height distribution for the sample of cells. This pulse-height distribution is analogous to a cell-size distribution and is especially useful to cell biologists who are studying the growth characteristics of individual cells, for it reveals for any selected phase of a population growth curve the relative numbers of cells of all sizes (and therefore ages) present in the population. Thus, with an electronic cell-gating instrument it is possible not only to rapidly enumerate the cells in the population but also to obtain measurements of the sizes of the cells. It would be difficult to overstate the value of this instrument to the cell biologist.

Flow Cytometry.   The use of flow cytometry for rapidly counting and analyzing cells was developed by M. J. Fulwyler and L. A. Herzenberg. Typically, the cells to be analyzed are first stained with a *fluorescent dye* (e.g., ethidium bromide, which combines with the nuclear DNA) or treated with a **fluorescent antibody** that binds to proteins in the cell surface. The cell suspension is then pumped through a small orifice and into the center of a narrow, horizontal channel in the ''flow cell assembly'' (Fig. 2–5). Pumped into the same channel through a second small opening at one end is ''sheath fluid,'' which serves to sweep the cell suspension away from the measuring orifice and out of the flow cell assembly through a third opening. A beam of light from a laser is directed at the measuring orifice where it excites the fluorescent material in each emerging cell causing the cell to emit a flash of fluorescent light. The fluorescence is detected by a system of photomultiplier tubes, which relay electrical signals to the instrument's analyzing system. The number of fluorescent light pulses detected by the photomultiplier tubes is directly related to the number of cells emerging from the measuring orifice. Moreover, the magnitude of each pulse varies in proportion to the quantity of fluorescent material associated with each cell, and this provides still another measurement parameter. For example, if the fluorescent pulses are produced by the cell's DNA, then not only will the cells be counted but a measurement is made of each cell's DNA content. Thus, one can readily follow the changes that take place in the amount of DNA per cell over the cell cycle. Using flow cytometry, thousands of cells can be enumerated and also analyzed in just a few minutes.

The Continuous Culture of Cells

When cells are cultured in a container of fixed volume, both the population density (i.e., the number of cells per unit volume of culture) and the population size (i.e., the total number of cells in the culture) increase at the same rate during the exponential phase of growth. If we consider a ''biological unit'' in such a culture as a cell plus some volume of the surrounding medium, then this unit undergoes continuous change as the population expands. The medium surrounding each cell is becoming successively depleted of nutrients, while metabolic waste products released by the cells accumulate in increasing amounts in the medium, and the cell itself may be increasing (or decreasing) in size. In multicellular organisms like man, the cells of the various tissues and organs are bathed in a body fluid that continuously provides fresh nutrients while carrying away waste products. In other words, in vivo a cell's surroundings stay more or less constant. This condition differs markedly from that of a batch culture but can be approximated experimentally by methods of **continuous culture** in which an effort is made to keep the biological unit constant.

A variety of techniques can be employed for the continuous culture of cells. One popular method involves the use of an instrument known as a **chemostat** (Fig. 2–6), introduced by A. Novick and L. Szilard. In the chemostat, fresh, sterile culture medium is continuously pumped into the culture chamber, while at the same time an equal volume of old medium and some of the cells are removed. Thus the total volume of the cell chamber does not change.

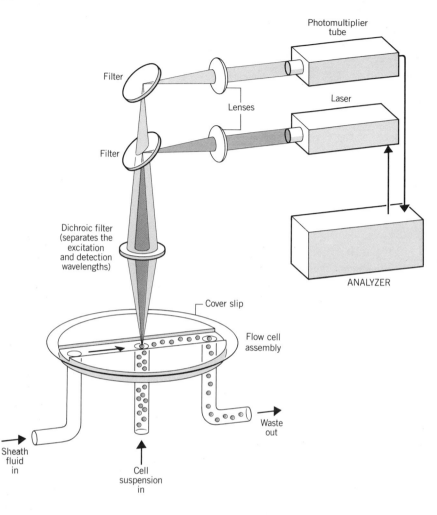

Figure 2–5
Basic components of a flow cytometer using fluorescence as a parameter of measurement (see text for details).

The rate of expansion of the cell population is regulated by the rate at which a chemical substrate essential for cell growth is added to the medium, and this is adjusted so that the population size and population density remain constant. Since cellular waste products are continuously diluted by the new medium added to the chamber, these substances do not attain toxic levels. With the chemostat it is possible to maintain the cell population in the exponential phase of growth for a greatly extended period of time. In certain continuous culture devices, constant cell population density is maintained by monitoring the turbidity of the suspension and appropriately adjusting the rates at which medium flows into and out of the culture chamber. These devices are known as **turbidostats.** Another method for keeping a volume of medium surrounding each cell chemically con-

stant involves the inoculation of small numbers of cells into very large volumes of growth medium. In this way, the growth and proliferation of thc cells will have little influence on the chemical composition of the medium for many generations. The continuous culture of cells offers numerous advantages over growth in fixed volumes, since a balanced, steady-state growth rate can be maintained and physiological studies can be carried out with cells that are little influenced by fluctuations in their environment.

The Growth of Individual Cells: The Cell Cycle

We have already seen that it is possible to determine the distribution of cell sizes (i.e., volumes) in a random pop-

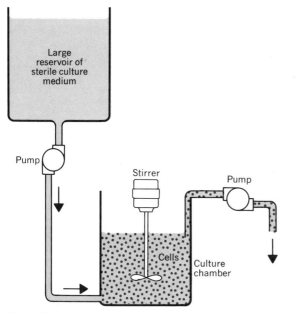

Figure 2–6
Simplified diagram of a chemostat. Sterile, cell-free culture medium enters the culture chamber at the same rate as medium containing cells is withdrawn. The population density of the culture remains constant and is determined by the rate at which a substrate or nutrient required for growth is added. Although the basic principles are represented in this diagram, most laboratory chemostats are considerably more complex.

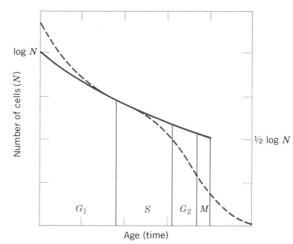

Figure 2–7
Theoretical distribution of cell ages in an ideal population of growing cells (solid curve). The areas under the curve represent the relative numbers of cells in each of the four phases of the cell cycle at any point in time. In a real population, some variation in cell cycle time is observed among the cells. The resulting deviation from the hypothetical age distribution is shown by the broken curve.

ulation of cells. Since cell size is related to cell age, an age distribution is also obtained. If cells were produced in a manner like the commercial production of small spherical objects (such as ball-bearings or marbles), then their sizes would be distributed normally about some mean size. However, in a random, exponentially expanding population of cells, the distribution of size or age is not normal. Instead, a distribution exists in which there are more young (small) cells than old (large) cells. In an *ideal* population (i.e., one in which a cell divides equally into two daughter cells, each of which grows to the size of the parent cell before dividing), the age or size distribution takes the form of the solid red curve seen in Figure 2–7. This curve indicates that there is an *exponential decrease* in the relative number of cells with age or size. This is so because the absolute number of cells entering division is constantly increasing in an exponential manner, so that each new generation of

cells is exponentially greater in number than the one that preceded it. It should be emphasized that the curve of Figure 2–7 is derived mathematically and applies the assumption that the cells have identical generation times.

The distribution described above probably does not exist in nature, since cells display variations in individual generation times and do not divide into progeny of exactly equal volume. Thus the distribution of cell size ranges beyond V and $2V$ (where $2V$ is the volume of the ideal parent cell, and V is the volume of each daughter cell). Furthermore, it has been shown that the small cells produced by unequal or early division do not, in turn, give rise to smaller progeny; instead, there is a random fluctuation about some inherited average size. A size distribution typical for a random population of cells is shown in Figure 2–8.

Although several possibilities exist for the rate of cell growth between divisions, most studies of cell volume distributions in either random or "synchronized" cultures (see below) indicate that cells grow either linearly or exponentially. For example, *Tetrahymena pyriformis* is believed to grow exponentially in volume between divisions, while *Chlorella ellipsoidea* appears to grow linearly. In

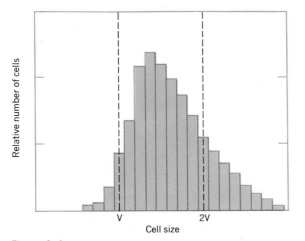

Figure 2–8
A typical distribution of cell sizes observed in a random culture of cells. The distribution is not "normal" but is skewed toward larger cell size. V and 2V are the initial and final sizes (i.e., volumes) of an ideal cell whose size doubles before equal division into two daughter cells occurs.

many cell cultures, the cell volume distribution remains constant throughout the exponential phase of population expansion, but in some cultures (such as cultures of the flagellate *Polytomella agilis* and the yeast *Schizosaccharomyces pombe*), the cell volume distribution shifts toward smaller size during the exponential phase. The latter observation indicates that in some cases the growth of individual cells occurs less rapidly than the growth (in numbers of cells) of the entire population.

The sizes attained by cells in culture also depend on other factors, such as the temperature at which the cells are maintained. In *T. pyriformis,* average cell size increases with increasing culture temperature between 28°C and 34°C. In *P. agilis,* cells cultured at lower temperatures have a greater average size than cells cultured at higher temperatures (in the range 9°C to 25°C). It is apparent that, depending upon the conditions of the culture, the events that result in cell division may occur either more or less rapidly than the events that result in individual cell growth.

Phases of the Cell Cycle

The various phases of the growth and reproduction of cells constitute what is called the **cell cycle** (Fig. 2–9), a complete cycle taking place in one generation time. The prin-

cipal signposts of the cell cycle are the replication of the nuclear DNA and its distribution among the progeny cells. The replication of the nuclear DNA occurs in that portion of the cell cycle known as the **interphase.** This was definitively shown for the first time in the classic experiments of A. Howard and S. Pelc in 1953. Using the technique of **autoradiography** (*see* Chapter 14), Howard and Pelc found that the incorporation of radioactive phosphrous into newly synthesized DNA occurred only during a short time interval within the interphase. The distribution of the replicated DNA and the physical division of the parent cell into two daughter cells occur in the "period of division" referred to as the **M phase.** The symbol *M* stands for "mitosis," the process of nuclear division familiar to even the beginning biology student and consisting of the **prophase, metaphase, anaphase,** and **telophase.** The entire cell cycle of most higher animal and plant cells is about 10 to 25 hours, of which only about one hour is spent in the *M* phase. (Procaryotic cells, which lack a nucleus and whose DNA is generally considerably smaller in content and is differently organized, may have a cell cycle of only 20 or 30 minutes.)

The replication of nuclear DNA occurs in a specific portion of the interphase called the **S phase** (*S* for "synthesis"). Some time elapses between the conclusion of the *M* phase of the parent cell and the onset of the *S* phase in the two daughter cells. This interval is known as the **G_1 phase** (**G** for "gap"). Completion of the *S* phase is usually not immediately followed by the *M* phase. The gap between the end of the *S* phase and the onset of the *M* phase is called the **G_2 phase.** The G_1 phase of the cell cycle is characterized by the synthesis and accumulation of RNA and protein in the cytoplasm and is accompanied by considerable cell growth. During the *S* phase, protein synthesis continues but predominantly in the cell nucleus, where additional **histones** are being made for close association with the newly synthesized DNA. By the end of the *S* phase, the cell nucleus contains two full complements of genetic material and enters the G_2 phase, in which some cytoplasmic growth resumes and the cell prepares for nuclear division.

The lengths of the G_1, *S*, G_2, and *M* phases vary among different types of cells, but variations between cells of the same type are usually very small. The greatest variation is seen in the G_1 phase. When the cell cycle is very long, most of the prolongation can be accounted for by a length-

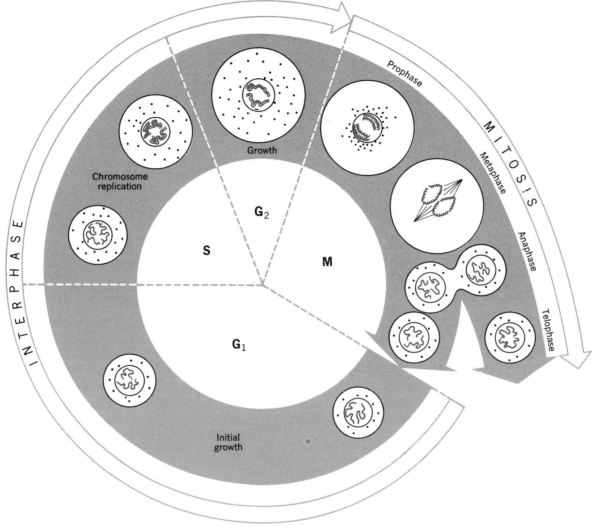

Figure 2–9
Various phases of the cell cycle, which generally lasts about 10 to 25 hours in higher animals and plants. In the diagram, the length of the *M* phase is deliberately exaggerated to show mitosis and cytokinesis.

ened G_1 phase. In contrast, in egg cells, where the cell cycle is very short, the G_1 phase is very brief or even nonexistent. The cells of animal embryos undergo a rapid sequence of divisions with very little intervening cell growth. The cell cycle may last only an hour with no G_1 phase and with DNA synthesis occurring during mitosis or soon after mitosis is completed. Thus the overall length of the cell cycle depends primarily on the length of the G_1

phase. The four areas under the solid curve of Figure 2–7 show for a random population the numbers of cells in each of the four phases of the cell cycle.

During the M phase, the nuclear contents undergo a series of changes and rearrangements. Prior to the onset of mitosis, the DNA is dispersed through the nucleus, but as the prophase of mitosis commences, the nuclear material condenses to form the clearly visible **chromosomes.** The dramatic nature of this condensation process can be appreciated when one considers that in a human cell, the DNA, which is 10–15 feet long in its dispersed state, condenses to form 46 duplicated chromosomes whose combined length is less than 1/25 of an inch. In the remaining phases of mitosis, the duplicated chromosomes are distributed to the daughter cells, leading to cell division or **cytokinesis** itself. These events are clearly depicted in the excellent interference optics photomicrographs of A. S. Bajer, reproduced in Figure 20–14. Mitosis is considered further in Chapter 20, which deals at length with the structure, organization, and functions of the cell nucleus. Additional discussion of this subject is therefore deferred until then.

In considering the various phases of the cell cycle, it is important to bear in mind that the stages are a part of a continuous process and that the cycle's subdivision into the G_1, S, and G_2 phases of interphase and the prophase, metaphase, anaphase, and telophase of mitosis is a matter of descriptive convenience only.

Synchronous Cell Cultures

The cells that are growing in a random culture represent a heterogeneous collection at various stages of the growth–division cycle or cell cycle. Some cells are dividing, some have just completed division, some are about to begin division, and so on. Any study of the progressive changes in either the chemical composition of the cells or cell morphology is difficult, if not impossible, when the cells are randomly distributed with respect to age. This problem could be avoided by studying the cycle of individual cells, but cells are usually too small for individual examination and analysis. To resolve the problem, some clever methods have been devised in which the cells in a culture are brought to the same stage of the cell cycle. Consequently, the entire population can be studied as though it were a single cell. This condition is known as **synchronous cell growth.**

The degree of synchrony that exists in a cell culture can be assessed by following changes in the culture's **mitotic index,** which is a measure of the fraction of cells undergoing division at any instant and is given by the formula

$$I_M = N_M/N \qquad (2\text{–}7)$$

where I_M is the mitotic index, N_M is the total number of

cells visibly in mitosis, and N is the total number of cells present.

During the exponential expansion of a cell culture, I_M remains constant; however, during synchronous growth, I_M changes from some minimal value (ideally 0) to some maximum value (ideally 1.0) during a short time interval. The efficiency of the synchronization procedure can be evaluated from a comparison of (1) the mitotic index, (2) the time required for the population to double during synchronous division, and (3) the generation time observed in a corresponding asynchronous (i.e., random) culture of the cells. A comparison of the growth curves for a batch culture and a synchronous culture is shown in Figure 2–10, and the corresponding mitotic indices are shown in Figure 2–11.

A variety of methods have been devised for inducing the synchronous growth of a cell culture; these fall into two major categories: *synchrony by induction* and *synchrony by selection*.

Synchrony by Induction

The most frequently employed methods for inducing synchrony involve *temperature cycles* (*temperature shocks*), *light cycles,* and *chemical manipulations*. When temperature is used to induce synchrony, the cells are subjected to alternating cold and warm periods. Little cell division occurs during the cold periods, but upon entry into the warm periods, cell division commences. For example, cultures of the flagellate protozoan *Polytomella agilis* can be synchronized by a repetitive temperature cycle of 22 hours at 9°C, followed by 2 hours at 25°C. Similar procedures have been successful with other protozoa, including *Astasia longa* and *Tetrahymena pyriformis*. Synchrony may also be induced by a rapid succession of very short cold and warm periods. Following this sequence, the cell population enters synchronous division.

Certain cultures of photosynthetic cells can be induced to divide synchronously by exposing them to alternating periods of dark and light, the population doubling with each light cycle.

If a specific substrate required for cell division is withheld from the culture medium for some time, the onset of division for all cells is delayed; the division of the population then follows the addition of the substrate to the culture medium. Agents that chemically inhibit cell division can be added to the culture. Cell growth continues but cell

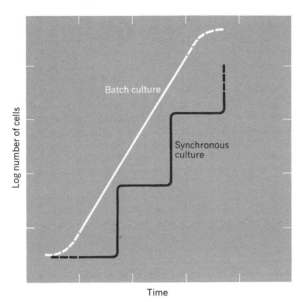

Figure 2–10
A comparison of the growth curves for idealized random (i.e., batch) and synchronous cell cultures.

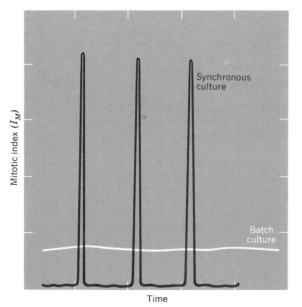

Figure 2–11
A comparison of mitotic indices during the growth of random (i.e., batch) and synchronous cell cultures.

division is arrested. An example of this is the action of **colchicine,** which arrests the cells at the metaphase of mitosis. If colchicine-treated cells are then transferred to a colchicine-free medium, inhibition is soon reversed and the cells all enter the G_1 phase. A general drawback to the use of colchicine and other drugs is the possibility of inducing abnormal behavior in the cells.

The "Double Thymidine Block." A popular procedure for inducing synchrony in cells is to cause the random population to accumulate at the beginning of the S phase by preventing the synthesis of DNA. One way to achieve this is through the addition of high concentrations of the nucleoside thymidine to the culture medium. The addition of the thymidine acts through a *feedback mechanism* within the cells (see Chapter 8) to prevent the production of other nucleosides that are needed for DNA synthesis. When the thymidine is first added to the culture, any cells that are in the S phase stop DNA synthesis immediately, thereby producing a culture in which some cells are halted at various stages of the S phase while the remaining cells are blocked at the entry into the S phase. After this, the block is lifted for an interval of time greater than the length of the S phase. This allows all cells to complete the S phase regardless of the point in the cell cycle where they were halted after the first application of the thymidine. Most of the cells in the population would now be at various stages of the G_2 phase. Thymidine is then applied again, and during this second exposure the cells continue through the cycle until they finish the G_1 phase. Therefore, as a result of the "double thymidine block" all of the cells in the culture are arrested at the start of the S phase.

These are but a few examples of the variety of chemical procedures in which cyclic manipulation of the culture medium's chemistry results in cell synchrony.

Synchrony by Selection

Selection techniques involve the mechanical isolation of cells of similar age from a random culture; these cells are then inoculated into fresh medium, where they grow and divide synchronously for some time. Among the methods frequently employed for mechanically isolating cells of similar age is **filtration.** In this procedure, a cell culture is passed through a filter that separates the larger cells from the smaller ones. The isolated smaller cells (which are also the younger cells of the population) are then inoculated into fresh medium to start a new culture.

Cells of similar age may also be isolated from a random culture by **sedimentation.** Since young cells are generally smaller than older cells, they sediment more slowly and may be isolated for subculture.

Another technique that is used in tissue culture is the "grow off" method in which the cells are adsorbed onto some surface (such as filter paper); during cell division one of the two daughter cells that is produced detaches from the surface and can be collected for subculturing.

Regardless of the procedure that is employed, the ultimate effect is to alter the random age distribution of the cells so that for several subsequent generations, all divisions occur over a short interval of time and all cells are at a similar stage of their cell cycle. Of the various procedures employed, synchrony by selection is to be preferred, since chemical or temperature variations are unnatural and may have undesirable effects upon the cell population. The synchrony achieved by any one of these methods is usually observed to decay after several generations of synchronous growth.

Culture Fractionation

Another way to study cells that are at the same stage of their cell cycle involves a technique called **culture fractionation.** This procedure avoids the potential problems of synchronization techniques where the inducing factor may itself alter or distort the normal events of the cycle. In culture fractionation, a random culture of cells (i.e., cells that are at various stages of the cell cycle) is *sorted* into subpopulations of varying cell size by sedimentation through a **liquid density gradient** (see Chapter 12) and is followed by the collection of the gradient and the entrained cells as a series of fractions. During sedimentation, the larger cells travel farther through the density gradient than the smaller cells. Cell sedimentation through the gradient may be achieved by simply using *gravity* (i.e., the cells fall through a stationary, cylindrical column of gradient) or may be accelerated using *centrifugal* techniques. The rationale

behind this approach is that a relationship exists between cell size and cell age (i.e., they are directly related to one another), so that the population is sorted into a linear sequence of fractions of increasing cell age. By studying and comparing chemical, physiological, or morphological properties of the cells in each of the separated fractions, specific effects occurring during the cell cycle can be assigned an age. Thus, changes taking place during cell growth may be followed even though a synchronous cell population is not being employed.

Culture fractionation requires the processing of very large quantities of material so that all of the collected fractions from the density gradient contain enough cells of the same age (i.e., size) for valid physical and/or chemical measurements. Sedimentation of large numbers of cells through a density gradient under the influence of the earth's gravity alone involves the use of large chambers called "Sta-Put" devices; centrifugal separations of large numbers of cells usually require "zonal rotors." Both of these laboratory techniques are discussed further in Chapter 12, along with other methods of particle separation and analysis.

Summary

All cells of multicellular animals and plants are derived initially from a single cell through growth and division. Mathematical models may be derived that describe the growth of cell populations, and proliferation may be studied experimentally using cell cultures. A number of alternative methods are used to culture cells and to measure population growth. The growth of a **random** or **batch** culture of cells characteristically includes **lag, exponential, stationary, and declining** phases. The extrapolation of observations made using an entire population of cells to the cycle of events occurring within an individual cell requires that the cell culture be **synchronous.** The cycle of individual cells (i.e., the *cell cycle*) takes a characteristic form and is divided into the G_1, S, G_2, and M phases. The lengths of these phases vary from one type of cell to another but are characterized by the same physiological events and culminate in the division of one cell into two.

References and Suggested Reading

Articles and Reviews

Edmunds, L. N., and Adams, K. J., Clocked cell cycle clocks. *Science 211,* 1002 (1981).

Hayflick, L., The cell biology of human aging. *Sci. Am. 242*(1), 58 (Jan. 1980).

Mattern, C. F. T., Bracket, F. S., and Olson, B. J., Determination of number and sizes of particles by electrical gating. *J. Appl. Physiol. 10,* 56 (1957).

Mazia, D., The cell cycle. *Sci. Am. 230*(1), 54 (Jan. 1974).

Pardee, A. B., Dubrow, R., Hamlin, J. L., and Kletzien, R. E., Animal cell cycle, in *Annual Reviews of Biochemistry,* Vol. 47 (E. E. Snell et al., Eds.), Annual Reviews Inc., Palo Alto, Calif., 1978.

Yanishevsky, R. M., and Stein, G. H., Regulation of the cell cycle in eukaryotic cells. *Int. Rev. Cytol. 69,* 223 (1981).

Books, Monographs, and Symposia

Adams, R. L. P., *Cell Culture for Biochemists,* Elsevier/North-Holland Biomedical Press, Amsterdam, 1980.

Dean, A. C. R., and Hinshelwood, C., *Growth, Function and Regulation in Bacterial Cells,* Oxford University Press, Oxford, 1966.

Mitchison, J. M., *The Biology of the Cell Cycle,* Cambridge University Press, New York, 1972.

Strehler, B. L., *Time, Cells and Aging* (2nd ed.). Academic Press, New York, 1977.

Zeuthen, E., *Synchrony in Cell Division and Growth,* Interscience, New York, 1964.

Part 2
MOLECULAR CONSTITUENTS OF CELLS

All cells are composed of water, proteins, polysaccharides, lipids, nucleic acids, salts, and minute quantities of a diversity of organic compounds such as vitamins and intermediates of cellular metabolism. The proteins, polysaccharides, lipids, and nucleic acids are relatively large molecules called **macromolecules** and are described separately in the following chapters. Water, ionized salts, and certain vitamins may complex with the macromolecules of the cell or they may remain free. The quantities of each of these classes of compounds vary widely from one type of cell to another and from one organism to another (Table 3–1). Water is the most common molecule, while polysaccharides and proteins are the most prevalent organic substances. In this chapter, some fundamentals of cellular chemistry and the properties of a number of common cellular constituents will be considered.

Standard Units of Measurement

One of the special concerns of any scientific endeavor is the effort to acquire accurate measurements in a form that can be understood and verified by others. To this end, cell biologists, like scientists in other disciplines, rely on the use of a number of *standard units of measure*. For many years this function has been fulfilled by the **metric system.** In the 1960s, in an effort to establish uniformity among all fields of science, an international conference on weights and measures recommended certain modifications of the metric system and gave it a new name. It is now called the *Systeme Internationale d'Unites* (i.e., *International System of Units*) and is abbreviated *S.I*. Table 3–2 lists the standard units of measure of the S.I. and presents their metric equivalents. For purposes of comparison, the corresponding equivalents of the *English* or *fps* (foot-pound-seconds) system are also given in the table.

Of special note for the cell biologist and biochemist is the recommended use of the *joule* (J), rather than the *calorie* (cal), as the standard unit of energy. In view of the fact that energy values expressed in calories are still to be found in so much of the biological literature, the caloric energy values along with their joule equivalents will often be cited.

Special Properties and Behavior of Water

Water is the essential solvent of life and is the main chemical constituent of all cells. The essential role played by water can be attributed to its unique physical and chemical properties, which collectively act to protect living systems and are necessary to the structure and function of cells. One of the unusual characteristics of water is that despite its low molecular weight it is a liquid at most environmental temperatures. Other molecules having similar molecular weights (e.g., ammonia and methane) are gases. In comparison with other solvents, water has high values for melting point, heat of fusion, boiling point, heat of vaporization, specific heat, and surface tension (see Table 3–3). Each of these properties serves to constrain major temperature fluctuations and to keep water in the liquid state. Water absorbs more heat energy per gram for every degree rise in temperature than other common solvents and therefore acts to moderate temperature changes. (Indeed, the calorie is specifically defined as the amount of energy required to raise the temperature of one gram of water from 14.5°C to 15.5°C.) Likewise, in order to convert one gram of water from the liquid to the vapor (gas) state at its boiling point, an additional 2259 J (540 cal) must be absorbed. This high heat of vaporization, together with the high surface tension of liquid water, tends to keep water in the liquid state. At the other temperature extreme, large amounts of energy (335 J or 80 cal per gram) must be lost

Table 3–1
**Chemical Composition of Some Representative Cells
(Percent of Total Wet Weight)**

Constituent	Cell or Tissue				
	Rat Liver	Rat Skeletal Muscle	Sea Urchin Egg	E. coli	Corn Seed
Water	69–72	76	77	73	13
Protein	16–22	21	16	19	9
Polysaccharide	2–10	0.6	1.4	1.1	73
Lipid	5	4–11	4.8	1.1	4
Nucleic acids	1–5	1–2	1	3.5	1
Salts	1	1	0.3	2.3	1

Table 3–2
**The Systeme Internationale (S.I., International System of Units) and
It's Metric and English Equivalents**

Measurement	S.I. Unit of Measure	Metric Equivalents	English System (fps) Equivalents
Distance	meter (m)	1 meter (m) 100 centimeters (cm)	3.28 feet (ft)
Mass	kilogram (kg)	1 kilogram (kg) 1000 grams (g)	2.205 pounds (1b)[a]
Time	second (sec)[b]	1 second (sec)	1 second (sec)
Force	newton (N) ($1 N = 1$ kg m/sec^2)	10^5 dynes (D) ($1 D = 1$ g cm/sec^2)	7.23 poundals ($= 7.23$ ft-lb/sec^2)
Energy	joule (J)	1 joule (J) 0.239 calories (cal) 10^7 ergs	23.73 ft-poundals
Volume	cubic meter (m^3)	1000 liters (l) 10^6 ml	35.3 cubic feet (cu ft)

[a] Strictly speaking, the pound is a unit of *weight*.
[b] Seconds may also be abbreviated using the symbol s.

in order for water to be converted from the liquid to the solid state.

Although water is often called the "universal solvent," not all substances dissolve in water. However, water does dissolve most salts and other ionic compounds, as well as nonionic polar compounds such as sugars, alcohols, and other molecules that contain hydroxyl, aldehyde, and ketone groups. Many substances that contain both polar and nonpolar groups (such as soaps, fatty acids, and glycerophosphatides) do not dissolve in water, but they do form **micelles.** A micellar arrangement is not a true solution but is a *suspension* or *dispersion*. The behavior of soap molecules in water is a good and also common example of micelle formation. Soap molecules, formed by the **saponification** of fatty acids (Chapter 6), consist of a long, nonpolar hydrocarbon chain terminating in a polar carboxyl group that is ionically bonded to a metal ion such as K^+ or Na^+. When dispersed in water, the soap molecules aggregate to form spherical clusters, called micelles, in which the polar carboxyl groups of the soap molecule are arranged at the surface of the sphere, where they form weak bonds with the surrounding water, while the non-polar hydrocarbon chains project inward (Fig. 3–1).

The special physical properties of water are founded in its molecular structure. In water, two hydrogen atoms are

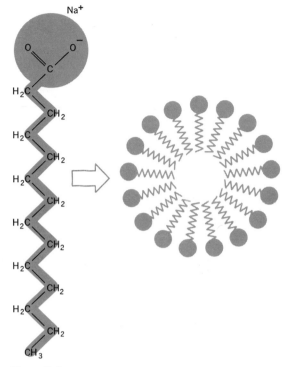

Figure 3–1
Soaps (fatty acids linked to metal ions) *coalesce* **in aqueous suspensions to form** *micelles.* **The polar ends of the soap molecules face the surrounding water, while the hydrocarbon chains are directed toward the center of the micelle.**

Table 3–3
Physical Properties of Several Common Solvents

Property	Water	Methanol	Acetone	Ammonium Hydroxide
Melting point (°C)	0	−99	−94	−78
Heat of fusion				
J/g	335	92	84	351
cal/g	80	22	20	84
Boiling point (°C)	100	64.5	56.5	−20
Heat of vaporization				
J/g	2259	1100	523	1330
cal/g	540	263	125	318
Specific heat				
J/g/°C	4.2	2.4	2.1	4.2[a]
cal/g/°C	1.0	0.566	0.506	0.999[a]
Surface tension				
(dynes/cm)	72.8	22.6	23.7	73.5
(N/m)	0.0728	0.0226	0.023	0.0735

[a] Value for a 50% solution in water.

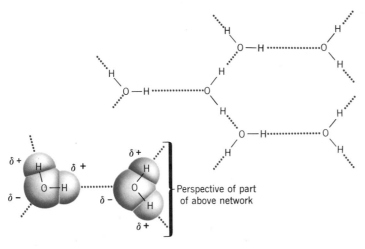

Figure 3–2
Water molecules behave like dipoles and form *hydrogen bonds* with each other. As a result, water molecules tend to arrange themselves into hexagonal lattices. Redrawn from Holum, *Fundamentals of General, Organic and Biological Chemistry*, pg. 123; copyright © 1978 by John Wiley & Sons, Inc.

each covalently bonded to a single oxygen atom. The oxygen atom, being more electrophilic than the hydrogens, attracts the shared electrons more strongly and produces an electrical asymmetry in the molecule. As a result, the oxygen atom takes on less than a full negative charge, technically called a *partial negative charge* and denoted by δ^-. The hydrogens become *partially positive* (Fig. 3–2). Because the bond angle between the hydrogen atoms is 104.5°, each water molecule acts as a *dipole* even though it has no net charge. Because of their dipole character, water molecules tend to form bonds with each other. The bonds linking neighboring water molecules are called **hydrogen bonds** and are formed by the attractions of the electronegative oxygen of one molecule for the electropositive hydrogens of two other molecules. Thus, each oxygen atom may bond to as many as four hydrogens, forming a *tetrahedron* about the oxygen (Fig. 3–3). The hydrogen bonding that occurs between water molecules in liquid water (Fig. 3–2 and 3–4) accounts in part for water's high heat of vaporization and surface tension.

When the temperature of water is lowered, the concomitant decrease in kinetic energy allows the water to become more dense until at 4°C, maximum density occurs. A continued decrease in kinetic energy allows more extensive and less transient formation of hydrogen bonds among the water molecules. This promotes the development of a lattice structure. As the freezing point of water is approached, more space develops between the molecules, and the density decreases; thus, freezing water and ice rise to the surface of the liquid.

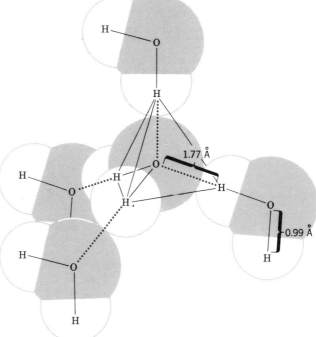

Figure 3–3
Tetrahedral arrangement of hydrogen atoms about oxygen in water. Two of the four hydrogen atoms are covalently bonded to the oxygen in water, while two additional hydrogens (from two other water molecules) form hydrogen bonds.

Ionic substances are readily dissolved in water because of the water molecules' dipolar character. When salts (or other polar compounds) are dissolved in water, the orientation of the water molecules with respect to each other is

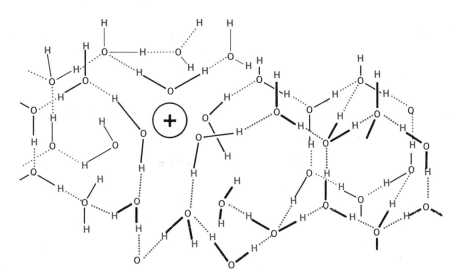

Figure 3–4
The lattice structure formed in water by hydrogen bonding is disturbed by the presence of ions, which act to attract some of the water molecules. In this illustration, the *cation* attracts the partially negative oxygen atoms of several water molecules.

disturbed. For example, sodium chloride (one of the most common salts found in cells and tissues) ionizes in water to form Na^+ and Cl^-; these ions attract the water molecules and disturb the water lattice. Figure 3–4 depicts the manner in which a positive ion disrupts the water lattice as some of the water molecules arrange themselves about the cation. Water molecules form *spheres of hydration* that *encage* either cations or anions (Fig. 3–5). Were it not for the formation of spheres of hydration about their ionic groups, many large molecules such as proteins would not be soluble in water. By encaging the ionized groups of proteins in layers of water, the groups are prevented from interacting with each other. As a general rule, ionized salts bind water more strongly than the polar groups of macromolecules, and this property is occasionally used to separate macromolecules from their solvent. For example, when salts are added to a protein solution, the salt ions form more extensive hydration spheres than the proteins. In effect, the redistribution of the water around the salt ions diminishes the amount of solvent available for the ionized groups of the proteins. The ionized groups of the proteins interact with each other, and a protein precipitate is formed. This method for separating proteins from their solvent is called "salting out" and is discussed more thoroughly in Chapter 13.

Water molecules also undergo ionization; the capacity of water to form ions is another one of its important biological properties. As we have already seen, in water each hydrogen atom is covalently bonded to one oxygen atom and forms a weak electrostatic bond with an oxygen atom of

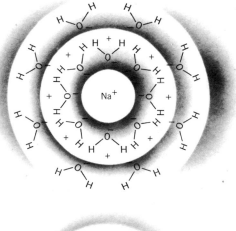

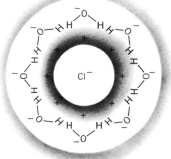

Figure 3–5
Hydration spheres **formed about cations (top) and anions (bottom) dissolved in water.**

another water molecule (i.e., the water molecules are hydrogen bonded to one another). With measurable frequency, the covalent bond linking a hydrogen to oxygen is broken and the hydrogen establishes a closer association with the oxygen to which it was electrostatically bonded.

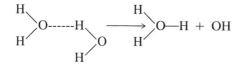

The water molecules have thus ionized, dissociating into a *hydroxide anion* (OH^-) and a *hydronium cation* (H_3O^+). At standard temperature and pressure, one liter of water contains 1.0×10^{-7} moles of OH^- and H_3O^+. Both the hydronium ions and the hydroxide ions form hydrogen bonds with other water molecules. The notation H_3O^+ is rarely used; instead H^+ (called a *hydrogen ion* or *proton*) is used more frequently, even though free protons do not generally exist per se but are bonded to water to form hydronium ions.

Salts, Ions and Gases

Salts are found in all cells and exist in an ionized form. The ions may be dissolved in the cell fluid or specifically bound to other molecules such as proteins or lipids. Salts have two broad roles in the cell. One role is osmotic, in that the total concentration of dissolved salts affects the *flux of water* across the cell's membranes. The second is the more specific role that certain ions play in contributing to the structure and function of cellular particles and macromolecules. The specific cellular roles of a number of ions are given in Table 3–4.

Gases that enter the cell from the environment or are produced by the metabolism of a cell dissolve in the cytoplasmic water. The most common gases in air are nitrogen (N_2), oxygen (O_2) and carbon dioxide (CO_2). At sea level, N_2 accounts for 78.03% of all gases, O_2 20.99%, and CO_2 0.03%. Both nitrogen and oxygen dissolve in water in their molecular form, but their solubility in water is low. For example, at 25°C and at 1 atmosphere pressure, only 2.83 ml of oxygen and 1.43 ml of nitrogen dissolve in 100 ml of distilled water. The behavior of carbon dioxide in water is quite different. Some of the dissolved carbon dioxide reacts with water, forming carbonic acid, which in turn dissociates into ions:

$$CO_2 + H_2O \rightleftharpoons \underset{\substack{\text{carbonic} \\ \text{acid}}}{H_2CO_3} \rightleftharpoons H^+ + \underset{\text{bicarbonate}}{HCO_3^-}$$

Most of the carbon dioxide in cells is in the bicarbonate and carbonate forms. CO_2 gas is more soluble in water than are O_2 and N_2. For example, 75.9 ml of CO_2 dissolves in 100 ml of water at 25°C and 1 atmosphere pressure.

Acids, Bases, pH, and Buffers

According to the Brønsted definitions, an **acid** is a *proton donor,* and a **base** is a *proton acceptor.* For example, when acetic acid is dissolved in water, a certain number of the acetic acid molecules dissociate as follows:

$$\underset{\text{(acetic acid)}}{CH_3COOH} + H_2O \rightleftharpoons \underset{\text{(acetate anion)}}{CH_3COO^-} + H_3O^+$$

In the reaction, a proton of acetic acid has been donated to water, producing an anion (acetate) and a cation (the hydronium ion). If the reaction is reversed, the acetate anion acts as a proton acceptor and therefore fits the criterion of a base. Acetic acid and acetate ions are referred to as a **conjugate acid–base pair.** Protons are attracted both to the conjugate base and to water. An acid that readily donates protons to water is called a *strong acid,* while an acid that does not readily donate protons to water is termed a *weak acid.* Each acid is characterized by its tendency to dissociate; the extent of dissociation may be determined from the acid's dissociation constant, K, defined by

$$K = \frac{[H^+] [A^-]}{[HA]} \qquad (3\text{--}1)$$

where for a given solution, $[HA]$ is the concentration of undissociated acid, $[A^-]$ is the concentration of conjugate base, and $[H^+]$ is the hydrogen ion concentration. The dissociation constant varies somewhat according to the temperature of the solution. Customarily in physiology and biochemistry, one refers to the *apparent dissociation constant, K',* of an acid. This value is based on the concentrations of reactants and products that can actually be measured following dissociation of the acid in water. Table 3–5 lists the apparent dissociation constants of some common acids.

The concentration of H^+ in cells varies widely. The

Table 3–4
Cellular Functions of Certain Ions

Element	Ionic form Present	Percent Total Weight	Function
Molybdenum	$MoO_4^=$	0.00001	Cofactor or activator of certain enzymes (e.g., nitrogen fixation, nucleic acid metabolism, aldehyde oxidation)
Cobalt	Co^{++}	0.00001	Constituent of vitamin B_{12}
Copper	Cu^+, Cu^{++}	0.0004	Constituent of plastocyanin and cofactor of respiratory enzymes
Iodine	I^-	0.0005	Constituent of thyroxin and other hormones
Boron	$BO_3^=$, $B_4O_7^=$	0.002	Activates arabinose isomerase
Zinc	Zn^{++}	0.003	Cofactor or activator of certain enzymes (e.g., carbonic anhydrase, carboxypeptidase)
Manganese	Mn^{++}	0.003	Cofactor or activator of certain enzymes (e.g., several kinases, isocitric decarboxylase)
Iron	Fe^{++}, Fe^{+++}	0.01	Constituent of hemoglobin and the cytochromes
Magnesium	Mg^{++}	0.1	Constituent of chlorophyll; activates ATPase
Sulfur	$SO_4^=$	0.1	Constituent of coenzyme A, biotin, thiamine, proteins
Phosphorus	$PO_4^=$, $H_2PO_4^-$	0.2	Constituent of lipids, proteins, nucleic acids, sugar phosphates
Calcium	Ca^{++}	0.5	Constituent of plant cell walls; matrix component of bone tissue; cofactor of coagulation enzymes
Potassium	K^+	1.0	Cofactor for pyruvate kinase and K^+-stimulated ATPases

range may exceed 0.1 to 10^{-10} M. Although these concentrations may not seem high, small changes in H^+ concentration have noticeable effects upon cell function. It is inconvenient to express the H^+ concentration in molar terms (such as 1×10^{-7} M), and the pH scale proposed by Sørensen is used instead. Accordingly,

$$pH = -\log_{10} [H^+] \qquad (3\text{–}2)$$

As noted earlier, at 25°C, the H^+ (and OH^-) concentration of pure water is 1×10^{-7} M and the solution is at neutrality. At this concentration,

$$pH = -\log_{10} [1.0 \times 10^{-7}] = 7.0 \qquad (3\text{–}3)$$

Table 3–6 shows the range of the pH scale and lists the pH values of a number of familiar liquids. In the laboratory, the pH of a solution is determined potentiometrically using an instrument called a *pH meter*. The instrument contains two electrodes that are immersed in the liquid sample; one of these is a *reference electrode* (whose electrical potential is constant), and the other is a *glass electrode* (whose electrical potential depends upon the pH of the sample). An *electrometer* measures the difference in electrical potential between the two electrodes, and this value is then converted

Table 3–5
Dissociation Constants for Some Common Acids

Acid	K'	pK'
Formic acid	1.78×10^{-4}	3.75
Acetic acid	1.74×10^{-5}	4.76
Propionic acid	1.35×10^{-5}	4.87
Lactic acid	1.38×10^{-4}	3.86
Succinic acid	6.16×10^{-5}	4.21
Phosphoric acid ($PO_4^=$)	7.25×10^{-3}	2.14
Monobasic phosphate ($H_2PO_4^-$)	6.31×10^{-8}	7.20
Dibasic phosphate ($HPO_4^=$)	3.98×10^{-13}	12.4
Carbonic acid	1.70×10^{-4}	3.77

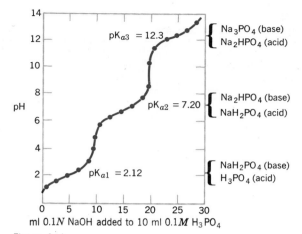

Figure 3–6
Titration curve of phosphoric acid (H_3PO_4) with sodium hydroxide (NaOH) showing the buffering plateaus.

to pH units and is displayed. Most pH meters are accurate to within 0.01 pH units.

A **buffer** solution is formed by mixing together a *weak acid* and its *conjugate base*. These solutions effectively slow the rate of pH change over a limited portion of the pH scale when a quantity of strong acid or base is added to the buffer. As is shown in Figure 3–6, titration of H_3PO_4 (phosphoric acid) with up to about 8 ml of NaOH causes a gradual increase in pH to about 2.5. In contrast, the next few ml of NaOH that are added bring about a rapid rise in pH to about 6.0. The gradual pH change observed upon

addition of the first 8 ml of base is the result of the conjugate acid–base buffering capacity of the H_3PO_4 and NaH_2PO_4 (i.e., the NaOH reacted with H_3PO_4, forming NaH_2PO_4 and H_2O); that is

$$Na^+ + OH^- + H^+ + H_2PO_4^- \rightarrow Na^+ + H_2PO_4^- + H_2O$$

A second buffering "plateau" is produced by the conjugate

Table 3–6
The pH Scale

H⁺ Concentration (M)	pH	Examples	Condition
1.0	0	0.1 M HCl	Acid
0.1	1	Human gastric fluid	
0.01	2	Orange juice	
0.001	3	Vinegar	
0.0001	4	Pineapple juice	
10^{-5}	5	Tomato juice	
10^{-6}	6	Cow's milk	
10^{-7}	7	Most body fluids; pure water	Neutral
10^{-8}	8	Sea water	Basic
10^{-9}	9	Soil water	
10^{-10}	10	Alkaline desert ponds	
10^{-11}	11		
10^{-12}	12	Limewater	
10^{-13}	13		
10^{-14}	14	1 M NaOH	

acid–base buffer system of NaH_2PO_4 and Na_2HPO_4. The third plateau is achieved through the actions of Na_2HPO_4 and Na_3PO_4.

Just as the hydrogen ion concentration can be expressed in terms of pH (or negative logarithms), the dissociation constant for a weak acid, K_a, can be similarly defined:

$$pK_a = -\log K_a \qquad (3\text{--}4)$$

The pK_a for each conjugate acid–base pair occurs at the *midpoint* of its buffering plateau. The most effective buffering range of an acid–base pair usually extends 0.5 pH units to each side of the pK_a value. The relationship between pH and pK_a is expressed in the Henderson–Hasselbach equation:

$$pH = pK_a + \log_{10}\frac{[A^-]}{[HA^-]} \qquad (3\text{--}5)$$

where the weak acid is represented by [HA] and its conjugated base by [A$^-$]. When the concentrations of the two are equal, as at the midpoint of a plateau, the last term of equation 3–5 becomes zero (i.e., $\log_{10} 1 = 0$), and therefore, $pH = pK_a$. The Henderson–Hasselbach equation can be used to calculate the pH if the concentrations of the acid and base are known. The equation is also useful for calculating the necessary quantities of acid and conjugate base that must be mixed together in order to prepare a buffer that has a particular pH. However, it should be pointed out that the equation is an approximation and is most accurate when almost equal amounts of the acid and base are mixed. The pK values of several buffers that are routinely used in cellular studies are listed in Table 3–7.

Chemical Bonds

A molecule is a stable union of two or more atoms. The formation of the bonds that exist between the atoms comprising a molecule requires less energy than that needed to keep the atoms apart. The breakage of the bonds that join atoms consumes energy. If a large amount of energy is required to break the bond, the bond is called a *strong bond*. If small amounts of energy are required, the bond is a *weak bond*. Four major types of chemical bonds may be identified: **covalent bonds, ionic bonds, hydrogen bonds,** and **hydrophobic bonds.** The energies of these bonds are given in Table 3–8. Covalent bonds are the bonds formed by the *sharing* of one or more electron pairs by the atoms comprising the molecule. These are generally the strongest

Table 3–7
Common Buffer Systems

Compound	pK_{a1}	pK_{a2}	pK_{a3}	pK_{a4}
Acetic acid	4.7			
Ammonium chloride	9.3			
Carbonic acid	6.4	10.3		
Citric acid	3.1	4.7	5.4	
Diethanolamine	8.9			
Ethanolamine	9.5			
Fumaric acid	3.0	4.5		
Glycine	2.3	9.6		
Glycylglycine	3.1	8.1		
Histidine	1.8	6.0	9.2	
Maleic acid	2.0	6.3		
Phosphoric acid	2.1	7.2	12.3	
Pyrophosphoric acid	0.9	2.0	6.7	9.4
Triethanolamine	7.8			
Tris-(hydroxymethyl) amino methane	8.0			
Veronal (sodium diethylbarbiturate)	8.0			
Versene (ethylenediaminotetraacetic acid)	2.0	2.7	6.2	10.3

bonds formed between atoms. While ionic bonds formed between oppositely charged atoms have a high energy level in crystals (such as in crystalline NaCl, KCl, etc; see Table 3–8), the ionic bonds that are formed in aqueous solution (the usual case for cells) have much lower energy levels and are considerably weaker than covalent bonds. As we have already seen, very weak bonds called hydrogen bonds occur when hydrogen is shared by two electronegative atoms. Hydrogen bonds most commonly occur between the oxygen atoms of water molecules, but they can also be formed between two nitrogen atoms and between nitrogen and oxygen (see also Chapters 4 and 7). Hydrophobic bonds are associations of molecules or parts of molecules that have nonpolar groups. They are not chemical bonds in the usual sense. The attractions of water molecules for one another is far greater than their attraction to hydrophobic groups and this causes the hydrophobic groups to aggregate and have minimum contact with the surrounding water. The association of the hydrophobic groups is stabilized by **van der Waals interactions**—weak attractions between all uncharged atoms or molecules that are in close proximity.

Coordination Compounds, Ligands, and Chelates

In addition to the types of bonds discussed above, oxygen and nitrogen can form a special kind of covalent association, called a **coordination bond,** with certain divalent and trivalent metal ions such as Ca^{++}, Co^{++}, Cu^{++}, Fe^{++}, Mg^{++}, Mn^{++}, Ni^{++}, and Zn^{++}. In this type of bond, all of the shared electrons that form the bond are donated by the nitrogen or oxygen atom. No electrons are donated by the metal. The energy of such bonds is about 251 kJ or 60 kcal per mole. Molecules that contain nitrogen or oxygen atoms capable of donating electron pairs to metals or metal ions are called **ligands.** If two or more electron pair donating atoms are present in the same molecule, the molecule is called a *chelate* or **chelating agent.** The metal bound by the chelating agent is said to be *sequestered*. Water may also coordinate metal ions, as in the hydrate forms of Mg, Ca, and Al salts. Figure 3–7 shows the chemical structure of cyanocobalamin (vitamin B_{12}) in which the cobalt ion forms coordination bonds (depicted as arrows) with several

Table 3–8
Energies of Chemical Bonds Determined at 25°C

Type of Bond		Kilojoules/Mole	Kilocalories/Mole
Covalent bonds			
Single bonds	O—H	428	102.4
	H—H	436	104.2
	C—H	338	80.9
	N—H	314	75.0
	C—O	1074	256.7
	C—C	602	144.0
	S—H	344	82.3
	C—N	749	179.0
	C—S	732	175.0
	S—S	426	101.9
Double bonds	C=O	497	118.9
	C=N	749	179.0
	C=C	602	144.0
	P=O	623	149.0
Ionic bonds	NaCl	408	97.5
	KCl	424	101.3
	NaI	304	72.7
Hydrogen bonds		8–42	33–175
Hydrophobic bonds (van der Waals attractions)		8	33

atoms. The oxygen-transporting protein hemoglobin (discussed in some detail in Chapter 4) is a prime example of an iron chelate. Some enzymes are activated by chelation of a metal (e.g., see cofactors of Table 9–3).

Special Compounds

Among the compounds that have special roles in cell metabolism are the **nucleoside phosphates** (e.g., adenosine triphosphate, ATP; adenosine diphosphate, ADP; and aden-

osine monophosphate, AMP), the **pyridine nucleotides** (e.g., nicotinamide adenine dinucleotide, NAD$^+$; and nicotinamide adenine dinucleotide phosphate, NADP$^+$), and the **vitamins.**

Nucleoside Phosphates

The metabolic activities of cells are accompanied by a complex series of energy transformations, and the principal intermediaries of these energy exchanges are the nucleoside

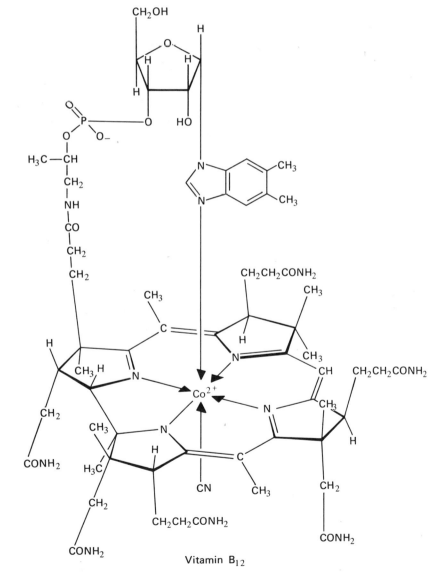

Vitamin B$_{12}$

Figure 3–7

Chemical structure of cyanocobalamin (vitamin B$_{12}$) in which the cobalt ion is *sequestered* by the tetrapyrrole. Redrawn from Brady and Holum, *Fundamentals of Chemistry*, pg. 704; copyright © 1981 by John Wiley & Sons, Inc.

phosphates. Although ATP (Fig. 3–8) serves as the immediate energy donor in most energy-consuming (i.e., **endergonic**) reactions, other nucleoside triphosphates such as guanosine triphosphate (GTP), uridine triphosphate (UTP), and cytidine triphosphate (CTP) serve in this capacity in certain instances. Energy-producing (i.e., **exergonic**) reactions are often accompanied by the synthesis of nucleoside triphosphates from the corresponding nucleoside diphosphate. The chemical formulae of the various nucleosides may be found in Chapter 7.

The high rate of turnover of ATP attests to its vital role in metabolism. For example, in humans a quantity of ATP equal to about one-half the total body weight is consumed each day. Of course, a corresponding amount of ATP must also be synthesized daily. An examination of the chemical structure of ATP (Fig. 3–8) sheds some light on why this molecule so readily serves as an energy source for endergonic cellular reactions. The three phosphate groups of ATP exist in an ionized state at physiological pH and the four negatively-charged oxygen atoms strongly repel one another. When ATP is hydrolyzed,

$$ATP^{-4} + H_2O \rightarrow ADP^{-3} + HPO_4^{-}$$

both of the products are negatively charged and continue to repel each other. The amount of energy that must be consumed in order to recombine these anions to reform ATP reflects the amount of energy that is made available during the hydrolysis. For many years, the bonds linking successive phosphate groups in ATP were represented by a "squiggle" (i.e., ~ P) and called "high energy bonds." Actually, there is nothing special about the bonds themselves and the amount of bond energy is not that great (31 kJ or 7.3 kcal per mole). What is important is that the phosphate groups have a very high transfer potential. The study of energy transfer and transformation is a special field of biochemistry called bioenergetics; this topic is discussed in Chapter 9.

Cyclic Nucleoside Monophosphates. AMP and GMP have cyclic forms called *cyclic AMP* (cAMP) (Fig. 3–9) and *cyclic GMP* (cGMP). Both cAMP and cGMP have been found to have a *regulatory* influence in a variety of cellular processes. cAMP is produced in the plasma membrane of cells and acts as an *intracellular transmitter* (sometimes called a "second messenger") by triggering certain

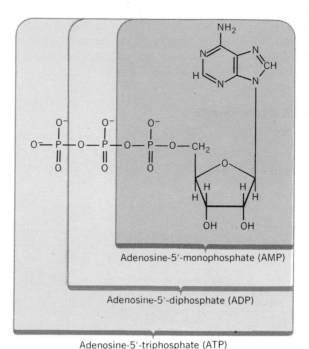

Figure 3–8
Chemical structure of ATP, ADP, and AMP.

metabolic reactions in the cell. The production of cAMP in the plasma membrane takes place as a response to compounds (usually **hormones**) that bind to specific receptor sites on the outside surface of the membrane (see Chapter 15). A good example of the regulatory role played by the cyclic nucleotides is the action of cAMP during periods of exercise in which great demands are made by the muscle tissues for free sugars. During exercise, the hormone *epinephrine* is released into the bloodstream and upon reaching the liver and muscles binds to the plasma membranes.

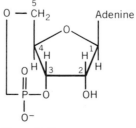

Figure 3–9
Chemical structure of *cyclic AMP* (adenosine-3′,5′-monophosphate).

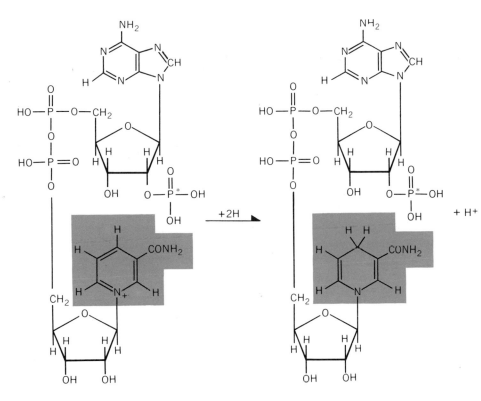

Figure 3–10
Acceptance of hydrogen by the coenzyme NADP⁺. The portion of the coenzyme formed by the vitamin niacin is shaded in color. In NAD⁺, the phosphate group (identified by the asterisks) is replaced by an OH group.

cAMP formed in the membranes in response to epinephrine binding then diffuses into the hyaloplasm. In the hyaloplasm, cAMP binds to and activates an enzyme called *protein kinase,* which in turn catalytically activates a second enzyme, *phosphorylase.* Phosphorylase initiates the breakdown of *glycogen* (Chapter 5), thereby making available the large numbers of free sugar molecules whose consumption fulfills the energy demands of muscular activity.

cAMP also influences *transcription* of nuclear DNA (see Chapters 11 and 20) where it binds to the *catabolite gene activator protein* (*CAP*), and in the case of the *lac* operon (Chapter 11) promotes the transcription of genes for specific cellular enzymes. Both cAMP and cGMP appear to be involved in the control of cell proliferation. Cells that are in the log phase of growth (see Chapter 2) contain smaller amounts of cAMP and higher amounts of cGMP than cells in other phases of the growth cycle. cAMP and cGMP affect the assembly of *microtubules* in cells (Chapter 23) and this would be expected to directly influence cell growth.

Pyridine Nucleotides

The pyridine nucleotides NAD⁺ (formerly called DPN⁺) and NADP⁺ (formerly TPN⁺) usually function as **coenzymes** for certain dehydrogenation reactions in cells. The hydrogens that are removed from compounds by various dehydrogenase enzymes are accepted by the pyridine nucleotides. Hydrogens are usually removed in pairs, with one hydrogen binding to the coenzyme as a hydride (Fig. 3–10) and the other carried as a hydrogen ion.

Vitamins

Vitamins are organic substances that are present in small amounts in all cells and are essential to certain phases of the cell's metabolism. Although vitamins are not used in large amounts by cells, the lack of a particular vitamin is usually accompanied by characteristic deficiency symptoms. Not all organisms are able to synthesize all of the

Table 3–9
Vitamins and Coenzymes

Vitamin	Examples of Associated Coenzymes	Function of Coenzyme and Enzyme
Vitamin A	11-*cis*-retinal	Visual cycle reactions
Thiamine (vitamin B_1)	Thiamine pyrophosphate	Decarboxylation reactions, oxidases, transketolases
Pyridoxine (vitamin B_6)	Pyridoxal phosphate	Amino acid metabolism (decarboxylation/transamination)
Cyanocobalamin (vitamin B_{12})	Cyanocobalamine	C—C, C—O, C—N bond cleavage
Vitamin D	1,25-dihydroxycholecalciferol	Metabolism of calcium and phosphate
Vitamin E	—	Antioxidant activity
Vitamin K	—	Synthesis of precursor enzyme for blood co-agulation (prothrombin)
Biotin	Biocytin	Carboxylation reactions
Folic acid	Tetrahydrofolic acid	Various reactions involving single carbon transfers
Lipoic acid	Lipoyllysine	Dehydrogenation, acyl group transfer
Niacin	Nicotinamide nucleotides (NAD^+, $NADP^+$)	Numerous oxidation/reduction reactions by dehydrogenases
Pantothenic acid	Coenzyme A (CoASH)	Acetyl group transfer
Riboflavin (vitamin B_2)	Flavin coenzymes (FMN, FAD)	Several oxidation/reduction reactions by de-hydrogenases

vitamins that are essential to their metabolism, and they therefore rely on an external dietary source. The role of most vitamins is closely allied to enzyme function, the vitamin comprising all or part of the nonprotein portion of the enzyme (called the **coenzyme**). Table 3–9 lists the more common vitamins, the coenzymes of which they are a part, and the role of the coenzyme and/or the associated enzyme. The vitamin *niacin* is required for the synthesis of the coenzymes NAD^+ and $NADP^+$ described above (Fig. 3–10). *Riboflavin* (vitamin B_2) is used in the synthesis of the coenzyme *flavin adenine dinucleotide* (*FAD*) which also serves as a hydrogen acceptor in dehydrogenation reactions. The vitamin *pantothenic acid* is used in the synthesis of *coenzyme A*. The vitamin constituents of these two coenzymes are shown in Fig. 3–11.

Unlike animals, which rely on an exogenous source of certain vitamins, plants are able to synthesize all of their required vitamins; most of the synthesis takes place in the plant's green tissues.

Summary

Although the quantities of specific chemical constituents vary from one type of cell or organism to another, all cells contain *proteins, polysaccharides, lipids,* and *nucleic ac-*ids. In addition to these macromolecules, all cells contain a wide variety of smaller compounds such as *water, salts, ions, metabolic intermediates,* and *vitamins.*

The physical and chemical properties of water make it a most suitable solvent for living systems. The boiling point, heat of vaporization, specific heat, and surface tension tend to keep water in the liquid state. Because of its polar character, water readily dissolves ionic and polar compounds and induces the formation of **micelles** by substances that are partly polar and partly hydrophobic. The association of water with other molecules (including other water molecules) usually takes the form of **hydrogen bonds.** These bonds have relatively low energy, yet they are vital to the molecular architecture of the cell.

Salts are usually present in ionic form and many of these are associated with macromolecules. Salt ions are also important to the osmotic behavior of cells and to buffering activities.

Some of the more important special compounds that are present in cells are *nucleoside phosphates, ligands,* and *vitamins.* Nucleoside phosphates are important sources of energy and also act to regulate specific metabolic reaction sequences in cells. Vitamins are often components of **coenzymes,** a coenzyme acting in concert with an enzyme to facilitate a specific chemical reaction.

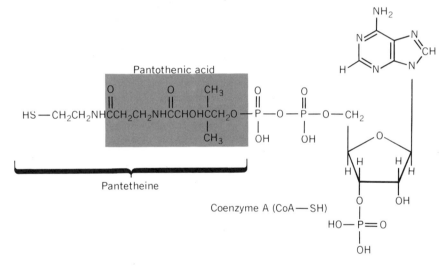

Figure 3–11

The coenzymes flavin adenine di-nucleotide (FAD) and coenzyme A. The vitamin component of each coenzyme is shaded in color.

Riboflavin

Flavin adenine dinucleotide (FAD)

Pantothenic acid

Pantetheine

Coenzyme A (CoA—SH)

References and Suggested Reading

Articles and Reviews

Ibers, J. A., and Holm, R. H., Modeling coordination sites in metallo-biomolecules. *Science 209,* 223 (1980).

Morton, B. A., The vitamin concept. *Vitam. Horm. 32,* 155 (1974).

Bitensky, M. W., and Gorman, R. E., Cellular responses to cyclic AMP. *Prog. Biophys. 26,* 409 (1973).

Books, Monographs, and Symposia

Brady, J. E., and Holum, J. R., *Fundamentals of Chemistry,* Wiley, New York, 1981.

Conn, E. E., and Stumpf, P. K., *Biochemistry* (4th ed.), Wiley, New York, 1976.

Greenberg, L. H., *Physics for Biology and Premed Students,* Saunders, Philadelphia, 1975.

Harrison, P. M., and Hoare, R. J., *Metals in Biochemistry,* Chapman and Hall, London, 1980.

Holum, J. R., *Elements of General and Biological Chemistry* (5th ed.), Wiley, 1979.

Holum, J. R., *Fundamentals of General, Organic, and Biological Chemistry,* Wiley, New York, 1978.

Neckers, D. C., and Doyle, M. P., *Organic Chemistry,* Wiley, New York, 1977.

Roberts, J. D., and Caserio, M. C., *Basic Principles of Organic Chemistry* (2nd ed.), W. A. Benjamin, Menlo Park, Calif., 1977.

Solomons, T. W. G., *Organic Chemistry* (2nd ed.), Wiley, New York, 1980.

Stryer, L., *Biochemistry* (2nd ed.), W. H. Freeman, San Francisco, 1981.

Chapter 4

THE CELLULAR MACROMOLECULES: PROTEINS

In most cells, over 90% of the total mass (excluding water) is represented by large molecules called **macromolecules.** The macromolecules vary in size from several hundred to several hundred million molecular weight units (i.e., daltons). Four major classes of macromolecules may be identified; these are the **proteins, polysaccharides, lipids,** and **nucleic acids** (the latter three classes are considered in Chapters 5, 6, and 7). The relative amounts of these substances in a "typical" cell are given in Table 4-1.

Of all macromolecules found in the cell, the proteins are the most chemically and physically diverse. The term **protein** was introduced to the biological and chemical literature in 1838 by the Dutch chemist Gerard Johannes Mulder (1802–1880), who recognized the primary importance of this substance in living matter (the word "protein" is derived from the Greek *proteios,* which means "of the first order" or "first rank"). Mulder's writings clearly indicate that he recognized the universality of protein in living things; however, he and his contemporaries believed protein to be a single substance; that is, all protein, regardless of its source, was essentially the same. Of course, this is not so, for today we recognize that there are a myriad of chemically different proteins. In fact, it has been variously estimated that a single human cell contains thousands of different proteins.

The proteins serve a variety of biological roles, but they may be divided functionally into two major classes: *structural proteins* and *dynamic proteins.* The *intracellular* structural proteins form the mechanical framework of the cell, such as *tubulin,* the *actin*-like proteins of *microtubules* and *microfilaments,* and certain of the membrane proteins. *Extracellular* structural proteins are found in multicellular organisms and also play a supportive role; included here are proteins such as *collagen* of skin, cartilage, and bone, and *keratin* in nails and hair. The dynamic proteins include the **enzymes,** which serve as catalysts in intracellular and extracellular metabolism; also included in this class are certain **hormonal** proteins (*insulin, thyroxin, erythropoietin,* etc.), certain *blood pigments* (*hemoglobin, hemocyanin,* etc.), and other proteins the roles of which are not fundamentally structural. Some proteins, such as actin, may serve both structural and dynamic functions in the cell.

Table 4–1
The Cellular Macromolecules

Substance	Percentage of Total Cell Weight
Water	65–85
Protein	10–25
Polysaccharide	1–5
Lipid	2–10
Nucleic acid	0.5–5

Proteins are sometimes also classified according to their molecular organization. Accordingly, there are the **fibrous** or threadlike proteins (such as *collagen, fibrin, actin, myosin,* etc.) and the more compact **globular** proteins (such as *hemoglobin, myoglobin,* the plasma proteins, and most enzymes). Although it is convenient to classify proteins according to their functional category, the student should recognize that all systems of classification are at best artificial.

The Amino Acids

The first intensive biochemical studies of proteins were those carried out by the German biochemist Emil Fischer around the turn of the century. In fact, Fischer received the first Nobel Prize award (in 1902) for his biochemical studies (see Table 1–1). Among Fischer's contributions to our early understanding of protein chemistry was the discovery that all proteins consist of chains of smaller units which he named "amino acids." It is now known that there are over 20 different amino acids that occur regularly as constituents of proteins and the size (i.e., molecular weight), shape, and function of proteins is determined by the number, type, and distribution of the amino acids present in the molecule. Proteins occur in a wide spectrum of molecular sizes from small molecules such as the hormone *adrenocorticotrophic hormone* (ACTH), which consists of 39 amino acids and has a molecular weight of 4500, to extremely large proteins such as the invertebrate blood pigment *hemocyanin,* which consists of 8200 amino acids and has a molecular weight greater than 900,000 (see Table 4–2 for additional examples).

Amino acids conform to the general formula shown in Figure 4–1. The alpha carbon atom of each amino acid is covalently bonded to four groups: (1) a hydrogen atom, (2) an amino group, (3) an acid group, and (4) a side chain called an R group. It is the specific chemical nature of the R group that distinguishes one amino acid from another. The amino acids that most frequently occur in proteins are listed in Table 4–3, where they are classified according to the chemical nature of the R group. By way of contrast, Table 4–4 lists the amino acids according to the functional properties of the R groups—properties that determine their specific contributions to protein structure, as will be discussed later. As seen in the tables, the amino acids aspartic acid and glutamic acid have amide configurations called asparagine and glutamine, which also occur regularly in proteins. In addition to the common amino acids there are also a number of rare amino acids. Beta-alanine (found in the vitamin pantothenic acid), gamma-aminobutyric acid (found in brain tissue), and ornithine, citrulline, and homoserine are amino acids that occur regularly as intermediates in metabolism, but these are rarely, if ever, incorporated into proteins.

The four groups that are attached to the alpha carbon atom of an amino acid are all different except in the case of glycine, where R is a hydrogen atom; therefore, all amino acids except glycine are optically active. The four groups may be considered to lie at the four corners of a regular tetrahedron, with the alpha carbon atom at the center. Accordingly, these groups may be arranged in either of two ways, yielding the L and D isomers (Fig. 4–2). Only the L forms of amino acids have been identified in proteins, although the D forms occur in some antibiotics.

Table 4–2

**Molecular Weight and Amino Acid Content
of Some Representative Proteins**

Protein	Number of Amino Acids	Molecular Weight
Adrenocorticotrophic hormone	39	4,500
Insulin	51	5,700
Ribonuclease	124	12,000
Cytochrome-c	140	15,600
Horse myoglobin	150	16,000
Trypsin	180	20,000
Hemoglobin	574	64,500
Urease	4,500	473,000
Snail hemocyanin	8,200	910,000

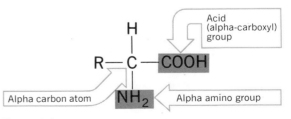

Figure 4–1

General chemical formula of an amino acid.

Table 4–3
The Common Amino Acids

Class	Name	R group	Class	Name	R group
Neutral amino acids	Glycine	H—	Basic amino acids (cont'd)	Lysine	NH_2—$(CH_2)_4$—
	Alanine	CH_3—	Aromatic amino acids	Phenylalanine	
	Valine	CH_3 CH— CH_3			
	Leucine	CH_3 CH—CH_2— CH_3		Tyrosine	
	Isoleucine	CH_3—CH_2—CH— CH_3			
	Serine	HO—CH_2—		Tryptophan	
	Threonine	OH CH_3—CH—			
Acidic amino acids	Aspartic acid	$HOOC$—CH_2—			
	Glutamic acid	$HOOC$—CH_2—CH_2—			
Amidic amino acids	Asparagine	NH_2 O=C—CH_2—	Sulfur-containing amino acids	Cysteine	HS—CH_2—
	Glutamine	NH_2 O=C—CH_2—CH_2—		Methionine	CH_3—S—$(CH_2)_2$—
Basic amino acids	Histidine	HC=C—CH_2— N NH CH	Secondary amino acids	Proline	H_2C—CH_2 H_2C CH—$COOH$ NH
	Arginine	NH_2 C—NH—$(CH_2)_3$— NH		Hydroxyproline	HO—CH—CH_2 H_2C CH—$COOH$ NH

The Peptide Bond

Proteins are composed of one or more chains of amino acids called **polypeptides.** The neighboring amino acids in the chain are linked together by **peptide bonds**; these bonds are formed, in effect, by the elimination of one molecule of water from the two amino acids sharing the bond, such that the alpha-carboxyl carbon atom of one amino acid is linked to the alpha-amino nitrogen atom of the neighboring amino acid (Fig. 4–3). This type of reaction is known as a **condensation reaction** or, in this particular case, a **dehydration synthesis.** As will be seen in Chapter 22, which deals with the mechanism by which cells synthesize their proteins, the polymerization of the amino acids to form these chains involves a number of intermediate steps. The amino and carboxyl groups that are not directly bonded to the alpha carbon atom (such as those that occur in the side chains of lysine, glutamic acid, aspartic acid, etc.) do not form covalent bonds with neighboring amino acids.

The two atoms that are involved in the peptide linkage and the four adjacent atoms lie in the same plane in space and form what is called a **planar amide group** (Fig. 4–4). Primarily through the pioneering work of L. Pauling and R. B. Corey, the specific interatomic distances and bond angles of the planar amide group are known; these are shown in Figure 4–5. In 1954, Pauling was awarded the Nobel Prize for his work on protein structure (Table 1–1). In the planar amide group, the C—N bond (i.e., the peptide bond) and the C—O bond assume a character between that of a double bond and single bond. This prevents rotation about either bond and results in the planarity of the group.

It should be noted that a planar group includes parts of two neighboring amino acids, and the alpha carbon atoms at each end of a planar group are also included as the ends of neighboring planar groups. The bond angle between planar units is 111°; this is the angle formed between the C—C bond of one planar group and the C—N bond of the next group. Note that this is an *intra*-amino acid bond angle! Figure 4–6 shows two successive planar groups arranged in such a manner that they are *coplanar* (i.e., both planar groups lie in the same plane). The alpha carbon atoms are identified simply by α. The important 111° angle formed between the C—C bond of one group (the upper group) and the C—N bond of the other (the lower group) is specifically identified. Clockwise rotation (when viewed from the position of the alpha carbon atom) of the upper

Table 4–4
Functional Classification of the Amino Acids

Hydrophilic amino acids
Acidic
Aspartic acid
Glutamic acid
Tyrosine
Neutral
Serine
Threonine
Cysteine
Asparagine
Glutamine
Basic
Arginine
Lysine
Histidine
Hydrophobic amino acids
Glycine
Alanine
Valine
Leucine
Isoleucine
Phenylalanine
Methionine
Tryptophan
Proline

planar group about the C—C bond sweeps out the angle called Ψ (psi). Clockwise rotation (again when viewed from the alpha carbon atom) of the lower planar group about the C—N bond sweeps about the angle termed φ (phi). It should be noted that the 111° angle between planar groups remains fixed regardless of the values of Ψ and φ. The flexibility of a chain of amino acids (i.e., a polypeptide) results from the rotations of successive planar groups about these bonds yielding the various values of Ψ and φ and does not result from the linkages *between* consecutive amino acids. This point cannot be overemphasized if the structures that can be assumed by polypeptides are to be properly understood. Figure 4–7 is a stereoscopic view showing two successive planar groups that are not coplanar.

The total length of two (or more) planar groups is maximized when Ψ and φ are zero. Chain length is reduced as the Ψ and φ angles are altered between 0° and 360°. If these angles are held constant within a stretch of polypep-

Figure 4–2
Stereoscopic diagrams of the arrangements of groups about the alpha carbon atom of L (upper stereo pair) and D (lower stereo pair) amino acids. *R,* side chain; *N,* amino group; *C,* carboxyl group; *H,* hydrogen; α, alpha carbon.

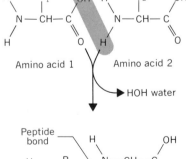

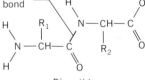

Figure 4–3
The elimination of water between two amino acids links them together through a peptide bond. This is an example of a *dehydration synthesis.*

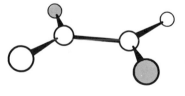

Figure 4–4
Stereoscopic diagram of the arrangement of atoms in a planar amide group. The alpha carbon atoms are shown in color.

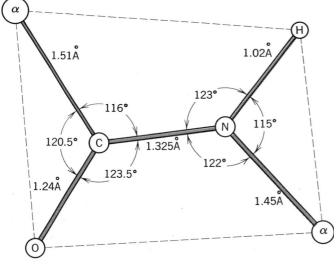

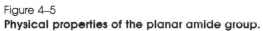

Figure 4–5
Physical properties of the planar amide group.

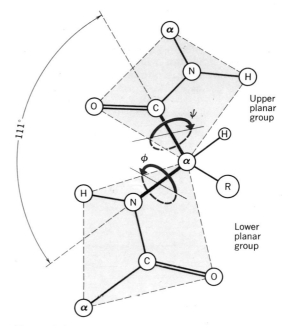

Figure 4–6
Two successive planar amide groups.

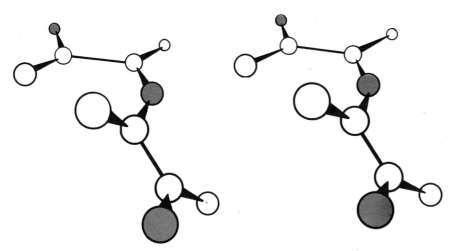

Figure 4–7
Stereoscopic diagram of two successive planar amide groups that are not coplanar. The alpha carbon atoms are shown in color.

tide, the series of planar groups naturally takes on the shape of a helix. In theory, all kinds of helices can be formed using various values for Ψ and ϕ. However, in actuality, most combinations of Ψ and ϕ are not possible, because they would result in the overlap of atoms in space. Much of the experimental data that have been accumulated to date indicate that most polypeptides contain many regions of helical coiling and few regions of complete extension. The relative lengths of helical or extended segments of a polypeptide are determined by the specific sequence of amino acids present in the polypeptide (see below).

Helical Polypeptides

Fundamental Properties of Helices

In strictly geometrical terms, helices can differ from one another in **direction, pitch,** and **diameter** (Fig. 4–8). The *direction* of a helix can be *right-handed* (i.e., rotation is *clockwise* as the helix moves forward along its linear axis) or *left-handed* (i.e., rotation is *counterclockwise*). The *pitch* is the amount of *linear translation* along the helix axis for each 360° rotation of the helix.

The helices of Figure 4–8 are geometric and more or less ideal, but helices that are formed by polypeptide chains necessarily deviate from the ideal by virtue of the complex arrangements of the constituent amino acids. For helical polypeptides, rules regarding direction and diameter remain the same and the pitch (p) is defined as the *product of the number of amino acids per turn of the helix* (n) and the *amount of linear translation per amino acid* (d). Hence, $p = (n)(d)$, where for right-handed helices n is a positive

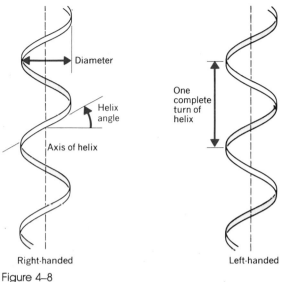

Figure 4–8
Fundamental properties of geometric helices.

number and for left-handed helices n is a negative number (see Fig. 4–9). Polypeptide helices can be *integral* (i.e., n is an integer) or *non-integral* (n is not an integer). In subsequent chapters, we will see that polysaccharides (Chapter 5) and nucleic acids (Chapter 7) also form helices with characteristic directions and dimensions.

The Hydrogen Bond

The stability of polypeptide helices results in part from the formation between different parts of the helix of a number of weak electrostatic bonds called **hydrogen bonds.** The

Figure 4–9
Examples of helical polypeptides having different numbers of amino acids per turn of the helix (i.e., different n values).

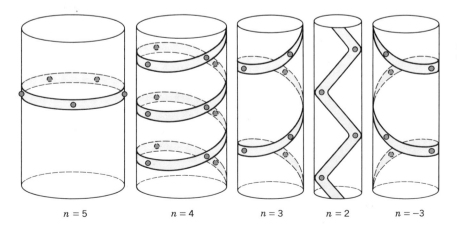

$n = 5$ $n = 4$ $n = 3$ $n = 2$ $n = -3$

attraction of atomic nuclei for orbital electrons (referred to as **electrophilia**) is not equal for all atoms; rather, it depends upon the number of protons in the atomic nucleus. For example, in a carbonyl group, the oxygen atom is more electrophilic than the carbon atom; as a result, the electron density is greater in the vicinity of the oxygen nucleus than the carbon nucleus. Consequently, there is a small negative charge associated with the oxygen atom and a small positive charge associated with the carbon atom. Similar *partial charges* (represented by δ^+ and δ^-) occur in amino and hydroxyl groups in which the hydrogen atoms are less electrophilic than nitrogen and oxygen. Weak electrostatic bonds may be formed between carbonyl oxygen atoms and amino hydrogen atoms or hydroxyl hydrogen atoms; e.g.,

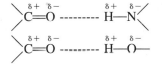

In polypeptide helices a number of such bonds are formed between the amino hydrogen atoms and the carbonyl oxygen atoms of amino acid residues that are separated by various distances along the polypeptide. Although individual hydrogen bonds are weak, their large numbers in a helix provide a significant stabilizing influence.

The Alpha Helix

The most commonly occurring helical structure in proteins, and the first whose properties were worked out, is the **alpha helix.** Principally through the work of Linus Pauling, it is known that this helix has 3.6 amino acids per turn, a pitch of 5.4 Å, and a resulting linear translation of 1.5 Å per amino acid. In the alpha helix, as in all polypeptide helices so far studied, the R groups of the amino acids are directed radially away from the axis of the helix. The helix is stabilized in part by the formation of hydrogen bonds; each bond occurs between the alpha carbonyl oxygen atom of one residue and the alpha amino hydrogen of another that is *four residues further along the polypeptide*. The hydrogen bonds bridge the carbonyl oxygens and amino hydrogens at the ends of sequences of 13 covalently bonded atoms in the polypeptide (see Figs. 4–10 and 4–11). The alpha helix is also referred to as the "3.6_{13}" helix, which reveals the value of n (i.e., $n = 3.6$) and also the nature

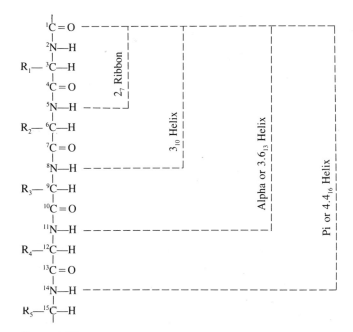

Figure 4–10
Hydrogen bonding patterns of the 2_7 ribbon, and 3_{10}, alpha, and pi helices.

of the intrahelical hydrogen bonding. Theoretically, alpha helices may be right-handed or left-handed, but only the right-handed forms have been identified in proteins. With the exception of proline, all of the various amino acids may be included internally in stretches of alpha helix. Because of the unusual nature of its side chain, proline is incompatible with alpha helical structure. Consequently, it is not unusual to find that a section of alpha helix ends (or begins) at a position occupied by proline. A stereoscopic view of a portion of a right-handed alpha helix is presented in Fig. 4–12.

Other Polypeptide Helices

Among other polypeptide helices that are found in proteins are the right-handed forms of the *pi* (or 4.4_{16}) helix, the 3_{10} helix, and a helixlike structure called the 2_7 ribbon (Table 4–5). Their hydrogen bonding patterns are compared with that of the alpha helix in Figure 4–10 (see also Table 4–5).

Molecular model building studies have shown that of the several hydrogen-bonded polypeptide helices that are the-

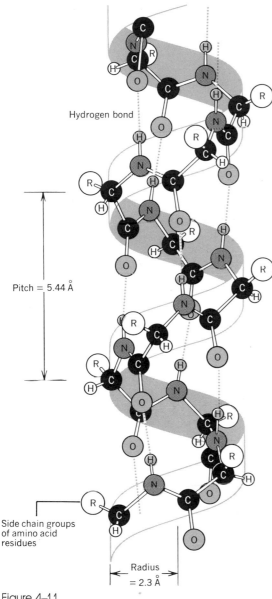

Hydrogen bond

Pitch = 5.44 Å

Side chain groups
of amino acid
residues

Radius
= 2.3 Å

Figure 4–11
Dimensions of the alpha helix.

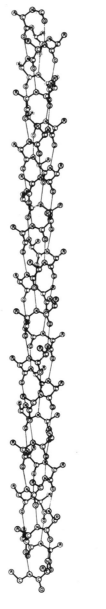

Figure 4–12
Stereoscopic diagram of an alpha helix. (Courtesy of C. K. Johnson, Oak Ridge National Laboratory.)

oretically possible, only the right-handed α-helix is strain free, with hydrogen bonds parallel to the helical axis. The other helices are formed only with varying degrees of internal strain. Thus it is not surprising that the α-helix is found in large quantities in *globular proteins,* whereas the others are rare and limited to short polypeptide segments.

Fibrous Proteins

The helices that have just been described are characteristic of globular proteins. While some fibrous proteins contain α-helices (e.g., *wool, keratin,* and *myosin*) or 3_{10} helices

Table 4–5
Structural Parameters of Right-Handed Helices

Name of Helix	n	p	d	H-bond Atoms
Alpha helix	3.6	5.4 Å	1.5 Å	13
Pi helix	4.4	3.5 Å	0.8 Å	16
3_{10} helix	3.0	6.0 Å	2.0 Å	10
2_7 ribbon	2.0	5.6 Å	2.8 Å	7

(e.g., *F-actin*), additional helices and some nonhelical repeating structures are characteristic of other fibrous proteins. Among these, we will consider those that occur in *collagen* and *silk*.

Collagen

The most commonly occurring protein among animals is the fibrous protein collagen, which is comprised principally of the three amino acids—*glycine, proline,* and *hydroxyproline.* A. Rich, F. H. C. Crick, and G. Ramachandran have shown that collagen is a rod-shaped molecule formed by the intertwining of three polypeptide chains. Each of the three chains is individually twisted to form a left-handed helix having three residues per turn (i.e., $n = -3$), and these are then twisted about one another to form a coiled-coil or "superhelix" that is *right-handed.* This arrangement is shown in Figure 4–13. Every third residue in each of the polypeptide chains forming the triple helix is glycine, whose side chain is small enough to project into the central *core* of the molecule. The three chains are held together in part through hydrogen bonding *between* proline and hydroxyproline residues in separate chains (i.e., "interchain hydrogen bonding"). This arrangement of the helices ac-

counts for the special properties of collagen, namely, its resistance to stretching and its tensile strength.

Silk

Pauling and Corey have shown that silk fibers are composed of a number of parallel, extended polypeptide chains. Neighboring chains are *opposite* in *polarity* and are held together by interchain hydrogen bonding, as shown in Figure 4–14. Successive planar groups of each of the polypeptide chains are not quite coplanar so that the overall structure takes on a "pleated" appearance and is referred to as an **antiparallel beta-pleated sheet.** (Pauling and Corey called it a "beta" structure because it was the second repeating polypeptide configuration that they worked out, the first being the alpha helix.) Since each polypeptide chain in silk is nearly fully extended, silk fibers tend to resist stretching, although lateral flexibility between parallel chains exists.

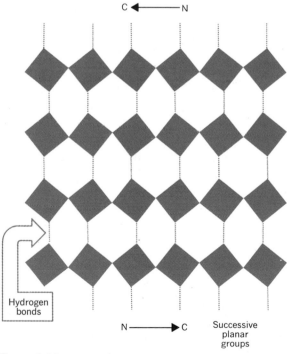

Figure 4–14
Structure of *silk*. Chains consisting of nearly coplanar planar amide groups form a *pleated-sheet* pattern linked together by hydrogen bonds.

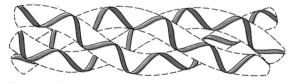

Figure 4–13
Structure of *collagen*. In collagen, three left-handed polypeptide helices (color) are intertwined to form a right-handed coiled-coil or *superhelix*.

Regions of antiparallel beta-pleated sheet structure have been observed in proteins that are characteristically globular (as in the immunoglobulins described later in this chapter and in enzymes that belong to the chymotrypsin family, see Chapter 8). In these proteins, the beta structures are formed from a single polypeptide chain with intrachain hydrogen bonding. Furthermore, parallel beta-pleated sheet structures (in which neighboring polypeptide chains have the same polarity) are also permissible.

Random Coils

It is not unusual for stretches of a polypeptide chain to twist and turn without rigorously conforming to a particular repeat structure. Such regions are called **random coils.** In many globular proteins, random coils account for a large percent of the total structure. Myoglobin and the globin chains of hemoglobin appear to be exceptions and contain extensive regions of helical organization (see below). Fibrous proteins contain little random coil and may therefore be considered more highly ordered in this regard.

Levels of Protein Structure

Primary Protein Structure

Four levels of protein structure may be delineated; these are called **primary, secondary, tertiary,** and **quaternary** structure. The primary structure of a protein is a specific and sequential delineation of the *covalent association of all of the amino acids in the protein.* So far, we have considered only the covalent peptide bond that links consecutive amino acids of the polypeptide chain. Another covalent bond called the **disulfide bridge** also occurs in many proteins. This bond is formed between the sulfur atoms of two cysteine residues by the elimination of the sulfhydryl hydrogen atom of each residue (Fig. 4–15). Disulfide bridges can occur between cysteine residues that are in the same polypeptide chain or in *different* polypeptide chains.

All of the alpha-amino and alpha-carboxyl groups of the amino acids in a polypeptide chain participate in the formation of peptide bonds except two, which are located at either end of the polypeptide chain. The end of the polypeptide that bears the amino acid with the free alpha-amino group is called the *N-terminus* (or α-amino end), while the end that has the amino acid with the free alpha-carboxyl

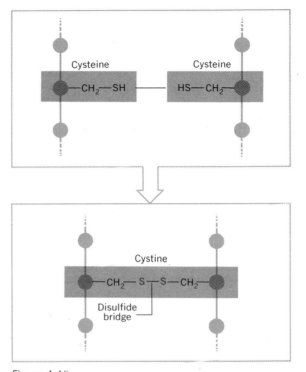

Figure 4–15
The *disulfide bridge*. Removal of hydrogen from the R groups of two cysteine residues links them together through a disulfide bridge and forms cystine.

group is called the *C-terminus* (or α-carboxyl end). Primary structure is usually delineated *beginning with the N-terminal amino acid* of each of the polypeptide chains of the protein (if indeed there is more than one polypeptide chain). The number, nature, and positions of any intrachain or interchain disulfide bridges are also specified. Some of these relationships are depicted in Figure 4–16. An internationally recognized set of abbreviations is used for each amino acid to assist in the description of all or part of the primary structure (Table 4–6).

Inherent Variety of Protein Primary Structures. The diversity of amino acids that may be included in proteins provides for an enormous number of different primary structures. Consider, for example, the mathematical variety that is possible in a polypeptide chain consisting of 61 amino acids (and this would be considered a relatively small protein). Each of the 61 residue positions can be occupied by any of 20 different amino acids. Therefore, altogether there

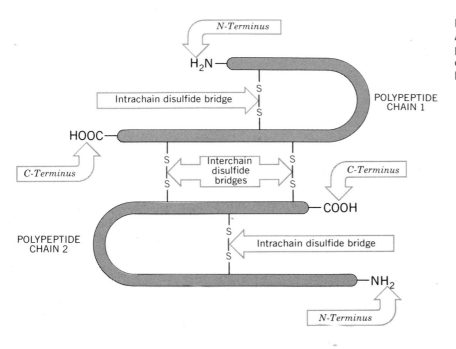

Figure 4–16
A hypothetical protein consisting of two polypeptide chains. Disulfide bridges occur within each chain and also covalently link the chains together.

Table 4–6
Standard Abbreviations for the Common Amino Acids

Amino Acid	Abbreviation	Amino Acid	Abbreviation
Alanine	ala	Leucine	leu
Arginine	arg	Lysine	lys
Aspartic acid	asp	Methionine	met
Asparagine	asn or asp-NH_2	Phenylalanine	phe
Cysteine	cys	Proline	pro
Glutamic acid	glu	Serine	ser
Glutamine	gln or glu-NH_2	Threonine	thr
Glycine	gly	Tryptophan	try
Histidine	his	Tyrosine	tyr
Isoleucine	ile	Valine	val

would be 20^{61} possible polypeptide molecules (i.e., 20^{61} different primary structures are possible). Now, $20^{61} = 2.3 \times 10^{79}$, and since it has been estimated that the entire universe contains 0.9×10^{79} atoms, there is greater potential variety in a polypeptide chain that is 61 amino acids long than there are atoms in the universe!

Secondary Protein Structure

In the description of a protein's primary structure, the three-dimensional shape or arrangement of the amino acid se-

quence making up the polypeptide chain (or chains) is not considered. In contrast, the description of a protein's secondary structure identifies (1) the position and extent of those regions of the polypeptide chain (or chains) that are twisted to form helices, (2) the nature or type of helix that is present, and (3) the position, extent, and nature of any nonhelical regions. For convenience, the various segments of each polypeptide chain can be assigned a specific nomenclature. Beginning at the N-terminus, the helical regions are denoted by the letters A, B, C, D, and so on, and the amino acids within each helix assigned numbers (e.g., C1, C2, C3, etc.). The interhelical regions of each chain are denoted by the letters of the adjoining helices (i.e., nonhelical regions AB, BC, CD, etc.) and the amino acids within these regions are also assigned numbers (i.e., BC1, BC2, BC3, etc.). The nonhelical region at the N-terminus (if indeed the N-terminus is not part of a helix) is denoted NA and its amino acids numbered consecutively (NA1, NA2, NA3, etc.). If there is a nonhelical segment at the C-terminus, it is identified on the basis of the last helix. For example, in a polypeptide chain containing eight helices (A through H), a nonhelical segment at the C-terminus would be identified as HC (and its amino acids numbered HC1, HC2, HC3, etc.) Using this type of nomenclature the specific position of any amino acid can be identified (see Fig. 4–17).

Tertiary Protein Structure

Tertiary protein structure refers to the manner in which the helical and nonhelical regions of a globular polypeptide are *folded back on themselves* to add yet another order of shape to the protein molecule. In globular proteins, it is the nonhelical regions that permit the folding. The folding of a polypeptide is not random but occurs in a specific fashion, thereby imparting certain steric properties to the protein. Well before the three-dimensional atomic structure of the first protein was worked out, W. Kauzmann anticipated the general principles that would govern the overall shape of a protein. Kauzmann predicted in 1959 that all polar groups in the protein would either interact with each other or be solvated by the surrounding water, and that considerations of entropy (see Chapter 9) would draw the nonpolar parts of the protein together in the molecule's interior. This kind of specific folding is achieved and maintained by a variety of interactions between one part of the polypeptide chain and another and between the polypeptide and neighboring molecules of water. The interactions include (1) **ionic bonds** or **salt bridges,** (2) **hydrogen bonds,** (3) **hydrophobic bonds,** and (4) **disulfide bridges.**

Ionic Bonds (Salt Bridges). In aqueous solutions, most amino acids occur in an ionized (or dissociated) state. For example, most molecules of glycine occur in the following form when they are dissolved in water:

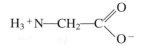

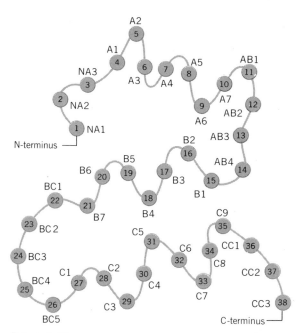

Figure 4–17

A hypothetical polypeptide illustrating the nomenclature used to describe amino acid positions in the chain's secondary structure.

In this form, a hydrogen ion (i.e., a proton) has been dissociated from the alpha-carboxyl group, while another has been removed from the surrounding water by the alpha-amino group. The resulting ion is called a **zwitterion** because it bears two different kinds of charge—positive and negative. Note that while having both kinds of charge, the glycine molecule has no *net* charge.

The acidic amino acid aspartic acid has the following zwitterionic form:

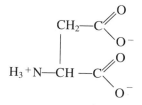

In this case, aspartic acid bears one positive charge and two negative charges and thus has a net negative charge (i.e., −1). Glutamic acid behaves in a similar manner.

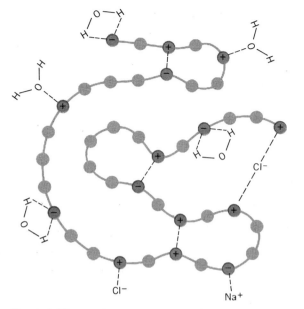

Figure 4–18
The ionized R groups and N- and C-terminals of a poly-peptide chain (color) can form various ionic bonds (electrostatic interactions) with the surroundings and with each other.

Finally, the basic amino acid lysine yields the following zwitterion in solution:

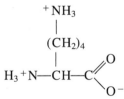

In this form, lysine carries two positive charges and one negative charge and has a net positive charge (i.e., $+1$).

In polypeptide chains, the alpha-amino and alpha-carboxyl groups of all of the amino acids except those that are at the N- and C-terminals are involved in peptide linkages. Therefore, except at the ends of the polypeptide chain, these groups are not ionized and contribute no charge to the polypeptide. However, the side chains of acidic and basic amino acids (as well as certain others) may contribute positive and negative charges along the length of the polypeptide if either conditions of local pH or the nature of the other side chains in that region of the tertiary structure allow dissociation or protonation. Electrostatic attraction between oppositely charged side chains of the amino acids of a polypeptide may bring these regions of the chain closer

together and stabilize their positions relative to one another. The bonds so formed are called **ionic bonds** or **salt bridges** (also **salt bonds**). It is also possible for ionized side chains of amino acids in the interior of the protein molecule to react with and bind water and in many proteins a certain quantity of water is permanently retained within the molecule by such interactions. Since salt ions (e.g., Na^+ and Cl^-) are also present in the surroundings of most proteins, these may also play a role in ionic bond formation between different ionized groups in the interior of the molecule. Ionic bonds also occur between charged side chains that project from the protein's surface and surrounding water and salt ions. The various kinds of ionic bonds are shown in Figure 4–18.

Hydrogen Bonds. Hydrogen bonds formed between alpha-amino hydrogen atoms and alpha-carboxyl oxygen atoms have already been discussed in connection with the stabilization of helices and parallel chains of the beta-pleated sheet structure. Hydrogen bonds can also be formed between undissociated carboxyl-containing side chains of the acidic amino acids and the amino groups of the basic amino acids lysine, tryptophan, and histidine. The hydroxyl groups of serine, threonine, and tyrosine may also participate in hydrogen bonding, as may the secondary carboxyl and amino groups of asparagine and glutamine. Although individually weak, these bonds collectively contribute to the stability of a specific tertiary structure.

Hydrophobic Bonds. A third class of interactions that stabilize tertiary protein structure are hydrophobic bonds. These are interactions between amino acids whose side chains are hydrophobic (e.g., aromatic amino acids, leucine, isoleucine, valine). The side chains of these amino acids are drawn together by their mutual hydrophobic properties becoming organized in such a manner as to have minimal contact with the surrounding water. Placed in close proximity to one another, juxtaposed atoms of separate side chains undergo **van der Waals interactions** with each other, resulting in the formation of weak bonds. Again, it is the large numbers of these interactions that impart stability to the structure. Figure 4–19 depicts the stabilization of a fold in a polypeptide by the hydrophobic association between two valine side chains.

Disulfide Bridges. Because they are covalent, disulfide bridges are the strongest bonds formed between one part of a polypeptide and another. The nature and formation of

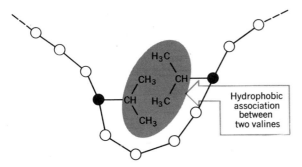

Figure 4–19
The hydrophobic side chains of amino acids such as leucine, isoleucine, valine, and phenylalanine interact with each other internally in the protein or in clefts at the protein's surface.

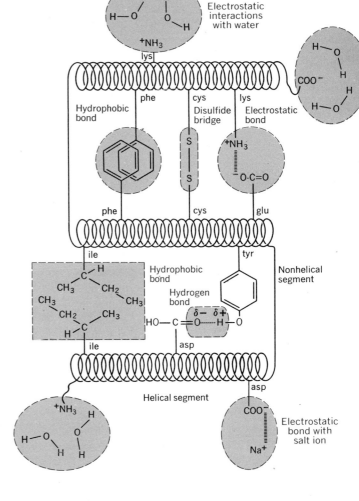

Figure 4–20
A hypothetical polypeptide illustrating the various types of interactions involved in maintaining a particular tertiary structure.

these bonds have already been discussed in connection with primary protein structure (see above). Such bonds can be formed between cysteine residues in different regions of a polypeptide (and also between cysteine residues in different polypeptide chains of a protein, see below). Where they occur, disulfide bridges contribute a considerable stabilizing influence to tertiary folding.

The four classes of bonds discussed above are depicted together in the generalized tertiary protein structure diagrammed in Figure 4–20. As you examine this diagram, it is important to note that bonds stabilizing tertiary folding may *simultaneously stabilize secondary structure*. For example, the disulfide bridge and the hydrophobic and electrostatic bonds that keep the top and middle helices of the

protein depicted in Figure 4–20 parallel to each other also serve to prevent the unwinding of these two helices. Thus, in a general sense, *specific interactions between one part of a protein and another can play a stabilizing role at more than one level of the protein's structure*.

Quaternary Protein Structure

Many proteins consist of more than one polypeptide chain. In proteins that are comprised of two or more polypeptide chains, the *quaternary structure* refers to the specific orientation of these chains with respect to one another and the nature of the interactions that stabilize this orientation. The individual polypeptide chains of the protein are usually

Table 4–7
Representative Subunit Composition of Some Proteins

Protein	Molecular Weight	Subunits		
		Number	Designation	Molecular Weight
Hemoglobin A (human)	64,500	4	Alpha chains (2)	15,700
			Beta chains (2)	16,500
Lactate dehydrogenase	135,000	4	A-chains (0 to 4)	33,600
			B-chains (4 to 0)	33,600
Immunoglobulin G	150,000	4	Light chains (2)	25,000
			Heavy chains (2)	50,000
Tryptophan synthetase (*E. coli*)	150,000	4	Alpha chains (2)	29,500
			Beta chains (2)	45,000
Aspartate transcarbamylase	306,000	12	C-chains (6)	34,000
			R-chains (6)	17,000
L-arabinose isomerase (*E. coli*)	360,000	6	(identical)	60,000
Apoferritin (iron-storage protein)	456,000	24	(identical)	19,000
Thyroglobulin	670,000	2	(identical)	335,000

referred to as its **subunits.** Table 4–7 lists some representative proteins that are comprised of subunits and gives their numbers, designations, and molecular weights. As can be seen from this sampling, proteins can contain either a small number of large subunits (e.g., *thyroglobulin*), a large number of small subunits (e.g., *apoferritin*), or any intermediate combination. Moreover, in some proteins the subunits are polypeptide chains whose primary structures are identical to each other (e.g., L-*arabinose isomerase*), while in others the subunits are different (e.g., *immunoglobulin G*).

The same classes of interactions that contribute to the stability of tertiary protein structure also serve to stabilize the quaternary association of subunits; namely, ionic bonds, hydrogen bonds, hydrophobic bonds, and disulfide bridges. As we shall see in Chapter 8, many cellular enzymes are comprised of subunits, and the resulting quaternary structure is of fundamental importance in the regulation of enzyme activity. The molecular weights of proteins comprised of subunits are often high enough for the molecules to be seen and studied by electron microscopy of negatively stained preparations. Electron microscopy thus provides additional information about quaternary structure, for it is often possible to discern the number and orientation of the protein's subunits. The subunit organization of the enzyme

L-*arabinose isomerase* is quite evident in the electron photomicrographs of Figure 4–21.

Among the groups of proteins whose quaternary structures have been extensively studied are the *hemoglobins* and *immunoglobulins*. Probably more is known about the chemistry, organization, and functions of members of these two groups than about any other proteins. Later in the chapter, we shall consider these two groups of proteins in considerable detail.

To this point we have been using the terms **primary, secondary, tertiary,** and **quaternary structure** exclusively in connection with proteins. However, corresponding levels of organization are recognized for *polysaccharides* and *nucleic acids* in which the order of their building blocks (i.e., sugars and nucleotides) and the coiling and folding of their chains can also be delineated.

Establishment of Secondary, Tertiary, and Quaternary Structure

Although it is reasonably clear just how a specific secondary, tertiary, and quaternary structure can be maintained, how such a specific arrangement is initially achieved is not completely understood. There is no evidence for the existence in cells of templates or special enzymes that function

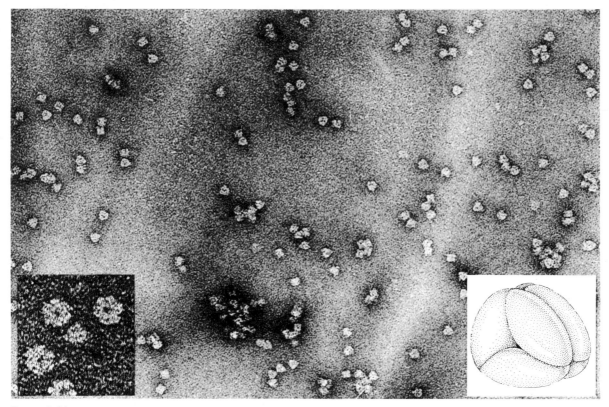

Figure 4–21

Molecules of the enzyme L-*arabinose isomerase* revealed by the technique of negative staining (see Chapter 1). The quaternary structure of the enzyme is formed from two stacks containing three subunits each. The arrangement is seen more clearly in the enlarged insert (the arrow identifies a second enzyme in the sample, *glutamine synthetase*) and in the diagram. (Courtesy of Drs. F. A. Eisreling and L. J. Wallace.)

in molding a particular three-dimensional shape. Some polypeptides (perhaps most) *spontaneously* assume their biologically active shape from among the many different alternatives that exist. As we have seen, polypeptide chains are quite flexible. Thus, free to twist and fold and subject mainly to kinetic constraints, polypeptides can progress through a variety of different shapes, progressively assuming the shape that is energetically most stable.

We shall see in Chapter 22 that a polypeptide chain is formed by the sequential addition of amino acids beginning at the chain's N-terminus. Encoded in the polypeptide's primary structure are critical sequences of amino acids that promote the progressive assumption of a specific tertiary structure by the polypeptide as the primary structure is being laid down. Such progressive folding would appreci-

ably reduce the total number of possible alternative shapes assumable by a polypeptide. In other words, once folding of a portion of the polypeptide proceeds to its thermodynamically stable state, certain alternative configurations made possible by continued elongation of the primary structure are no longer kinetically accessible. For proteins comprised of two or more polypeptide chains and possessing quaternary structure, it has been suggested that the polypeptide subunits may randomly encounter one another, automatically orienting themselves and combining in such a manner as to establish the proper (i.e., biologically active) quaternary structure. Some of the experimental evidence supporting these ideas follows.

The side chains of the acidic and basic amino acids are often dissociated (or protonated), thereby creating negative and positive sites along the polypeptide; these charged sites can be coordinated and stabilized by water molecules, which are also partially polar, or by dissolved ions. Consequently, segments of a polypeptide containing such charged residues would naturally tend to orient themselves at the outer surface of the protein where they could face and interact with the surrounding aqueous environment. The side chains of other amino acids such as tyrosine, glutamine, and so forth, which are electrically neutral on the whole, contain atoms of nitrogen and oxygen that are partially polar due to the differential electrophilia of these atoms. Therefore, water would also be attracted to these groups, although not to the same degree as fully charged side chains. By interacting with electrically charged surface groups, the water molecules minimize the strength of the electric fields surrounding these groups and in so doing stabilize the molecule's structure.

Amino acids such as leucine and phenylalanine are nonpolar and have hydrophobic side chains. Hydrophobic groups disturb the haphazard arrangement of water molecules, tending to impart order to the water (as occurs, for example, in ice crystals). Any increase in order makes a system potentially less stable by reducing the system's entropy (which is a measure of the disorder of a system, see Chapter 9). Since entropy tends to be maximized, the hydrophobic side chains are usually turned away from the surrounding water and face each other internally.

Accordingly, among the many proteins that have been extensively studied, most of the amino acids with polar side chains reside at the protein's outer surface, while the hydrophobic side chains are either confined to the interior of the molecule or form crevices in its surface so as to have minimal contact with the surrounding water and ions.

Anfinsen's Studies of the Enzyme Ribonuclease.   The notion that, acting in concert, the specific primary structure of a polypeptide and the innate properties of the side chains of its amino acids induce the polypeptide to spontaneously assume its biologically active tertiary structure originated with the studies of the protein **ribonuclease** begun in the mid-1950s by C. B. Anfinsen; later (1972) Anfinsen was awarded the Nobel Prize for this most definitive work. Ribonuclease is an enzyme produced in large amounts by cells in the pancreas; from here it is conveyed through ducts to the duodenal section of the small intestine, where the enzyme acts to degrade RNA in ingested food. Ribo-

nuclease contains 124 amino acids, has a molecular weight of about 12,000 and consists of a single polypeptide chain. Anfinsen identified four disulfide bridges in the protein (i.e., there are four cystine positions), suggesting that the enzyme is highly folded (Fig. 4–22). As is the case with nearly all enzymes, the ability of ribonuclease to carry out its enzymatic activity depends upon the maintenance of the appropriate three-dimensional structure. In concentrated solutions of beta-mercaptoethanol in urea, the disulfide bridges of the protein are broken, yielding cysteine residues where cystine existed before. The resultant unfolding of the protein is accompanied by a loss of enzyme activity. The enzyme is said to be "denatured." If the beta-mercaptoethanol and urea are removed by dialysis and the denatured ribonuclease is reacted with oxygen, the four disulfide bridges re-form, and essentially all of the enzymatic activity is recovered (see Fig. 4–22). Corresponding observations have been made with other proteins; that is, they are capable of spontaneously reestablishing their biologically active ter-

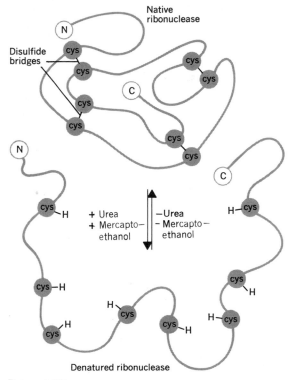

Figure 4–22

Denaturation and spontaneous renaturation of the enzyme ribonuclease.

tiary structure after having undergone extensive molecular disorganization. In these proteins, the three-dimensional structure that is crucial to biological function is directed by the primary structure, and in the appropriate environment one particular configuration is overwhelmingly favored energetically over other possibilities. Polypeptides possess sufficient flexibility to explore various configurations progressively assuming the most stable of these.

Spontaneous Formation of Quaternary Structure. Studies of a number of different proteins comprised of two or more polypeptide chains support the concept of spontaneous assumption of a specific and functional quaternary structure from the separate subunits. Among those proteins in which the phenomenon is readily demonstrated is the blood pigment hemoglobin. Normal adult human hemoglobin, called hemoglobin A, may be represented as follows:

$$\alpha_2\beta_2$$

This notation indicates that the molecule is a tetramer, consisting of four polypeptide chains—a pair of alpha chains and a pair of beta chains (Fig. 4–23). When a hemoglobin solution is made weakly alkaline or weakly acidic, the hemoglobin tetramers undergo a stepwise dissociation into subunits; that is

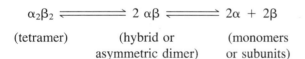

$$\alpha_2\beta_2 \rightleftharpoons 2\,\alpha\beta \rightleftharpoons 2\alpha + 2\beta$$

| (tetramer) | (hybrid or asymmetric dimer) | (monomers or subunits) |

As indicated in the above reaction, the dissociation of the tetramer into monomers or subunits is reversible; that is, upon restoration of the normal pH, fully functional hemoglobin molecules are sequentially re-formed. This observation supports the view that intrinsic properties of the individual subunits of a protein are sufficient to promote the assumption of the biologically active quaternary structure. In the case just cited, it seems that the subunits seek out one another in solution and complex to form the biologically functional tetramer.

Conjugated Proteins

Conjugated proteins are proteins that contain nonprotein constituents or **prosthetic groups.** The prosthetic groups

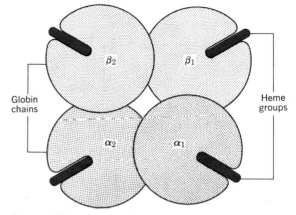

Figure 4–23
Relative positions and arrangement of the four globin chains and heme groups of hemoglobin.

are *permanently* associated with the molecule, usually through covalent and/or noncovalent linkages with the side chains of certain amino acids. Conjugated proteins can be subdivided into three major classes: (1) **chromoproteins,** (2) **glycoproteins,** and (3) **lipoproteins.**

Chromoproteins

The chromoproteins are a heterogeneous group of conjugated proteins related to each other only in that they all possess *color.* The *hemoglobins, myoglobins,* and other heme-containing proteins such as the *cytochromes* and *hemerythrins* belong to this group. The prosthetic groups of the chromoproteins, such as the heme groups of hemoglobin and the cytochromes, are bound to the polypeptide portion of the molecule through a combination of covalent and noncovalent bonds. We shall examine the interaction between the hemes and polypeptide chains of hemoglobin in some detail later in the chapter.

Glycoproteins

Glycoproteins (from the Greek *glykys* meaning "sweet") are proteins containing varying amounts of *carbohydrate* (the chemistry of carbohydrates is dealt with in the next chapter). A number of important proteins fall in this category, including many of the blood plasma proteins and a large number of enzymes and hormones. The surfaces (i.e., plasma membranes) of most cells also contain quantities of glycoproteins, and these molecules serve there as antigenic determinants and as receptor sites. Virtually all the carbohydrate present in red blood cells occurs as membrane glycoproteins. Although over 100 different sugars (or monosaccharides) are known, only about nine occur as regular constituents of glycoproteins (Table 4–8).

The amount of carbohydrate present in glycoproteins varies from less than 1% to more than 85% (Table 4–9). For example, in egg white ovalbumin (molecular weight 45,000), there is only one monosaccharide per molecule, while in mucin (a secretion of the submaxillary gland having a molecular weight of about 1 million), about 800 monosaccharides are present. The carbohydrate moieties of glycoproteins are usually bound to the protein through covalent bonds with either asparagine, threonine, hydroxylysine, serine, or hydroxyproline (see Fig. 4–24). The carbohydrate bonded at each site of the protein may consist of a single monosaccharide unit (as in Fig. 4–24) or a linear or branched chain of several monosaccharides (called an oligosaccharide), as depicted in Figure 4–25.

Lipoproteins

Lipid-containing proteins are called *lipoproteins*. This class includes some of the blood plasma proteins and also a large number of membrane proteins. The lipid content of lipoproteins is often very high, accounting for as much as 40 to 90% of the total molecular weight of the complex. In lipoproteins, the amount of lipid present markedly affects

Table 4–8
Carbohydrates Occurring as Regular Constituents of Glycoproteins

Glucose
Galactose
Mannose
Fucose
Acetylglucosamine
Acetylgalactosamine
Acetylneuraminic acid
Arabinose
Xylose

Table 4–9
Carbohydrate Content of Glycoproteins

Glycoprotein	Percentage of Carbohydrate	Function
Ovalbumin	1	Hen-egg food reserve
Follicle-stimulating hormone (FSH)	4	Hormone
Fibrinogen	5	Blood coagulation protein
Transferrin	6	Iron transport protein of blood plasma
Ceruloplasmin	7	Copper transport protein of blood plasma
Glucose oxidase	15	Enzyme
Peroxidase	18	Enzyme
Luteinizing hormone	20	Hormone
Haptoglobin	23	Hemoglobin-binding protein of blood plasma
Erythropoietin	33	Hormone
Mucin	50–60	Mucus secretion
Blood-group glycoproteins	85	Unknown

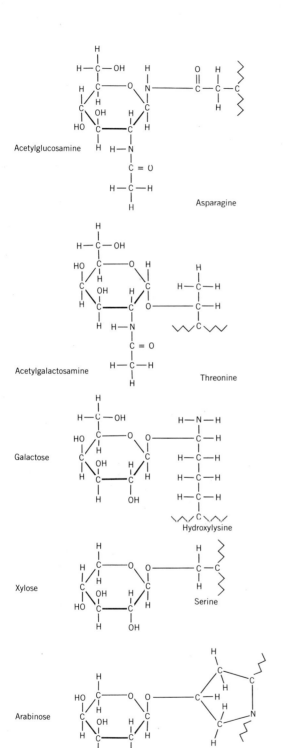

Acetylglucosamine

Asparagine

Acetylgalactosamine

Threonine

Galactose

Hydroxylysine

Xylose

Serine

Arabinose

Hydroxyproline

Figure 4–24
Covalent bonds formed between carbohydrates and the side chains of specific amino acids in *glycoproteins*.

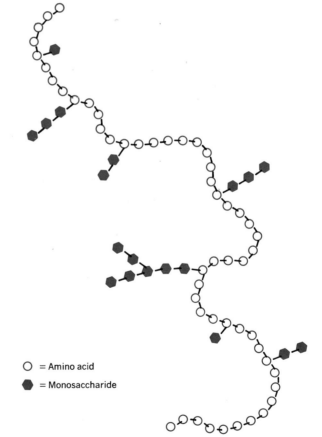

○ = Amino acid

⬡ = Monosaccharide

Figure 4–25
The carbohydrate portion(s) of glycoproteins may be short or long chains of sugars and may be branched.

the *density* of the molecule, and this property is often used as the basis for lipoprotein classification. Whereas uncomplexed proteins have a density of about 1.35, lipoproteins vary in density down to 0.9 (i.e., a lipoprotein may be less dense than water).

The interactions between the lipid and protein portions of a lipoprotein usually involve similar functional groups. For example, the hydrophobic portions of fatty acids, sterols, glycerides, and the like (see Chapter 6) form van der Waals interactions with the hydrophobic side chains of the nonpolar amino acids. Covalent bonds are believed to occur between the phosphate moieties of certain phospholipids and the hydroxyl-containing side chains of amino acids like serine. Lipoproteins are discussed further in connection with membrane structure in Chapter 15.

Nucleoproteins

In eucaryotic cells, specific proteins called **nucleoproteins** are found intimately associated with nuclear DNA. Also, in procaryotes as well as eucaryotes, ribonucleoprotein complexes (i.e., protein plus RNA) occur. These proteins are not usually classified with the conjugated proteins, since the nucleic acids involved cannot be regarded as prosthetic groups. Two types of proteins have been identified in nucleoproteins, the **histones** and **nonhistones.** Histones have a rather restricted amino acid composition (containing about 25% arginine and lysine) and are quite similar in all plant and animal cells. Their highly basic nature accounts for the close associations they form with the nucleic acids and lends credence to the notion that they are involved in the tight packing of DNA molecules during the condensation of chromatin to form chromosomes. The nonhistones are considerably more heterogeneous in amino acid composition and have acidic properties. There is much evidence to implicate the nonhistones in the regulation of gene expression. The histones and nonhistones are considered further in conjunction with the discussion of nuclear organization and function in Chapter 20.

In the remaining parts of this chapter, we shall be concerned with the structural properties of two specific proteins, the hemoglobins and the immunoglobulins. These proteins have received particularly intense study during the past few decades—studies that have revealed some rather startling facts about (1) the relationship between structure and biological activity, and (2) the diversification of structure in functionally related proteins within a species of organism and among different species.

Vertebrate Hemoglobins

Our understanding of the structure of the hemoglobin tetramer and the individual globin polypeptide chains has advanced rapidly during the past 15 years, primarily because information about this molecule has been obtained by a variety of independent methods of investigation. These include (1) *X-ray crystallographic* studies of hemoglobin crystals, providing atomic resolution on the order of just a few Angstroms; (2) biochemical studies of amino acid sequences in isolated globin chains; (3) determination of the nucleotide sequences in globin chain genes (from these sequences, the primary structures of the globin chains can be inferred); (4) *electron spin resonance (ESR)* studies of the heme groups and their manner of interaction with the globin chains; (5) radioactive tracer studies of the mechanism of globin chain biosynthesis; and (6) physiological studies of the behavior of the molecule. Most notable among those scientists who have contributed to our present understanding of the structure and function of hemoglobin is M. F. Perutz, whose work with this protein began more than 45 years ago. In 1962, Perutz, together with his colleague J. C. Kendrew, was awarded a Nobel Prize for his studies of hemoglobin and a closely related protein, *myoglobin.*

The following discussion of the anatomy of hemoglobin molecules will illustrate a number of general features that are characteristic of protein structure and function. There is, however, a more far-reaching lesson to be learned from these studies of hemoglobin. Hemoglobin, like many other proteins (especially enzymes), is widespread in nature and the primary structure of the molecule is now known for many different species of organism. Although differences in composition do exist, there are overriding similarities among all of the hemoglobins. These similarities are not merely chemical curiosities; rather, *they provide us with an insight into the nature of evolution as it proceeds at the molecular level.*

General Structure of the Hemoglobins

The hemoglobins are conjugated globular proteins; that is, they contain some nonprotein parts. The function of hemoglobin is to transport oxygen in the blood from the lungs to the tissues (see below), and therefore hemoglobin molecules can exist in two states: oxygenated and unoxygenated. In humans, the most common type of hemoglobin is hemoglobin A (usually abbreviated HbA) which consists of 574 amino acids and has a molecular weight of 64,500. Its secondary, tertiary, and quaternary structure is typical of all higher vertebrate hemoglobins. The protein portion of the molecule, called globin, is comprised of four polypeptide chains, each of which is also globular in shape. The four globin chains consist of two identical pairs; two alpha chains (141 amino acids each) and two beta chains (146 amino acids each). The nonprotein portion of hemoglobin consists of four heme groups—one associated with each of the four globin chains.

Cysteine residues are not uncommon among the various hemoglobins; for example, there are six in HbA. Yet, despite the fact that disulfide bridges between cysteine residues play an important role in the maintenance of secondary, tertiary, and quaternary structure in many proteins, there are no such linkages in the hemoglobins. Instead, the tertiary structure of each chain and the quaternary structure of the whole tetramer are maintained by noncovalent linkages.

Hemoglobin molecules are highly symmetric (Fig. 4–23). The molecule can be divided into two identical halves, each consisting of an αβ dimer (called a "hybrid" or "asymmetric" dimer). The complete tetramer is similar to a flattened sphere (i.e., an oblate ellipsoid), having a maximum diameter of about 55 Å. The four polypeptide chains are arranged in such a manner that unlike chains have numerous stabilizing interactions, whereas like chains have few. That is, each alpha chain has many contacts with both betas, and each beta chain has many contacts with both alphas; however, alpha–alpha and beta–beta contacts are few. The like chains face each other across an axis of symmetry such that rotation of any chain 180° about this axis would make it congruent with its identical partner. A cavity about 25 Å long and varying in width from about 5 to 10 Å passes through the molecule along the axis. This arrangement is seen in Figure 4–26.

The interior of each subunit contains many amino acids with hydrophobic side chains; these form hydrophobic bonds with one another. Amino acids whose side chains are ionized predominate at the surface of the subunits and the molecule as a whole. Amino acids with nonpolar side

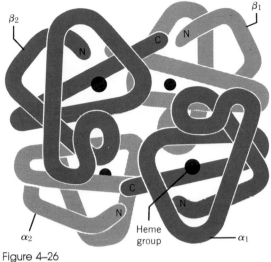

Figure 4–26
The hemoglobin molecule.

chains may be found along a polar segment at the surface, but in such cases the side chains are buried in a crevice formed in the molecule's surface and thereby have little contact with the surrounding water. As noted above, contacts between like subunits are rare. In the oxygenated molecule, ionic bonds (salt bridges) occur between the polar amino and carboxyl terminals of the respective chains (see Fig. 4–26) and also between a few other residues. Contacts between unlike subunits are much more numerous and include 17 to 19 hydrogen bonds between about 35 amino acids of each hybrid dimer—that is, between α_1 and β_1 and between α_2 and β_2. Fewer bonds occur between α_1 and β_2 and between α_2 and β_1. No covalent bonds link any of the subunits together.

Association of the Globin Chains with Heme and the Heme with Oxygen

All of the globin chains have the same fundamental secondary structure (Fig. 4–27). The helical regions are α-helix although there may be small excursions into pi and 3_{10} configurations. The heme group that is associated with each globin chain consists of an organic portion and an iron atom (Fig. 4–28). The organic part is called *protoporphyrin IX* and is formed from four *pyrrole rings* linked together by methene bridges. All of the atoms of this tetrapyrrole lie in a single plane. Four *methyl*, two *vinyl*, and two

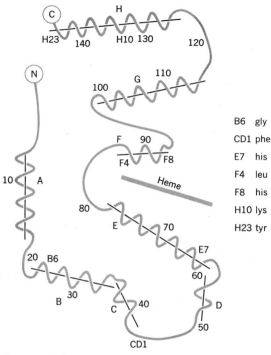

B6   gly
CD1   phe
E7   his
F4   leu
F8   his
H10   lys
H23   tyr

Figure 4–27
Secondary structure of the beta chain of human hemoglobin. The highly conserved amino acid positions discussed in the text are specifically noted.

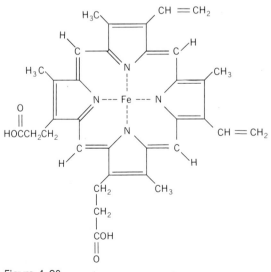

Figure 4–28
Chemical formula of heme.

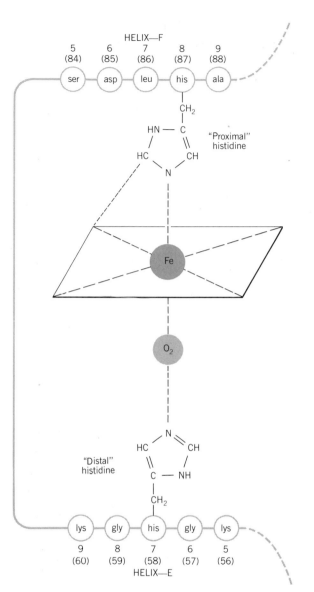

Figure 4–29
The iron atom of the heme group (the parallelogram) forms a bond with the side chain of the "proximal" histidine F8 (amino acid 87 in human alpha chains and 92 in beta chains) and also interacts with the "distal" histidine E7 (amino acid 58 in alpha chains and 63 in beta chains). During oxygenation, molecular oxygen is bound to the iron on the distal side of the heme group. In the unoxygenated state, repulsion occurs between the proximal histidine and one of the pyrrole nitrogen atoms (dotted line). Only the alpha chain is depicted here.

Figure 4–30
Stereoscopic pair of diagrams showing the tertiary structure of the alpha globin chains of hemoglobin. (Courtesy of A. Dickerson and I. Geis.)

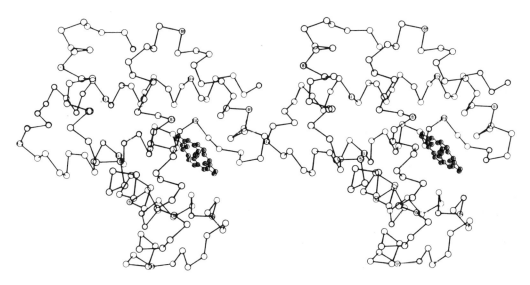

propionic acid groups are attached to the tetrapyrrole as side chains. The iron atom is linked to each pyrrole ring and depending upon whether or not the hemoglobin is oxygenated, the iron lies either in the same plane as the tetrapyrrole or just above it. Each globin chain envelopes its heme group in a deep cleft (Fig. 4–27), exposing only the propionic acid side chains to the surrounding water.

Within each cleft, the heme groups interact with the R groups of 16 amino acids from seven segments of the globin chain. Except for the iron atom, the portion of the heme group that is enclosed in the cleft is strongly hydrophobic and most of the bonds that are formed with the heme involve hydrophobic side chains of amino acids such as leucine, valine, and phenylalanine. Each iron atom can form up to six bonds, four of which occur with the pyrrole nitrogen atoms. The side chains of two *histidine* residues (amino acids E7 and F8 in both the alpha and beta chains) that are positioned on opposite sides of the plane of the tetrapyrrole also interact with the iron atom, but only one of these (called the "proximal" histidine, F8) forms a bond with the iron (Fig. 4–29). The "distal" histidine (E7) contacts the heme group but does not form a bond with the iron. Instead, the sixth bond is formed with molecular oxygen. In the unoxygenated state, the bond between iron and the proximal histidine "lifts" the iron atom out of the plane of the tetrapyrrole, but when oxygen enters the cleft on the distal side of the heme group and combines with the iron atom, the iron is drawn back into the plane. The

tertiary structure of an alpha globin chain together with its heme group is depicted stereoscopically in Figure 4–30.

In addition to human hemoglobin, hemoglobins from about 30 other species have been completely analyzed, including those of the chimpanzee, gorilla, monkey, rabbit, mouse, horse, pig, cow, dog, sheep, camel, goat, llama, kangaroo, echidna, chicken, turtle, frog, carp, and lamprey. Several vertebrate myoglobins (conjugated oxygen-binding proteins closely related to the hemoglobins and consisting of a single polypeptide chain, also called globin, and a heme group), including those of the sperm whale, porpoise, seal, dolphin, and horse, have also been analyzed. In all of these molecules, there is a remarkable similarity in the secondary and tertiary structures of the globin chains. Since the primary structures of these globins are known, it is possible to analyze the relationship between their common secondary and tertiary structures and their respective primary structures.

When two different chains are compared and found to have the same amino acid occupying the same position of the secondary structure (e.g., position B6, E7, and so on), then the two chains are said to be "homologous" at that position. Such comparisons reveal an interesting relationship between the extent of homology of the globin chains of various species and the evolutionary relatedness of the organisms. Moreover, certain positions are invariably occupied either by a specific amino acid or by a limited number of different amino acids, implying that a high degree of conservation has occurred. Such conservation is clearly related to the constraints that must be satisfied for the assumption and maintenance of a specific and common tertiary structure among the various globin chains.

Highly Conserved Amino Acid Positions.   Among all of the hemoglobin (exclusive of the abnormal hemoglobins) and myoglobin molecules so far studied, there are seven positions that are invariably occupied by the same amino acid in all or *almost* all globin chains (Table 4–10 and Fig. 4–27). Seven is believed to be too small a number to *impose* or maintain a specific secondary and tertiary globin chain structure. However, it is acknowledged that these residues play an important part in the *maintenance* of the globin chain's characteristic shape. For some of these positions, the role is readily explained; for example, amino acids E7 and F8 are the **distal** and **proximal** histidine residues (Table 4–8).

There are also five globin chain positions occupied by amino acids whose side chains are invariably *ionized*. Three of these are invariably basic amino acids (positively charged side chains) and the other two are invariably acidic amino acids (negatively charged side chains). The positions containing basic amino acids (i.e., lys or arg) are B12, E5, and H10; the positions containing acidic amino acids (i.e., asp or glu) are A4 and B8.

There are 33 positions in the interior of the globin chains that do not contact the surrounding environment; among these, 30 positions are invariably occupied by amino acids with *nonpolar* side chains. There are also 10 positions in crevices at the surface of the molecule invariably occupied by nonpolar amino acids, Thus, more than one-fourth of the amino acid positions in globin are invariably nonpolar. These positions may be occupied by any of several amino acids, or in some instances the amino acid may be specific. For example, position AB1 may be ala, gly, or ser; F3 may be ala, ser, thr, gln, lys, or pro; in contrast, F4 is nearly always leu. It is generally agreed that the invariably nonpolar positions, together with the nonpolar portions of the heme group, play a major role in determining the shape of the globin chain. Internally, the hydrophobic interactions between side chains form the skeletal *framework* on which the shape is founded. The substitution through genetic mutation of a single polar amino acid in a position that is invariably nonpolar usually is sufficient to disrupt the normal organization of the polypeptide and render it biologically nonfunctional. At the surface of the molecule, ionic and hydrogen bonds between residues of the same chain, between residues of different chains, and between residues and the surrounding water and ions also contribute to the stability and universal shape of the molecules.

Function and Action of Hemoglobin: Cooperativity in Proteins

The function of hemoglobin is to reversibly bind molecular oxygen; that is

$$Hb \;+\; 4\,O_2 \; \underset{\text{in tissues}}{\overset{\text{in lungs}}{\rightleftharpoons}} \; Hb(O_2)_4$$

unoxygenated   oxygen   oxyhemoglobin
hemoglobin

The manner in which oxygen is bound by and released from hemoglobin reveals yet another characteristic of protein (especially enzyme) action, namely that of **cooperativity.** Hemoglobin is contained within the red blood cells (erythrocytes) and is oxygenated as the blood circulates through the capillary networks of the lungs. Upon leaving the lungs, virtually every hemoglobin molecule is combined with four molecules of oxygen. In this state, the hemoglobin is said to be 100% *saturated.* Later, when the blood is circulated to the other body tissues, hemoglobin releases its bound oxygen. The amount of oxygen that is released is determined by the concentration of dissolved oxygen gas (i.e., the *partial pressure*) in the surrounding plasma and body fluid. In muscle, for example, it would not be unusual

Table 4–10
Highly Conserved Globin Chain Positions

Position	Amino Acid	Role
Invariant		
CD1	phe	interaction with heme
F8	his	contact with iron atom of heme
Almost invariant		
B6	gly	permits close contact of helices B and E
E7	his	contact with heme
F4	leu	interaction with heme
H10	lys	uncertain
H23	tyr	uncertain

for the percent saturation of hemoglobin to fall to 40% or lower as the released oxygen diffuses from the erythrocytes into the plasma and then into the muscle tissue. As noted earlier, the closely related and structurally similar oxygen binding protein myoglobin (Mb), which acts to temporarily store oxygen in certain muscle tissues, consists of a single polypeptide chain and heme group. Consequently, although hemoglobin can reversibly bind four molecules of oxygen, *myoglobin binds only one*. Figure 4–31 compares the oxygen association/dissociation curves of hemoglobin and myoglobin.

Referring to Figure 4–31, as the oxygen partial pressure rises between 0 and 20 mm Hg, myoglobin rapidly combines with oxygen and quickly approaches complete saturation. The association/dissociation curve takes the form of a *rectangular hyperbola*. The behavior of hemoglobin is considerably different and is much more complex. At low oxygen partial pressure, the affinity of hemoglobin for oxygen is considerably less that that of myoglobin. For example, at 20 mm Hg partial pressure, hemoglobin is only about 21% saturated. However, between 20 and 60 mm Hg oxygen partial pressure, the affinity of hemoglobin for oxygen is greatly increased and the hemoglobin approaches saturation. Above 60 mm Hg, hemoglobin binds only small quantities of additional oxygen. Unlike the myoglobin curve, the oxygen association/dissociation curve of hemoglobin is *sigmoid*. From a physiological standpoint, the unique oxygen-binding characteristics of hemoglobin are crucial. The partial pressure of oxygen in the blood leaving the lung capillaries is generally in excess of 100 mm Hg, but by the time the blood reaches the capillaries of the various body tissues, the partial pressure has fallen to 80 mm Hg. An examination of the curve of Figure 4–31 shows that in this interval hemoglobin will have released only a small percentage of its oxygen. However, in the tissue capillaries, where the oxygen partial pressure often falls below 40 mm Hg, hemoglobin releases much of its bound oxygen. In actively exercising muscle, where the partial pressure may drop to 20 mm Hg, still greater quantities of oxygen would be released. Thus, within the range of 60 to 20 mm Hg (a range within which most of the tissues of the body operate), a relatively small decrease in oxygen partial pressure is accompanied by a quantitative release of hemoglobin-bound oxygen.

In contrast, myoglobin retains nearly all of its bound oxygen even at a partial pressure of 20 mm Hg. Indeed, the greater affinity for oxygen displayed by myoglobin at

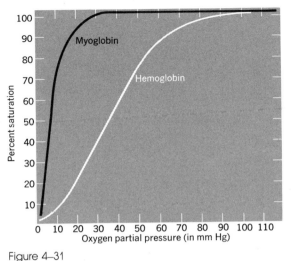

Figure 4–31
Oxygen association/dissociation curves of hemoglobin and myoglobin.

all partial pressures provides for the efficient transfer of oxygen from the blood to the myoglobin-containing musculature. If the hemoglobin oxygen association/dissociation curve was hyperbolic (like myoglobin's) instead of sigmoid, then inadequate amounts of oxygen would be released from the blood to the tissues, leading to asphyxiation even during moderate exercise. The difference between the behavior of these closely related proteins results from the *heme–heme interactions* that are possible in the tetrameric hemoglobin molecule but which cannot occur in the monomeric myoglobin. The heme–heme interactions exhibited by hemoglobin belong to a general class of interactions that are possible in proteins that are comprised of two or more subunits; the phenomenon is known as **cooperativity.**

Cooperativity in Hemoglobin.   The complex behavior of hemoglobin, which is precisely what is required of an efficient oxygen-transporting system, may be attributed to its quaternary structure, for when hemoglobin is dissociated into its subunits, the separate subunits exhibit the oxygen-binding behavior of myoglobin (i.e., a hyperbolic curve). In the intact tetramer, the various subunits exhibit cooperativity.

In the unoxygenated state, the hemoglobin molecule is said to be in the "T" (i.e., *tense*) state, whereas oxygenation leads to the "R" (i.e., *relaxed*) state. In the T-structure, the iron atom of each heme group is pulled out of the plane of the tetrapyrrole and toward helix F by its inter-action with the side chain of each proximal histidine. Bind-ing of oxygen draws the iron atom into the plane and pulls the histidine side chain toward the heme group; hence, a small change in tertiary structure occurs. Now, since each globin chain is bonded to neighboring subunits, *changes taking place in one subunit will induce changes in another* by altering the nature of the subunit associations. In partic-ular, during oxygenation, salt bridges linking subunits to-gether are broken, leading to the R-state. The changes in tertiary and quaternary structure that accompany the oxy-genation and deoxygenation of hemoglobin have been sus-pected for some time, for it has long been known that crystals formed by oxyhemoglobin and deoxyhemoglobin are quite different.

The binding of the first oxygen molecule to unoxygenated hemoglobin involves one of the alpha subunits; while this occurs slowly, the configurational change in this subunit brings about a change in the accompanying subunit (i.e., the beta member of the asymmetric dimer). The result is a more rapid binding of the second oxygen molecule by the beta subunit. The changes that have taken place in one asymmetric dimer induce changes in the other half of the molecule (i.e., the other asymmetric dimer), including a 15° rotation of the two dimers with respect to one another. Although this movement accelerates the binding of the third oxygen molecule, the fourth oxygen is bound less rapidly.

Effects in which one or more subunits of a protein alter the structure of other subunits in such a way as to modify the protein's behavior are called *cooperative effects*. In the case of hemoglobin, it may now be understood why the oxygen association/dissociation curve is sigmoid and not hyperbolic, for the individual alpha and beta subunits of the tetramer *do not operate independently* of one another in the binding of oxygen but instead influence each other. The importance of cooperative effects in proteins cannot be overstated, for we shall see in Chapter 8 that cooperativity is also manifested by a number of enzymes whose affinities for their substrates can be modulated through subtle changes in the individual subunits of the enzyme.

Evolution of Proteins

Ontogeny and Phylogeny of Hemoglobin

Functionally and structurally similar proteins exist in di-verse animal and plant species. Many of these are believed to have a common evolutionary origin. The picture of protein evolution is probably clearest in the case of the hemoglobins, where it is possible to make accurate com-parisons of hemoglobin structure in a large number of different animal species. Moreover, in humans (and also in other vertebrates), different kinds of hemoglobin molecules are present in the blood at different stages of development, and this ontogeny also sheds light on the phylogenetic picture. In addition to hemoglobin A, several other nor-mally occurring hemoglobins are present in humans. In the very early stages of embryonic development, a hemoglobin called *Gower-1* appears. This hemoglobin consists of two alpha-like chains called zeta (ζ) chains and two beta-like chains called epsilon (ϵ) chains; the Gower-1 tetramer is thus represented as $\zeta_2\epsilon_2$.

Beginning at about the eighth week of development, the ζ and ϵ chains are gradually replaced by the adult α chain and also two different beta-like chains that are designated gamma^G (γ^G) and gamma^A (γ^A); the two types of gamma chains are identical except that in γ^G the amino acid glycine occurs at position H13, while in γ^A the position is occupied by alanine. Alpha chain production, once begun, continues throughout embryonic, fetal, and adult life. The next hemo-globin to appear is called *Gower-2* and consists of two alpha and two epsilon chains (i.e., $\alpha_2\epsilon_2$). This is quickly followed by the appearance of hemoglobin *Portland*, which consists of two zeta and two gamma chains (i.e., $\zeta_2\gamma_2$). Both Gower-2 and Portland are soon replaced by *hemoglo-bin F* (designated HbF and called *fetal hemoglobin*), con-sisting of two alpha and two gamma chains (i.e., $\alpha_2\gamma_2$). HbF predominates in the blood through the remainder of fetal development. Beginning soon after the twelfth week of development, the adult beta chains begin to appear and there is a progressive increase in HbA (i.e., $\alpha_2\beta_2$) and an accompanying decrease in HbF. Shortly before birth, yet another beta-like chain appears called the delta chain (δ); together with alpha chains, the delta chains form a hemo-globin known as A$_2$ (i.e., $\alpha_2\delta_2$). By about six months after birth, little if any HbF can be found in the blood. At this time, about 98% of the hemoglobin is HbA and the re-maining 2% is HbA$_2$. In a normal individual, this ratio persists through adult life. The various human hemoglobins and their subunit compositions are listed in Table 4–11. Their differential temporal appearance is shown in Figure 4–32.

A consistent feature of the human hemoglobins is that two members of each tetramer are always alpha-like chains

Table 4–11
The Human Hemoglobins

	Symbol	Chains
Embryonic hemoglobins		
Portland	P	$\zeta_2\gamma_2$
Gower–1	G_1	$\zeta_2\varepsilon_2$
Gower–2	G_2	$\alpha_2\varepsilon_2$
Fetal hemoglobin	F	$\alpha_2\gamma_2$
Adult hemoglobins		
Hemoglobin A	A	$\alpha_2\beta_2$
Hemoglobin A_2	A_2	$\alpha_2\delta_2$

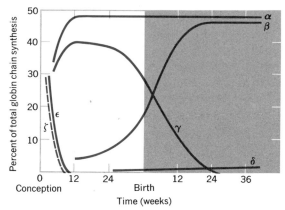

Figure 4–32
Ontogeny of the alpha (α), beta (β), gamma (γ), delta (δ), epsilon (ε), and zeta (ζ) globin chains of human hemoglobin (see text for explanation).

and the other two are beta-like. Conditions similar to this have been found in all other mammals studied so far and also in other vertebrates. Except in the case of the lowest vertebrates (e.g., lampreys and hagfishes), hemoglobin molecules are always tetramers and are formed from one pair of alpha-like chains and one pair of beta-like chains. When adult organisms have two or more hemoglobins, they share a common alpha-like chain. Transition from one developmental stage to another, as in the metamorphosis of amphibians, is accompanied by a change in hemoglobin in which one pair of chains (i.e., the alpha-like chains) is conserved. In the lowest vertebrates, hemoglobin is a monomer (i.e., one globin chain and one heme group), and in this regard is similar to all of the vertebrate myoglobins. All of these observations play a critical role in the devel-

opment of the current concept concerning the evolution of the hemoglobin molecule.

The primary structures of the alpha-like and beta-like chains of the human hemoglobins reveal a high degree of **homology.** An examination of Table 4–12 reveals that there are 84 differences between the alpha and beta chains, implying that there are 62 homologies; that is, nearly one-half of all positions in each chain are occupied by the same amino acid. A common evolutionary origin for the alpha and beta chains is suggested in view of their extensive homology.

Table 4–12
Differences Between Human Globin Chains

		Alpha-like		Beta-like				
		Alpha	Zeta	Beta	Gamma^G	Gamma^A	Delta	Epsilon
Alpha-like	Alpha	—	44	84	89	90	85	76
	Zeta	44	—	75	78	78	77	72
Beta-like	Beta	84	75	—	39	40	10	24
	Gamma^G	89	78	39	—	1	41	18
	Gamma^A	90	78	40	1	—	42	17
	Delta	85	77	10	41	42	—	40
	Epsilon	76	72	24	18	17	40	—

Still higher degrees of homology exist among the alpha-like and among the beta-like chains (Table 4–12). The alpha and zeta chains have the same amino acids at 97 positions, with only 44 differences. The beta and gamma[G] chains have the same amino acids at 107 positions, with only 39 differences. Beta and epsilon chains have only 24 differences; beta and delta chains have only 10 differences; and so on. Thus, the beta, gamma, delta, and epsilon chains are more closely related to each other than any of them are to the alpha-like chains; the alpha and zeta chains are more closely related to each other than either one is to the beta-like chains.

Just as striking are the similarities between human alpha chains and the alpha chains of other mammals and between human beta chains and the beta chains of other mammals. Table 4–13 lists these differences for a number of mammals in order of diminishing phylogenetic relationship to man. No differences exist between the primary structures of human and chimpanzee alpha and beta globin chains. Gorilla and human globin chains differ in only two positions. Although the more distantly related mammals reveal increasingly large numbers of differences, it is consistently observed that *the alpha chains or beta chains of the various species are more closely related than are the alpha and beta chains within a species* (e.g., human alpha chains and kangaroo alpha chains have a higher degree of homology than do human alpha and human beta chains). The latter class of observations is basic to contemporary theory on hemoglobin evolution. It is interesting (but not surprising) that a common revelation of these hemoglobin studies is that *species taxonomically more closely related have greater similarities in the primary structures of their globin chains*.

The secondary and tertiary structure of myoglobin is almost identical to that of the globin chains of hemoglobin, and its primary structure reveals an especially high degree of homology with the alpha-like globin chains. In view of their similarity, myoglobin and hemoglobin are believed to have a common evolutionary origin. This notion is also supported by the observations that the hemoglobin of the lowest vertebrates (e.g., lampreys and hagfishes) consists of a single polypeptide chain and not a tetramer.

On the basis of the numerous studies of vertebrate myoglobins and hemoglobins, it is generally agreed that all of the globin chains have a common evolutionary origin. The proposed scheme for the evolution of human myoglobin

Table 4–13
Differences Between Human Globin Chains and Those of Other Mammals

Mammal	Alpha Chain	Beta Chain
Chimpanzee	0	0
Gorilla	2	2
Monkey	4	8
Rabbit	18	18
Pig	18	17
Horse	18	25
Kangaroo (marsupial)	27	40
Echidna (monotreme)	36	31

and hemoglobin is shown in Figure 4–33. According to this scheme, the *structural genes* (see Chapters 20 and 21) for the various globin chains arose through a series of gene duplications, and subsequently these genes underwent separate and independent evolution. The degree of similarity that exists between the primary structures of the globin chains (Table 4–12) is used to establish the relative phylogenetic positions of the structural genes. Thus, the greater degree of homology between the beta and delta globin chains than between the beta and gamma chains is presumably the consequence of the more recent duplication of the primitive beta/delta gene, and so on.

Based upon the extent of homology between the primary structures of polypeptide chains from different species and the presumed frequencies with which gene mutations occur, it is possible to estimate the age (and even the primary structure) of the ancestral polypeptide. Such studies are encompassed in a field of biology called molecular paleogenetics. Phylogenetic relationships between species suggested on the basis of protein similarities are remarkably consistent with those that were predicted many decades ago using more conventional taxonomic criteria.

Immunoglobulins

One of the body's most important defense mechanisms against infection is the production by the reticuloendothelial tissues of a class of proteins called **antibodies** or **immunoglobulins.** These proteins circulate in the bloodstream, where they make up part of the "gamma globulin" fraction of blood plasma (see Chapter 13). The production of the immunoglobulins is stimulated by chemical substances,

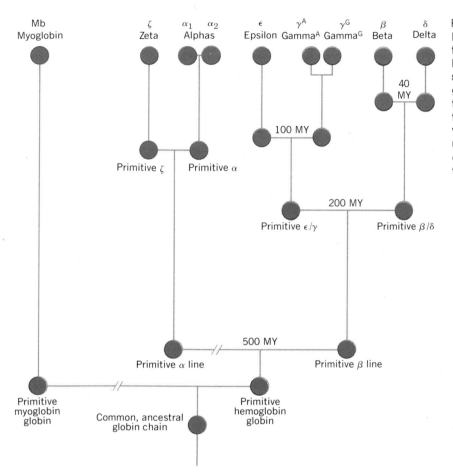

Figure 4–33

Evolution of the globin chains found in the human hemoglobins and in myoglobin. Beginning with a common, ancestral structural gene and a corresponding globin chain, a series of gene duplications occurred. Following each duplication, the product genes underwent individual diversification through mutation, ultimately producing the multiplicity of different normal globin chains existing today. (MY = millions of years.)

called *antigens,* present in or released from the infecting agent (e.g., constituents of bacterial membranes or the coats of viruses) and recognized as being foreign or alien to the body. This response is called the "immune response." Although antigens can belong to virtually any chemical category, they usually are proteins, polysaccharides, or nucleic acids. The immunoglobulin molecules that are produced during an immune response have the capacity to bind to the antigens, the reaction between the two being highly specific; that is, each type of immunoglobulin that is produced in the body reacts with a particular antigen and no other. A major source of the antibodies that circulate in the blood are the "plasma cells" derived by the differentiation of a class of white blood cells known as *B-lymphocytes.*

The binding of antibodies to antigens that are in the surface of invading microorganisms (such as bacteria) trig-

gers a series of reactions that leads to the destruction of the foreign cells. These reactions involve a host of other blood plasma proteins collectively referred to as **complement** and which are normally present in the plasma (i.e., unlike the immunoglobulins, they are not synthesized in response to the appearance of an antigen). Attachment of the immunoglobulin molecules to the surface antigens of the infecting cells is followed by the sequential binding and activation of the complement proteins. Acting in a manner similar to digestive enzymes, the complement proteins create holes in the bacterial surface. This produces a "Donnan effect" (see Chapter 15) in which ions and water enter the foreign cell, causing it to swell and eventually rupture (i.e., lyse). In effect, the immunoglobulins by binding to the surface antigens have served to identify the cells that are to be attacked and destroyed by complement. Each antibody has more than one antigen-binding site. Therefore, when free antigen molecules are encountered in the body, they can be cross-linked to form a precipitate. When antibodies react simultaneously with antigens in the surfaces of separate cells, the cells are **agglutinated;** that is, they are clumped together into small masses. Precipitated antigens and agglutinated cells are ultimately engulfed and disposed of by *phagocytic* white blood cells. The immune response appears to be of rather recent evolutionary origin, since immunoglobulin production is characteristic only of vertebrates.

Immunoglobulin Structure

The human body is capable of synthesizing more than a million different kinds of immunoglobulin molecules, each capable of reacting with a different antigen, but all of them appear to share the same fundamental quaternary structure. During the early stages of an infection, the response to the antigen involves the production of a specific class of immunoglobulin known as *immunoglobulin M* or *IgM,* having a molecular weight of about 1,000,000. Later, the amount of IgM gradually declines as another class or "isotype" of immunoglobulin called *IgG* appears. Although other immunoglobulins are also present in blood (see Table 4–14), IgG represents the most abundant form (as much as 80%). IgG has been more extensively studied than the other immunoglobulins and much of the description that follows is based upon results obtained from studies of IgG.

IgG molecules are composed of four polypeptide chains of two different kinds—a pair of identical high-molecular-weight chains, called "heavy" or **H chains,** and a pair of identical lower-molecular-weight chains called "light" or **L chains.** The *L* chains have a molecular weight of 20,000 to 25,000 and consist of about 214 amino acids. The *H*

chains have a molecular weight of 50,000 to 55,000 and contain about 450 amino acids. In associating with each other, the four chains form a molecule whose quaternary structure resembles that of a letter "Y" (Fig. 4–34). Each arm of the Y contains a complete *L* chain and part of an *H* chain, while the leg of the Y contains the remaining parts of the *H* chains. Near its C-terminus, each *L* chain is linked to an *H* chain by a disulfide bridge, and two additional disulfide bridges link the *H* chains together.

Each of the four polypeptide chains that form an immunoglobulin is divided into separate regions called "domains." There are *two* domains in the *L* chains and *four* in the *H* chains. Within each of the domains, folding of the polypeptide chain produces two parallel planes each containing several segments with folded beta structure (Fig. 4–35). In each folded beta structure, neighboring stretches of the polypeptide have opposite polarity (i.e., they are antiparallel). The two parallel planes in each domain are held together by disulfide bridges and also by van der Waals interactions between hydrophobic side chains of amino acids in each sheet.

In each of the arms of the immunoglobulin, the *H* and *L* chains associate to establish a globular quaternary structure maintained principally by van der Waals reactions between hydrophobic side chains of amino acids in each chain; some hydrogen bonding between chains also occurs.

Antigen and Complement Binding Sites

Studies employing digestive enzymes such as trypsin and papain have revealed which portions of an immunoglobulin molecule combine with the antigens and which part binds complement. These digestive enzymes characteristically

Table 4–14
Immunoglobulins

Class	Light Chains	Heavy Chains	Tetramer	Molecular Weight	Sedimentation Coefficient
IgG	k or λ	γ^a	$k_2\gamma_2$ or $\lambda_2\gamma_2$	150,000	7S
IgA	k or λ	α	$(k_2\alpha_2)_n$ or $(\lambda_2\alpha_2)_n$	180,000 to 500,000	7S, 10S, 13S
IgM	k or λ	μ	$(k_2\mu_2)_5$ or $(\lambda_2\mu_2)_5$	950,000	18S to 20S
IgD	k or λ	δ	$k_2\delta_2$ or $\lambda_2\delta_2$	175,000	7S
IgE	k or λ	ε	$k_2\varepsilon_2$ or $\lambda_2\varepsilon_2$	200,000	8S

[a] There may be several different γ chains.

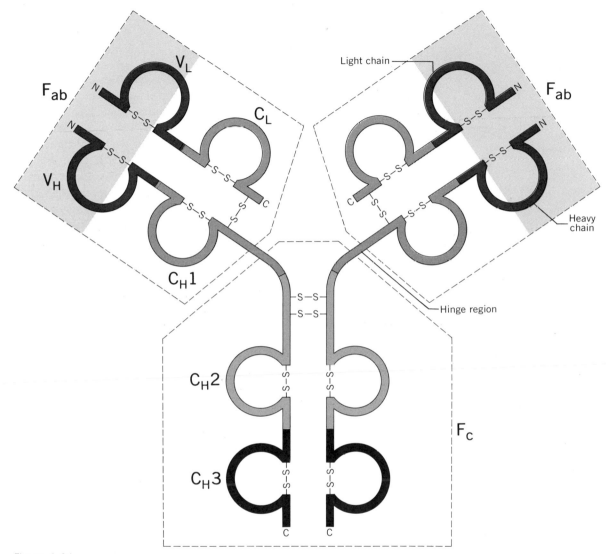

Figure 4–34
Fundamental architecture of an immunoglobulin. Each molecule is formed from four polypeptide chains (two "heavy" chains and two "light" chains) linked together by disulfide bridges. Each chain contains variable (color) and constant (gray shading) domains. Enzymatic digestion of the immunoglobulins for subsequent chemical analysis cleaves the molecule in the "hinge" region and produces the F_{ab} and F_c fragments. The area of the immunoglobulin that binds the antigen is lightly shaded. (See the text for details.)

Planes of the
beta-pleated sheets

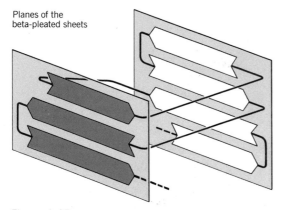

Figure 4–35
Within each domain of an immunoglobulin, the polypeptide chain creates two parallel planes or sheets. In each sheet, adjacent segments of the polypeptide are arranged in the folded beta structure and are antiparallel. In this diagram, segments of the polypeptide that connect successive stretches of beta structure are shown as thin lines, while the beta structures are depicted as broad arrows. Note that adjacent segments of the beta structure are not always formed by the continuation of the same region of the polypeptide chain.

cleave immunoglobulins in a region (called the "hinge" region) located on the C-terminal side of the disulfide bridge that links each L chain to an H chain (see Fig. 4–35). Two types of fragments are produced by such enzymatic cleavage, one of which retains the ability to bind the antigen (called the F_{ab} fragment). Each of the two F_{ab} fragments that is produced consists of a complete L chain and the N-terminal half of an H chain. The other type of fragment consists of the C-terminal halves of both H chains and is called the F_c fragment because it can be *crystallized*. The antigen appears to be bound near the ends of the two F_{ab} units (the shaded areas in Fig. 4–34). The angle between the F_{ab} and F_c units is variable because of the flexibility of the hinge regions of the molecule. As a result, the overall shape of an immunoglobulin can vary between that of a letter "Y" and a letter "T." This flexibility is unquestionably related to the immunoglobulin's capacity to bind to the antigens. The complement proteins associate with the F_c fragment. Immunoglobulins contain a small amount of carbohydrate (i.e., the immunoglobulins are glycoproteins). The carbohydrate portion consists of a short chain of sugars covalently bonded to an asparagine residue and sandwiched along the interface between the F_{ab} and F_c units.

Origin and Evolution of the Immunoglobulins

Homologous and Variable Domains

Among the immunoglobulins, certain domains of both H and L chains are homologous or constant (i.e., essentially the same from one immunoglobulin to another) and certain domains are variable (i.e., they are the basis of the differences among the immunoglobulins). Each L chain has one *variable* and one *constant* domain; each H chain has one variable and three constant domains. The variable domains occur near the N-terminals of the polypeptide chains and together create an antigen-binding site that is unique to that immunoglobulin molecule. The variable domain of each light chain is designated V_L and the constant domain C_L. The variable domain of a heavy chain is designated V_H and the constant regions are called C_H1, C_H2, and C_H3.

Not all light chains have the same constant regions; instead, there are two different classes of light chains, designated kappa (κ) and lambda (λ). About 40% of the amino acid positions of κ and λ chains are homologous; this degree of homology is similar to that exhibited by the alpha and beta chains of hemoglobin (see above). There are five different types of heavy chains, and these are designated γ, α, μ, δ, and ε. In the heavy chains, the variable region is about the same size as in the light chains but the constant region is about three times as long. Various combinations of the different light and heavy chains give rise to the five classes (or "isotypes") of immunoglobulins listed in Table 4–14.

As we have already noted, the N-terminal regions of the L and H chains have different primary structures; these differences are thought to reflect the ability of different immunoglobulins to react with different antigens. The variable regions of the L chains (V_L) are homologous with the variable regions of the H chains (V_H). Moreover, the constant regions of the H chains (C_H1, C_H2, and C_H3) are similar to one another and are also homologous with the constant region of the L chains (C_L).

The internal homologies between various parts of the immunoglobulins suggest that these polypeptides, like those that form the various hemoglobins, have a common ancestral origin. During evolution, gene duplications followed by separate diversification of the multiple genes probably produced all of the immunoglobulin genes that are present today in the human genome. However, unlike the hemoglobin saga in which duplications and diversification produced the seven globin structural genes that are usually expressed, hundreds of different genes for the various regions of the *L* and *H* chains have arisen.

Basis of Immunoglobulin Diversity. For many years, the prevailing view with regard to the proteins of a cell was that each different polypeptide chain was the product of a separate structural gene. For example, the seven expressed globin chain genes of man give rise to the seven alpha- and beta-like chains, which in different combinations produce all of the human hemoglobins. The immunoglobulin picture is considerably more complex. The immune system of one individual is capable of producing more than a million different immunoglobulins, but this is not the consequence of the existence of a corresponding number of structural genes for the *H* and *L* chains. The basis of so great an immunoglobulin diversity is the rather astounding finding that *H* and *L* polypeptide chains are the products of the combined contributions of three different kinds of genes. These are called the *C* (for *constant*), *V* (for *variable*), and *J* (for *joining*) genes. The *C* genes encode the constant regions of the *H* and *L* chains; the *V* genes encode most of the variable region of each *L* and *H* chain; and the *J* genes encode the remaining short segment of the *V* region. For the *L* chains, the constant regions are encoded by one or two genes, whereas about eight different genes encode the *C* region of the *H* chains. Several hundred different *V* genes and about five *J* genes encode the variable regions of the *H* and *L* chains. Some very recent findings suggest that there may be a fourth type of immunoglobulin gene; it has been tentatively called a ''D'' (for ''diversity'') gene.

During embryonic development and also during an immune response, the differentiation of antibody-producing cells is accompanied by the *translocation* of chromosomal segments (see Chapter 20), thereby bringing various *V, J,* and *C* genes together. The combined *V, J,* and *C* genes serve for the construction of one complete *H* or *L* chain. The existence of a large number of different constant region genes for the *H* chains accounts for the greater variety of heavy chain classes seen in Table 4–14. The diversity of immunoglobulins is therefore partly explained by the very large number of possible combinations of *V, J,* and *C* genes. Additional diversity is the result of irregular translocations of the *J* genes, ''point mutations'' during the replication of the genetic material, and other chromosomal phenomena which in a more general sense serve to increase genetic variation in all living things—not just in antibody-produc-

ing cells. These mechanisms are considered further in Chapters 20 and 21.

Summary

Of the four major classes of macromolecules, the **proteins** are the most diverse and complex, being composed of one or more chains of **amino acids.** About 20 different amino acids occur in proteins and are covalently linked together within each chain by **peptide bonds.** The specific sequence of covalent associations of all the amino acids in the protein is called the **primary structure.** Each polypeptide chain may contain one or more regions twisted to form *helical* structures stabilized by **covalent, hydrophobic, electrostatic,** and/or **hydrogen** bonds. This is called the **secondary** level of protein structure. A more compact structure may be achieved by folding the polypeptide chain in the nonhelical regions; this establishes the polypeptide's **tertiary** structure. In proteins that have two or more polypeptides, the specific orientation of the chains with respect to one another is called **quaternary** structure.

The secondary, tertiary, and quaternary structure of proteins may be established during assembly of the primary structure or spontaneously following completion of synthesis. Hydrophobic and hydrophilic amino acids are specifically distributed in relation to the final tertiary and quaternary structure of the molecule.

Proteins containing **prosthetic groups** are called **conjugated proteins** and include **chromoproteins, glycoproteins,** and **lipoproteins.** Chromoproteins are chemically heterogeneous but are related in that they possess color. In glycoproteins the prosthetic groups consist of one or more branches of sugar chains covalently linked to the polypeptides. In the lipoproteins, the prosthetic groups consist of lipids linked to the protein through either covalent bonds or hydrophobic interactions.

Among all proteins studied, the structure, action, and evolution of the *hemoglobins* and *immunoglobulins* are best understood. The alpha-like and beta-like globin chains of hemoglobin are encoded in separate genes descended from a commmon ancestor through gene duplication and mutation. In marked contrast, the heavy and light chains of the immunoglobulins are encoded by hundreds of genes belonging to three different classes. Variable recombination of members of each class of genes ultimately provides the incredible diversity of primary structures that characterizes the immunoglobulins within a species.

References and Suggested Reading

Articles and Reviews

Dayhoff, M. O., Computer analysis of protein evolution. *Sci. Am. 221*(1), 86 (July 1969).

Dickerson, R. E., The structure and history of an ancient protein. *Sci. Am. 226*(4), 58 (April 1972).

Doolittle, R. F., Similar amino acid sequences: chance or common ancestry? *Science 214*, 149 (1981).

Edelman, G. M., The structure and function of antibodies. *Sci. Am. 223* (2), 34 (Aug. 1970).

Efstratiadis, A., Posakony, J. W., Maniatis, T., Lawn, R. M., O'Connell, C. O., Spritz, R. A., DeRiel, J. K., Forget, B. G., Weissman, S. M., Slightom, J. L., Blechl, A. E., Smithies, O., Baralle, F. E., Shoulders, C. C., and Proudfoot, N. J., The structure and evolution of the human β-globin gene family. *Cell 21*, 653 (1980).

Eyre, D. R., Collagen: Molecular diversity in the body's protein scaffold. *Science 207*, 1315 (1980).

Fraser, R. D. B., Keratins. *Sci. Am. 221*(2), 86 (Aug. 1969).

Gale, R. E., Clegg, J. B., and Huehns, E. R., Human embryonic haemoglobins Gower-1 and Gower-2. *Nature 280*, 162 (1979).

Gross, J., Collagen. *Sci. Am. 204*(5), 120 (May 1961).

Kendrew, J. C., The three-dimensional structure of a protein molecule. *Sci. Am. 205*(6), 96 (Dec. 1961).

Kitchen H., and Boyes S. (eds.), Hemoglobin: Comparative molecular biology models for the study of disease. *Ann. N. Y. Acad. Sci. 241* (1974).

Perutz, M. F., Electrostatic effects in proteins. *Science 201*, 1187 (1978).

Perutz, M. F., The hemoglobin molecule. *Sci. Am. 211*(5), 64 (Nov. 1964).

Perutz, M. F., Haemoglobin structure and respiratory transport. *Sci. Am. 239*(6), 92 (Dec. 1978).

Perutz, M. F., Structure and function of haemoglobin. I. A tentative atomic model of horse oxyhaemoglobin. *J. Mol. Biol. 13*, 646 (1965).

Perutz, M. F., X-ray analysis, structure and function of enzymes. *Eur. J. Biochem. 8*, 455 (1969).

Perutz, M. F., Structure and mechanism of haemoglobin. *Br. Med. Bull. 32*, 195 (1976).

Perutz, M. F., Kendrew, J. C., and Watson, H. C., Structure and function of haemoglobin. II. Some relations between polypeptide chain configuration and amino acid sequence. *J. Mol. Biol. 13*, 669 (1965).

Perutz, M. F., Muirhead, H., Cox, J. M., and Goaman, L. C. G., Three-dimensional Fourier synthesis of horse oxyhaemoglobin at 2.8 Å resolution: The atomic model. *Nature 219*, 131 (1968).

Poljak, R. J., Correlations between three-dimensional structure and function of immunoglobulins. *C. R. C. Crit. Rev. Biochem. 5*, 435 (1978).

Schroeder, W. A., The hemoglobins. *Ann. Rev. Biochem. 32*, 301 (1963).

Sharon, N., Glycoproteins. *Sci. Am. 230*(5), 78 (May 1974).

Tanford, C., The hydrophobic effect and the organization of living matter. *Science 200*, 1012 (1978).

Uy, R., and Wold, F., Posttranslational covalent modification of proteins. *Science 198*, 890 (1977).

Wood, W. G., Haemoglobin synthesis during human fetal development. *Br. Med. Bull. 32*, 282 (1976).

Zuckerkandl, E., The evolution of hemoglobin. *Sci. Am. 212*(5), 110 (May 1965).

Books, Monographs, and Symposia

Anfinsen, C. B., *The Molecular Basis of Evolution*, Wiley, New York, 1963.

Ayala, F. J., *Molecular Evolution*, Sinauer Assoc., Sunderland, Mass., 1976.

Bunn, H. F., Forget, B. G., and Ranney, H. M., *Human Hemoglobins*, Saunders, Philadelphia, 1977.

Dickerson, R. E., and Geis, I., *The Structure and Action of Proteins*, Harper & Row, New York, 1969.

Fox, J. L., Zdenek, D., and Blazej, A., *Protein Structure and Evolution*, Marcell Dekker, New York, 1975.

Gottschalk, A., Glycoproteins and glycopeptides, in *Comprehensive Biochemistry*, Vol. 8 (M. Florkin and E. A. Stotz, Eds.), Elsevier, Amsterdam, 1963.

Ingram, V. M., *The Hemoglobins in Genetics and Evolution*, Columbia University Press, New York, 1963.

Ingram, V. M., *Biosynthesis of Macromolecules* (2nd ed.), W. A. Benjamin, Menlo Park, Calif., 1972.

Masters, C. J., and Holmes, R. S., *Haemoglobin, Isoenzymes and Tissue Differentiation*, North-Holland, Amsterdam, 1975.

McConkey, E. H., *Protein Synthesis*, Vol. 1, Marcel Dekker, New York, 1971.

McGilvery, R. W., *Biochemical Concepts*, Saunders, Philadelphia, 1975.

Russel, T. R., Brew, K., Faber, H., and Schultz, J., *From Gene to Protein: Information Transfer in Normal and Abnormal Cells*, Academic Press, New York, 1979.

Schulz, G. E., and Schirmer, R. H., *Principles of Protein Structure*, Springer-Verlag, New York, 1979.

Stryer, L., *Biochemistry*, (2nd ed.), W. H. Freeman, San Francisco, 1981.

Timasheff, S. N., and Fasman, G. D., *Structure and Stability of Biological Macromolecules*, Marcel Dekker, New York, 1969.

Wold, F., *Macromolecules: Structure and Function*, Prentice-Hall, Englewood Cliffs, N.J., 1971.

Chapter 5
THE CELLULAR MACROMOLECULES: POLYSACCHARIDES

Polysaccharides are complex carbohydrates that play important and diverse roles in the physiology of cells and tissues; some (e.g., cellulose and chitin) serve as structural and supportive extracellular elements, while others (e.g., starch and glycogen) are stored within cells as reserve energy sources. Carbohydrates are composed of simple sugar molecules called **monosaccharides,** many of which conform to the general formula $C_nH_{2n}O_n$. In some monosaccharides, nitrogen and sulfur are also present. The simple sugars are linked together in chains to form carbohydrates of varying size. Carbohydrates consisting of two simple sugars are called **disaccharides;** those containing three simple sugars are called **trisaccharides.** The term **oligosaccharides** encompasses the disaccharides and trisaccharides and other sugars with up to 10 monosaccharide units. The polysaccharides contain very large numbers (often many thousands) of simple sugar units. More than 200 different naturally occurring monosaccharides have been identified but only a small number of these occur as regular constituents of the polysaccharides. The chemical structures of a number of common sugars have been known since 1900, mainly as a result of the work of Emil Fischer. For his work on carbohydrates, Fischer received the first Nobel Prize. Fischer's contributions to our early understanding of protein chemistry have already been noted (Chapter 4).

Monosaccharides

The most commonly occurring monosaccharides are either **aldoses** or **ketoses** and contain three to six carbon atoms forming an unbranched chain; those sugars containing three carbons are **trioses,** those with four carbons **tetroses,** those with five carbon atoms **pentoses,** and those with six carbon atoms **hexoses** (Figs. 5–1 and 5–2). Each of the carbon atoms of the chain is assigned a number beginning with the end that is closest to the carbon bearing the *double-bonded* oxygen. In the aldoses, the number 1 carbon (called the "ultimate" carbon) forms the double bond with oxygen, whereas in the ketoses, it is usually the "penultimate" carbon. The middle carbon atom (i.e., carbon number 2) of glyceraldehyde is asymmetric; that is, it has four chemically different groups bonded to it, and it therefore has two **optical isomers.** The two isomers of glyceraldehyde are shown in Figure 5–3 and are identified as the D and L forms. By chemical convention, the hydroxyl group of the asymmetric carbon is to the right in the D form and to the left in the L form. All of the aldoses may be considered to be derivatives of glyceraldehyde, while the corresponding ketoses are derivatives of dihydroxyacetone. D or L notation is determined by similarity of the second carbon atom to glyceraldehyde and is independent of whether the sugars

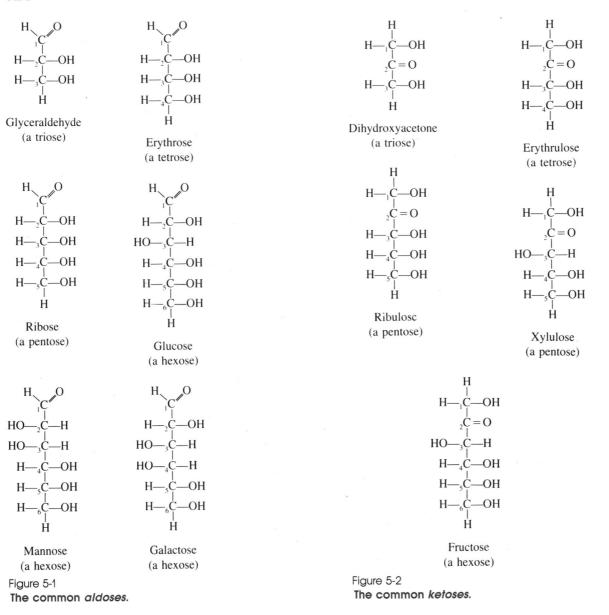

Glyceraldehyde
(a triose)

Erythrose
(a tetrose)

Ribose
(a pentose)

Glucose
(a hexose)

Mannose
(a hexose)

Galactose
(a hexose)

Figure 5-1
The common *aldoses*.

Dihydroxyacetone
(a triose)

Erythrulose
(a tetrose)

Ribulose
(a pentose)

Xylulose
(a pentose)

Fructose
(a hexose)

Figure 5-2
The common *ketoses*.

are dextrorotatory or levorotatory. The most common aldoses and ketoses found in nature (e.g., those in Figs. 5–1 and 5–2) are the D forms.

The number of possible isomers of a monosaccharide is determined by the number, n, of asymmetric carbon atoms present in the chain. Glucose ($n = 4$) has 2^4 or 16 possible isomers. In spite of the variety of possible monosaccharide isomers, only a few different forms occur naturally in any abundance. For example, there are only three naturally occurring isomers of glucose; these are (1) D-glucose, (2) D-mannose, and (3) D-galactose (see Fig. 5–1).

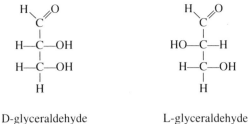

D-glyceraldehyde   L-glyceraldehyde

Figure 5-3
D and L forms of *glyceraldehyde.*

Pyranoses and Furanoses

Some of the monosaccharides can occur in either of two structural configurations: *open-chain* and *cyclic* (or *ringed*). Cyclic forms that contain a six-member ring are called **pyranoses** and those that contain a five-member ring are called **furanoses.** Both the open-chain and cyclic configurations are in reversible equilibrium with one another in aqueous solution, although the cyclic forms are much more prevalent. As an example, the formation of the cyclic structures of D-glucose is shown in Figure 5–4. The aldehyde group of carbon atom number 1 reacts with the hydroxyl group of carbon number 5 to produce a six-member ring containing oxygen. The formation of the ring form of glucose makes carbon number 1 asymmetric and thereby increases the number of possible isomers. The two ring forms of D-glucose are called the α and β anomers. In aqueous solutions of glucose, the β anomer accounts for about 63% of the molecules present, the α form 36%, and the open chain form about 1%.

The three-dimensional arrangement of the constituent atoms of the ring structure of monosaccharides is more readily visualized when the structure is depicted in the **Haworth projection** (after W. H. Haworth) (Fig. 5–5). In this projection, the members of the ring are arranged in a plane that is perpendicular to the plane of the page. Groups to the right of the carbon skeleton in the open-chain configuration are depicted below the plane of the ring, and those to the left of the carbon skeleton extend above the plane of the ring. Certain bonds between members of the ring are deliberately shaded more heavily in order to assist the viewer's perception of the three-dimensional perspective.

The Haworth projection of glucose (and other pyranoses) in which the members of the ring are all depicted as lying in the same plane is not entirely accurate. While carbon

Figure 5-5
Haworth projection of glucose.

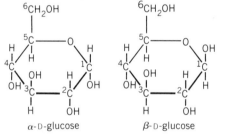

α-D-glucose   β-D-glucose

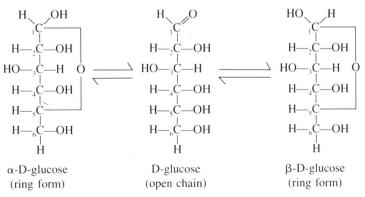

α-D-glucose   D-glucose   β-D-glucose
(ring form)   (open chain)   (ring form)

Figure 5-4
Open-chain and cyclic forms of D-glucose.

Figure 5-6
Stereoscopic views of the "boat" (*a*) and "chair" (*b*) forms of aldohexoses and the complete α anomer (*c*). The oxygen atom of the ring is shown in color.

(a)

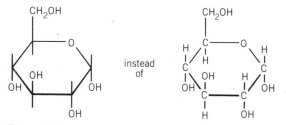

(b)

(c)

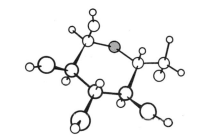

instead of

Figure 5-7
For simplicity and convenience the carbon atoms and associated hydrogens of the ring may be omitted from the Haworth projection.

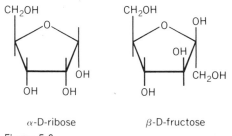

α-D-ribose   β-D-fructose

Figure 5-8
Cyclic forms of D-ribose and D-fructose.

atoms 2, 3, and 5 and the oxygen atom lie in the same plane (to form the corners of a square), carbon atoms 1 and 4 lie either slightly above or below the plane. This gives rise to the so-called *boat* and *chair* forms of glucose (Fig. 5–6), of which the chair form is energetically favored. Other cyclic aldohexoses also exhibit boat and chair forms. For simplicity and convenience, the carbon atoms of the ring and their associated hydrogens are often omitted from the Haworth representation (Fig. 5–7). The cyclic forms of two other common monosaccharides, α-D-ribose and β-D-fructose are shown in Figure 5–8. It should be noted that while both glucose and fructose are hexoses, glucose (an aldohexose) forms a pyranose whereas fructose (a keto-hexose) forms a furanose.

Disaccharides

Disaccharides consist of two ringed monosaccharides. The bonds that unite neighboring monosaccharides are called **glycosidic bonds** and are formed by the condensation of a hydroxyl group of carbon atom number 1 of one monosaccharide with the hydroxyl group of either the number 2, 4, or 6 carbon atom of another. The formation of the common

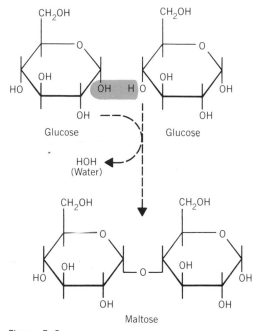

Figure 5–9
Condensation of two molecules of glucose to form the disaccharide *maltose*.

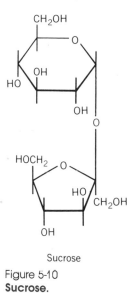

Sucrose

Figure 5-10
Sucrose.

Lactose

Figure 5-11
Lactose.

disaccharide *maltose* from two molecules of glucose is shown in Figure 5–9. In maltose, the oxygen bridge is formed between the number 1 carbon atom of one α-D-glucose unit and the number 4 carbon atom of the other. The bond formed is referred to as an α1→4 glycosidic bond.

Another important disaccharide is *sucrose* (i.e., ordinary "table" sugar) and is formed by the condensation of α-D-glucose and β-D-fructose (Fig. 5–10). Milk contains the disaccharide *lactose,* which consists of the hexoses β-D-galactose and β-D-glucose (Fig. 5–11). In lactose, the glycosidic bond is of the beta variety; i.e., β1→4 (compare with maltose).

Polysaccharides

The **polysaccharides** (or *glycans*) are composed of long chains of sugars and can be divided into two main functional groups: the *structural* polysaccharides and the *nutrient* polysaccharides. The structural polysaccharides serve primarily as extracellular or intercellular supporting ele-

ments. Included in this group are *cellulose* (found in plant cell walls), *mannan* (found in yeast cell walls), *chitin* (in the shells of arthropods and the cell walls of some fungi), *hyaluronic acid* (in connective tissue) and the *peptidoglycans* of bacteria. The nutrient polysaccharides serve as reserves of monosaccharides and are in continuous metabolic turnover. Included in this group are *starch* (plant cells and bacteria), *glycogen* (animal cells), and *paramylum* (in certain protozoa).

On chemical bases, the polysaccharides can be divided into two broad classes: the **homopolysaccharides** and the **heteropolysaccharides.** In the homopolysaccharides, all of the constituent sugars are the same. Included in this class are cellulose, starch, glycogen, and paramylum. In the heteropolysaccharides, the constituent sugars may take different forms; included here is hyaluronic acid. Some polysaccharides are unbranched (i.e., linear) chains whose structure may be ribbon-like or helical (usually a left-handed spiral). Other polysaccharides are branched and assume a globular form. A description of some of the more important polysaccharides follows.

Cellulose

Cellulose is not only the most common of all polysaccharides, it is the most abundant organic substance in the living world. Indeed, it is estimated that more than half of all organic carbon is present as cellulose. Cellulose is the major component of plant cell walls, where it plays a structural role. Cellulose is an unbranched polymer of glucose in which the neighboring monosaccharides are joined by β1→4 glycosidic bonds (Fig. 5–12). Chain lengths vary from several hundred to several thousand glucosyl units. Successive pyranose rings in cellulose are rotated 180° relative to one another so that the chain of sugars takes on a "flip-flop" appearance.

In plant cell walls, large numbers of cellulose molecules are organized into cross-linked, parallel *microfibrils* (Fig. 5–13), whose long axis is that of the individual chains. The cellulose microfibrils may be coated with *hemicellulose* (a smaller polysaccharide formed from xylose, mannose, or galactose). The plant cell wall is made up of a series of layers of these microfibrils, with the walls of neighboring cells cemented together by *pectin* (a polymer of *galact-uronic acid*). Vast amounts of cellulose may be deposited between neighboring cells as is vividly seen in the scanning electron photomicrographs of wood shown in Figure 5–13. Cellulose has also been identified in the cell walls of algae and some fungi.

Chitin

Chitin is an extracellular structural polysaccharide found in large quantities in the body covering (*cuticle*) of arthropods and in smaller amounts in sponges, mollusks, and annelids. Chitin has also been identified in the cell walls of most fungi and some green algae. The chemical structure of chitin is closely related to that of cellulose; the difference is that the hydroxyl group of each number 2 carbon atom is replaced by an acetamido group. Hence, chitin is an unbranched polymer of *N-acetylglucosamine* (Fig. 5–14) containing several thousand successive aminosugar units linked by β1→4 glycosidic bonds.

Hyaluronic Acid

Hyaluronic acid is an unbranched heteropolysaccharide containing repeating disaccharides of *N-acetylglucosamine* and *glucuronic acid* (Fig. 5–15). Glucuronic acid is linked to *N*-acetylglucosamine in each disaccharide by a 1→3 glycosidic bond, but successive disaccharides are 1→4 linked. Hyaluronic acid is found in connective tissue, in the synovial fluid of joints, in the vitreous humor of the eyes, and also in the capsules of bacteria.

Figure 5-12
Small segment of a cellulose molecule.

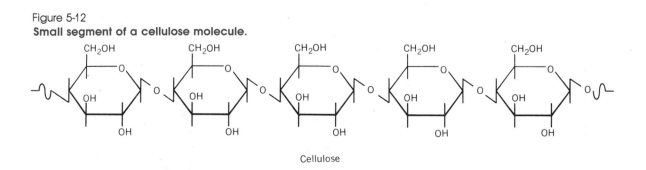

Cellulose

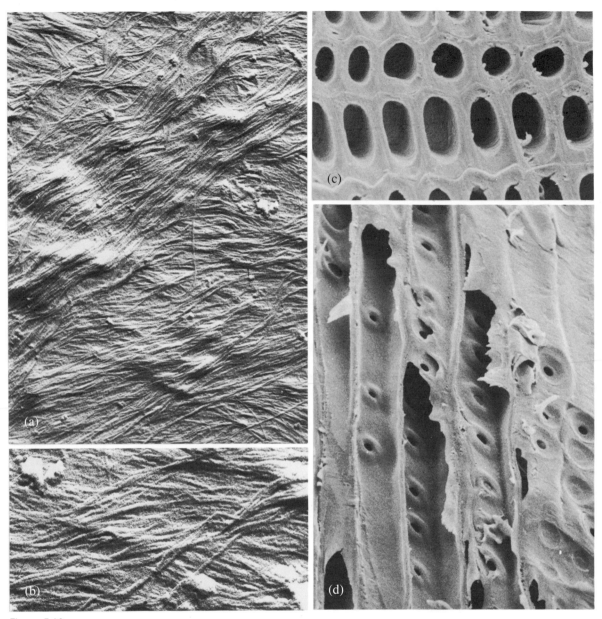

Figure 5-13
(*a*) Transmission electron photomicrograph of cellulose microfibrils in the cell wall of cotton cells. As the wall develops the microfibrils are laid down as successive layers; within a layer the orientation of the microfibrils is consistent but may vary from one layer to the next. Magnification 15,000×. (*b*) Higher magnification view (30,000×). (*c*) Scanning electron photomicrograph of pine wood cut perpendicular to the cells' long axis. Note the thickness of the cell walls (magnification 350×). (*d*) Longitudinally cut pine wood (magnification 350×). This tissue is younger than that shown in (*c*) and the cell walls are thinner; note the bordered pits through which materials can flow between adjacent cell surfaces. (Photomicrographs courtesy of Dr. J. H. M. Willison.)

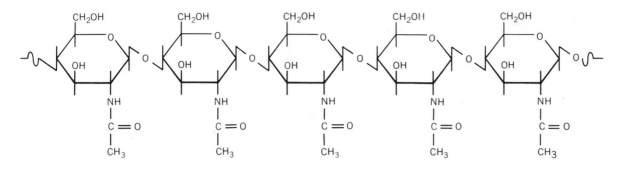

Chitin

Figure 5-14
Small segment of a chitin molecule.

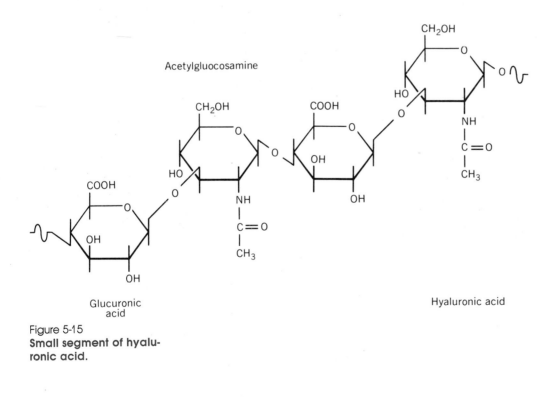

Acetylgluocosamine

Glucuronic
acid

Hyaluronic acid

Figure 5-15
Small segment of hyaluronic acid.

Inulin

Inulin is an unbranched nutrient polysaccharide found in the bulbs of such plants as artichokes, dahlias, and dandelions. It consists of repeating *fructose* units in $\beta 2 \rightarrow 1$ linkage (Fig. 5–16).

Glycogen

Glycogen is a branched nutrient homopolysaccharide containing *glucose* in $\alpha 1 \rightarrow 4$ and $\alpha 1 \rightarrow 6$ linkages. It is found in nearly all animal cells and also in certain protozoa and algae. In view of its ubiquitous occurrence, it is an ex-

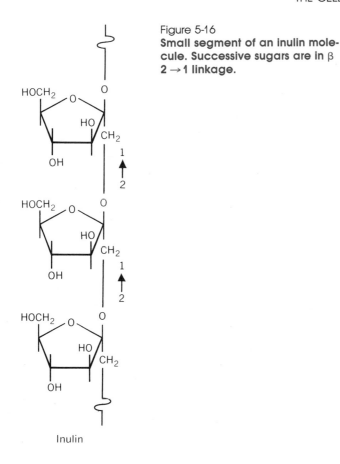

Figure 5-16

Small segment of an inulin molecule. Successive sugars are in β 2 → 1 linkage.

Inulin

glycosidic bonds. This yields the "bush"- or "tree"-like structure depicted in Figure 5–18. It should be noted that in a glycogen molecule there is only one glucose unit whose number 1 carbon atom bears a hydroxyl group. All of the other 1-OH groups are involved in $\alpha1\rightarrow4$ and $\alpha1\rightarrow6$ glycosidic bonds. The single free 1-OH group is called the "reducing end" of the molecule and is noted by the letter *R* in Figure 5–17. In contrast, numerous "nonreducing ends" are present (i.e., free 4-OH and 6-OH groups) at the terminals of the outermost chains.

In Figure 5–18 the individual glucose units are represented by circles and the branch points (i.e., the $\alpha1\rightarrow6$ linkages) by heavier connections. In this model of glycogen, a number of different kinds of chains may be distinguished. *A chains* are attached to the molecule by a single $\alpha1\rightarrow6$ linkage (chains of open circles in Fig. 5–18). *B chains* (gray circles) bear one or more A chains. Each glycogen molecule contains only one *C chain* (colored circles), and this is the chain that ends in the free reducing group. *Exterior chains* are those portions of individual chains between the nonreducing end groups and the outermost branch points. Finally, those parts of individual chains between branch points are called *interior chains*. The exterior chains of glycogen are usually six to nine glucosyl units long, while interior chains contain only three to four glucosyl units. Approximately 8 to 10% of all glycosidic linkages are the $\alpha1\rightarrow6$ type.

Glycogen molecules are often sufficiently large to be studied by electron microscopy. Although the molecules can be seen in transmission electron photomicrographs of osmium tetroxide-fixed tissues (especially liver, see Fig. 1–13), morphologic studies of glycogen are usually carried out with material that has been negatively stained with phosphotungstic acid and osmium tetroxide (Fig. 5–19).

Starch

Starch is a nutrient polysaccharide found in plant cells, protists, and certain bacteria and is similar in many respects to glycogen. (In fact, glycogen is often referred to as "animal starch.") Starch usually occurs in cells in the form of granules visible by both light and electron microscopy. In plant cells (such as potato or corn), these granules may be several micrometers in diameter, whereas in microorganisms their diameter may be only 0.5 to 2 μm. Starch granules contain a mixture of two different polysaccharides,

tremely important polysaccharide and has been the object of numerous extensive studies. In man and other vertebrates, glycogen is stored primarily in the liver and muscles and is the principal form of stored carbohydrate. In an unstarved animal, as much as 10% of the liver weight may be glycogen. Glycogen undergoes an almost continuous biosynthesis and degradation, especially in liver tissue. Liver glycogen serves as a reservoir for glucose under starvation conditions (it may be almost completely depleted during 24 hours of starvation) and during muscular exertion. Glycogen is quickly resynthesized from newly ingested carbohydrate.

Glycogen molecules exist in a continuous spectrum of sizes, with the largest molecules containing many thousands of glucosyl units. A small portion of a glycogen molecule is shown in Figure 5–17. The glucosyl units that are linked by $\alpha1\rightarrow4$ glycosidic bonds are organized into long chains; the chains are interconnected at branch points by $\alpha1\rightarrow6$

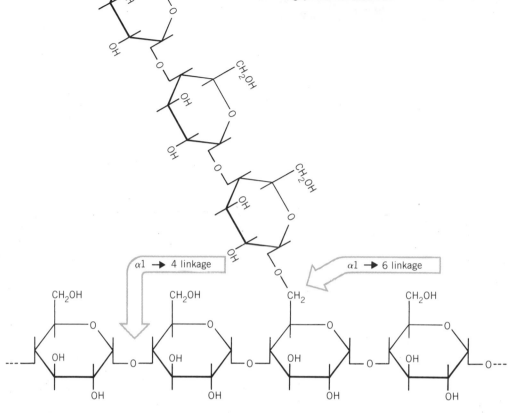

Figure 5-17
A portion of a glycogen molecule. Glucosyl units are linked by α 1 → 4 gly-cosidic bonds to form long chains, and these chains are interconnected by α 1 → 6 glycosidic bonds.

α1 → 4 linkage

α1 → 6 linkage

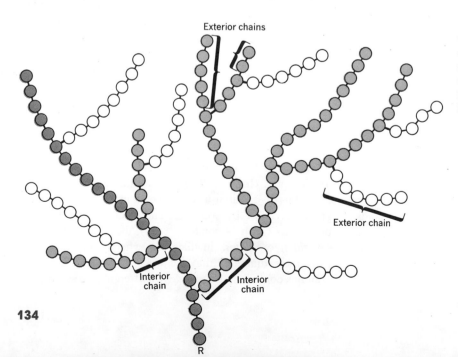

Exterior chains

Exterior chain

Interior chain

Interior chain

R

Figure 5-18

"Bush"- or "tree"-like structure of the gly-cogen molecule. A chains are shown by open circles, B chains are gray circles, and the C chain is shown in color. The reducing end of the molecule is de-noted by the letter *R*.

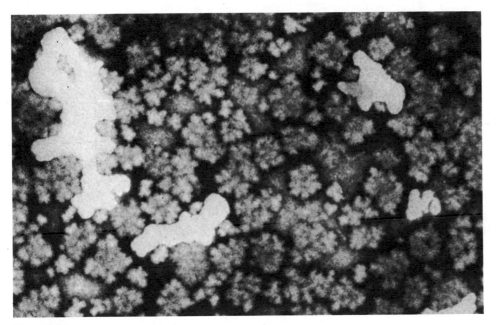

Figure 5-19
Electron photomicrograph of negatively stained glycogen particles (rosettelike structures) from liver tissue magnified 80,000×.

amylose and *amylopectin,* and the relative amounts of these two polysaccharides vary according to the source of the starch.

The amylose component of starch is an unbranched α1→4 polymer of glucose and may be several thousand glucosyl units long. The polysaccharide chain exists in the form of a *left-handed* helix (Fig. 5–20). The familiar blue color that is produced when starch is treated with iodine is believed to result from the coordination of iodide ions in the interior of the helix.

Amylopectin is a branched polysaccharide containing α1→4 and α1→6 linked glucosyl units; in this respect it is similar to glycogen. However, amylopectin has a more open structure with fewer α1→6 linkages and longer chain lengths. Some of the characteristics of amylopectin and glycogen are compared in Table 5–1.

Other Polysaccharides

In addition to the polysaccharides already described, several others should be briefly reviewed. *Mannan,* a homopolymer of mannose, is found in the cell walls of yeast and is also stored intracellularly in some plants. In yeast, the mannan has a branched structure, whereas in plants it is a linear molecule. *Paramylum* is a nutrient homopolysaccharide stored intracellularly as large granules in certain protozoa (e.g., *Euglena*). It consists of unbranched chains of glucosyl units in 1→3 linkage. *Dextran* is a branched homopolymer of glucose produced by certain bacteria. However, unlike glycogen and amylopectin, the glycosidic linkages vary from one species to another and may be 1→6, 1→4, 1→3, or 1→2. Some of the properties of the various polysaccharides are summarized in Table 5–2.

Proteoglycans, Glycoproteins, and Glycolipids

Polysaccharides also occur in covalent combination with proteins and lipids. In the **proteoglycans,** the protein portion of the molecule represents only a small portion of the

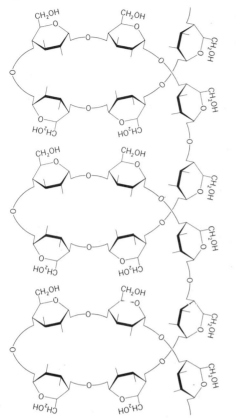

Figure 5-20
Left-handed helix formed by the amylose polysaccharide.

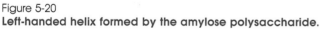

Table 5–1
Some Chemical Properties of Amylopectin and Glycogen

	Amylopectin	Glycogen
Molecular weight	10^7	10^6–10^9
Average chain length	20–25	10–14
Percentages 1 $\longrightarrow$ 6 linkages	4–5	8–10
Exterior chain length	12–17	6–9
Interior chain length	5–8	3–4

to account for more than half of the total molecular weight. Glycoproteins serve diverse roles in cells and tissues and include certain *enzymes, hormones,* and *antibodies* (see also Chapter 4).

Glycolipids are covalent combinations of carbohydrate and lipid. The carbohydrate portion may be a single monosaccharide or a linear or branched chain. Glycolipids are components of most cell membranes; their chemistry is considered in somewhat greater detail in Chapter 6.

Summary

The **polysaccharides** are macromolecules composed of individual sugar units called **monosaccharides.** Among these, the most common is the hexose *glucose.* Although monosaccharides may contain three or more carbon atoms and are derivatives of *glyceraldehyde* or *dihydroxyacetone,* the six-carbon forms occur most often in polysaccharides. Individual sugars may take the open-chain configuration or may form ringed structures; only the ringed **pyranoses** and **furanoses** occur as regular constituents of polysaccharides. From a functional standpoint, polysaccharides may be divided into *nutrient* and *structural* types. Nutrient polysaccharides such as *glycogen* and *starch* are often stored in cells as discrete granules visible either by light or electron microscopy. The most abundant structural polysaccharide, *cellulose,* serves as the major supportive element in the cell walls of higher plants. Polysaccharides may be formed from unbranched chains of sugars (as in cellulose and the *amylose* component of starch) or from branched chains (as in glycogen and the *âmylopectin* component of starch). All sugars are the same type in the **homopolysaccharides** (e.g., cellulose, glycogen, and starch) but vary in the **hetero-**

total molecular weight. These macromolecules are also known as **mucopolysaccharides.** Probably the most important of the proteoglycans is *chondroitin* found extracellularly in cartilage and bone. The polysaccharide portion of chondroitin consists primarily of chains of *glucuronic acid-acetylglucosamine* disaccharides. In some of the acetylglucosamine units, hydroxyl groups are replaced by sulfate groups. Several dozen polysaccharide chains may be present, each covalently linked to the protein portion of the molecule through the hydroxyl group of serine residues.

The cell walls of certain bacteria contain *peptidoglycans.* These are heteropolymers of *N*-acetylglucosamine and *N*-acetylmuramic acid in which short peptides are linked to the polysaccharide backbone.

In the **glycoproteins,** the carbohydrate portion consists of much shorter chains, which are often branched. The chains may be few in number or there may be so many as

Table 5–2
Polysaccharides

Class	Name	Source	Composition	Linkages
Structural polysaccharides	Cellulose	Plant cell walls	Glucose (beta linkage)	Unbranched 1 ⟶ 4
	Mannan	Yeast cell walls	Mannose (beta linkage)	Branched 1 ⟶ 2, 1 ⟶ 3 and 1 ⟶ 6
	Chitin	Arthropod shells, fungal cell walls	Acetylglucosamine and glucuronic acid (beta linkage)	Unbranched 1 ⟶ 4
	Hyaluronic acid	Synovial fluid (joints), subcutaneous tissue	Acetylglucosamine and glucuronic acid (beta linkage)	Unbranched 1 ⟶ 3 and 1 ⟶ 4
	Peptidoglycans	Bacterial cell walls	Acetylglucosamine and Acetylmuramic acid	Unbranched 1 ⟶ 4
Nutrient polysaccharides	Inulin	Artichokes, dandelions	Fructose (beta linkage)	Unbranched 2 ⟶ 1
	Paramylum	Certain protozoa (i.e., *Euglena*)	Glucose (beta linkage)	Unbranched 1 ⟶ 3
	Glycogen	Certain protozoa (i.e., *Tetrahymena*) and most animals	Glucose (alpha linkage)	Branched 1 ⟶ 4 and 1 ⟶ 6
	Starch: Amylopectin	Plant cells and some protozoa (e.g., *Polytomella*)	Glucose (alpha linkage)	Branched 1 ⟶ 4 and 1 ⟶ 6
	Amylose		Glucose (alpha linkage)	Unbranched 1 ⟶ 4

polysaccharides. Polysaccharides may also occur in covalent combination with proteins, short peptides, and lipids, forming **proteoglycans, glycoproteins, peptidoglycans, and glycolipids.**

References and Suggested Reading

Articles and Reviews

Manners, D. J., The molecular structure of glycogens. *Adv. Carbohyd. Chem. 13,* 261 (1957).

Manners, D. J., Enzymatic synthesis and degradation of starch and glycogen. *Adv. Carbohyd. Chem. 17,* 371 (1962).

Sharon, N., Carbohydrates. *Sci. Am. 243*(5) 90 (Nov. 1980).

Books, Monographs, and Symposia

Aspinall, G. O., *Polysaccharides,* Pergamon Press, Oxford, 1970.

Conn, E. E., and Stumpf, P. K., *Outlines of Biochemistry* (4th ed.), Wiley, New York, 1976.

Ginsberg, V., and Robbins, P., (eds.) *Biology of Carbohydrates.* Wiley-Interscience, New York, 1981.

Lehninger, A. L., *Biochemistry* (2nd ed.), Worth, New York, 1975.

McGilvery, R. W., and Goldstein, G., *Biochemistry, A Functional Approach* (2nd ed.), Saunders, Philadelphia, 1979.

Rees, D. A., *Polysaccharide Shapes,* Halsted Press, New York, 1977.

Sharon, N., *Complex Carbohydrates,* Addison-Wesley, London, 1975.

Stryer, L., *Biochemistry* (2nd ed.), W. H. Freeman, San Francisco, 1981.

Chapter 6
THE CELLULAR
MACROMOLECULES: LIPIDS

Lipids are a chemically heterogeneous collection of molecules that are only sparingly soluble in water but are highly soluble in nonpolar, organic solvents such as acetone, benzene, chloroform, and ether. It is because of their similar solubility properties that they are usually placed together in one class. Like the carbohydrates, lipids serve two major roles in cells and tissues: (1) they occur as constituents of certain structural components of cells, particularly membranous organelles; and (2) they may be stored within cells as reserve energy sources. The most common lipids include the **fatty acids, neutral fats, phospholipids,** *(glycerophosphatides, plasmalogens,* and *sphingolipids),* **glycolipids, terpenes,** and **steroids.** When they occur independently, several of these lipids are not macromolecular in structure; however, because they have similar properties and frequently combine with each other and with other kinds of molecules to form macromolecular complexes, they are considered in this section.

Cellular Distribution of Lipids

Table 6–1 shows the distribution of certain lipids in the four major cell fractions that are obtained when liver tissue is homogenized and fractionated by a technique called *differential centrifugation* (see Chapter 12). The distribution

is believed to be fairly representative, at least of animal cells. It should be noted that the mitochondrial and microsomal fractions are richest in lipid. Most of the lipid in these fractions is phospholipid (i.e., glycerophosphatides, etc.) serving as structural constituents of the mitochondrial and microsomal membranes where it is combined with membrane protein (see Chapter 15). Most of the lipids that are found in the nonsedimenting, soluble phase of the cell (i.e., the cytosol) are neutral fats and are in rapid metabolic turnover. The neutral fat content of the cytosol fractions of tissues that are specialized for fat storage (e.g., adipose tissue) is much higher. Procaryotic cells contain less lipid than eucaryotic cells and lack steroids. In bacteria, more than 90% of all of the cellular lipid is associated with the plasma membrane.

Fatty Acids

The fatty acids are an important and also abundant group of lipids. In cells, the fatty acids only sparingly occur freely; instead, they are esterified to other components and form the saponifiable cell lipids (see below). A fatty acid molecule may be either **saturated** or **unsaturated.** The saturated fatty acids consist of long hydrocarbon chains terminating in a carboxyl group and conform to the follow-

Table 6–1
Lipid Distribution in Animal Cells

Fraction	Total Lipid (% dry weight)	Percent of Total Lipid		
		Neutral Fats	Phospholipids[a]	Terpenes and Steroids
Nuclei	16	3	90	5
Mitochondria	21	1	90	6
Microsomes[b]	32	0	90	6
Cytosol	7	70	30	4

[a]Glycerophosphatides, plasmalogens, and sphingolipids

[b]The microsomal fraction includes fragments of the ER, ribosomes, and other tiny cellular inclusions

ing general formula:

$$CH_3—(CH_2)_n—COOH$$

In nearly all naturally occurring fatty acids, n is an even number from 2 to 22. In the saturated fatty acids most commonly found in animal tissues, n is either 12 (i.e., *myristic acid*), 14 (i.e., *palmitic acid*), or 16 (i.e., *stearic acid*). In unsaturated fatty acids, at least two but usually no more than six of the carbon atoms of the hydrocarbon chain are linked together by double bonds. The two most common unsaturated fatty acids are *oleic acid* and *linoleic acid,* depicted in Figure 6–1 along with the saturated fatty acid stearic acid. In all of the common saturated and unsaturated fatty acids, the long hydrocarbon chain is *unbranched.* Unsaturated fatty acids predominate in the *saponifiable* lipids of higher plants and in animals that live at low temperatures. Lipids in the tissues of animals inhabiting warm climates contain larger quantities of saturated fatty acids. Some animals, especially mammals, are unable to synthesize certain fatty acids and therefore require them in their diets. The essential fatty acids, like linoleic acid, must be obtained from plant material. Fatty acids found in bacterial cells are usually of short chain length and, when unsaturated, have only one double bond.

Fatty acid molecules contain both hydrophilic and hydrophobic parts. In their dissociated states (shown in the formulas of Fig. 6–1), the carboxyl ends of the molecules are mildly soluble in water, while the long hydrocarbon chains repel water. In neutral solutions, salts of the fatty acids form small spherical droplets or **micelles** in which the dissociated carboxyl groups occur at the surface, and the hydrophobic carbon chains project toward the center. (This arrangement is shown in Figure 3–1.) As noted above, most of the fatty acids that are found in cells and tissues are esterified to other molecules. Since the ester linkages can be hydrolyzed by the addition of alkali, these lipids are called *saponifiable lipids.*

Saponifiable Lipids

Neutral Fats

The **neutral fats, triglycerides,** or **triacylglycerols** are esters of glycerol and three fatty acids and have the general formula shown in Figure 6–2. In this formula, n, n' and n'' may be the same number or different numbers. Usually, the fatty acid that is esterified to the first carbon atom of glycerol (i.e., the top carbon in Fig. 6–2) is saturated, whereas the fatty acid esterified to the middle carbon atom is unsaturated. Either saturated or unsaturated fatty acids are found at the third carbon position. Unlike the fatty acids, the neutral fats are entirely nonpolar. Neutral fats, which are often deposited in cells as potential sources of chemical energy, represent the major type of stored lipid; they occur as droplets in the cytoplasm and most of the lipid that is recovered in the soluble phase or cytosol of disrupted cells takes this form. The lipid droplets stored in abundance in muscle cells are clearly seen in Fig. 6–3.

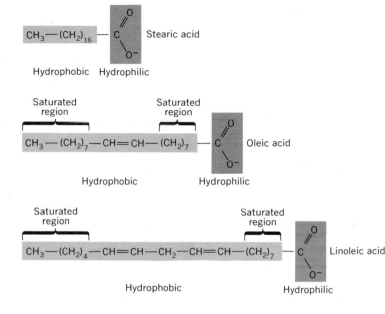

Figure 6-1
The *saturated* fatty acid *stearic acid* and the *unsaturated* fatty acids *oleic* and *linoleic* acid.

Glycerophosphatides

The major members of this group of lipids are derivatives of *phosphatidic acid*. Phosphatidic acid is similar to a triglyceride except that one of the fatty acids is replaced by a phosphate group (Fig. 6–4). Phosphatidic acid and its derivatives are present in cell membranes where they play an active role in membrane function (Chapter 15) in addition to serving as structural constituents (Fig. 6–3). The most common derivatives of phosphatidic acid are *phosphatidyl choline* (also known as *lecithin*), *phosphatidyl ethanolamine* (also known as *cephalin*), *phosphatidyl serine,* and *phosphatidyl inositol* (Fig. 6–5). Free rotation about the single bonds of the glycerol backbone allow the hydrophilic phosphate and its derivatives to face away from the hydrophobic portion, as shown in the structural formulas of Figure 6–5. Consequently, glycerophosphatides (and other lipids discussed below) are **amphipathic**—that is, one end of the molecule is strongly *hydrophobic* (i.e., the end containing the hydrocarbon chains), while the other end is *hydrophilic* due to the charged nature of the dissociated phosphate group and other substituents.

In the presence of water, glycerophosphatides can behave in various ways (Fig. 6–6). For example, they can form

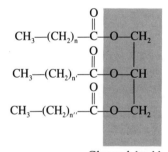

Glycerol backbone

Figure 6–2
Generalized formula for a *neutral fat (triglyceride)*.

monolayers (i.e., *monomolecular layers*) on the water with their polar ends projecting into the water and their hydrophobic ends directed away from the water. Also, like fatty acids, glycerophosphatides can form *micelles* with the polar ends of the molecules at the micelle's surface and the hydrocarbon chains projecting toward the center. However, of special significance is the tendency of glycerophospha-

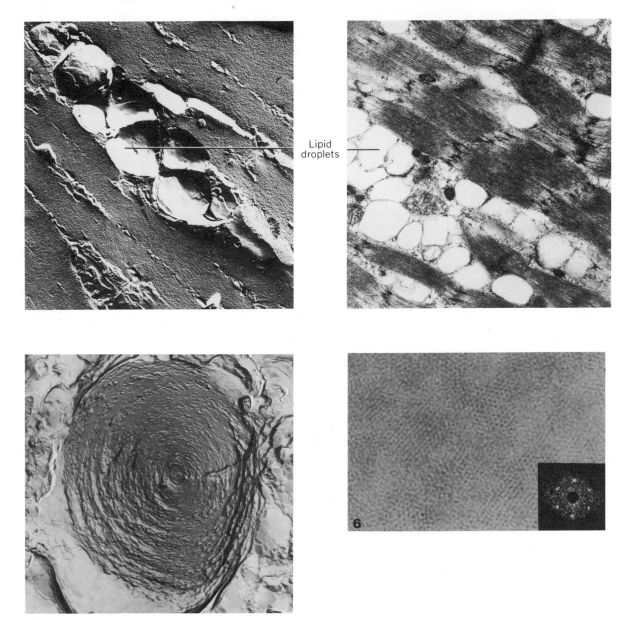

Lipid
droplets

Figure 6-3
Lipids in cells. *Top.* Fat droplets in striated muscle tissue. Freeze-fracture (left) and
transmission (right) electron micrographs. (Courtesy A. P. Bland and C. J. Roberts;
copyright © 1981 by Pergamon Press Ltd.; *Micron 12,* 163.) *Bottom.* Cardiolipin, an
acidic phospholipid commonly found in the inner membranes of mitochondria
and chloroplasts. *Left,* freeze-fractured granule revealing concentric layers; *right,*
crystalline lipid structure and X-ray diffraction pattern (*insert*). (Courtesy Drs. P.
Arslan, M. Beltrame, and U. Muscatello; copyright © 1980 by Pergamon Press Ltd.;
Micron 11, 118.)

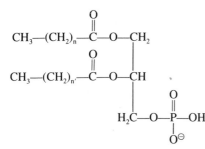

Figure 6-4
Phosphatidic acid.

tides to form *bilayers* or *bimolecular sheets* (Fig. 6–6). Bilayers formed predominantly from glycerophosphatides are believed to be fundamental to the structure of the plasma membrane, the membranes of the endoplasmic reticulum (see Chapter 15), and the membranes of vesicular organelles.

Hydrophobic interactions between the tails of the glycerophosphatides are the most important forces promoting the formation of bilayers. The assembly of a bilayer is spontaneous so long as there is adequate lipid present; the spontaneous nature of bilayer formation is reminiscent of the automatic folding of protein molecules that takes place due to the interaction of hydrophobic amino acid side chains (Chapter 4).

Liposomes.   When aqueous suspensions of glycerophosphatides are subjected to rapid agitation using ultrasound (i.e., *insonation*), the lipid disperses in the water and forms **liposomes** or *lipid vesicles* (Fig. 6–7). These liposomes consist of a spherical bilayer of the glycerophosphatide molecules enclosing a small volume of the aqueous medium. If small molecules and ions are initially dissolved in the water, some of these will be enclosed by the liposomes. The ability of different substances inside or outside the liposomes to traverse the bilayer can be studied experimen-

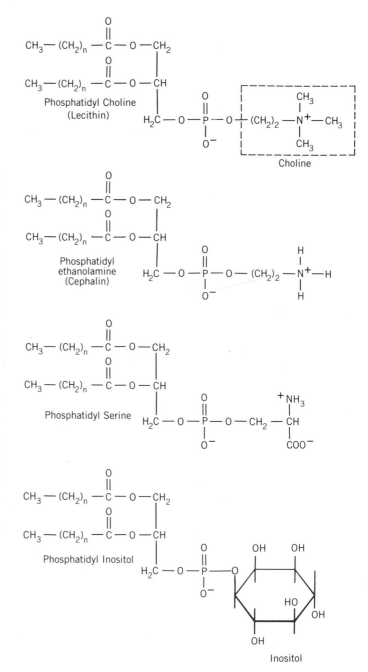

Figure 6-5
Phosphatidyl choline (lecithin), phosphatidyl ethanolamine (cephalin), phosphatidyl serine, and phosphatidyl inositol.

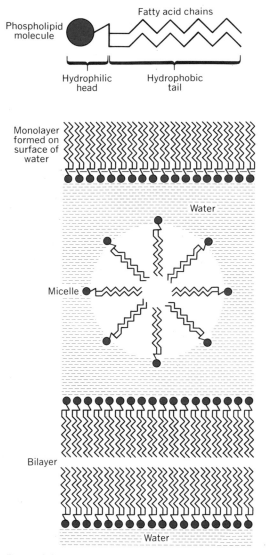

Figure 6-6

In the presence of water, glycerophosphatides can form *monolayers* **on the water's surface,** *micelles,* **or** *bimolecular sheets* **(***bilayers***).**

tally. When this is done, it is found that liposomes exhibit many of the permeability properties of natural cellular membranes. This subject is pursued in Chapter 15.

Plasmalogens

Plasmalogens are a special class of lipids especially abundant in the membranes of nerve and muscle cells and also characteristic of cancer cells. A plasmalogen is similar to a glycerophosphatide except that the fatty acid at the number 1 position of the glycerol is replaced by an unsaturated ether (Fig. 6–8).

Sphingolipids

Sphingolipids are derivatives of *sphingosine,* an amino alcohol possessing a long, unsaturated hydrocarbon chain (Fig. 6–9a). The *myelin sheath* surrounding many nerve cells is particularly rich in the sphingolipid *sphingomyelin* (Fig. 6–9b). In sphingomyelin, the amino group of the sphingosine skeleton is linked to a fatty acid and the hydroxyl group is esterified to phosphorylcholine.

Glycolipids

The outer surface of the plasma membrane of most cells is studded with short chains of sugars. These sugars are either parts of **glycoproteins** (see Chapter 4) or they are attached to membrane lipids, thereby forming **glycolipids.** Glycolipids in the plasma membrane play vital roles in *immunity, blood group specificity,* and *cell–cell recognition.* The carbohydrates that are found in glycolipids vary from individual monosaccharides (e.g., glucose and galactose) to short, branched or unbranched oligosaccharides. The lipid portion of a glycolipid is similar to sphingosine with the amino group of the sphingosine skeleton acylated by a fatty acid (as in sphingomyelin) and the hydroxyl group associated with the carbohydrate. The simplest glycolipids are the *cerebrosides* which, as their name suggests, are abundant in brain tissue, where they are found in the myelin sheaths and may account for as much as 20 percent of the sheath's dry weight. Small amounts of cerebrosides are also found in kidney, liver, and spleen cells. Usually the sugar of a cerebroside is either glucose or galactose (Fig. 6–10). In the *gangliosides* (also associated with nerve tissue), the carbohydrate portion of the molecule consists of a chain of sugar molecules usually including glucose, galactose, and neuraminic acid. One particular ganglioside, called G_{M2}, may accumulate in the lysosomes of brain cells because of a genetic deficiency that results in the failure of the cells to produce a lysosomal enzyme that degrades this ganglioside. The condition is known as *Tay-Sachs* disease and frequently leads to paralysis, blindness, and retarded development.

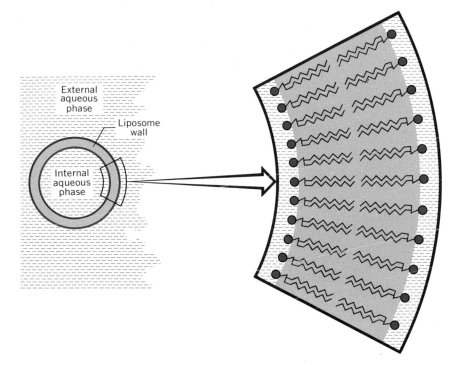

Figure 6-7
Liposomes are small spherical bodies the surface of which is formed by a lipid bilayer. Liposomes show some permeability properties that are similar to those of natural membranes enclosing cells.

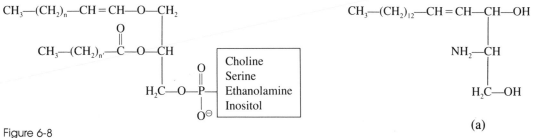

Figure 6-8
Generalized formula for a *plasmalogen*.

(a)

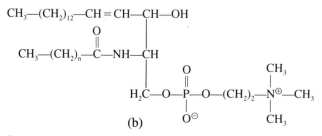

(b)

Figure 6-9
The sphingolipids *sphingosine (a)* and *sphingomyelin (b)*.

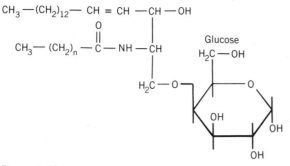

Figure 6–10
A *glycolipid.* The molecule shown is a cerebroside and contains only one sugar, in this instance glucose. *Cerebrosides* may also contain galactose. In *gangliosides*, the hexose is replaced by a chain of sugars.

Nonsaponifiable Lipids

The nonsaponifiable lipids (also known as simple lipids) are lipids that do not contain fatty acids as constituents. Two major classes of nonsaponifiable lipids are the **terpenes** and the **steroids.**

Terpenes

The terpenes include certain fat-soluble vitamins (such as *vitamins A, E,* and *K*), carotenoids (pigments of plant cells that are involved in photosynthesis), and certain coenzymes

such as coenzyme Q (ubiquinone). All of the terpenes are synthesized from various numbers of a five-carbon building-block called an **isoprene** unit (Fig. 6–11). The isoprene units are bonded together in a head-to-tail organization. Two isoprene units form a *monoterpene,* four form a *diterpene,* six a *triterpene,* and so on. The monoterpenes are responsible for the characteristic odors and flavors of plants (e.g., *geraniol* from geraniums, *menthol* from mint, and *limoneme* from lemons).

Steroids

The steroids are an especially important class of lipids consisting of a system of fused cyclohexane and cyclopentane rings. All are derivatives of **perhydrocyclopentanophenanthrene,** which consists of three fused cyclohexane rings (in a nonlinear arrangement) and a terminal cyclopentane ring (Fig. 6–12a).

Steroids have widely different physiological properties and these are determined by the nature of the chemical groups attached to the basic skeleton. Some steroids are **hormones** (e.g., *estrogen, progesterone, corticosterone*) and affect cellular activities by influencing gene expression (Chapters 11 and 20). Some steroids are **vitamins** (e.g., vitamin D_2) and influence the activities of certain cellular enzymes. Other steroids are regular constituents of membranes, where they influence membrane structure, perme-

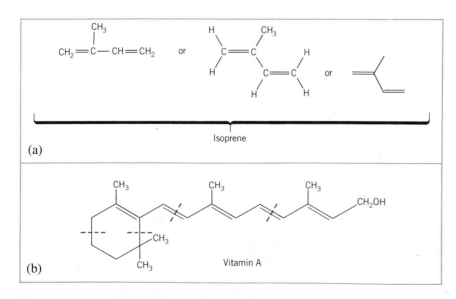

Figure 6-11
(*a*) *Isoprene,* the building block of the *terpenes;* (*b*) *vitamin A* (*retinol*). The ends of neighboring isoprene units are indicated by the broken lines.

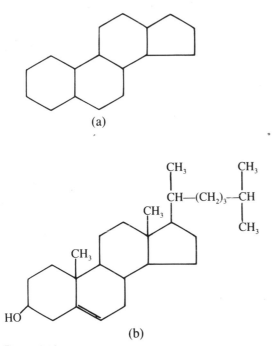

(a)

(b)

Figure 6-12
Perhydrocyclopentanophenanthrene (a) and one of its derivatives, *cholesterol (b)*.

ability, and transport. Probably the best known steroid and an important constituent of the plasma membranes of many animal cells is *cholesterol* (Fig. 6–12*b*).

Summary

The lipids are a heterogeneous collection of molecules soluble in nonpolar solvents that play two major roles in cells and tissues: they are sources of reserve energy and structural constituents of cellular membranes. The simplest lipids, the **saturated** and **unsaturated** fatty acids, consist of unbranched hydrocarbon chains terminating in carboxyl groups. In the **neutral fats,** the carboxyl groups of fatty acids are esterified to the three carbon atoms of *glycerol* and represent the most abundant form of stored lipid. Neutral fats are in rapid metabolic turnover. **Glycerophosphatides,** in which one of the fatty acid chains of a neutral fat is replaced by the polar phosphate group or a phosphate derivative, are **amphipathic** and, as such, play an important part in the structure and organization of cell membranes. The most physiologically diverse lipids are the **terpenes,** (formed from **isoprene** units) and the **steroids** (formed from a series of fused, ringed hydrocarbons). Some terpenes function as vitamins and coenzymes. Steroids serve as hormones, as vitamins, and as constituents of the plasma membranes of many cells.

References and Suggested Reading

Articles and Reviews

Capaldi, R. A., A dynamic model of cell membranes. *Sci. Am. 230*(3), 26 (Mar. 1974).

Bell, R. M., and Coleman, R. A., Enzymes of glycolipid synthesis in eukaryotes. *Annu. Rev. Biochem. 49,* 459 (1980).

Fieser, L. F., Steroids. *Sci. Am. 192*(1), 52 (Jan. 1955).

Fox, C. F., The structure of cell membranes. *Sci. Am. 226*(2), 30 (Feb. 1972).

Hadley, N. F., Cuticular lipids of plants and arthropods. *Biol. Rev. 56,* 23 (1981).

Hokin, L. E., and Hokin, M. R., The chemistry of cell membranes. *Sci. Am. 213*(4), 78 (Oct. 1965).

Kirscher, H. W., and Heftman, E., Symposium: Sterol analysis. *Lipids 15,* 697 (1980).

Lindgren, F. T., Symposium: High density lipoproteins. *Lipids 14,* 504 (1979).

Morrell, P., and Norton, W. T., Myelin. *Sci. Am. 242*(5), 88 (May 1980).

Rosenthal, M. D., Accumulation of neutral lipids by human skin fibroblasts. *Lipids 16,* 173 (1981).

Rothman, J. E., and Lenard, J., Membrane asymmetry. *Science 195,* 743 (1977).

Books, Monographs, and Symposia

Conn, E. E., and Stumpf, P. K., *Outlines of Biochemistry* (4th ed.), Wiley, New York, 1976.

Lands, W. E. M., and Crawford, C. G., *Enzymes of Biological Membranes* (A. Mantonosi, Ed.), Vol. 1, Plenum, New York, 1976.

Lehninger, A. L., *Biochemistry* (2nd ed.), Worth, New York, 1975.

McGilvery, R. W., *Biochemistry, A Functional Approach* (2nd ed.), Saunders, Philadelphia, 1979.

Stryer, L., *Biochemistry* (2nd ed.), W. H. Freeman, 1981.

Chaper 7
THE CELLULAR MACROMOLECULES: NUCLEIC ACIDS

The nucleic acids were discovered over 100 years ago, but their role in genetics and in the control of cellular activity has been elucidated only during the past 50 years. This is probably due in part to the great emphasis placed on the study of proteins during the first half of this century and to the mistaken belief by many scientists during this period that the proteins were endowed with the genetic information of the cell. This error is understandable in view of the fact that the composition and organization of proteins is so diverse, whereas the chemical nature of the nucleic acids (and also their structure) is so much more regular and restricted. Consequently, it was difficult to reconcile the great diversity of life with the fundamental chemical similarities manifested by all the nucleic acids. We begin our consideration of the nucleic acids by describing some of the major observations and discoveries that ultimately led biologists to recognize that these molecules and not proteins were intimately involved in the transmission of genetic information and in the determination and control of the activities of the cell. In this respect, this chapter will differ from the previous chapters on cellular macromolecules, which were almost entirely chemically oriented.

Cellular Roles of the Nucleic Acids
The Discovery of DNA

Friedrich Miescher (Swiss, 1844–1895) is credited with the discovery of nucleic acid in 1869. He isolated nuclei from white blood cells present in pus, using dilute solutions of hydrochloric acid to dissolve away other cell structures and then added the protein-digesting enzyme *pepsin* to further degrade residual cell protein adhering to the nuclei. Nuclei isolated in this manner were then extracted with alkali, and the chemical composition of the extract was analyzed. In view of the unique chemical composition of the extract (which differed markedly from protein), Miescher called this material "nuclein." By comparing the chemical analyses reported by Miescher with those carried out more recently, it is clear that Miescher's nuclein was, in fact, the nucleic acid DNA (deoxyribonucleic acid). The term **nucleic acid** was introduced 20 years following Miescher's discovery by another biochemist, Richard Altmann.

Miescher also worked with the sperm cells of salmon, which contain particularly large nuclei (more than 90% of the cell mass is accounted for by the nucleus). In addition to isolating nuclein from the sperm nuclei, he used acid to extract an organic material having an unusually high nitrogen content. He called this substance "protamine." The protamines are proteins containing an unusually rich content of lysine and arginine residues. We now recognize that extracts of cell nuclei also contain proteins called *histones* and that these are intimately associated with the nuclear DNA (see Chapter 20). The function of the cell nucleus was unknown in Miescher's time, and while Miescher was convinced of the fundamental importance of nuclein, especially during fertilization, it was not until 60 years after his discovery that a series of experiments was carried out that established the genetic role of the nucleic acids. Also during this period, it was shown that two types of nucleic acid occur in cells: deoxyribonucleic acid (DNA) and ribonucleic acid (RNA). Most notable among the experiments that established the genetic role of the nucleic acids were those on (1) the *transformation of bacterial types* and (2) *virus reproduction.*

Transformation of Bacterial Types

Two types of pneumonia bacteria (*Diplococcus pneumoniae*) exist and are readily distinguished by the appearance of their colonies when cultured on agar plates; they are called ''smooth'' (*S*) and ''rough'' (*R*) types. The *S* type (which is the normal, *virulent* kind) is enclosed within a polysaccharide capsule and gives rise to smooth, shiny colonies. In contrast, the *R* type is noninfective (i.e., nonvirulent), is unable to synthesize the polysaccharide capsule, and gives rise to granular (rough-appearing) colonies. In 1928, F. Griffith showed that it was possible to transform the rough bacteria into the smooth type. He simultaneously injected small numbers of live *R* bacteria and large numbers of heat-killed *S* bacteria into mice, many of which subsequently died. When these mice were examined, they were found to contain live *S* bacteria. Since the *S* bacteria originally injected into the mice were incapable of reproducing, Griffith concluded that some of the *R* bacteria must have been *transformed* into the *S* type in the presence of the dead *S* cells. These observations were subsequently confirmed by a number of investigators who also ruled out the possibility of contamination of the inoculum by a few live *S* cells or the mutation of some of the *R* cells into the *S* type following injection.

In 1932, J. L. Alloway showed that similar transformations were possible in vitro, for when an extract of *S* cells was added to a culture of *R* cells, some of the latter were permanently transformed into the viable *S* type. Therefore, there appeared to be a substance in *S* cells that was capable of bringing about an inheritable change in the *R* cells; this unknown material was termed the **transforming principle.**

The transforming principle was finally identified in 1944 by O. T. Avery, C. M. MacLeod, and M. McCarty. Although crude extracts of heat-killed *S* cells were found to contain protein, polysaccharide, lipid, and nucleic acid, the removal of the protein, polysaccharide, and lipid by a combination of chemical procedures, including enzymatic hydrolysis, chloroform extraction, and alcohol fractionation, resulted in a product that retained the transforming activity. However, when the product was treated with the enzyme *deoxyribonuclease* (which degrades DNA), the capacity to transform *R* cells was lost. This evidence, together with chemical analyses, showed that the transforming principle was DNA. These experiments have been repeated many times since then, and similar transformations have been demonstrated in other bacterial species, in yeast, and in mammalian cells.

Virus Reproduction

As noted in Chapter 1, viruses are composed principally of protein and nucleic acid. Depending on the type of virus, the nucleic acid is either DNA or RNA. The protein forms a coat around the head and tail structures of the particle and encloses the core of nucleic acid. In 1952, A. D. Hershey and M. W. Chase conducted a series of experiments to determine whether the viral protein or nucleic acid was required for virus reproduction. Hershey and Chase employed the bacterium *E. coli* and the T2 virus (a DNA-containing virus that infects *E. coli*) in two sets of experiments. In one set of experiments, *E. coli* was cultured in a medium containing the radioactive isotope of sulfur, ^{35}S (the principles of radioactive tracer experiments are discussed in Chapter 14). During the growth of the bacterial population, ^{35}S was incorporated into the cells. The culture was then infected with the T2 *bacteriophage,* and during the reproduction of the bacteriophage within the host cells, bacterial ^{35}S was used in the synthesis of phage protein (i.e., it was incorporated into cysteine and methionine residues). Nucleic acids do not contain sulfur. Following lysis of the bacterial cells, the ^{35}S-labeled viruses were collected and used to infect *E. coli* cultured on media devoid of ^{35}S. A Waring blender was then used to separate by physical agitation what was left of the attached viruses from the surfaces of the bacterial hosts. A comparison of the radioactive sulfur content of the bacteria and freed viruses revealed that nearly all the ^{35}S remained with the viruses and had not entered the bacterial cytoplasm.

In other experiments, *E. coli* was cultured in media containing the radioisotope ^{32}P prior to infection with the phage. The nucleic acids contain phosphorus, and therefore ^{32}P was incorporated into newly synthesized viral DNA.

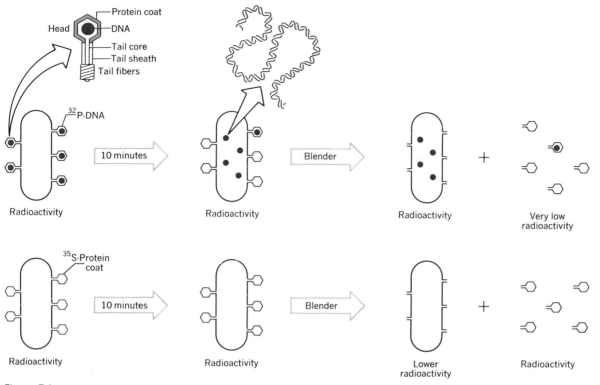

Figure 7-1

The experiments of Hershey and Chase. Parts shown in color represent portions of the virus that have been labeled with radioactive isotope (see text for details).

When labeled viruses were used to infect additional bacteria and the blender again was employed to shake off the attached viruses, it was found that the ^{32}P radioactivity was within the infected cells. The two sets of experiments are depicted in Figure 7–1. Hershey and Chase concluded that it was the DNA of the virus that entered the host cell during infection and that DNA was required for the reproduction of genetically identical virus particles by the metabolic machinery of the host cell. It should be noted that some viruses contain RNA instead of DNA; in these cases, it is the viral RNA that enters the host cell during infection.

The observations of Hershey and Chase also explained earlier findings by T. F. Anderson and R. M. Herriott that the T2 phage loses its ability to reproduce when distilled water is added to a suspension of the virus particles prior to their addition to a bacterial culture. Even though such viruses are still able to attack the bacterial host, the sudden osmotic shock created by exposure to distilled water causes

them to empty their nucleic acid content into the suspending medium.

Tobacco mosaic virus (TMV), the virus that infects tobacco leaves, contains RNA instead of DNA. In the early 1950s, H. Fraenkel-Conrat separated the RNA and protein components of TMV and found that the RNA separately injected into the tobacco leaves could infect the plant and produce new virus particles containing both RNA and protein. Fraenkel-Conrat also found that when protein isolated from one strain of TMV was mixed with RNA isolated from another strain, a reconstituted "hybrid" virus was produced that retained the ability to infect the tobacco leaves. The new viruses produced following infection by the hybrid were isolated and their RNA and protein components separated and analyzed. It was found that the type of protein in the virus coat was identical to that of the strain used as the source of RNA (see Fig. 7–2); that is, it was *not* the same as the protein of the hybrid viruses causing

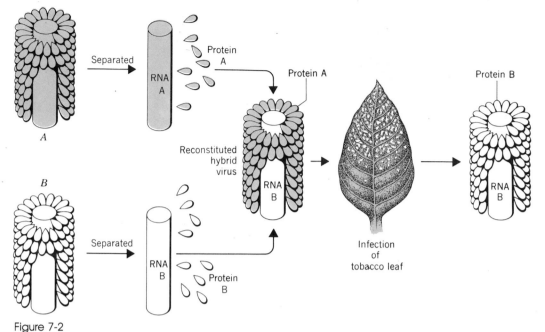

Figure 7-2
Fraenkel-Conrat's reconstitution experiments demonstrating that RNA and not protein is the genetic material in tobacco mosaic virus (see text for details).

the infection. These observations support earlier conclusions concerning the genetic role of the nucleic acids and also prove that these molecules alone contain information that determines the specific nature of newly synthesized protein.

Although the most direct evidence supporting the notion that DNA (or in the case of some viruses, RNA) is the genetic material was obtained using microbial systems, observations supporting this idea were also made with higher organisms. Based on numerous studies of the mechanism of fertilization, including those by O. Hertwig (1865), H. Fol (1877), E. Strasburger (1884), A. Weismann (1892), and E. B. Wilson (1895), it was generally acknowledged by the turn of the century that the chromosomes contained within the cell nucleus were concerned with the transmission of heredity. Therefore, the development of staining reactions (such as the Feulgen reaction) that are specific for DNA and the subsequent microscopic localization of the DNA in the chromosomes implicated DNA as the genetic material. Further evidence was provided during the 1940s from quantitative chemical analyses of DNA present in measured quantities of cells (i.e., numbers, dry

weight, etc.). These analyses revealed that the amount of DNA per cell was more or less constant within the various tissues of an organism. Moreover, it was also found that the total quantity of DNA present in the cell nucleus was related to its *ploidy* (i.e., the number of complete sets of chromosomes). The gametes (sperm and egg cells) of an organism were shown to have one-half as much DNA as the diploid somatic cells. This is precisely what would be expected if DNA served as the genetic material.

Although RNA appears to be the genetic material of some virus particles, there is no evidence that it plays a similar role in cells. Instead, RNA serves as an intermediary between the genetic information of DNA and the expression of this information as the synthesis of the cell's enzymes and other proteins (Chapter 22). In eucaryotic cells, most of the DNA is present in the nucleus, although certain organelles (mitochondria and chloroplasts) also contain DNA (see Chapters 16 and 17). Before we explore just how the relationship between the two nucleic acids can be effected and the mechanism by which DNA serves as the genetic material, it is first necessary to consider the chemistry of these two macromolecules.

Table 7–1
Chemical Constituents of DNA and RNA

	DNA	RNA
Purines	Adenine	Adenine
	Guanine	Guanine
Pyrimidines	Cytosine	Cytosine
	Thymine	Uracil
Pentose	2-Deoxyribose	Ribose
	Phosphoric acid	Phosphoric acid

Composition and Structure of the Nucleic Acids

As we have already noted, there are two major classes of nucleic acids: DNA and RNA. Both are composed of unbranched chains of subunits called **nucleotides,** each of which contains (1) a nitrogenous base (either a **purine** or **pyrimidine**), (2) a **pentose,** and (3) **phosphoric acid.** In RNA, the pentose is *ribose,* whereas in DNA it is *2-deoxyribose.* Both DNA and RNA contain the purine nitrogen bases *adenine* (abbreviated A) and *guanine* (G) and the pyrimidine *cytosine* (C), but in DNA a second pyrimidine is *thymine* (T), while in RNA it is *uracil* (U). A number of other nitrogenous bases have been identified in DNA and RNA, but these occur much less frequently. The phosphoric acid component of each nucleotide is, of course, chemically identical in both nucleic acids. These relationships are summarized in Table 7–1, and the corresponding chemical formulas are shown in Figure 7–3.

The pentose of each nucleotide unit is simultaneously bonded through its number 1 carbon atom to the nitrogen base (forming a *nucleoside*) and through its number 5 carbon atom to phosphoric acid. The structures of the four deoxyribonucleotides of DNA together with the specific numbering system used to identify each constituent atom are shown in Figure 7–4. Successive nucleotides of DNA and RNA are joined together by phosphodiester linkages involving the 5'-phosphate group of one unit and the 3'-hydroxyl group of the neighboring unit (Figure 7–5).

The ''backbone'' of a nucleic acid molecule is formed by the repeating sequence of pentose and phosphate groups, and this is the same in all molecules. What distinguishes one DNA (or RNA) molecule from another is the specific *sequence* of purine and pyrimidine bases present in the chain of nucleotides and the *total number* of nucleotides (i.e., the size of the molecule).

Structure of DNA

At one time, it was believed that the four purines and pyrimidines of DNA occurred in approximately equal amounts in the molecule. However, the studies of E. Chargaff and others in the late 1940s showed that this was not the case. Instead, they found that the relative amounts of the nitrogen bases varied between species but were constant within a species. The constancy noted within a species was maintained regardless of the tissue or organ from which the DNA was isolated. Furthermore, the relative amounts of the nitrogen bases were similar in closely related species and quite different in unrelated species. Chargaff also made the following extremely important finding. *Regardless of the species used as the source of DNA, the molar ratios of adenine and thymine were always very close to unity, and the same was true for guanine and cytosine.* No such constant relationship could be demonstrated for any other combination of nitrogen base pairs. This implied that for some reason, every molecule of DNA contained equal amounts of adenine and thymine and also equal amounts of guanine and cytosine.

Using chemical information of this sort, together with the results of X-ray crystallographic studies of DNA, J. D. Watson, F. H. C. Crick, M. H. F. Wilkins, and R. Franklin proposed a model for the structure of DNA in the early 1950s. They suggested that a molecule of DNA consists of two helical polynucleotides wound around a common axis to form a right-handed ''double helix.'' In contrast to the arrangement of amino acid side chains in helical polypeptides (where the side chains are directed radially away from the helix axis), the purine and pyrimidine bases of each polynucleotide chain were directed inward toward the center of the double helix so that they faced each other.

On the basis of stereochemical studies, Watson and Crick further suggested that the only possible arrangement of the nitrogen bases within the double helix that was consistent with its predicted dimensions was that in which a purine always faced a pyrimidine, for the diametric distance between the two polynucleotide chains is too small to accommodate two juxtaposed purines. Which purine was matched with which pyrimidine became clear from a consideration of which pairs would be able to form the hydrogen bonds

Nucleic acids are composed of repeating subunits called nucleotides.
Each nucleotide is composed of three units.

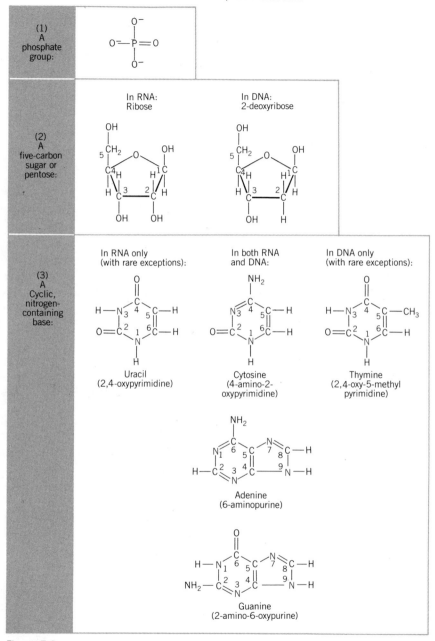

Figure 7-3
Structural formulas of the constituents of nucleic acids. When the pentoses are present in nucleosides, nucleotides, or nucleic acids, the five carbon atoms are numbered 1′, 2′, 3′, 4′, and 5′, respectively, so that they may be distinguished from the carbons of the nitrogen bases. (Modified from Gardner and Snustad, *Principles of Genetics*, 6th ed., p. 76. Copyright © 1981 by John Wiley and Sons, Inc.)

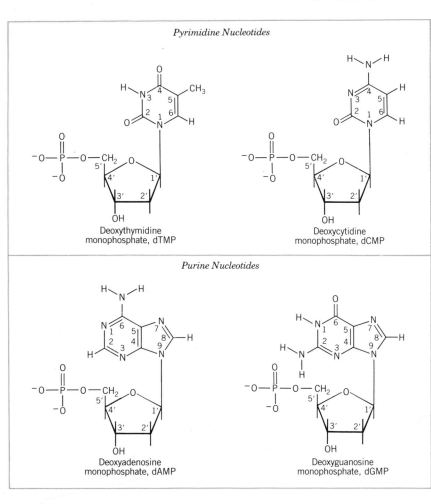

Figure 7-4

The four *deoxy*ribonucleotides of DNA. RNA contains similar *ribo*nucleotides, which contain the pyrimidines uracil and cytosine and the purines adenine and guanine. (Modified from Gardner and Snustad, *Principles of Genetics,* 6th ed., p. 77. Copyright © 1981 by John Wiley and Sons, Inc.)

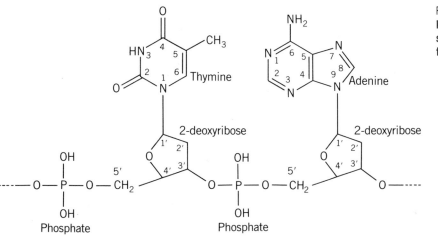

Figure 7-5

Portion of one strand of a DNA molecule showing the phosphodiester linkage between successive nucleotides.

necessary to stabilize the double-helical structure. Accordingly, Watson and Crick concluded that *adenine must be matched with thymine and guanine with cytosine.* This

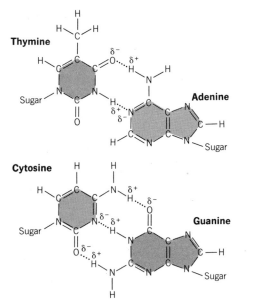

conclusion was, of course, in agreement with the chemical findings of Chargaff (see above)—in fact, Chargaff's data may have been critical to the development of Watson and Crick's proposals. The manner in which hydrogen bonds are formed between adenine and thymine and between guanine and cytosine is shown in Figure 7–6. Although individually weak, the great number of these bonds contributes appreciably to the stability of the double-helical structure. In addition, the double helix is stabilized by hydrophobic bonds between neighboring nitrogen bases of each polynucleotide chain.

Certain other features of the structure of DNA should be noted. The two polynucleotide chains that make up the molecule are **antiparallel.** That is, beginning at one end of the molecule and progressing toward the other, successive nucleotides of one chain are joined together by $3' \rightarrow 5'$ phosphodiester linkages, whereas the complementary nucleotides of the other chains are joined by $5' \rightarrow 3'$ phosphodiester linkages. This antiparallel arrangement is depicted diagrammatically in Figure 7–7. The two polynucleotides are twisted around one another in such a way as to produce two helical grooves in the surface of the molecule; these are called the *major* and *minor* grooves and are shown in Figure 7–8, together with some of the physical dimensions of the double helix.

Figure 7-6
Formation of hydrogen bonds between thymine and adenine and between cytosine and guanine. Note that three bonds are formed between C and G and only two between T and A.

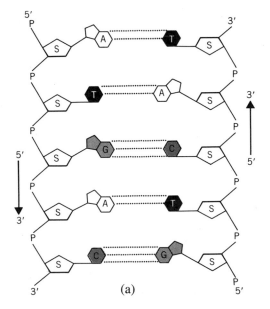

(a)

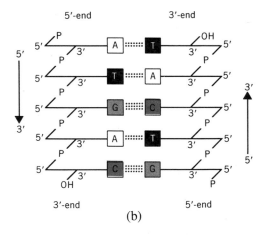

(b)

Figure 7-7
Molecular structure of DNA showing the pentose sugar (S)—phosphate (P) backbones of the polynucleotide chains and their antiparallel nature. (*a*) Molecular structure; (*b*) schematic abbreviated representation, emphasizing the opposite polarity of the complementary strands.

Replication of DNA

One of the intrinsic properties of the genetic material is its capacity for replication. The manner in which DNA satisfies this requirement is apparent from the nitrogen base pairing required in the model. Since the sequence of bases in one polynucleotide chain automatically determines the sequence of bases in the other, it is clear that one-half of a molecule (i.e., one of the two helixes) contains all the information necessary for constructing a whole molecule. For example, if we know that the sequence of bases along one polynucleotide chain of DNA is A T G A C and so on, then the complementary sequence in the other chain must be T A C T G, and so on, since A bonds to T and C bonds to G (see above). Therefore, if the double helix were unwound, each separate polynucleotide chain could act as a template for the production of a new, complementary chain. The result would be two identical double helices where there was only one before. Of course, one-half of each new double helix would be represented by one of the original polynucleotide chains. The basic features of this process are shown in Figures 7–9 and 7–10. A detailed description of the mechanism by which the replication of DNA occurs, together with its experimental verification, is given in Chapter 21.

Denaturation and Renaturation of DNA

DNA is readily denatured by extremes of temperature and pH. The denaturation takes the form of an unwinding of the double helix as hydrogen bonds between complementary bases are disrupted. This form of DNA denaturation is referred to as *melting* and produces separate DNA strands. Solutions of DNA absorb ultraviolet light (UVL) having a wavelength of 260 nm. When the temperature of a native solution of DNA is elevated, the resulting melting is accompanied by an increase in UVL absorption. This **hyperchromic effect** occurs because the purines and pyrimidines of separated strands can absorb more light energy than when they are part of a double helix. Some viruses contain a single-stranded form of DNA (see below), and since this form does not exhibit the hyperchromic effect, it is readily distinguished from double-stranded forms.

When DNA is melted thermally, denaturation begins in regions of the double helix that are rich in A—T base pairs and progressively shifts to regions of greater and greater G—C content. This is because the two hydrogen bonds holding each A—T pair together can be broken more easily (hence, at a lower temperature) than the three hydrogen bonds holding each G—C pair together. A quantitative measure of the change in UVL absorbance that takes place as the temperature of a DNA solution is slowly elevated is

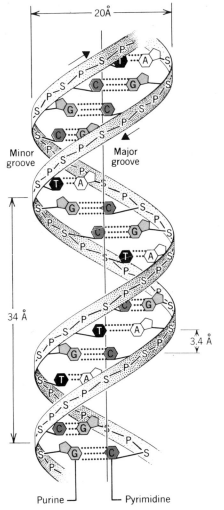

Figure 7-8

Diagram of DNA double helix showing complementary base-pairing, major and minor surface grooves, and certain molecular dimensions. The double helix has a diameter of 20 Å. Each complete turn of a helix accounts for 34 Å of linear translation and each nucleotide 3.4 Å. (Hence, there are 20 nucleotides per helical turn). The usual symbols for the bases are used, with S for sugar and P for phosphate.

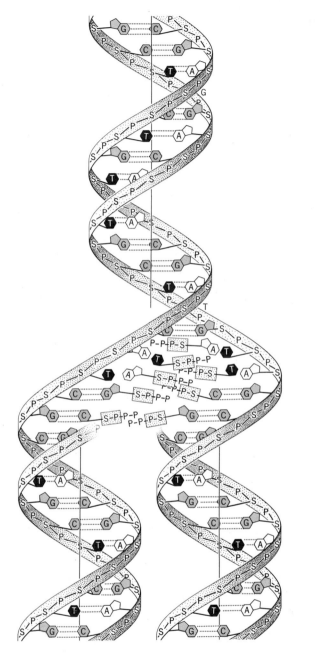

Figure 7-9
Model for replication of DNA. The replication of the DNA molecule could be effected if the double helix unwound and the separated chains acted as templates for the production of two new complementary chains. In this figure, the original double helix has partially unwound, and two complementary chains are being formed in this region by the addition of nucleotide units.

called a *melting curve* (Fig. 7–11). The point in the melting curve at which the change from double-stranded to single-stranded DNA is half complete is called the T_m value and is characteristic of a particular source of DNA. The species specificity that is characteristic of DNA melting curves reflects differences in the G—C and A—T compositions of different kinds of DNA.

Thermally denatured DNA can be renatured by lowering the temperature of the solution. Separated strands recombine to form double helices as hydrogen bonds between complementary bases are re-formed. This *reannealing* can be monitored as a decrease in UVL absorption by the DNA solution. The capacity for denatured DNA to reanneal can be used to assess the size of an organism's **genome** (the total number of genes, see Chapters 20 and 21) and the *complexity* of the DNA that is present.

When reannealing studies are to be performed, the isolated DNA is first broken by shearing force into lengths of several hundred to several thousand nucleotide pairs. The double-helical DNA is then thermally denatured yielding single strands. A known concentration of the single-stranded DNA is then incubated at the reannealing temperature (usually about 25 degrees below the T_m), and the reannealing rate determined from the rate of change in UVL absorbance. A large genome reanneals more slowly than a small genome because there are a greater number and variety of DNA fragments. Thus, each fragment takes a longer time to "seek out" and anneal with its complementary partner. The kinetics of DNA renaturation (Fig. 7–12) is represented by a curve relating the percentage of reassociated fragments to the "C_0t number" [i.e., the concentration of DNA in moles of nucleotides per liter (C_o) times the reaction time (t) in seconds]. As seen in Fig. 7–12, viral DNA (curve *c*) reanneals more rapidly than procaryote DNA (curve *d*), and the latter reanneals more quickly than eucaryote DNA (curve *e*). The discovery of a eucaryotic

Figure 7-10
Replication of the base sequence of DNA. The complementary base sequence of a portion of a DNA molecule is shown at the top of the figure. The complementary bases in this region are separated as the double helix unwinds and serve as templates for the production of two new polynucleotide chains (middle). Finally, two complete DNA molecules are produced with identical base sequences (bottom). One-half of each of these DNA molecules is represented by one newly produced chain (color) and by one original chain.

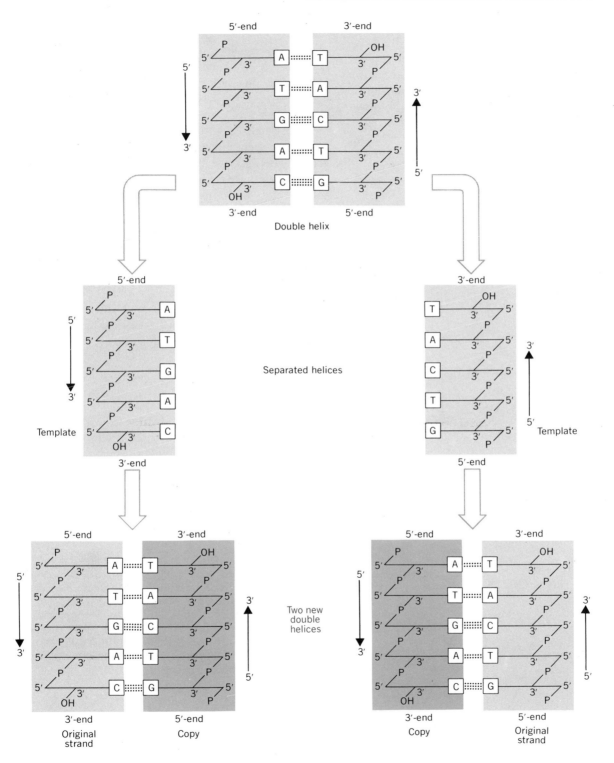

Double helix

Separated helices

Template

Template

Two new
double
helices

Original
strand

Copy

Copy

Original
strand

DNA fraction in mammalian cells that reanneals unexpectedly rapidly (curve *b*) revealed for the first time the existence in the genomes of higher organisms of repetitive DNA sequences (see Chapter 22). In Figure 7–12, curve *a* shows the reannealing rate of a solution containing a mixture of synthetically produced polyuridylic acid (a nucleotide chain with only uracil bases) and polyadenylic acid strands. Even though uracil is not usually found in DNA, like thymine it can form hydrogen bonds with adenine and does so in many RNA molecules (see below).

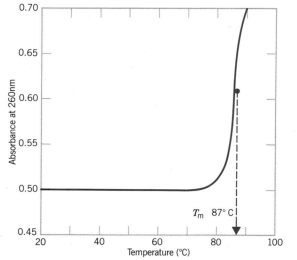

Figure 7-11

A typical thermal denaturation curve (or *melting curve*) for DNA showing the *hyperchromic effect* and the evaluation of the T_m of the nucleic acid. (Redrawn from Conn and Stumpf, *Outlines of Biochemistry*, 4th ed., p. 121. Copyright © by John Wiley and Sons, Inc.)

Z-DNA

Until quite recently, it was presumed that all naturally occurring double-helical DNA conformed to the right-handed Watson-Crick model. However, in 1979, A. Rich confirmed earlier observations reported by F. M. Pohl and T. Jovin that a left-handed form of DNA also exists. As in the right-handed DNA, the two helices are held together by complementary base pairing and the strands are antiparallel. Because the two sugar–phosphate backbones of the molecule trace a zigzag course around the axis of the helices (Fig. 7–13) this left-handed DNA has been called **Z-DNA.**

While the structure of Z-DNA originally proposed by Rich was based upon studies of DNA crystals produced in the laboratory, Z-DNA has recently been identified in polytene salivary gland chromosomes of the vinegar fly, *Drosophila.* The Z-form coexists with the right-handed form (called **B-DNA**) in the same molecule. At the present time, it is speculated that switches in DNA helicity may be related to biological functions such as selective gene expression (see Chapters 11 and 21).

"Single-Stranded" DNA

Although in nearly every case so far studied DNA consists of two polynucleotide chains twisted about one another to form a double helix, it is now apparent that in a few bacterial viruses (i.e., the φX174 and S13 *E. coli* phages), DNA exists as a single polynucleotide chain. This was initially suspected when chemical analyses of the nitrogen base contents of these viral DNAs revealed that the amounts

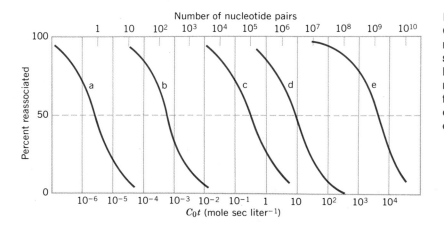

Figure 7-12

C_0t curves for different types of DNA. The kinetics of reannealing is a function of the size and complexity of the genome. (*a*) Mixture of poly U and poly A fragments; (*b*) repetitive DNA fraction from mice; (*c*) DNA from T4 phage of *E. coli*; (*d*) *E. coli* DNA; and (*e*) nonrepetitive DNA fraction from a calf.

Figure 7-13

Double helical DNA can exist in both *left-handed* (Z-DNA) and *right-handed* (B-DNA) forms. In the left-handed form, the phosphate—sugar backbone of each nucleotide (shown by heavy black and red lines) follows a zigzag pattern around the axis of the helices.

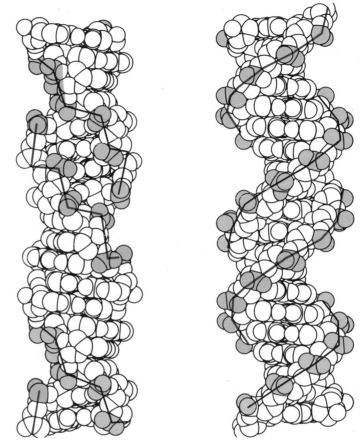

of adenine and thymine, as well as guanine and cytosine, were not equal. During reproduction of these viruses, the single-stranded DNA (referred to as the '' + strand'') is injected into the host bacterial cell, where it acts as a template for the reproduction of a complementary polynucleotide chain (called the '' − strand''); these two polynucleotides combine to form a conventional double helix, which then serves as the template for the production of additional + strands. The newly produced + strands are then enclosed in the viral protein coats to form new virus particles.

Structure of RNA

As noted earlier, RNA and DNA differ chemically in two notable ways: in RNA, *ribose* is the pentose (not deoxyribose as in DNA) and the pyrimidine *uracil* occurs in place of thymine (Fig. 7–14). Early analyses of the nitrogen base contents of RNAs from various sources revealed that the molar ratios of these bases were quite different from those of DNA and were also quite variable. On this basis, it was concluded that RNA occurs as a single polynucleotide chain. This contention has more recently been supported by other physicochemical studies, but it should be noted that there are some viral RNAs that are double stranded. Although only one polynucleotide chain is usually present, RNA does possess regions of double-helical coiling where the single chain loops back upon itself. These regions are stabilized by the formation of hydrophobic bonds between neighboring bases (as in DNA) and also by the formation of hydrogen bonds between opposing units of guanine and cytosine and between adenine and *uracil*. In RNA, A and U can form two hydrogen bonds similar to the two bonds formed between A and T in DNA.

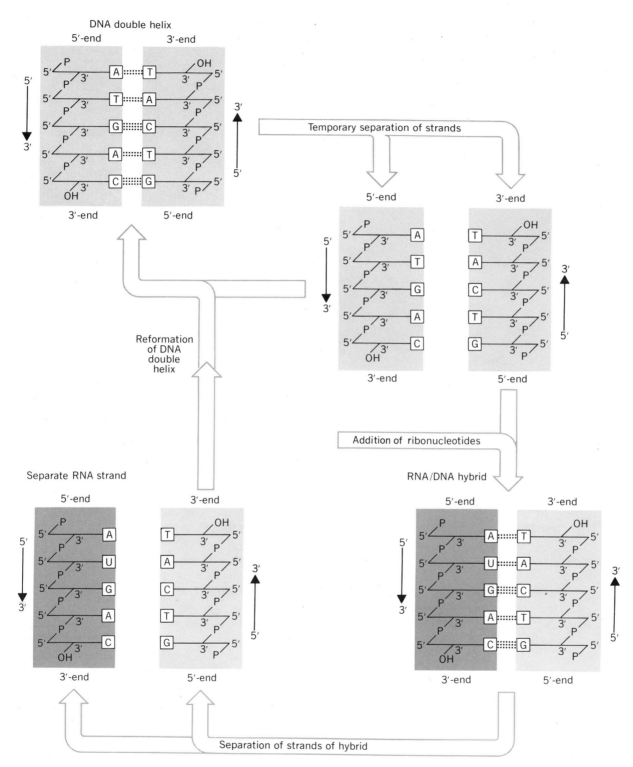

Figure 7-14
Transcription of the base sequence of one of the helices of DNA into the complementary base sequence of RNA.

Synthesis of RNA

Except perhaps in the case of the reproduction of certain RNA viruses (see below), the synthesis of RNA appears to be directed by DNA. The formation of the RNA polynucleotide takes place using the base sequence along only *one* of the two deoxyribonucleotide helices of DNA (producing a temporary RNA-DNA hybrid) and resulting in the release of a single, complementary polyribonucleotide chain in which the base uracil occurs in place of thymine (Figure 7–14).

Replication of DNA and RNA Viruses

The viruses may be subdivided into two classes: (1) viruses whose genetic complement consists of DNA, and (2) viruses whose genetic complement consists of RNA. In cells infected by DNA viruses, the infecting viral DNA is **replicated,** forming new viral DNA which is then **transcribed** into RNA; this RNA is then **translated** into viral protein (Fig. 7–15*a*). The newly produced viral DNA and viral proteins combine in the assembly of new, complete virus particles that are released upon lysis of the host cell. A *latent state* can also be established in which the viral DNA is incorporated into the host cell's genome, being replicated and distributed along with the host cell's native DNA, until it is transcribed once again into additional viral RNA and then into viral proteins (see Chapter 1).

For most RNA viruses (poliomyelitis, influenza, common cold, etc.), DNA involvement is essentially bypassed. For example, during the infection of a cell with polio virus, the single-stranded RNA (called a " + strand") enters the host cell, where it acts as a template for the synthesis of complementary " − strands." The latter are then employed in the proliferation of new + strands, and these are translated into viral proteins (Fig. 7–15*b*).

The mechanism just described is varied in several other viruses where the RNA is either double stranded (e.g., reovirus, in which only one of the two rRNA strands produced during replication is transcribed) or where the infecting single RNA strand is complementary (rather than identical) to the newly produced viral RNAs that are to be translated into viral proteins (e.g., Sendai virus, Newcastle disease virus).

It has recently become clear that yet another mechanism exists in the case of the RNA tumor viruses (e.g., Rous

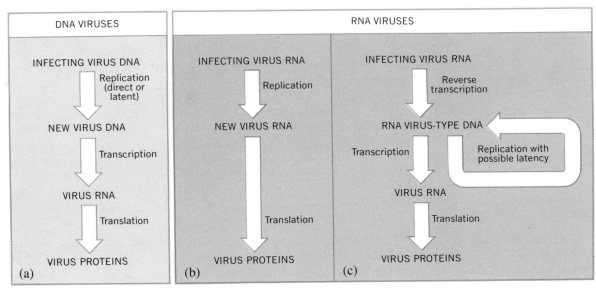

Figure 7-15
Replication of DNA and RNA viruses.

sarcoma virus). These viruses do not transfer information from RNA to RNA, but rather from RNA to DNA and then to new RNA. The viral RNA is employed as a template for the synthesis of DNA by the infected cell (a phenomenon that has come to be called "reverse transcription"). Some of the resulting "viral DNA" may be incorporated into the genome of the host cell, establishing what is called the *provirus* state.

The provirus state has been suggested as the basis of a number of different RNA virus-induced and DNA virus-induced cancers. According to this view, one or more of the provirus genes—which are normally repressed by the host cell—may become derepressed and cause the production of an *oncogenic* (cancer-causing) substance that alters the cell's normal properties or behavior. Such a change may be delayed for a number of generations, depending upon the period of latency.

During the 1960s and early 1970s, the so-called central dogma of molecular biology was the orderly and *unidirectional* flow of information encoded in the base sequences of a cell's DNA to RNA and then to protein; that is,

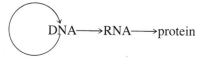

The discovery of reverse transcription by certain RNA viruses in which the information of RNA is passed on to DNA has necessitated a reexamination of that dogma and raised the question of whether or not a similar interaction between RNA and DNA might normally occur in cells (i.e., cells not infected by viruses) under specific conditions (e.g., during cellular differentiation). The central dogma might more appropriately be represented as

Types of Cellular RNA

Cells contain three major functional types of RNA: **ribosomal RNA** (abbreviated rRNA), **messenger RNA** (mRNA), and **transfer RNA** (tRNA). All of these are transcribed from DNA and are engaged in mediating the genetic message of DNA by participating in the synthesis

of the cell's proteins, as discussed in Chapter 22. It has already been noted that RNA occasionally serves as the genetic material of viruses.

Of the cellular RNAs, rRNA is the most abundant, accounting for up to 85% of the total RNA of the cell. Only three or four different kinds of rRNA are present in cells, and these are confined for the most part to the cell's ribosomes. mRNA accounts for about 5 to 10% of the cell's RNA and is much more heterogeneous with respect to size and nitrogen base content than the rRNAs. This results from the relationship (see below) between the chain lengths and base sequences of mRNAs and the variable sizes and primary structures of polypeptides synthesized in a cell. Most mRNA occurs in the cytoplasm, where it transiently combines with ribosomes during protein synthesis.

About 10 to 20% of the cell's RNA is tRNA. All tRNA molecules are similar in size and typically contain 75 nucleotide units. In spite of these similarities, a single cell may contain about 60 species of tRNA differing in their base sequences. Because most of the tRNA is recovered in the cytoplasmic (i.e., soluble) phase of disrupted cells following centrifugation, tRNA is also called *soluble* RNA (sRNA). In view of its small size and relative ease of isolation, tRNA has been more extensively studied than the other two ribonucleic acids, and the specific primary, secondary, and tertiary structures of many tRNAs have already been determined. tRNAs contain moderate amounts of unusual nucleotides such as *ribothymidine, dihydrouridine, pseudouridine,* and *methylguanosine.* These are formed by modification of the four common RNA bases and play a crucial role in establishing the unique spatial organization of these molecules.

Although the mechanisms of DNA replication and protein synthesis are considered in depth in subsequent chapters, it is appropriate that a brief accounting of the functional relationships among the nucleic acids and between nucleic acids and proteins be made at this time. Inheritable information is encoded in the various nitrogen base sequences possessed by the cell's DNA, and by the processes of *transcription* and *translation,* these base sequences are employed to specify the primary structures of all proteins produced by the cell. Most important among these proteins are the enzymes that catalyze and regulate the myriad of chemical reactions characterizing the cell's metabolism. Therefore, the information of DNA confined essentially to the cell nucleus manifests itself primarily in the cytoplasm as the synthesis of a unique assemblage of proteins. The replication of DNA that precedes mitotic cell division and the equal distribution of the duplicated DNA among the progeny cells provides for the passage of complete sets of information from one generation of cells to another. In addition to serving as templates for their own replication, the nucleotide sequences of DNA are used during transcription to produce complementary base sequences of RNA. The resulting RNAs then serve as intermediaries in translating the original message into protein. Of paramount importance in this process are the mRNA molecules whose base sequences directly determine the *primary structures* of the polypeptides. These mRNA molecules leave the nucleus of the cell following their synthesis and attach in the cytoplasm to one (or several) ribosomes. The rRNA of each ribosome is believed to play a role in this attachment. tRNA molecules also produced in the nucleus enter the cytoplasm, where they combine with specially activated amino acids (distinct tRNAs exist for each species of amino acid). The resulting complexes, directed by the base sequences of mRNA attached to the ribosomes, sequentially deposit their amino acids in the growing polypeptide chains. Further details of this process are examined in Chapter 22.

Summary

Although **nucleic acids** were discovered and chemically characterized more than 100 years ago, their genetic role was not appreciated until the phenomenon of bacterial transformation and the mechanisms of virus reproduction were understood. The two major nucleic acids, DNA and RNA, are composed of unbranched chains of subunits called **nucleotides,** each nucleotide containing **phosphate,** a **pentose,** and a **nitrogenous base.** Specific differences exist in the nucleotides found in DNA and in RNA. In DNA, two **antiparallel** polynucleotide chains are twisted around one another to form a **double helix,** the helices being held together in part through hydrogen bonds between juxtaposed nitrogenous bases. Constraints imposed by the known shape and organization of the double helix and the nitrogenous bases capable of forming such stabilizing bonds suggest that the base sequence in one polynucleotide necessarily determines the sequence of bases in the other. Therein lies the capacity for replication. DNA serves as the genetic material in all cells and in many viruses.

Most RNAs are formed by a single polynucleotide twisted about itself in certain regions, thereby forming a periodic double-helical structure. In certain viruses, RNA serves as the genetic material, but in procaryotic and eucaryotic cells, RNA serves three primary functions. The *ribosomal* RNAs are the most abundant and serve as functional components of the cell's ribosomes during protein synthesis. The *messenger* RNAs contain the sequence of nucleotides that specify the primary structures of the cell's proteins. *Transfer* RNAs function in the transport of amino acids to the messenger RNA-ribosome complex during protein synthesis or *translation*. The cell's RNAs are produced by transcription of its DNA.

References and Suggested Reading

Articles and Reviews

Baltimore, D., Viruses, polymerases and cancer. *Science 192,* 632 (1973).

Bauer, W. R., Crick, F. H. C., and White J. H., Supercoiled DNA. *Sci. Am. 243*(1), 118 (July 1980).

Bishop, J. M., Oncogenes. *Sci Am. 246*(3), 80 (Mar. 1982).

Britten, R. J., and Kohne, D. E., Repeated segments of DNA. *Sci. Am. 222*(4), 24 (April 1970).

Brown, D. D., The isolation of genes. *Sci. Am. 229*(2), 20 (Aug. 1973).

Campbell, A. M., How viruses insert their DNA into the DNA of the host cell. *Sci. Am. 235*(6), 102 (Dec. 1966).

Chan, H. W., Israel, M. A., Garon, C. F., et al., Molecular cloning of polyoma virus DNA in *Escherichia coli:* plasmid vector system. *Science 203,* 883 (1979).

Crick, F. H. C., The structure of the hereditary material. *Sci. Am. 194*(4), 54 (Oct. 1954).

Fiddes, J. C., The nucleotide sequence of a viral DNA. *Sci. Am. 237*(6), 54 (Dec. 1977).

Grobstein, C., The recombinant-DNA debate. *Sci. Am. 237*(1), 22 (July 1977).

Holley, R. W., The nucleotide sequence of a nucleic acid. *Sci. Am. 214*(2), 30 (Feb. 1966).

Kolata, G., Z-DNA: from the crystal to the fly. *Science 214,* 1108 (1981).

Kornberg, A., The synthesis of DNA. *Sci. Am. 219*(4), 64 (Oct. 1968).

Mirsky, A. E., The discovery of DNA. *Sci. Am. 218*(6), 78 (June 1968).

Reddy, V. B., Thimmappaya, B., Dhar, R., et al., The genome of simian virus 40. *Science 200,* 494 (1978).

Sanger, F., Determination of nucleotide sequences in DNA. *Science 214,* 1205 (1981).

Simons, K., Garoff, H., and Helenius, A., How an animal virus gets into and out of its host cell. *Sci. Am. 246*(2), 58 (Feb. 1982).

Sinsheimer, R. L., Single-stranded DNA. *Sci. Am. 207*(1), 109 (July 1962).

Stent, G. S., The multiplication of bacterial viruses. *Sci. Am. 188*(5), 36 (May 1953).

Temin, H. M., RNA-directed DNA synthesis. *Sci. Am. 226*(1), 24 (Jan. 1972).

Wang, A. H. J., Quigley, G. J., Kolpak, F. J., Crawford, J. L., van Boom, J. H., van der Marel, G., and Rich, A., Molecular structure of a left-handed double helical DNA fragment at atomic resolution. *Nature 282*, 680 (1979).

Wang, A. H. J., Quigley, G. J., Kolpak, F. J., van der Marel, G., van Boom, J. H., and Rich, A., Left-handed double helical DNA: variations in the backbone conformation. *Science 211*, 171 (1981).

Books, Monographs, and Symposia

Braun, W., *Bacterial Genetics,* Saunders, Philadelphia, 1953.

Goldstein, L., *The Control of Nuclear Activity,* Prentice-Hall, Englewood Cliffs, N.J., 1967.

Haggis, G. H., Michie, D., Muir, A. R., et al., *Introduction to Molecular Biology,* Wiley, New York, 1965.

Lehninger, A. L., *Biochemistry,* Worth, New York, 1975.

Levine, R. P., *Genetics,* Holt, Rinehart and Winston, New York, 1962.

Luria, S. F., Darnell, J. E., Baltimore, D., and Campbell, A., *General Virology* (3rd ed.), Wiley, New York, 1978.

McGilvery, R. W., *Biochemical Concepts,* Saunders, Philadelphia, 1975.

Stewart, P. R., and Letham, D. S. (Eds.), *The Ribonucleic Acids,* Springer-Verlag, New York, 1973.

Stryer, L., *Biochemistry* (2nd ed.), W. H. Freeman, San Francisco, 1981.

Taylor, J. H., *Selected Papers on Molecular Genetics,* Academic Press, New York, 1965.

Watson, J. D., *Molecular Biology of the Gene* (3rd ed.), W. A. Benjamin, Menlo Park, Calif., 1976.

Woodward, D. O., and Woodward, V. W., *Concepts of Molecular Genetics,* McGraw-Hill, New York, 1977.

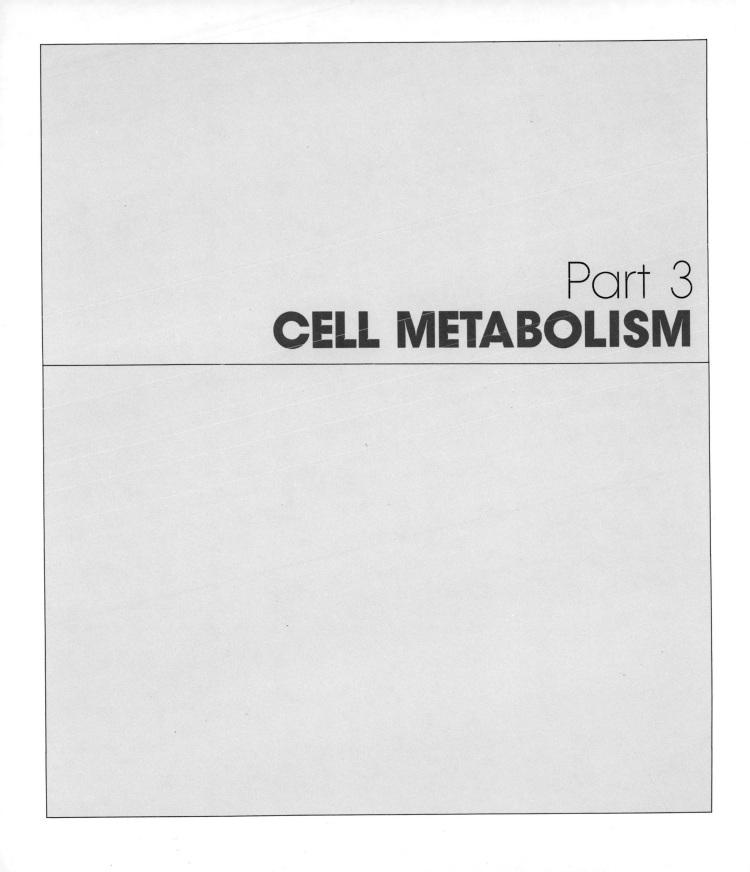

Part 3
CELL METABOLISM

The metabolism of the cell is characterized by a myriad of simultaneously occurring chemical reactions. Nearly all of these reactions are catalyzed by a special class of proteins called **enzymes.** It has been estimated that an average cell contains about 3000 different enzymes. In the absence of these enzymes, the cellular reactions would proceed at much slower, in most cases negligible rates. Enzymes integrate the cellular chemical reactions and provide the order without which the complex processes of life would not be possible. Most enzymes are characterized by a high degree of specificity; that is, they will catalyze one particular chemical reaction but not another. Enzyme molecules themselves are not consumed in the chemical reactions that they catalyze; instead, they are made available over and over again to repeatedly catalyze their reactions.

More than a thousand different enzymes have been specifically identified, and of these, several hundred have already been isolated or crystallized. Most of the enzymes responsible for such diverse functions as alimentary digestion, blood coagulation, muscle contraction, carbohydrate and fat metabolism, and nucleic acid and protein biosynthesis are known. Before considering the properties of enzymes and the mechanisms of enzymic catalysis, some general features of chemical reactions will be reviewed, including the various levels of **molecularity** and the associated **reaction kinetics.**

Molecularity of Chemical Reactions

Chemical reactions may be classified according to their **level of molecularity;** for example, *monomolecular* reactions, *bimolecular* reactions, *trimolecular* reactions, and so on. A monomolecular reaction may be written

$$A \rightleftharpoons P \tag{8–1}$$

and involves the conversion of one molecular species (i.e., *A*) to another (i.e., *P*) without the addition or removal of atoms. Instead, an intramolecular reorganization of the molecule occurs. The interconversion of the α and β isomers of glucose via the open-chain intermediate is an example of this type of reaction (see Chapter 5). In the case just cited, the intramolecular changes are spontaneous and do not require catalysis by an enzyme. In contrast, the conversion of glucose-6-phosphate to fructose-6-phosphate during glycolysis (Fig. 8–1) is an example of an enzyme-catalyzed monomolecular reaction. Enzymes catalyzing intramolecular reorganizations are termed **isomerases.**

Monomolecular reactions are relatively uncommon in cells. Much more numerous are bimolecular reactions and reactions of higher order, involving two or more reactants and/or two or more products; for example,

$$A + B \rightleftharpoons P \tag{8–2}$$

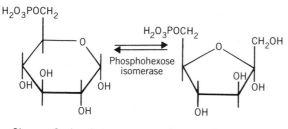

Glucose–6–phosphate   Fructose–6–phosphate

Figure 8-1
Reaction catalyzed by an *isomerase*. During monomolecular reactions, the atoms are rearranged, but none is added or removed.

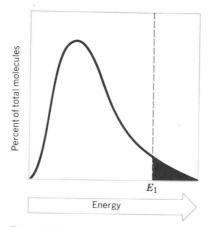

Figure 8-2
Maxwell-Boltzmann distribution of molecular kinetic energies. Not all molecules in solution have the same kinetic energy; instead, these energies are distributed as shown with those molecules with greater amounts of energy to the right. Only those molecules having some minimal kinetic energy (shown beyond E_1) may participate in a particular chemical reaction.

$$A \rightleftharpoons P_1 + P_2 \qquad (8–3)$$

$$A + B \rightleftharpoons P_1 + P_2 \qquad (8–4)$$

$$A + B + C \rightleftharpoons P \qquad (8–5)$$

Among the major classes of enzymes catalyzing bimolecular and higher order reactions are the **hydrolases, dehydrogenases, decarboxylases,** and **transferases.** Hydrolases catalyze reactions in which water is either added to or removed from the reactant(s). Dehydrogenases oxidize compounds by catalyzing the removal of hydrogen atoms. Decarboxylases remove carbon dioxide from carboxylic acids, and transferases remove reactive groups from one compound and transfer them to another. Numerous examples of these and other enzyme-catalyzed cellular reactions will be encountered in subsequent reading.

Reaction Kinetics

To begin our consideration of the kinetics of chemical reactions, consider the reaction described by the equation 8–2 in which A and B are the reactants and P is the single product of the reaction. One prerequisite for the reaction to occur is the ''collision'' of molecules A and B. However, a collision alone is not sufficient to guarantee the formation of molecule P. In addition, molecules A and B must possess some minimal amount of kinetic energy at the instant of collision and must be oriented with respect to one another at that instant such that the chemical bonds strained by the collision allow the shifts of orbital electrons necessary for the formation of the product.

The probability of a reaction occurring between A and B is determined in part by the probability of finding molecules A and B in the same region of space at the same time (i.e., this would constitute a collision). The probability of finding molecule A in the specific region of space is directly proportional to its concentration, and the same is true for molecule B. Therefore, the probability of finding *both* molecules A and B in the same region of space at the same instant in time is equal to the product of the probabilities of each being there. That is,

$$\text{Probability of collision} \sim [A]\,[B] \qquad (8–6)$$

where [A] is the concentration of molecule A and [B] is the concentration of molecules B in the solution. The same line of reasoning leads us to conclude that the rate of the chemical reaction, that is, the change in the concentration of A and B with time as they are converted to P (i.e., $d[A]/dt$ or $d[B]/dt$), would be directly proportional to the *product* of [A] and [B].

In a solution of any molecular species, not all the molecules will have the same kinetic energy. Instead, the energies are distributed as shown in Figure 8–2. This distribution, known as a **Maxwell-Boltzmann** distribution, means that different molecules in the solution have different kinetic energies. This variation results from random collisions of the molecules with one another during which some molecules gain kinetic energy while others lose kinetic energy. For each chemical reaction, there is a minimum kinetic energy that the participating molecules must have in order for the chemical reaction to occur. As Figure 8–2 shows, only a small percentage of the molecules in solution have this energy at any instant in time (those to the right of E_1).

Even though a collision might involve two molecules with the requisite kinetic energies, it is still necessary that they be oriented with respect to one another in a specific manner at the instant of collision in order for the reaction to proceed. Thus, only a small percentage of the collisions involving molecules with the required kinetic energy will result in the appropriate chemical reaction. Hence, the actual rate at which the reaction takes place is equal to some constant, k_1, times the probability of a collision. This constant, called a *rate constant,* accounts for both molecular orientation and kinetic energy considerations. Thus,

$$\text{Rate of reaction} = \frac{d[A]}{dt} = \frac{d[B]}{dt} = k_1[A][B] \qquad (8\text{–}7)$$

The rate of the reverse reaction is

$$\frac{d[P]}{dt} = k_2[P] \qquad (8\text{–}8)$$

Once a reversible reaction has reached equilibrium, no *net* change in the concentrations of the reactants or the product(s) occurs. Therefore, at this time

$$\frac{d[A]}{dt} = \frac{d[B]}{dt} = \frac{d[P]}{dt} \qquad (8\text{–}9)$$

and

$$k_1 [A] [B] = k_2 [P] \qquad (8\text{–}10)$$

Consequently,

$$\frac{k_1}{k_2} = \frac{[P]}{[A][B]} = K \qquad (8\text{–}11)$$

where K is the **equilibrium constant** for the reaction and describes the relative concentrations of all reacting molecular species at equilibrium.

Effect of Enzyme on Reaction Rate

Two general mechanisms are involved in enzyme catalysis. First, the presence of the enzyme increases the likelihood that the potentially reacting molecular species will encounter each other with the required orientations in space. This occurs because the enzyme has a high affinity for the reactants (more appropriately referred to as the **substrates**) and temporarily binds them. The association of substrates with an enzyme is not arbitrary; instead the substrates are bound to the enzyme in such a manner that each substrate is oriented with respect to the other in precisely the manner required for the reaction to occur. Second, the formation of temporary bonds (mostly noncovalent) between the enzyme and substrate forces a redistribution of electrons within the substrate molecules, and this redistribution imposes a *strain* upon specific covalent bonds within the substrates—a strain that culminates in bond breakage. Biochemists refer to the introduction of bond strains in a substrate by association with an enzyme as ''substrate activation.'' The net effect of this is to greatly increase the percentage of molecules in the population that at any instant are sufficiently reactive toward each other. This rather brief and general accounting of enzyme function is expanded later in the chapter as the specific structures and actions of certain representative enzymes are considered.

For a reversible enzyme-catalyzed reaction such as

$$A + B \overset{E}{\rightleftharpoons} P \qquad (8\text{–}12)$$

where E is the enzyme (and can catalyze the reaction in both directions),

$$\frac{d[A]}{dt} = \frac{d[B]}{dt} = k_1[A][B]e \qquad (8\text{–}13)$$

In this equation, e is the ''enzyme factor''—a *factor* that accounts for the increase in reaction rate through catalysis—and is proportional to the concentration of the enzyme in solution. That is, if the reaction proceeds at some rate at a given enzyme concentration, then the reaction would proceed twice as rapidly at twice the enzyme concentration (Fig. 8–3).

For the reverse reaction,

$$\frac{d[P]}{dt} = k_2[P]e \qquad (8\text{–}14)$$

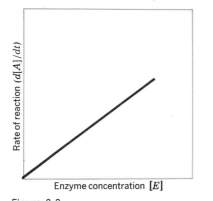

Figure 8-3
Relationship between enzyme concentration and the rate of an enzyme-catalyzed reaction.

Table 8–1
Turnover Numbers of Some Enzymes

Enzyme	Turnover Number[a]
Carbonic anhydrase	3.6×10^6
Acetylcholinesterase	1.5×10^6
Urease	1.0×10^6
Amylase	1.0×10^5
Lactic dehydrogenase	6.0×10^4
Chymotrypsin	6.0×10^3
Lysozyme	3.0×10^1

[a]Moles of substrate converted to product per minute per mole of substrate-stained enzyme.

It can be shown using equations 8–13 and 8–14 that the equilibrium constant for the reaction (i.e., k_1/k_2) is not altered by the presence of the enzyme. Thus, *the enzyme greatly affects (i.e., increases) the rate at which equilibrium is achieved, but it does not alter the respective equilibrium concentrations of the various molecular species.*

The marked effect of enzymes on reaction rates can be clearly appreciated by considering the **turnover number** of an enzyme. The turnover number is the number of moles of substrate converted to product per minute per mole of enzyme (when the enzyme is fully saturated with substrate). For example, the hydrolytic enzyme *urease,* which catalyzes the conversion of urea to ammonia, has a turnover number of about 10^6. In the absence of urease, the hydrolysis of urea is several orders of magnitude slower. One molecule of *amylase,* an enzyme involved in the breakdown of starch and other polysaccharides, can hydrolyze about 10^5 glycosidic bonds per minute, but *no* detectable hydrolysis occurs in the absence of amylase. The turnover numbers of most enzymes fall between 10^2 and 10^6 (Table 8–1).

The Kinetics of Enzyme Action

In most instances, the association of the enzyme with the substrate is so fleeting that the complex is extremely difficult to detect. Yet, as early as 1913, L. Michaelis and M. L. Menten postulated the existence of this transient complex. On the basis of their observations with the enzyme *invertase,* which catalyzes the hydrolysis of sucrose to glucose and fructose, they proposed that enzyme-catalyzed reactions were characterized by a sequence of phases that involves (1) the formation of a complex (ES) between the enzyme (E) and the substrate (S); (2) the modification of the substrate to form the product or products (P), which briefly remain associated with the enzyme (EP); and (3) the release of the product or products from the enzyme; that is,

$$E + S \rightleftharpoons ES \rightleftharpoons EP \rightarrow E + P \qquad (8\text{–}15)$$

These events are more conventionally described by the equation

$$E + S \underset{k_2}{\overset{k_1}{\rightleftharpoons}} ES \overset{k_3}{\longrightarrow} E + P \qquad (8\text{–}16)$$

In this equation, it is assumed that the combination of enzyme and substrate is reversible; k_1 is the rate constant for the formation of ES (the dimensions of the rate constant are *seconds*$^{-1}$), and k_2 is the rate constant for the dissociation of ES. After the ES complex is formed, S is converted to P with the rate constant k_3. As long as the concentration of P remains negligible (as it would be at the outset of catalysis or if P is in some manner quickly removed from the system), then it is not necessary to consider the reverse flux from $E + P$ to EP to ES.

Figure 8–4 depicts the relationship that exists between substrate concentration and the rate at which reaction products appear for enzyme-catalyzed reactions of this type. The curve describes the *initial rate of product formation* at a fixed enzyme concentration when the substrate concen-

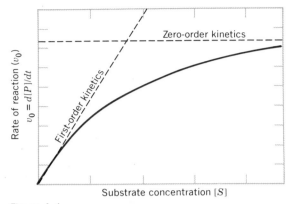

Figure 8-4
Relationship between substrate concentration and rate of enzyme catalyzed reaction. Curve describes *initial* rate of product formation (*P*) when the enzyme concentration is held constant and the substrate concentration [*S*] is increased on successive trials. At low substrate concentration, the reaction flux (v_o) follows first-order kinetics. As the substrate concentration is increased, reaction rate approaches a maximum value at which time zero-order kinetics is followed.

tration is varied in successive trials. At low concentrations of substrate, the initial velocity of the reaction (i.e., v_o) is directly proportional to the substrate concentration (i.e., follows first-order kinetics). However, as the substrate concentration is increased, the reaction velocity levels off, approaching a maximum value. At this high substrate concentration, the reaction velocity is limited by the amount of time required to convert bound substrate (i.e., *ES*) to product (*P*) and free enzyme (*E*). The curve now follows zero-order kinetics. The curve of Figure 8–4 is called a Michaelis-Menten curve, and for an idealized enzyme catalyzed reaction is a *rectangular hyperbola*.

Michaelis and Menten are also credited with the first mathematical study of the relationship between substrate concentration and reaction rates. They introduced two particularly useful mathematical expressions which for any enzyme relate [*S*] to *V* (velocity) and permit quick comparisons of various enzyme-catalyzed reactions; the two expressions are now called the **Michaelis-Menten constant** and the **Michaelis-Menten equation.** They are derived as follows:

Let [*S*] = the concentration of free substrate, *S;* (we may assume that the amount of available substrate is

so great that the amount combined with the enzyme may be ignored in comparison. Hence, total substrate concentration and the free substrate concentration are the same);

$[E]_T$ = the total concentration of enzyme, *E;*

$[E]$ = the concentration of free enzyme (that not complexed with substrate);

$[ES]$ = the concentration of enzyme-substrate complex (since the total amount of enzyme present is assumed to be very small in comparison with the total amount of substrate, a significant proportion of the total enzyme may be involved in the *ES* complex. Hence, separate terms for the total and free enzyme concentrations are warranted);

$[P]$ = the concentration of product, *P.*

The total enzyme concentration is given by

$$[E]_T = [E] + [ES] \qquad (8–17)$$

For equation 8–16, the initial rate of product formation, v_o, is given by

$$v_o = \frac{d[P]}{dt} = k_3\,[ES] \qquad (8–18)$$

The change in the concentration of *ES* with time is equal to the rate at which *ES* is formed minus the rate at which it is being eliminated. That is,

$$\frac{d\,[ES]}{dt} = k_1\,[E]\,[S] - (\,k_2\,[ES] + k_3\,[ES]\,) \qquad (8–19)$$

Therefore, by simplifying,

$$\frac{d\,[ES]}{dt} = k_1\,[E]\,[S] - (k_2 + k_3)\,[ES] \qquad (8–20)$$

If sufficient substrate is available, then $d[ES]/dt$ would be zero, since for every *P* leaving the complex, a new *S* would enter it. Under such conditions,

$$k_1[E][S] = (k_2 + k_3)\,[ES] \qquad (8–21)$$

and by rearrangement

Table 8–2
Some Representative K_M Values

Enzyme	Source	Substrate	K_M
Sucrase	Intestine	Sucrose	$2 \times 10^{-2}\ M$
Urease	Soybeans	Urea	$2.5 \times 10^{-2}\ M$
Catalase	Liver	Hydrogen peroxide	$2.5 \times 10^{-2}\ M$
Carbonic anhydrase	Blood	CO_2	$9 \times 10^{-3}\ M$
Chymotrypsin	Pancreas	Peptides	$5 \times 10^{-3}\ M$
Phosphatase	Bone	Glycerophosphate	$3 \times 10^{-3}\ M$
Hexokinase	Liver	Glucose	$1.5 \times 10^{-4}\ M$
Lysozyme	Egg white	Hexa-N-acetylglucosamine	$6 \times 10^{-6}\ M$

$$\frac{k_2 + k_3}{k_1} = \frac{[E][S]}{[ES]} \tag{8-22}$$

The ratio of constants given in equation 8–22 may be set equal to a new constant, K_M, which is the Michaelis-Menten constant. Thus, the Michaelis-Menten constant is a constant that relates the steady-state concentrations of enzyme, enzyme-substrate complex, and substrate.

Each enzyme-catalyzed reaction reveals a characteristic K_M value, and this value is a measure of the tendency of the enzyme and the substrate to combine with each other. In this sense, the K_M value is an index of the affinity of the enzyme for its particular substrate. Some representative K_M values are given in Table 8–2. It is to be stressed that *the greater the affinity of an enzyme for its substrate, the lower the K_M value*. This is because the K_M value is numerically equal to the substrate concentration at which half of the enzyme molecules are associated with substrate (i.e., in the *ES* form).

The relationships derived above are based on reactions in which a single substrate molecule is bound to the enzyme. However, they also apply in situations where more than one substrate is bound to the enzyme, as long as the concentrations of all but one substrate species are held constant or are not rate-limiting (i.e., present in large excess). For example, even the pioneering studies of Michaelis and Menten involved a bimolecular enzyme-catalyzed reaction. They employed the enzyme *invertase*, which forms a complex with one molecule of sucrose and one molecule of water and releases glucose and fructose as products. However, in this reaction (as in nearly all hydrolyses), the concentration of water remains virtually unaltered during the course of the reaction.

K_M values are seldom determined using the relationship given in equation 8–22, since the enzyme-substrate concentration [ES] cannot easily be measured. Some additional algebraic manipulations may be carried out in order to convert equation 8–22 into a more useful form. Since $[E] = [E]_T - [ES]$, then

$$K_M = \frac{[E][S]}{[ES]} = \frac{([E]_T - [ES])[S]}{[ES]} \tag{8-23}$$

$$K_M[ES] + [S][ES] = [E]_T[S] \tag{8-24}$$

$$[ES](K_M + [S]) = [E]_T[S] \tag{8-25}$$

and

$$[ES] = \frac{[E]_T[S]}{K_M + [S]} \tag{8-26}$$

Since $v_o = d[P]/dt = k_3[ES]$, then

$$v_o = \frac{k_3[E]_T[S]}{K_M + [S]} \tag{8-27}$$

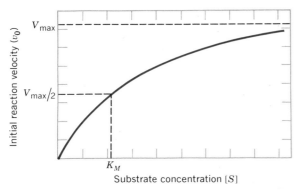

Figure 8-5
Relationship between substrate concentration [S] and re-action rate (v_o). When the rate of reaction is one-half its maximum initial value (V_{max}), the substrate concentration is numerically equal to the Michaelis-Menten constant (K_M). All such curves are rectangular hyperbolas. Consequently, the lower the K_M value, the lower the substrate concentration at which the reaction proceeds at a maximum rate.

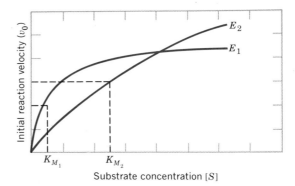

Figure 8-6
Relationship between substrate concentration [S] and re-action rate (v_o) for two enzymes having the same substrate. Enzyme 1 (E_1) has a lower K_M and attains maximum rate of reaction at lower substrate concentration. Enzyme 2 (E_2) has a higher K_M value and produces a maximum reaction rate at higher concentration of substrate. As a result, the reaction catalyzed by enzyme 1 is favored at low substrate levels, while the reaction catalyzed by enzyme 2 is favored at high substrate concentrations.

As seen in Figure 8–4, the initial reaction velocity increases with substrate concentration until the enzyme concentration becomes limiting, and at that time, the initial reaction velocity approaches a maximum value (i.e., V_{max}). This occurs because all or nearly all the enzyme is maintained in the ES form. Thus, $[ES]$ will equal $[E]_T$, and $k_3 [E]_T$ corresponds to V_{max}.

By substituting V_{max} for $k_3 [E]_T$ in equation 8–27, we obtain

$$v_o = \frac{V_{max} [S]}{K_M + [S]} \qquad (8\text{–}28)$$

Equation 8–28 is the Michaelis-Menten equation. From this equation, it may be seen that when the substrate concentration is numerically equal to the K_M value of the enzyme, then the reaction velocity is equal to one-half of the maximum value. For example, if $[S]$ and K_M are both equal to 3, then equation 8–27 simplifies to

$$v_o = \frac{V_{max} (3)}{3 + 3} = \frac{V_{max}}{2} \qquad (8\text{–}29)$$

Consequently, the Michaelis-Menten constant for an enzyme may be determined from the substrate concentration at which the reaction velocity proceeds at one-half its maximum value (Fig. 8–5).

From the above, we may conclude that the maximum velocity of an enzyme-catalyzed reaction depends on the affinity between the enzyme and its substrate (i.e., the K_M value). The higher the K_M value, the lower the affinity of the enzyme for the substrate; and the smaller the K_M value, the greater the affinity. In many instances, the same substrate may enter either of several different enzyme-catalyzed reactions occurring in cells. Which of the alternative reactions predominates in the cell depends on the K_M values of the respective enzymes and the concentration of available substrate. At very low substrate concentrations, the specific reaction catalyzed by the enzyme with the lowest K_M will dominate, whereas at higher substrate concentrations, the reaction catalyzed by the enzyme having the greatest K_M value can dominate (if enough of the enzyme is present). Therefore, which of several different metabolic pathways is actually followed by the initial substrate in the course of a series of enzyme-catalyzed reactions may be regulated by controlling the amount of available substrate. This and other metabolic regulatory mechanisms are discussed in Chapter 11. The relationship between reaction velocity and substrate concentration for two enzymes that act on the same substrate is depicted in Figure 8–6.

In order to obtain the K_M value of an enzyme experimentally, it is necessary to determine v_o for a series of substrate concentrations. In practice the evaluation of K_M from a plot similar to that in Figure 8–6 is difficult, since the precise value of V_{max} cannot be determined. This is because the curve is a rectangular hyperbola and approaches V_{max} asymptotically. In 1934, H. Lineweaver and D. Burk introduced a different form of the Michaelis-Menten equation (8–28) that simplifies the determination of K_M from experimentally obtained values for [S] and v_o. In this form, the acquired data are used to construct a straight line rather than a hyperbola. By taking the inverse of equation 8–28 we obtain,

$$\frac{1}{v_o} = \frac{K_M}{V_{max}\,[S]} + \frac{[S]}{V_{max}\,[S]} \qquad (8\text{--}30)$$

or

$$\frac{1}{v_o} = \frac{K_M}{V_{max}}\frac{1}{[S]} + \frac{1}{V_{max}} \qquad (8\text{--}31)$$

The terms are now arranged in the form of the general equation of a straight line, $y = ax + b$, in which a is the slope of the line and b is the intercept on the y-axis. Thus, for equation 8–31, $1/v_o$ serves as the y-axis and $1/[S]$ as the x-axis. Consequently, the y-intercept will be $1/V_{max}$, the slope will be K_M/V_{max}, and the x-intercept will be $-1/K_M$.

The graph showing this relationship (known as a **double reciprocal plot**) is presented in Figure 8–7.

Effects of Inhibitors on Enzyme Activity

The interaction between the substrate and the enzyme takes place in a particular region of the enzyme molecule called the **active site** (discussed later). In many instances, compounds other than the normal substrate for a particular enzyme-catalyzed reaction may bind to the enzyme's active site, and this has a significant effect on the kinetics of the normal reaction. One possible consequence of this phenomenon is the inhibition of normal enzyme activity; such compounds are therefore called **enzyme inhibitors.** (Usually, the inhibitor is unaltered by its interaction with the enzyme.) In some instances, the normal substrate (S) and the inhibitor (I) *compete* with each other for the active site of the enzyme; the manner in which this affects the normal kinetics of the reaction is shown in Figure 8–8. V_{max} is not altered by the presence of a competitive inhibitor, but the K_M value is elevated. As can be seen in Figure 8–8, the effect of the inhibitor is maximal at low substrate concentration (i.e., when $1/[S]$ is large) and minimal at high substrate concentration (i.e., when $1/[S]$ approaches 0.)

A classical example of this form of inhibition is the competition between succinic acid and malonic acid for the enzyme *succinic acid dehydrogenase.* In this instance, competition between these two compounds for the active site of the enzyme is understandable in view of their marked chemical similarity (Fig. 8–9). Succinic acid is the normal substrate for the enzyme and, in the absence of the inhibitor, is converted to fumaric acid.

Enzyme inhibition can also be *noncompetitive* in that the binding of the inhibitor to the enzyme cannot be reversed by increasing the concentration of the normal substrate. A common example of negative inhibition is the action of

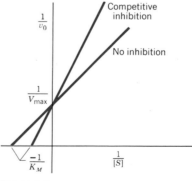

Figure 8-8
Effect of competitive inhibitor on normal enzyme kinetics. V_{max} is not altered, but K_M value is increased. At low substrate concentrations, the effect of inhibitor on reaction rate is marked but is completely reversed as substrate concentration is increased and 1/[S] approaches 0.

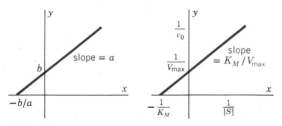

Figure 8-7
Lineweaver-Burk *double reciprocal plot* of 1/v_o and 1/[S] may be used to construct a line that may easily be extrapolated backward to find the K_M value for the enzyme.

(a)

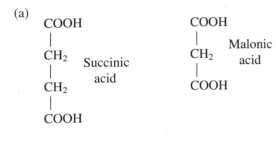

(b)

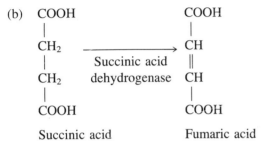

Succinic acid   Fumaric acid

Figure 8-9
(a) Chemical formulas of succinic acid and malonic acid; (b) Malonic acid is a competitive inhibitor of *succinic acid dehydrogenase.* Note the chemical similarity between the true substrate (succinic acid) and the inhibitor.

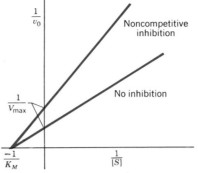

Figure 8-10
Effect of a noncompetitive inhibitor on normal enzyme kinetics. The K_M value is not altered, but V_{max} is reduced. Noncompetitive inhibition can be reduced but not reversed by increasing the substrate concentration.

heavy metals such as mercury on the active sites of enzymes containing a reactive sulfhydryl (i.e., $-SH$) group. In effect, the presence of the inhibitor prevents some percentage of the enzyme present from participating in normal catalysis. As a result, the maximum reaction velocity is depressed, even though the K_M value remains the same (Fig. 8–10).

The inhibition of enzyme activity by competitors also occurs naturally in cells and serves in the feedback regulation of certain metabolic pathways. A special form of feedback inhibition involves **allosterism,** a topic that will be considered at length later in the chapter.

Mechanics of Enzyme Catalysis

Enzymes are proteins, and therefore their capacity for catalysis is intimately related to a specific tertiary or quaternary molecular structure. If the tertiary or quaternary structure of an enzyme is altered, a loss of enzyme activity usually follows. Thus, environmental factors that modify

protein structure also influence enzyme activity. Key environmental factors that can affect enzyme activity are pH and temperature. As discussed at some length in Chapter 4, the polar side chains of certain amino acids form electrostatic bonds with each other and with surrounding ions and water molecules; these interactions contribute in part to the specific tertiary and quaternary structure of the protein. Whether or not a particular amino acid side chain bears a charge is determined in part by the pH of the protein's environment. As the pH is lowered (i.e., the concentration of H^+ is increased), groups that may be negatively charged, such as the secondary COO^- of aspartic acid and glutamic acid and the O^- of tyrosine, become protonated, thereby neutralizing these negative charges. At the same time, some secondary amino groups, such as those of lysine and arginine, may accept additional protons, thereby imparting charge to these formerly neutral side chains. In contrast, as the pH is elevated (i.e., the concentration of OH^- is increased), positively charged side chains dissociate protons and are thereby neutralized, while the loss of protons from secondary COOH and OH groups renders these groups negative. Some of these relationships are depicted in Figure 8–11.

In addition to playing important parts in the maintenance of a specific tertiary or quaternary molecular structure, the polar side chains of some of the amino acids in the enzyme (i.e., those in the active site) may be involved in binding the substrate to the enzyme and thereby introduce bond strains into the substrate molecule. Consequently, most enzymes can operate only within a narrow pH range and

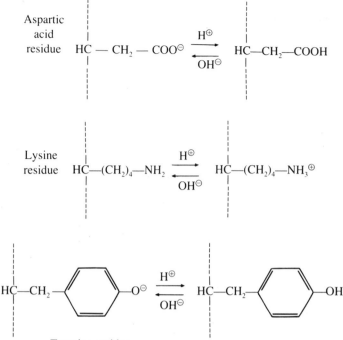

Aspartic acid residue

Lysine residue

Tyrosine residue

Figure 8-11
Influence of a change in pH on the side chains of aspartic acid, tyrosine, and lysine.

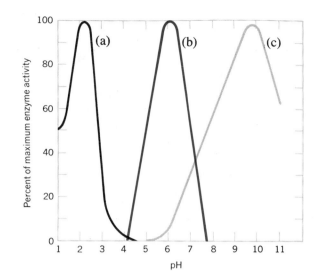

Figure 8-12
Effect of pH on enzyme activity. (*a*) *Pepsin,* (*b*) *glutamic acid decarboxylase,* and (*c*) *arginase.*

Table 8–3
Optimum pH Values for Certain Enzymes

Enzyme	Optimum pH
Pepsin	2.0
Glutamic acid decarboxylase	5.9
Urease	6.7
Salivary amylase	6.8
Pancreatic lipase	7.0
Trypsin	9.5
Arginase	10.0

display a pH optimum (Fig. 8–12). The pH optima of enzymes vary over a broad range of values; several examples are given in Table 8–3. On either side of the pH optimum, enzyme activity declines as the configuration of the protein is altered and/or its affinity for the substrate is correspondingly decreased.

Temperature also influences enzyme activity, and most enzymes display a temperature optimum close to the normal

temperature of the cell or organism possessing that enzyme. Accordingly, the temperature optima of plant cell enzymes and enzymes of poikilothermic animals inhabiting cold regions of the earth are usually lower than enzymes of homeothermic animals.

If the temperature is elevated far above the optimum, enzyme activity decreases. This is the result of an alteration of the enzyme's structure (an unraveling process called **denaturation**). Most enzymes are irreversibly denatured if maintained at temperatures above 55°C to 65°C for an extended period of time.

The Active Site

The formation of the enzyme-substrate complex is not a random process. This was recognized as long ago as 1894, when Emil Fischer postulated that an enzyme allows only one or a few compounds to fit onto its surface. This is the ''lock-and-key'' hypothesis according to which the enzyme and its substrate have a complementary shape. The specific substrate molecules (and prosthetic groups, if any) are

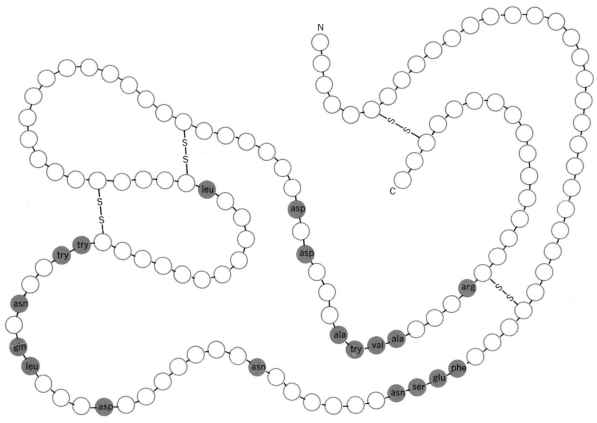

Figure 8-13
Structure of lysozyme showing distribution of amino acids forming the active site.

bound to a specific region of the enzyme molecule called its **active site.** The active site of an enzyme is formed by a number of amino acid residues whose side chains have two principal roles: (1) they serve to *attract* and *orient* the substrate in a specific manner within the site (such amino acids are called **contact residues** and contribute in large degree to substrate specificity); and (2) they participate in the *formation of temporary bonds* with the substrate molecule, bonds that polarize the substrate, introduce strain into certain of its bonds, and trigger the catalytic change (such amino acids are termed **catalytic residues**).

The contact and catalytic residues that make up the active site may be located in widely separated regions of the enzyme's primary structure, but as a result of stabilized polypeptide chain folding, they are brought into the appropriate juxtaposition. This is exemplified by the enzyme lysozyme (Fig. 8–13), in which the amino acids that form

the active site in the folded structure (shaded residues) are widely separated in the primary structure. The bonds formed between a substrate and the amino acid side chains forming the active site may be either covalent or noncovalent.

Figure 8–14 depicts the interaction between enzyme and substrate according to the lock-and-key model. The substrate has polar (i.e., △ and △) and nonpolar (Ⓗ, hydrophobic) regions and is attracted to and associates with the active site that is complementary in both shape and charge distribution (Fig. 8–14a, b). Positive, negative, and hydrophobic regions of the active site are created by the side chains of the contact residues, which align the substrate for interaction with the site's catalytic residues (Ⓐ and Ⓑ). Following catalysis (Fig. 8–14c), the products are released from the active site (Fig. 8–14d), thereby freeing the enzyme for another round of catalysis. The lock-and-key

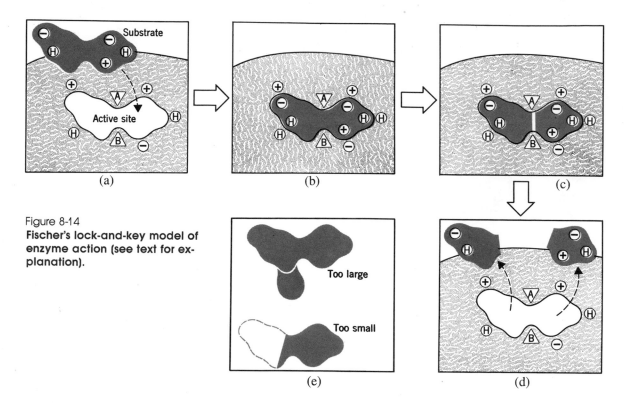

Figure 8-14
Fischer's lock-and-key model of enzyme action (see text for explanation).

model of enzyme catalysis accounts for enzyme specificity, since compounds that lack the appropriate shape or are too large or too small (Fig. 8–14e) cannot be bound to the active site.

Although the lock-and-key model accounts for much of the substrate specificity data, certain observations about enzyme behavior do not fit or are difficult to explain using this model. For example, there are a number of instances in which compounds other than the true substrate bind to the enzyme even though they fail to form reaction products. Furthermore, for many enzyme-catalyzed reactions, substrates are bound to the active site in a specific temporal order. In the 1960s, Daniel Koshland proposed the "induced-fit" theory of enzyme action according to which the active site of the enzyme does not initially exist in a shape that is complementary to the substrate but is induced to assume the complementary shape as the substrate becomes bound. As Koshland put it, the active site is induced to assume complementary shape "in much the same way as a hand induces a change in the shape of a glove." Thus, according to this model, the enzyme (or its active site) is *flexible*.

The induced-fit model is depicted diagrammatically in Figure 8–15. The active site and substrate initially have different shapes (Fig. 8–15a) but become complementary upon substrate binding (Fig. 8–15b). The shape change places the catalytic residues in position to alter the bonds in the substrate (Fig. 8–15c), following which the products are released (Fig. 8–15d) and the active site returns to its initial state. Although molecules that are larger or smaller than the true substrate or that have different chemical properties may nonetheless be bound to the active site, none succeed in inducing the proper alignment of catalytic groups, and no catalysis occurs (Fig. 8–15e, f). The induced-fit model explains the effects of certain competitive and noncompetitive inhibitors of enzyme action.

Before proceeding further, it should be acknowledged that some enzyme-catalyzed reactions are adequately explained by the lock-and-key model, so that a flexible active site is not a strict requirement for catalysis. By the same token, the possession of a flexible active site does not imply that just any molecule may become bound to the enzyme.

A change in the shape of the active site of an enzyme can also be induced by binding at sites on the enzyme's

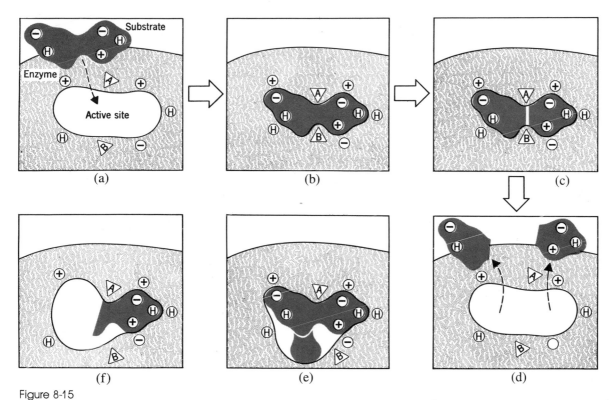

Figure 8-15
Koshland's induced-fit model of enzyme action (see text for explanation).

surface that are far removed from the active site. In such a case, the change is transmitted through the enzyme molecule from the site of binding to the active site. Such changes may either decrease or increase the enzyme's activity. The latter phenomenon will be discussed later in the chapter in connection with enzyme **cooperativity** and **allosterism.**

Extensive studies during the past decade have revealed the precise atomic structure of a number of important enzymes including *lysozyme, ribonuclease, chymotrypsin, trypsin, carboxypeptidase, papain, elastase,* and *thrombin.* A consideration of some of these reveals certain generalizations about the organization of enzyme molecules and provides insight into the mechanisms by which catalysis is achieved.

Lysozyme

Lysozyme is an enzyme produced by both animal and plant tissues and was first to have its complete three-dimensional

structure revealed (by C. C. F. Blake, D. C. Phillips, and A. C. T. North, in 1965). The enzyme cleaves polysaccharide chains found in the cell walls of certain bacteria by hydrolyzing the glycosidic bonds between neighboring hexosyl residues. Egg white lysozyme has a molecular weight of about 14,600 and consists of a single polypeptide chain of 129 amino acids. The enzyme is oval in shape with a deep cleft across its midline that divides the molecule into two parts. The shape of lysozyme and the orientation of the substrate are depicted stereoscopically in Figure 8–16. The polypeptide contains three short helical regions and a segment arranged in the form of a beta-pleated sheet. All the polar residues are located on the enzyme's surface, while nearly all uncharged residues are buried internally. Hydrogen bonds between one portion of the polypeptide chain and another are pronounced and appear to be vital in sustaining the active tertiary structure of the molecule.

The cleft at the center of the molecule contains the active site and is studded with hydrophobic groups that probably form van der Waals bonds with the substrate. The bacterial

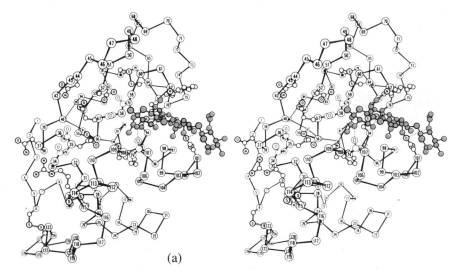

(a)

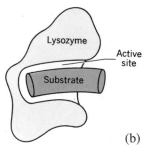

(b)

Figure 8-16
(*a*) Stereoscopic diagram of lysozyme. (Courtesy of R. E. Dickerson and I. Geis.) (*b*) Diagram illustrating how the binding of the substrate occurs in a deep cleft that forms the active site of lysozyme.

polysaccharide that acts as substrate for the enzyme consists of chains of *N-acetylglucosamine* (NAG) and *N-acetylmuramic acid* (NAM) in which the two sugars alternate. In the bacterial cell wall, these chains are cross-linked by short polypeptides. Six hexosyl units of the substrate polysaccharide are simultaneously bound at the active site of lysozyme. A change in tertiary structure, which takes the form of a narrowing and deepening of the cleft in the enzyme, accompanies substrate binding and forces a modification of the conformation of one of the six hexosyl groups (i.e., NAM) bound at the active site. This conformational change involves a distortion of the "chair" form normally assumed by this sugar and primes this region of the substrate for catalytic alteration. It is the glycosidic bond between this group and its neighbor that is broken during catalysis. Of the 19 or more amino acids that compose the active site, only two have been identified as catalytic; these are asp (amino acid no. 52) and glu (amino acid no. 35). The remaining 17 residues align the substrate through hydrogen bonds and nonpolar interaction.

Asp 52 carries a dissociated carboxyl group (i.e., negative charge), but glu 35, being surrounded by the nonpolar side chains of other residues, is protonated. The sequence of events occurring during catalysis is depicted in Figure 8–17*a, b,* and *c* in which only three of the six hexosyl units bound to the active site (i.e., numbers 3, 4, and 5) are shown. The binding of the substrate to the enzyme and the conformational changes that occur in both as a result are followed by an attack by a hydrogen ion of the undisso-

ciated COOH group of glu 35 on the oxygen atom forming the bridge between hexosyl units 4 and 5 (Fig. 8–17*a*). This, of course, is the specific region of the substrate placed under strain during binding. The glycosidic bond is broken, leaving carbon atom no. 1 of hexosyl unit 4 with a positive charge (i.e., a carbonium ion) (Fig. 8–17*b*). The carbonium ion is stabilized by the formation of a temporary ionic bond with the dissociated side chain of asp 52 (Fig. 8–17*b*). The dissociation of a neighboring molecule of water provides a hydroxyl ion for attack upon the carbonium ion and a hydrogen ion for glu 35 (Fig. 8–17*c*). Once the reaction is completed, the two resulting polysaccharide fragments leave the active site.

Ribonuclease

Ribonuclease is produced by the pancreas and secreted into the small intestine where it catalyzes the hydrolytic digestion of polyribonucleotide chains. The enzyme consists of a single 124 amino acid polypeptide having a molecular weight of about 13,700. The reaction catalyzed by ribonuclease, shown in Figure 8–18, involves the hydrolytic cleavage of the ester linkage between the 3′-phosphate group of a pyrimidine nucleotide and the 5′ position of the adjacent nucleotide.

Like lysozyme, the active site of ribonuclease lies in a cleft in the enzyme's surface. The cleft is lined by a number of positively charged side chains of lysine and arginine residues, which probably act as contact residues by forming

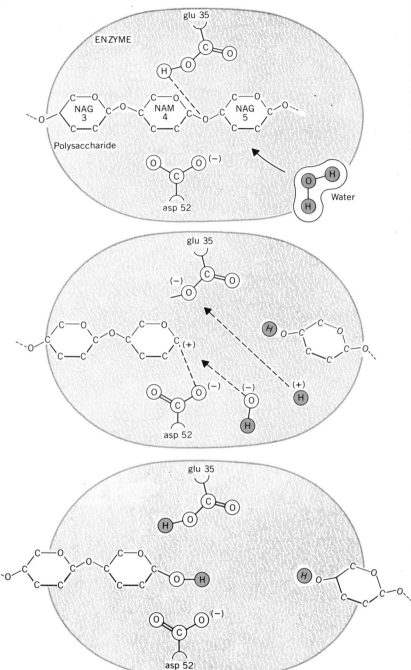

Figure 8-17
**Proposed mechanism for the action of lyso-
zyme.** Only three of the five or six glycosyl
units of the polysaccharide that are bound
to the active site are shown (numbered 3, 4,
and 5). Hydroxyl groups, other groups, and
hydrogen atoms of the pyranose rings have
been omitted for clarity. Only the side
chains of the two catalytic residues are
shown. See text for details of the catalytic
process.

ionic bonds with the negatively charged phosphate groups
of the ribonucleic acid backbone (Fig. 8–18). Projecting
into the cleft from either side are two histidine residues that
are believed to hydrolyze the RNA substrate by general
acid–base catalysis in which the P—O bond linking the
ribose of one nucleotide with the ribose of a neighboring
nucleotide is broken. Ribonuclease is depicted stereoscopically in Figure 8–19.

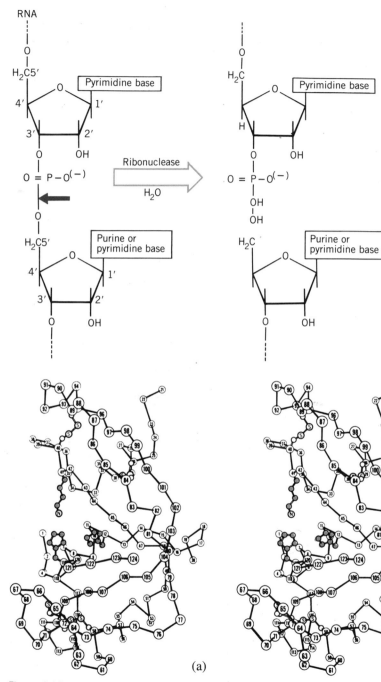

Figure 8–18
The chemical reaction catalyzed by ribonuclease. The enzyme cleaves the P—O bond (heavy arrow) on the far side of the phosphate group linking the 3′C of a pyrimidine nucleoside with the 5′C of the adjacent nucleoside.

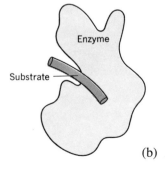

Figure 8-19
(a) Stereoscopic view of ribonuclease. Histidine and lysine residues of the active site are shown in color. (Courtesy of R. E. Dickerson and I. Geis.) (b) Diagram showing the general outline of the enzyme and the position of substrate in the cleft containing the active site.

Chymotrypsin

The digestive enzyme chymotrypsin is derived from the inactive precursor polypeptide *chymotrypsinogen* produced in the pancreas and containing 245 amino acids. The activation of chymotrpysinogen involves the preliminary cleavage of four of its peptide bonds, resulting in three separate polypeptide chains cross-linked by two disulfide bridges; two dipeptides released in the activation process do not become a part of the functional enzyme (see Fig. 22–40). Chymotrypsin digests alimentary protein by hydrolyzing peptide bonds on the carboxyl side of amino acids with large hydrophobic side chains (i.e., phenylalanine, tyrosine, and tryptophan residues).

Unlike lysozyme and ribonuclease, the active site of chymotrypsin is not in a cleft but resides in a shallow dish-shaped cavity on the surface of this roughly spherical enzyme. In the active site is a hydrophobic pocket that is believed to act as the receptor of the hydrophobic side chain of the substrate amino acid residue and thereby imparts specificity to this enzyme. In addition to these and other contact residues, the active site contains three catalytic residues: asp 102, his 57, and ser 195. Figure 8–20 depicts the series of stages that are believed to be involved in the action of chymotrypsin. The imidazole group of histidine attracts the hydroxyl hydrogen atom of the serine, thereby rendering the serine oxygen strongly nucleophilic and particularly reactive toward amides and esters (Fig. 8–20a). Binding of the substrate polypeptide to the enzyme (Fig. 8–20b) is followed by formation of a covalent bond between the serine oxygen and the alpha-carboxyl carbon atom of the substrate residue (Fig. 8–20c). This is followed by donation of the serine proton to the alpha-amino nitrogen atom of the neighboring substrate residue, the breakage of the peptide bond, and the release of the first product (Fig. 8–20d).

At this stage, a molecule of water enters the reaction (Fig. 8–20e), donating a hydrogen ion to serine and a hydroxyl group to the alpha-carboxyl carbon (Fig. 8–20f). Dissociation of this carboxyl group (Fig. 8–20g) leaves the carboxyl group negatively charged. Repulsive forces between this group and the negatively charged aspartic acid residue of the active site facilitate separation of the enzyme and final product (Fig. 8–20h).

The behavior of the catalytic serine residue is critical to the function of chymotrypsin. Other proteases, including trypsin, elastase, collagenase, thrombin, fibrinolysin, and some lysosomal proteases, share a similar overall tertiary structure and active site, and they cleave peptide bonds via the identical action of a catalytic serine residue. The different substrate specificities of these "serine proteases" may be attributed to the nature and location of the contact residues in their active sites. In view of their remarkable similarity, the serine proteases are believed to have a common evolutionary origin.

The accumulated information on lysozyme, ribonuclease, and the serine proteases, as well as several other enzymes, permits certain generalizations to be made concerning enzyme structure and action. The active site of the enzyme nearly always resides in a depression in its surface, and the binding of the substrate to the active site is followed by conformational change in the enzyme, the substrate, or both. The binding and/or the conformational changes that follow place a strain on certain bonds in the substrate and render them particularly susceptible to chemical attack. The catalysis itself may be facilitated by as few as one or two amino acid side chains of the enzyme which, because of their spatial arrangement in the active site and the polar or hydrophobic nature of their surroundings, are especially reactive.

Cofactors

Enzymes are composed of one or more polypeptide chains. However, there are a number of cases in which nonprotein constituents called **cofactors** must be bound to the enzyme (in addition to the substrate) in order for the enzyme to be catalytically active. In these instances, the exclusively protein portion of the enzyme is called the **apoenzyme.** Three kinds of cofactors may be identified: **prosthetic groups, coenzymes,** and **metal ions.**

Prosthetic groups are organic compounds and are distinguished from other cofactors in that they are permanently bound to the apoenzyme. For example, in the peroxisomal enzymes *peroxidase* and *catalase,* which catalyze the breakdown of hydrogen peroxide to water and oxygen, *heme* is the prosthetic group and is a permanent part of the enzyme's active site (see also Chapter 19).

Coenzymes are also organic compounds, but their association with the apoenzyme is transient, usually occurring only during the course of catalysis. Furthermore, the same coenzyme molecule may serve as the cofactor in a number

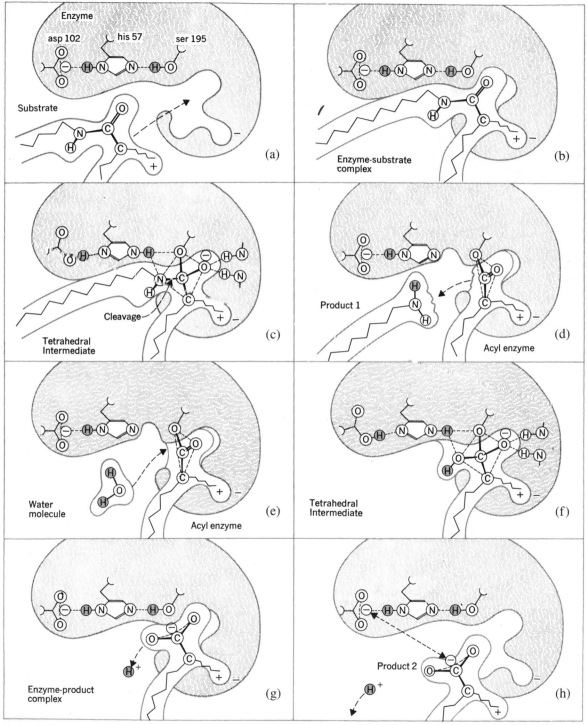

Figure 8-20
Stages involved in proteolysis by chymotrypsin. See text for details of the catalytic process.

Table 8–4
Some Enzymes That Require Cofactors

Cofactor	Enzyme	Reaction
Prosthetic groups		
Heme	Catalase	$2 H_2O_2 \longrightarrow 2 H_2O + O_2$
Heme	Peroxidase	$2 H_2O_2 \longrightarrow 2 H_2O + O_2$
Flavin adenine dinucleotide (FAD)	Succinic dehydrogenase	Succinic acid $\longrightarrow$ fumaric acid
Coenzymes		
Flavin mononucleotide (FMN)	Some dehydrogenases	Removal of hydrogen atoms
Thiamine pyrophosphate (TPP)	Some decarboxylases	Removal of CO_2
Nicotinamide adenine dinucleotide		
(NAD)	Some dehydrogenases	Removal of hydrogen atoms
Lipoic acid	Some decarboxylases	Removal of CO_2
Metal ions		
Zn^{++}	Carboxypeptidase	Hydrolysis of proteins
Zn^{++}	Carbonic anhydrase	$CO_2 + H_2O \leftrightarrow H_2CO_3$
Cu^{++}	Ascorbic acid oxidase	Ascorbate $\leftrightarrow$ dehydroascorbate
Cu^{++}	Uric acid oxidase	Uric acid $\longrightarrow$ allantoin + CO_2 + H_2O_2
Mg^{++}	Hexokinase	Glucose + ATP $\longrightarrow$ glucose phosphate

of different enzyme-catalyzed reactions. In general, coenzymes not only assist enzymes in the cleavage of the substrate but also serve as temporary acceptors for one of the products of the reaction. The essential chemical components of many coenzymes are *vitamins*. For example, the coenzymes *nicotinamide adenine dinucleotide* (NAD) and *nicotinamide adenine dinucleotide phosphate* (NADP) contain the vitamin *niacin; coenzyme A* contains *pantothenic acid; flavin adenine dinucleotide* (FAD) contains *riboflavin* (i.e., vitamin B_2); *thiamine pyrophosphate* contains *thiamine* (i.e., vitamin B_1), and so on. The chemistry of some of these cofactors was discussed in Chapter 3.

A number of enzymes require *metal ions* for their activity. The metal ions form **coordination bonds** (see Chapter 3) with specific side chains at the active site and at the same time form one or more coordination bonds with the substrate. The latter assist in the polarization of the substrate bonds to be cleaved by the enzyme. For example, zinc is a cofactor for the proteolytic enzyme *carboxypeptidase* and forms coordination bonds with the side chains of two histidines and one glutamic acid residue at the active site. A fourth bond is formed between zinc and the alpha-carboxyl group of the substrate amino acids, and it is here that the cleavage of the peptide occurs. Table 8–4 contains

a list of some of the cofactor-requiring enzymes.

The observation that catalytic activity is lost when an enzyme is stripped of its cofactor testifies to the crucial role played by these atoms or molecules. The role is diverse:

1. In some cases, the cofactor completes the active site of the enzymes or modifies it in such a manner that substrate binding can ensue.

2. The cofactor acts as a donor of electrons or atoms to the substrate and following the reaction is returned to its former state.

3. Cofactors may also serve as temporary recipients of either one of the reaction products or simply an electron or proton, again being recycled to its former state some time after the main reaction is completed.

4. Finally, the cofactor, together with the side chains of residues at the active site, may serve to polarize the substrate and prime it for catalytic alteration.

Isoenzymes

Occasionally, several different enzyme molecules, all of which catalyze the same chemical reaction, have been isolated from a single tissue. Such families of enzymes are called **isoenzymes** or **isozymes** (see also Chapter 11). Among the various isoenzymes the *lactic dehydrogenases* have been most extensively studied, and five different forms have been identified. All are composed of four polypeptide chains of two types called *M* and *H* subunits. Thus the lactic dehydrogenase isoenzymes may take any one of the following forms: M_4, M_3H, M_2H_2, MH_3, or H_4. Similar arrangements are believed to exist for other groups of isoenzymes.

Isozymes should be distinguished from **allelozymes,** which are multiple forms of single polypeptide enzymes resulting from variations in a single allelic or structural gene pair. Although allelozymes act on the same substrate, and in this regard are similar to isozymes, isozymes result from combinations of polypeptide chain products of two or more separate pairs of structural genes.

Zymogens

A number of enzymes arise from an inactive form called a **zymogen** or **proenzyme.** Several of the alimentary digestive enzymes belong to this group including *pepsin, trypsin,* and *chymotrypsin.* The conversion of the zymogen to the active enzyme involves the preliminary cleavage of one or more of the zymogen's peptide bonds, followed occasionally by removal of a portion of the original protein molecule. This phenomenon may best be understood by considering a few examples.

Pepsin, the major protein-digesting enzyme of the stomach, is synthesized in the form of the precursor polypeptide *pepsinogen* (molecular weight 42,000). Upon entering the acidic gastric juice, the pepsinogen molecule is hydrolyzed at several positions to yield a number of small peptides and the active enzyme pepsin (molecular weight of 35,000). The activation of pepsinogen can also be carried out by pepsin itself.

Trypsin, another proteolytic digestive enzyme, is produced in the pancreas as the zymogen *trypsinogen* and secreted into the duodenum (the anterior portion of the small intestine). In the duodenum, another enzyme, *enterokinase,* catalyzes the removal of six amino acids from the *N-terminus* of trypsinogen, thereby yielding trypsin. Trypsin itself can convert trypsinogen molecules to more trypsin.

The enzyme chymotrypsin also participates in the alimentary digestion of protein and is produced in the pancreas as the inactive proenzyme *chymotrypsinogen.* The activation of this zymogen involves a series of peptide bond cleavages catalyzed by trypsin already present in the duodenum and also by chymotrypsin itself. These cleavages split the chymotrypsinogen molecule into three polypeptides that remain interconnected in the activated enzyme through disulfide bridges that were part of the molecule's primary structure.

The activation of the proenzyme *trypsinogen* is shown in Figure 8–21 and illustrates some of the general features of zymogen activation. The active site of the proenzyme is devoid of binding and/or catalytic activity until peptide bonds of the zymogen are broken. Following this peptide cleavage, the remaining portion of the molecule is reorganized with a consequent unmasking of the active site, which can now bind and act on the substrate.

Not only proenzymes but also other proteins may be activated by a preliminary proteolysis. For example, during the coagulation of blood, the formation of the matrix of the clot is brought about through a cascade of proteolytic activations that finally convert inactive, soluble protein monomers (fibrinogen) in blood plasma to the active, polymerizable form, which then produces the insoluble protein threads (fibrin). Some protein hormones are also synthesized as inactive precursors that are activated only upon peptide cleavage (e.g., the conversion of *proinsulin* to *insulin*).

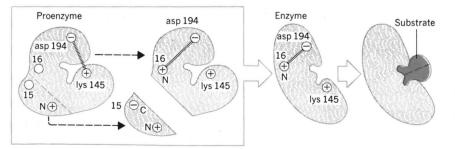

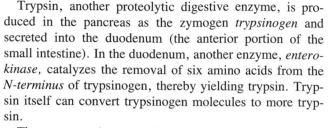

Figure 8–21

Trypsinogen activation involves removal of the peptide. Thus amino acid 16 of trypsinogen becomes the N-terminus of trypsin and interacts ionically with asp 194. This alters the orientation of lys 145 (which was previously neutralized by asp 194) generating the site for substrate binding and eventual cleavage (dashed line).

The Regulation of Enzyme Activity

Literally thousands of different enzyme-catalyzed reactions characterize the metabolism of a cell, and these are organized into a number of interconnected (branching and converging) pathways. Some of the more important or universal pathways are considered in Chapter 10. It is clear that the orderly functioning of a cell or organism demands that controls be placed on these reactions so that specific metabolic pathways (and, therefore, specific cell functions) are active or operative at certain times and inactive at others. One way in which such regulation can be achieved is by *altering the activity of specific enzymes under specific circumstances.*

Consider as an example the following situation. Suppose that in the course of five enzyme-catalyzed reactions, substrate A is converted to end product F; that is,

$$A \xrightarrow{E_1} B \xrightarrow{E_2} C \xrightarrow{E_3} D \xrightarrow{E_4} E \xrightarrow{E_5} F$$

where E_1, E_2, E_3, and so on are the enzymes involved in the pathway. If the end product F is able to bind to enzyme E_1 and in so doing render the enzyme inactive (i.e., unable to catalyze the conversion of A to B), then the synthesis of end product F can be regulated, for under conditions of low F concentration, the metabolic pathway leading to F would be functioning, while at high concentrations of F, E_1 inhibition by the end product would put a halt to the pathway leading to F. The phenomenon in which the end product of a metabolic pathway can regulate its own production by inhibition of this sort is called **feedback inhibition.**

Many metabolic pathways are characterized by a number of branch points. In the example below, substrate A may be converted either to end product F or end product I. In such a pathway, the enzyme affected by feedback inhibition would occur at the branch point; that is, enzyme E_3 would be inhibited by end product F. Consequently, although the conversion of A to F would be inhibited, the formation of I would continue. The continued synthesis of I would not be possible if feedback inhibition occurred before the branch point.

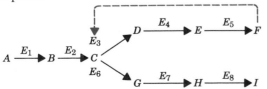

That the phenomenon of feedback inhibition actually occurs in cells was definitively established in the 1950s. One of the earliest discovered examples of feedback inhibition and a classic case is the effect of cytidine triphosphate (CTP, one of the final products formed during pyrimidine synthesis) on the enzyme *aspartate transcarbamylase* (Fig. 8–22). The biosynthesis of CTP begins with the interaction of aspartic acid and carbamyl phosphate to form carbamyl aspartate. The reaction is catalyzed by aspartate transcarbamylase (ATCase) and is followed by four more reactions, culminating in the formation of CTP. When the free CTP concentration is low (i.e., when CTP is being consumed in the cell's metabolism), the formation of carbamyl aspartate proceeds uninhibited. However, when the CTP level rises, carbamyl aspartate synthesis is inhibited.

It is now clear that feedback inhibition is only one of several ways in which enzyme activity can be regulated. Not only end products but also other metabolites may be bound to an enzyme and in so doing alter its activity. The binding can take place at the active site or at other sites on the enzyme's surface. Indeed, in instances where the enzyme molecule has more than one active site (see below), the binding of substrate at one site can influence subsequent substrate binding at another. The various mechanisms that regulate enzyme activity will be considered in the following section.

Number of Polypeptide Chains and Number of Binding Sites of an Enzyme

Many enzymes, including lysozyme and ribonuclease, are composed of a single polypeptide chain. Others consist of two or more chains. For example, the enzyme *glycogen phosphorylase* consists of two polypeptide chains; *fumarase* (a Krebs cycle enzyme) contains four polypeptide chains; and *aspartate transcarbamylase* (involved in the metabolic pathway leading to the synthesis of cytidine triphosphate) consists of 12 polypeptide chains. Each of the enzyme's constituent polypeptide chains is referred to as a *subunit,* the subunits being held together by disulfide bridges, electrostatic interactions, hydrogen bonds, van der Waals interactions, etc. (Chapter 4).

Enzymes may have two separated, *functionally different* binding sites. One type of site, the active site, binds the substrate of the enzyme and possesses catalytic activity, while the other type of site, the *allosteric (allo,* "other" + *steric,* "space") or *regulatory* site, lacks catalytic ac-

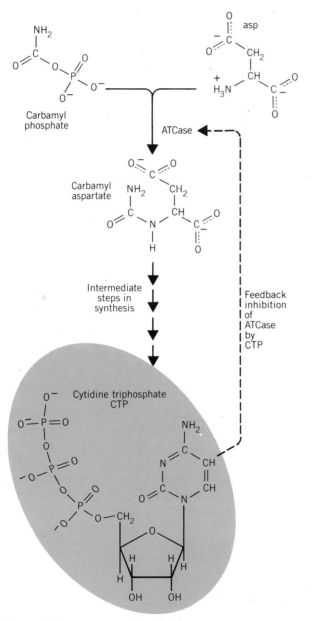

Figure 8-22
Feedback inhibition of *aspartate transcarbamylase* by cytidine triphosphate.

tivity and binds an *effector* molecule. Such enzymes are usually referred to as **allosteric enzymes.** Depending upon the enzyme, active and allosteric sites may be on the same polypeptide or on separate subunits. Effector molecules that inhibit enzyme activity (as in feedback inhibition) are called *negative* effectors. In some cases, binding of an effector molecule enhances enzyme activity, and these molecules are known as *positive* effectors. A single enzyme may possess regulatory sites capable of binding either negative or positive effectors; these effectors may compete with each other for the same regulatory site or be bound at separate regulatory sites.

In some enzymes, two or more subunits may each possess a catalytic site, and substrate binding to the catalytic site on one subunit may influence substrate binding at another catalytic site. Such enzymes exhibit **cooperativity**— a phenomenon already discussed in connection with the reversible oxygenation of hemoglobin (see Chapter 4). Enzymes that exhibit cooperativity and allosteric enzymes do not obey conventional Michaelis-Menten kinetics; the consequences of this are discussed later in the chapter.

Model for Allosteric Enzyme Function

A simple scheme depicting the influence of positive and negative effectors on allosteric enzyme activity is given in Figure 8–23, in which substrates and effectors and active and allosteric binding sites are represented as geometric shapes. The active and regulatory sites of the enzyme (in the absence of bound effectors) are depicted as circular areas. The binding of a positive effector (hexagon) induces a conformational change in the enzyme molecule at the regulatory site (symbolized as a change from circular to hexagonal shape). The conformational shape change is transmitted through the molecule (zigzag arrow) until it reaches and alters the active site in such a way that the substrate is bound more readily than in the absence of such activation. In Figure 8–23, the alteration of the active site is represented as a change from circular to square shape. Binding of a negative effector (diamond) induces a different type of conformational change in the regulatory site (circular to diamond shape), which is transmitted through the enzyme's structure such that the resulting change at the active site (circular to triangular shape) prevents subsequent substrate binding. The negative effector may, of course, be the end product of the pathway involving this enzyme.

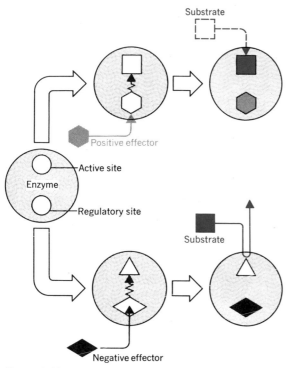

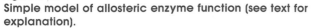

Figure 8-23
Simple model of allosteric enzyme function (see text for explanation).

The behavior of aspartate transcarbamylase, discussed earlier in connection with feedback inhibition, serves as a good example of the mechanism depicted in Figure 8–23. Although aspartate transcarbamylase activity is inhibited by the end product CTP (i.e., CTP is a negative effector), ATCase activity is stimulated by ATP (i.e., ATP acts as a positive effector). It appears that CTP and ATP compete for the same allosteric site of the enzyme.

The model of Figure 8–23 deliberately fails to stipulate whether the allosteric enzyme is composed of one or more than one polypeptide chain, whether more than one active and regulatory site is present, or whether the sites are on the same or on different subunits. Although all of these possibilities exist, most allosteric enzymes consist of several polypeptide subunits with active and regulatory sites on *separate* subunits. In the case of ATCase, there are 12 polypeptide chains—six catalytic (C) chains (molecular weights of 34,000) and six regulatory (R) chains (molecular weights of 17,000). Each C chain contains one active site for aspartate and each R chain one regulatory site. Table

8–5 lists some of the allosteric enzymes, their substrates, and negative and positive effectors.

Cooperativity in Enzymes

When enzymes contain more than one active site, the binding of a substrate molecule to the first site may influence substrate binding to a second site. Binding of the second substrate may influence binding of a third, and so on. This phenomenon is called **cooperativity.** The influence may be positive in that binding of the first substrate molecule facilitates binding of subsequent substrate molecules (called "positive cooperativity") or the influence may be negative in that binding of a second or subsequent substrate molecule occurs less readily than binding of the first (called "negative cooperativity"). In a sense, the substrate itself is acting as either a positive or negative effector for the enzyme. These relationships are depicted in Figure 8–24. Cooperative effects are not restricted to enzymes but are observed with other proteins. Earlier (Chapter 4), we considered the positive cooperativity that exists among the globin chains of hemoglobin, a cooperativity that facilitates successive binding of oxygen molecules to the alpha and beta globin chains (i.e., positive cooperativity).

Both cooperative effects involving active sites on neighboring subunits of an enzyme and true allosteric effects involving regulatory sites may occur in a single enzyme molecule. A case in point is that of *cytidine triphosphate synthetase,* an enzyme involved in nucleic acid metabolism and consisting of four subunits. Two of these subunits contain active sites that bind the substrate glutamine, and the other two have regulatory sites that bind GTP. When glutamine is bound to the active site of one subunit, a conformational change transmitted through the enzyme to another subunit renders the latter's active site unable to bind glutamine (i.e., negative cooperativity). On the other hand, when GTP is bound to the regulatory site of one subunit, this has a positive effect on glutamine binding but it negatively affects GTP binding at the other regulatory site. In other words, the effector GTP serves to activate catalysis but to inhibit further GTP binding.

Allosterism, Cooperativity, and Michaelis-Menten Enzyme Kinetics

As noted earlier, allosteric enzymes and enzymes that exhibit cooperative effects do not display conventional

Table 8–5
Some Allosteric Enzymes

Enzyme	Substrate	Negative Effector	Positive Effector
Aspartate transcarbamylase	Aspartic acid and carbamyl phosphate	CTP	ATP
Threonine deaminase	Threonine	Isoleucine	Valine
Acetolactate synthetase	Pyruvic acid	Valine	—
Isocitric dehydrogenase	Isocitric acid	Alpha-ketoglutaric acid	Citric acid
Phosphofructokinase	Fructose-6-P and ATP	ATP	3'-AMP, 5'-AMP
Phosphorylase b	Glucose-1-P, glycogen, and phosphate	ATP	5'-AMP
Glycogen synthetase	Uridine diphosphoglucose	—	Glucose-6-P
Homoserine dehydrogenase	Homoserine and aspartic semialdehyde	Threonine	Isoleucine and methionine

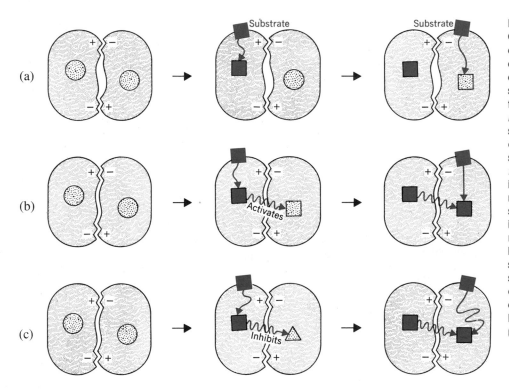

Figure 8-24
Cooperativity in enzymes. The enzyme depicted here consists of two subunits. In (a), binding of the substrate to the active site of one subunit is without effect on the other subunit; i.e., *noncooperativity*. In (b), substrate binding to one subunit activates the site of the other subunit (i.e., *positive cooperativity*), thereby facilitating binding of the second substrate molecule (indicated by the straight path to the active site in contrast to the somewhat more devious path shown in (a). In (c), binding of the first substrate molecule to the active site of one subunit produces a conformational change in the enzyme that inhibits substrate binding at the other active site (i.e., *negative cooperativity*).

Michaelis-Menten kinetics. Figure 8–25 compares the Michaelis-Menten curves for enzymes exhibiting noncooperativity, positive cooperativity, and negative cooperativity. Curve 8–25a shows the normal hyperbolic binding pattern exhibited by most enzymes. In this example, an 81-fold increase in substrate concentration is required in order to elevate enzyme activity from 10% to 90% of its maximum level. Curve 8–25b depicts the sigmoid pattern characteristic of positive cooperativity. Here, only a nine-fold increase in substrate concentration elevates enzyme activity from 10% to 90% of maximum. Note the resemblance of this curve to that for hemoglobin oxygenation (Chapter 4).

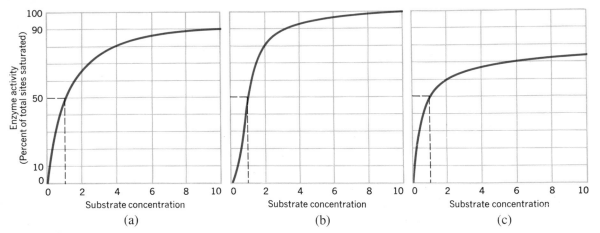

Figure 8–25

Michaelis-Menten curves (*a*) for an enzyme that does not exhibit cooperativity, (*b*), an enzyme exhibiting positive cooperativity, and (*c*), an enzyme exhibiting negative cooperativity.

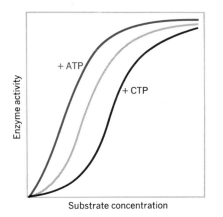

Figure 8-26

Effects of CTP (feedback inhibitor, negative effector) and ATP (positive effector) on the Michaelis-Menten curve of the positively cooperative enzyme *aspartate transcarbamylase*.

Negative cooperativity produces the curve shown in 8–25*c*. Although the curve appears hyperbolic, it actually is not. An increase in substrate concentration greater than 6000-fold would be required to elevate the enzyme activity from 10% to 90% of maximum. The curves have been drawn so that a substrate concentration of 1 unit corresponds to 50% of maximum enzyme activity.

As noted earlier, enzymes may exhibit cooperativity and *also* be affected by the binding of effectors to regulatory

sites. ATCase, the enzyme discussed above, is a good example; it is negatively affected by CTP, positively affected by ATP, and shows positive cooperativity among its substrate binding sites. Figure 8–26 shows the effects of positive and negative effectors on ATCase.

Summary

Regardless of the level of *molecularity,* nearly all reactions occurring in cells are catalyzed by the class of proteins called **enzymes.** Enzymes increase the likelihood that potentially reacting molecules (**substrates**) will encounter each other with the necessary orientation in space and activate their substrates by imposing ''strains'' upon certain bonds. Although greatly increasing reaction rates, enzymes do not alter equilibrium concentrations. In the presence of increasing amounts of substrate, reactions approach a limiting velocity, the kinetics of which may be used to determine the enzyme's **Michaelis-Menten constant;** this is a measure of the tendency of the enzyme and substrate to combine with each other.

The catalytic properties of enzymes are intimately related to the enzyme's primary, secondary, tertiary, and quaternary structure, and changes in structure are often accompanied by loss of catalytic activity. Consequently, most enzymes operate within narrow ranges of pH and temperature. Association of the substrate takes place in the enzyme's **active**

site, which contains **contact** and **catalytic** amino acids. The active site nearly always resides in a depression or cavity in the enzyme's surface, and the binding of substrate is followed by a conformational change in the enzyme and/ or the substrate. Some enzymes require a nonprotein component called a **cofactor** in order to be catalytically active. Cofactors may be permanently bound to the enzyme (i.e., **prosthetic groups**) or may form transient associations (e.g., **coenzymes** and **metal ions**). In some organisms, several different enzyme molecules catalyze the same chemical reaction; such families of enzymes are called **isozymes.** Certain enzymes possess an inactive form called a **zymogen** or **proenzyme.**

In addition to the active site, some enzymes possess an **allosteric** or **regulatory** site. The binding of a positive *effector* to the allosteric site increases the activity of the enzyme, whereas the binding of negative effector decreases enzyme activity. Changes in the levels of activity of allosteric enzymes are fundamental to the control of many metabolic pathways through feedback mechanisms. Enzymes composed of two or more polypeptide chains may contain two or more active sites, and the binding of substrate to one site often influences binding of additional substrate to other sites. Such influence, which can be positive or negative, is called **cooperativity.**

References and Suggested Reading

Articles and Reviews

Anderson, C. M., Zucker, F. H., and Steitz, T. A., Space-filling models of kinase clefts and conformation changes. *Science 204,* 375 (1979).

Bell, R. M., and Koshland, D. D., Covalent enzyme-substrate intermediates. *Science 172,* 1253 (1971).

Hammes, G. G., and Wu, C.-W., Regulation of enzyme activity. *Science 172,* 1205 (1971).

Koshland, D. E., Correlation of structure and function in enzyme action. *Science 142,* 1533 (1963).

Koshland, D. E., Protein shape and biological control. *Sci. Am 229*(4), 52 (Oct. 1973).

Koshland, D. E., and Neet, K. E., The catalytic and regulating properties of enzymes. *Annu. Rev. Biochem. 37,* 359 (1968).

Neurath, H., Protein-digesting enzymes. *Sci. Am 211*(6), 68 (Dec. 1964).

Perutz, M. F., X-ray analysis, structure and function of enzymes. *Eur. J. Biochem. 8,* 455 (1969).

Phillips, D.C., The three-dimensional structure of an enzyme molecule. *Sci. Am 215*(5), 78 (Nov. 1966).

Stoud, R. M., A family of protein-cutting proteins. *Sci. Am 231*(1), 74 (July 1974).

Books, Monographs, and Symposia

Bernhard, S. A., *The Structure and Function of Enzymes,* W. A. Benjamin, New York, 1968.

Boyer, P. D. (Ed.), *The Enzymes* (3rd ed.), Academic Press, New York, 1970.

Cornish-Bowden, A., *Fundamentals of Enzyme Kinetics.* Butterworths, London, 1979.

Dickerson, R. E., and Geis, I., *The Structure and Action of Proteins,* Harper & Row, New York, 1969.

Ferdinand, W., *The Enzyme Molecule.* Wiley, New York, 1976.

Fersht, A., *Enzyme Structure and Mechanism,* W. H. Freeman, San Francisco, 1977.

Foster, R. L., *The Nature of Enzymology.* Halsted Press, New York, 1980.

Lehninger, A. L., *Biochemistry,* Worth, New York, 1975.

McGilvery, R. W., *Biochemical Concepts,* Saunders, Philadelphia, 1975.

Snell, F. M., Shulman, S., Spencer, R. P., and Moos, C., *Biophysical Principles of Structure and Function,* Addison-Wesley, Reading, Mass., 1965.

Stryer, L., *Biochemistry* (2nd ed.), W. H. Freeman, San Francisco, 1981.

Chapter 9
BIOENERGETICS

In the study of physiology and biochemistry, cells are often thought of as tiny machines in which all events may be explained in terms of either chemical reactions, fluid dynamics, electrical fluxes across partitions, or the absorption or emission of light. In other words, cellular activities, regardless of their level of complexity, are ultimately founded on the known laws of physics and chemistry. Those laws of physics and chemistry that are fundamental to an understanding of cellular activities, especially the production and consumption of energy during cell metabolism, will be considered in this chapter.

Energy and Metabolism

The **metabolism** of a cell is characterized by a myriad of chemical reactions in which energy is either consumed, produced, or converted (i.e., *transduced*) from one form into another. Metabolism can be subdivided into two broad categories: **catabolism** and **anabolism.** During catabolic reactions (or reaction sequences, see below), molecules are broken down by the cell into simpler forms; during anabolism, complex molecules are formed from simpler ones. The catabolic and anabolic reactions that proceed in cells are accompanied by energy changes and it is the study of these changes that constitutes the field of **bioenergetics.**

Consider, for example, an anabolic process such as the synthesis of new membranes within the cell. Such biosynthesis requires (consumes) energy, and this energy must ultimately be obtained from the cell's environment in some form. Within the cell, the energy (perhaps in a new form) is consumed in order to "drive" the cell's membrane-synthesizing processes. Whole organisms and individual cells may be assigned to different groups according to the nature of the materials that they must acquire from their surroundings in order to support their metabolic needs. Most plant cells (i.e., those that contain chlorophyll) and many different kinds of bacteria require only CO_2, H_2O (or H_2S), simple nitrogenous compounds like NH_3, and a few vitamins from their environment in order to fulfill their minimum metabolic needs. These cells or organisms are called **autotrophs.** With the exception of trace amounts of certain vitamins, they can live and grow in the complete absence of an exogenous supply of organic materials. (Indeed, some autotrophs do not even need an external source of vitamins.) When an autotroph can utilize light as a source of energy, it is called a **photoautotroph.** Other autotrophs can obtain their energy from the oxidation of inorganic substances such as ammonium ions (i.e., NH_4^+), ferrous iron (i.e., Fe^{++}), or elemental sulfur (S). This kind of autotroph is called a **chemoautotroph** (see Table 9–1 for examples).

All animal cells (and certain plant cells and most bacteria) depend upon an external source of organic compounds and specific vitamins for their metabolism and are therefore called **heterotrophs.** Some heterotrophs (e.g., a few algae

Table 9–1
Comparison of Autotrophs and Heterotrophs

Energy Source	Carbon Source	
	CO_2	Organic Substances
Light	Photoautotrophs	Photoheterotrophs
	Plants	A few algae
	Most algae	Some purple and
	Purple bacteria	green bacteria
	Green bacteria	Some cyanobacteria
	Most cyanobacteria	
Oxidizable chemicals	Chemoautotrophs	Chemoheterotrophs
	Some bacteria	Animals
		Most bacteria
		Fungi
		Protozoa

Source: Adapted from T. D. Brock, *Biology of Microorganisms* (3rd ed.), 1979, p. 207. Courtesy of Prentice-Hall, Englewood Cliffs, N.J.

and bacteria) can also use light as an energy source and are called **photoheterotrophs.** However, most heterotrophs require organic compounds as both a source of energy and as raw materials for the synthesis of intracellular components; such heterotrophs are called **chemoheterotrophs** (Table 9–1). Energy that is derived by the catabolism of organic materials is used to meet anabolic needs.

The primary sources of energy and raw materials for heterotrophic metabolism are polysaccharides, lipids, and proteins. Organisms that remove these macromolecules from their environment break them down in the successive catabolic stages of metabolism. As these compounds are chemically degraded, the chemical energy that is inherent in their molecular structure is both released in the form of heat and used to create the bonds that form new molecules, as in the attachment of free (inorganic) phosphate to ADP to form ATP (Fig. 9–1). The ultimate primary products of catabolism are NH_3, CO_2, and H_2O.

Although autotrophic organisms can use CO_2, H_2O, and small nitrogenous compounds from their environment, these small compounds do not by themselves contain enough extractable chemical energy to sustain the organisms. Consequently, autotrophs also absorb energy in the form of light and, using the light energy, synthesize simple organic acids from CO_2 and water (i.e., photosynthesis), phosphorylate ADP to form ATP, and synthesize amino acids from the organic acids using incorporated NH_3 (a

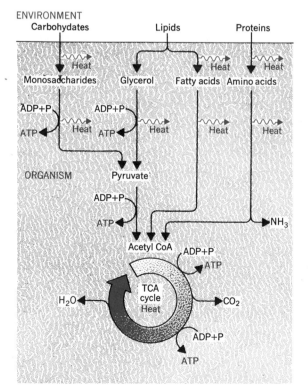

Figure 9-1
The flow of materials and energy during catabolism. Some of the energy is released as heat and some is used for the synthesis of ATP.

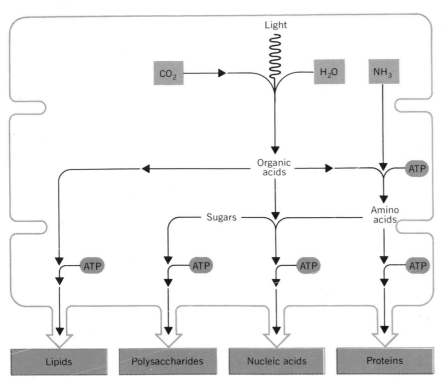

Figure 9-2
The flow of materials and energy during anabolism. ATP serves as a source of energy during many stages of anabolism.

process called *amination*). Drawing from the pool of ATP as a source of chemical energy, and using these simpler molecules, the cell then synthesizes complex molecules such as proteins, polysaccharides, nucleic acids, and lipids (Fig. 9–2). All of these metabolic activities are accompanied by the loss of some unusable chemical energy as heat.

Autotrophic organisms not only possess the enzymes for the anabolic processes just described but also the catabolic enzyme systems similar to those of heterotrophs. Accordingly, they too can produce ATP by breaking down polysaccharides, lipids, and proteins. Like autotrophs, heterotrophs have ATP-dependent anabolic enzyme systems that synthesize macromolecules, but most heterotrophs are unable to carry out photosynthesis. Compounds are cycled *within* cells and also *between* cells and *between* whole organisms, as depicted in Figure 9–3. Each transition is accompanied by a specific energy change.

The Laws of Thermodynamics

Thermodynamics is the study of energy changes—that is, the conversion of energy from one form into another. Such changes conform to two laws called the *first and second laws of thermodynamics*.

The First Law of Thermodynamics

The first law is concerned with the conversion of energy *within* a "system," where a system is defined as a body (e.g., a cell or an organism) and its surroundings. This law, which applies to both biological and nonbiological systems, states the following:

Energy cannot be created or destroyed but can be converted from one form into another: during such a conversion, the total amount of the energy of the system remains constant.

This law applies to all levels of organization in the living world; it applies to organisms, cells, organelles, and the individual chemical reactions that characterize metabolism. In practice, it is difficult to measure the energy possessed by cells (i.e., to limit the "system" to an individual cell), since energy may escape into the environment surrounding

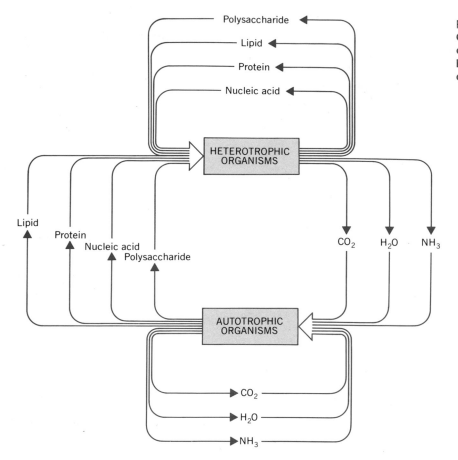

Figure 9-3
Cycles that characterize the interchange of materials within and between heterotrophic and autotrophic cells and organisms.

the cell during the measurement. Similarly, energy may be acquired by the cell from its environment; for example, a photosynthesizing cell absorbs energy from its environment in the form of light. The acquisition or loss of energy by a cell (or anything else) should not be confused with the destruction or creation of energy, which according to the first law of thermodynamics cannot occur.

From a biological viewpoint, the first law of thermodynamics indicates that at any given moment a cell possesses a specific quantity of energy. This energy may take any of several forms; it may be (1) *potential* (e.g., the energy of the bonds that link atoms together in a molecule or the pressure–volume relationships within the cell as a whole or within membrane-enclosed intracellular components); (2) *electrical* (e.g., the distribution of different amounts of electrical charge across cellular membranes); or (3) *thermal*

(i.e., the temperature-dependent constant and random motions of molecules and atoms). According to the first law, these forms of energy may be interconverted; e.g., some of the cell's potential energy can be converted into electrical or thermal energy, but in doing this the cell does not create or destroy energy. For example, when a cell breaks down polysaccharide to ultimately form CO_2 and H_2O, some of the potential energy present in the carbohydrate is conserved as potential energy by the cell by phosphorylating ADP. The ATP so produced represents a new energy source (and also one that is of greater immediate utility for the cell). However, not all of the energy of the original carbohydrate is conserved as potential energy; some of it becomes thermal and is transferred to the surroundings as heat. It is important to recognize that none of the energy is destroyed, and it should be possible to account for all of the energy originally present in the polysaccharide in other forms within the system (i.e., in the ATP that is produced and in the heat that is released).

The Second Law of Thermodynamics

The first law of thermodynamics tells us that the total energy of an isolated system consisting of a cell (or organism) and its surroundings is the *same* before and after a series of events or chemical reactions has taken place. What the first law does not tell us is the *direction* in which the reactions proceed. This problem can be illustrated using a simple example. Suppose we place a small cube of ice in a liter of hot water, seal the combination in an insulated container (e.g., a vacuum bottle), and allow the system (i.e., the ice and the water) to reach an equilibrium. In such a system, we would not be surprised to find that the ice melts and that this is accompanied by a decrease in the temperature of the water. When we later examine the system, we find that we are left only with water (no ice) and that the water is at a reduced temperature. The flow of heat, which is thermal energy, from the hot water to the ice thereby causing the ice to melt is *spontaneous* and the energy that is "lost" by the water is "gained" by the melting ice so that the total energy of the system remains the same. We certainly would not expect ice to form spontaneously in a sealed system that contains warm water, *even though such an eventuality is not prohibited by the first law.* Consequently, the important lessons to be learned from this illustration are that *energy changes have direction and may be spontaneous.*

In order to anticipate the spontaneity of a reaction and predict its direction, one must take into account a function called **entropy**. Entropy is a measure of the degree of *randomness* or *disorder* of a system, the entropy increasing with increasing disorder. Accordingly, the second law of thermodynamics states:

In all processes involving energy changes within a system, the entropy of the system increases until an equilibrium is attained.

In the illustration that was presented above, the highly ordered distribution of energy (i.e., large amounts of energy in the hot water and smaller amounts of energy in the ice)

was lost as the ice melted to form water. In the resulting warm water, the energy was more randomly and uniformly distributed among the water molecules.

The units of entropy are J/mole, indicating that entropy is measured in terms of concentration. When equal numbers of moles of a solid, liquid, and gas are compared at the same temperature, the solid has less entropy than the liquid, and the liquid has less entropy than gas (the gaseous state is the state of greatest disorder). Entropy can be thought of as the energy of a system that is of no value for performing work (i.e., it is *not* "useful" energy). For example, the catabolism of sucrose or other sugars by a cell is accompanied by the formation of energy-rich ATP. Although superficially it may appear as though useful energy has increased in the form of the ATP gained by the cell, the total amount of useful energy has actually decreased and the amount of unavailable energy increased. It is true that some of the potential energy of the sugar has been converted to potential energy in the form of ATP, but some has also been converted to thermal energy, which tends to raise the temperature of the cell and therefore its entropy. Suggestions that cells can decrease entropy by carrying out photosynthesis are misleading. Although it is true that during photosynthesis cells convert molecules with very little potential energy (CO_2 and H_2O) into larger molecules with considerably more potential energy (sugars) and that there is an accompanying decrease in the entropy of the cell, energy in the form of light was absorbed from the cell's environment. Since the light energy consumed during photosynthesis is a part of the whole system (i.e., the cell and its surroundings), it is clear that there has actually been an overall decrease in useful energy and an increase in entropy (see Fig. 9–4).

The entropy change during a reaction may be quite small. For example, when sucrose undergoes hydrolysis to form the sugars glucose and fructose, much of the potential energy of the original sucrose is present in the resulting glucose and fructose molecules. Changes in entropy are extremely difficult to measure, but the difficulty can be circumvented by employing two other thermodynamic functions: **enthalpy** or heat content (denoted H) and **free energy** (denoted G). The change in enthalpy (ΔH) is a measure of the total change in energy that has taken place, whereas the change in free energy (ΔG) is the change in the amount of energy available to do work. Changes in

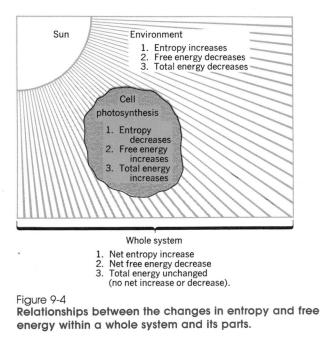

Figure 9-4
Relationships between the changes in entropy and free energy within a whole system and its parts.

entropy (ΔS), enthalpy, and free energy are related by the equation

$$\Delta G = \Delta H - T \Delta S \qquad (9\text{--}1)$$

in which T is the absolute temperature of the system.

The change in free energy can also be defined as the *total* amount of free energy in the products of a reaction minus the *total* amount of free energy in the reactants; that is,

$$\Delta G = G_{(products)} - G_{(reactants)} \qquad (9\text{--}2)$$

A reaction that has a negative ΔG value (i.e., the sum of the free energy of the products is less than that of the reactants) will occur *spontaneously;* a reaction for which the ΔG is zero is at *equilibrium;* and a reaction that has a positive ΔG value will not occur spontaneously and proceeds only when energy is supplied from some outside source.

ΔG, ΔG°, and $\Delta G^{\circ\prime}$ Values

The hydrolysis of sucrose,

$$\text{sucrose} + H_2O \rightarrow \text{glucose} + \text{fructose}$$

has a negative ΔG value, and therefore when sucrose is added to water, there is the spontaneous conversion of some of the sucrose molecules to glucose and fructose. However, the reverse reaction

$$\text{glucose} + \text{fructose} \rightarrow \text{sucrose} + H_2O$$

has an equal *but positive* ΔG value and therefore does not occur without an input of energy. Hence, special attention must be paid to the direction in which the reaction is written (i.e., the direction of the arrow) and the sign of the ΔG value. If 5 moles of sucrose are mixed with water, the formation of glucose and fructose will take place spontaneously and the ΔG may be determined; this value is, of course, greater than if 4 or 2 moles of sucrose are used. Thus, ΔG values are dependent upon the amounts and concentrations of reactants and products. More uniform standards of reference that have been established by convention are the *standard free energy changes* ΔG° and $\Delta G^{\circ\prime}$. ΔG° represents the change in free energy that takes place when the reactants and products are maintained at 1.0 molar concentrations (strictly speaking, 1.0 molal) during the course of the reaction and the reaction proceeds under standard conditions of temperature (25°C) and pressure (1 atmosphere) and at pH 0.0. The $\Delta G^{\circ\prime}$ value is a much more practical term for use with biological systems in which reactions take place in an aqueous environment and at a pH that usually is either equal or close to 7.0. The $\Delta G^{\circ\prime}$ value is defined as the standard free energy change that takes place at pH 7.0 when the reactants and products are maintained at 1.0 molar concentrations (Table 9–2).

The changes in standard free energy are independent of the route that leads from the initial reactants to the final products. For example, glucose can be converted to carbon dioxide and water either by combustion in the presence of oxygen or through the actions of cellular enzymes. Changes in standard free energy are the same, regardless of the method that is used; thus, the value of the standard free energy change provides no information about the reaction sequence by which the change has taken place. By the same token, the values obtained for changes in standard free energy tell us nothing about the *rate* at which the changes have taken place.

The $\Delta G^{\circ\prime}$ value can be calculated from the equilibrium

Table 9–2
Free Energy: Symbols, Definitions and Explanations

Symbol	Definition	Explanation
G	Free energy (also called Gibbs free energy)	The maximum energy that can be derived from a particular molecule capable of doing work under conditions of constant temperature and pressure.
ΔG	Change in free energy	The change in free energy that takes place during a chemical reaction (equal to the free energy of the products of the reaction minus the free energy of the reactants.)
ΔG°	*Standard* free energy change[a]	An expression more frequently used in physical chemistry than in biochemistry. The change in free energy that takes place under the following standard conditions: reactants and products are maintained at 1.0 molal concentrations; the temperature is 25°C; the pressure is 1 atm; and the pH is 0.0.
$\Delta G^{\circ\prime}$	*Standard* free energy change[a]	Expression used in biochemistry and physiology. Change in free energy under the following standard conditions: reactants and products are maintained at 1.0 molar concentrations; the temperature is 25°C; the pressure is 1 atm; and the pH is 7.0.
$\Delta G^\prime$	*Standard* free energy change[a]	Expression rarely used anymore; same as ΔG° but not necessarily at pH 0.0 The pH value must be specified.
$\Delta G\ddagger$	Free energy of activation	A measure of the activation energy of a chemical reaction; does not provide information about the direction or equilibrium of the reaction. (See Chapter 8.)

[a]If the reaction is independent of pH, then $\Delta G^\circ = \Delta G^{\circ\prime} = \Delta G^\prime$

constant, $K^\prime_{eq}$ of a reaction using the following relationship:

$$\Delta G^{\circ\prime} = - R T \ln K^\prime_{eq} \qquad (9\text{–}3)$$
$$= - 2.303 R T \log_{10} K^\prime_{eq} \qquad (9\text{–}4)$$

where R is the gas constant (8.314 J/mole/degree), T is the absolute temperature (in degrees Kelvin), and $K^\prime_{eq}$ is the equilibrium constant. Table 9–3 lists a number of $\Delta G^{\circ\prime}$ values for common reactions.

For the reaction

$$A + B \longrightarrow C + D$$

the equilibrium constant is defined as follows:

$$K^\prime_{eq} = \frac{[C]\,[D]}{[A]\,[B]} \text{ at equilibrium} \qquad (9\text{–}5)$$

where $[A]$ and $[B]$ are the concentrations of the reactants, and $[C]$ and $[D]$ are the concentrations of products. If the equilibrium constant is 1.0, then the $\Delta G^{\circ\prime}$ value equals zero. If the equilibrium constant is *greater* than 1.0, then the $\Delta G^{\circ\prime}$ value is negative (e.g., -11.41 kJ/mole for a $K^\prime_{eq}$ value of 100), and the reaction is said to be **exergonic** (i.e., "energy releasing") because it proceeds sponta-

Table 9–3
**Standard Free Energy Changes of Common
Biochemical Reactions at pH 7.0 and 25°C**

Reaction	$\Delta G^{o'}$	
	kJ mol^{-1}	kcal mol^{-1}
Hydrolysis:		
Acid anhydrides:		
Acetic anhydride $+ H_2O \rightarrow$ 2 acetate	-91.2	-21.8
Pyrophosphate $+ H_2O \rightarrow$ 2 phosphate	-33.4	-8.0
Esters:		
Ethyl acetate $+ H_2O \rightarrow$ ethanol $+$ acetate	-19.7	-4.7
Glucose-6-phosphate $+ H_2O \rightarrow$ glucose $+$ phosphate	-13.8	-3.3
Amides:		
Glutamine $+ H_2O \rightarrow$ glutamate $+ NH_4^+$	-14.2	-3.4
Glycylglycine $+ H_2O \rightarrow$ 2 glycine	-9.2	-2.2
Glycosides:		
Sucrose $+ H_2O \rightarrow$ glucose $+$ fructose	-29.3	-7.0
Maltose $+ H_2O \rightarrow$ 2 glucose	-16.7	-4.0
Esterification:		
Glucose $+$ phosphate $\rightarrow$ glucose-6-phosphate $+ H_2O$	$+13.8$	$+3.3$
Rearrangement:		
Glucose-1-phosphate $\rightarrow$ glucose-6-phosphate	-7.11	-1.7
Fructose-6-phosphate $\rightarrow$ glucose-6-phosphate	-1.67	-0.4
Elimination:		
Malate $\rightarrow$ fumarate $+ H_2O$	$+3.14$	$+0.75$
Oxidation:		
Glucose $+ 6O_2 \rightarrow 6CO_2 + 6H_2O$	$-2,870$	-686
Palmitic acid $+ 23O_2 \rightarrow 16CO_2 + 16H_2O$	$-9,782$	-2338

neously in the direction written when starting with unimolar concentrations of reactants and products. When the K'_{eq} value is *less* than 1.0, the $\Delta G^{o'}$ value is positive (e.g., 5.71 kJ/mole for a K_{eq} of 0.1), and the reaction is said to be **endergonic** (i.e., "energy consuming") because it does not proceed spontaneously in the direction written when starting with unimolar concentrations of reactants and products.

Calculations of $\Delta G^{o'}$ values are usually based on experimental measurements of isolated reactions, that is, with reactions that take place *independently* of other reactions and that are not associated with cells. ΔG^o and $\Delta G^{o'}$ values do not provide information about the free energy changes of reactions as they might take place in cells or under conditions in which the concentrations of reactants and products, pH, and so on, may change. This may be dramatically illustrated by considering the following example.

At pH 7.0 and at 25°C, the equilibrium constant for the reaction

dihydroxyacetone phosphate $\rightarrow$ glyceraldehyde-3-phosphate

is 0.0475. Therefore, using equation 9–3

$$\Delta G^{o'} = -2.303\ R\ T \log_{10} K'_{eq}$$
$$= -2.303\ (8.314 \text{J/mole/degree})\ (298) \log_{10} (0.0475)$$
$$+7.55\ \text{kJ/mole}$$

The positive value indicates that this reaction does not proceed spontaneously in the direction written. However, in cells, this reaction is but one of a series of reactions in a metabolic pathway called *glycolysis* (Chapter 10). Other reactions of glycolysis that occur *prior* to this one and which have negative $\Delta G^{o'}$ values produce additional substrate (i.e., dihydroxyacetone phosphate) and reactions with negative $\Delta G^{o\prime}$ values that occur *after* this step remove the product (glyceraldehyde-3-phosphate). As a result, the reaction proceeds in the direction written under the conditions specified above, *even though the $\Delta G^{o\prime}$ value is positive.* This example illustrates the important point that *the $\Delta G^{o\prime}$ value for a specific biological reaction cannot be used to predict reliably whether or not that particular reaction is actually taking place within the cell.*

Coupled Reactions

Cells commonly employ either of two mechanisms in order to cause endergonic reactions to take place. One mechanism, illustrated above by the glycolysis example, is to create reactant and product concentrations that are markedly different from equilibrium values. This can be brought about either through the production of additional reactants or through the removal of products by drawing them into other cellular reactions. For example, the reaction

$$\text{glucose-6-phosphate} \xrightarrow{\text{isomerase}} \text{fructose-6-phosphate}$$

has a $\Delta G^{o\prime}$ value of 1.67 kJ/mole, and the isolated reaction would not be expected to occur spontaneously when reactants and products occur in equimolar concentrations. However, in cells, there is an enzyme that catalyzes the conversion of fructose-6-phosphate to fructose-1,6-diphosphate

$$\text{fructose-6-phosphate} \xrightarrow[\text{ATP\ ADP}]{\text{fructokinase}} \text{fructose-1,6-diphosphate}$$

and the $\Delta G^{o\prime}$ value of this reaction is -14.22 kJ/mole. In the parlance of bioenergetics, the equilibrium of this reaction "lies far to the right." In cells, the two reactions are *coupled;* that is, by removing the product of the first reaction, its concentration is shifted away from its equilibrium value, allowing more product to be formed, and therefore both reactions proceed sequentially.

The *total* free energy change for a sequence of coupled reactions is equal to the sum of the free energy changes of the individual reactions of the sequence. In the example just given, the $\Delta G^{o\prime}$ for the sequence of reactions is $-14.22 + 1.67 = -12.55$ kJ/mole.

A second type of coupling occurs when exergonic and endergonic reactions are catalyzed by the same enzyme. In cells, the following two reactions

$$\text{ATP} + \text{H}_2\text{O} \longrightarrow \text{ADP} + \text{phosphate}$$
$$(\Delta G^{o\prime} = -30.65 \text{ kJ/mole})$$

$$\text{glucose} + \text{phosphate} \longrightarrow \text{glucose-6-phosphate} + \text{H}_2\text{O}$$
$$(\Delta G^{o\prime} = +13.86 \text{ kJ/mole})$$

are catalyzed (and coupled) by the enzyme *glucokinase;* that is,

$$\text{glucose} + \text{ATP} \xrightarrow{\text{glucokinase}} \text{glucose-6-phosphate} + \text{ADP}$$
$$(\Delta G^{o\prime} = -16.79 \text{ kJ/mole})$$

The exergonic reaction is thus coupled to the endergonic reaction by *glucokinase,* and some of the free energy yielded by the exergonic reaction is consumed in order to drive the endergonic reaction.

Intracellular Phosphate Turnover

In all cells, most of the major energy changes and reaction couplings involve nucleotides. Although the cell draws upon a number of different nucleotide pools,

$$\text{ATP} \rightarrow \text{ADP} + \text{phosphate}$$

exchanges are the most common. Other exchanges, such as

$$\text{ATP} \rightarrow \text{AMP} + \text{pyrophosphate}$$

occur, for example during lipid metabolism, as do

$$\text{CTP} \rightarrow \text{CDP} + \text{phosphate}$$

reactions. Uridine triphosphate (UTP) is utilized in polysaccharide synthesis and guanosine triphosphate (GTP) is consumed during protein synthesis. The deoxy derivatives of the nucleotides, dATP, dGTP, dTTP, and dCTP, are used during DNA synthesis.

Regeneration of the nucleoside triphosphates is catalyzed by a relatively nonspecific nucleoside diphosphate kinase:

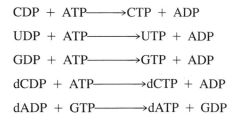

$$CDP + ATP \longrightarrow CTP + ADP$$

$$UDP + ATP \longrightarrow UTP + ADP$$

$$GDP + ATP \longrightarrow GTP + ADP$$

$$dCDP + ATP \longrightarrow dCTP + ADP$$

$$dADP + GTP \longrightarrow dATP + GDP$$

The total concentration of all nucleotides in cells that are metabolizing in a steady state is relatively constant, and most of the adenylate is present in the form of ATP. When cellular activity increases, the level of ATP decreases and the level of ADP + AMP increases. The changes that take place in the levels of these compounds cause an increase in reaction sequences, such as those of glycolysis (Chapter 10), which generate ATP. Conversely, when cellular activity decreases, the transient rise in ATP acts to slow down the ATP-generating systems. The regulation of ATP synthesis (discussed in Chapter 11) is achieved through feedback control of allosteric enzymes by ATP, ADP, and AMP.

Although ATP hydrolysis is a major source of free energy in cells, cells do not form substantial reservoirs of ATP. The amount of ATP present in a cell lasts for only a short time during periods of increased cellular activity. Most cells contain additional pools of compounds that can quickly be consumed in order to generate additional ATP. For example, striated muscle, smooth muscle, and nerve cells contain reservoirs of *phosphocreatine*. When the ATP levels in these cells undergo a marked decline, the enzyme *creatine kinase* is activated and catalyzes the reaction

$$\text{phosphocreatine} + ADP \longrightarrow \text{creatine} + ATP$$

$$(\Delta G^{o\prime} = -12.6 \text{ kJ/mole})$$

Phosphoarginine and polymetaphosphate play a similar role in invertebrate and bacterial cells, respectively.

Redox Reactions and Redox Couples

The movements of electrons between cellular **reductants** and **oxidants** represent another form of energy transfer in cells. A reductant (or *reducing agent*) is a substance that loses or donates electrons to another substance; the latter substance is the oxidant (or *oxidizing agent*). Conversely, an oxidant is a substance that accepts electrons from another substance, the latter being the reductant. Reactions that involve the movement of electrons between reductant and oxidant are called **redox reactions.**

Different chemical substances have different potentials for donating or accepting electrons. The tendency of hydrogen to dissociate

$$H_2 \rightleftharpoons 2H^+ + 2e^-$$

thereby releasing electrons is used as a standard against which the tendencies of other substances to release or accept electrons is measured. The electron donor (e.g., H_2 in the above reaction) and the electron acceptor (e.g., $2H^+$ in the above reaction) are called a **redox couple** or **half cell.** The tendency of any chemical substance to lose or gain electrons is called the **redox potential** and is measured in volts (v). Measurements are made using an electrode that has been standardized against the H_2–$2H^+$ couple whose redox potential is set at 0.0 under standard conditions (pH 0.0, 1 M [H^+], 25°C, and 1 atmosphere pressure). This potential is noted by the symbol E_o. For biochemical reactions that normally occur at pH 7.0, the redox potential of the H_2–$2H^+$ couple is -0.421 v; standard redox potentials at pH 7.0 are noted by the symbol $E_o{}'$. The $E_o{}'$ values of a number of biologically important redox couples are given in Table 9–4.

Any substance with a more positive $E_o{}'$ value than another has the potential for oxidizing that substance (i.e., removing electrons from the substance with the more negative $E_o{}'$ value). The greater the difference in redox potentials, the greater the energy changes involved. The change in standard free energy, $\Delta G^{o\prime}$, is related to $E_o{}'$ as follows

$$\Delta G^{o\prime} = n\mathscr{F}\Delta E_o{}'$$

where n is the number of electrons exchanged per molecule, $\mathscr{F}$ is the *Faraday* (96,406 J/volt) and $\Delta E_o{}'$ is the difference in redox potential between the more positive and the more negative members of the redox couple. For example, the oxidized form of cytochrome c oxidizes the reduced form of cytochrome b by removal of 2 electrons. The difference between the redox potentials of the two is

$$+0.254 - (+0.030) = +0.224 \text{ v}$$

Therefore,

Table 9–4
Standard Redox Potentials at pH 7.0 and 25°–37° C

Reductant	Oxidant	E_o' (volts)
Pyruvate	Acetate $+ 2H^+ + 2e^-$	$- 0.70$
Acetaldehyde	Acetate $+ 2H^+ + 2e^-$	$- 0.58$
H_2	$2H^+ + 2e^-$	$- 0.42^a$
$NADH_2$	$NAD^+ + 2H^+ + 2e^-$	$- 0.32$
Ethanol	Acetaldehyde $+ 2H^+ + 2e^-$	$- 0.197$
Lactate	Pyruvate $+ 2H^+ + 2e^-$	$- 0.185$
Succinate	Fumarate $+ 2H^+ + 2e^-$	$- 0.031$
Ubiquinol	Ubiquinone $+ 2H^+ + 2e^-$	$+ 0.10$
2 Cytochrome $b_{(red)}$	2 Cytochrome $b_{(ox)} + 2e^-$	$+ 0.030$
2 Cytochrome $c_{(red)}$	2 Cytochrome $c_{(ox)} + 2e^-$	$+ 0.254$
2 Cytochrome $a_{3(red)}$	2 Cytochrome $a_{3(ox)} + 2e^-$	$+ 0.385$
H_2O	$\frac{1}{2} O_2 + 2H^+ + 2e^-$	$+ 0.816$

$^aE_o = 0.0$ volts.

$$\Delta G^{o\prime} = -2 \ (96{,}406 \ J/v)(0.224 \ v)$$
$$= -43.19 \ kJ \ (per \ mole \ of \ each \ cytochrome)$$

In cells, the energy changes during redox reactions may be coupled to the synthesis of ATP; this is discussed in Chapter 16.

Light and the Transduction of Energy

The conversion of energy from one form into another is called **transduction.** An example of transduction is the conversion of light energy (*electromagnetic radiation*) into potential chemical bond energy in such seemingly diverse processes as *photosynthesis* and *vision*. The conversion of chemical bond energy into light energy (i.e., transduction in the reverse direction) occurs in *bioluminescence* (e.g., the emission of light by fireflies and by certain oceanic microorganisms, Fig. 9–5).

Electromagnetic radiations consist of more than just the spectrum of colors perceived by the human eye. Each of the radiations of the *electromagnetic spectrum* (Table 9–5) can be thought of as a stream of moving packets of energy called **photons.** The stream of photons has the character of a *wave* (Fig. 9–6) in which the distance from any point on the wave to an equivalent point on the neighboring wave is the **wavelength.** Among the differences between the radiations of the electromagnetic spectrum are differences

in wavelength. At one extreme are the *long* waves, such as radio and television waves, which have wavelengths up to 1000 m. At the opposite end of the spectrum are the X rays and gamma rays, which have wavelengths of 0.1×10^{-9} to 0.1×10^{-12} m. Most of the biologically important wavelengths are shorter than 10^{-6} m. In air, each photon travels at the same velocity, namely 3×10^8 m sec^{-1}.

The energy of radiation is said to be *quantized;* that is, it has specific values or **quanta.** Quantum energy is related to wavelength by *Planck's Law:*

$$q = hv = \frac{hc}{\lambda}$$

$$(9–7)$$

where q is the quantum energy in Joules, h is Planck's constant (6.624×10^{-34} J sec), v is the frequency of the radiation, c is the velocity of light, and λ is the radiation's wavelength (in meters). It may be seen from this equation that radiations of shorter wavelength possess more energy than radiations of longer wavelength. It should be noted that there are a number of forms of radiation that are not electromagnetic but are particulate. Particulate radiations include the *alpha* rays and *beta* rays that are emitted from the nuclei of radioactive isotopes. These radiations are considered in Chapter 14, which deals with the uses of radioisotopes as *tracers* of cellular metabolism.

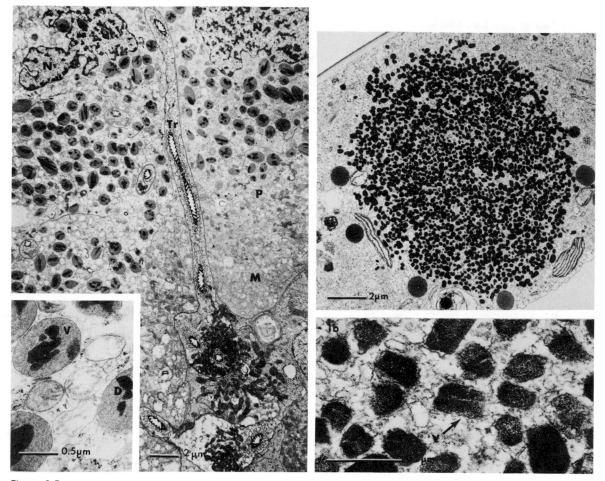

Figure 9-5
Left, Light-emitting cells (*photocytes*) of the firefly, *Photinus greeni. P,* photocyte; *N,* nucleus; *M,* mitochondrion; *V,* light-emitting vesicle (enlarged in lower left insert); *D,* dense inclusions; *Tr,* tracheole. (Courtesy of J. F. Case and K. Linberg; copyright © 1980 by Academic Press; *Intl. Rev. Cytol. 68,* 173.) *Right,* Light-emitting vesicles (v) of the bioluminescent dinoflagellate *Pyrocystis fusiformis.* (Courtesy of Beatrice Sweeney; copyright © 1980 by Academic Press; *Intl. Rev. Cytol. 68,* 173.)

When radiation passes through a substance, some (perhaps all) of the energy of the radiation is absorbed by certain electrons of the atoms and molecules comprising the substance. The energy of each photon is either *totally* transferred to an electron or *none* of the photon energy is transferred. That is, with regard to a specific electron, either *all or none* of the photon energy is absorbed. Moreover,

the *Einstein-Stark* law of photochemical equivalence demands that the absorption of *one* quantum of light energy results in the activation of *one* atom (or *one* molecule). The absorption of light energy by an atom or molecule often involves a shift of an electron from one orbital to another. Each electron possesses energy, the amount of which is determined by the location of the electron orbital in space

Table 9–5
Wavelengths of the Radiations of the Electromagnetic Spectrum

Type of Radiation	Photon Energy (joule $\times 10^{-19}$)	Wavelength (nm[a])
Television waves	2.0×10^{-10}–2.0×10^{-8}	10^9–10^{11}
Radar waves	2.0×10^{-8}–2.0×10^{-5}	10^6–10^9
Radio waves	2.0×10^{-11}–2.0×10^{-5}	10^6–10^{12}
Infrared (far)	0.005–0.99	2×10^3–4×10^5
Infrared (near)	0.99–2.5	780–2×10^3
Red light	2.5–3.2	620–780
Orange light	3.2–3.4	590–620
Yellow light	3.4–3.6	545–590
Green light	3.6–4.1	490–545
Blue light	4.1–4.6	430–490
Violet light	4.6–5.1	390–430
Ultraviolet (long)	5.1–6.6	300–390
Ultraviolet (short)	6.6–9.9	200–300
Ultraviolet (very short)	9.9–132.4	15–200
X-rays (soft)	99.3–2.0×10^4	0.1–20
X-rays (hard)	2.0×10^4–3.9×10^6	0.0005–0.1
Gamma rays	1.4×10^4–19.9×10^6	0.0001–0.14

[a] 1 nm (nanometer) = 10^{-9} m (meter); 1nm = 1 mμ (millimicron); 1 nm = 10 Å (angstrom).

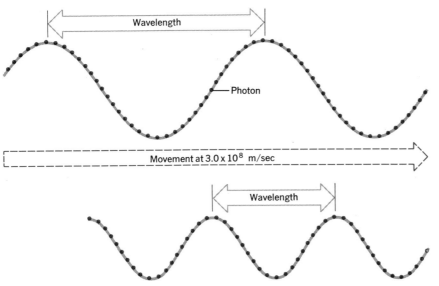

Figure 9-6
Wavelike character of electromagnetic radiation such as light. The energy of the light may be thought of as a stream of discrete packets or quanta called *photons*.

and the speed at which the electron moves. Absorption of light energy either raises an electron to an orbital of higher energy or accelerates the electron in its orbit. Electrons may orbit in pairs within an orbital, the members of the same orbital spinning in opposite directions. Most atoms at their lowest energy level (i.e., the *ground state*) have all of their electrons paired in this fashion. When the absorption of light energy raises an electron to a higher, unoccupied orbital, the electron may continue to spin in the direction opposite to its former partner or it may spin in the

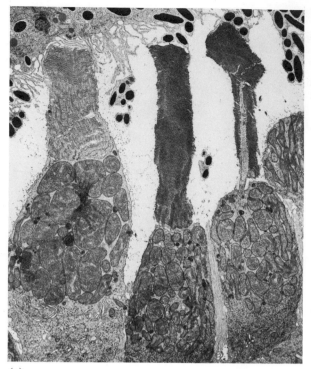

(a)

Figure 9–7

Rod cells of the vertebrate eye. (*a*) Transmission electron photomicrograph of retina tissue showing the parallel, closely packed intracellular membranes (disks) of the outer rod segment and the connecting cilium that leads to the inner rod segment. (Courtesy of Dr. S. Fisher. Copyright © 1976 by Academic Press. *J. Ultra. Res. 55*, 114.) (*b*) Diagram of a typical rod cell.

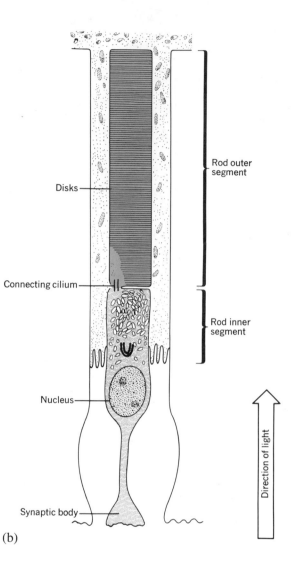

Disks

Connecting cilium

Nucleus

Synaptic body

Rod outer segment

Rod inner segment

Direction of light

(b)

same direction as its former partner. Since the energy of radiation varies with wavelength, and since the transfer of energy takes place in quantum units, it is evident that only certain wavelengths will excite specific atoms in a molecule. If the quantum transfer is great enough, the electron may altogether escape from the atom, and the atom (or molecule) would therefore be *ionized*. Radiations of very short wavelengths (e.g., gamma rays and X rays) cause appreciable ionization when they pass through cells and tissues.

Atoms containing electrons that have been boosted to higher orbitals are intrinsically unstable and frequently return (in less than 10^{-9} seconds) to their ground state as the electrons move again to their former orbitals. The gain in energy that boosts an electron to a higher orbital is called a *primary reaction,* while the release of energy during the

return of an electron to a lower orbital is called a *secondary reaction*. During the secondary reaction, the energy that is released may be transferred to neighboring molecules or may take the form of light (i.e., fluorescence) or heat. The transfer of energy to neighboring molecules during a secondary reaction underlies such biological mechanisms as vision (discussed briefly below) and photosynthesis (considered at length in Chapter 17).

Vision

The **photoreceptor** cells of vertebrate eyes (called **rods** and **cones**) contain a number of pigment molecules (*caro-*

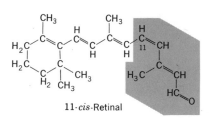

11-*cis*-Retinal

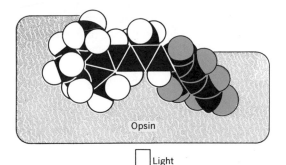

Opsin

Light

Opsin

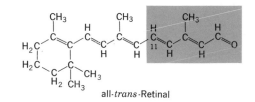

all-*trans*-Retinal

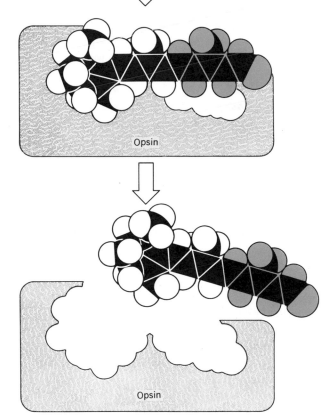

Opsin

Figure 9-8
Chemical changes in rhodopsin induced by light.

tenoids) that are excited by light. Among these, *rhodopsin,* a pigment in the rod cells, has been well studied. The pigment consists of a protein called *opsin* and *retinal,* a compound related to vitamin A. The rhodopsin molecules are localized in the outer segment of the rod cells (Fig. 9–7) in disks formed by stacks of intracellular membranes. When light energy of the appropriate wavelength is absorbed by rhodopsin, the retinal changes from the *cis* to the *trans* isomer (Fig. 9–8). This change alters the bonding between the retinal and the opsin, causing the eventual release of the retinal from the protein. The separation of these compounds in turn alters the permeability of the outer segment disk membranes to cations such as Na^+, Ca^{++}, and K^+ and brings about a change in electrical potential across the rod cell's plasma membrane. As a result, a transmitter substance is released by the inner rod segment, and this serves as a chemical stimulus for nerve cells that conduct impulses from the rods to the brain.

Other Kinds of Transductions

A number of additional examples of energy transductions will be encountered in succeeding chapters of the book, but a few examples will be briefly previewed here. Cells frequently utilize potential chemical energy such as that in ATP in order to move *permeable* solutes through their membranes in a direction that is *against* the solute's concentration gradient. These vital cellular transport mechanisms are treated more fully in Chapter 15, which deals with the structure and functions of the cellular membranes. The solute concentration gradient that is created by a transport mechanism may itself serve as a potential source of energy. Mitochondria generate an ion gradient across their membranes, which is believed to serve as an energy source for the phosphorylation of ADP. The process is discussed in detail in Chapter 16.

In nerve cells, there is the selective transport of Na^+ and K^+ across the plasma membrane against the concentration gradients of these permeable ions, and this serves to create an electrical potential across the membrane. The establishment of this potential underlies the ability of nerve cells to propagate impulses (see Chapter 24). The potential chemical energy inherent in ATP is used in muscle cells to bring about the sliding of protein filaments past one another during contraction. Muscle contraction is therefore a vivid example of a transduction in which chemical energy is transformed to mechanical energy. This mechanism is more fully described in Chapter 24.

Summary

The **metabolism** of a cell can be subdivided into two categories: **catabolism** (the breakdown of molecules into simpler forms, accompanied by the release of energy), and **anabolism** (the synthesis of complex molecules from simpler ones, accompanied by the consumption of energy). **Bioenergetics** is the study of all of the energy changes that takes place in living systems. Depending upon the nature of the materials that must be obtained from their surroundings, cells (and organisms) are classified as **autotrophs** (**photoautotrophs** or **chemoautotrophs**) or **heterotrophs** (**photoheterotrophs** or **chemoheterotrophs**).

Metabolic reactions that yield energy are called **exergonic** reactions, while reactions that consume energy are called **endergonic** reactions. In most instances the energy of exergonic reactions is used to attach phosphate to ADP, thereby forming ATP. Energy changes conform to the first two *laws of thermodynamics.*

> *First law: Energy cannot be created or destroyed but can be converted from one form into another.*
> *Second law: In all processes involving energy changes within a system, the entropy of the system increases until equilibrium is achieved.*

Reactions are said to be *coupled* when the energy of the exergonic reaction drives the endergonic reaction. **Redox reactions,** in which electrons are transferred from one compound (the **reductant**) to another (the **oxidant**), are also accompanied by energy exchanges.

Conversion of energy from one form to another is called **transduction.** Light energy occurs in units called **quanta;** the energy of a quantum is inversely proportional to the **wavelength** of the light. Light energy is converted to chemical energy in such biological mechanisms as photosynthesis and vision.

References and Suggested Reading

Articles and Reviews

Florkin, M., and Stotz, E. H., Bioenergetics, in *Comprehensive Biochemistry,* Vol. 22, American Elsevier, New York, 1967.

Ingraham, L. L., and Pardee, A. B., Free energy and entropy in metabolism, in *Metabolic Pathways* (D. M. Greenberg, Ed.), Vol. 1, Academic Press, New York, 1967, p. 2.

Lipmann, F., Metabolic generation and utilization of phosphate bond energy, *Adv. Enzymol. 18,* 99 (1941).

Sweeney, B. M., Intracellular source of bioluminescence. *Intern. Rev. Cytol. 68,* 173 (1980).

Books, Monographs, and Symposia

Atkinson, D. E., *Cellular Energy Metabolism and Its Regulation,* Academic Press, New York, 1977.

Blum, H. F., *Time's Arrow and Evolution* (3rd ed.), Harper, New York, 1968.

Lehninger, A. L., *Bioenergetics,* W. A. Benjamin, Menlo Park, Calif., 1972.

Lehninger, A. L., *Biochemistry,* Worth, New York, 1975.

Stryer, L., *Biochemistry* (2nd ed.), W. H. Freeman, San Francisco, 1981.

Wall, F. T., *Chemical Thermodynamics,* W. H. Freeman, San Francisco, 1965.

Chapter 10
CELL METABOLISM

The **metabolism** of cells includes (1) all the individual chemical reactions, (2) the sequences of these reactions, (3) the interrelationships that exist among the reaction sequences, and (4) the various mechanisms that regulate the reactions. As we have already seen (Chapter 9), individual reactions may be energy producing (the exergonic reactions), or they may be energy consuming (endergonic reactions). Commonly, the *primary* reactants or substrates are converted into final products by means of a sequence of reactions, each reaction enzymatically catalyzed and requiring a product of the prior reaction as the substrate.

The overall reaction sequences may be classified as **catabolic** (i.e., degradative) if the ultimate products of the reaction sequence are considered to be subunits or parts of the initial substrate. Alternatively, if the products are a result of the combining of two or more different substrates, the sequence is considered to be **anabolic** (i.e., synthetic). A sequence of reactions is usually referred to as a *metabolic pathway*. Some metabolic pathways are common to all living organisms or cells; several of these are considered in this chapter. Some pathways are especially active, receiving as substrates the products of a variety of other, less active pathways. Alternatively, these *central pathways* may feed substrates into a number of other, less active pathways.

Figure 10–1 diagrammatically shows some of the relationships between the catabolic and anabolic pathways followed by the major groups of cellular compounds. Intermediates in the breakdown of carbohydrates can be diverted to lipid synthesis or to the formation of nitrogenous compounds such as nucleotides and amino acids. Lipids in microbial and plant (and to a limited extent animal) cells can be converted into carbohydrates and nitrogen compounds. Likewise, nitrogen compounds, once denitrified, can be converted into lipids or carbohydrates. All of these compounds may be further degraded, their catabolism acting as sources of energy for ATP synthesis or to provide compounds that serve as a source of reducing power (e.g., NADH and NADPH) to be used in cellular anabolic reactions.

It is possible to identify specific sites within a cell where particular metabolic pathways are operative. For example, the enzymes necessary for the tricarboxylic acid (Krebs) cycle reactions are located in the mitochondria; the primary reactions of steroid synthesis are associated with the smooth endoplasmic reticulum; fatty acids are oxidized by reaction sequences in mitochondria and are synthesized in the cytosol; the reaction sequences that successively break down sugars to form pyruvic acid or lactic acid also take place in the cytosol; proteins are synthesized by the cell's ribosomes.

The intermediates, as well as the end products of a pathway, may be drawn off and used in other pathways. For example, during the breakdown of carbohydrates, a large number of intermediate compounds are formed before

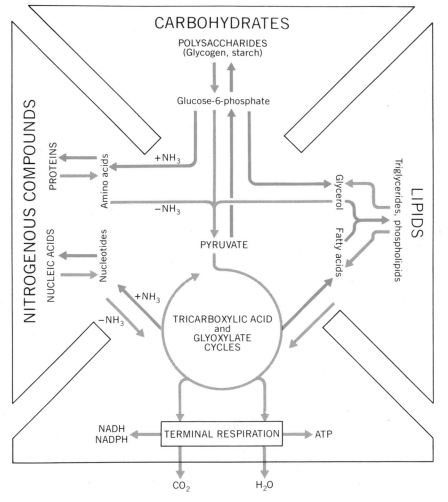

Figure 10-1
The major pathways of metabolism. By using these pathways, the cell is able to interconvert essentially all of the major compounds. Dark arrows represent degradative or *catabolic* pathways and color arrows are biosynthetic or *anabolic* pathways.

the ultimate products, CO_2 and H_2O, are formed. Some of these intermediates may be diverted from the catabolic process and used in the formation of fatty acids. Other intermediates may be used in the formation of amino acids. A number of mechanisms are used by cells to regulate the activity of metabolic pathways. These *metabolic regulatory mechanisms* are discussed in the next chapter.

In this chapter, some of the major pathways common to cells are discussed. For convenience the major pathways of carbohydrate metabolism will be reviewed first, followed by those of lipid and nitrogen metabolism. Although they are described separately, it should be remembered that in the cell many pathways are operative at the same time, and one pathway may influence the rate and direction of reactions of another pathway.

The terms used to identify metabolic intermediates vary somewhat in the chemical and biological literature. For example, *pyruvic acid* is sometimes identified as *pyruvate, lactic acid* as *lactate,* and so on. In situ, many acids dissociate, the anion (e.g., pyruvate) often being the more common form. The chemical formula may be presented as the acid or the anion.

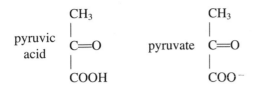

Also variable are the names assigned to phosphorylated compounds. The position of the phosphate group in the molecule is identified either at the beginning or at the end of the compound's name. Hence, *3-phosphoglyceraldehyde* and *glyceraldehyde-3-phosphate* refer to the same chemical substance.

Revealing the Individual Steps of Metabolic Pathways

Our current understanding of cellular metabolic pathways has been assembled from a large number of individual observations and experiments. Initially the existence of a pathway is identified by observing that the consumption of certain reactants leads to the accumulation of certain products. For example, the consumption of sugar and the production of carbon dioxide and alcohol during **alcoholic fermentation** is a pathway that has been known for centuries; the overall reaction is:

$$C_6H_{12}O_6 \longrightarrow 2\ CO_2 + 2\ C_2H_5OH \quad (10\text{--}1)$$
$$\text{(glucose)} \qquad\qquad\qquad \text{(ethanol)}$$

By quantitative chemical analysis, it is possible to measure the amount of sugar consumed and the amount of CO_2 and ethanol produced. While such an analysis reveals that all of the carbon in the sugar may be accounted for in these two products (i.e., no other carbon-containing products are formed), it does not tell us anything about the number or kinds of individual steps that may characterize the overall reaction. Also not revealed are *coupled reactions* such as the formation of ATP during fermentation. The individual

steps of a pathway may be revealed using *marker* and *tracer* techniques and through the isolation and identification of specific enzymes associated with the pathway.

Marker and Tracer Techniques

A frequently used technique for identifying the steps in a pathway is to follow the metabolism of a substrate molecule in which one of the atomic positions has been "labeled" with a radioactive isotope (Table 14–1) or in rare instances a "heavy" isotope (such as deuterium, ^2H, or heavy nitrogen, ^{15}N). Such a labeled substrate is said to be a *tracer* because specialized techniques are available to identify these labeled compounds. The tracer is then made available to the cell and metabolism allowed to proceed. After a short period of time, metabolism is halted by either rapidly dropping the temperature of the cells or by adding metabolic inhibitors (Chapter 8). The cells are then broken open and potential chemical intermediates of the metabolic pathway are isolated. This procedure usually involves centrifugal fractionation of the cells followed by some combination of extraction and chromatographic procedures; these techniques are discussed at length in Chapters 12 and 13. The intermediates peculiar to the pathway being studied can be distinguished from other chemical substances present on the basis of their isotope content (e.g., by their radioactivity) and can then be chemically identified.

Because in a pathway such as

$$\text{substrate*} \rightarrow B^* \rightarrow C^* \rightarrow D^* \rightarrow E^* \rightarrow \text{product*} \quad (10\text{--}2)$$

the reaction forming B occurs before the reaction forming C, which in turn occurs before the reaction forming D (and so forth), the amount of time that elapses between the addition of the tracer to the cells and the time at which metabolism is halted can be used to reveal the correct sequence. In this type of tracer experiment, the labeled substrate is made available at a designated time, and the reactions are allowed to proceed for only a short interval so that the labeled substrate can participate in no more than the first one or two reactions of the pathway. As a result, only a few of the intermediates formed will contain the label. Isolation and analysis of the labeled intermediates is then carried out. The experiment is repeated using another sample of cells and labeled substrate but is allowed to proceed for a longer period of time before being halted. As a result an additional (labeled) intermediate may be identified. The experiment is repeated again and again, each time revealing new intermediates of the pathway.

When it is suspected that after one or two reactions a substrate may split into two or more different products,

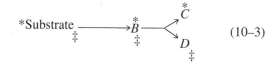

$$*\text{Substrate} \xrightarrow[\ddagger]{} *B \xrightarrow[\ddagger]{} \begin{array}{c} *\overset{*}{C} \\ \nearrow \\ \searrow \\ D \\ \ddagger \end{array} \qquad (10\text{–}3)$$

then the atoms at various positions in the substrate may be labeled so that the products entering each branch may be identified and followed. In the above diagram, the labeled positions are represented by the symbols * and ‡. Because some intermediates are not formed in sufficient amounts to permit efficient extraction and identification, nonlabeled intermediates may be added to the fractionated cells. The nonlabeled intermediate mixes with the labeled intermediate and thereby provides sufficient ''carrier'' for extraction and analysis; this technique is known as **isotopic trapping.**

Another method used to verify the position of an intermediate in a pathway is by *overloading*. For example, suppose that the position of intermediate C in reaction sequence 10–2 is uncertain. To clarify the role of C, two parallel experiments are performed. In one experiment, the substrate is labeled and the amount of label appearing in the product is then followed as described earlier. In the second experiment, an excess amount of intermediate C is added to the cells along with the labeled substrate. If C is indeed an intermediate in the pathway, then the amount of label recovered in products formed beyond the step that produces C will be greatly diminished. This is because quantities of both labeled and unlabeled C will be incorporated into the subsequent intermediates.

Enzyme Techniques

Since nearly all metabolic reactions in cells are catalyzed by enzymes, it should be possible to identify an enzyme for each metabolic reaction. In enzyme studies of this type, cells are disrupted and a cell-free **homogenate** prepared. After adding selected substrates and cofactors to the homogenate, the presence of a particular enzyme can be dem-

onstrated by measuring the rate of disappearance of the substrate of that enzyme, or the rate of appearance of a specific product. The homogenate can be fractionated by centrifugation into its components, thereby isolating the mitochondria, cell membranes, ribosomes, other organelles, or the cytosol. Each of these fractions can be tested for enzymatic activity so that the native location of the reactions in intact cells may be established.

Use of Enzyme Inhibitors and Enzyme-Deficient Cells

Yet another approach to determining metabolic reaction sequences is through the use of specific enzyme inhibitors or mutant cells that fail to produce a specific enzyme. Numerous enzyme inhibitors are known that block specific reaction steps. In the conversion of substrate A to product E, that is,

$$A \xrightarrow{1} B \xrightarrow{2} C \xrightarrow{3} D \xrightarrow{4} E \qquad (10\text{–}4)$$

an inhibitor that blocks the enzyme catalyzing reaction 2 would also prevent the formation of final product E. However, even in the presence of the inhibitor, one should still be able to make the following two observations: (1) intermediate B should continue to be formed from A, since step 1 is not inhibited; in fact, B may even be found to accumulate in excess; and (2) addition of an exogenous source of C or D should be accompanied by the continued production of final product E. The use of an inhibitor for one of the reactions of a metabolic pathway thus provides a means for identifying the intermediate *before* the block. Inhibitors can also be used to test for suspected intermediates that come *after* the block.

In a similar manner, a mutant cell that lacks the genetic information for producing a specific enzyme may serve as a test organism for studying the steps of the metabolic pathway involving that enzyme.

Carbohydrate Metabolism

The individual steps in the catabolism of carbohydrates are known in great detail. For the most part, cells break down carbohydrates by similar metabolic pathways whether they are plant cells, animal cells, or bacterial cells. The central pathway, **glycolysis,** found in most cells is outlined in Figure 10–2; however, there are alternative pathways for oxidizing carbohydrates.

Sugars are cleaved from polysaccharides such as starch and glycogen by the action of *phosphorylase* enzymes. The *phosphorolysis* introduces inorganic phosphate and produces glucose-1-phosphate as shown in the upper half of Figure 10–3. Disaccharides such as sucrose and maltose are hydrolyzed by *saccharases* without the addition of phosphate, thereby releasing their constituent monosaccharides. These sugars are phosphorylated by enzymatic reaction with ATP. The function of this phosphorylation is twofold: (1) it increases the energy content of the molecule, and (2) it introduces a charged moiety into the molecule, making it relatively impermeable to membranes and therefore unlikely to diffuse out of the cell or into an organelle. The phosphorylated sugars produced from either of these sources enter the enzymatic reactions outlined in Figure 10–2, producing the intermediate pyruvate.

Glycolysis

The major features of glycolysis are as follows:

1. The sugars are first doubly phosphorylated. In the case of monosaccharides such as glucose, fructose, mannose, and the like, 2 moles of ATP per mole of monosaccharide are utilized. Sugars that are derived from glycogen and starch require only 1 mole of ATP per mole of glucose-equivalent, since the first inorganic phosphate is acquired during the phosphorolysis of the polysaccharide.

2. The six-carbon sugar diphosphate is split by *aldolase* (reaction 10–8 in Fig. 10–3), producing 2 three-carbon units, glyceraldehyde-3-phosphate and dihydroxyacetone phosphate; the latter subsequently forms a second mole of glyceraldehyde-3-phosphate (10–9).

3. A major oxidation and phosphorylation of the substrate is catalyzed by *glyceraldehyde-3-phosphate dehydrogenase.* Two moles of hydrogen are removed per mole of substrate and reduce 2 moles of the coenzyme NAD^+. In the same reaction, inorganic phosphate is incorporated into the acid (10–10).

4. In the final steps of glycolysis, the intermediates are dephosphorylated by reaction with ADP. For each mole of monosaccharide oxidized to pyruvate, 2 moles of ATP are consumed and 4 moles of ATP are produced, resulting in

a *net* production of 2 moles of ATP. Note that when the original sugar molecule is derived from glycogen or starch by phosphorolysis, there is a net production of 3 moles of ATP per mole of glucose.

Pyruvate does not accumulate in very large amounts in cells. Instead, it is converted into other products (Fig. 10–4). The enzymes that act on pyruvate vary among different kinds of organisms and with the nature of the environment. The more common fates of pyruvate are (1) its fermentation to ethanol and carbon dioxide in cells such as yeast, (2) its anaerobic conversion into lactate in cells such as muscle cells, and (3) its conversion into acetate in the mitochondria of cells and organisms living under aerobic conditions.

Anaerobic Respiration and Fermentation

The reactions of glycolysis have no specific requirement for oxygen. Oxidation reactions do occur, such as the removal of two hydrogens from glyceraldehyde-3-phosphate, and NAD^+ is reduced to NADH, but oxygen per se is not consumed. Under *anaerobic* conditions (i.e., in the absence of oxygen), pyruvate may be reduced to a variety of different compounds. Alcoholic fermentation (reactions 10–19 and 10–20 in Fig. 10–4) is a common pathway in microorganisms and is of industrial importance. In these two steps, 1 mole of CO_2 is removed from each mole of pyruvate (i.e., 2 moles of CO_2 per mole of monosaccharide) and NADH is reoxidized to NAD^+, thereby producing ethanol.

The stoichiometric relationships and the cyclic involvement of NAD^+ are especially important in this pathway. During the oxidation of each mole of glyceraldehyde-3-phosphate (reaction 10–10), one mole of NAD^+ is reduced to form NADH; and during the conversion of acetaldehyde to ethyl alcohol (reaction 10–20) a mole of NADH is oxidized to form NAD^+. Therefore, the levels of NAD^+ and NADH are unaffected by the conversion of glyceraldehyde-3-phosphate to ethyl alcohol. The quantities of these coenzymes in cells are very small. Therefore, if the NAD^+ reduced in reaction 10–10 was not reoxidized, this central pathway would soon be blocked at the glyceraldehyde-3-phosphate step by the lack of NAD^+, and the pathway would necessarily have to cease.

Like NAD^+ and NADH, ATP and ADP are cycled between the ATP-requiring reactions in the early steps of glycolysis and the ATP-producing reactions in the later

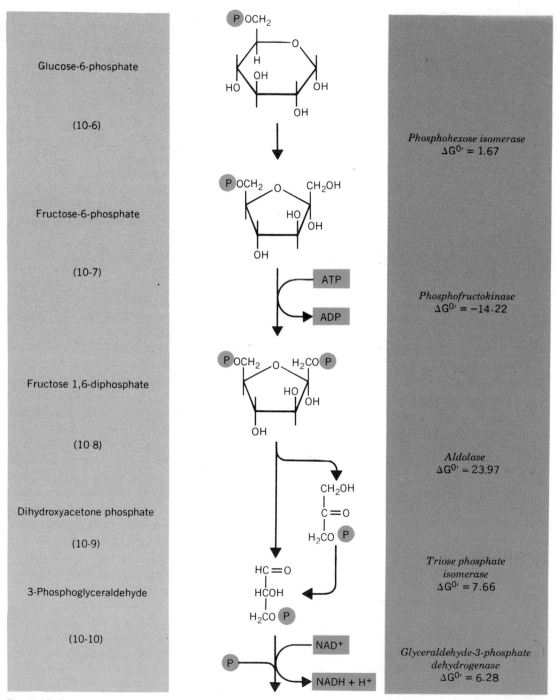

Glucose-6-phosphate

(10-6)

Fructose-6-phosphate

(10-7)

Fructose 1,6-diphosphate

(10-8)

Dihydroxyacetone phosphate

(10-9)

3-Phosphoglyceraldehyde

(10-10)

Phosphohexose isomerase
$\Delta G^{0\prime} = 1.67$

Phosphofructokinase
$\Delta G^{0\prime} = -14.22$

Aldolase
$\Delta G^{0\prime} = 23.97$

Triose phosphate
isomerase
$\Delta G^{0\prime} = 7.66$

Glyceraldehyde-3-phosphate
dehydrogenase
$\Delta G^{0\prime} = 6.28$

Figure 10–2

Glycolysis. Although individual reactions are reversible, reversal of the entire pathway does not normally occur via this sequence of steps (see Figure 10-8). The $\Delta G^{0\prime}$ values given in this figure are in kJ/mole and apply to each reaction in the direction written. The names of the intermediates appear on the left; the names of the enzymes appear on the right. $\textcircled{P}$ = phosphate group.

Figure 10–2
continued

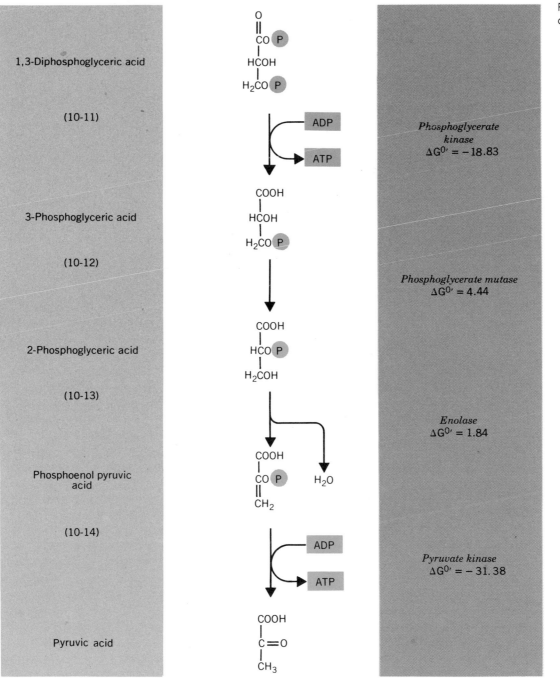

1,3-Diphosphoglyceric acid

(10-11)

3-Phosphoglyceric acid

(10-12)

2-Phosphoglyceric acid

(10-13)

Phosphoenol pyruvic
acid

(10-14)

Pyruvic acid

*Phosphoglycerate
kinase*
$\Delta G^{0\prime} = -18.83$

Phosphoglycerate mutase
$\Delta G^{0\prime} = 4.44$

Enolase
$\Delta G^{0\prime} = 1.84$

Pyruvate kinase
$\Delta G^{0\prime} = -31.38$

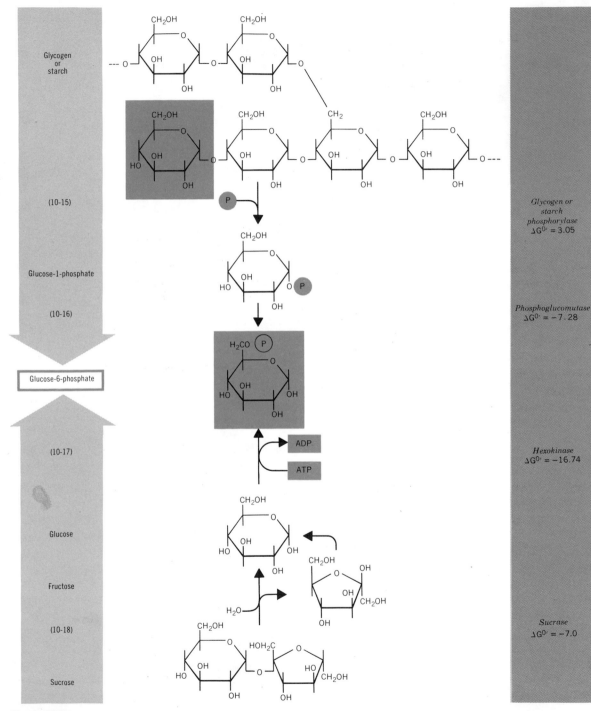

Figure 10-3
Conversion of polysaccharides and oligosaccharides into phosphorylated monosaccharides (e.g., glucose-6-phosphate) as a preliminary to their catabolism via the glycolytic pathway. Ⓟ = phosphate group.

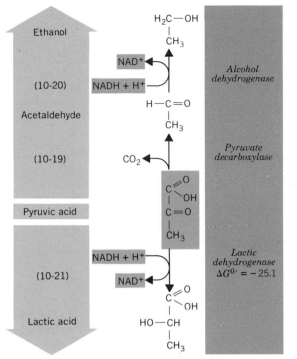

Figure 10-4
The anaerobic conversion of pyruvic acid (pyruvate) to ethanol and to lactic acid (lactate).

Oxidation of Pyruvate

Except in the case of the *strict anaerobes* (organisms that *cannot* live in the presence of oxygen), when oxygen is present, most cells will oxidize pyruvate rather than reduce it to lactate, ethanol, or other compounds.

Energetically, there is an advantage for the cell to oxidize the pyruvate to CO_2 and water rather than to reduce it to lactate. In the production of lactate, there is an overall change in standard free energy ($\Delta G^{o'}$) of -123.2 kJ (-29.44 kcal) per mole of glucose and a net production of 2 moles of ATP. However, in the breakdown of a mole of glucose through glycolysis and further oxidation to CO_2 and water, the $\Delta G^{o'} = -2870$ kJ (or -686 kcal) per mole of glucose, and a net of 36 (or 38, see below) moles of ATP are produced. Therefore, pyruvate still has a significant amount of stored chemical energy that the aerobic cell can extract.

All of the enzymes that bring about the complete oxidation of pyruvate in eucaryotes are found in the mitochondria, either in the matrix or associated with the mitochondrial membranes. The various stages of pyruvate oxidation—(1) formation of acetyl-CoA, (2) the tricarboxylic acid cycle (Krebs cycle), and (3) electron transport and oxidative phosphorylation—are so closely associated with the structure of the mitochondrion that a detailed description of these reactions is deferred to Chapter 16. The reactions are summarized in Figure 16–19.

Other Pathways of Carbohydrate Catabolism

Pentose Phosphate Pathway

An alternative oxidative pathway for the oxidation of sugars is the **pentose phosphate pathway** (also called the **hexose monophosphate shunt** or **phosphogluconate pathway**). This pathway occurs in plants and in most animal tissues. However, its activity in comparison with glycolysis is usually lower and varies considerably from tissue to tissue. The pathway serves as the primary mechanism for (1) converting hexoses into those pentoses necessary for the synthesis of nucleotides and nucleic acids, (2) degrading pentoses so that they may be catabolized in the glycolytic pathway, and (3) generating reduced pyridine nucleotides (e.g., NADPH) to be used for synthetic reactions such as fatty acid synthesis, steroid synthesis, and amino acid syn-

steps. Cells contain small pools of ATP, ADP, and AMP; when needed, these compounds are drawn from the pools and later returned to the pools. During the early stages of glycolysis, 2 moles of ATP are consumed per mole of glucose, whereas during the later stages 4 moles of ATP are produced. The early stages proceed by drawing ATP from the pool, while the latter reactions return ATP to the pool. However, there is a net generation of ATP from glycolysis that is then available for other energy-requiring reactions within the cell.

Another common fate of pyruvate that occurs in the absence of oxygen is its conversion to lactate. This is a normal process in active muscle cells that are not receiving adequate amounts of oxygen and in many plant and bacterial cells that live under anaerobic conditions. In a single reaction, pyruvate is converted to lactate (reaction 10–21). During the reaction, one mole of NADH is converted to NAD^+. This stoichiometrically balances NADH production during the earlier glyceraldehyde oxidation step.

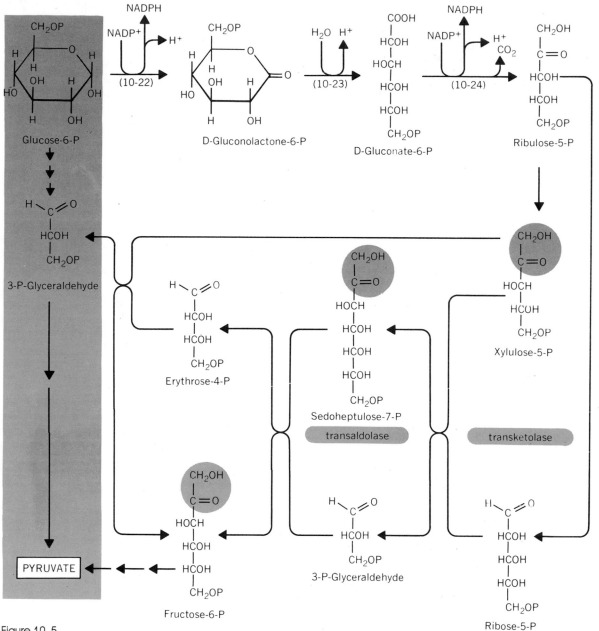

Figure 10–5

The *pentose phosphate pathway* (*hexose monophosphate shunt; phosphogluconate pathway*) The 2-carbon fragment transferred during the *transketolase* reaction is shaded in color. The 3-carbon fragment transferred during the *transaldolase* reaction includes the same two carbons transferred earlier. $\circledP$ = phosphate group.

thesis. In animal tissues, the latter functions occur extensively in the liver, in the mammary glands, and in the cortex of the adrenal glands. It has been found that 20% of the sugar metabolized in the mammary glands proceeds via the pentose phosphate pathway. In heart or skeletal muscle, little synthesis of fatty acids (or other related substances) occurs, and therefore it is not surprising that the pentose phosphate pathway is not active in these tissues.

Figure 10–5 illustrates the major reaction steps of the pentose phosphate pathway. Reactions 10–22 and 10–24 (Fig. 10–5) describe the two oxidative steps that produce the NADPH required in the synthetic reactions. Reaction 10–24 is also the decarboxylation step that converts the hexose to a pentose. Through the action of an **isomerase** and an **epimerase,** three pentose phosphates may be formed: ribose-5-phosphate, ribulose-5-phosphate, and xylulose-5-phosphate. These compounds may be drawn off into nucleoside biosynthesis; however, if they continue through the pentose phosphate pathway, they are acted on by **transketolases** and **transaldolases** as seen in Figure 10–5. The transketolase transfers a 2-carbon fragment from xylulose-5-phosphate to ribose-5-phosphate, thereby producing sedoheptulose-7-phosphate and glyceraldehyde-3-phosphate. The transaldolase catalyzes the transfer of a 3-carbon fragment from sedoheptulose-7-phosphate to glyceraldehyde-3-phosphate, thus forming erythrose-4-phosphate and fructose-6-phosphate. Because of the action of these enzymes 3-, 4-, 5-, 6-, and 7-carbon sugar phosphates are interconvertible, and intermediates of the pentose phosphate pathway may be channeled into glycolysis and from there to the mitochondria where the tricarboxylic acid cycle and terminal respiration produce CO_2 and H_2O. Figure 10–6 shows the major reaction steps for the route of hexoses through the pentose phosphate pathway and into the glycolytic pathway.

In plant cells, the enzymes (and therefore the reactions) of the pentose phosphate pathway are also found in the stroma of chloroplasts and are used there to regenerate pentoses in photosynthesis (Chapter 17).

The Glyoxylate Pathway

The glyoxylate pathway is essentially a bypass of the CO_2-evolving steps of the tricarboxylic acid cycle; in fact the pathway is often called the "glyoxylate bypass." The enzymes for this bypass are commonly found in plants but are generally absent in the tissues of animals. The enzymes are localized in organelles called glyoxysomes (see Chapter 19). Figure 10–7 summarizes the key reactions of the bypass. Two moles of acetate are required for each turn of the pathway rather than the 1 mole required in the tricarboxylic acid cycle. The first mole of acetate (acetyl CoA) condenses with oxalacetate to form citrate, which is then converted to isocitrate. The inducible enzyme isocitric lyase then catalyzes the conversion of isocitrate to succinate and

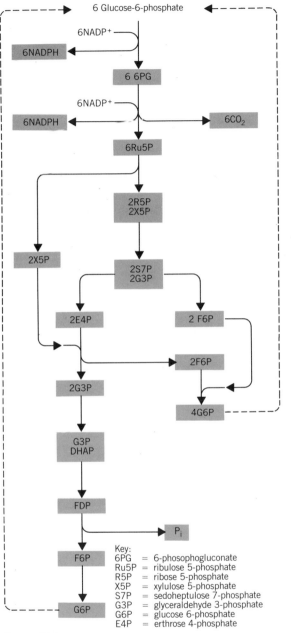

Figure 10-6

Balance sheet and diagram illustrating the complete oxidation of one molecule of glucose-6-phosphate to carbon dioxide via the pentose phosphate pathway.

Key:
6PG = 6-phosophogluconate
Ru5P = ribulose 5-phosphate
R5P = ribose 5-phosphate
X5P = xylulose 5-phosphate
S7P = sedoheptulose 7-phosphate
G3P = glyceraldehyde 3-phosphate
G6P = glucose 6-phosphate
E4P = erthrose 4-phosphate

glyoxylate. The glyoxylate reacts with a second mole of acetate (acetyl CoA) to produce malate, which is converted back to oxalacetate.

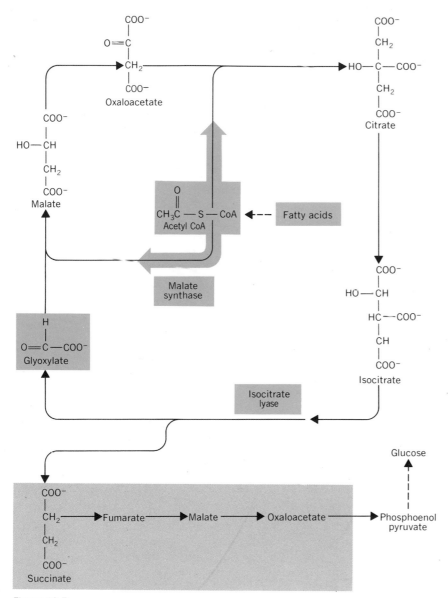

Figure 10-7

The glyoxylate pathway (glyoxylate bypass). Acetyl CoA, derived from the acetate that is released from the breakdown of compounds such as fatty acids, is consumed in two steps in the cycle.

This bypass set of reactions enables plants and microorganisms to transform acetyl CoA derived from the breakdown of fatty acids into carbohydrate. The excess succinate produced by the glyoxylate pathway can be channeled back up the glycolytic pathway for the formation of sugar and subsequently polysaccharides. In order to make the synthesis of sugar energetically possible by reversing the steps of glycolysis, certain bypass reactions are employed. These are discussed below.

Gluconeogenesis

Glucose can be synthesized by reversing most of the reactions of glycolysis. The initial reactions used in order to start various substrates back along the glycolytic pathway differ among different cells and tissues. Also, the carbohydrates that are finally produced vary with conditions and with cell types. Photosynthetic organisms can reduce CO_2 through a reversal of the glycolytic reactions after initially ''fixing'' the CO_2 using ribulose-diphosphate (Chapter 17). Liver cells can regenerate glucose starting with lactate by reversing the glycolytic reactions. Almost all cells can transaminate or deaminate key amino acids thereby producing tricarboxylic acid cycle intermediates, and by converting these into phosphoenol pyruvate, glycolysis may then be reversed to produce glucose. Plant and bacterial cells can oxidize fatty acids to acetyl CoA and, via the glyoxylate pathway, form tricarboxylic acid intermediates; these can then be converted to phosphoenol pyruvate and thereby initiate the reversal of glycolysis. Figure 10–8 diagrams the major initial reactions and features of gluconeogenesis.

In the reversal of glycolysis, there are three reactions that occur in place of those found in the catabolic sequence:

1. Pyruvate is not directly converted to phosphoenol pyruvate (the energetics of that particular reaction are not favorable; $\Delta G^{o\prime} = 31.4$ kJ/mole (or 7.5 kcal/mole). Instead, the pyruvate is converted first to tricarboxylic acid cycle intermediates within the mitochondria, which then pass into the cytosol, where *phosphoenolpyruvate carboxykinase,* together with GTP, converts these to phosphoenolpyruvate ($\Delta G^{o\prime} = 4.2$ kJ/mole or 1.0 kcal mole).

2. The dephosphorylation of fructose-1,6-diphosphate is not coupled to ATP by the catabolic sequence enzyme (*phosphofructokinase*) but instead is acted upon by *fructose diphosphatase* to yield fructose-6-phosphate and inorganic phosphate.

3. In some tissues such as liver and kidney the glucose-6-phosphate may be broken down to glucose and inorganic phosphate by *glucose-6-phosphatase* instead of hexokinase.

Synthesis of Glycogen and Starch

Glycogen and starch are highly branched polymers of glucose (see Chapter 5). The syntheses of these two polysaccharides proceed by similar reactions in both animals and plants, although the enzymes that catalyze the syntheses are different and so too are certain intermediates. In both pathways, (Fig. 10–9), the first reaction is the conversion of glucose-6-phosphate to glucose-1-phosphate. In the case of glycogen synthesis, the glucose-1-phosphate reacts with UTP to form uridine diphosphoglucose (UDPG), whereas in plants glucose-1-phosphate reacts with ATP to form adenosine diphosphoglucose (ADPG). In both cases, the resulting nucleoside diphosphoglucose serves as a glucose donor to a preexisting polysaccharide. The glucose is added to the end of one of the outer chains of the polysaccharide via an α 1→4 glycosidic linkage. The chain accepting the additional sugar residue is called a ''primer.'' Repeated rounds of sugar addition continuously add to the growing glycogen or starch molecule.

Branch points are created in glycogen and the amylopectin component of starch (see Chapter 5) by separate enzymes called **branching enzymes.** These enzymes remove segments of outer chains containing a number of α 1→4 linked glucose units and then reattach them to the polysaccharide molecule via α 1→6 linkages.

Lipid Metabolism

Triglycerides

The synthesis of fatty acids and their ultimate incorporation into triglycerides follows a pathway that is significantly different from that which results in the catabolic breakdown of these compounds. In effect, the linear fatty acid molecule is assembled two carbon units at a time by the consumption of acetate in the form of acetyl CoA (Fig. 10–10). The addition of each 2-carbon unit to the chain involves six steps.

1. The **priming reaction,** in which acetyl CoA is bound first to the *nonenzymatic acyl carrier protein* (ACP) and then transferred to the enzyme *ACP-acyltransferase.*

2. The **malonyl reaction,** in which a second acetyl CoA converted to malonyl CoA by incorporation of HCO_3^- is bound to ACP.

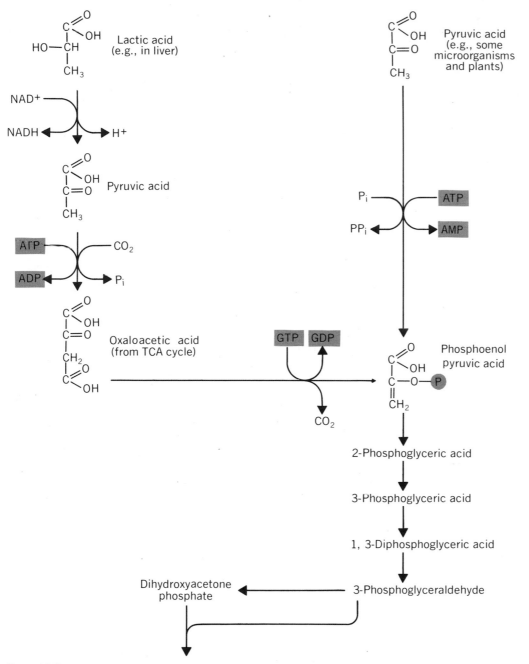

Figure 10-8
Gluconeogenesis. **Procaryotic and eucaryotic cells use various pathways in order to convert pyruvic acid and intermediates of the tricarboxylic acid cycle into glucose-6-phosphate, glucose, and polysaccharides. Some of these reactions are shown here.**

Figure 10–8
continued

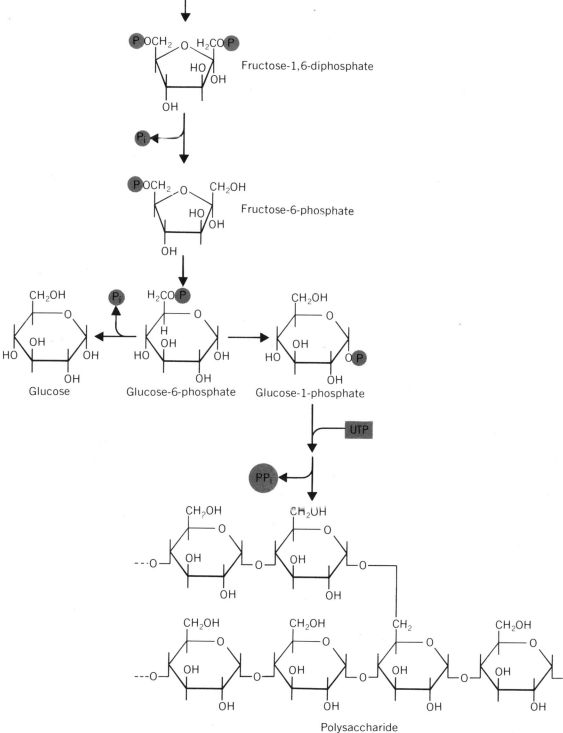

Fructose-1,6-diphosphate

Fructose-6-phosphate

Glucose

Glucose-6-phosphate

Glucose-1-phosphate

UTP

Polysaccharide
(glycogen or starch)

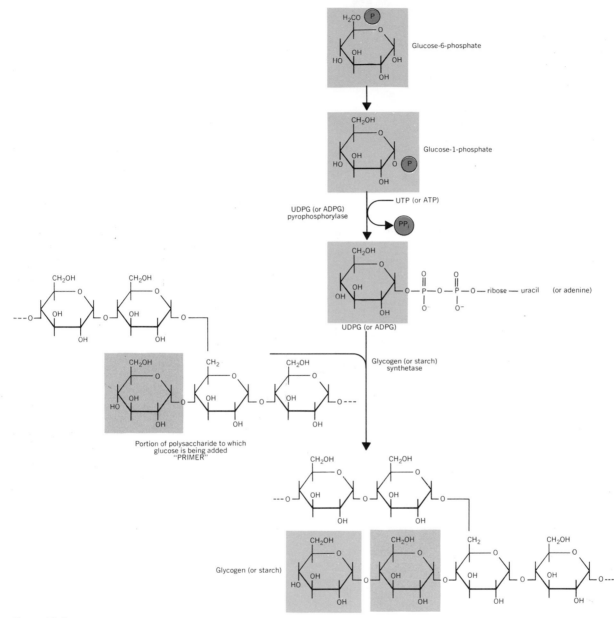

Figure 10-9
Pathways employed in animals and plants for incorporating glucose (shaded in color) into polysaccharides. UDPG, uridine diphosphoglucose; ADPG, adenine diphosphoglucose.

3. The **condensation reaction,** in which the malonyl group loses CO_2, becoming an acetyl group which then attaches to the acetyl group on the enzyme.

4. A **first reduction step** involving NADPH.

5. A **dehydration step**

6. A **second reduction step** involving NADPH

The resulting fatty acid, which is four carbons in length,

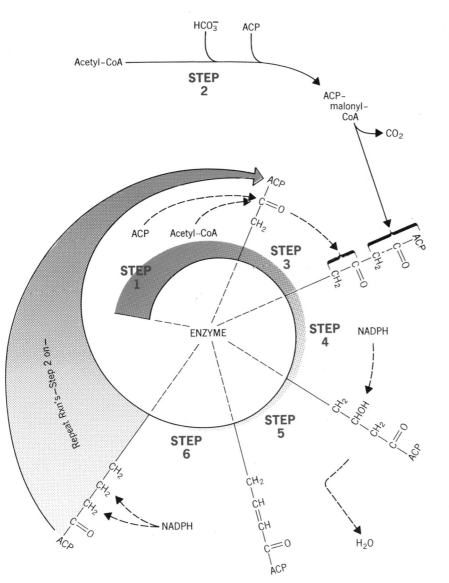

Figure 10-10
**Pathway for the synthesis of fatty acids
(see text for explanation).**

can now serve as the primer, and the sequence is repeated from step 2 onward. Each cycle adds two carbon units to the growing chain. Desaturation of the fatty acids occurs by additional enzymatic steps. The acetyl CoA used during the synthesis is generated primarily in the mitochondria by decarboxylation of pyruvate, by oxidative degradation of some amino acids, or by oxidation of other fatty acids. The mitochondrial acetyl CoA must be converted into other molecules in order to leave the mitochondria and enter the cytosol. In the cytosol, the acetyl CoA is reformed and incorporated into the fatty acids. Triglycerides are formed by the condensation of fatty acyl CoA molecules and either dihydroxyacetone-phosphate (produced by the reactions of glycolysis) or glycerol-3-phosphate (formed through the reduction of dihydroxyacetone phosphate by NADH).

Degradation of triglycerides occurs initially in the cytosol by hydrolysis to glycerol and fatty acids; the glycerol then enters the glycolytic pathway. Fatty acids must be activated

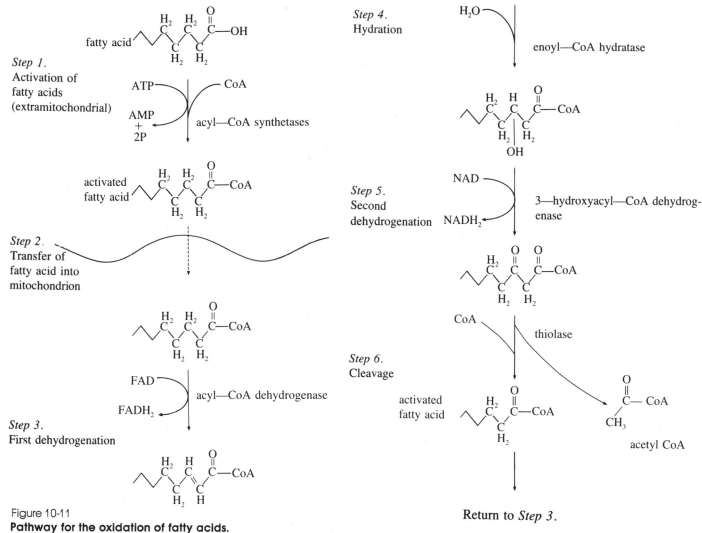

Figure 10-11
Pathway for the oxidation of fatty acids.

and transported into the matrix of the mitochondria, where they undergo the degradative and oxidative steps outlined in Figure 10–11. The sequential repetition of steps 3 to 6 (Fig. 10–11) removes two carbon units at a time from the fatty acid until breakdown is complete and only acetyl CoA or malonyl CoA remains. This process is known as β-*oxidation,* since it is the bond at the second carbon atom that is cleaved. Other oxidations are known; for example, α-oxidation occurs in some germinating seeds producing CO_2, ω-oxidation occurs in the liver of mammals, but the role of this process is not understood.

The synthesis and degradation of most other lipids such as phospholipids and sterols are also known in detail but will not be considered here. Synthesis of cholesterol occurs by the progressive buildup of 2-carbon fragments (i.e., acetyl CoA units) and involves some 25 enzymatic steps.

Nitrogen Metabolism

The metabolism of cells requires a variety of nitrogen-containing compounds. Central to the requirement for nitrogen is the formation of amino acids. Amino acids are necessary not only for the synthesis of proteins but also as the primary source of nitrogen in the synthesis of the nucleotide building blocks of nucleic acids. Not all cells or organisms are able to synthesize all of the amino acids necessary to their existence. For example, the cells of human tissues are able to make only 10 of the 20 amino acids required in the synthesis of human proteins. The other 10 are referred to as the *essential amino acids* and are obtained from plant or animal tissues in the diet. Nitrogen for the synthesis of the nonessential amino acids of humans and other higher animals is obtained from ammonium ions. In most organisms the enzyme systems for utilizing nitrate, nitrite, or atmospheric nitrogen are not present. However, many microbes, including those in symbiotic relationship with leguminous plants, and a few plants can use atmospheric nitrogen. Most higher plants can make the amino acids needed for protein synthesis by using ammonia, nitrate, and nitrite as sources of nitrogen.

Each of the 20 different amino acids required for protein synthesis is synthesized by a specific system of enzymes. Several utilize intermediates of the glycolytic pathway and the tricarboxylic acid cycle. The degradation of amino acids involves pathways different from the synthetic pathways but the products of deamination are usually further metabolized by the glycolytic or the tricarboxylic acid cycle reactions.

Functions of Metabolic Pathways

The sequences of reactions that form the metabolic pathways serve a number of functions. First, and most obvious, is the synthesis of an end product needed by the cell or organism. This product may be used as part of the structure of the cell, or in cases where it is secreted from the cell, it may be incorporated into an extracellular structural part of an organism. The proteins in the extracellular matrix of cartilage and the pectins of the middle lamellae of plant cell walls are examples of such secretions. The end product may also function in the cell as a regulatory agent for other reactions. Hormones are such end products. End products may also be stored as reserves, such as starch, glycogen, and certain lipids.

Second, a metabolic pathway may function to provide energy-rich compounds such as ATP, GTP, and UTP and these may be used in other energy-requiring reactions. Metabolic pathways rarely yield ATP, GTP or UTP as end products of reaction sequences; instead they are produced at one or more of the intermediate steps of a pathway. For example, in glycolysis, ATP is produced at two intermediate steps (Fig. 10–2, reactions 10–11 and 10–14). Other compounds such as NADH and NADPH are also products of intermediate reactions and may be utilized directly in other reactions of the same or a different pathway or may be oxidized to provide energy in the form of ATP.

Third, intermediates of a metabolic pathway may be drawn upon and used as substrates for other metabolic sequences. For example, the tricarboxylic acid cycle (Krebs cycle) is a sequence of reactions whose net effects are the oxidation of acetyl CoA and the production of CO_2, NADH, and GTP. However, many of the intermediates of the Krebs cycle are drawn off and utilized for other purposes. Acetyl CoA itself is used in lipid synthesis, and α-ketoglutarate and oxalacetate are used in amino acid synthesis. Metabolic reactions that serve in both a catabolic (energy-producing) and an anabolic (biosynthetic) function are called **amphibolic** pathways. Amphibolic pathways, which may have their intermediates diverted to other reaction sequences, frequently include specialized reactions that replenish these intermediates. The specialized enzymatic reactions are called **anaplerotic** reactions. The "Wood-Werkman" reaction in bacteria in which carbon dioxide combines with pyruvate to form oxalacetate is an example of such an anaplerotic mechanism. The glyoxylate pathway described earlier is also an anaplerotic mechanism, effectively forming the tricarboxylic acid cycle intermediates, succinate and malate.

Calculations of Energy Change

For most of the major metabolic pathways of the cell the specific enzymatic reactions in which ATP, GTP, or other related compounds are formed or consumed are known. By inspecting metabolic charts, such as those of Figures 10–2, 10–3, and 10–4, one can calculate the number of moles of ATP consumed and/or produced and the *net* change for any reactant-to-product sequence. For example, when 1 mole of sucrose is converted to pyruvate, 4 moles of ATP are consumed,

$$
\begin{array}{ll}
& \text{2 moles in reaction 10–7} \\
& \underline{\text{2 moles in reaction 10–17}} \\
\text{Total} = & \text{4 moles (consumed)}
\end{array}
$$

and 8 moles of ATP are produced :

$$
\begin{array}{ll}
& \text{4 moles in reaction 10–11} \\
& \underline{\text{4 moles in reaction 10–14}} \\
\text{Total} = & \text{8 moles (produced)}
\end{array}
$$

Therefore, in the glycolytic oxidation of 1 mole of sucrose, there is a net production of 4 moles of ATP. These ''paper calculations'' reveal the theoretical amounts of ATP expected from the glycolytic sequence of reactions. In laboratory experiments, these numbers are approached but are not always obtained. This may be due to the fact that laboratory conditions are not always optimal, intermediates may be drawn into reactions that are not part of the pathway being evaluated (this is especially true in experiments using intact cells), cofactors may not be present in the proper concentration, and so on.

The *efficiency* of a metabolic pathway is usually determined by an analysis of the change in free energy. For example, in the conversion of glucose to lactate (Figs. 10–2, 10–3, 10–4), the enzymatic sequence may be considered to be composed of the exergonic reaction

$$
\underset{\text{(glucose)}}{C_6H_{12}O_6} \longrightarrow \underset{\text{(lactate)}}{2\ C_3H_6O_3}
$$

and the endergonic reaction

$$
2\ ADP + 2P_i \longrightarrow 2\ ATP + H_2O
$$

The standard free energy change ($\Delta G^{o\prime}$) for the catabolic reactions is determined by summing the values for each step.

Note. The $\Delta G^{o\prime}$ values given for a reaction as shown in a metabolic chart are for the reaction *as written*, including the *direction* written; for example

$$
\text{glycerate-3-P} \longrightarrow \text{glycerate-2-P}
$$
$$
\Delta G^{o\prime} = 4.44 \text{ kJ or } 1.06 \text{ kcal}
$$

and

$$
\text{glycerate-2-P} \longrightarrow \text{glycerate-3-P}
$$
$$
\Delta G^{o\prime} = -4.44 \text{ kJ or } -1.06 \text{ kcal}
$$

If both an exergonic reaction and an endergonic reaction are written together, for example

$$
\text{phosphoenol pyruvate} \xrightarrow{\;\;ADP \quad ATP\;\;} \text{pyruvate}
$$
$$
\Delta G^{o\prime} = -23.9 \text{ kJ or } -5.7 \text{ kcal}
$$

then the $\Delta G^{o\prime}$ given is the sum of the two separate reactions; that is

$$
\text{phosphoenol pyruvate} \longrightarrow \text{pyruvate} + P_i
$$
$$
\Delta G^{o\prime} = -54.4 \text{ kJ or } -13.0 \text{ kcal}
$$

and

$$
ADP + P_i \longrightarrow ATP
$$
$$
\Delta G^{o\prime} = 30.5 \text{ kJ or } 7.3 \text{ kcal}
$$

In the case of the catabolic reactions of Figures 10–2, 10–3, and 10–4 which form lactate, the summation would be:

Reaction Number	$\Delta G^{o\prime}$ (kJ/mole)*	
10–17	+ 13.81	
10–6	+ 1.67	
10–7	+ 16.32	
10–8	+ 23.97	
10–9	+ 7.66	
10–10	+ 12.56	(i.e., + 6.28 × 2)
10–11	− 98.74	(i.e., − 49.37 × 2)
10–12	+ 8.88	(i.e., + 4.44 × 2)
10–13	+ 3.68	(i.e., + 1.84 × 2)
10–14	− 123.84	(i.e., − 61.92 × 2)
10–21	− 50.20	(i.e., − 25.10 × 2)
Total	− 184.24	

* Excluding the $\Delta G^{o\prime}$ attributed to ATP→ADP + P_1 or ADP + P_1→ATP

A summation of the net ADP/ATP reactions would be:

Reaction Number	$\Delta G^{o\prime}$ (kJ/mole)	
10–17	− 30.54	
10–7	− 30.54	
10–11	+ 61.08	(i.e., 30.54 × 2)
10–14	+ 61.08	(i.e., 30.54 × 2)
Total	+ 61.08	

The reactions of glycolysis and the fermentation of glucose to lactate thus produce much more free energy than they consume in the formation of ATP. The efficiency of these pathways would therefore be $61.08/184.24 = 0.332$ or 33.2%.

Attempts to measure the energetics of glycolysis and fermentation in intact cells (i.e., in vivo) have produced striking results. In red blood cells (which are ideal in such studies, since this cell derives most of its energy from glycolysis and from metabolism to lactate), the efficiency is about 53%; this is much higher than that predicted above. To determine this in vivo efficiency, the steady-state concentrations of all the glycolytic intermediates are measured, and from these values, the actual equilibrium constants are calculated and ΔG values are determined. The ΔG values (rather than the $\Delta G^{o\prime}$ values) reveal the greater efficiency of red blood cells. Skeletal muscle cells also reveal an efficiency level higher than that anticipated from calculations and summations of the type carried out above. Therefore while $\Delta G^{o\prime}$ values provide figures for easy comparison under defined conditions (i.e., 1 molar concentrations of reactants and products, pH 7.0 and 25°C), differences in substrate concentration, pH, temperature and other factors may produce variations in efficiency under natural conditions, that is, in vivo.

Summary

Cells break down (i.e., **catabolize**) compounds by sequences of enzyme catalyzed reactions to obtain usable chemical energy or to form smaller molecules that can be used to build (i.e., **anabolize**) other, larger molecules. The reaction sequences, called **metabolic pathways,** are frequently associated with specific organelles, the enzymes that catalyze the reactions compartmentalized between or within the membranes of the organelle.

In general, carbohydrates, lipids, and proteins can be catabolized to yield energy and a pool of small molecules, and these may either be used in the synthesis of new macromolecules or excreted from the cell. The products of catabolism in one organelle may be transported to other organelles for further catabolism or anabolism. Reaction sequences and their intracellular location have been revealed using **tracer techniques,** by studies of *enzyme activity* and *inhibition,* and through the use of *mutant* cells.

The most common metabolic pathways in cells are:

1. Glycolysis—the catabolism of monosaccharides to pyruvate

2. Fermentation—the catabolism of monosaccharides to products such as ethanol and CO_2 in the absence of air (anaerobic conditions)

3. Tricarboxylic acid (Krebs) cycle—the oxidation of pyruvate and its catabolism to CO_2 and water in the presence of oxygen

4. Pentose phosphate pathway (phosphogluconate pathway or hexose monophosphate shunt)—a pathway for formation and/or catabolism of pentoses

5. Glyoxylate pathway—the conversion of acetyl CoA to carbohydrate

6. Gluconeogenesis—the synthesis of glucose from simple organic acids

7. Glycogen and starch synthesis—the incorporation of sugars into polysaccharides

8. Fatty acid synthesis—the pathway leading to the formation of fatty acids and triglycerides from acetyl CoA and glycerol

9. β-oxidation—the pathway for breakdown of fatty acids

10. Amino acid synthesis—the mechanism for formation of amino acids to be incorporated into proteins

---

References and Suggested Reading

Articles and Reviews

Hinkle, P. C., and McCarty, R. E., How cells make ATP. *Sci. Am.* 238(3), 104 (Mar. 1978).

Krebs, H. A., The history of the tricarboxylic acid cycle, *Perspect. Biol. Med. 14,* 154 (1970).

Villar-Palasi, C., and Larner, J., Glycogen metabolism and glycolytic enzymes. *Annu. Rev. Biochem. 39,* 639 (1970).

Books, Monographs, and Symposia

Atkinson, D. E., *Cellular Energy Metabolism and Its Regulation,* Academic Press, New York, 1977.

Axelrod, B., Other pathways of carbohydrate metabolism, in *Metabolic Pathways* (D. M. Greenberg, Ed.), Vol. 1, Academic Press, New York, 1967, p. 272.

Dagley, S., and Nicholson, D. E., *Metabolic Pathways,* Wiley, New York, 1970.

Larner, J., *Intermediary Metabolism and Its Regulation,* Prentice-Hall, Englewood Cliffs, N. J., 1971.

Lehninger, A. L., *Biochemistry,* Worth Publishers, New York, 1975.

Stanbury, J. O., Wyngaarden, J. B., and Fredickson, D. S., *The Metabolic Basis of Inherited Disease,* McGraw-Hill, New York, 1972.

Stryer, L., *Biochemistry* (2nd ed.), W. H. Freeman, San Francisco, 1981.

The diverse metabolic reactions and reaction sequences in cells have been briefly described and outlined in the preceding three chapters. From these descriptions, it is clear that a substrate can be enzymatically converted into a great variety of intermediates and products. Although there are a number of essentially unidirectional reactions, the metabolic network of reversible and cyclical reaction sequences provides for the possible conversion of almost any metabolite into any other metabolite. In a superficial comparison, one could visualize the metabolic pathways as a branching and connecting network of water pipes in which the water can be caused to flow between any two points in the network under suitable conditions. For example, simply an *excess* of water (metabolite) in one part of the system could cause flow to the other parts of the system. In vivo and in vitro experiments have indicated that the "flow" of metabolites through metabolic pathways is not as free and uncontrolled as this pipeline–network analogy suggests. Metabolic conversions are controlled or regulated by a variety of mechanisms that channel metabolites into needed compounds or into stable reserve products and prevent energetically wasteful conversions.

Cells have evolved a diverse set of regulatory mechanisms. Individual reactions may be controlled by one or more processes from simple mass action to complex hormonally controlled enzyme systems. The more common of these processes are described in this chapter.

Regulation by Mass Action

For any reversible reaction, such as

$$A + B \rightleftharpoons C + D \qquad (11\text{–}1)$$

in which A and B are reactants and C and D are products, the ratio of the concentrations of products $[C]$ and $[D]$, and reactants $[A]$ and $[B]$ at equilibrium is

$$\frac{[C]\,[D]}{[A]\,[B]} = K'_{eq}$$

At equilibrium concentrations, the reaction rate in one direction equals the reaction rate in the opposite direction, resulting in no net change in the concentration of reactants or products. The value of the equilibrium constant depends upon temperature and pressure. The **standard equilibrium constant,** K'_{eq}, is the equilibrium constant at 25°C, 1 atm pressure, and pH 7.0. If the concentration of any one of the components of the reaction is altered, then the concentration of at least one other component must also change to satisfy the constancy of the K'_{eq} value. For reaction 11–1, if the K'_{eq} value is greater than 1.0, then the equilibrium of the reaction lies to the right as written; and if the ratio is less than 1.0, the equilibrium lies to the left as written.

In a sequence of reactions (e.g., see Fig. 11–1), each

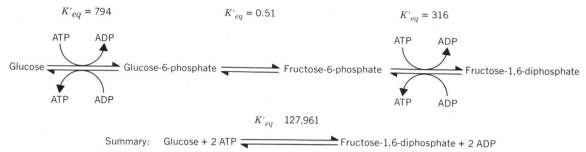

Figure 11-1
Example of a linear metabolic pathway (see text for discussion). Note that the K'_{eq} value for the summary reaction is the product of the individual K'_{eq} values of each reaction.

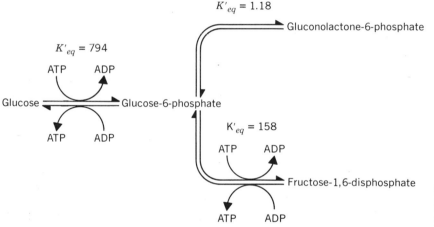

Figure 11-2
Example of branching in a metabolic pathway (see text for discussion).

reaction may affect the other reactions. For the reaction sequence of Figure 11–1, the equilibrium of the intermediate reaction, in which glucose-6-phosphate is converted to fructose-6-phosphate, lies to the left, whereas in the other two reactions, the equilibrium lies well to the right. If the entire sequence of reactions is allowed to come to equilibrium, then the concentrations of all of the molecules present will satisfy each of the three K'_{eq} values. If additional glucose and ATP are added, the entire sequence of reactions will proceed to the right even though the K'_{eq} value of the middle reactions is less than 1.0. This occurs because the production of glucose-6-phosphate by the first reaction temporarily disturbs the equilibrium that existed. The middle reaction thus proceeds to the right to a greater extent than it proceeds to the left, resulting in a net increase in fructose-6-phosphate. Some of this fructose-6-phosphate is then converted to fructose-1,6-diphosphate by the third

reaction. The rate at which the net change in concentrations of reactants and products occurs and the direction in which the reaction proceeds is the result of *mass action*.

Which of the two alternative metabolic paths is taken at a *branch point* (Fig. 11–2) is controlled to some degree by the law of mass action. Although the equilibrium lies to the right for each of the reactions of Figure 11–2, the equilibrium is "further" to the right in the reaction that produces fructose-1,6-diphosphate, and more product would be formed by reactions along that branch than the branch that forms gluconolactone-6-phosphate. However, during glucose metabolism as it takes place in situ, both fructose-1,6-diphosphate and gluconolactone-6-phosphate are intermediates in a more extensive sequence of reactions than is shown in Fig. 11–2 (see Figs. 10–2 and 10–3), and the K'_{eq} values of subsequent reactions may affect the direction of the reactions differently in situ.

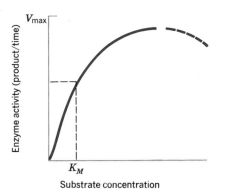

Figure 11-3
Effect of substrate concentration on enzyme activity.

Table 11–1
Maximum Activity of Glycolytic Enzymes in Mouse Brain Tissue under Anoxic Conditions

Enzyme (in Order of Glycolysis)	V_{max} (mmoles/kg/min)
Hexokinase	15.2
Phosphoglucoisomerase	154.0
Phosphofructokinase	26.7
Aldolase	7.6
Glyceraldehyde-3-phosphate dehydrogenase	96.0
Phosphoglycerate kinase	750.0
Phosphoglycerate mutase	145.0
Enolase	36.0
Pyruvate kinase	95.0
Lactic dehydrogenase	129.0

Source: Copyright © American Society of Biological Chemists, Inc., *J. Biol. Chem. 239,* 31 (1964).

Regulation by Enzyme Activity

Regulation of metabolism is most commonly achieved at the cellular level by altering the *activities* of enzymes or by altering the *number* of enzyme molecules present. Regulation through altered enzyme activity can be brought about by (1) *changes in substrate concentration,* (2) *allosteric effectors,* and (3) *irreversible and reversible covalent bond modification in the enzyme molecules.*

Substrate Concentration Effects

As described in Chapter 8, enzyme activity as measured by the rate of product formation increases as a hyperbolic function as the substrate concentration is raised (Fig. 11–3) until a maximum reaction velocity (V_{max}) is achieved. Excessively high substrate concentrations may actually reduce enzyme activity. Each enzyme, subject to experimental conditions, has a characteristic V_{max}, as exemplified in Table 11–1 for the glycolytic enzymes in brain tissue. If the substrates present in brain tissue were to occur in excess, one would expect that the reaction catalyzed by aldolase (which has the lowest V_{max}) would be the rate-limiting reaction in the sequence. However, in vivo studies indicate that enzymes are rarely saturated by substrate. Under typical conditions, *hexokinase, phosphoglucoisomerase,* and *aldolase* operate in the presence of substrate concentrations equal to or somewhat greater than the K_M value (Michaelis-Menten constant). Small changes in substrate concentration

do not significantly alter the rate of metabolism in this part of the glycolytic pathway. Of greater regulatory importance for these reactions is the amount of enzyme present. The last six glycolytic enzymes in brain tissue operate at substrate levels that are significantly below the K_M value. Therefore, small changes in substrate concentration would be more likely to alter the level of activity of these enzymes. Under anoxic conditions the rate of production of lactate is not appreciably increased by supplementing the tissue with glucose. However, supplements of compounds such as glycerol, which enters the glycolytic pathway below the level of aldolase, causes an increase in lactate production.

The *affinity* between an enzyme and its substrate affects the rate of enzyme-catalyzed reactions. For example, at branch points of metabolic pathways, two enzymes compete for the same substrate. The reaction catalyzed by the enzyme with the lower K_M value is favored at low substrate concentrations, while the reaction catalyzed by the enzyme with the higher K_M value is favored at high substrate concentrations (Fig. 11–4). Thus, when present in low concentrations, a substrate may be channeled primarily into one pathway, while the major direction of metabolism may shift to other pathways at higher substrate concentrations.

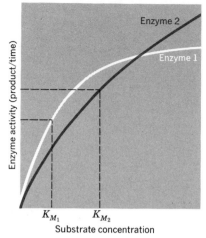

Figure 11-4
Relationship between substrate concentration and enzyme activity for two enzymes that act on the same substrate but have different K_M values.

Allosteric Effectors

The regulatory effects exercised by substrates through mass action generally influence *all* reactions in metabolic pathways; therefore, these control mechanisms are not very specific. There are, however, a number of mechanisms by which specific reactions in a pathway can be regulated. In one such mechanism enzymes catalyzing specific reactions of the pathway are influenced by the type and amount of certain **regulatory metabolites** that are present. These enzymes are called **allosteric enzymes** because their catalytic activity is modified when specific metabolites, called **allosteric effectors,** are bound to a site on the enzyme *other than the active site* (see Chapter 8). If the metabolite has an inhibitory effect, it is called a *negative* effector; if its effects are stimulatory, then it is called a *positive* effector. Some of the more common allosteric effectors are listed in Table 11–2.

Feedback Inhibition. Inhibitory allosteric effects are caused when the product of a reaction sequence binds to an allosteric site of an enzyme that catalyzes a reaction at the beginning of a linear pathway or at the branch point of a branched pathway (Fig. 11–5a, b). Binding of the effector to the enzyme causes a change in the configuration of the enzyme (called an *allosteric transition*) and this reduces

Table 11–2
Common Regulatory Metabolites

Glucose-6-phosphate
Fructose-1,6-diphosphate
1,3-diphosphoglycerate
Citrate
Acetyl CoA
AMP, ADP, ATP
Cyclic AMP, cyclic GMP
GTP, UTP, etc.
NAD, $NADH_2$
Fatty acids
Amino acids

the catalytic activity of the enzyme. Thus, as the end product of a reaction sequence begins to accumulate, more and more of the enzyme catalyzing the early stage of the reaction pathway leading to that product is rendered inactive. In this way, the rate of formation of the end product is dramatically reduced. Where branches occur in a metabolic pathway, the end product of one branch usually affects the enzyme at the branch point leading to that product (Fig. 11–5b). This type of inhibition is called **feedback inhibition** or **end-product inhibition.**

Feedback Stimulation. Positive effectors generally act in one of two ways—by **feedback stimulation** or by **feedforward stimulation.** In feedback stimulation, the effector may be the end product of one branch of a pathway that combines with and stimulates the enzyme catalyzing the reaction at the branch point leading into the *other* pathway (Fig. 11–5c). In other words, the reactions leading to the synthesis of the positive effector are attenuated, while the reactions that lead to the synthesis of an alternative product are enhanced.

Feedforward Stimulation. In feedforward stimulation, the initial substrate of a pathway (or an early intermediate in the pathway) stimulates the activity of an enzyme further along the metabolic pathway.

When an allosteric enzyme is affected by one effector it is said to be *monovalent.* However, some allosteric enzymes may bind two or more effectors and are termed *polyvalent.* A single allosteric enzyme may possess sites that combine with either negative or positive effectors, and these effec-

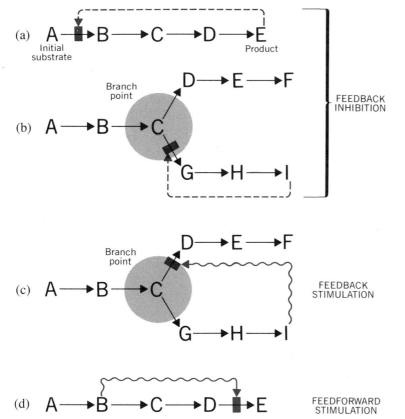

Figure 11-5
Feedback inhibition, feedback stimulation, and feedforward stimulation in linear and branched metabolic pathways. The small colored rectangles represent the allosteric enzymes; broken colored arrows indicate inhibition, while wavy colored arrows indicate stimulation.

tors may compete with each other for the same allosteric site or be bound at different sites on the enzyme. The kinetics of allosteric enzyme activity are discussed in Chapter 8.

Control of Metabolic Processes by Allosteric Enzymes

Alternative Pathways for Glycogen Synthesis and Degradation

An excellent example of allosteric enzyme regulation of metabolic processes is provided by the interrelationship in animals between the metabolic pathways that result in (1) the *synthesis* of glycogen from glucose, and (2) the *oxidation* of glucose to CO_2 and water. Nearly all of the energy-consuming processes in the body proceed at the expense of ATP, and much of this ATP is derived through the oxidation of glucose. During periods of elevated activity (e.g., exercise), glycogen is broken down to yield glucose, which then enters the metabolic pathway converting it to CO_2 and water, with consequent generation of ATP. In contrast, during periods of rest or low energy demand, absorbed glucose is converted to glycogen. Three of the enzymes involved in glucose metabolism are allosteric; these are *phosphofructokinase* (an enzyme required in the series of reactions that convert *glucose-6-phosphate* to CO_2

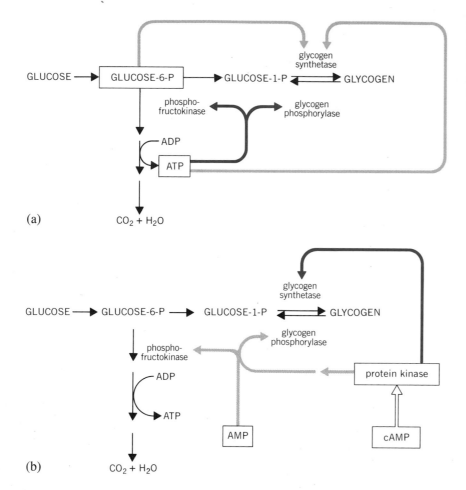

Figure 11-6
Allosteric regulation of glycogen synthesis (*glycogenesis*) and glycogen degradation (*glycogenolysis*). (*a*) Glycogenesis; (*b*) glycogenolysis. Colored arrows show negative action of effector; gray arrows show positive action of effector.

and water), *glycogen synthetase* (involved in the incorporation of *glucose-1-phosphate* into glycogen), and *glycogen phosphorylase* (which removes glucose as glucose-1-phosphate from glycogen during glycogen catabolism).

When ATP levels are high and no major consumption of energy is taking place in the body, glucose is diverted into glycogen (i.e., "glycogenesis" predominates). This is achieved because ATP acts as a negative effector of phosphofructokinase and glycogen phosphorylase and as a positive effector, along with glucose-6-phosphate, of glycogen synthetase (Fig. 11–6*a*).

When the ATP level falls (e.g., during exercise) and there is an increased demand for ATP, glycogen synthesis is halted as absorbed glucose is directly consumed in the production of ATP and additional glucose is made available through the catabolism of glycogen (i.e., "glycogenolysis"). This pathway is activated by the positive effects

on phosphofructokinase and glycogen phosphorylase of the ATP precursor, AMP. The hormone *epinephrine* (adrenalin), secreted into the bloodstream during periods of great activity, also has an effect on these metabolic pathways in muscle and in liver.

When epinephrine in the bloodstream reaches the muscles, it binds to the surface of the muscle cells and promotes the synthesis of cyclic AMP (cAMP) by the enzyme *adenylcyclase*. The cAMP then allosterically activates a second enzyme (*protein kinase*), which ultimately activates glycogen phosphorylase but inactivates glycogen synthetase (Fig. 11–6*b*).

The pathways just described illustrate the mechanisms for turning allosteric enzymes on and off. In the absence of such mechanism, both pathways might simultaneously be active so that their effects cancel one another—a most unproductive state! Allosterism thus provides a basis for

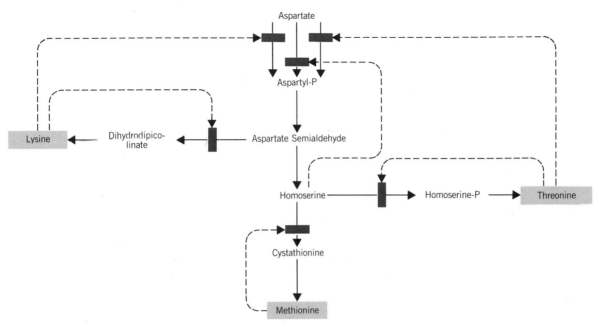

Figure 11-7
Allosteric feedback control of amino acid synthesis in *E. coli* cells. Small colored rectangles are allosteric enzymes. Dashed arrows indicate feedback inhibition.

regulating the levels of activity of related metabolic pathways.

The Regulation of Amino Acid Synthesis

Escherichia coli provides a clear example of control of divergent metabolic pathways by feedback inhibition. An outline of the metabolic pathways for the synthesis of three amino acids is shown in Figure 11–7. Lysine, methionine, and threonine are each synthesized from aspartate, and each may be utilized in protein synthesis. Without metabolic controls, the consumption or utilization of any one of these amino acids would stimulate the pathways and cause unneeded synthesis of the unused amino acids as well as the one utilized. Such an unregulated system would consume vital resources and energy; both factors could have survival implications to the organism and evolutionary consequences to the species. However, in *E. coli,* the allosteric regulatory mechanisms are most effective. The accumulation of each amino acid produces a feedback inhibition of the first enzyme in the specific branch of the pathway leading to the synthesis of that amino acid. In Figure 11–7, this negative effect is shown by dashed lines. Moreover, an additional level of regulation is achieved through effects on the enzyme *aspartokinase,* which catalyzes the phosphorylation of aspartate. This enzyme exists in three forms or *isozymes,* as described later in this chapter and in Chapter 8. The presence of the three isozymes is symbolized in Figure 11–7 by using three separate arrows to show the conversion of aspartate to aspartylphosphate. One of the isozymes is specifically and completely inhibited by threonine; the second (which is present only in small amounts) is specifically inhibited by homoserine; and the third isozyme is specifically inhibited by lysine. In addition, the latter isozyme is *repressed* by lysine. (*Repression* is a regulatory mechanism that reduces the *number* of enzyme molecules in the cell and is discussed later in this chapter.)

Covalent Bond Modification of Enzyme Activity

A number of enzymes are produced in an *inactive* or *zymogen* form and must be modified through cleavage of certain covalent bonds in order to become active. The modification may be irreversible, as is the case with hydrolytic modification of zymogens such as *pepsinogen* (to form *pepsin*)

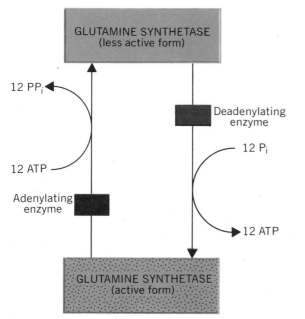

Figure 11-8

Activation and deactivation of the enzyme glutamine synthetase by adenylation and deadenylation. The less active form of the enzyme has 12 molecules of AMP covalently bonded to 12 tyrosine residues; their removal by the deadenylating enzyme activates the enzyme.

and *trypsinogen* (to form *trypsin*), which were discussed in Chapter 8. Other enzymes may be reversibly covalently activated and deactivated. The reversible activation and deactivation of *glutamine synthetase* is a well-studied example (Fig. 11–8). This enzyme catalyzes the conversion of glutamate to glutamine,

$$\text{glutamate} + NH_3 \longrightarrow \text{glutamine}$$
$$\text{ATP} \qquad \text{ADP} + \text{phosphate}$$

and the transfer of a glutamyl group to hydroxylamine,

$$\text{glutamine} + NH_2OH \longrightarrow \gamma\text{-glutamyl-NHOH} + NH_3$$

Glutamine synthetase is inactivated by a reversible adenylation in which 12 molecules of AMP become covalently bonded to the hydroxyl groups of 12 tyrosine residues of the enzyme. The inactivation is catalyzed by *adenylating enzyme,* which uses ATP as the source of adenylate and releases pyrophosphate. Reactivation is brought about

through the action of *deadenylating enzyme,* which, in the presence of inorganic phosphate, cleaves the bonds linking the AMP molecules to the inactive form of the enzyme, thereby producing ADP (Fig. 11–8).

Regulation by Number of Enzyme Molecules

Isozymes

In a number of instances, an enzyme that catalyzes a specific reaction may exist in multiple forms in a cell or organism. These multiple forms or **isozymes** are encoded by different genes, have different amino acid complements, and can therefore be separated from one another by electrophoresis and chromotographic procedures. Although they catalyze the same reactions, the isozymes usually have different K_M and V_{max} values and catalyze the reaction more or less effectively depending upon the reaction conditions. One of the most exhaustively studied of the known isozymes is *lactic dehydrogenase,* which catalyzes one of the terminal reactions in glycolysis; that is,

$$\text{pyruvate} + NADH + H^+ \rightleftharpoons \text{lactate} + NAD^+$$

In rats and in a number of other vertebrates, this enzyme is present in five forms. Each of the five forms has a molecular weight of about 134,000 daltons, consists of four polypeptide chains of about 33,500 daltons each, and catalyzes the same reaction. Each of the four polypeptides may be of two types, usually referred to as *M* and *H.* In rat skeletal muscle tissue, the predominant form of the isozyme contains four polypeptides of the M type; in contrast, in rat heart muscle, the predominant form contains four polypeptides of the H type. The other isozyme forms are made up of three H and one M, two H and two M, and one H and three M polypeptides (usually abbreviated H_3M, H_2M_2, and HM_3). Some of the latter forms predominate in other tissues, although each tissue has some of each isozyme (Fig. 11–9).

The M_4 isozyme is prevalent in embryonic tissue and in skeletal muscle tissue. It has a low K_M and a high V_{max}. It is well adapted to these tissues, which are frequently deprived of oxygen and must depend upon the energy made available during the breakdown of glucose to lactate. Similarly, the H_4 isozyme is well suited to the metabolism of

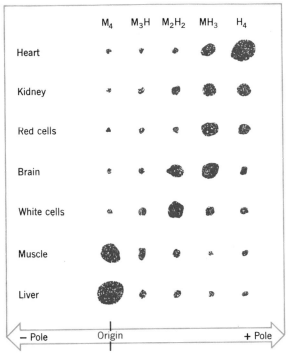

	M_4	M_3H	M_2H_2	MH_3	H_4
Heart					
Kidney					
Red cells					
Brain					
White cells					
Muscle					
Liver					

− Pole   Origin   + Pole

Figure 11-9

***Lactic dehydrogenase* isozymes. The spots show the relative amounts of each isozyme as they appear when extracts of different tissues are separated by two-dimensional electrophoresis (see Chapter 13).**

heart muscle. This isozyme has a high K_M and a low V_{max} and is inhibited by excess pyruvate. Heart muscle is primarily aerobic, converting pyruvate to CO_2 and H_2O rather than to lactate. The lactic dehydrogenase enzyme is most active during emergency conditions when the oxygen supply is low.

Regulation of Enzyme Synthesis (Controlled Gene Expression)

Another method used by cells to regulate metabolism is by altering the types and numbers of enzymes that are manufactured through the integrated actions of the cell's genetic and protein synthesizing apparatuses. This mechanism is also referred to as **controlled gene expression.** Although the topics of gene expression and protein synthesis are treated in detail in Chapters 21 and 22, a brief description here of the basic mechanism will set the foundation for

understanding the metabolic control mechanisms that are described below.

The primary structures of all cellular enzymes are encoded in the cell's DNA or genes. To produce these enzymes, the DNA is first *transcribed* into messenger RNA (mRNA), and the mRNA is subsequently *translated* by the cell's ribosomes into polypeptides that take on the enzymatic properties. This two-step process (transcription followed by translation) provides two additional levels at which cellular metabolism can be regulated. There can be **transcriptional control** in which there is *selective* transcription of specific genes, and there can be **translational control** operating at the level of mRNA-ribosome interaction. Most of our current understanding of metabolic regulation exercised at these levels stems from research conducted with procaryotic organisms, for the regulation of gene expression in eucaryotic cells is much more complex.

Constitutive and Inducible Enzymes

Studies using procaryotes indicate that enzymes fall into two categories with respect to their occurrence and numbers in cells. Those that appear to always be present and that occur in relatively constant concentrations are called **constitutive enzymes.** For example, the enzymes of glycolysis are usually constitutive. These enzymes are the products of genes that are continuously expressed. The other type of enzyme, called an **inducible enzyme,** is found lacking in cells or is present only in small amounts. However, upon introduction of a specific metabolite, usually a substrate, the concentration of the enzyme quickly increases. The metabolite that initiates the appearance of the enzyme is called an **inducer.** Inducible enzymes are the products of genes that are selectively expressed; these genes are referred to as inducible genes.

One of the first inducible enzymes to be intensively studied was β-*galactosidase*. Wild-type *E. coli* cells metabolize glucose and will metabolize only the glucose even if lactose, another sugar, is also present. The enzymes of glucose metabolism are constitutive and are always present in the cell, while the enzyme needed to initiate lactose metabolism, β-galactosidase, is present in only minor amounts—according to one study, no more than five copies per cell. If wild-type *E. coli* cells are placed in a growth medium containing only lactose as the carbon source, they are at first unable to utilize this disaccharide (Fig. 11–10).

However, within a few minutes the cells respond to the presence of lactose by synthesizing β-galactosidase, which hydrolyzes the lactose to glucose and galactose; these sugars are then metabolized by glycolysis. The lactose acted as an inducer of the β-galactosidase, and in cells grown on lactose thousands of copies of this enzyme are present.

A number of β-galactosides besides lactose are able to act as inducers; these include methyl β-galactoside and allolactose. Actually, the presence of these inducers initiates the synthesis of not one but three *enzymes* in *E. coli:* (1) β-*galactoside permease,* an enzyme formed in the plasma membrane that promotes the transfer of galactosides into the cell; (2) β-*thiogalactoside transacetylase,* the specific action of which is not understood, and (3) β-*galactosidase,* the key enzyme for initiating lactose metabolism. When, as in this case, induction can be brought about by a single agent and results in the appearance of several enzymes, the process is known as **coordinate induction.** The appearance of inducible enzymes is an example of metabolic regulation through transcriptional control.

Repressible Enzymes

The presence of a specific substance may inhibit the continued production of a specific enzyme or a sequence of enzymes in a metabolic pathway; this process is called *enzyme repression* (or in the case of the repression of a sequence of enzymes, *coordinate repression*). Like enzyme induction, enzyme repression represents regulation at the transcriptional level. *E. coli* cells growing in a medium in which NH_4^+ is the only nitrogen source contain all of the enzyme systems necessary to synthesize amino acids from organic acids. However, if one of the amino acids is added exogenously, the synthesis of certain enzymes in the metabolic pathway leading to that amino acid will be repressed, and with continued growth of the culture these enzymes soon become diluted out.

In another bacterium, *Salmonella typhimurium,* a family of genes encode nine enzymes that catalyze the reactions of a pathway leading to the synthesis of the amino acid histidine. Histidine is an important constituent of proteins, and continuous production of protein in this bacterium depends upon the availability of histidine. The histidine must either be synthesized by the pathway that involves these nine enzymes or must be supplied exogenously. When histidine is added to a culture of *S. typhimurium,* the production of the histidine biosynthetic enzymes is halted (Fig.

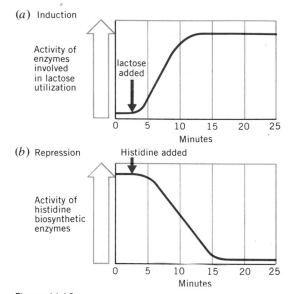

Figure 11-10
(a) *Induction* **of the synthesis of enzymes required for utilization of lactose by** *E. coli;* **(b)** *Repression* **of the synthesis of enzymes required for the synthesis of histidine in** *Salmonella typhimurium* **(see text for discussion). (From Gardner and Snustad,** *Principles of Genetics,* **6th ed., p. 400. Copyright © 1981 by John Wiley and Sons, Inc.)**

11–10). The genes for the histidine enzymes are said to have been *repressed.* If the bacteria are then transferred to a medium lacking histidine, the genes are *derepressed* and the histidine enzymes soon reappear. Enzyme repression is a metabolic control mechanism acting at the level of transcription and should not be confused with feedback inhibition (discussed earlier in the chapter).

The Operon

In 1961, F. Jacob and J. Monod proposed the **operon model** to explain the genetic basis of enzyme induction and repression in procaryotes. A few years later (1965), these two investigators were awarded the Nobel Prize for their incisive work. Although Jacob and Monod's original operon model applied specifically to the regulation of the genes for lactose metabolism in *E. coli,* additional findings by these two noted scientists as well as the work of many others have revealed the mechanisms by which other operons function.

The components of an operon are illustrated in Figure 11–11*a.* The operon consists of a series of **structural genes** (denoted by SG1, SG2, SG3), a segment of DNA called an **operator** adjacent to the structural genes, and next to the operator, an additional DNA segment called a **promoter.** Elsewhere on the chromosome, there is a segment of DNA called a **regulator gene** and next to this is a DNA segment that functions as a promoter for the regulator gene.

In the absence of an inducer (e.g., lactose in an *E. coli* culture) *RNA polymerase* binds to the promoter of the regulator gene and progresses along the DNA transcribing the regulator gene. The RNA produced is then translated into a **repressor protein,** which binds to the operator of the operon. Binding of the repressor to the operator (see the top half of Fig. 11–11*b*) prevents RNA polymerase from attaching to this segment of the DNA. As a result, the structural genes cannot be transcribed and no mRNA or enzyme is formed. It is now known that the repressor prevents the binding of the RNA polymerase by steric hindrance.

When Jacob and Monod conducted their studies in the 1960s, the existence of the promoter and the mechanism of repressor function were not known. Today it is recognized that operons also regulate enzyme induction (Fig. 11–11*b*) and repression (Fig. 11–11*c*). In the case of the operon model for the induction of enzymes (Fig. 11–11*b*), the inducer (e.g., lactose) combines with the repressor to form a complex that cannot bind to the operator. Therefore, RNA polymerase can associate with the *operon promoter* and proceeds along the DNA transcribing it into mRNA, which in turn is translated into polypeptides and enzymes. In the operon model for enzyme repression (Fig. 11–11*c*), the repressor cannot bind to the operator unless it first complexes with a **corepressor.** In the example presented earlier in which histidine-synthetic enzymes in *S. typhimurium* are repressed by histidine that is supplied exogenously, histidine would serve the role of the corepressor. In the absence of the corepressor, RNA polymerase associates with the operon promoter, and transcription and translation occur. The mRNA transcripts of **multigenic operons** (i.e., operons containing two or more structural genes) contain the code for several polypeptides and are said to be **polycistronic.** Therefore, there is the **coordinate expression** of all of these genes of the operon.

The Lac Operon

The progressive unraveling of the molecular organization and function of the *lac operon* is a classic study in physi-

ology and genetics. Jacob and Monod began their studies of this operon in *E. coli* in the early 1960s; since then, they and many others have continued the study, so that today it is one of the best understood regulatory systems (Fig. 11–12).

In *E. coli,* the lac operon is inducible, consisting of three structural genes of known lengths, designated *z, y,* and *a,* which respectively code for β-galactosidase, β-galactoside permease, and β-galactoside transacetylase. The location and nucleotide sequences are also known for the promoter (*P*) and operator (*O*) segments of the operon. The regulator gene consists of two contiguous portions of known lengths—an *i*-gene and the promoter of the *i*-gene *(P(i))* (see Fig. 11–12). The *i*-gene codes for a polypeptide consisting of 360 amino acids; four of these polypeptides combine to form a biologically active tetramer that functions as a repressor of the operator gene. The repressor binds to the operator and, in so doing, prevents RNA polymerase from associating with the promoter of the operon.

The normal inducer of the lac operon is *allolactase,* which is produced from lactose by β-galactosidase (Fig. 11–13). A few copies of this enzyme are present in *E. coli* cells even in the uninduced state. When allolactose combines with the repressor, a steric change occurs that causes the repressor to be released from the operator. As a result, transcription of the structural genes by RNA polymerase begins and the three enzymes quickly appear in the cells.

The promoter of the lac operon contains two active components that regulate transcription. One is the RNA polymerase binding site, and the second is the **catabolite activator protein** (*CAP*) binding site. This CAP site functions to prevent transcription of the lac operon when sufficient glucose is present. Today, a number of operons are known in addition to the lac operon. Some of these are listed in Table 11–3.

Catabolic Repression

Catabolic repression is a specific type of repression of enzyme production in which a metabolite such as glucose acts to repress the formation of enzymes that would allow the catabolism of *other,* related metabolites. For example, when *E. coli* cells are cultured in a medium that is rich in glucose, the glucose represses the formation of β-galactosidase even if lactose (an inducer of this enzyme) is added to the medium. Glucose is even known to repress the production of enzymes that formerly were thought to be

Figure 11-11

Operon model for the regulation of gene expression in procaryotic cells. (*a*) Essential genetic components; the operon consists of one or more *structural genes* (SG1, SG2, SG3) and the adjoining *promoter* and *operator* sequences. (*b*) Mode of regulation of gene expression in an *inducible* operon. *PR*, promoter for the regulator gene; *R*, regulator gene; *PO*, promoter for the operon; *O*, operator, (*c*) Mode of regulation of gene expression for a *repressible operon* (see text for explanation). (From Gardner and Snustad, *Principles of Genetics*, 6th ed., p. 402-403. Copyright © 1981 by John Wiley and Sons, Inc.)

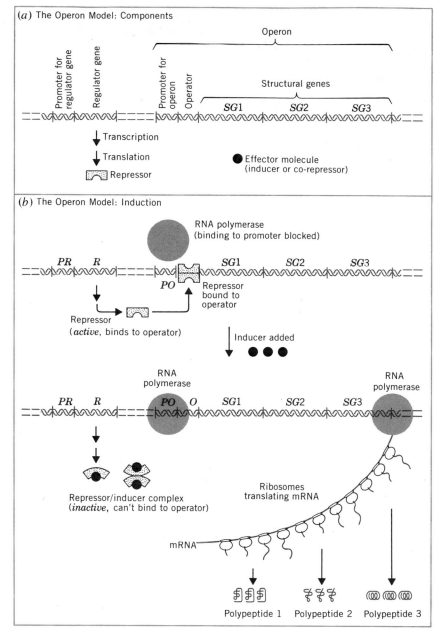

Figure 11-11 continued

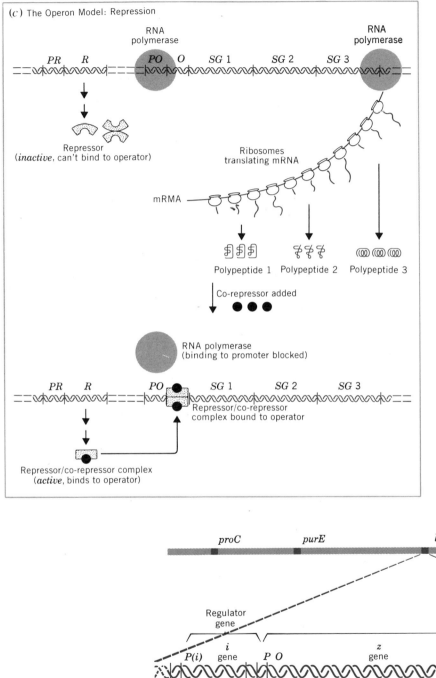

(c) The Operon Model: Repression

RNA polymerase

Repressor
(*inactive*, can't bind to operator)

Ribosomes
translating mRNA

mRMA

Polypeptide 1   Polypeptide 2   Polypeptide 3

Co-repressor added

RNA polymerase
(binding to promoter blocked)

Repressor/co-repressor
complex bound to operator

Repressor/co-repressor complex
(*active*, binds to operator)

Figure 11-12

The *lac* operon is but one of several operons known to exist in the *E. coli* chromosome (upper part of figure). The *lac* operon is inducible and consists of three structural genes (z, y, and a) and the promoter (P) and operator (O) sequences adjoining the z-gene. The regulator gene (i) is contiguous with the operon in this particular case. The regulator gene has its own promoter (P(i)). The numbers below each of the genes are their approximate lengths (in nucleotide pairs). (From Gardner and Snustad, *Principles of Genetics*, 6th ed., p. 404. Copyright © 1981 by John Wiley and Sons, Inc.)

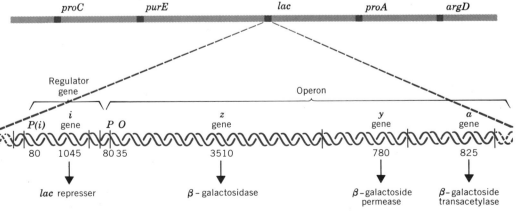

Regulator
gene

Operon

lac represser   β – galactosidase   β –galactoside
permease   β –galactoside
transacetylase

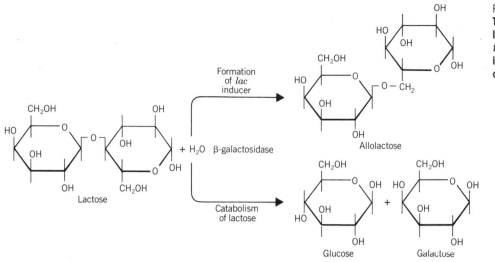

Figure 11-13
The two reactions that are catalyzed by the enzyme β-*galactosidase*. Allolactose produced in the upper reaction is an inducer of the *lac* operon.

Formation of *lac* inducer

Catabolism of lactose

Lactose + H₂O β-galactosidase

Allolactose

Glucose + Galactose

Table 11–3
Some More Common Operons in Procaryotes

Operon	Number of Enzymes	Function
lac	3	Hydrolysis of lactose and transport of galactose
his	10	Histidine synthesis
gal	3	Galactose ⟶ UDP-glucose
leu	4	Leucine synthesis
trp	4	Tryptophan synthesis
ara	4	Metabolism and transport of arabinose
pyr	5	Aspartate ⟶ UMP

constitutive. A common phenomenon in bacteria is the suppression of aerobic respiration and electron transport by high glucose concentrations, *even in the presence of ample oxygen*. Under these conditions, the cells utilize the glycolytic and fermentative pathways.

The manner in which the catabolite brings about the effect is only partially understood. In the case of catabolite repression by glucose in bacteria, it appears that glucose affects the amount of cyclic AMP (cAMP) present in the cells. When the concentration of glucose is high, the concentration of cAMP is low; and low levels of glucose are accompanied by high concentrations of cAMP. It is possible that glucose affects the synthesis of cAMP, which is formed from ATP by *adenylcyclase* (Fig. 11–14). cAMP is necessary for CAP to bind to the promoter site. cAMP binds to the CAP, and once the cAMP-CAP complex binds to the promoter (Fig. 11–15), then RNA polymerase can attach to the promoter and begin transcription.

For example, in the case of the lac operon, when the glucose level is high (even if lactose is also present), the cAMP is low, and therefore the cAMP-CAP complex is not available to bind to the promoter and allow transcription to start. However, in the absence of glucose and in the presence of lactose (which forms a complex with the repressor), cAMP is plentiful and is available to combine with CAP so that transcription proceeds. The rate of lac operon transcription in the absence of glucose is 50 times as great as in the presence of glucose.

Translational Control

Regulatory mechanisms usually function at the beginning of a sequence of events rather than at the end. For example, allosteric feedback control is achieved by regulating the enzyme that catalyzes the first reaction in a sequence of reactions. Likewise, control of the number of enzyme molecules produced in a cell is most commonly brought about at the transcriptional level rather than at the level of translation. Regulatory mechanisms conserve cell energy and prevent the buildup of materials that will not be used.

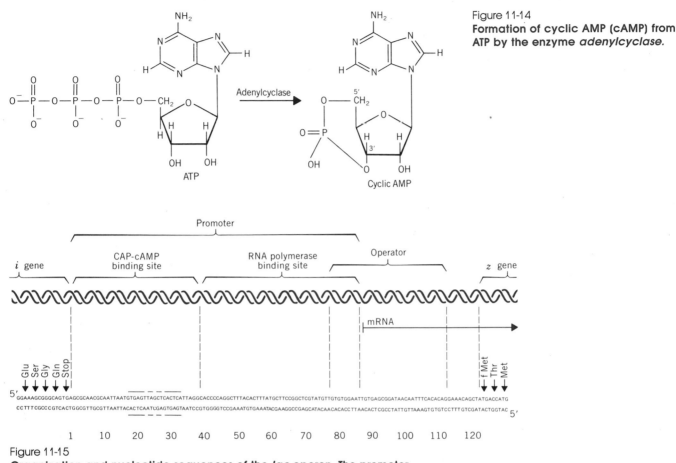

Figure 11-14
Formation of cyclic AMP (cAMP) from ATP by the enzyme *adenylcyclase*.

Figure 11-15
Organization and nucleotide sequences of the *lac* operon. The promoter consists of two components; the site that binds the cAMP-CAP complex and the RNA polymerase binding site. The adjoining segments of the repressor (*i*) and β-*galactosidase* (*z*) structural genes are also shown. The horizontal arrow (labeled mRNA) shows the position at which the transcription of the operon begins (i.e., what will be the 5′ end of the mRNA). The numbers at the bottom of the figure give the distances (in nucleotide pairs) from the end of the *i*-gene. (From Gardner and Snustad, *Principles of Genetics*, 6th ed., p. 408. Copyright © 1981 by John Wiley and Sons, Inc.)

Therefore, *translational control* mechanisms would be expected to be rare, and at the present time there is little evidence for very much control being exercised at this level. Indirect evidence suggests that translational control of enzyme synthesis may occur for enzymes that are part of the same operon. For example, since the three structural genes in the lac operon are controlled by the same operator, one would expect that equal quantities of the three enzymes would be produced. However, induced cells contain more copies of one enzyme than another. This could be explained if the ribosome detaches from mRNA before completing the translation of the entire polycistronic message.

The economic impact of the use of control mechanisms is reflected also in the synthesis of ribosomal RNA (rRNA). When *E. coli* cells are cultured in a medium that is deficient in amino acids, they are unable to synthesize proteins. As a consequence, the cells also stop producing rRNA—a mechanism called **stringent control.** Mutants of *E. coli* do exist, however; they continue to synthesize rRNA under these conditions and are called **relaxed mutants.** Paper chromatographic analysis (see Chapter 13) of nucleotide extracts of the stringent cells and relaxed mutants reveals that there are two spots in the chromatogram of stringent cells that are absent in the chromatogram of the relaxed mutants. Initially these spots were called "magic spot I" and "magic spot II," but they have now been identified as *guanosine-5'-diphosphate-2'-(or 3')-diphosphate (ppGpp)* and *guanosine-5'-triphosphate-2'-(or 3')-diphosphate (pppGpp),* respectively. The studies of Cashel and Gallant and their colleagues indicate that these nucleotides trigger the events that turn off rRNA synthesis when the amino acids needed by the cells for protein synthesis are lacking.

Regulation of Enzyme Production in Eucaryotic Cells

As might be expected, the regulation of enzyme production in eucaryotic cells is much more complex than in procaryotic cells. Eucaryotic cells contain a number of chromosomes instead of the single chromosome found in procaryotes. Moreover, eucaryotic cell chromosomes are diploid (see Chapter 20), and at times polyploid. The DNA of the chromosomes is usually supercoiled and highly folded, and the chromosomes themselves are physically separated from the cytoplasmic ribosomes by the nuclear envelope. Undoubtedly, these factors make control of enzyme synthesis more complex, even though transcription and translation involve what is basically the same mechanism as in procaryotes. However, structural genes for a group of enzymes that might constitute an inducible component such as an operon are not found adjacent to one another on a chromosome and may, in fact, be distributed among different chromosomes.

Enzyme induction does occur in primitive eucaryotic organisms such as yeast and *Neurospora,* but operons are rare in higher eucaryotes. In all but a few cases, the mRNAs of higher eucaryotes contain the coding sequence of only one structural gene (i.e., they are **monogenic**). The induction process in the fungi is slower, and the change in concentration of enzymes is not as great as in the procaryotes. In yeast, the induction of β-galactosidase takes much longer than in *E. coli;* also, the increase in enzyme activity is 10-fold and not 1000-fold. The enzyme *tryptophan-2,3-oxygenase* found in the liver cells of vertebrates is inducible, but induction requires many hours.

Enzyme Induction by Hormones

Enzyme induction in procaryotic cells is usually triggered by a potential metabolite (such as the induction of β-galactosidase by lactose). As noted above, this form of enzyme induction occurs in some eucaryotic cells as well; however, in higher animals there also is a highly developed control system in which *hormones* act as *messengers* that coordinate the biosynthetic activities of cells. This coordination may take the form of activating (or deactivating) enzymes that *already* exist in the cell or the effect may be on gene expression, resulting in the production of additional enzymes. The term *messenger* is appropriate, since the hormones are produced and secreted into the bloodstream by one tissue of the body and travel via the blood to other tissues upon which they have an effect.

From a chemical point of view, hormones are quite diverse. Some hormones (e.g., *thyroxin* and *epinephrine*) are small molecules derived from amino acids; others are proteins (e.g., *insulin* and *erythropoietin*) or steroids (e.g., *progesterone* and *cortisol*). Hormones regulate the production of enzymes through either of two mechanisms depending upon the size and chemical nature of the hormone molecule. The steroid hormones enter the **target** cell by passing through the plasma membrane and into the cytoplasm. In the cytoplasm, the hormone combines with a receptor molecule and the complex then migrates to the cell nucleus, where it has a direct impact on the expression of certain genes (Fig. 11–16). In vitro studies using chick oviduct cells have shown that the hormone *estrogen* enters the cytoplasm to form a hormone-receptor complex that migrates to the cell nucleus and induces the increased rate of transcription of the genes for *lysozyme* and *ovalbumin.*

Other hormones (e.g., insulin, epinephrine, and thyroxin) cannot pass through the plasma membrane of the target cell. Instead, the hormone binds to a specific receptor protein on the membrane's surface. Following this, either one or both of the following occur: (1) the hormone, together with the membrane receptor, is *internalized;* and/or (2) binding of the hormone serves to activate the enzyme *adenylate cyclase* in the membrane, thereby leading to the increased production of cellular cAMP. cAMP (sometimes cGMP) acts as a *second messenger* (the hormone being the first), coordinating the cellular activities. Internalization takes the form of an infolding of the membrane to form a vesicle containing the hormone and the receptor. The vesicle subsequently fuses with a Golgi body or lysosome, the

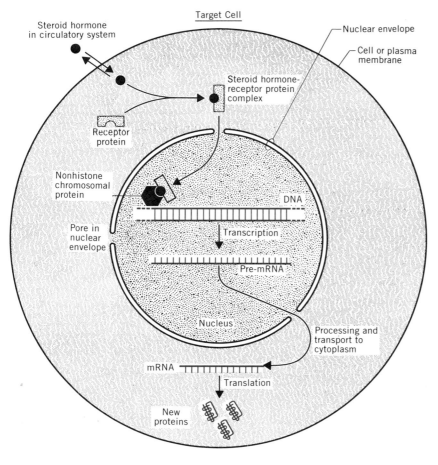

Target Cell

Steroid hormone in circulatory system

Nuclear envelope

Cell or plasma membrane

Steroid hormone-receptor protein complex

Receptor protein

Nonhistone chromosomal protein

DNA

Pore in nuclear envelope

Transcription

Pre-mRNA

Nucleus

Processing and transport to cytoplasm

mRNA

Translation

New proteins

Figure 11-16

Metabolic regulation via steroid hormones. The small steroid hormone penetrates the cell's plasma membrane and forms a complex with a receptor protein. The hormone-receptor complex enters the cell nucleus, where it affects the transcription of DNA. (From Gardner and Snustad, *Principles of Genetics,* **6th ed., p. 404. Copyright © 1981 by John Wiley and Sons, Inc.)**

enzymes of which (see Chapters 18 and 19) presumably alter the receptor-hormone complex, resulting ultimately in the formation of a product that enters the cell nucleus and affects gene expression. The direct effects of cAMP on gene expression in procaryotes have already been discussed. Although a similar role in eucaryotes remains speculative, cAMP unquestionably affects the activities of enzymes that already exist in the cell. Some of these effects are direct. For example, cAMP directly activates a number of **protein kinases** (enzymes that phosphorylate proteins). Other effects of cAMP are more appropriately considered indirect, in that they are the result of protein phosphorylation by the activated kinases.

Calcium Ion and Calmodulin Regulation

In the past few years, it has become quite clear that calcium ions play a very important role in the regulation of certain cellular activities. For example, such diverse phenomena as cell motility, muscle cell contraction, chromosome movement, endocytosis, exocytosis, and glycogen catabolism are influenced by the level of Ca^{++} in the cell. Some of the effects of calcium ions are mediated through a specific calcium-binding protein called **calmodulin** (Fig. 11–17) found in nearly all eucaryotic cells. Calmodulin is a small protein (M.W. 16,720) consisting of a single polypeptide chain containing 149 amino acids. Especially interesting is the finding that its primary structure is essentially the same in all species studied, indicating that the protein has undergone little change in the course of evolution.

Each molecule of calmodulin binds four calcium ions and its common occurrence in eucaryotes suggests that it may have some universal regulatory function along with Ca^{++}. When Ca^{++} is made available to cells and is bound by calmodulin, the complex formed influences cellular ac-

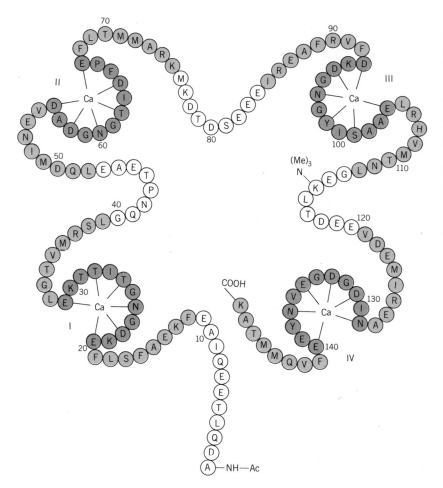

Figure 11-17
The *calmodulin* molecule. Calmodulin chelates Ca⁺⁺, the complex having a regulatory function in many cellular processes. The primary structure of this small protein (indicated here using a one-letter code) has undergone very little change in the course of evolution. A, ala; D, asp; E, glu; F, phe; G, gly; H, his; I, ile; K, lys; L, leu; M, met; N, asn; P, pro; Q, gln; R, arg; S, ser; T, thr; V, val; *and* Y, tyr. (Modified from Cheung, W. Y., *J. Cyclic Nucleo. Res. I,* 71; copyright © 1981 Raven Press. Courtesy of W. Y. Cheung.)

tivity in various ways. For example, the Ca⁺⁺-calmodulin complex may serve to directly activate certain enzymes. Such a stimulatory effect has been demonstrated for *phosphorylase kinase,* an enzyme that controls glycogen catabolism (Table 11–4). The effect of the Ca⁺⁺-calmodulin complex on cell metabolism may be slow and indirect. For example, the complex acts to stimulate *adenylate cyclase* activity in the plasma membranes of cells, resulting in the production of cAMP. The role of cAMP as a second messenger influencing cell metabolism has already been noted (see above). These are but two examples of the manner in which calcium ions and this protein act to regulate cell metabolism; other examples are given in Table 11–4.

Z-DNA

The discovery of a left-handed form of DNA (called **Z-DNA,** see Chapter 7) that coexists with the right-handed form (i.e., **B-DNA**) in cells has led to the proposition that the alternating handedness of the DNA may in some way be involved in the regulation of gene expression in eucaryotic cells. This notion awaits further experimental investigation.

Repression in Eucaryotes

Unlike procaryotic cells, the cells of higher animals and plants undergo extensive *differentiation,* forming tissues

Table 11–4
Effect of Calmodulin-Ca^{++} Complexes on Cell Enzymes

Enzyme	Effect
Adenylate cyclase	Activates enzyme and thereby regulates the rate of cyclic AMP synthesis
Cyclic nucleotide diesterase	Activates enzyme and thereby regulates the rate of cyclic nucleotide degradation
Myosin kinase	Activates enzyme and thereby regulates cell movement
Phosphorylase kinase	Activates enzyme and thereby regulates glycogen breakdown
NAD kinase (plants)	Activates enzyme and thereby regulates the levels of coenzyme present
Enzymes envolved in Ca^{++} pumps	Regulates transport of Ca^{++} across the plasma membrane and the sarcolemma
Tubulin	Disassembles this protein, thereby leading to the disassembly of microtubules

Table 11–5
Relative Concentrations of Calf-Thymus Chromatin Components

DNA	100
Histone	114
Nonhistone protein	33
RNA	7

that have specific and limited physiological roles. Yet most differentiated cells of higher plants and animals contain *complete* genomes. This implies that large segments of the genome go unexpressed. RNA-DNA saturation hybridization experiments (see Chapter 20) indicate that 6 to 30% of the genome is expressed. In some cases, the repression is reversible. For example, when differentiated carrot root cells are isolated and grown in tissue culture, new differentiated cells are produced, including cells that are characteristic of vascular tissue, storage tissue, and epidermal tissue. However, in many differentiated cells, such as nerve and muscle cells of higher animals, large portions of the genome are permanently repressed.

The mechanism of gene repression in eucaryotic cells is not yet clear. The chromosomes of eucaryotic cells contain large quantities of proteins and RNA in addition to DNA (Table 11–5), and over the years, these have been considered potential repressors. However, their chemical properties do not adequately support such a contention. For example, Stedman and Stedman proposed as long ago as the 1940s that the histones might act as gene repressors. How-

ever, there are only five major classes of histones in eucaryotic cells, and they occur in about equal amounts, with little variation between different tissues of an organism or between species. It would therefore appear that histone function is more fundamental (see Chapter 20). If the histones do have repressor activity, then the effect is nonspecific. Electrophoretic analysis (Chapter 13) of the nonhistone proteins in chromatin reveals that they occur in much greater variety than the histones. However, their diversity is insufficient to support the contention that they play a role in gene repression.

The Britten-Davidson Model of Gene Regulation in Eucaryotes

Although several models have been proposed to explain gene regulation in eucaryotes, none has been documented with evidence that would give it the degree of certainty that surrounds the Jacob-Monod operon model for procaryotic cells. However, the model proposed by R. Britten and E. Davidson in 1969 has attracted a great deal of attention. According to this model (Fig. 11–18), the nuclear chromosomes contain DNA sequences called **sensor** genes that recognize various cellular substances such as metabolic inducers (substrates), hormone-receptor complexes, or regulatory nucleotides (e.g., ppGpp). When the inducer enters the nucleus, it binds to the sensor and promotes the transcription of an adjacent **integrator** gene whose product is a specific **activator** RNA. The activator RNA can attach to appropriate DNA sequences that constitute **receptor sites** on either the same or a different chromosome. Presumably, the function of the activator would be analogous to that of the cAMP-CAP complex described earlier in the chapter for procaryotes. The binding of the activator to a receptor site promotes transcription of adjacent structural genes.

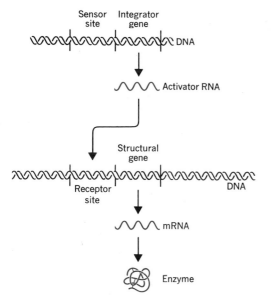

Figure 11-18

Model of regulation of gene expression in eucaryotic cells. Binding of an inducer to a sensor site causes transcription of adjacent integrator gene. The transcript (activator RNA) then binds to the receptor site adjacent to a structural gene on the same or different chromosome. Transcription and translation of the structural gene ensue.

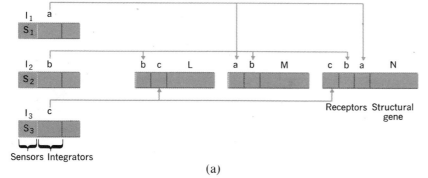

Figure 11-19

Britten-Davidson models for the regulation of gene expression in eucaryotic cells. In the upper model (a), there is a redundancy of receptor genes; in the lower model (b), the redundancy occurs in the integrator genes (see text for explanation).

(a)

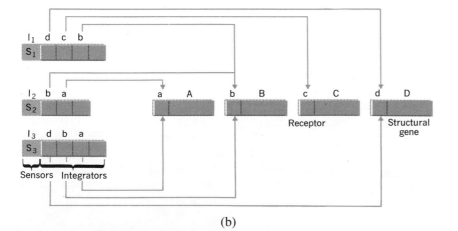

(b)

After undergoing some modification called "processing," the mRNA transcript leaves the nucleus and is translated into protein. (Processing and translation of mRNA are discussed in Chapter 22.)

A number of elaborate modifications of the basic model (Fig. 11–19) can explain the variations in gene expression and differentiation in eucaryotes. For example (Fig. 11–19a), a number of different sensor genes (S_1, S_2, and S_3), upon binding different inducers (I_1, I_2, and I_3), promote the formation of different activator RNA molecules (a, b, and c) by their companion integrator gene. The presence of *multiple* receptor sites for each structural gene (i.e., L, M, and N) would imply that different combinations of structural genes would be transcribed, depending upon binding of the various activators to their respective receptor sites. The binding of an activator to any one receptor would trigger transcription of the adjacent structural gene.

In an alternative model (Fig. 11–19b), transcription of structural genes in various combinations results from the binding of a specific inducer. In this variation, the sensors initiate activator synthesis in a number of adjacent integrator genes, and each activator then associates with *one* receptor. The models have a number of interesting possibilities and are supported by the observation that a large proportion of the DNA in eucaryotic cells (e.g., 40% in calf thymus gland cells) consists of repeated nucleotide sequences that are too small to be structural genes; these, perhaps, are receptor sequences.

Compartmentalization

A final regulatory mechanism in eucaryotic cells is the physical separation and isolation of groups of enzymes within membranous boundaries; that is, specific groups of enzymes are *compartmentalized* within the cellular organelles. For example, in both plant and animal cells the enzymes of the tricarboxylic acid or Krebs cycle are physically separated from those of glycolysis because the former are confined within mitochondria (Chapter 16), while the latter are present in the cytosol. In plant cells, the enzymes of the dark reactions of photosynthesis are physically isolated within chloroplasts (Chapter 17). The enzymes of the glyoxylate cycle are compartmentalized in microbodies

(Chapter 19), while many of the cell's powerful hydrolytic enzymes are restricted within the lysosomes.

Many of these compartmentalized enzymes act on substrates or employ cofactors that are produced by enzymes that are restricted to other parts of the cell. Regulation of the transport of these compounds across cellular membranes from one cell compartment to another affords yet another level at which the control of metabolism can be exercised.

Summary

A variety of metabolic regulatory mechanisms control the direction and rate of individual enzyme-catalyzed reactions in cells and entire metabolic pathways. These mechanisms include the following:

1. **Regulation by mass action:** The rate and direction of a reaction is changed by the addition of substrates or the removal of products.

2. **Regulation of enzyme activity:** The rate is altered by inhibition or activation of an enzyme.

3. **Regulation by allosteric effectors:** Metabolites attach to enzymes at a site other than the active site and in so doing stimulate or inhibit the enzyme's catalytic activity.

4. **Regulation through the modification of covalent bonds in the enzyme:** Many cellular enzymes possess inactive forms called *zymogens,* which are activated only after certain amino acids or peptides are removed.

5. **Regulation by isozymes** or multiple forms of an enzyme: The various forms exhibit different activities under different conditions.

6. **Regulation through the selective expression of certain genes:** In procaryotes, this form of regulation takes the form of inducible and repressible *operons.*

7. **Regulation by compartmentalization:** In eucaryotic cells, the enzymes peculiar to specific metabolic pathways are isolated within specific intracellular organelles.

References and Suggested Reading

Articles and Reviews

Cheung, W. Y., Calmodulin plays a pivotal role in cellular regulation. *Science 207,* 19 (1980).

Crawford, I. P., and Stanffer, G. V., Regulation of tryptophan biosynthesis. *Annu. Rev. Biochem. 49,* 163 (1980).

Chock, P. B., Rhee, S. G., and Stadtman, E. R., Interconvertible enzyme cascades in cellular regulation. *Annu Rev. Biochem. 49,* 813 (1980).

Klec, C. B., Cronch, T. H., and Richman, P. G., Calmodulin. *Annu Rev. Biochem. 49,* 489 (1980).

Maniatis, T., and Ptashne, M., A DNA operator-repressor system. *Sci. Am. 234*(1), 64 (June 1976).

O'Malley, B. W., and Schrader, W. T., The receptors of steroid hormones. *Sci. Am. 234*(4), 32 (Feb. 1976).

Books, Monographs, and Symposia

Atkinson, D. E., *Cellular Energy Metabolism and Its Regulation,* Academic Press, New York, 1977.

Conn, E. E., and Stumpf, P. K., *Outline of Biochemistry* (4th ed.), Wiley, New York, 1976.

Gardner, E. J., and D. P. Snustad, *Principles of Genetics* (6th ed.), Wiley, New York, 1981.

Larner, J., *Intermediary Metabolism and Its Regulation,* Prentice-Hall, Englewood Cliffs, N.J., 1971.

Lehninger, A., *Biochemistry,* Worth Publishers, New York, 1975.

Vogel, H. J. (Ed.), *Metabolic Pathways,* Vol. 5, *Metabolic Regulation,* Academic Press, New York, 1971.

Whelan, W. J., *Biochemistry of Carbohydrates,* Butterworths, London, 1975.

Part 4
TOOLS AND METHODS OF CELL BIOLOGY

Chapter 12
FRACTIONATION OF TISSUES AND CELLS

Much of our current knowledge concerning the structure, chemical composition, and function of the various cell organelles has been obtained following the isolation of these components from cells using various **fractionation** procedures. Such studies involve three major phases. In the first, the tissue or cell suspension is *disrupted* in order to release the cell components. The second phase involves the *sorting* of the cell components into *fractions,* such that the members of any single fraction are the same but differ from the members of any other fraction. The final phase consists of an *examination and analysis* of the separated cell fractions.

Methods for Disrupting Tissues and Cells

Various methods have been devised for disrupting tissues and suspended cells, but the method of choice is usually the procedure that causes *minimal damage* to the released cell constituents. Most physical procedures are based on the effects of **shearing** forces, and because the released cell parts undergo rapid deterioration at room temperature, these procedures are usually carried out at low temperature and in cold buffer solution. Among the older methods used is grinding the sample with a *mortar* and *pestle,* often with the aid of abrasive materials such as sand, alumina, or ground glass. This procedure has several disadvantages, including the loss of some cell constituents by adsorption

to the abrasive and the necessity of removing the abrasive material either before or during the fractionation procedure.

Shearing forces adequate to disrupt most cells and tissues may also be obtained using a blender in which steel blades rapidly rotate through the cell or tissue suspension. The product of this and similar techniques is called an **homogenate.** Tissues may also be homogenized by placing them, along with cold buffer solution, in a cylindrical glass tube fitted with a glass or Teflon plunger. As the plunger is driven down the tube (generally by hand), the tissue and buffer are forced upward through the narrow space between the wall of the tube and the plunger. The shearing forces so generated are usually sufficient to disperse the tissue after several up-and-down strokes. This is the method of choice for the disruption of soft tissues, such as liver, brain, and kidney. One of the more rigorous methods for disrupting cells is by the use of a **pressure cell,** which consists of a steel cylinder and close-fitting steel piston. The piston is pushed into one end of the cylinder using a press or hydraulic jack, and the sample is forced out of the cylinder through a narrow opening at the other end. The size of this opening may be controlled by a needle valve. Bacteria and other microorganisms enclosed within a tough cell wall are frequently disrupted using this approach.

Cells may also be disrupted by **insonation** (ultrasound) using a sonifier. In this procedure, the probe of the sonifier is immersed in the cell suspension and caused to vibrate in the fluid (usually at about 20,000 cycles per second). These ultrasonic vibrations produce a number of effects in the fluid that act collectively to cause the disruption of the cells. Shock waves (alternate compressions and rarefactions) arising from the tip of the probe create turbulent flow of the fluid in which the cells are suspended and may disrupt the cells. The shock waves also cause **cavitation** of the fluid; that is, microscopic bubbles are formed in the fluid near the tip of the probe, and these rapidly stream away from the probe along with some of the fluid. The friction created between this stream and the suspended cells also contributes to their disruption. Some of these bubbles disintegrate into still smaller bubbles that travel away from the probe like miniature projectiles; when these impinge upon the cells, the shear force may disrupt them.

Cells can also be disrupted by chemical means. For example, enzymes that specifically degrade the components of the cell wall or plasma membrane may be added to the tissue or cell suspension. Alternatively, proteolytic or lipolytic agents (such as certain *detergents*) that dissolve the membrane may be used. Some cells are sufficiently fragile that they may be disrupted by successive freezing and thawing. Erythrocytes (red blood cells) and certain other cells can be broken by the **osmotic pressure** created within them when they are placed in a hypotonic salt solution or distilled water.

After the cells have been disrupted, the goal is to separate and isolate the structures that have been released from them. Since cellular organelles and other constituents vary in size, shape, and density, they settle through the liquid in which they are suspended at different rates. Consequently, disrupted cells are more often fractionated by some form of **centrifugation** than by any other method. Indeed, centrifugation has become one of the most widely employed procedures in cellular research and one of the most important tools of the cell biologist.

Centrifugation

If a container is filled with a suspension of particles of varying size and density, the particles will gradually settle to the bottom of the container under the influence of gravity. The rate at which settling occurs can be greatly increased by increasing the gravitational effect upon the particles. This is the rationale behind the use of centrifugation. A tube containing a suspension of particles (e.g., a tissue

homogenate) is placed in the rotor of a centrifuge and then is rotated at high speed. The resulting acceleration greatly increases the force acting on the suspended particles, causing their more rapid sedimentation to the bottom of the tube along paths that are *perpendicular* to the axis of rotation (i.e., along radii of the circle being swept out by the rotating tube; see Fig. 12–1).

Although the practical application of centrifugal force can be traced back more than a thousand years, the modern era of centrifugal methodology began in the early years of this century when the focal point of biochemistry was the study of the chemical and physical properties of protein solutions and other **colloids.** The discovery by Emil Fischer in 1902 that proteins consisted of long chains of amino acids turned out to be the main impetus behind a dramatic surge in the development of centrifugal methods. Fischer and most other notable chemists of the period believed that individual protein species (myoglobin, hemoglobin, etc.) were **polydisperse;** that is, a given protein molecule was believed to occur in various sizes (i.e., various polypeptide chain lengths). That this is not the case was shown by Theodor Svedberg, who developed the first ''analytical ultracentrifuges'' (see below), in which the movements of sedimenting particles can be followed optically. Svedberg

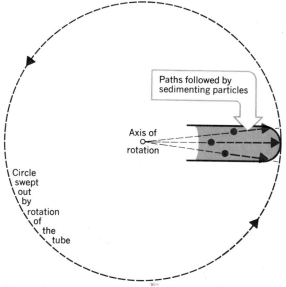

Figure 12-1
Paths followed by particles during centrifugation. The centrifuge tube is being viewed from a position above the plane of rotation.

showed that a given protein type is not polydisperse but is *homogeneous*, with all molecules of that protein having the same, well–defined molecular weight. In 1926, Svedberg received the Nobel Prize in recognition of his pioneering centrifugal studies of proteins and other colloids.

Theory of Centrifugation

The force acting on a particle during centrifugation is given by the equation

$$F = m\omega^2 x \qquad (12–1)$$

where m is the mass of the particle, ω (omega) is the angular velocity of the spinning rotor in radians per second (one unit of circumference contains 2π radians), and x is the distance of the particle from the axis of rotation.

Usually, the value given for the force applied to particles during centrifugation is a relative one; that is, it is compared with the force that the earth's gravity would have on the same particles. It is called **relative centrifugal force** or RCF and is given by

$$\mathrm{RCF} = \frac{F_{centrifugation}}{F_{gravity}} = \frac{m\omega^2 x}{mg} = \frac{\omega^2 x}{g} \qquad (12–2)$$

where g is acceleration due to gravity and equals 980 cm/sec/sec.

Since g is a constant, then

$$\frac{\omega^2}{g} = \frac{\left[\dfrac{(2\pi)(\mathrm{rpm})}{60}\right]^2}{980} \qquad (12–3)$$

In equation 12–3, the term $(2\pi/60)^2/980$ is also a constant and is equal to 1.119×10^{-5}.

Thus the equation for relative centrifugal force (i.e., eq. 12–2) simplifies to

$$\mathrm{RCF} = 1.119 \times 10^{-5}(\mathrm{rpm})^2(x) \qquad (12–4)$$

Consequently, in order to determine the relative centrifugal force in effect during centrifugation it is necessary to measure the revolutions per minute (rpm) of the rotor and the distance between the sample and the axis of rotation. Since

RCF is the ratio of two forces, it has no units. However, it is customary to follow the numerical value of the RCF with the symbol g. This indicates that the applied RCF is a multiple of the earth's gravitational force.

Sample Problem. What is the *maximum* relative centrifugal force applied when red blood cells are sedimented at 1000 rpm in a rotor of maximum sample radius equal to 10 cm?

$$\mathrm{RCF}_{max} = 1.119 \times 10^{-5}(1000)^2(10)$$
$$= 1.119 \times 10^2 \text{ or } 112 \ g$$

The sedimentation of particles by centrifugation is, in effect, a method for concentrating them; therefore, one of the major physical forces opposing such concentration is **diffusion.** In the case of the sedimentation of cells or subcellular organelles (nuclei, mitochondria, etc.), the effects of diffusion are essentially nil. However, when centrifugation is used to sediment much smaller particles (such as cellular proteins, nucleic acids, or polysaccharides), the effects of diffusion become significant.

Sedimentation Rate and Sedimentation Coefficient

The RCF or "g force" applied to particles during centrifugation may readily be calculated using equation 12–4 and is independent of the physical properties of the particles being sedimented. However, a particle's **sedimentation rate** at a specified RCF depends upon the properties of the particle itself. Also, since the RCF varies directly with x, it is clear that the sedimentation rate *changes* with changing distance from the axis of rotation. (For particles settling under the influence of the earth's gravity alone, the sedimentation rate is constant.) The instantaneous sedimentation rate of a particle during centrifugation is determined by three forces: (1) F_C (i.e., the centrifugal force), (2) F_B, the **buoyant force** of the medium, and (3) F_f, the **frictional resistance** to the particle's movement. For a spherical particle P of volume V and density ρ_P sedimenting through a liquid medium of density ρ_M,

$$F_C = m_P\omega^2 x \qquad (12–5)$$

The buoyant force is given by

$$F_B = (m_P/\rho_P)\rho_M\omega^2 x \qquad (12–6)$$

where (m_P/ρ_P) is the volume of the particle and (m_P/ρ_P) (ρ_M) is the mass of the liquid medium displaced by the particle. The frictional force resisting the sedimentation of a sphere through a liquid at 1 cm/sec is given by **Stokes' law,** according to which

$$f = 6\pi\eta r \qquad (12\text{--}7)$$

In this equation, η is the viscosity of the medium and r is the radius of the spherical particle. For a particle sedimenting at a rate other than 1 cm/sec, the frictional resistance would be

$$F_f = f(dx/dt) = 6\pi\eta r(dx/dt), \qquad (12\text{--}8)$$

where dx/dt is the instantaneous sedimentation rate.

If the sedimentation of a particle during centrifugation is viewed as a series of small steplike sedimentation rate increases, then the sedimentation rate *between* increases results from the balance of F_C, F_B, and F_f. That is, at any instant the sedimentation rate results from the fact that

$$F_C = F_B + F_f$$

Accordingly,

$$m_P\omega^2 x = [(m_P/\rho_P)\rho_M\omega^2 x] + [6\pi\eta r(dx/dt)] \qquad (12\text{--}9)$$

Substituting the volume of a sphere (i.e., $\frac{4}{3}\pi r^3$) times its density for its mass into equation 12–9, we obtain

$$(\tfrac{4}{3}\pi r^3)(\rho_P)(\omega^2 x) = (\tfrac{4}{3}\pi r^3)(\rho_M)\omega^2 x + 6\pi\eta r(dx/dt) \qquad (12\text{--}10)$$

By factoring and transposing, we then obtain

$$\tfrac{4}{3}\pi r^3(\rho_P - \rho_M)\omega^2 x = 6\pi\eta r(dx/dt) \qquad (12\text{--}11)$$

Now, the **sedimentation coefficient,** s, is the rate at which a particle sediments under conditions of unit acceleration; that is,

$$s = (dx/dt)/\omega^2 x \qquad (12\text{--}12)$$

By transposing terms in equation 12–11 and solving for s as defined in equation 12–12, we obtain

$$s = \frac{\tfrac{4}{3}\pi r^3\,(\rho_P - \rho_M)}{6\pi\eta r} = \frac{2r^2\,(\rho_P - \rho_M)}{9\eta} \qquad (12\text{--}13)$$

Equation 12–13 is very important to your understanding of particle sedimentation and should be carefully examined, for the equation shows that those properties of a particle that determine its rate of sedimentation during centrifugation are *radius* and *effective density* (i.e., the difference between the density of the particle and the density of the liquid through which the particle is sedimenting). Since the sedimentation rate varies in proportion to the square of the particle's radius (but is a first–order function of the particle's density) it is clear that particle *size* is "more influential" than particle density in determining sedimentation rate.

From equation 12–13, it may also be noted that two particles having different sizes or different densities can have similar sedimentation coefficients. On the other hand, two particles with either similar size or similar density can have markedly different sedimentation coefficients.

The Analytical Ultracentrifuge

The sedimentation coefficient of a particle may be determined experimentally in an instrument known as an **analytical ultracentrifuge.** The rotor that spins in this ultracentrifuge typically contains two compartments. Into one

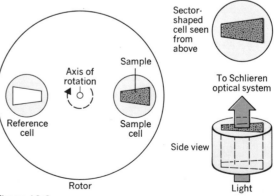

Figure 12-2
Diagram of an analytical ultracentrifuge rotor (viewed from above) and the sector-shaped sample and reference cells.

compartment is placed a "reference cell" containing sample–free solvent, while the other receives a cell containing the sample to be analyzed. The interior of each cell is *sector shaped* and bounded above and below by parallel quartz windows to permit light from below the rotor to pass through the reference and sample during rotation (Fig. 12–2). As the particles sediment, *boundaries* are formed at the trailing edges of each particulate species. Changes in the distance between each boundary and the axis of rotation are measured as a function of time by the instrument's optical system. These measurements are then used to calculate the respective sedimentation coefficients of the particles in the suspension. The sharpness of the boundaries that are formed serves as an index of the heterogeneity of the sample. As noted earlier, the first analytical ultracentrifuges were designed and built by T. Svedberg in the 1920s.

The sedimentation coefficients of many cellular macromolecules such as proteins and nucleic acids fall in the range 1×10^{-13} seconds to 200×10^{-13} seconds (i.e.,

Table 12–1
Some Representative Sedimentation Coefficients

Particle	Sedimentation Coefficient, s (in Svedberg Units, S)
Proteins	
Cytochrome c	1.7
Myoglobin	1.8
Pepsin	2.8
Hemoglobin	4.1
Nucleic acids	
Transfer RNA	4
T_7 bacteriophage DNA	30
Ribosomal RNA (eucaryotes)	5, 5.8, 18, and 28
Viruses	
Tobacco mosaic virus (TMV)	180
Simian virus 40 (SV 40)	240
Influenza virus	700
Subcellular components	
Nucleosomes	11
Ribosomes	70 to 80
Plasma membranes	up to 10^5
Lysosomes	4×10^3 to 2×10^4
Peroxisomes	4×10^3
Mitochondria	1×10^4 to 7×10^4
Chloroplasts	10^5 to 10^6
Nuclei	10^6 to 10^7

the dimensions of s are seconds). For convenience, a unit called the **Svedberg unit,** abbreviated S, is used to describe sedimentation coefficients and is equal to the constant 10^{-13} seconds. Thus, most cellular proteins have sedimentation coefficients between 1 and 200 S. The sedimentation coefficients of a number of cell constituents are listed in Table 12–1.

Calculation of the sedimentation coefficient from data collected with the analytical ultracentrifuge involves the following amplification of the relationship of equation 12–12. If a boundary is x_1 centimeters from the axis of rotation at time t_1 and x_2 centimeters at time t_2, then equation 12–12 may be solved by integration as follows. First, by transposition,

$$s \, (dt) = \frac{1}{\omega^2} \, (dx/x) \qquad (12\text{–}14)$$

Integrating between the limits set above, we obtain

$$s \int_{t_1}^{t_2} dt = \frac{1}{\omega^2} \int_{x_1}^{x_2} dx/x \qquad (12\text{–}15)$$

and

$$s \, (t_2 - t_1) = \frac{1}{\omega^2} \, (\ln x_2 - \ln x_1) = \frac{1}{\omega^2} \left(\ln \frac{x_2}{x_1} \right) \qquad (12\text{–}16)$$

Therefore,

$$s = \frac{1}{\omega^2 \, (t_2 - t_1)} \ln \frac{x_2}{x_1} \qquad (12\text{–}17)$$

Sample Problem. A solution containing a single particulate species is accelerated in the analytical ultracentrifuge at 60,000 rpm. At time t_1, the boundary between the trailing edge of the sedimenting particles and the axis of rotation is 6.0 cm, while at t_2, 60 minutes later, the boundary is 6.8 cm from the axis. What is the sedimentation coefficient of the particles?

$$\omega = \frac{60,000 \, (2\pi)}{60} \text{ radians per second}$$

$$= 6,280 \text{ radians per second}$$

$$s = \frac{1}{(6,280)^2 (60) \, (60)} \, 2.3 \log_{10} (6.8/6.0)$$

$$\frac{1}{1.42 \times 10^{11}} \, 2.3 \, (0.053)$$

$$= 8.6 \times 10^{-13} \text{ or } 8.6 \, S$$

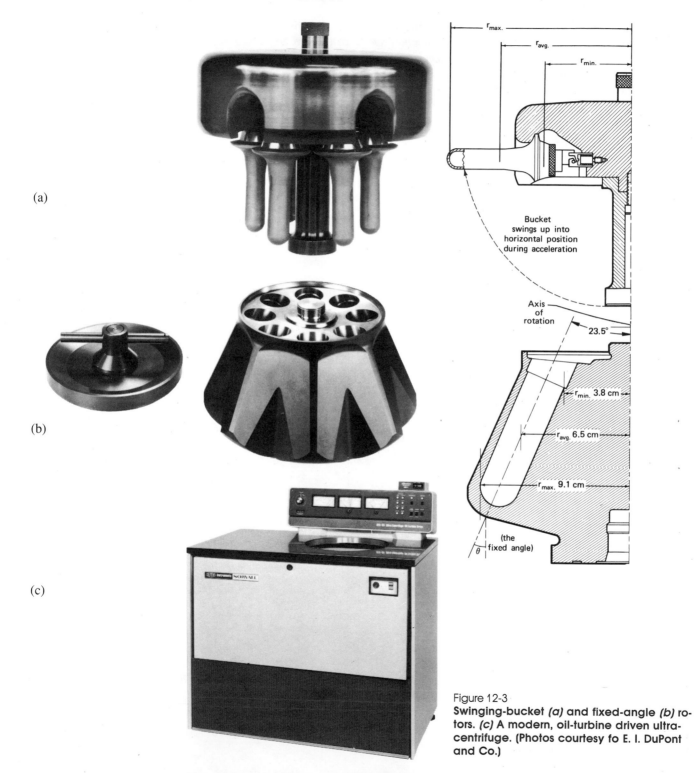

(a)

(b)

(c)

Figure 12-3
Swinging-bucket *(a)* **and fixed-angle** *(b)* **rotors.** *(c)* **A modern, oil-turbine driven ultra-centrifuge. (Photos courtesy fo E. I. DuPont and Co.)**

As its name implies, the analytical ultracentrifuge is an instrument for analysis and does not physically separate the multiple components of a mixture from one another. Furthermore, the amount of material that may be studied is quite limited as a result of the small size of the sample cell.

Preparative Centrifugation

The counterpart to analytical centrifugation is **preparative centrifugation,** which provides for the isolation of cell components for further analysis. Two basic types of centrifuge rotors are regularly employed for conventional preparative centrifugation; these are the **swinging-bucket rotor** and the **fixed angle rotor** (Fig. 12–3). The swinging-bucket rotor consists of a series of metal buckets attached to the central *harness* of the rotor. The samples to be centrifuged (previously placed in centrifuge tubes) are inserted into the buckets, which swing upward from a vertical position to a horizontal position during acceleration of the rotor (Fig. 12–3). During deceleration, the buckets return to the vertical position and the centrifuge tubes are removed. In a fixed angle rotor, the centrifuge tubes are maintained at a constant angle (usually 15° to 45° from verticality) throughout centrifugation (Fig. 12–3). As a result, the radially sedimenting particles quickly strike the tube wall, where convection currents rapidly carry them to the bottom of the tube.

Differential Centrifugation

During centrifugation, particles sediment through the medium in which they are suspended at rates related to their size, shape, and density. Differences in the sedimentation coefficients of the various subcellular particles provide the means for their effective separation. **Differential centrifugation,** a technique introduced to cellular research in the early 1940s by the noted biologist and Nobel–laureate Albert Claude, is one of the classical procedures for isolating subcellular particles and involves the stepwise removal of classes of particles at increasing RCF.

The material to be fractionated is subjected first to low–speed centrifugation in order to sediment the largest (or densest) particles present. Following this, the unsedimented material (called the **supernatant**) is transferred to another tube and centrifuged at a higher speed to sediment particles of somewhat smaller size (and/or lower density). The sequence is repeated several times until all particles have been sedimented; each sediment is then used for further experimentation and analysis. The procedure regularly employed for the differential fractionation of liver tissue may serve as a convenient example of this method and is shown diagrammatically in Figure 12–4.

The removed liver tissue is homogenized in cold buffer and centrifuged for 10 minutes at 700 g. This is usually sufficient to sediment all the cell nuclei to the bottom of the centrifuge tube, thereby providing the **nuclear fraction.** Depending upon the effectiveness of the homogenization procedure, some unbroken cells and large cell fragments may also be recovered in this fraction. The overlying supernatant (called the **nuclear supernatant**) is removed and transferred to another tube for a second centrifugation at 20,000 g for 15 minutes. This sediments nearly all the mitochondria (i.e., the *mitochondrial fraction*). Again, the supernatant (i.e., **mitochondrial supernatant**) is removed and is subjected to a third centrifugation at 105,000 g for 60 minutes. This causes the sedimentation of a fraction called **microsomes,** which includes ribosomes and small fragments of intracellular membranes. The **microsomal supernatant** is referred to as the **soluble phase** of the cells, or **cytosol,** and includes soluble proteins, soluble nucleic acids, soluble polysaccharides, lipid droplets, and other small particles. In the procedure just described the liver tissue is separated into four major fractions.

Differential centrifugation has several major disadvantages. Since the homogenate is initially distributed uniformly throughout the centrifuge tube, the first particles sedimented will necessarily be contaminated with all other constituents of the homogenate. This effect is shown in Figure 12–5. As the smaller particles are sedimented, they in turn will be contaminated by even smaller particles. In fact, the only particles to be obtained in relatively pure form will be those that sediment most slowly. In the example given above, the initial nuclear fraction would contain some mitochondria, microsomes, and cytosol; the mitochondrial fraction obtained in the second centrifugation would be contaminated with microsomes and cytosol, and so on. These disadvantages may not be serious in some instances, since the major subcellular particles of a tissue homogenate have sedimentation values that differ from one another by one or more orders of magnitude. However, serious difficulties arise if the particles to be separated have similar sedimentation coefficients. This problem is illus-

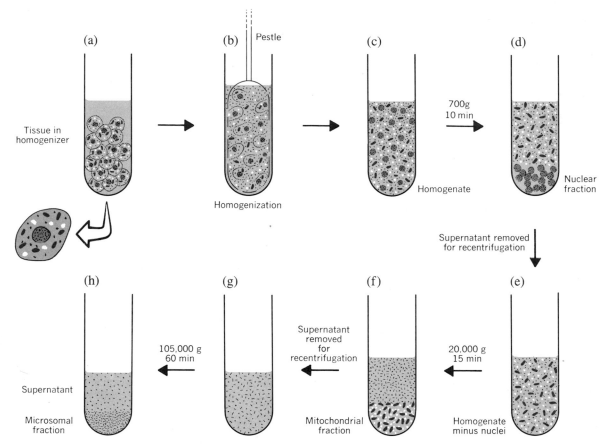

Figure 12-4
Fractionation of liver tissue by differential centrifugation: nuclei (pale color), mitochondria (solid color), lysosomes (white), microsomes (black), and cytosol (gray).

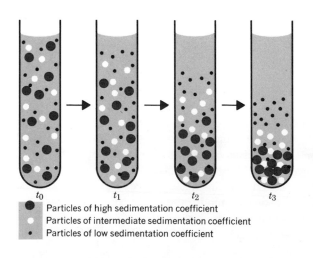

Particles of high sedimentation coefficient
Particles of intermediate sedimentation coefficient
Particles of low sedimentation coefficient

Figure 12-5
Impurities of sediments produced by differential centrifugation. At time t_0, all particles are distributed uniformly through the centrifuge tube. During centrifugation (times t_1, t_2, and t_3), all particles sediment at rates that are determined by their sedimentation coefficients. Although the particles that have the highest sedimentation coefficient quickly reach the bottom of the tube, cross-contamination by small quantities of particles having lower sedimentation coefficients is unavoidable.

trated in Fig. 12–4*f,* in which the mitochondrial fraction also contains the cells' lysosomes.

Density Gradient Centrifugation

The resolution achieved during centrifugation can be greatly improved if the mixture of particles is confined at the outset to a narrow zone at the top of the centrifuge tube and the particles then permitted to sediment from this position. Initial stability under these conditions can be obtained only if the particles are layered onto a **density gradient,** that is, a column of fluid of increasing density. The technique, known as density gradient centrifugation, was introduced to cellular research in 1951 by M. K. Brakke.

Density gradients may be **stepwise** *(discontinuous)* or **continuous.** A step gradient is prepared by successively layering solutions of decreasing density in the centrifuge tube. Continuous density gradients are prepared by mixing dense and light solutions in varying proportions at a controlled rate and delivering the mixture to the tube in a continuous stream. A simple procedure for producing a ''linear'' density gradient is shown in Figure 12–6. In this procedure, the light solution (in the right cylinder) flows into and is mixed with the dense solution (in the left cylinder). The mixture is then delivered to the centrifuge tube. Using this method, a density gradient is formed that decreases linearly as a function of volume between the limiting densities originally present in the two cylinders. Gradients with other shapes (hyperbolic, exponential, etc.) may be prepared using containers of various noncylindrical shapes or gradient–generating devices specifically manufactured for this purpose. Solutes used to provide solutions for density gradients are selected on the basis of their solubility in water and their compatibility with cells and subcellular organelles. Sucrose, Ficoll (a copolymer of sucrose and epichlorohydrin), and Percoll (a colloidal form of silica) are the most popular, since they have minimal detrimental effects on organelles and provide densities up to about 1.3 g/cm³. Cesium chloride and other dense salts are used when the limiting density must be considerably higher (i.e., up to 1.9 g/cm³). The rate at which particles sediment through a density gradient is given by equation 12–13. Since the quantity $(\rho_P - \rho_M)$ changes as the particles sediment deeper into the gradient, *s* also changes. If a particle reaches a position in the gradient where the particle's density and the gradient's density are the same (i.e.,

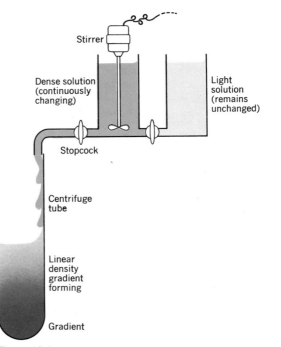

Figure 12-6
Density gradients for centrifugation may be produced in many ways. Shown here is the production of a "linear" density gradient, using a two-chamber device. The light solution flows into the chamber containing the dense solution and is rapidly mixed there. The mixture is then delivered to the centrifuge tube. The density of the liquid in the mixing chamber decreases *linearly* as the volumes of the two chambers progressively diminish.

$\rho_P = \rho_M$), then *s* becomes zero, and no further sedimentation of the particle occurs. If a particle suspension is overlaid with a gradient whose maximum density is greater than that of the particles (i.e., $\rho_M > \rho_P$), then during centrifugation the particles will *rise* through the gradient, for when ρ_M is greater than ρ_P, *s* becomes a negative term.

Rate Sedimentation.   When the densest region of a density gradient (i.e., the liquid at the bottom of the centrifuge tube) is less dense than the particles being sedimented, all particles will eventually reach the bottom of the tube. However, if the duration of centrifugation is carefully limited, this will not occur; instead, the particles will be distributed through the density gradient *in order of their sedimentation coefficient.* Fractionations carried out in this manner are called **rate separations** and are based upon the combined contributions of particle size and particle density. The rate

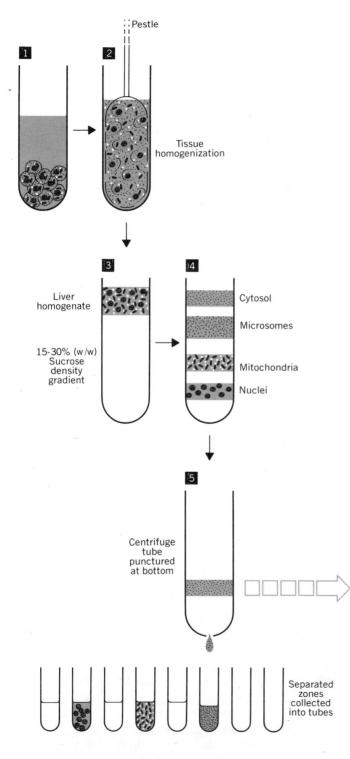

Figure 12-7
Rate separation of a liver homogenate by density gradient centrifugation. The tissue homogenate is carefully layered onto the surface of a 15 to 30% w/w sucrose density gradient and is centrifuged at 10,000 *g* for 20 min. The homogenate is thus separated into four major fractions: *nuclear, mitochondrial, microsomal,* and *soluble phase* (i.e., *cytosol*). Following centrifugation, the separated fractions can be collected by puncturing the bottom of the centrifuge tube and permitting the gradient to drip into a series of test tubes.

fractionation of a liver tissue homogenate is depicted in Figure 12–7. The four major cell fractions (nuclear, mitochondrial, microsomal, and cytosol) are separated during a single centrifugation.

Isopycnic Sedimentation.   When the densest region of a density gradient is denser than the particles to be sedimented, no particles will reach the bottom of the tube regardless of how long centrifugation is carried out. Instead, the particles will sediment through the gradient until they reach their *isodense or* **isopycnic** position. Fractionations carried out on the basis of particle density alone are called isopycnic separations. It is to be noted that this implies that small, dense particles with low sedimentation coefficients sediment further through such a density gradient than large, light particles with high sedimentation coefficients (Fig. 12–8).

In rate separations, both particle size and particle density determine the final positions of particles in the gradient. However, of the two parameters, size is more influential, since *s* is a second–order function of particle radius but a first–order function of particle density (examine equation 12–13). In isopycnic separations, particle density alone determines final position in the density gradient.

Maximum resolution of particle mixtures by density gradient centrifugation is achieved when both rate and isopycnic centrifugations of the sample are carried out in sequence. First, using rate centrifugation, the particle mixture is separated into fractions of similar particle sedimentation coefficient; then, collected fractions are subfractionated into classes of equal density by isopycnic centrifugation. This technique is known as two–dimensional centrifugation and provides for the separation of cell components of similar sedimentation coefficient or similar density. (It is extremely unlikely that two different particles will have similar sedimentation coefficients *and* similar

Rate centrifugation
in shallow
density gradient

Isopycnic centrifugation
in steep
density gradient

● Particles of large size and intermediate density
● Particle of intermediate size and low density
• Particle of small size but high density

Figure 12-8

Comparison of rate and isopycnic centrifugation. During rate centrifugation, the particles become distributed through the gradient in an order related to their respective sedimentation coefficients. Consequently, a large but not very dense particle may sediment further through the gradient in a shorter period of time than a small, very dense particle. In contrast, during isopycnic centrifugation, particles become distributed through the gradient in order of their respective densities (i.e., independent of particle size differences).

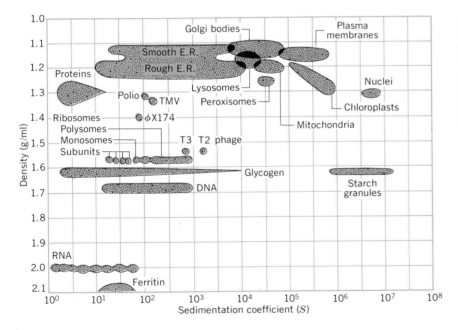

Figure 12-9

Densities and sedimentation coefficients of a number of biological particles and substances. The banding densities of the organelles are based upon their behavior in sucrose density gradients, whereas the banding densities of the viruses and macromolecules are based upon results using CsCl salt gradients.

densities.) Figure 12–9 compares the sedimentation coefficients and densities of a variety of cellular components. After centrifugation is completed, the separated zones may be recovered from the density by puncturing the bottom of the centrifuge tube and allowing the gradient to elute (Fig. 12–7).

In certain instances, it is not necessary to layer the sample to be fractionated onto a preformed density gradient. Instead, the sample and the density gradient solutions are mixed and placed in the centrifuge tube. During centrifugation at high speed, the density gradient forms automatically within the tube, and the particles migrate (upward and/or downward) to their isopycnic positions. This procedure is known as **equilibrium isopycnic centrifugation** and is regularly carried out in using solutions of CsCl or other heavy salts (Fig. 12–10).

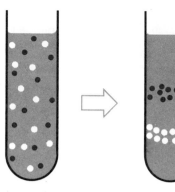

Figure 12-10
Equilibrium isopycnic centrifugation. During centrifugation, the liquid in which the particles were initially (and uniformly) suspended automatically forms a density gradient. The families of particles either sediment or float to their respective isopycnic positions as the gradient is being formed.

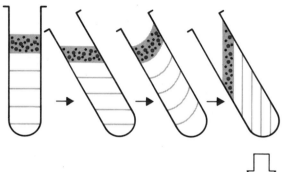

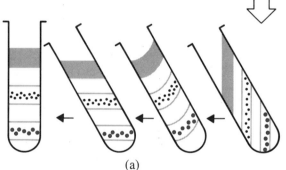

(a)

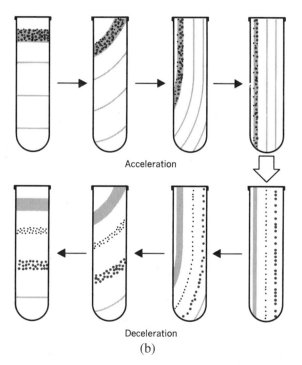

Acceleration

Deceleration

(b)

Figure 12-11
(a) **Isopycnic centrifugation using a fixed-angle rotor.** *(b)* **Isopycnic centrifugation in a vertical tube rotor. In both cases, the axis of rotation is to the left of the tube, and the density gradient is caused to reorient between the vertical and radial position during rotor acceleration and deceleration.**

Although density gradient centrifugation is usually carried out in swinging–bucket rotors, isopycnic (but not rate) separations can also be performed in fixed angle rotors (Fig. 12–11a). When the gradient with sample layered above is placed in the inclined position in the rotor, the sample zone and equal density planes within the gradient form ellipses. During acceleration of the rotor, each isodense elliptical layer becomes a small section of a *paraboloid of revolution* with focal point on the axis of rotation. The family of paraboloids becomes increasingly steep as rotor acceleration continues and eventually approaches verticality; the phenomenon is called **gradient reorientation.**

Particles in the gradient sediment radially until they encounter the sloping tube wall; from there, they are carried down through the gradient by centrifugal force and convection until they reach the isodense region of the gradient, where they form layers. Rotor deceleration reorients the gradient again so that the separated particles form horizontal layers.

A recent modification of the fixed–angle approach to isopycnic centrifugation involves the use of **vertical tube rotors** in which the fixed angle is reduced to 0° (i.e., the centrifuge tubes are maintained in the vertical position throughout; Fig. 12–11*b*). Here, too, the gradient is reoriented during rotor acceleration and deceleration.

Zonal Centrifugation

Density gradient centrifugation using tubes is the most widely employed technique for separating cells and cell organelles and for isolating cellular macromolecules. However, although it is one of the cell biologist's most valuable tools, it is not without disadvantages, since the amount of material that can be fractionated in a single tube is so small. When large quantities of sample must be fractionated (in order to isolate sparse organelles such as lysosomes or peroxisomes), a very large number of tubes and gradients is needed. The lack of sector shape in a centrifuge tube causes "wall effects" (Fig. 12–12), including adsorption to the tube wall, premature sedimentation down the tube, and convection disturbances. These two problems served as the impetus for the development in the 1960s of **zonal rotors,** a technological advancement pioneered by N. G. Anderson.

A zonal rotor consists of a large cylindrical chamber subdivided into a number of sector–shaped compartments by vertical septa (or vanes) that radiate from the axial core to the rotor wall. The entire chamber is used during centrifugation and is loaded with a single density gradient, each sector–shaped compartment serving as a large centrifuge tube. The large chamber capacity of these rotors (typically 1 to 2 liters) eliminates the need for multiple runs and multiple density gradients. Two basic forms of the zonal rotor are used for tissue fractionation; these are (1) **dynamically unloaded (rotating–seal) rotors** and (2) **reorienting gradient ("reograd") rotors.** The similarity in appearance of these rotors (Fig. 12–13) is misleading, since the basic principles of operation are substantially different.

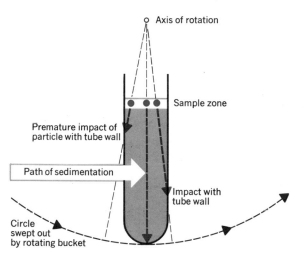

Figure 12-12
Wall effects produced in centrifuge tubes. The paths followed by sedimenting particles are perpendicular to the axis of rotation. Consequently, particles that are initially near the center of the sample zone have an uninterrupted path leading to the bottom of the tube. However, particles that are initially located near the margins of the sample zone will encounter the wall of the centrifuge tube before reaching the bottom. This can result in particles sticking to the tube wall or more rapid sedimentation to the bottom of the tube as a result of convective effects.

Dynamically Unloaded Zonal Rotors

Operation of a dynamically unloaded zonal rotor is depicted schematically in Figure 12–14. Two fluid lines connect the center and edge of the rotor chamber with a rotating–seal assembly that permits loading the density gradient (and sample) into the rotor *while it is spinning*. The center fluid lines open into the rotor chamber through the core section, while the edge lines pass radially through each of the septa and open at the rotor wall.

The zonal rotor is filled with the density gradient while rotating at low speed. The light end of the gradient is loaded first through the edge lines and is followed by the denser mixture (Fig. 12–14*a*). The dense end of the gradient gradually displaces the lighter fluid toward the core of the rotor. Addition of a dense "cushion" forces some of the light end of the gradient out of the rotor (Fig. 12–14*b*). The sample to be fractionated is introduced through the center lines, thereby displacing some of the cushion out of the edge lines (Fig. 12–14*c*). Additional light fluid (called

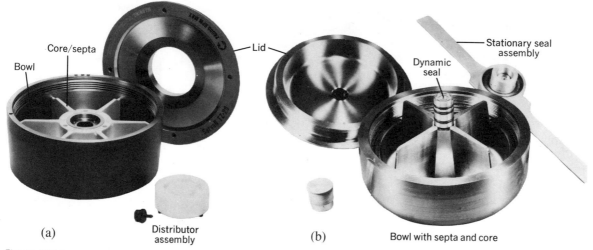

Figure 12-13

(a) A reorienting gradient (*reograd*) zonal rotor (TZ-28) and *(b)* dynamically unloaded zonal rotor (B-29). The bowl of the TZ-28 rotor is unloaded through lines in the distributor, core, and septa after the rotor has slowly decelerated to rest and the gradient has been reoriented. The B-29 rotor is unloaded while the rotor is spinning; this is achieved by using a static and rotating-seal assembly that provides for continuous communication with the core and wall regions of the rotor bowl through channels in the core and the septa.

overlay) is then pumped into the center line to push the sample clear of the core region (Fig. 12–14*d*). Now the upper (stationary) portion of the seal assembly is removed and the rotor is accelerated to a higher speed for separation of the particles in the sample (Fig. 12–14*e*). The separated particles form a series of concentric cylindrical zones in the rotor bowl. Following particle separation, the rotor is decelerated to a lower speed, the static seal reinserted, and the entire gradient displaced through the center lines by pumping dense fluid through the edge line (Fig. 12–14*f*). The eluting gradient may be monitored and collected in tubes for subsequent analysis.

Reograd Zonal Rotors

The operation of reograd zonal rotors (developed in the 1960s by P. Sheeler and J. R. Wells) is depicted in Figure 12–15. Unlike dynamically unloaded zonal rotors, the gradient and sample can be loaded into the reograd rotor either at rest through the septa lines (which communicate with the bowl floor) or while spinning using the core lines. Unloading is *always* carried out with the rotor at rest. If the rotor is loaded at rest, then the gradient is reoriented from the vertical to the radial position during acceleration (Fig. 12–15*a, b*). During centrifugation, different particles in the sample sediment to form a family of concentric cylindrical zones in the radial gradient (Fig. 12–15*d*), but during rotor deceleration, they are reoriented to form horizontal layers (Fig. 12–15*e, f*). The density gradient and entrained particles are then withdrawn from the stationary rotor through the septa lines and collected as fractions for further analysis.

Because the chambers of zonal rotors are sector shaped, detrimental wall effects are minimized or eliminated entirely. Zonal rotors have greater capacities than swinging–bucket rotors, and, therefore, larger quantities of material can be fractionated during a single centrifugation. In order to fractionate a comparable amount of tissue or cells using conventional rotors, several successive centrifugations would be required and separate density gradients would have to be prepared for each tube.

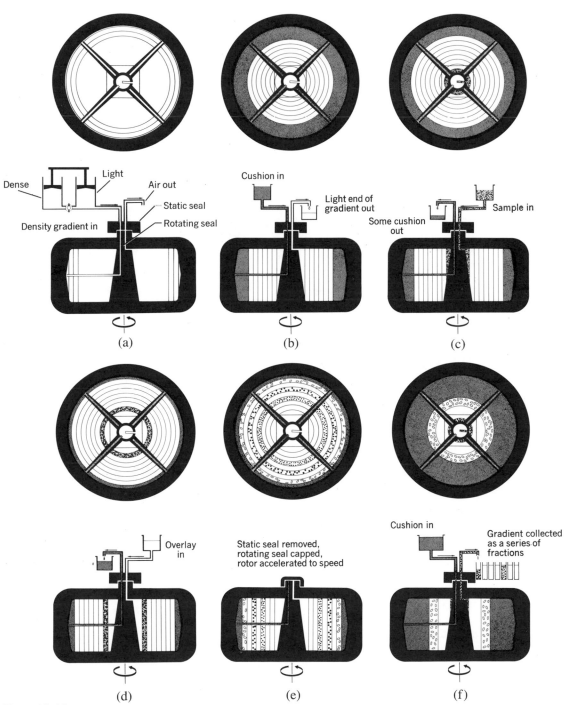

Figure 12–14
Operation of a dynamically unloaded zonal rotor such as the B-29 shown in Figure 12-13. Lines drawn inside the rotor's sector-shaped chambers depict hypothetical planes of equal density. The "cushion" supporting the dense end of the gradient is shown in color (see the text for additional details).

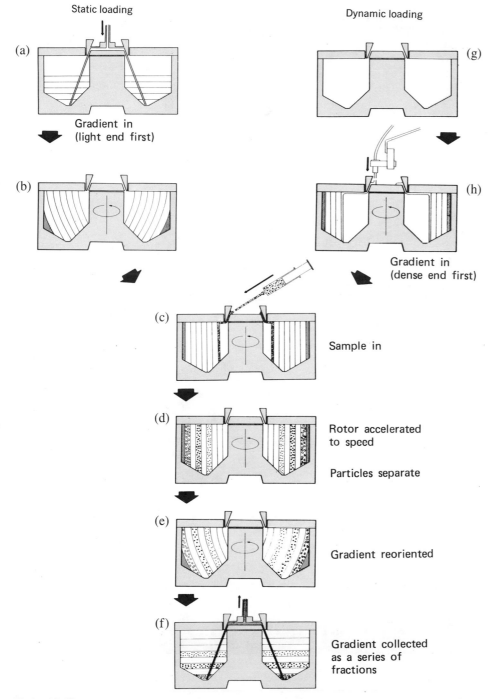

Static loading

Dynamic loading

(a)

Gradient in
(light end first)

(b)

(g)

(h)

Gradient in
(dense end first)

(c) Sample in

(d) Rotor accelerated
to speed

Particles separate

(e) Gradient reoriented

(f) Gradient collected
as a series of
fractions

Figure 12-15
**Operation of a reograd zonal rotor such as the TZ-28 shown in Figure 12-13.
Lines drawn inside the rotor chambers depict hypothetical planes of equal
density. The cushion is shown in color (see the text for additional details).**

Examination and Analysis of Separated Cell Fractions

The final phase of any study involving the fractionation of cells or tissues is the *examination* and *analysis* of the separated fractions so that the morphology, chemical composition, organization, and function of the isolated components may be revealed or better understood. A host of procedures can be employed in such studies, but they will not be pursued here, since they are more properly considered individually in conjunction with closely allied topics in other chapters of the book. For example, the isolation of the plasma membrane–containing fraction (or fractions) from a tissue homogenate might be followed by morphological or biochemical analyses involving electron microscopy, chemical characterization and isolation of the membrane constituents, determination of component function in membrane transport, and so on. These topics are dealt with in Chapter 1, 13, and 15, which specifically consider the principles of electron microscopy, chemical fractionation procedures, and plasma membrane structure and function. Indeed, a considerable portion of this book is devoted to the presentation of the results of examination and analysis of isolated cell components and to the integration of those results into a functional whole.

Methods for Separating Whole Cells

Throughout this chapter, we have been concerned with methods used to separate and isolate particles released from disrupted cells and tissues. Until quite recently, little attention was directed to a related problem—namely, how to separate *whole viable* cells from one another when the tissue or culture being studied is heterogeneously composed. For example, an organ such as the liver is composed of many different types of cells, including hepatic cells, Kupffer cells, connective tissue cells, smooth muscle cells, blood cells, and so on. Therefore, a homogenate of liver tissue contains subcellular particles from diverse kinds of cells. Even a culture of the same cell type may be heterogeneous with regard to cell ages (see Chapter 2) and therefore is representative of a broad spectrum of morphological characteristics or physiologic activity. The development of methods for separating different types of cells present in a tissue has lagged behind technological advances in the area

of subcellular fractionation. In the following section, we will examine some of the problems associated with whole–cell separations and some of the more important methods that have evolved to effect such separations.

Tissue Disaggregation

If the tissue to be fractionated consists of suspensions of individual cells (e.g., cultures of microorganisms, some tissue culture, blood cells, and certain tumors), the problem of whole–cell separation is far less difficult than when the cells comprise a solid tissue (such as liver, kidney, brain, etc.). It is therefore not surprising that, to date, most efforts directed toward whole–cell separations have involved natural suspensions of cells as the starting material. Some success has been obtained with solid tissues by employing chemical agents that induce tissue disaggregation—primarily *digestive enzymes* and *chelating agents*. These materials weaken the connections between neighboring cells, making it possible to mechanically disperse the tissue into individual cells without appreciable cell breakage. The tissue may be so treated after its removal from the animal, although *perfusion* of the organ with a solution of the disaggregating agent prior to the organ's excision is more often preferred.

Once the tissue has been reduced to a suspension of individual cells, fractionation into subpopulations follows. If the goal of the experiment is to isolate a particular cell subpopulation for further study, then the remaining cells in the suspension may be *selectively* destroyed or removed by chemical means. For example, the leucocytes of blood may be separated from the erythrocytes by selective destruction of the erythrocytes using *osmotic* or *chemical lysis*. Purification of a particular subpopulation of cells may also be achieved by taking advantage of *differential cell agglutinability* in the mixed population. Simply *freezing* and *thawing* a suspension of cells may differentially lyse specific subpopulations. In general, chemical procedures cause some changes in *all* the cells in the mixed population, so that the method of choice is more generally one that achieves a separation by mild physical means. Among the most popular of the latter methods are **adherence and filtration, conventional and zonal centrifugation, centrifugal elutriation, unit gravity separation, countercurrent distribution, electrophoresis,** and **fluorescence–activated cell sorting.**

Adherence and Filtration

Separations of cells using differences in adherence phenomena or filtration properties are among the oldest physical procedures used. Some cells readily adhere to glass beads, nylon wool, glass wool, and so on and may be separated from nonadhering cells by passing the cell suspension through a hollow glass column packed with these materials. Success has also been obtained by coating glass or plastic beads with antibodies, antigens, or haptens so that cells will be differentially adsorbed to the beads on the basis of chemical interactions between the cell membrane and the coating material. Sieves of varying pore diameter can also be used to separate populations of cells on the basis of differences in cell diameter.

Conventional and Zonal Centrifugation

Because of their relatively large size (i.e., in comparison with organelles and macromolecules), whole cells sediment quite rapidly. Consequently, attempts to fractionate suspensions of cells using centrifugation involve rotation at low rpm (i.e., small RCF) for short periods of time (typically less that 500 g for a few minutes). As with subcellular centrifugal fractionations, greatest resolution is obtained using density gradients in which the mixture of cells in the starting zone is separated into subpopulations on the basis of differences in average cell size and/or density. Most cells behave like miniature osmometers, so that strict attention must be paid to the selection of gradient solute. Salts are rarely used to prepare density gradients for cell separations because of their deleterious osmotic effects. Large, impermeable, and biologically inert polymers such as Percoll and Ficoll are the more frequent choices.

Significant wall effects accompany particle sedimentation in fixed–angle rotors; therefore, conventional approaches to the centrifugal separation of cells involve swinging–bucket rotors. Since some minimum rpm (and therefore minimum RCF) must be attained before the buckets reach the horizontal position, special swinging–bucket rotors in which the minimum and maximum radii are particularly small (i.e., the buckets are close to the axis of rotation) are used in order to maintain the low centrifugal force required. Although offering the advantage of greatly increased sample size, dynamically unloaded zonal rotors are rarely used because of the continuing cell sedimentation that occurs during the extended dynamic unloading period. Reograd zonal rotors can be used, however, since gradient unloading is carried out at 1 g (i.e., with the rotor at rest).

Centrifugal Elutriation (Counter–Streaming Centrifugation)

Centrifugal elutriation is an ingenious technique pioneered in the 1940s by P. E. Lindahl and brought to its present state of the art principally through the work of C. R. McEwen. In centrifugal elutriation, a suspension of cells is pumped into a specially designed rotor chamber through a marginally located entry port (Fig. 12–16), and this is followed by a continuous flow of suspending medium. Centrifugal sedimentation of the cells is opposed by the centripetal flow of the suspending medium. Both effects vary in magnitude across the radial dimension of the rotor chamber, since (1) the centrifugal force increases with distance from the rotor axis, and (2) the rate of centripetal liquid flow varies according to the cross–sectional area of the chamber (the area increases exponentially as the liquid travels toward the rotor axis). Depending upon its initial position in the chamber and its sedimentation coefficient, each cell will either sediment radially under the centrifugal force or be carried centripetally by the liquid flow. As a result, the cells migrate through the chamber to positions where these two forces cancel one another. Some cells (e.g., the smallest ones) may be swept from the rotor chamber entirely, while others form a zone within the rotor chamber and can be collected for further study. Centrifugal elutriation has been successfully applied to separations of blood cells, algae, yeasts, and other cells in culture.

Unit Gravity Separation

The separation of particles on the basis of sedimentation rate differences may not necessitate centrifugation if the particles are sufficiently large. For example, whole cells sediment fairly quickly even at 1 g (i.e., at unit gravity). Unit gravity procedures have been used effectively to separate different types of blood cells, tissue culture cells, and populations of microorganisms into subpopulations. The separation is achieved by layering the mixture of cells onto the top of a stationary density gradient and allowing the cells to settle through the gradient for some period of time. The gradient and separate cell populations are then collected

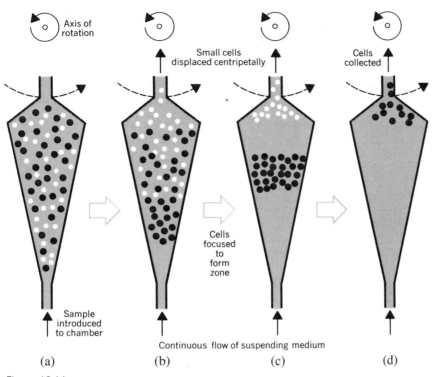

Axis of rotation

Small cells displaced centripetally

Cells collected

Sample introduced to chamber

Cells focused to form zone

Continuous flow of suspending medium

(a) (b) (c) (d)

Figure 12-16

Centrifugal elutriation. (a) A sample containing a mixture of cells is introduced into the separation chamber through the chamber's centrifugal edge while the rotor is spinning. (b) The combined effects of centrifugal force and centripetal liquid flow displace the cells. (c) Small cells are swept from the rotor while the larger cells are focused into a narrow zone. (d) The focused cells may then be displaced from the rotor chamber and collected. The walls of the separation chamber take the shape of an exponential horn (not shown here).

as a series of fractions. Devices used to separate cells in density gradients at unit gravity are called "sta–put" devices and vary from simple cylindrical chambers to more elaborate apparatus having moving, conical end caps. The principle is illustrated in Figure 12–17.

Not only is it possible to separate heterogeneous mixtures of cells using this simple approach, but a population of a single cell type (e.g., a cell culture) can be fractionated according to cell age when age and size are related. In this way, events that occur during successive phases of the *cell cycle* (see Chapter 2) can be studied by examining the cells present in different collected fractions.

A number of methods for separating and isolating the molecular constituents of cells are considered in Chapter 13. Some of these methods have been appropriately mod-

ified and applied with varying degrees of success to separations of different kinds of cells that make up a tissue; **countercurrent distribution** and **electrophoresis** will be mentioned briefly here.

Countercurrent Distribution

Cells may be separated from one another on the basis of differences in their partition between two immiscible liquids. Naturally, if the cells are to be separated without undue damage, the milieu selected must be compatible with the cells with respect to ionic composition, concentration, osmotic pressure, and so on. This demand significantly restricts the selection of liquids, especially in comparison with the range of choices available when countercurrent

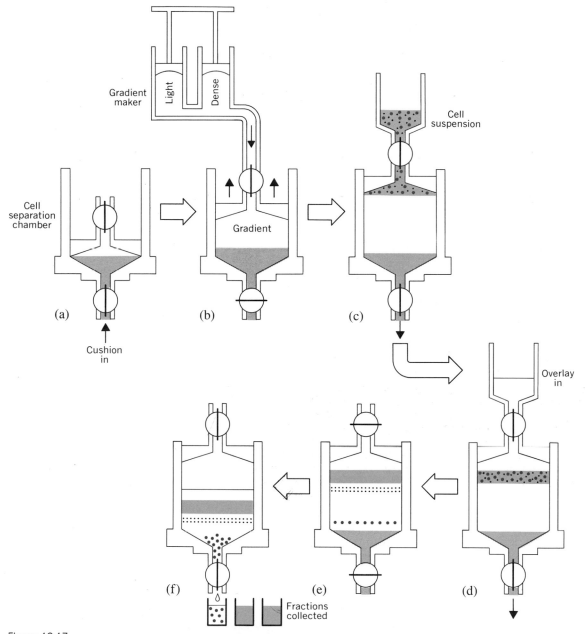

Figure 12-17
The separation of cells (or particles of comparable size) by unit gravity using a "sta-put" chamber. (a) A cushion is introduced into the bottom of the chamber, and this is followed (b) by introduction of the density gradient. (c) Once the gradient is loaded, the sample is carefully layered onto its surface, followed by an overlay (d). The particles separate from one another as they settle through the gradient under the influence of gravity and form a series of different zones. (e). (f). The density gradient and separated cells are then collected as a series of fractions.

distribution is employed for molecular separations (see Chapter 13). Greatest success has been obtained with phases consisting of polyethylene glycol and aqueous solutions of dextran (polyglucose in which most glycosidic linkages are 1→6). The technique has been especially fruitful in separations of different microorganisms and different blood cells.

Electrophoresis

Electrophoresis is one of the most popular methods used for separating different molecular species, especially proteins (see Chapter 13). However, electrophoresis can be used to separate whole cells. Cell separations using this technique are based on the fact that the plasma membranes of cells contain ionized groups (e.g., proteins, sialic acid, short carbohydrate chains) that impart a net electrostatic charge to the cell surface. Different types of cells possess different net charges so that when they are placed in the appropriate conductive medium and subjected to an electrical current, they will migrate through the medium at different rates. Hence, they become separated into subpopulations that can be collected for further study. Various chemical substances may be applied to the cells in order to selectively alter their normal surface charge distribution and assist in their electrophoretic separation. Electrophoresis has been applied successfully to separations of microorganisms, blood cells, ascites tumor cells, HeLa cells, and other cells in culture.

Fluorescence–Activated Cell Sorting

Fluorescent dyes such as **fluorescein** can react with and bind to the surfaces of cells; the type and quantity of dye bound varies for different kinds of cells. This differential property has been used for years to *visually* distinguish different types of cells in a mixed population and very recently has been employed in an elegant instrument that physically separates the cells. The instrument, known as a **fluorescence–activated** (or **"multiparameter"**) **cell sorter** and depicted diagrammatically in Figure 12–18, has a complex history, but its development may be credited to the combined contributions of M. J. Fulwyler, L. A. Herzenberg, R. G. Sweet, W. A. Bonner, and H. R. Hulett. The fluorescein–treated suspension of cells is mixed with electrolyte solution ("sheath fluid") and forced downward

through a tiny nozzle vibrating at 40,000 cycles per second. The vibrations of the nozzle break the emerging stream into uniform droplets approximately equal in number to the frequency of vibration (i.e., 40,000 droplets per second). The population density of the original cell suspension and the flow rate are adjusted so that each droplet contains no more than one cell (indeed, most droplets contain no cells).

Just prior to droplet formation, the stream is illuminated with an argon–ion laser beam that excites the fluorescent material in the cell surfaces. Two detectors respectively measure the amount of fluorescent light and the volume of the cell and trigger an electrical pulse that charges each cell–containing droplet as it is formed. The amount and sign of the electrostatic charge borne by the droplet depend upon the size of the entrained cell and the number of fluorescein molecules bound to its surface. These charge parameters can be selected by the operator and effectively divide the droplets into *three* classes: positively charged, negatively charged, and uncharged. The droplets then pass between two electrostatic plates; the charged droplets are appropriately deflected as they pass through the field between these plates, while uncharged droplets continue on their original course. Finally, the three droplet streams are collected in reservoirs. The left and right streams contain different populations of cells, while the undeflected center stream consists primarily of empty droplets, unwanted cells, and debris. The fluorescence–activated cell sorter can separate about 5000 cells per second.

Harvesting Cells and Subcellular Components: Continuous–Flow Centrifugation

This discussion of methods for the differential isolation and separation of cells and subcellular components would not be complete without a description of the use of centrifugation for *harvesting* large quantities of particulate material from large–volume suspensions. The technique is generally known as **continuous–flow centrifugation.**

The most common application of continuous–flow centrifugation is the harvesting of bacteria, algae, protozoa, and other cells grown in multiliter cultures as a preliminary to chemical, physiological, or morphological analysis. However, the technique is also frequently employed (1) to collect cell–free culture media prior to the isolation and assay of cellular excretion products such as enzymes, vi-

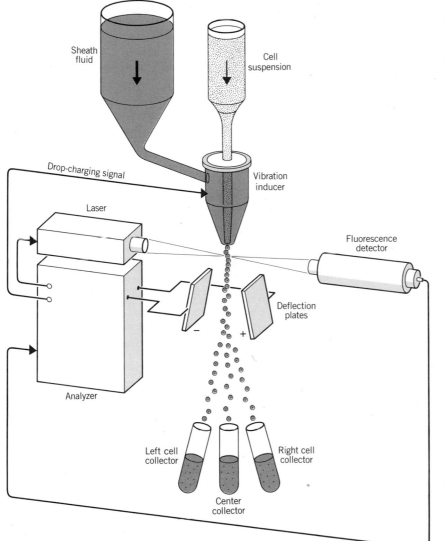

Figure 12-18
**The major components of a fluores-
cence-activated (or multiparameter)
cell sorter (see the text for additional
details.)**

tamins, and hormones; (2) to separate blood plasma from whole blood; (3) to remove the larger subcellular components such as nuclei, chloroplasts, and mitochondria from large volumes of tissue homogenates; and (4) to collect precipitates from large volumes of aqueous suspensions.

During continuous–flow centrifugation, the suspension of particles is introduced into the spinning centrifuge rotor as a continuous, uninterrupted stream. As the suspension passes through the rotor, particles are sedimented out of the stream and are trapped and concentrated within specific rotor chambers, while the clarified supernatant leaves the rotor and is collected separately. Continuous–flow centrifuge rotors thereby eliminate the need for a series of batch separations when very large volumes of particle suspensions must be processed. When processing is completed, the rotor is simply decelerated and opened, and the trapped cells or other particles removed. Although cells or other particles present in multiliter volumes can be harvested using conventional rotors, that approach is far less efficient. Even the largest conventional rotors generally accommodate only a few liters of suspension, so that a succession of spins is necessary when larger volumes of material must be handled. Equally important, the increased size and weight of these rotors restricts their maximum operating speeds and may necessitate extended centrifugation time in order to ensure total particle "cleanout." Since at any instant, continuous–flow rotors contain only a small fraction of the total volume of material to be centrifuged, they may be quite small. Thus, in addition to eliminating the need for successive runs, continuous–flow rotors can be operated at much higher speeds (hence greater RCF), providing more rapid and efficient particle cleanout.

The RCF experienced by the particles as they enter the collection chambers of the rotor depends on rotor speed and causes the particles to sediment at specific rates. If the rate of particle sedimentation is *greater* than the rate at which the surrounding liquid moves toward a centripetal exit port in the chamber, then the particles become trapped in the rotor. However, if the sedimentation rate is *less* than the rate of centripetal flow, the particles are carried toward the exit ports and out of the rotor. Usually flow rate and rotor speed are selected to provide maximum cleanout of the particle suspension. However, for heterogeneous populations of particles, the flow rate and rotor speed can often be adjusted so that a *differential fractionation* of the particles is achieved. That is, depending on their sizes, shapes, and densities, some particles will be trapped in the rotor, while others are conducted out of the rotor with the supernatant.

Summary

Tissue fractionation begins with the disruption of the cells and the preparation of a subcellular particle suspension or homogenate. Cells are most frequently disrupted using the shear forces generated by special *grinders, blenders, pressure cells,* or *insonators.* Chemical procedures involving lytic agents or osmotic pressure are also used in certain tissues.

Once the tissue homogenate is prepared, the method of choice for separating subcellular organelles and particles is **centrifugation.** Centrifugal fractionation may involve one or a combination of different approaches. In **differential centrifugation,** gross differences in the **sedimentation rates** of the subcellular particles are used to produce a series of particulate sediments at successively higher "*g* forces." Different families of particles may also be separated on the basis of size and/or density differences using the **density gradient** approach. A variety of centrifugal devices are used to effect the purification of subcellular particles, including conventional swinging–bucket and fixed–angle rotors and the more sophisticated vertical tube and zonal rotors.

Heterogeneous mixtures of very large particles, such as whole cells, may be fractionated by **centrifugal elutriation** or simply by **unit gravity** sedimentation. **Electronic sorting, countercurrent distribution,** and **electrophoresis** are alternative methods. When particularly large quantities of cells or particles are to be harvested, **continuous–flow** centrifugal procedures can be employed.

References and Suggested Reading

Articles and Reviews

Brakke, M. K., The origins of density gradient centrifugation. *Fractions No. 1,* 1 (1979).

de Duve, C., Tissue fractionation—past and present. *J. Cell Biol. 50,* 20 (1971).

de Duve, C., and Beaufay, H., A short history of tissue fractionation. *J. Cell Biol. 91,* 293 (1981).

Grabske, R. J., Separating cell populations by elutriation. *Fractions No. 1,* 1 (1978).

Herzenberg, L. A., Sweet, R. G., and Herzenberg, L. A., Fluorescence–activated cell sorting. *Sci. Am. 234,* 108 (1976).

Horan, P. K., and Wheeless, L. L., Quantitative single cell analysis and sorting. *Science 198,* 149 (1977).

Pretlow, T. G., Weir, E. E., and Zettergren, J. G., Problems connected with the separation of different kinds of cells. *Int. Rev. Exp. Pathol. 14,* 91 (1975).

Rembaum, A., and Dreyer, W. J., Immunomicrospheres: Reagents for cell labelling and separation. *Science 208,* 364 (1980).

Shortman, K., Physical procedures for the separation of animal cells. *Annu. Rev. Biophys. Bioengineering 1,* 93 (1972)

Books, Monographs, and Symposia

Anderson, N. G. (Ed.), *The Development of Zonal Centrifuges and Ancillary Systems for Tissue Fractionation and Analysis* (National Cancer Institute Monograph 21), U.S. Dept. Health, Education and Welfare, U.S. Govt. Printing Office, Washington, D.C., 1966.

Birnie, G. D., and Rickwood, D., (Eds.), *Centrifugal Separations in Molecular and Cell Biology,* Butterworths, London, 1978.

Cutts, J. H., *Cell Separation Methods in Hematology,* Academic Press, New York, 1970.

Hinton, R., and Dobrota, M., *Density Gradient Centrifugation,* North–Holland, Amsterdam, 1976.

Rickwood, D., (Ed.), *Centrifugation: A Practical Approach,* Information Retrieval, London, 1978.

Sheeler, P., *Centrifugation in Biology and Medical Science,* Wiley–Interscience, New York, 1981.

Chapter 13

THE ISOLATION AND CHARACTERIZATION OF CELLULAR MACROMOLECULES

Over the years, a number of sophisticated analytical and preparative techniques have been developed for separating, analyzing, and isolating the various macromolecular constituents of cells and tissues. Various forms of **electrophoresis, chromatography, gel filtration,** and **ultracentrifugation** are now in routine use in most laboratories engaged in molecular biological studies and have greatly increased our understanding of the chemistry and properties of the cellular macromolecules. Most of the methods that are used to separate and isolate different members of a class of macromolecules simultaneously provide information concerning their chemistry because parameters such as *molecular size, shape, density,* net and absolute *charge, differential solubility,* and so forth are used as the basis for the separation.

Nearly all of the techniques that are used for separating and isolating macromolecules require that these substances initially be in a soluble state. Consequently, macromolecules present in extracellular fluids such as plasma, lymph, and hormonal and digestive secretions are most easily isolated and have been the subject of the most intense studies to date (albumin, globulins, pepsin, trypsin, insulin, ACTH, etc.). However, if the component to be isolated is normally a constituent of the soluble phase of the cell or cytosol, then centrifugal isolation of this phase from disrupted cells quickly provides the starting material. Greater difficulty is encountered when macromolecules are to be isolated from particulate cell components such as nuclei, mitochondria, cellular membranes, and ribosomes, where the molecule may be an integral part of the organelle's structure.

In Chapter 12, a number of physical methods were described for disrupting cells. Often, more vigorous or more extensive applications of the same procedures to whole cells or to isolated organelles will also free some of the constituent macromolecules. For example, extended sonification or homogenization of mitochondrial suspensions solubilizes many of the mitochondrial enzymes and other constituents, and nucleic acids may be released from isolated nuclei under similar conditions. Chemical procedures may also be used to extract or solubilize the desired class of macromolecules. Lipids are often extracted from whole cells or isolated organelles using organic solvents such as chloroform–methanol mixtures or acetone. Proteins may be extracted from membranous elements using dissociating agents (such as urea and mercaptoethanol), chelating agents (such as ethylenediamine tetraacetic acid, EDTA) or organic detergents (such as sodium deoxycholate, sodium dodecyl sulfate, and Triton X–100). Whatever the method used, once a soluble mixture of macromolecules is obtained, these may then be separated from one another and isolated using one or a combination of methods.

In this chapter, a number of methods that are either in routine use by cell biologists for isolating and characterizing cellular macromolecules (proteins, carbohydrates, lipids, and nucleic acids) or of significant historical interest will be considered. Most of these methods (see Table 13–1) rely upon differences in molecular size, shape, electrostatic charge, solubility, or biological activity to effect the separation.

Salting In and Salting Out

Among the oldest methods for fractionating and isolating mixtures of macromolecules (especially proteins) are chemical procedures that differentially alter a molecule's solubility. Among other factors, a protein is maintained in the dissolved (i.e., solubilized) state by the interaction of its charged groups with water (which is partially polar) and

with salt ions in the solution (see Chapter 4). The salt concentration (or **ionic strength**) of the protein solution significantly and differentially influences the solubility of the proteins present. For example, many proteins are insoluble in pure (i.e., salt–free) water, whereas addition of small quantities of salt renders these molecules soluble. This effect is believed to be due to an interaction between the salt ions and certain charged groups of the protein that would otherwise react with each other, resulting in insolubility. Even proteins that are soluble in distilled water may be dissolved in much greater quantities by the simultaneous addition of small amounts of salts. This phenomenon is known as **salting in.**

If the salt concentration of a protein solution is successively increased, a point is eventually reached at which some of the proteins begin to precipitate. Further addition of salt results in greater precipitation. Precipitation occurs

Table 13–1
Methods for the Isolation and Characterization of Cellular Macromolecules

Method	Principal Impelling Factor(s)	Principal Retarding or Opposing Factor(s)	Separation Depends Primarily Upon
Countercurrent distribution	Mechanical	Solubility	Differential partition
Dialysis/ultrafiltration	Osmotic effects, concentration gradients, hydrodynamic force	Molecular sieve effects	Molecular size
Ultracentrifugation	Centrifugal force	Friction, bouyancy, diffusion	Molecular size, shape, effective density
Electrophoresis			
Moving-boundary	Electrostatic force	Friction, diffusion	Molecular ionic properties
Zone	Electrostatic force	Friction, diffusion, molecular sieve effects	Molecular ionic properties
Discontinuous	Electrostatic force, Kohlrausch function	Friction, diffusion, molecular sieve effects	Molecular ionic properties and molecular size
Immunoelectrophoresis	Electrostatic force	Diffusion, molecular sieve effects	Molecular ionic properties, biological activity
Isoelectric focusing	Electrostatic force	Diffusion	Molecular ionic properties
Paper chromatography	Hydrodynamic force	Association/dissociation effects, diffusion	Adsorption/partition differences
Thin-layer chromatography	Hydrodynamic force	Association/dissociation effects, diffusion	Adsorption/partition differences
Ion-exchange chromatography	Hydrodynamic force	Electrostatic forces	Molecular ionic properties
Affinity chromatography	Hydrodynamic force	Molecular affinity	Biological activity
Gel filtration	Hydrodynamic force	Molecular sieve effects	Molecular size
Gas chromatography	Gas pressure	Diffusion	Adsorption/partition differences

because spheres of hydration formed around the salt ions effectively "remove" the water molecules necessary to hydrate certain surface charges of the protein (see Chapter 3). Protein–protein interactions begin to dominate over protein–solvent interactions with the resulting precipitation of the proteins. In effect, the proteins are being "squeezed out of solution." The phenomenon is called **salting out.** The effect of salt concentration on the solubility of hemoglobin is shown in Figure 13–1.

Because different proteins are salted out at different salt concentrations, solutions containing mixtures of different proteins may be fractionated using this approach. Among the most popular salts used is ammonium sulfate, $(NH_4)_2SO_4$, because of its high solubility in water and its high ionic strength and because it does not irreversibly denature most proteins. The *ionic strength* (**I**) of a salt solution is given by

$$I = \frac{\Sigma M_n Z_n^2}{2} \qquad (13\text{–}1)$$

where M is the molar concentration of each ionic species present and Z is the electrostatic charge of each ion. Thus, the ionic strength of a 1.0 M NaCl solution is $[(1)(1)^2 + (1)(1)^2]/2$ or 1.0, whereas the ionic strength of a 1.0 M $(NH_4)_2SO_4$ solution is $[(2)(1)^2 + (1)(2)^2]/2$ or 3.0. The ammonium sulfate solution has three times the ionic strength of the NaCl solution because of the different concentrations of ions and different numbers of charges per ion that result from dissociation; i.e.,

$$1\ M\ NaCl \longrightarrow 1\ M\ Na^+ + 1\ M\ Cl^-$$

$$1\ M\ (NH_4)_2SO_4 \longrightarrow 2\ M\ NH_4^+ + 1\ M\ SO_4^=$$

To fractionate a protein solution using ammonium sulfate, the salt concentration of the solution is serially increased, each increment precipitating another group of proteins, which are then removed (usually by filtration or centrifugation) before the next addition of salt.

Isoelectric Precipitation

The distribution of polar groups in proteins and nucleic acids is influenced by pH. For a given protein, a pH may be identified at which there are equal amounts of positive

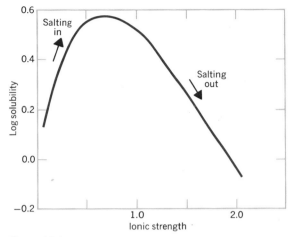

Figure 13-1
Salting in and ***salting out*** of carbonmonoxyhemoglobin using K_2SO_4.

and negative charge, and this is known as the **isoelectric point** (or **isoelectric pH**). Most proteins are *least* soluble and many proteins are *insoluble* at the isoelectric point, so that their removal from solution is most easily achieved by first adjusting the pH of the solution to the isoelectric point. Electrostatic charge distribution and the isoelectric properties of proteins and other macromolecules are considered in detail later in the chapter in connection with electrophoresis.

Dialysis and Ultrafiltration

Semipermeable membranes, such as those prepared from *cellophane* or *collodion* (cellulose nitrate), may be used to separate solutes on the basis of molecular weight differences. The technique, called **dialysis,** was first described in 1861 by T. Graham. In dialysis, the solute mixture is placed in a bag formed from tubular sheets of the semipermeable membrane, and the bag is immersed in an aqueous medium (e.g., distilled water). Molecules larger than the pores of the membrane are confined to the tubing, while smaller molecules diffuse into the surrounding liquid (Fig. 13–2). Semipermeable membranes can be treated chemically or physically in order to alter the sizes of the pores so that solutes of varying molecular weight are rendered permeable. Generally, dialysis is used with unmodified membranes to quickly separate low–molecular–weight solutes (i.e., molecular weight of less than 5000) such as

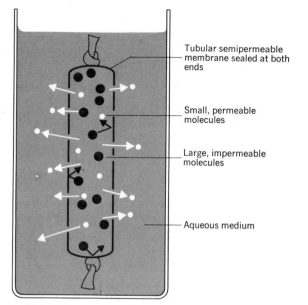

Figure 13-2
Dialysis using a semipermeable membrane. Despite the concentration gradient, the large molecules are unable to diffuse through the pores of the membrane.

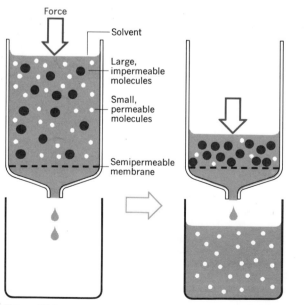

Figure 13-3
Ultrafiltration through a semipermeable membrane using hydrodynamic force.

Table 13–2
Relationship Between Molecular Weight and Permeability to Modified Cellophane Membranes

Solute	Molecular Weight	$T_{1/2}$ (min)[a]
Tryptophan	204	4
Bacitracin	1,422	15
Cytochrome *c*	12,000	60
Ribonuclease	13,600	120
Lysozyme	14,000	138
Trypsin	20,000	240
Chymotrypsin	25,000	300
Pepsin	35,500	4,800

[a] $T_{1/2}$ is the amount of time required for one-half of the solute to permeate the membrane.

salts, sugars, and amino acids from proteins and polysaccharides present in the solution. However, high–molecular–weight solutes do *differentially* penetrate membranes having large pore sizes (Table 13–2).

In **ultrafiltration,** force is used to drive the smaller molecules *along with solvent* through the semipermeable membrane. As a result, not only are permeable and impermeable molecules separated but the impermeable species is also simultaneously *concentrated* (Fig. 13–3).

Ultracentrifugation

Centrifugation can be employed not only for the separation of cells, subcellular organelles, and other particulate constituents of cells but also for molecular separations. Since its initial development in the 1920s by T. Svedberg, the **analytical ultracentrifuge** (described in Chapter 12) has been used repeatedly to evaluate the heterogeneity or purity of molecular constituents extracted from cells and to estimate molecular sizes on the basis of sedimentation rate. Preparative separations (as opposed to analytical studies) became possible with the development of ultracentrifuges capable of spinning rotors that could hold large amounts of sample and which could attain speeds producing an RCF in excess of 500,000 *g*. Using forces of this magnitude, true separations and isolations of molecular constituents of cells in rate or isopycnic gradients have become routine.

The principles of analytical and preparative centrifugation were considered at length in Chapter 12 and will not be pursued here. However, it should be noted that certain

across the tube, the protein molecules migrated toward one or the other electrode. Since different proteins migrate at different rates, a number of *interfaces* or *boundaries* formed between the leading (and trailing) edge of each protein type and the remaining mixture. A Schlieren optical system recorded the number of boundaries formed and their rates of migration. The moving–boundary technique was used for the analysis but not for the fractionation of complex mixtures of proteins, since the proteins in the mixture were not really separated from each other.

Zone Electrophoresis

Zone electrophoresis offers a number of important advantages over moving–boundary electrophoresis. In zone electrophoresis, the separation of the different proteins in a sample is in fact realized. For this reason, the technique can be *both* analytical and preparative. Generally, zone electrophoresis also yields greater resolution of the protein components of a mixture. In zone electrophoresis, the proteins are separated in a semisolid or porous supporting medium such as filter paper or various types of gels (polyacrylamide, cellulose acetate, hydrolyzed starch, etc.); these usually take the form of narrow sheets, slabs, or columns.

The principles of zone electrophoresis may be illustrated by considering the filter paper technique as an example. The strip of filter paper is saturated with the buffer/electro-

Table 13–3

Isoelectric Points of Some Proteins

Protein	Isoelectric pH
Lysozyme	11.0
Cytochrome *c*	10.6
Ribonuclease	9.5
Normal human hemoglobin	7.1
Myoglobin	7.0
Horse hemoglobin	6.9
Transferrin	5.9
Fibrinogen	5.8
Insulin	5.4
Beta lactoglobulin	5.1
Urease	5.0
Plasma albumin	4.8
Egg albumin	4.6
Haptoglobin	4.1
Pepsin	1.0

lyte solution to be used and is tautly suspended between two (inner) baths (Fig. 13–6). The inner baths communicate with two additional (outer) baths containing the electrodes; the connections may be achieved using baffles or wicks saturated with the buffer/electrolyte. The electrodes are not inserted directly into the inner baths because **electrolysis** would dramatically alter the pH of the bath and also the filter paper strip. By confining the electrolysis to the outer

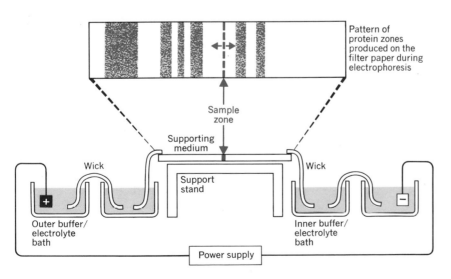

Figure 13-6

Essential components of a zone electrophoresis assembly (see text for details).

baths, the pH of the inner baths and the filter paper remains constant. The mixture of proteins to be separated is applied as a narrow zone perpendicular to the long axis of the filter paper.

When an electrical potential is applied, the proteins in the sample zone migrate through the filter paper toward the appropriate electrode. In so doing, they form a number of discrete zones distributed along the length of the paper (Fig. 13–6). Following separation, the individual zones or bands can be visualized using special stains. Alternatively, the paper strip may be cut into a number of sections and the proteins in each eluted and collected for further study.

Since the medium through which the proteins migrate is porous, the size of the protein influences its electrophoretic mobility—smaller proteins migrating more rapidly than larger proteins of equal net charge. Consequently, the sieving effect of the supporting medium results in a separation based upon both molecular size and charge. Polyacrylamide gels are especially effective for zone electrophoretic separations, since the sieving effect may be varied by changing the concentration of acrylamide used to prepare the gel (i.e., the greater the acrylamide concentration, the smaller the pore size in the gel).

An interesting modification of zone electrophoresis is ''diagonal electrophoresis'' or *two–dimensional* electrophoresis. In this approach, the sample is separated first on a rectangular sheet of filter paper or polyacrylamide gel at high pH; then, after the supporting medium is rotated 90° and equilibrated at low pH, the separated zones are electrophoresed again. The result is a better resolved distribution of the components in the original sample, since it is highly unlikely that two components would have the same electrophoretic mobility at grossly different pH's.

Disc or Discontinuous Electrophoresis.   Discontinuous electrophoresis is a specialized form of zone electrophoresis developed in the early 1960s by L. Ornstein and B. Davis. In view of its widespread use and importance as a research tool, it will be considered separately. The procedure provides the highest degree of resolution so far attainable by electrophoretic methods. This is a consequence of the extreme thinness of the sample zone. Disc electrophoresis is carried out in either columns or sheets of polyacrylamide gel, the pore sizes of which may be accurately controlled by regulating the concentration of acrylamide in the gel. Consequently, the gel acts like a molecular sieve, and separation is based on both the charge and the size of the macromolecule.

In order to separate two proteins by conventional methods of zone electrophoresis, it is necessary to permit migration to continue until one of the proteins has traveled at least one starting zone thickness further than the other. For example, suppose that the sample zone was 1 mm thick and contained two different proteins. If one of the proteins had no electrophoretic mobility at all under the pH conditions employed, then the other would have to migrate *at least* 1 mm in order for separation of the two proteins to occur. If both proteins migrate in the same direction but at different velocities, then an even greater distance must be traversed by the protein of higher electrophoretic mobility. Since the sharpness of the zone occupied by each protein decreases with time owing to diffusion in the gel, it is most desirable to begin with the narrowest starting zone possible and to minimize the duration of electrophoresis. In disc electrophoresis, the sample zone undergoes a preliminary concentration during which its thickness is reduced from one or more centimeters to just a few hundredths of a centimeter; as a result, high resolution is achieved during very brief runs. The preliminary concentration also makes it possible to analyze samples too small or too dilute to be studied using other methods.

The ability to concentrate and narrow the starting zone during disc electrophoresis is based on a phenomenon first noted and described by F. Kohlrausch in 1897. He showed that if two solutions of ions having significantly different electrophoretic mobilities are layered over one another (e.g., slow ion above, fast ion below) and subjected to an electrical field, the boundary between the two ionic species would be sharply maintained as the ions migrate. That is, under these conditions the ions of lower electrophoretic mobility migrate at the same rates as the faster ions. This may be explained as follows: The velocity at which an ion migrates in an electrical field is determined by the product of its electrophoretic mobility and the applied voltage gradient (volts/cm). Therefore, ions of low mobility can migrate as rapidly as ions of high mobility if their mobility–voltage products are equal. In the example cited above, at the instant voltage is applied, the trailing edge of the fast ions at the boundary moves away from the leading edge of the slow ions, resulting in the temporary formation of a zone of reduced conductivity and increased field strength. The increased field strength in this zone accelerates the

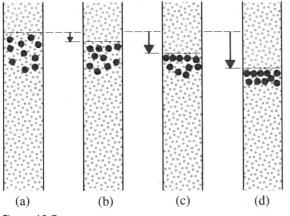

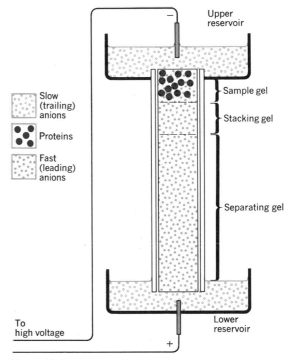

Figure 13-7
Concentration of ions of intermediate electrophoretic mobility (color) at the boundary between fast (small, closed circles) and slow (small, open circles) ions. (a) Before an electrical potential is applied, the ions of intermediate mobility occupy a region below the slow ion–fast ion boundary. (b) When an electrical potential is applied, the boundary moves downward and an increased voltage gradient is formed behind. (c) and (d) The increased voltage gradient accelerates both the slow ions and the ions of intermediate mobility, causing the latter to be concentrated into a narrow zone between the trailing edge of the fast ions and the leading edge of the slow ions. This phenomenon underlies the formation of extremely thin starting zones during disc electrophoresis of proteins.

Figure 13-8
Essential components of discontinuous electrophoresis (see text for details).

slow ions to keep up with the fast ions, thereby creating a steady state in which the mobility–voltage products of the two ions remain constant.

Taking this a step further, consider the situation in which a region below the fast ion–slow ion boundary contains a number of ions of intermediate electrophoretic mobility (Fig. 13–7a). When an electrical potential is applied, the increased voltage gradient behind the downward–moving boundary accelerates both the ions of low mobility and those of intermediate mobility. The mobility–voltage product of the latter, however, will be somewhat greater, so that these ions will be swept up by the boundary and form a zone of continuously decreasing thickness between the fast and slow ions (Fig. 13–7b, c, d). If instead of a single ionic species of intermediate mobility, a mixture of ions is placed in the region below the slow ion–fast ion boundary, these would be concentrated into narrow zones stacked one above the other in order of decreasing mobility. In disc electrophoresis, conditions are chosen such that the proteins

in the sample have mobilities intermediate to two specially selected fast and slow ions.

Disc electrophoresis gels are cylindrical blocks or flat sheets of polyacrylamide suspended between two reservoirs of buffer/electrolyte (Fig. 13–8). Usually, the polyacrylamide gel is divided into three regions called the **sample gel, stacking gel,** and **separating gel.** The sample gel contains the mixture of proteins to be separated and is prepared using low concentrations of acrylamide so that pore sizes are large and do not influence the rates of migration of different size proteins. The stacking gel is similar to the sample gel but lacks the proteins. The sample gel, stacking gel, and reservoirs have the same pH (usually 8.3). The separating gel differs from the other two regions in that it has a higher pH (usually 9.5) and is prepared with greater concentrations of acrylamide; this results in smaller pore sizes and provides the sieving effect. All three regions of the polyacrylamide gel contain the fast (leading) ion, whereas the buffer/electrolyte contains the slow (trailing) ion. In most instances, Cl⁻ serves as the fast ion and glycine (NH_2–CH_2–COO^-) as the slow ion; other amino acids or weak acids may also serve as the slow ion. At pH 8.3, nearly all proteins have an electrophoretic mobility between those of Cl⁻ and glycine.

When an electrical potential is applied across the gel, the voltage gradient causes the chloride ions to migrate down from the top of the gel. The increased field strength created immediately behind the trailing edge of the chloride ions accelerates the slower glycine ions to keep pace. As the Cl⁻–glycine boundary moves downward, it overtakes the more slowly migrating proteins, and they too are accelerated by the increased field strength behind the boundary. As a result, by the time the trailing edge of the chloride ions reaches the end of the stacking gel, all the proteins in the original sample have been concentrated into a series of contiguous thin zones.

Once the separating gel is reached, the change in pH dramatically alters the electrophoretic mobility of the trailing ion. In the case of glycine, the degree of ionization is very low at pH 8.3 but is several times greater at pH 9.5. As a result, the glycine ions now overtake each of the protein zones and catch up with the trailing edge of the Cl⁻ (the mobility of Cl⁻ is unaffected by the pH change), and the new interface then migrates rapidly through the remainder of the separating gel. The proteins, on the other hand, are physically retarded by the smaller pores in the separating gel, and since each protein zone is now in a uniform electrical field, these separate from one another strictly on the basis of net charge and size difference (just as in conventional zone electrophoresis).

It is apparent from the foregoing discussion that disc electrophoresis differs markedly from conventional zone electrophoresis and is a clever and unusual technique providing extraordinary resolution (Fig. 13–9). The method has been widely accepted since its introduction and has been applied in diverse cell studies, especially in the analysis of the protein components of isolated cell organelles. An example of this is given in Figure 13–10, which shows more than 30 different protein components separated from the plasma membranes of liver cells.

SDS–Gel Electrophoresis.

SDS (sodium dodecyl sulfate) is an ionic detergent that has been used for many years to solubilize membrane proteins so that they can be separated either electrophoretically or by other methods. Since mem-

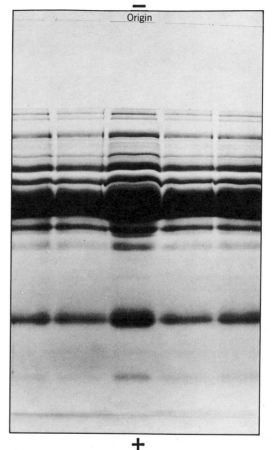

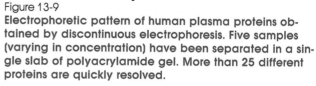

Figure 13-9
Electrophoretic pattern of human plasma proteins obtained by discontinuous electrophoresis. Five samples (varying in concentration) have been separated in a single slab of polyacrylamide gel. More than 25 different proteins are quickly resolved.

brane proteins are not usually soluble in aqueous solutions, the SDS is also incorporated into the separating medium (e.g., the electrophoresis gel); this prevents the proteins from forming an intractable precipitate once they enter the gel. SDS dissociates oligomeric proteins into their subunits and forms ionic bonds with the polypeptides. Therefore, the SDS alters the protein's native electrostatic charge and adds to its mass. As a result, when electrophoresis is carried out in small–pore polyacrylamide gels containing SDS, the proteins migrate through the gel at rates determined principally by molecular size (the smaller polypeptides traveling further through the gel than the larger polypeptides). If in

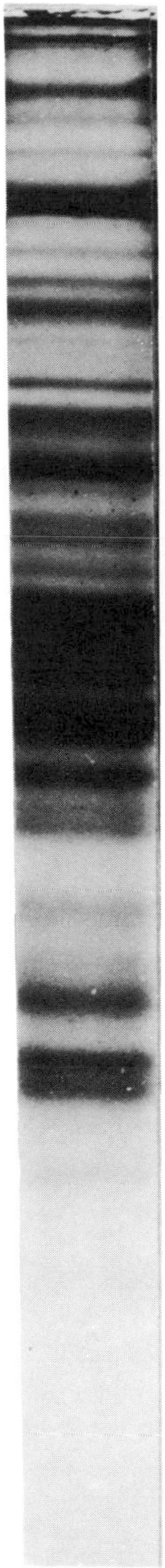

Figure 13-10
Discontinuous electrophoretic separation of proteins extracted from isolated plasma membranes of liver cells.

two neighboring *tracks* or *lanes* of the same gel, proteins of known molecular weights are electrophoresed alongside a sample containing proteins of unknown molecular weights, the latter's molecular weights can be estimated by comparing their final positions with those of the "markers." Hence, SDS–gel electrophoresis is a powerful tool for both separating and sizing macromolecules.

The Maxam–Gilbert Technique. In recent years, the SDS–gel electrophoresis method has been extended to the separation of polynucleotide fragments obtained by enzymatic cleavage of DNA and RNA. Not only are the polynucleotide fragments separated on the basis of size differences, but the *order* of the nitrogenous bases in the original DNA or RNA molecule is revealed. The technique is known by the names of its developers, A. M. Maxam and W. Gilbert.

To illustrate this ingenious and most valuable procedure, suppose that the base sequence of the following polynucleotide is to be determined:

$$5'\text{–CTACGTAG–}3'$$

After their $5'$ ends are labeled with radioactive phosphate, the polynucleotides are subjected to enzymatic cleavage using **endonucleases** that introduce breaks on the $5'$ side of specific bases. The amounts of the endonucleases and the lengths of the exposure times permit only *one cleavage per polynucleotide*. As a result, the following radioactive fragments are formed:

1. Cleavage at **A** yields **$5'$–CT–$3'$** and **$5'$–CTACGT–$3'$** fragments

2. Cleavage at **T** yields **$5'$–C–$3'$** and **$5'$–CTACG–$3'$** fragments

3. Cleavage at **G** yields **$5'$–CTAC–$3'$** and **$5'$–CTACGTA–$3'$** fragments

4. Cleavage at **C** yields **$5'$–CTA–$3'$** fragments

Other fragments are also formed, but they are not radioactive because they do not contain the original $5'$ ends.

When the four families of fragments are electrophoresed in neighboring lanes of an SDS–polyacrylamide gel, the resulting pattern of radioactive fragments revealed by *autoradiography* (Chapter 14) takes the form shown in Figure 13–11. Fragment **$5'$–CTACGTA–$3'$** produced by cleavage at G migrates the most slowly because it is the largest fragment. Therefore, lane G has the slowest band; migrat-

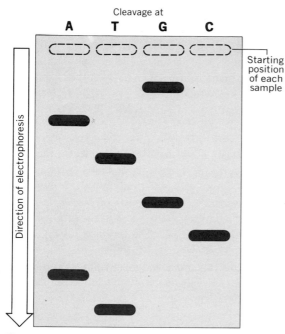

Figure 13-11

Maxam-Gilbert procedure for determining the base sequence of polynucleotides. The colored zones show the final positions of the *radioactive* fragments (see text for details).

ing more rapidly is a band in the A lane, then the T lane, and so on. Reading in order beginning at the top of the pattern, we obtain **GATGCAT,** which is the base sequence of polynucleotide beginning at its 3′ end. Using the Maxam–Gilbert technique, the base sequences in fragments containing 150 nucleotides are readily determined.

Immunoelectrophoresis

Immunoelectrophoresis, a technique developed in the 1950s by C. A. Williams and P. Grabar, combines the principles of electrophoresis and immunochemistry in order to separate and identify antigens and antibodies. In this technique, a portion of the sample of antigens to be analyzed is first injected into an experimental animal (usually a horse or cow). After a period of time during which the animal produces antibodies against the injected antigens, the immunoglobulin–rich blood serum is collected and prepared as the *antiserum*. The remaining portion of the sample to be analyzed is then subjected to conventional zone electrophoresis in a rectangular block of agar or other supporting

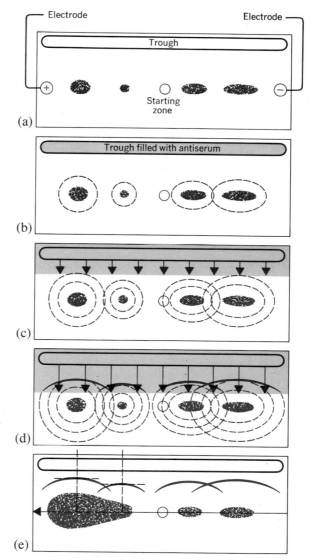

Figure 13-12

Immunoelectrophoresis. (*a*) Mixture of proteins separated by zone electrophoresis. (*b*) and (*c*) Trough filled with antiserum followed by diffusion of protein zones and antibodies through the gel. (*d*) Antibody–antigen reaction forms lines of precipitate (arcs) where the two diffusing fronts meet. (*e*) By drawing a perpendicular line from the widest point of the arc to the path of electrophoretic migration, the center of concentration of each protein can be determined.

medium. This distributes the various proteins in the original sample into a linear series of zones (Fig. 13–12*a*). A trough cut in the agar block parallel to the direction of electropho-

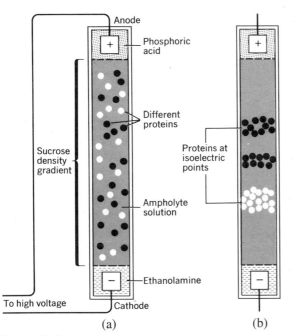

Figure 13-13
Essential features of isoelectric focusing. In (a) three proteins having different isoelectric points are shown distributed through the ampholyte-sucrose density gradient. When an electrical potential is applied, the ampholytes migrate to their isoionic points establishing a pH gradient. This is followed (b) by electrophoresis of proteins to their isoelectric positions in the gradient, forming a series of narrow zones.

retic migration is then filled with the antiserum. The protein zones in the agar block diffuse radially and eventually encounter the front of inwardly diffusing antibody, where the resulting antibody–antigen reaction produces a fine, arc–shaped line of precipitation (Fig. 13–12d). Each line of precipitation corresponds to a *pair* of precipitants (a protein in the sample plus an antibody from the antiserum); crossed arcs indicate that all the reactants are immunochemically different proteins.

Isoelectric Focusing

Isoelectric focusing is a form of electrophoresis used to determine the isoelectric pH's of proteins and to separate proteins from one another on the basis of differences in their isoelectric points. The method was developed in the 1960s by H. Svensson and O. Vesterberg and is extremely sensitive; proteins having isoelectric pH's that differ by as little as 0.0025 units may be resolved. In the preparative form of isoelectric focusing, a glass column is filled with a solution of synthetic charged ampholytes (aliphatic aminocarboxylic acids of various molecular weights) having a broad and continuous range of isoionic points. The ends of the column contain the electrodes and electrode solutions: a strong organic base for the cathode (usually ethanolamine) and a strong acid for the anode (usually phosphoric acid). The distribution of constituents in the column is stabilized by the presence of a sucrose density gradient that also prevents convection during electrophoresis. The mixture of proteins to be separated may initially be confined to a specific region of the column or uniformly distributed through the density gradient (Fig. 13–13).

When a potential is applied to the electrodes, the ampholytes migrate in the column to positions where they are equally attracted by both electrodes. Thus, they arrange themselves in order of their isoelectric points, the most acidic ampholytes being located near the anode and the most basic near the cathode. An examination of the pH in different regions of the column at this time would reveal that a pH gradient has been established in which pH increases in progressing from the anode to the cathode. Depending upon the pH of the surrounding solution and the nature of the amino acid side chains present in the molecule, each protein assumes a characteristic net charge. Proteins close to the anode will bear a net positive charge and be repelled by that electrode, while the same proteins near the cathode carry net negative charges and are repelled by that electrode. As a result, most proteins will migrate away from the electrodes and toward the center of the column. As this occurs, the proteins pass through the pH gradient, and the charges on their amino acid side chains are altered. Proteins migrating away from the anode become less and less positive as they pass through regions of increasing pH, whereas proteins migrating away from the cathode become less negative (see Fig. 13–13). Eventually, the net charge of each protein becomes zero as that region of the pH gradient corresponding to the isoelectric pH is reached, and at that time the electrophoretic migration of the protein ceases. Consequently, the proteins become distributed through the gradient as a series of narrow zones in order of their isoelectric points (see Fig. 13–13). Once the migration of all proteins terminates, the contents of the column may be drained and the separated proteins collected as a series of fractions and further studied. The isoelectric point of each protein is determined from the pH profile of the collected gradient.

For analytical applications, the liquid density gradient is replaced by a solid medium such as polyacrylamide gel. Several different samples, including standards of reference, can be focused in neighboring lanes of the gel.

A description of all electrophoretic methods currently in use is beyond the scope of this discussion. However, it should be recognized that other forms of electrophoresis such as *density gradient* electrophoresis and *continuous–flow* electrophoresis are also regularly employed. In density gradient electrophoresis, the separation of proteins is carried out in a glass column filled with buffer and stabilized by a density gradient. Continuous–flow electrophoresis provides for the continuous application of small volumes of sample onto the supporting medium, coupled with the continuous removal of the proteins as separation is achieved. Both zone electrophoresis and isoelectric focusing have continuous flow formats. With this technique it is possible to process very large volumes of starting material.

Although all methods of electrophoresis have proven extremely valuable for the separation and identification of proteins, certain limitations of the technique should be noted. The resolution of a protein mixture into a number of discrete zones does not guarantee that all the different proteins present in the original sample have been separated, for two or more different proteins having similar sizes may also have the same net charge under a given set of conditions and display the same electrophoretic mobilities (see Fig. 13–5). These proteins would not be resolved by zone electrophoresis. By the same token, two different proteins may have the same isoelectric pH and would not be separated by isoelectric focusing. Therefore, in order to evaluate the effectiveness of a separation and to determine the purity of the separated fractions, it is necessary to carry out electrophoresis under a variety of pH conditions or with a variety of supporting media or even to apply a combination of altogether different methods in the analysis.

Countercurrent Distribution

A brief, preliminary description of countercurrent distribution was given in Chapter 12 in connection with the separation of whole cells in mixed populations (e.g., blood). However, the technique lends itself more directly to the separation of various molecular species in solution; and because the physicochemical principles in effect during countercurrent distribution are the foundation of many other separation procedures (paper chromatography, thin–layer chromatography, ion–exchange chromatography, affinity chromatography, gas chromatography, etc.), this method will be treated in some detail.

It has been known for a century or more that many solutes differentially distribute or *partition* themselves between the separated phases of two immiscible liquids. Solutes having gross differences in their physical properties may therefore be separated by one or a few simple extraction procedures (e.g., using separatory funnels). Reasoning that the separation of a complex mixture of solutes on the basis of their partition differences could be made considerably more effective using a *cascade* of several dozen, several hundred, or even several thousand discrete extraction steps, L. C. Craig in the late 1940s and early 1950s developed a number of special instruments to achieve this. These instruments and their more modern counterparts are generally known as Craig apparatuses and the technique as **countercurrent distribution.**

The fundamental type of Craig apparatus consists of a series or *train* of chambers or "cells" divided into upper and lower compartments that house the respective immiscible liquid phases (e.g., butanol/water, phenol/water, propanol/water). The upper phase is usually the **mobile** phase, and the lower phase is the **stationary** phase. Equilibration of a dissolved solute between the phases within a chamber is achieved by agitation, so that one phase becomes finely dispersed in the other (thereby greatly increasing the interfacial surface area between the two phases). Once equilibration is achieved, the phases are allowed to separate. The mobile phase is then transferred to the next chamber of the train and mixed with a fresh volume of stationary phase. Simultaneously, a fresh volume of mobile phase is added to the first chamber. Agitation, equilibration, separation, and transfer are then serially repeated for the entire train of chambers.

A solute dissolved in a given pair of immiscible liquids will partition itself in a characteristic way—a specific solute concentration ending up in each phase. The **partition coefficient,** *P*, is given by

$$P = \frac{\text{concentration of solute in upper phase}}{\text{concentration of solute in lower phase}} = \frac{(S)_U}{(S)_L} \quad (13–3)$$

For example, if a solute has a partition coefficient of 1.0 in a butanol/water mixture, it will be *equally* divided between the butanol and the water phases once equilibration and separation

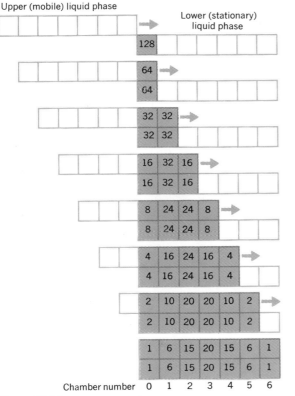

Upper (mobile) liquid phase

Lower (stationary) liquid phase

Chamber number 0 1 2 3 4 5 6

Figure 13-14
Countercurrent distribution of 128 molecules of a hypothetical solute having a P value of 1.0 in a train of seven chambers. At the outset, 128 molecules are dissolved in the lower half of chamber 0 and then equilibrated with an equal volume of mobile phase; this results in the equal partition of the solute in the two solvents. The upper phase of chamber 0 is then transferred and mixed with the lower phase of chamber 1, while fresh upper phase is added to chamber 0. Equilibration produces a new partition in chambers 0 and 1. The process is repeated until all seven chambers (i.e., C_0 through C_6) of the train have been used, yielding a final distribution of the solute. In practice, from several dozen to several thousand partition steps may be employed.

P \ C	0	1	2	3	4	5	6
0.2	33	40	20	5	1	0	0
0.5	9	26	33	22	8	2	0
1.0	2	9	23	31	23	9	2
2.0	0	2	8	22	33	26	9
3.0	0	0	3	13	30	36	18

Rounded off to the nearest whole percent.

Figure 13-15
Percentage distributions in each chamber of a seven-chamber train of five solutes having various P values. Attention is drawn (color shading) to those chambers containing the peak concentration of each solute.

have been achieved. Note that the solute is soluble in *both* liquids but partitions itself between the two in a characteristic way.

Figure 13–14 depicts the countercurrent distribution of 128 molecules of a hypothetical solute having $P = 1.0$ when carried through a train consisting of seven sets of chambers. In this example, the peak of the solute distribution in the train occurs in chamber number 3, which contains 40 molecules (or 31%) of the original solute. For a train consisting of $(C + 1)$ chambers (numbered C_0, C_1, C_2, C_3, etc.), each containing phases

of equal volume, the fraction F of solute in compartment C_n at the conclusion of the countercurrent distribution is given by

$$F = \frac{C!P^n}{n!\,(C-n)!\,(P+1)^C} \qquad (13\text{--}4)$$

where n is the number of transfers needed to reach chamber C_n (and is equal to the chamber number). For the example given in Figure 13–14, the fraction of solute in chamber C_3 would be $6!(1.0)^3/3!(6-3)!(2.0)^6$, which equals 31%, as already noted. Figure 13–15 tabulates the final distributions of a mixture of five solutes having different P values through a train of seven chambers and illustrates how the peak concentrations of each solute would be found in different chambers in the train. Although it may be noted that in this example considerable overlap exists among the chambers, it should be borne in mind that the number of chambers that make up the train has deliberately been kept low in order to simplify the example. Usually, a train consists of at least several dozen chambers. Accordingly, Figure 13–16 shows the resolution attainable for solutes having P values of 0.5 and 2.0 when the train consists of 30 chambers. It can be shown that the best resolution of two solutes, A and B, is achieved when $P_A \times P_B = 1.0$. The number of chambers occupied by a particular solute as a function of the total number of chambers present in the train rapidly diminishes as n increases. Hence, it is clear why trains in which hundreds or even thousands of transfers are possible may be used to provide maximum resolution of the solutes present in the original mixture.

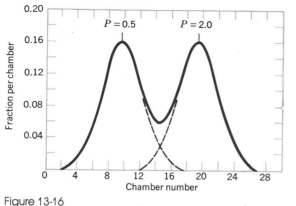

Figure 13-16

Countercurrent distribution of solutes having *P* values of 0.5 and 2.0 in a train consisting of 30 chambers.

The principles and mathematical relationships given above for a series of discrete partition transfers during countercurrent distribution apply equally well to seemingly quite distinct physical separation methods such as the various forms of chromatography described later in the chapter. In chromatography, the partition steps occur on numerous particles of tightly hydrated media such as starch, agarose, cellulose, silica, and paper fibers. This tightly bound *solvent* serves as the stationary phase, while a mobile phase containing solvent and the solute mixture to be fractionated flows by. This *continuous* process may be thought of as an infinite sequence of discrete, microscopic partition steps. Each microscopic step would be analogous to the stages shown in Figure 13–14, the upper mobile phase representing a continuously flowing solvent, and the lower stationary phase, the stationary, hydrated supporting medium. Although equilibrium is not truly achieved at each "step," the total number of steps is so great that the solute mixture is effectively fractionated.

Countercurrent distribution is particularly effective when dealing with molecules having molecular weights below 5000 (peptides, oligonucleotides, phospholipids, etc.), but the technique is also used successfully with larger solutes. The first protein isolated and purified using countercurrent distribution was the hormone insulin, which has a relatively low molecular weight (6500). However, ribonuclease (M. W. 13,600), lysozyme (M. W. 14,000), albumin (M. W. 68,000), and the α and β chains (M. W. 16,000) of human hemoglobin have also been isolated using this approach. Little success is obtained with polysaccharides

because of the problem of finding an organic phase providing useful partition coefficient differences. Ribonucleic acids are readily separated using countercurrent distribution; indeed, R. W. Holley's pioneering analysis of the primary structure of transfer RNA (Chapter 22) was carried out using alanyl–tRNA isolated and purified by countercurrent distribution. Lipids, especially steroids, are well suited for separation using this approach because of their greater solubility in organic solvents and the wide choice of solvent combinations that can be used.

Paper Chromatography

Paper chromatography is a technique in which a mixture of solutes is separated into discrete zones on a sheet of filter paper on the basis of (1) differences in solute partition between a stationary aqueous phase tightly bound to the cellulose fibers of the paper and a mobile organic liquid phase passing through the sheet by capillary action (i.e., liquid–liquid partition) and (2) differences in solute *adsorption* to the cellulose fibers and dissolution in the mobile liquid (i.e., solid–liquid partition). Although the separation is based upon a combination of both phenomena, liquid–liquid partition differences are the more significant. The principles of the technique were set down in the 1940s by R. Consden, A. H. Gordon, A. J. Martin, and R. L. Synge and are not unlike those that are in effect during countercurrent distribution.

In practice, a rectangular sheet of filter paper is saturated with the aqueous phase and allowed to air dry, and the sample is applied near the end of the sheet as a narrow zone. The sheet is then suspended in a closed chamber in which the air has been saturated with the vapors of the mobile organic phase, the edge of the paper immersed in a bath containing the mobile phase (Fig. 13–17a). Capillary action causes the mobile phase to slowly percolate through the paper from one end (the end containing the mixture to be separated) to the other. Movement of the liquid may be downward (descending chromatography) or upward (ascending chromatography). The solute mixture differentially partitions itself between the flowing solvent and the stationary phase time and time again as the solvent front advances toward the edge of the paper. Usually, the solvent front is allowed to migrate through the paper until it has almost reached the other end, at which time the sheet is removed and dried and the solute zones located by the

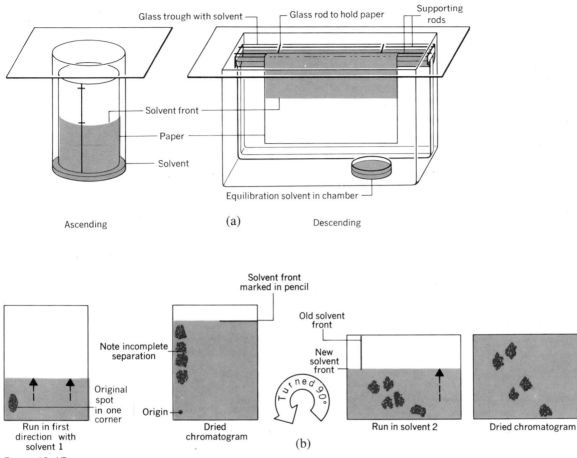

Figure 13–17

Paper chromatography. (a) Essential features of an ascending (left) and descending (right) chromatographic apparatus. (b) Stages in *two-dimensional* chromatography.

appropriate chemical or physical means. However, in certain instances where the solutes trail far behind the solvent front, it is desirable to allow the solvent to run off the edge of the paper sheet (descending chromatography only) so that maximum resolution of the solutes is achieved.

The rate of movement of a solute during paper chromatography is usually expressed as a dimensionless term R_f, where R_f is the *ratio* of the distance traveled by the solute to the distance traveled by the solvent front. Naturally, the R_f can be calculated only in those instances when the solvent is not allowed to leave the end of the paper sheet.

Greater resolution of the solutes may be obtained using **two-dimensional paper chromatography** (Fig. 13–17b); after chromatography using a particular solvent system in one direction along the paper sheet, the sheet is dried and rotated 90°, and another solvent system is used to chromatograph the solutes a second time. In this manner, solutes not fully separated by partition in the first solvent may be completely separated using the second solvent. Paper chromatography can also be combined with zone electrophoresis to provide two-dimensional analysis of a solute mixture, a technique known as "fingerprinting" (see Chapter 22).

Thin–Layer Chromatography

Thin–layer chromatography (abbreviated TLC) is an especially valuable method for rapidly separating unsaturated and saturated fatty acids, triglycerides, phospholipids, ster-

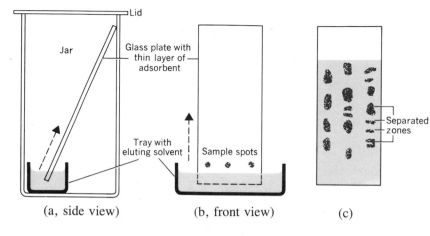

Figure 13-18
Thin-layer chromatography. (*a*) and (*b*) Glass plate containg thin layer of adsorbent and applied samples is immersed in tray containing shallow layer of solvent. Capillary action causes solvent to ascend through adsorbent (arrows) which separates sample spots into series of zones (*c*).

(a, side view)   (b, front view)   (c)

oids, peptides, nucleotides, and numerous other biological substances. In effect, TLC is a modification of paper chromatography in which the sheets of filter paper are replaced by glass plates covered with a thin, uniform layer of adsorbent. The essential features of the technique may be described as follows. An aqueous slurry of the selected adsorbent is uniformly spread over a glass plate to produce a thin layer and is then dried. Following this, the sample (usually prepared in a volatile solvent) is applied near one end of the long axis of the plate as a spot or thin line. When the sample has dried, the plate is supported vertically so that the end near the sample zone is immersed in a tray containing a shallow layer of the eluting solvent (usually an organic solvent of low polarity). Capillary action causes the solvent to ascend slowly through the layer of adsorbent, and as in paper chromatography, the solutes become distributed along the plate on the basis of differential partition between the stationary and mobile phases (Fig. 13–18). Although ascending TLC is the most common, descending and horizontal separations may also be carried out. TLC separations are very rapid, rarely exceeding 20 to 30 minutes.

After the separation has been achieved the glass plate is removed from the tray of solvent and allowed to dry. Zones containing colored substances can be detected directly, and others may be identified if they contain compounds that fluoresce when exposed to ultraviolet light. Many zones may also be rendered visible by spraying the plate with certain reagent dyes or stains. The adsorbent may also be scraped off various regions of the plate and the separated molecules eluted from the adsorbent particles. Table 13–4 lists some frequently used adsorbents and their applications.

Table 13-4
Adsorbents Used for Thin-Layer Chromatography

Adsorbent	Materials Separated
Silica gel	Amino acids, polypeptides, fatty acids, steroids, phospholipids, glycolipids, plasma lipids
Alumina	Amino acids, steroids, vitamins
Kieselguhr	Oligosaccharides, amino acids, fatty acids, triglycerides, steroids
Celite	Steroids
Cellulose powder	Amino acids, nucleotides
Hydroxylapatite	Polypeptides, proteins
Polyethylenimine	Nucleotides, oligonucleotides

Ion–Exchange Column Chromatography

Proteins and other macromolecules may also be separated by the technique known as ion–exchange chromatography. The separation is carried out in tall glass columns packed with grains of a synthetic ion–exchange **resin** (polymers to which numerous ionizable groups have been chemically added). Resins bearing negative charges are called **cation exchangers** and positively charged resins **anion exchangers.** As a solution of ions is passed through the column, the ions compete with each other for the charged sites on the resin. Consequently, the rate of movement of any ion through the column depends on its affinity for the resin sites, its degree of ionization, and the nature and concentration of competing ions in the solution. The differential rates of movement of ions through the column is the basis

Table 13-5
Some Commonly Used Ion Exchangers

	Polymer	Functional (Ionic) Group
Anion exchangers		
Diethylaminoethylcellulose	Cellulose	$-O-(CH_2)_2-N^+H-(C_2H_5)_2$
Dowex-2	Polystyrene-divinyl-benzene	$-N^+-(CH_3)_2C_2H_5OH$
Polylysine-kieselguhr (PLK)	Polylysine	$-(CH_2)_4-NH_3^+$
Amberlite IRA-400	Polystyrene-divinyl-benzene	$-N^+(CH_3)_3$
Cation exchangers		
Carboxymethylcellulose	Cellulose	$-O-CH_2-COO^-$
Hydroxylapatite	—	$-PO_4^{\equiv}$
Amberlite XE-64	Polymethacrylate	$-COO^-$
Sephadex SE	Dextran	$-(CH_2)_2-SO_3^-$

for protein and nucleic acid separations, since these molecules possess a variety of positively and negatively charged groups. Some ion exchangers used for protein and nucleic acid separations are listed in Table 13–5.

Among the resins most widely used for protein separations are diethylaminoethyl cellulose (DEAE–cellulose, an anion exchanger) and carboxymethylcellulose (CM–cellulose, a cation exchanger) developed by E. A. Peterson and H. A. Sober. These exchangers, like many others, are produced by reacting uncharged, high–molecular–weight polymers with ionizable compounds (Fig. 13–19). It is generally believed that the interaction between a resin and a protein involves the formation of a number of electrostatic bonds between the charged sites on the resin particle and oppositely charged dissociated side chains of certain amino acids. Since a number of bonds are formed, proteins are more firmly bound than singly charged substances. For example, at the appropriate alkaline pH, DEAE–cellulose could form a number of bonds with the negatively charged side chains of aspartic acid, glutamic acid, and tyrosine residues present in a protein. This also implies that the affinity of a protein for the resin can be altered by a change in pH. A reduction in pH sequentially suppresses the formation of bonds between the protein and DEAE–cellulose as hyrodgen ions bind to the negative side chains of the protein and displace the resin. A similar result would be effected by the addition of salt ions, which would compete with proteins for the resin sites (Fig. 13–20).

The separation of proteins by ion–exchange chromatography is carried out as follows. The resin is suspended in a buffer or salt solution (called the starting solvent), and the resulting slurry is used to fill the chromatographic column. At this time, the charges on the resin are neutralized by ions in the solvent. The sample (also dissolved in the starting solvent) is applied at the top of the column as a narrow zone. Depending upon the pH and salt concentration of the solvent, certain proteins present in the sample will form electrostatic bonds with sites on the resin (displacing solvent ions formerly bound at those sites), while the remaining proteins remain in the solvent. If a volume of the starting solvent is now passed through the column, the unbound proteins will be carried away with the flow and elute at the base of the column. In order to displace other proteins from the sample zone, it is necessary to change to another solvent whose pH or ionic strength will alter the degree of ionization of the amino acid side chains or compete more effectively with the proteins for the charged sites of the resin. In the case of proteins adsorbed to DEAE–cellulose or other anion exchangers, this can be accomplished by decreasing the pH and increasing the salt concentration of the solvent. If the second solvent differs only slightly from the first, only a small group of proteins will be released from the sample zone. During their descent through the column, transient bonds will be formed with the resin many times as the proteins and solvent ions compete for the resin sites. Depending upon the relative affinities of the proteins for the resin and the solvent ions, different proteins may reach the bottom of the column at different times. Thus, changing the solvent not only releases an additional group of proteins but also separates

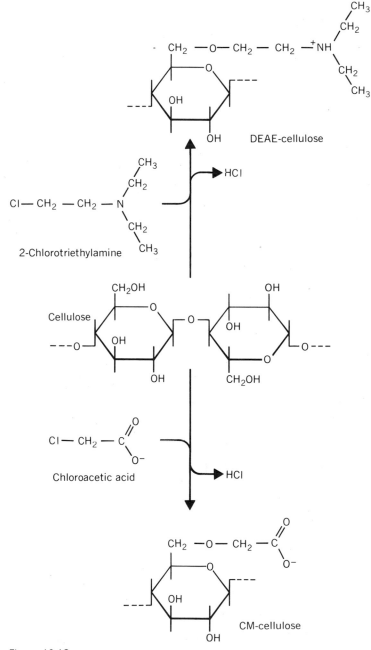

Figure 13-19
Synthesis of DEAE-cellulose and CM-cellulose by reacting cellulose with 2-chlorotriethylamine or chloroacetic acid.

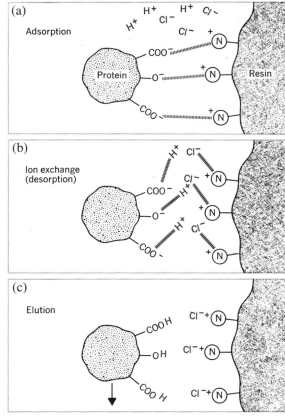

Figure 13-20
Desorption of protein from an anion-exchange resin by reducing the pH of the solvent. In (*a*) a theoretical protein molecule is adsorbed to the resin by forming electrostatic bonds between certain dissociated amino acid side chains and oppositely charged sites of resin. (N^+ represents any of several quaternary nitrogen groups associated with anion-exchangers; see Table 13-5). As the pH of the solvent flowing through the column is reduced, hydrogen ions compete more effectively for the dissociated amino acid side chains and break the bonds with the resin (*b*). Anions in the solvent (in this case Cl^-) bind to the resin, while the released protein is carried away with the flow of solvent (*c*). A similar effect is produced by increasing the salt concentration of the solvent. Generally, altering the pH and increasing the salt concentration are carried out simultaneously so that a combination of interactions results in protein desorption.

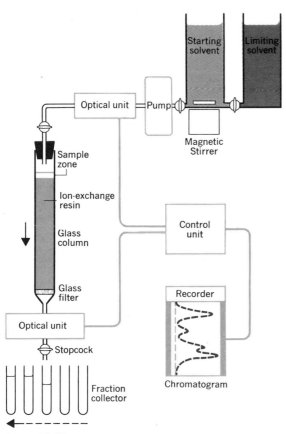

Figure 13-21
Diagram of components used in ion-exchange column chromatography. Gradient elution method is shown. The solvent is pumped through the optical unit of an ultraviolet light absorption monitor before entering the column. Flow out of the column is also monitored and the two absorbances compared in order to identify that portion of the effluent containing proteins. The absorbances are plotted on a strip-chart recorder. Peaks in the tracing (called a chromatogram) correspond to protein zones emerging from the column.

them into a series of zones during their passage through the column. The sequential addition of a series of solvents of different pH and salt concentration, called **stepwise elution,** eventually displaces and separates all the proteins in the sample zone. Alternatively, the proteins may be eluted by passing a solvent of continuously changing pH and ionic strength through the column—a method known as **gradient elution.** The effluent from the column is collect as a series of fractions (Fig. 13–21).

The conditions required for protein desorption depend on the number of bonds formed between the protein and the resin and on the nature of the ionized amino acid side chains. Therefore, two proteins having the same *net* surface charge density (and, consequently, similar electrophoretic mobilities) might be desorbed under quite different conditions and emerge from the column at different times if their *absolute* surface charges differ. For example, under a given set of conditions, one protein may bear eight negative and five positive amino acid side chains and have a net charge of -3; another may have four negative and one positive side chain and be more weakly bound to the resin (desorbing earlier) but would have a similar electrophoretic mobility (i.e., its net charge would also be -3). For the same reason, two proteins desorbed under the same conditions and emerging from the column together may have quite different electrophoretic mobilities. Consequently, the rechromatography of a given protein fraction or its further examination by electrophoresis is usually recommended in order to evaluate the purity of isolated components.

Although the discussion of ion–exchange chromatography has centered around protein separations, it should be recalled that mixtures of different nucleic acids may also be resolved using this technique. Polylysine–kieselguhr columns have been particularly successful for chromatographic separations of both ribonucleic acids and deoxyribonucleic acids.

Affinity Chromatography

Affinity chromatography is a novel form of column chromatography in which the molecules (principally proteins and nucleic acids) to be isolated from the sample under study are retarded in their passage through the column by their specific *biological* reaction with the column matrix. It is the biological nature of the interaction between the sample and the column matrix that distinguishes this form of chromatography from others. The specificity may take several forms including (1) an antigen–antibody reaction, (2) the interaction of an enzyme with its substrate, allosteric effector, or coenzyme, and (3) hydrogen bonding between complementary polynucleotides.

The column is packed with porous gel particles (usually agarose, cross–linked dextrans, or cellulose) to which ligands having a high affinity for specific biological com-

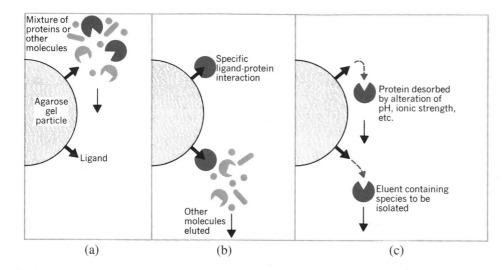

Figure 13-22
Affinity chromatography (see text for discussion).

ponents have been covalently coupled. When the sample is applied to the column, only the constituents having a high affinity for the ligand are bound, while other components are rapidly eluted. The ultimate desorption and isolation of the bound species are achieved by significantly altering the pH and/or ionic strength of the eluent (Fig. 13–22).

Affinity chromatography has been used with great success for the isolation of specific immunoglobulins from antisera. This is achieved by first coupling the gel with specific antigenic materials. When antiserum is applied to the column, specific immunoglobulins react with the antigen–coated gel particles and are retained by the column, while other species are eluted. Gels to which the hormone insulin has been covalently bound have been used to isolate insulin–binding receptor sites of cell membranes. Specific cellular enzymes may be isolated by passing a tissue extract through a column in which the gel particles have previously been derivitized with the enzyme's substrate or specific cofactor or competitive inhibitor. Messenger RNA can be isolated from tissue homogenates using agarose impregnated with polyuridylic acid (polyU); mRNA's affinity for the column matrix results from the occurrence of sequences of polyA in many eucaryotic mRNAs (Chapter 22).

Gel Filtration (Molecular Sieving)

In 1959 J. Porath and P. Flodin introduced **gel filtration,** a chromatographic method for separating molecules on the basis of size (i.e., molecular weight) differences. The procedure is carried out using glass columns packed with *nonionic,* porous gel particles. Since the gels do not possess ionic groups, separation of a mixture of macromolecules does not involve the temporary formation of bonds with the gel. The most commonly used gels are **cross–linked dextrans** (produced by reacting dextran with epichlorohydrin or acrylamide) in which the degree of cross–linkage determines the average pore size of the gel. Gel particles may be produced with a variety of pore sizes.

In gel filtration, the dry gel particles are first swollen by hydration in water or buffer and then packed into a glass column. The volume of the column is effectively divided into two phases: the solvent surrounding the gel particles (solvent phase) and the solvent within the gel particles (gel phase). Whether or not a given solute molecule can pass between phases depends on the molecular weight of the solute and on the pore sizes of the gel. Any solute molecules larger than the largest pore size cannot pass into the gel phase and are said to be above the **exclusion limit** of the gel. Smaller molecules are able to penetrate the gel phase to varying degrees.

If a mixture of molecules of various sizes is placed at the top of the column and is followed by the passage of solvent through the column, molecules above the gel exclusion limit will pass between the gel particles and emerge from the bottom of the column most rapidly. Molecules below the exclusion limit will pass through both the solvent and gel phases and emerge later. Therefore, molecules in

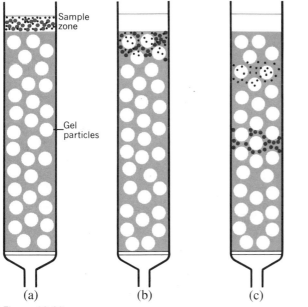

Figure 13-23

Separation of molecules by gel filtration. (a) Glass column packed with gel particles and overlaid with sample zone containing molecules of various sizes (small and large dots). (b) Molecules with molecular weights in excess of the gel exclusion limit percolate between the gel particles, while smaller molecules may enter the gel phase. (c) As a result, molecules reach the bottom of the column in order of decreasing size.

Table 13–6
Gel Filtration Media and Their Fractionation Ranges[a]

Type of Gel	Fractionation Range (Molecular Weight Units)
Epichorohydrin cross-linked dextrans	
G–10	0–700
G–15	0–1,500
G–25	1,000–5,000
G–50	1,500–30,000
G–75	3,000–80,000
G–100	4,000–150,000
G–150	5,000–300,000
G–200	5,000–600,000
Acrylamide cross-linked dextrans	
S–200	5,000–250,000
S–300	10,000–1,500,000
S–400	10,000–2,000,000
S–500	40,000–20,000,000
S–1000	500,000–100,000,000
Agarose	
2B	70,000–40,000,000
4B	60,000–20,000,000
6B	10,000–4,000,000

[a]The upper limit of the fractionation range is the gel particles' *exclusion limit*

the original sample are quickly separated into two different populations—molecules above and those below the gel exclusion limit (Fig. 13–23). However, if two molecules are below the gel exclusion limit but one is larger than the other, the larger molecule will pass through the column more rapidly, since it will spend less time in the gel phase than the smaller molecule (i.e., the chances that a sufficiently large gel pore will be encountered during descent through the column are less for larger molecules below the exclusion limit than for smaller molecules). Hence, even molecules below the gel exclusion limit undergo a separation and emerge from the column in order of decreasing size.

Cross–linked dextran and agarose gels may be obtained with a variety of exclusion limits (Table 13–6) and are used routinely for protein, nucleic acid, polysaccharide, and lipid separations. In most cases, a single solvent may be used to effect the separation, since molecules are not bound to the gel.

In addition to its usefulness for separating different size macromolecules, gel filtration is also used to estimate molecular weights. As in ion–exchange chromatography and electrophoresis, the determination of the purity of separated components may require additional analyses, since *different* molecules may have the same size and pass through the column at the same rate.

Gas Chromatography

Gas chromatography is a special form of column chromatography in which a gas is used as the mobile phase (instead of a liquid) and either a liquid or a solid is used as the stationary phase. When a liquid is used as the stationary phase, the technique is called **gas–liquid chromatography** (GLC), and separations are based primarily on differences in the partition of the molecules in the sample between the stationary liquid and the moving gas. In **gas–solid chromatography** (GSC), separations result from the differential

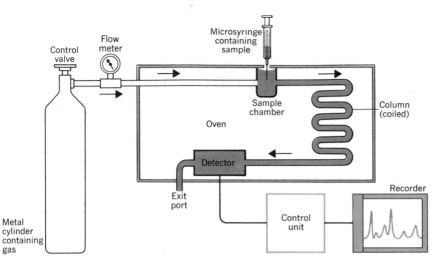

Figure 13-24
**Basic components of a gas chromato-
graph (see text for details).**

adsorption of sample molecules to the stationary solid phase as they are carried through the column by the gas. Of the two methods, GLC is, by far, the method most often employed.

The basic components of the gas chromatograph are the source of gas, the sample introduction chamber, the chromatographic column, the detector, and the recorder (Fig. 13–24). The gas (usually nitrogen, carbon dioxide, helium, or argon) is contained within a high–pressure cylinder connected to the column through metal tubing. A valve, pressure gauge, and flowmeter are used to accurately regulate the flow of gas. The sample is introduced into the flow of gas using a microsyringe and needle inserted through a self–sealing diaphragm in the sample chamber. The chamber itself is enclosed within a heating block so that the sample (if it is not already in a gaseous form) will immediately be vaporized upon introduction into the chamber and will be swept into the column by the gas. Gas chromatograph columns are made of glass, copper, or stainless steel tubing and are also enclosed in an oven; since they may be several feet long, they are often twisted to form a spiral.

The selection of packing material for the column depends upon whether the separation is to be based upon partition or adsorption. For GLC, the column is packed with an inert solid such as kieselguhr that is impregnated and lightly coated with a liquid of low volatility (so that it will not be eluted from the column at the operating temperature used). For GSC, the stationary phase is an adsorbent such as

charcoal or silica gel. Different molecules in the sample will be carried through the column at different rates, depending on their adsorption or partition characteristics, and will emerge from the end of the column at different times. Located near the exit of the column and also housed within an oven is the detector that monitors the composition of the emerging gas and relays electrical signals proportional to the amounts of separated components to a strip–chart recorder. The separated components are thus recorded as a series of peaks in the chromatogram tracing. The most common form of detector is the **flame ionization** chamber in which the components are successively mixed with hydrogen and air and burned in a high–voltage field. Migration of the ionized fragments in this field creates a current registered by the recorder. Most chromatographic separations are analytical, and the technique is so sensitive that only minute quantities of sample are required. However, separated components may also be collected as a condensate in tubes as they emerge from the heated column and are rapidly cooled.

Gas chromatography may be used to separate lipids, oligosaccharides, and amino acids after their preliminary conversion to volatile derivatives (many lipids may be chromatographed without conversion). Gas chromatography has been used with great success for the separation of different fatty acids. Depending upon whether a nonpolar or a polar liquid phase is used, fatty acids may be separated according to boiling point, size or degree of saturation.

The methods described in this chapter for separating,

isolating, and studying macromolecules and their constituents are widely used in cellular research. Students interested in pursuing this subject can find a more comprehensive discussion of these methods in the books and articles listed at the end of the chapter, together with descriptions of other separation methods that are used less often but are also important.

Summary

Over the years, a variety of analytical and preparative techniques have been developed for separating, analyzing, and isolating the macromolecular constituents of cells and tissues. Older methods such as **salting in, salting out,** and **isoelectric precipitation** relied on differences in the solubility of the molecular species under investigation. Differences in molecular size are used to achieve separations in such diverse approaches as **dialysis, ultrafiltration, gel filtration** (molecular sieving), and **ultracentrifugation.** Differences in the net electrostatic charges of molecules are used in various types of **electrophoretic** separations, including **moving–boundary** electrophoresis, **zone** electrophoresis, and **disc** electrophoresis. Biological activities of the molecules are used to advantage in such techniques as **immunoelectrophoresis** and **affinity chromatography.** Many **chromatographic** techniques utilize *partition* differences to achieve separations, the most popular formats being **thin–layer** chromatography, **anion** and **cation exchange** chromatography, and **gas** chromatography. Because such molecular parameters as size, charge, and solubility are used as the basis for effecting a separation, information concerning the unique physical and chemical properties of the species under investigation is simultaneously acquired.

References and Suggested Reading

Articles and Reviews

Bishop, R., Current major application areas in electrofocusing. *Sci. Tools* 26(1), 2, 1979.

Cooper, A. R., and Atzinger, D. P., Aqueous gel permeation chromatography. *Am. Lab.*, p. 13 (Jan. 1977).

Ettre, L. S., Pioneers in chromatography, Part I. *Am. Lab.*, p. 85 (Oct. 1978).

Ettre, L. S., Pioneers in chromatography, Part I. *Am. Lab.*, p. 120 (Nov. 1978).

Gray, G. W., Electrophoresis. *Sci. Am. 185*(6), 45 (Dec. 1951).

Haglund, H., Isoelectric focusing in natural pH gradients—a technique of growing importance for fractionation and characterization of proteins. *Sci. Tools 14,* 17 (1967).

Laurent, T. C., and Killander, J., A theory of gel filtration and its experimental verification. *J. Chromatography 14,* 317 (1964).

Peterson, E. A., and Sober, H. A., Chromatography of proteins I. Cellulose ion–exchange adsorbents. *J. Am. Chem. Soc. 78,* 751 (1956).

Sober, H. A., Gutter, F. J., Wyckoff, M. M., and Peterson, E. A., Chromatography of proteins II. Fractionation of serum proteins on anion–exchange cellulose. *J. Am. Chem. Soc. 78,* 756 (1956).

Vesterberg, O., Isoelectric focusing. *Am. Lab.*, p. 13 (June 1978).

Whipple, H. E. (Ed.), Gel electrophoresis. *Ann. N. Y. Acad. Sci.,* Vol. 71, (1964).

Williams, C. A., Immunoelectrophoresis. *Sci. Am. 202*(3), 130 (March 1960).

Books, Monographs, and Symposia

Bobbit, J. M., *Thin–Layer Chromatography,* Reinhold, New York, 1963.

Braver, J. M., Pesce, A. J., and Ashworth, R. B., *Experimental Techniques in Biochemistry,* Prentice–Hall, Englewood Cliffs, N.J., 1974.

Dean, J. A., *Chemical Separation Methods,* Van Nostrand Reinhold, New York, 1969.

Determann, H., *Gel Chromatography,* Springer–Verlag, New York, 1968.

Gel Filtration: Theory and Practice, Parmacia Fine Chemicals, Uppsala, Sweden, 1981.

Krugers, J., and Keulemans, A. I. M., *Practical Instrumental Analysis,* Elsevier, Amsterdam, 1965.

Lehninger, A. L., *Biochemistry* (2nd ed.), Worth, New York, 1975.

Lowe, C. R., and Dean, P. D. G., *Affinity Chromatography,* Wiley, London, 1974.

Lowe, C. R., *An Introduction to Affinity Chromatography,* North–Holland, Amsterdam, 1979.

Mangold, H. K., and Schmid, H. H. O., Thin–layer chromatography (TLC), in *Methods of Biochemical Analysis,* Vol. XII (D. Glick, Ed.), Wiley (Interscience), New York, 1964.

Morris, C. J. O. R., and Morris P., *Separation Methods in Biochemistry,* Pitmann and Sons, London, 1964.

Newman, D. W., *Instrumental Methods of Experimental Biology,* Macmillan, New York, 1964.

Porath, J., Cross–linked dextrans as molecular sieves, in *Advances in Protein Chemistry,* Vol. XVII (C. B. Anfinsen, Ed.), Academic Press, New York, 1962.

Randerath, K., *Thin–Layer Chromatography,* Academic Press, New York, 1963.

Righetti, P. G., and Drysdale, J. W., *Isoelectric Focusing,* North–Holland, Amsterdam, 1976.

Stock, R., and Rice, C. B. F., *Chromatographic Methods,* Chapman and Hall, London, 1967.

Chapter 14

RADIOACTIVE ISOTOPES
AS TRACERS IN CELL BIOLOGY

Isotopes are chemical elements that have the same **atomic number** (i.e., the number of protons in the nucleus of the atom) but different **atomic masses** (i.e., the sum of the number of protons and neutrons in the nucleus). Certain isotopes are unstable and undergo spontaneous nuclear changes (called **transmutations**) accompanied by the emission of **particulate** and sometimes also **electromagnetic** radiations. These atoms are said to be radioactive and are called *radioisotopes* or *radionuclides*. Radioactive atoms may readily be detected by instruments sensitive to their radiations. Generally, an organism cannot distinguish between the stable and radioactive forms of the same element so that both are metabolized in an identical manner. It is for this reason that radioisotopes have proven extremely useful to biologists, since these elements may conveniently be employed as *tracers*. That is, the fate of a given element (or molecule) in an organism or individual cell may be studied by introducing the radioactive form of that element and following the uptake and subsequent localization of the radioactivity. Some of the radioisotopes frequently employed as tracers are listed in Table 14–1.

If the radioisotope is initially a part of a molecule, then the fate of all or part of that molecule may similarly be followed. The use of radioisotopically labeled compounds is particularly desirable when the compound to be administered is a normal constituent of the cell or organism and would be impossible to distinguish from stable molecules already present. Many organic and inorganic compounds of biological interest may now be obtained that have one or more specific atomic positions occupied by radioisotopes. Because of the extremely high sensitivities of many radiation detectors, the quantities of radioisotopes employed in tracer studies can be kept small enough to preclude significant damage to cell constituents by the radiation.

Advantages of the Radioisotope Technique

Results obtained from experiments involving the use of radioisotopes are quantitative, since the amount of radioactivity present and available for detection is directly proportional to the radioisotope content. Moreover, numerous biological studies carried out routinely using radioisotopes can be performed only with great difficulty or are virtually impossible without them. Some examples may be cited to illustrate the value of the radioisotope technique.

Determination of Molecular Fluxes Under Conditions of Zero Net Exchange

The movement of a particular ionic or molecular species between different tissues of an organism or between a cell and its surroundings is often in **dynamic equilibrium.** That is, although a continuous exchange of a given substance

Table 14–1
Some Radioisotopes Frequently Used as Tracers

| Isotope | Radiation Emitted | | Half-life | Used as a Tracer of |
	Beta Particles	Gamma Rays		
^{3}H	Yes	No	12.3 years	Virtually any organic compound
^{14}C	Yes	No	5570 years	Virtually any organic compound
^{24}Na	Yes	Yes	15 hours	Salt metabolism, exchanges across membranes
^{32}P	Yes	No	14.3 days	Nucleic acid metabolism, phospholipid metabolism, salt metabolism
^{35}S	Yes	No	87.2 days	Protein metabolism
^{36}Cl	Yes	No	300,000 years	Salt metabolism
^{42}K	Yes	Yes	12.5 hours	Salt metabolism, exchanges across membranes
^{45}Ca	Yes	No	164 days	Salt metabolism, bone deposition
^{59}Fe	Yes	Yes	45.1 days	Heme synthesis, hemoglobin synthesis
^{131}I	Yes	Yes	8.1 days	Proteins

between one region and another takes place, no *net* transfer of material occurs. Alternatively, the concentration of a given substance within a tissue or cell may remain fairly constant as a result of the balanced biosynthesis and degradation of that material. These situations cannot be easily detected or studied by routine chemical analyses. The continuous flux of Na^+ and K^+ across the plasma membrane of the erythrocyte is a typical example of this type of equilibrium (see Chapter 15). The concentration of K^+ within the mammalian erythrocyte is much higher than in the surrounding blood plasma, while the reverse is true for Na^+. These large concentration differences are sustained in spite of the continuous passage of Na^+ and K^+ across the plasma membrane in both directions. Consequently, this dynamic steady state is not revealed by chemical analysis and, prior to the utilization of radioisotopes of sodium and potassium to study this situation, the concentration differences were interpreted as the consequence of the impermeability of the erythrocyte membrane to Na^+ and K^+. When the radioisotopes of sodium and potassium, ^{24}Na and ^{42}K, were used to label the plasma surrounding the red cells, the passage of radioactivity into the erythrocytes was observed, eventually approaching an equilibrium identical to the concentration equilibrium determined chemically. These studies unequivocally demonstrated that the plasma membrane of the red cell is readily permeable to Na^+ and K^+ and that a continuous exchange of these two ions between the cell and the surrounding plasma takes place. Observations of this kind provided great impetus to the development of

concepts concerning the continuous metabolic turnover of cellular materials present in constant concentrations and the **active transport** (Chapter 15) of materials across the plasma membrane against concentration gradients.

In addition to providing qualitative information, radioisotopes may be used to accurately measure the *rates* at which the metabolic turnover of materials within a cell or tissue occurs, as well as the rates of exchange of materials across the cell membrane. Prior to the availability of radioisotopes, no satisfactory methods were available to measure these rates. In the case of the erythrocyte, it has been shown that about 2% of the cell's Na^+ and K^+ are exchanged with the plasma each hour.

Simplification of Chemical Analyses

Depending on the nature of the chemical analyses required, quantitative determinations of the distribution of a given element or compound in different tissues of the body or in different parts of an individual cell may be very difficult. This difficulty may be compounded if certain components to be analyzed contain only trace amounts of the substance to be measured. Since radioisotope-containing compounds are not distinguished from their stable (nonradioactive) counterparts by an organism, the use of labeled compounds in studies of this sort can greatly simplify and reduce the number of analyses. Two requirements must be fulfilled: (1) the administration of the labeled compound must quickly be followed by its uniform distribution among nonradioactive molecules of the same kind that are already present in the system, and (2) once a uniform distribution is attained, the **specific activity** of the substance in question must be determined in one of the several components (i.e., tissues, cell fractions) to be analyzed. Specific activity is defined here as *the quantity of radioactive element or compound per unit mass of total element or compound* (the units of measurement are described later). If the labeled compound does become uniformly distributed in the system, then the specific activity will be equal throughout. Consequently, all subsequent measurements of the quantity of that compound may be made simply by measuring the amount of radioactivity present in the tissue or cell part to be analyzed. In nearly every instance, this is far easier than performing a series of quantitative chemical measurements of the element or compound in question.

The rates at which different elements or compounds are incorporated by individual cells or by the body tissues, together with a determination of the specific loci of the deposition, can also be conveniently determined using radioisotopes. For example, cells may be incubated in a medium containing labeled material or, in the case of measurements in whole animals, the material may be introduced into the bloodstream. In these cases, the specific activity of the element or compound is determined before it is made available for incorporation and the subsequent rate of incorporation determined from either the rate of disappearance of radioactivity from the medium or bloodstream or from the rate of appearance of radioactivity in the cells or tissue. The specific locus of deposition or utilization within the cell may be determined from separate radioactivity measurements of *fractionated* tissue (see Chapter 12) or by *autoradiographic* procedures (to be described later) combined with light or electron microscopy. Again, it should be noted that specific quantitative chemical analyses of the material being incorporated by the cells are unnecessary when the radioisotope technique is applied.

"Isotope Dilution" Methods

A common problem confronting physiologists and one that especially lends itself to the radioisotope technique is the determination of the *total* quantity or volume of a given material in the body or in a cell when quantitative isolation of that material for analysis is not possible. Determinations of total circulating blood volume, erythrocyte mass, chloride space, exchangeable sodium, and so on fall in this category. The manner in which such unknown quantities are measured using radioisotopes may be exemplified by considering the following generalized case. Suppose that material X occurs in a system (i.e., a tissue, a cell) in unknown abundance and a labeled form of this material, X*, is either available commercially or can be prepared experimentally, having a known specific activity, SA, given in this instance by the equation

$$SA = \frac{(X^*)}{(X^* + X)} \tag{14-1}$$

A measured quantity of the labeled material is introduced into the system and permitted to equilibrate thoroughly with the stable material already present. Following this, a small quantity of the material is withdrawn from the system and its specific activity determined. Since the X* originally introduced was uniformly mixed with unlabeled X already present, its specific activity will have been reduced in direct proportion to the total amount of X present in the system. Consequently, the total amount of X originally present can be determined from the relationship

$$X_T = X^* \left(\frac{SA_1}{SA_2} - 1 \right) \tag{14-2}$$

where

X_T = the total amount of X originally present in the system,

X^* = the quantity of labeled material added,

SA_1 = the original specific activity of the material added, and

SA_2 = the specific activity of the sample removed for analysis.

The isotope dilution technique is a fundamental part of *radioimmunoassay* procedures (discussed later in the chapter).

Precursor–Product Relationships

Radioisotopes have been widely applied to study precursor–product relationships, for in many cases the introduction of a labeled compound (i.e., **precursor**) into a system is soon followed by its chemical conversion into another form (i.e., **product**). For example, soon after ^{32}P-labeled phosphate is introduced into the bloodstream, radioactivity appears in a variety of tissue phospholipids. In a similar manner, the introduction of radioactive iron is soon followed by the appearance of radioactivity in liver ferritin and also in newly synthesized hemoglobin of red blood cells maturing in the bone marrow. In other words, radioisotopically labeled compounds may be used to determine which of a variety of alternative metabolic pathways is followed by a compound by examining the radioisotope content of the alternative metabolic products. If several different pathways are open to the precursor, then the extent to which each is followed under a variety of experimental conditions may be determined from the respective radioisotope contents of the products.

The formation of an end product often involves a number of intermediate precursors and products; that is,

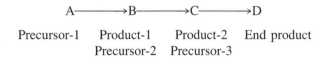

A————→B————→C————→D

Precursor-1   Product-1   Product-2   End product
                Precursor-2   Precursor-3

Before radioisotopes were available to study such metabolic pathways, it was often difficult to determine which com-

pound was the immediate precursor of each product. However, if each intermediate has a single precursor (as above), then the precise sequence can be determined using labeled material by following the change in specific activity of each intermediate compound with time. This may be done because *the specific activities of an intermediate and its immediate precursor are equal when the intermediate's specific activity reaches a maximum value* (Fig. 14–1). By degrading the intermediate and final products of a series of chemical reactions and determining the distribution of radioisotope among the constituent atoms, it is often possible to determine the fate of each atom in the pathway.

Properties of Radioactive Isotopes

Most radiations emitted by radioisotopes are the result of changes in the unstable atomic nuclei. Whether or not a given atomic nucleus is stable depends in turn upon the numbers of neutrons (N) and protons (Z) that it contains. The relationship between nuclear stability and the neutron:proton composition of the nucleus is shown graphically in Figure 14–2. It should be noted that for the lighter elements, nuclear stability exists when $N \cong Z$, whereas in the stable heavier elements, the number of neutrons exceeds the number of protons, with the allowable neutron excess increasing with atomic number. Isotopes with N and Z numbers outside of the stable region shown in Figure 14–2 undergo spontaneous changes in which neutrons and pro-

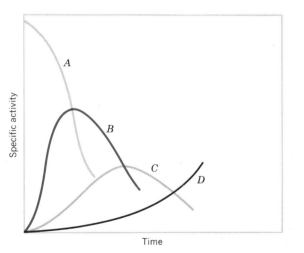

Figure 14–1
Relationship between the specific activities of precursors and products in the chemical pathway A → B → C → D following application of a pulse of radioactive tracer.

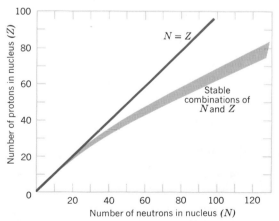

Figure 14–2
Relationship between stability of the atomic nucleus and the number of neutrons and protons present.

tons are interconverted. These nuclear transmutations are accompanied by the emission of particulate and sometimes also electromagnetic radiation.

Types of Radiation Emitted by Radioisotopes

The most common types of nuclear radiations are **alpha particles**, positive and negative **beta particles**, and **gamma rays.** Alpha particles, which consist of two protons and two neutrons and are therefore identical to helium nuclei, are emitted by radioisotopes of high atomic number (e.g., uranium, polonium, thorium, and radium) and are rarely used as tracers in biological studies.

Positive beta particles, also called **positrons**, are emitted from nuclei in which the N:Z ratio is lower than that which is stable. The nuclear transmutation involves the conversion of a proton into a neutron, positron, and neutrino, the latter two being ejected from the nucleus. The positron possesses a unit positive charge and is equal in mass to an electron, while the neutrino has neither mass nor charge. The isotope $^{11}_{6}C$ *decays* by positron emission (the numbers that precede the elements' symbol are its atomic mass [the superscript] and its atomic number [the subscript]):

$$^{11}_{6}C \longrightarrow {^{11}_{5}}B \ + \ _{(+1)}^{0}e \ + \ \text{neutrino}$$
$$\text{(positron)}$$

In view of the similarity between beta particles and electrons, the symbol ''e'' is often used to describe the beta particle. Very few positron-emitting radioisotopes are employed as biological tracers.

Nearly all radioisotopes used as tracers by biologists emit negative beta particles (**negatrons**). It has become customary to drop the term ''negative'' so that the expression ''beta particle'' is understood to imply ''negative beta particle.'' Although technically incorrect, this terminology is widely used and will be employed here. Beta particles are emitted from nuclei in which the N:Z ratio is above that which is stable. The nuclear change involves the conversion of a neutron to a proton with the resulting ejection of a beta particle and neutrino. ^{14}C and ^{32}P may serve as examples of this form of decay:

$$^{14}_{6}C \longrightarrow {^{14}_{7}}N \ + \ _{(-1)}^{0}e \ + \ \text{neutrino}$$
$$\text{(beta particle)}$$

and

$$^{32}_{15}P \longrightarrow {^{32}_{16}}S \ + \ _{(-1)}^{0}e \ + \ \text{neutrino}$$

The emission of beta particles from the nuclei of ^{3}H, ^{14}C, ^{32}P, ^{35}S, ^{36}Cl, and ^{45}Ca atoms changes the N:Z ratio to a stable value and lowers the energy content of the nucleus to the ground state. The energy lost by the nucleus in this process is divided between the beta particle and the neutrino. However, for many radioisotopes (e.g., ^{24}Na, ^{59}Fe, and ^{131}I), intranuclear changes resulting in the emission of beta particles and neutrinos produce nuclei with stable N and Z combinations but with energy levels that are still above the ground state. In these instances, the excess energy is eliminated and the ground state is attained by the emission of one or more gamma rays; for example,

$$^{59}_{26}Fe \longrightarrow {^{59}_{27}}Co \ + \ _{(-1)}^{0}e \ + \ \text{neutrino} \ + \ \text{gamma ray(s)}$$

Unlike alpha and beta radiation, gamma radiation is not particulate but is *electromagnetic* (see Chapter 9).

Energy of Radiation and Its Interaction with Matter

Radiation energy is measured in **electron volts** (abbreviated *ev*), one electron volt being the kinetic energy acquired by an electron in a potential difference of one volt. The beta particles and gamma rays emitted by radioisotopes that are often used as tracers have energies ranging from about 1×10^4 to 4×10^6 ev or 0.01 to 4.0 Mev (1 Mev equals 1 million electron volts). This energy range may be compared with the energy of chemical bonds, which is of the order of a few electron volts. The kinetic energy acquired by a beta particle during transmutation of the parent atom's nucleus can vary from 0 Mev up to some maximum value, E_{max}, which is a characteristic of the particular radioisotope. As an example, the energy spectrum for ^{32}P beta particles is shown in Figure 14–3. It should be noted that the *maximum* energy of a ^{32}P beta particle is 1.71 Mev but that the *average* energy is much less (actually, 0.70 Mev). Although the value for E_{max} varies, the *shape* of the energy curve is similar for all radioisotopes.

In order to reduce the energy level of the radioisotope nucleus to the ground state, a specific quantity of energy, Q, must be given off as radiation during the nuclear change.

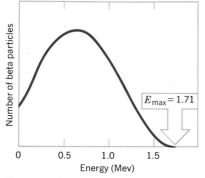

Figure 14–3
Energy spectrum of beta particles emitted by ^{32}P.

For isotopes emitting only beta particles, Q equals E_{max}, and the energy of the neutrino accounts for the difference between E_{max} and the actual kinetic energy acquired by the beta particle. For example, if the transmutation of a particular ^{32}P atom results in the emission of a 1.20 Mev beta particle, then the accompanying neutrino would have an energy of 0.51 Mev (i.e., 1.20 Mev + 0.51 Mev = 1.71 Mev = E_{max}). If the beta particle had escaped the nucleus with the maximum possible energy, then no neutrino would have been emitted. Table 14–2 lists the maximum beta particle energies of some radioisotopes.

In contrast to beta particles, gamma rays are emitted at specific energy levels. Thus, all gamma rays emitted by ^{42}K have an energy of 1.5 Mev (Table 14–2). The Q value of an isotope emitting both beta and gamma radiation is equal to the sum of the energies possessed by the beta particle, neutrino, and subsequent gamma ray(s). Occasionally, as in the case of ^{24}Na, the emission of a beta particle may be followed by the emission of two (or more) successive gamma rays in order to reduce the nuclear energy level to the ground state. It may also be noted from Table 14–2 that the decay of some radioactive isotopes (e.g., ^{42}K, ^{59}Fe, and ^{131}I) proceeds along two or more alternative pathways; that is, beta particles having more than one E_{max} may be emitted, and since the Q value is constant, these are necessarily followed by gamma rays of different energies.

Beta particles and gamma rays interact with matter by ionizing and exciting atoms in their path. In the case of beta particles, ionization results from the repulsion of orbital electrons by the negatively charged particle. As a result, beta particles tracking through matter produce a *wake* of electrons and positively charged ion (called **ion pairs**). The number of ion pairs produced per unit path length, called the **specific ionization,** is not constant but increases as the beta particle slows down. Also, the path

Table 14–2
Energies of Beta Particles and Gamma Rays Emitted by Radioisotopes Frequently Used as Tracers

| Isotope | Energy (Mev) | | Q (Mev) |
	Beta Particle (E_{max})	Gamma Ray	
^{3}H	0.0176		0.0176
^{14}C	0.154		0.154
^{24}Na	1.39 plus	1.37 & 2.75	5.51
^{32}P	1.71		1.71
^{35}S	0.167		0.167
^{36}Cl	0.714		0.714
^{42}K	3.60		3.60
	or 2.10 plus	1.50	3.60
^{45}Ca	0.254		0.254
^{59}Fe	0.27 plus	1.29	1.56
	or 0.46 plus	1.10	1.56
^{60}Co	0.31 plus	1.17 & 1.33	2.81
^{131}I	0.61 plus	0.36	0.97
		or 0.28 & 0.08	
	or 0.34 plus	0.63	0.97

of the beta particle is not linear but is quite erratic as a result of its repulsion by the orbital electrons of atoms with which it interacts. High-energy beta particles may traverse the linear equivalent of 1 to 2 m in air. Since gamma radiation is electromagnetic, its probability of interacting with matter is less than that for beta particles. Consequently, gamma rays have a much lower specific ionization and a much longer and also linear path length.

Half-life

The number of atoms in a sample of radioisotope that disintegrate during a given time interval decreases logarithmically with time and is unaffected by chemical and physical factors that normally alter the rates of chemical processes (temperature, concentration, pressure, etc.). Radioactive decay is therefore a classical example of a first-order reaction. A convenient term used to describe the rate of decay of a radioisotope is the **physical half-life**, T_p—that is, the amount of time required to reduce the amount of radioactive material to one-half its previous value. Each radioactive isotope decays at a characteristic rate and therefore has a specific half-life (see Table 14–1). For example, the amount of radioactivity arising from a sample of ^{59}Fe is reduced to one-half its original value in 45.1 days, to one-fourth in 90.2 days, to one-eighth in 135.3 days, and so on. The amount of decay occurring in the course of a tracer experiment must be taken into account when radioisotopes of short physical half-life such as ^{24}Na, ^{32}P, ^{42}K, ^{59}Fe, ^{125}I, and ^{131}I are used. Of course, this is not a problem in tracer experiments employing ^{3}H and ^{14}C (Table 14–1) since the length of the experiment is insignificant in comparison with the half-life.

When radioisotopes are used in in vivo experiments of extended duration, the turnover rate of the element in the body (or in the cell) must also be considered, for the rate of decrease of radioactivity will be a function of *both* radioactive decay and metabolic turnover. In these instances, a more useful term is the **effective half-life**, T_e, which is the amount of time required to reduce the radioisotope content of the body (or cell) to one-half its original value by the *combined* effects of decay and turnover; it is determined using the relationship

$$T_e = \frac{T_b \times T_p}{T_b + T_p} \qquad (14\text{–}3)$$

where

T_e = effective half-life,
T_p = physical half-life, and
T_b = **biological half-life** and is defined as the normal amount of time required for the turnover of one-half of the body content of a given element (radioactive or nonradioactive).

The physical, biological, and effective half-lives of several elements are compared in Table 14–3. It should be noted that T_e can never be greater than T_p and that the slower the rate of turnover of an element, the closer T_e approaches T_p.

On the basis of their biological and physical half-lives, their loci of deposition within the body, the types of radiation emitted, and the energy of the radiation, radioisotopes may be classified according to their degree of hazard to the investigator if accidentally ingested (Table 14–4).

Detection and Measurement of Radiation

The selection of instruments for the detection and measurement of radioisotopes is based primarily on the type and

Table 14–3
Physical Biological and Effective Half-lives of Some Radioisotopes

Element	Radio-isotope	Half-life		
		Physical	Biological	Effective
Hydrogen	^{3}H	12.3 yr	19 days	19 days
Carbon	^{14}C	5,570 yr	180 days	180 days
Phosphorus	^{32}P	14.3 days	3 yr	14.1 days
Calcium	^{45}Ca	164 days	73 yr	163 days
Iron	^{59}Fe	45.1 days	3.4 yr	43.5 days
Cobalt	^{60}Co	5.3 yr	8.1 days	8.1 days
Iodine	^{131}I	8.1 days	156 days	7.7 days
Radium	^{226}Ra	1,620 yr	104 days	104 days

Table 14–4
Relative Internal Hazard of Selected Beta- and Gamma-Emitting Isotopes

Degree of Hazard	Radioisotope
Slight hazard	^{24}Na ^{42}K ^{64}Cu
Moderately dangerous	^{3}H ^{14}C ^{22}Na ^{32}P ^{35}S ^{36}Cl ^{59}Fe ^{60}Co
Very dangerous	^{45}Ca ^{55}Fe ^{90}Sr

energy of the emitted radiation. The most commonly used detectors are: (1) **Geiger-Müller counters,** which are employed primarily with isotopes emitting beta particles of intermediate or high energy (E_{max} above 0.2 Mev) and which may also be used at low efficiency for the measurement of gamma radiation: (2) **solid scintillation counters,** which are generally employed with gamma-ray-emitting isotopes: and (3) **liquid scintillation counters,** which are used with isotopes emitting low-energy beta particles (E_{max} below 0.2 Mev).

Geiger-Müller Counters

The most widely used instrument for the detection and measurement of radiation is the Geiger-Müller (or G-M) counter. The detector itself, called a G-M tube, consists of a cylinder several inches long containing two electrodes and filled with a readily ionizable inert gas such as helium or argon. The insulated metallic internal surface of the cylinder serves as the cathode, and a narrow wire passing down the center of the tube serves as the anode (Fig. 14–4). One end of the G-M tube is covered by a thin material such as mylar plastic or mica and is called the **end-window.** The anode and cathode terminals at the other end of the tube are connected to a source of high voltage and a **scaler,** a device that simply counts electrical pulses.

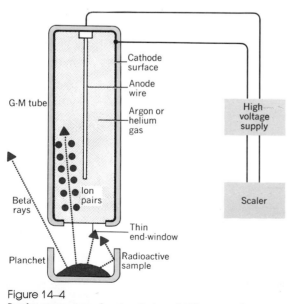

Figure 14–4

Basic components of a Geiger-Müller counter.

When a radioactive sample (usually deposited on a small metal disk called a **planchet**) is placed near the end-window, radiation enters the G-M tube, ionizing some of the gas molecules and forming a number of ion-pairs (i.e., positively charged argon or helium atoms and electrons). If a sufficiently high electrical potential is applied to the electrodes, the ion-pairs will migrate toward the appropriate electrode. During this migration, the ions collide with and ionize additional gas molecules, so that the passage of a single beta particle through the gas results in a large number of ions being collected at the electrodes. These events produce an electrical pulse that is recorded by the scaler as a *count.* Ideally, each ionizing ray entering the G-M tube is registered as a count and the amount of radioactivity is expressed as *counts per minute* (cpm).

Since radiation is emitted in all directions from a radioactive source, it is apparent that only a small percentage of the rays arising from the sample are directed toward the end-window. Therefore, even if all the rays entering the G-M tube are detected and counted, the cpm recorded for the sample is only a fraction of the true *rate* of disintegration (i.e., *disintegrations per minute, dpm*) of the isotope. This does not pose a serious problem when the *relative* isotope contents of a number of samples are to be determined (this is generally the case) and if constant *geometric* conditions are maintained for each sample (distance of the sample from the end window, volume of sample, etc.).

For some radioisotopes such as ³H, ¹⁴C, ³⁵S, ⁴⁵Ca, and others of low E_{max}, much or all of the energy of the emitted beta particles may be dissipated before the ray enters the ionizing gas. For example, the energy may be expended within the sample itself (called **self-absorption**), in the air between the sample and the end-window, or in the material of the end-window. Even with radioisotopes emitting beta particles of high E_{max} (Table 14–2), beta particles in the low region of the energy spectrum may go undetected. Because of the low specific ionization of gamma rays and the low density of the gas in the G-M tube, gamma rays may pass through the tube without causing ionizations and therefore go undetected. For these reasons, Geiger-Müller counters are usually not suitable for the detection and measurement of radioisotopes emitting gamma rays or beta particles of low E_{max}.

Even when no radioactive sample is placed below the end-window of the G-M tube, a small count is recorded. This is known as the *background* count and results from

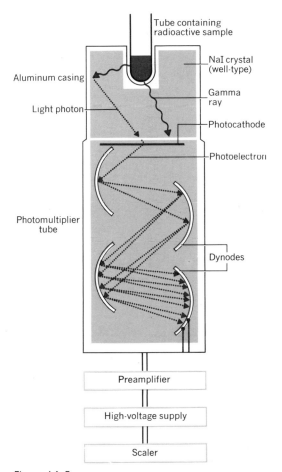

Tube containing
radioactive sample

NaI crystal
(well-type)

Aluminum casing

Gamma
ray

Light photon

Photocathode

Photoelectron

Photomultiplier
tube

Dynodes

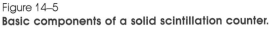

Preamplifier

High-voltage supply

Scaler

Figure 14–5
Basic components of a solid scintillation counter.

cosmic radiation (primarily gamma rays), naturally occurring radioisotopes in laboratory materials (such as ^{40}K in glass and naturally occurring ^{14}C and ^{3}H in organic compounds), radioactive samples left in the vicinity of the detector, and electronic ''noise'' within the components of the counting system. Therefore, the background count must always be subtracted from the count obtained for a radioactive sample. The magnitude of the background count may be reduced by placing lead shielding around the detector so that much of the cosmic radiation and radiations from other sources are absorbed before reaching the detector.

The total amount of radioisotope present in a sample at any instant may be determined from its rate of disintegration, that is its dpm; the basic unit of measurement is the **curie (Ci)** and is defined as that quantity of radioisotope

undergoing 2.22×10^{12} dpm. (Note that the curie content of a radioactive sample decreases exponentially with time at a rate determined by the physical half-life of the radioisotope.) In most tracer experiments, the quantity of radioisotope used is generally at the millicurie (mCi) or microcurie (μCi) level. The curie content of a labeled compound is generally provided at the time of purchase so that the efficiency of the counting system may be determined by comparing the recorded dpm of an aliquot of the isotope with its known dpm. Generally, this value is 10% or less for G-M counters but is much higher in solid and liquid scintillation systems. Once the efficiency of the counting system is known, then the specific activity of a radioactive sample (which we may now define as the *number of curies per unit mass of element or compound*) collected during the course of a tracer experiment may be calculated from its observed counting rate and composition.

Automated G-M counters of modest efficiency are available in which fresh ionizing gas is continuously supplied to the detector and which also permit the planchet containing the sample to seal to the end of the detector. This eliminates the end-window together with air and end-window absorption. These ''windowless gas-flow counters'' may also be equipped with automatic sample (i.e., planchet) changers. G-M counters are effectively employed in tracer experiments involving ^{24}Na, ^{32}P, ^{36}Cl, and other ''hard beta'' emitters but generally are not used with ^{3}H and ^{14}C, which emit ''soft beta'' rays.

Solid Scintillation Counters

Solid scintillation counters are used to detect and measure radioisotopes emitting gamma rays. The detector (Fig. 14–5) consists of a large crystal of thallium-activated sodium iodide and a photomultiplier tube encased in an aluminum housing; the latter is interfaced with a preamplifier, a source of high voltage, and a scaler. The radioactive sample to be counted is placed either against the end of the detector containing the crystal or, for greatly improved counting efficiency, into a well-shaped opening drilled into the crystal's surface (Fig. 14–5). Because of its high density, the crystal absorbs much of the energy of the gamma rays, causing excitation of electrons of atoms composing the crystal and raising them to higher energy orbitals. As these electrons return again to their lower energy orbitals, flashes of light or **scintillations** are emitted, proportional in number

Table 14–5
Chemical Substances Measured by Radioimmunoassay (RIA)

Steroids
    Aldosterone
    Testosterone
    Estradiol
    Digitalis
Peptides
    Antidiuretic hormone (ADH, vasopressin)
    Bradykinin
    Secretin
    Gastrin
Proteins
    Adrenocorticotrophic hormone (ACTH)
    Glucagon
    Insulin
    Calcitonin
    Somatotrophic hormone (STH, growth hormone)
    Parathyroid hormone
    Thyroglobulin
Other
    Cyclic AMP
    Morphine
    Prostaglandins

to the number and energy of the gamma rays exciting the crystal. The light photons are converted by the photomultiplier tube into electrical pulses of corresponding magnitude and frequency and these are relayed to the scaler. Since the magnitude of the electrical pulses produced is proportional to the energy of the gamma rays, and since gamma rays are monoenergetic, the inclusion of the appropriate circuitry in the counting system (i.e., a pulse height analyzer, see Chapter 2) allows different gamma-ray-emitting isotopes to be distinguished.

In contrast to G-M counters, few or no problems involving self-absorption and end-window absorption are incurred when solid scintillation methods are used with gamma-ray-emitting isotopes. However, the use of constant geometry and lead shielding around the detector to reduce the magnitude of the background count is important.

Radioimmunoassay.   At the present time, one of the most common uses of solid scintillation counting is in certain **radioimmunoassay** (RIA) procedures. The RIA technique is a tracer procedure used to measure trace amounts of any substance with antigenic properties. Depending upon the specific activity of the labeled compound, quantities of antigen that are as low as 10 *picograms* (i.e., 10 millionths of a microgram) may be detected. Prior to the development of RIA procedures, the assay of biological substances that did not possess enzymatic activity or some other easily measurable property were extremely difficult. However, since many different kinds of chemical substances can elicit an antibody response, they can be reliably quantitated by the RIA procedure. The sensitivity of the RIA technique equals or surpasses all chromatographic and spectrophotometric methods currently in use. The technique has been especially useful for assaying hormones (Table 14–5).

The RIA procedure is depicted schematically in Figure 14–6. A sample containing the antigen to be quantified is mixed with a second sample containing radioisotopically labeled antigen of very high specific activity. To the mixture is added antiserum containing antibody previously prepared against the antigen (see Chapter 4). The amount of antibody added must be limiting so that antigen is always in excess (even when no unlabeled antigen is present). The resulting antigen–antibody reaction produces a complex in which the amount of radioisotope present is *inversely* proportional to the original quantity of unlabeled antigen in the sample being assayed. In the final stage of the radioimmunoassay, the antigen–antibody complex is separated from free antigen by precipitation.

In the diagram in Figure 14–6, the antiserum contains enough antibody to bind 12 antigen molecules. If 12 labeled and 12 unlabeled antigen molecules are present, then only one-half of the labeled antigens can be bound. If 24 unlabeled antigen molecules are present (as in the diagram), then only one-third of the labeled antigens will be bound; and so on. Thus, the radioisotope content of the isolated antigen–antibody complexes can be used as a measure of the amount of unlabeled antigen present in the original sample. When the antigen is all or part protein or peptide, the labeled form is prepared by iodination with either ^{125}I or ^{131}I. The iodine is incorporated into the benzene ring of tyrosine residues. Both of these radioactive tracers emit gamma rays. In radioimmunoassays, radioactive tracers of iodine are preferred to those of hydrogen and carbon because 557 atoms of ^3H and 261,672 atoms of ^{14}C must be incorporated into each antigen molecule to yield the same dpm as one atom ^{131}I or seven atoms of ^{125}I. Although ^{131}I emits more energetic and more easily detected gamma rays, ^{125}I is usually the isotope of choice because of its longer half-life (60 days vs. 8.1 days).

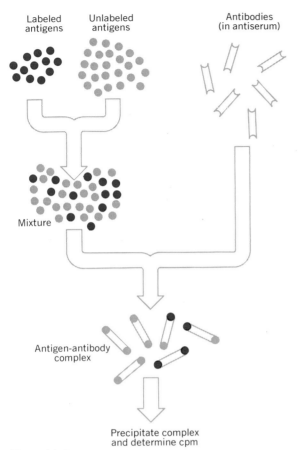

Figure 14–6
The radioimmunoassay (RIA) procedure. A mixture containing a known quantity of radioactively labeled antigens and an unknown quantity of unlabeled antigens is reacted with a limiting amount of antibody and the resulting antigen–antibody complexes are then isolated. The quantity of unlabeled antigen can be detrmined from the radioisotope content of the isolated complexes. The technique is highly sensitive.

Liquid Scintillation Counters

Except perhaps in the case of the radioimmunoassay of proteins, ^3H and ^{14}C are used more extensively than any other radioisotopes in tracer experiments. This is not surprising in view of the ubiquitous occurrence of hydrogen- and carbon-containing compounds in cells. The energies of the beta particles emitted by these isotopes are very low (especially tritium), so that it is only with great difficulty that they can be accurately measured using gas ionization methods, even in windowless counters. The great demand for a sensitive and accurate method for measuring ^3H and

^{14}C, as well as other soft beta emitters, led in the 1950s to the development of the **liquid scintillation** technique.

Before considering this important technique in detail, the basic mechanism involved in the detection process will briefly be summarized. The radioactive sample to be counted is mixed in a glass or plastic vial with a "scintillation fluid." Certain molecules dissolved in the fluid indirectly absorb the energy dissipated by the beta particles and respond by scintillating. The resulting photons of light are detected by photomultiplier tubes that relay proportional electrical pulses to the scaling units. Certain advantages of the procedure are immediately apparent, for if the sample is thoroughly mixed with the scintillation fluid, no self-absorption occurs. Moreover, since the technique is employed almost exclusively with weak beta-emitting isotopes, the energy of *all* beta rays is dissipated within the fluid, making it possible, at least in theory, to detect most if not all of the nuclear disintegrations.

The scintillation fluid contains three major components: (1) an organic **solvent** constituting the bulk of the mass: (2) a **primary solute** or **fluor**; and (3) a **secondary solute** or **fluor.** The most commonly used solvents and fluors are listed in Table 14–6. The energy of the beta particle is transferred to the solvent, causing ionization or excitation of solvent molecules. The latter effect is more important, since in returning to the nonexcited ground state, energy is transferred to the primary fluor molecules, causing their excitation. In returning to the ground state, the excited primary fluor emits photons of near ultraviolet light. The photocathodes of most photomultiplier tubes are not sensitive to photons having wavelengths in this region, and it is the role of the secondary fluor to absorb the light energy from the primary solute and reemit it as light of longer wavelength. (Secondary solutes are sometimes called "wave shifters.") For example, the light emitted by PPO has a wavelength maximum at 365 nm, whereas POPOP emits light having a wavelength maximum at 420 nm, close to the maximum sensitivity region of most photomultiplier tubes.

The size of the electrical pulse produced by the photomultiplier tube is proportional to the energy dissipated by the beta particle in the liquid scintillation fluid. Consequently, a pulse height spectrum is produced having a profile almost identical to that of the beta particle energy spectrum. In the case of ^3H and ^{14}C, the sizes of many of the pulses produced are so low that spurious low-amplitude pulses arising spontaneously within the photomultiplier tube

Table 14–6
Solvents and Fluors Used for Liquid Scintillation Counting

	Fluorescence Maximum (nm)
Solvents	
Toluene	
Methoxybenzene	
1,4-Dioxane	
Xylene	
Primary fluors	
PPO (i.e., 2,5-diphenyloxazole)	365
PBD (i.e., 2-phenyl-5-[4-biphenyl]-1,3,4-oxadiazole)	364
Naphthalene	325
Anthracene	405
Secondary fluors	
POPOP (i.e., 1,4-bis [5-phenyloxazolyl-2′] benzene)	420
DMPOPOP (i.e., 1,4-bis [4-methyl-5-phenyloxazol-2′] benzene)	430
BBOT (i.e., 2,5-bis [5-tert-butylbenzoxazolyl-2′] thiophene)	438

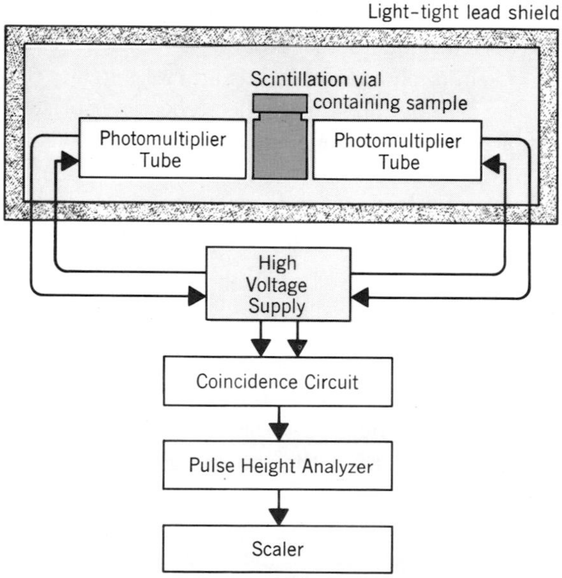

Figure 14–7
Basic components of a liquid scintillation counter.

itself (i.e., "noise") result in a false high count. To minimize this problem, in most liquid scintillation counters the sample vial is placed between the faces of *two* photomultiplier tubes connected through a pulse height discriminator to a *coincidence circuit* (Fig. 14–7) which permits a count to be registered only when *simultaneous* pulses are produced by the tubes. Therefore, a noise signal from one photomultiplier tube is not recorded unless another noise signal is generated by the other tube at the same time (actually, within 10^{-7} seconds of each other). Since noise signals are random, coincidental noise pulses are few in number.

As in G-M and solid scintillation counting, the background count must be subtracted from each recorded count. In liquid scintillation counting, the background count is derived from a number of sources, including coincidental tube noise, naturally occurring radioisotopes in the vial and scintillation fluid (most glass contain small amounts of ^{40}K, and organic solvents contain small amounts of ^{3}H and ^{14}C), cosmic radiation, and Cerenkov radiation.

Most liquid scintillation counters contain pulse height discriminators that provide for the rejection of pulses above and below a selected size range. This feature is particularly valuable in "double-label" experiments (experiments in which two different radioisotopes are used). Since the spec-

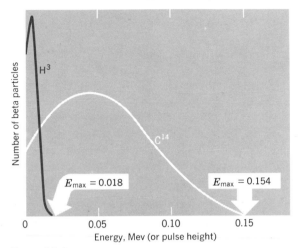

Figure 14–8
Energy or pulse height distributions of ³H and ¹⁴C. The energy and pulse height distributions differ so markedly that pulses arising only from ¹⁴C in a sample containing both isotopes may readily be obtained by adjusting the pulse height position of the discriminator.

trum of pulse heights produced is related to the beta particle energy spectrum of the radioisotope, the discriminator settings may be adjusted to distinguish pulses arising from each of the isotopes present in the sample. This is particularly easy in the case of samples that contain both ³H and ¹⁴C, since their beta particle energy spectra differ so markedly (Fig. 14–8).

Autoradiography

Any discussion of the uses of radioisotopes as tracers in cell biology would be incomplete without at least a brief description of **autoradiography.** The materials and equipment used in this technique differ significantly from those employed in the methods previously described. In autoradiography, the biological sample containing the radioisotope is placed in close contact with a sheet or film of photographic emulsion. Rays emitted by the radioisotope enter the photographic emulsion and expose it in a manner similar to visible light. After some period of time (usually several days to several weeks), the film is developed and the location of the radioisotope in the original sample determined from the exposure spots on the film. Unlike the methods described earlier, autoradiography is generally not employed as a quantitative technique (although it can be

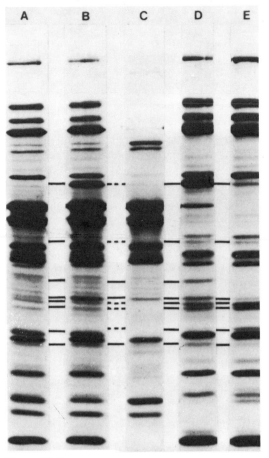

Figure 14–9
Autoradiogram of an electrophoresis gel. ¹⁴C-labeled proteins from whole T4 phage and from virus subfractions were separated in polyacrylamide and the gel covered with a sheet of film to expose the protein bands. (A) Whole phage; (B) heads; (C) necks; (D) extended tails; (E) tails. (Courtesy of Drs. D. Coombs and F. A. Eiserling.)

under certain conditions); instead, it is used to determine the specific region of *localization* of a radioactive tracer. The method is most often used with histological sections to determine the precise location of a labeled compound in a tissue or in a cell and may be applied either at the light microscope or electron microscope level. Autoradiography has also been successfully employed in conjunction with electrophoresis, chromatography (thin-layer chromatography, paper chromatography, etc.), and other molecular fractionation methods (see Chapter 13) for identifying zones containing labeled compounds (Fig. 14–9).

Since the main goal of autoradiography is to determine the precise location of the tracer, the degree of resolution obtained in the autoradiograph is of primary importance. Resolution depends on (1) the type of photographic emulsion used, (2) the distance between the radioactive sample and the emulsion, and (3) the nature of the emitted radiation. A variety of photographic emulsions are available varying in sensitivity and resolution according to the size and concentration of their silver halide grains—the least sensitive films generally offering the highest degree of resolution. Emulsions can also be obtained that are particularly sensitive to specific types of radiation. Since radiation is emitted in all directions from the radioactive source, the greater the distance between the film and the source, the more diffuse the resulting image. It is therefore very important to use very thin samples and place them in very close contact with the film. The highest degree of resolution is obtained with radioisotopes that emit rays of short path length and that have a high specific ionization. Alpha particles provide high resolution, but radioisotopes emitting this type of radiation are rarely useful in biological experiments. Autoradiographs become increasingly diffuse as the E_{max} of beta emitter increases, but ^{3}H, ^{14}C, ^{35}S, and ^{45}Ca do yield good resolution. Gamma rays are inefficient as a result of their extremely high ranges and low specific ionizations.

Autoradiography may be employed with large but thin slices of tissues or organs (gross autoradiography) or with smaller pieces sectioned and prepared for light or electron microscopy. For example, the deposition of calcium in bone has been studied using ^{45}Ca by cutting thin, flat, longitudinal slices through bone and placing them against large sheets of photographic film. When used in conjunction with light microscopy, the paraffin-embedded tissue sections are first mounted on slides that are then coated with photographic emulsion and stored in a lighttight, usually lead-lined box (in order to minimize background exposure that results from the effects of cosmic radiation and other sources of radiation). Several days or weeks later, after photochemical development, the sections are conventionally stained to better visualize the biological material and the distribution of radioisotope determined microscopically from the location of dark exposure spots (called ''grains'') on the section (Fig. 14–10).

Autoradiography may also be used with thin sections of tissues prepared for electron microscopy (Fig. 14–11). This procedure involves preliminary examination and photog-

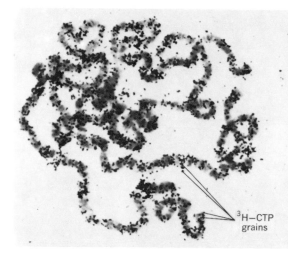

Figure 14–10

Autoradiogram of chromosomes from the polytene nucleus of a *Drosophila* salivary gland cell. The dark exposure spots or "grains" reveal the incorporation of ^{3}H-labeled cytidine triphosphate into actively transcribing regions of the chromosomes. The heavily labeled regions were specifically induced to transcribe by incubating the salivary gland at 37°C in medium containing the tracer. (Courtesy of Dr. J. Lee Compton.)

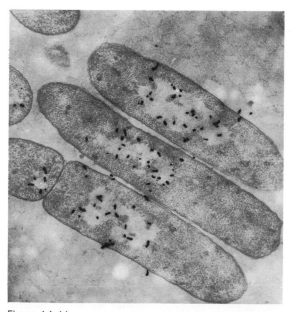

Figure 14–11

Autoradiogram of thin section through dividing *E. coli* cells. The bacteria were cultured on a medium containing ^{3}H-labeled thymidine, which was incorporated into the DNA of the nucleoids. (Courtesy of Dr. F. A. Eiserling.)

raphy of the thin section followed by coating the grid with a very thin layer of photographic emulsion containing silver halide crystals of particularly small size. After several days, the grid is developed and again examined and photographed with the electron microscope in order to identify those regions of the original section that contain clusters of metalic silver grain and therefore contain the radioactive tracer.

Summary

Certain atomic *isotopes* are unstable and emit particulate (and sometimes electromagnetic) radiations. A number of radioactive isotopes, especially ^{3}H, ^{14}C, ^{32}P, ^{35}S, ^{59}Fe, and ^{131}I, are particularly valuable to cell biologists as **tracers** of metabolism and function. With radioisotopes, it is possible to follow ionic or molecular fluxes under conditions of zero net exchange, to simplify chemical analyses, to measure pool sizes, and to determine precursor–product relationships.

The most common types of radiations are **alpha particles, beta particles,** and **gamma rays.** Only negative beta-particle-emitting isotopes and gamma-ray-emitting isotopes are normally used by biologists. The energy acquired by a beta particle during **nuclear transmutation** may vary over a specific range, whereas gamma rays have discrete energy values. Beta particles and gamma rays interact with matter by ionizing or exciting atoms in their path, and it is this property that is used in instruments designed to detect and measure the quantity of radioisotope present in a sample. The simplest and most common detector is the **Geiger-Müller counter,** used principally with tracers emitting beta particles of intermediate or high energy. **Solid scintillation counters** are used with isotopes emitting gamma rays, and **liquid scintillation counters,** with isotopes emitting weak beta rays. ^{3}H and ^{14}C are used as tracers more often than any other isotopes, and since these emit very weak beta particles, the liquid scintillation approach is a particularly important technique in cell biology. In the typical liquid scintillation counter, the sample is mixed with a scintillation fluid in a glass vial. The energies of the beta rays emitted by the sample are ultimately converted to flashes of light, which are detected by photomultiplier tubes and counted.

Autoradiography is a special modification of the radioisotopic tracer technique most often used in conjunction with light and/or electron microscopy. Biological samples (usually tissue sections) containing the tracer are placed in contact with special photographic films or emulsions. The emitted rays expose the film, producing *grains,* the numbers and distribution of which within a cell or tissue provide quantitative and qualitative information about the movement and/or localization of metabolic intermediates or products.

References and Suggested Reading

Articles and Reviews

Joftes, D. L., Radioautography, principles and procedure. *J. Nucl. Med. 4,* 143 (1963).

Ussing, H. H., Life with tracers. in *Annual Review of Physiology,* Vol. 42 (I. S. Edelman and S. G. Schultz, eds.), Annual Reviews, Inc., Palo Alto, 1980.

Books, Monographs, and Symposia

Chard, T., *An Introduction to Radioimmunoassay and Related Techniques.* North-Holland, Amsterdam, 1978.

Chase, G. D., and Rabinowitz, J. L., *Principles of Radioisotope Methodology* (3rd ed.) Burgess, Minneapolis, 1967.

Cooper, T. G., *The Tools of Biochemistry,* Wiley-Interscience, New York, 1977.

Finlayson, J. S., *Basic Biochemical Calculations: Related Procedures and Principles,* Addison-Wesley, Reading, Mass., 1969.

Fox, B. W., *Techniques of Sample Preparation of Liquid Scintillation Counting.* North-Holland, Amsterdam, 1976.

Freifelder, D., *Physical Biochemistry,* W. H. Freeman, San Francisco, 1976.

Friedlander, G., Kennedy, J. W., and Miller, J. M., *Nuclear Radiochemistry* (2nd ed.), Wiley, New York, 1964.

Glasstone, S., *Sourcebook on Atomic Energy* (3rd ed.), Van Nostrand, Princeton, N.J., 1967.

Gude, W. D., *Autoradiographic Techniques,* Prentice-Hall, Englewood Cliffs, N.J., 1968.

Mann, W. B., Ayers, R. L. and Garfinkel, S. B., *Radioactivity and Its Measurement.* Pergamon, Oxford, 1980.

Overman, R.T., and Clark, H. M., *Radioisotope Techniques,* McGraw-Hill, New York, 1960.

Tait, W. H., *Radiation Detection.* Butterworths, London, 1980.

Wang, C. H., and Willis, D. L., *Radiotracer Methodology in Biological Science,* Prentice-Hall, Englewood Cliffs, N.J., 1965.

Part 5
STRUCTURE AND FUNCTION OF THE MAJOR CELL ORGANELLES

Chapter 15
THE PLASMA MEMBRANE
AND THE ENDOPLASMIC RETICULUM

One of the most striking features of cellular organization is the pervasive use of membranes. Not only are cells bounded at their surface by the **plasma membrane,** but the membranous organelles compartmentalize specific cellular functions and activities. Especially in eucaryotic cells, the cytoplasmic matrix itself is pervaded by membrane-bordered channels that comprise the **endoplasmic reticulum** or **ER.** In subsequent chapters of the book, the structural organization and specific functions of the major cellular organelles are examined. In this chapter, we focus on the chemical composition of the plasma membrane and the membranes that form the endoplasmic reticulum and consider the mechanisms by which the transport of materials across these membranes can be achieved.

The *plasma membrane* delimits the cell, physically separating the cytoplasm and cellular organelles from the surrounding cellular environment. This implies that all substances either entering or exiting the cell must pass *through* the plasma membrane. Although the plasma membrane may play a passive role in the exchange of very small molecules between the cell and its surroundings, the flux of many substances is actually facilitated by continuous molecular changes within the membrane. In many instances, transport through the membrane is achieved by the active participation of carrier molecules within the membrane and incurs the expenditure of large amounts of metabolic energy. The cellular ingestion (or excretion) of some materials is associated with gross movements and separations of fragments of the membrane from the main body. Stages of this activity can be seen and studied with the electron microscope (sometimes also the light microscope).

The membranes of the endoplasmic reticulum divide much of the internal cell volume into two phases. The space *within* the channels of the ER comprises the **lumenal** phase of the cell or the **intracisternal space,** while the space *between* channels constitutes the **hyaloplasmic** phase or **cytosol.** The plasma membrane and the membranes of the endoplasmic reticulum have many properties in common. This is not surprising in view of the fact that portions of the plasma membrane are derived from the endoplasmic reticulum (see below). Both membranes are in a continuous state of flux as new components are added to the membranes and others are lost or degraded. Thus, the rather static appearance of these membranes in electron photomicrographs belies their dynamic nature. Moreover, periodic continuities exist between the plasma membrane and the ER with the result that the external cellular milieu is at times continuous with the intracisternal volume. Much of our current knowledge of cellular membranes has been acquired through studies of the plasma membrane since plasma membranes are more easily isolated from cells than ER membranes. To be sure, important differences between these membranes do exist; however, it is becoming increasingly apparent that their chemical composition, organization, and functions are similar. Indeed, in procaryotes (which lack an endoplasmic reticulum), the plasma membrane plays many of the roles of the endoplasmic reticulum of eucaryotic cells.

Early Studies on the Chemical Organization of the Plasma Membrane

Among all animal and plant cells, none has been more extensively studied than the mammalian erythrocyte or red blood cell. The erythrocyte has long been the favorite of investigators studying the plasma membrane because relatively pure membrane preparations are so easily obtained. The mature erythrocyte contains no nucleus, mitochondria, ribosomes, endoplasmic reticulum or other organelles; instead, this highly specialized cell consists essentially of a concentrated (semicrystalline) solution of hemoglobin encased in a membrane. Because of the cell's simplicity, its membranes are easily separated from other cytoplasmic constituents (primarily hemoglobin) by centrifugation following osmotic cell lysis.

Results obtained using erythrocytes have frequently been extrapolated to all cells. This is unfortunate because the erythrocyte is not a typical cell, and it is therefore unlikely that the chemical composition and organization of its limiting membrane are representative. Indeed, with increased interest in studying the plasma membranes of other cells and with the advent of methods for isolating these membranes, our knowledge of the plasma membrane has expanded rapidly in recent years. During this time, it has become increasingly obvious that many of the properties of the erythrocyte membrane are unique. For historical perspective, however, we will begin by considering the early studies of the red blood cell membrane.

Existence of Lipid in the Membrane

As early as 1899, E. Overton recognized that the boundary of animal and plant cells was ''impregnated'' by lipid material. Overton's conclusions were based on exhaustive studies of the rates of penetration of more than 500 different chemical compounds into animal and plant cells. In general, compounds soluble in organic solvents entered the cells more rapidly than compounds soluble in water. These differences were attributed to the ''selective solubility'' of the membrane; that is, lipid soluble materials would pass into

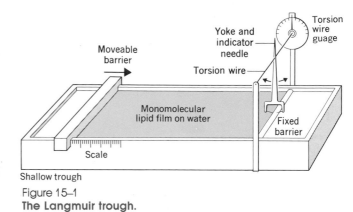

Figure 15–1
The Langmuir trough.

the cell by dissolving in the corresponding lipid elements that made up the membrane. Overton suggested that cholesterol and lecithins might be among the lipid constituents of the plasma membrane, a suggestion that was later substantiated chemically. The pioneering studies of Overton at the turn of the century set the stage for Gorter, Grendel, Cole, Danielli, Harvey, Davson, and others who attempted to determine the specific manner in which the lipid might be organized within the membrane.

The Langmuir Trough

One of the most valuable instruments used to study the behavior of lipid films is the **Langmuir trough** (Fig. 15–1). If lipid containing hydrophilic groups (such as the carboxyl groups of fatty acids or the phosphate groups of phospholipids) is dissolved in a highly volatile solvent and several drops are then carefully applied to the surface of the water, the lipid spreads out to form a thin, monomolecular film in which the hydrophilic parts of each molecule project into the water surface while the hydrophobic parts are directed up, away from the water (Fig. 15–2). In 1917, I. Langmuir introduced a clever technique for measuring the specific minimum surface area occupied by a monomolecular film of lipid and the force necessary to compress all the lipid molecules into this area. His device, known as the Langmuir trough or Langmuir film balance, has been used extensively over the past several decades in connection with physical measurements of membrane lipids.

The apparatus (shown in Fig. 15–1) consists of a shallow trough coated with a nonwettable material so that it can be filled with water to a level slightly higher than its edges.

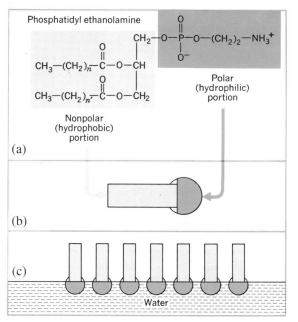

Figure 15–2

Formation of a monomolecular lipid film on water. Phosphatidyl ethanolamine *(a)* represents a typical lipid molecule possessing polar (hydrophilic) and nonpolar (hydrophobic) regions. These regions of the phospholipid molecule are depicted diagramatically in *(b)*. When spread on water, the hydrophilic parts of each lipid project into the water surface, while the hydrophobic parts are directed up, away from the water *(c)*.

A bar is placed across the width of the tray and is used to sweep dirt and dust from the water. A second bar, called the "fixed barrier" (which, in fact, is slightly movable), floats on the water surface and is connected to a torsion balance. After the surface of the water is swept clean, a third bar, called the "movable barrier," is placed across the clean region of the trough, and a known quantity of dissolved lipid is added to the space between it and the fixed barrier. The lipid spreads out over the water surface, forming a layer one molecule thick (Fig. 15–2). The movable barrier is then slowly pushed toward the fixed barrier, thereby compressing the lipid until it forms a continuous and rectangular monomolecular film. The force exerted by the film on the fixed barrier can be measured using the torsion wire gauge. From the area occupied by the lipid and the amount initially added to the trough, it is possible to calculate the surface area occupied by a single lipid

molecule. Langmuir himself employed this device to study the behavior of the surface films formed by a variety of organic compounds, and for this work he received the Nobel Prize in Chemistry in 1932. Others have applied the same technique in specific studies of membrane lipids.

Gorter and Grendel's Bimolecular Lipid Leaflet Model

In 1925, E. Gorter and F. Grendel published the results of their studies on the organization of lipid in the membrane of the red blood cell. Their studies were carried out using blood from a variety of mammals, including dogs, sheep, rabbits, guinea pigs, goats, and humans, and all yielded essentially the same results. The lipid present in accurately measured quantities of washed red blood cells was extracted with acetone and the acetone was then evaporated, leaving the lipid as a residue. This residue was redissolved in benzene and spread in a Langmuir trough to form a tightly packed monomolecular layer. The surface area occupied by the extracted lipid was then measured.

Gorter and Grendel determined the numbers of red cells present in each sample of blood analyzed and estimated the *total* surface area of the cells by multiplying the cell number by the average surface area per cell. (The surface area of the erythrocyte was estimated using the relationship proposed by Knoll that for red blood cells the surface area $= 2d^2$, d being the diameter of the cell determined microscopically. This estimate was subsequently shown to be in error.) By dividing the total surface area occupied by a monomolecular layer of membrane lipid extracted from these cells by the total cell surface area, the number of lipid layers present in the membrane was obtained. The value varied between 1.8 and 2.2, leading Gorter and Grendel to propose that the cell membrane was formed by a *bimolecular* lipid sheet. They further suggested that the polar ends of the lipid molecules of one layer were directed outward (from the cell) toward the surrounding plasma, while the polar ends of the lipid molecules forming the other layer were directed inward toward the cell hemoglobin. Thus the nonpolar, hydrophobic ends of the lipid molecules would face one another (Fig. 15–3).

In the past 10 years, the results of Gorter and Grendel have been reexamined under improved conditions by a number of investigators. The validity of Gorter and Grendel's bimolecular lipid leaflet model depends on the as-

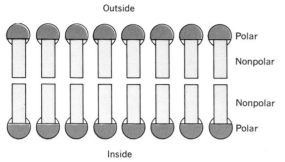

Outside

Polar

Nonpolar

Nonpolar

Polar

Inside

Figure 15–3
Bimolecular lipid leaflet model for the structure of the red blood cell membrane proposed by Gorter and Grendel.

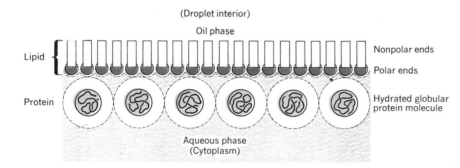

(Droplet interior)

Oil phase

Lipid

Nonpolar ends

Polar ends

Protein

Hydrated globular protein molecule

Aqueous phase (Cytoplasm)

Figure 15–4
Danielli and Harvey's 1935 model of the protein–lipid bilayer formed at the interface between a cell oil droplet and the aqueous cytoplasm.

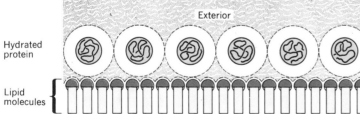

Exterior

Hydrated protein

Lipid molecules

Figure 15–5
Danielli-Davson membrane model (1935).

Lipoid

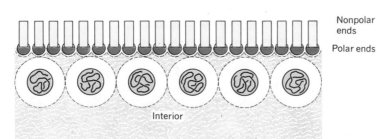

Nonpolar ends

Polar ends

Interior

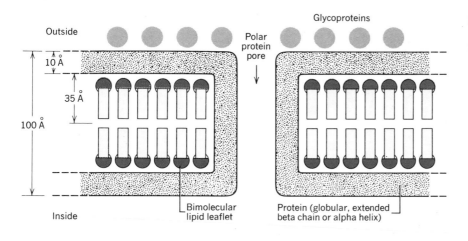

Figure 15–6
Modified Danielli-Davson membrane model (1950).

sumptions that (1) all the erythrocyte lipids are in the plasma membrane, (2) all of this lipid was extracted using their acetone procedure, and (3) the average surface area of the cells was accurately estimated. The first assumption has been verified—all the erythrocyte lipid *is* in the plasma membrane. However, it is now clear that Gorter and Grendel extracted only 70 to 80% of the total lipid. This error would seriously alter the predicted ratio of *lipid film area to cell surface area* if it were not for the fact that Gorter and Grendel also *underestimated* the red blood cell surface area by a comparable amount. The two errors canceled each other out, so that the ratio of 2:1 is still obtained.

The Danielli-Davson Membrane Model

A consistent observation made for plasma membranes that was not explained by the bimolecular lipid leaflet model was the very low surface tension of the cell membrane. In 1935, J. F. Danielli and E. N. Harvey proposed that oil droplets and other lipid inclusions in cells were bounded at their surfaces by an organized layer of the lipid and a layer of protein. It was postulated that the protein, which consisted of a monomolecular layer of hydrated molecules, faced the aqueous cytoplasm and simultaneously interacted with the polar portions of the lipid layer. The nonpolar portions of the lipid layer faced the hydrophobic oil phase of the droplet interior (Fig. 15–4). In this structure, the natural surface activity of the protein would account for the low interfacial tension of the droplet membrane. Shortly thereafter, Danielli and H. Davson suggested that the plasma membrane itself might be composed of two such lipid–protein bilayers—one facing the interior of the cell

and the other facing the external milieu. This arrangement is shown in Figure 15–5. Danielli and Davson proposed that such a membrane would exhibit *selective permeability,* being capable of distinguishing between molecules of different size and solubility properties and also between ions of different charge.

By the early 1950s, several modifications were made in the Danielli-Davson membrane model. Using the polarizing microscope, F. O. Schmitt showed that in the erythrocyte membrane, lipid molecules were oriented with their long axis perpendicular to the membrane surface; in contrast, the protein molecules were oriented tangentially. Accordingly, layers of polypeptides arranged in the pleated sheet configuration or as alpha helices were substituted for the hydrated globular molecules of the original model. It was also suggested that glycoproteins might be adsorbed to the outer membrane surface, thereby accounting for the antigenic properties of cell membranes. Pores in the membrane were presumed to be formed by periodic continuities (bridges) between the outer and inner protein layers. The modified Danielli-Davson membrane model is shown in Figure 15–6. In this arrangement, the association between the surface proteins and the bimolecular lipid leaflet would be maintained primarily by electrostatic interactions between the polar ends of each lipid molecule and charged amino acid side chains of the polypeptide layers. Either electrostatic or van der Waals bonds could bind other groups to the outer protein surface.

It should be kept in mind that the model shown in Figure 15–6 was formulated *before* the plasma membrane was first seen, since the use of the electron microscope to study the organization of the plasma membrane began around 1957.

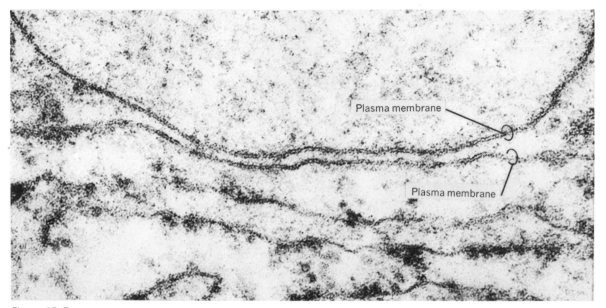

Figure 15–7
Trilaminar appearance of the plasma membrane. (Electron photomicrograph courtesy of R. Chao.)

Nevertheless, even the thickness of the membrane was estimated to be about 100 Å, based on the known lengths of extended phospholipid molecules (about 35 Å) and the thickness of *several* layers of pleated sheet polypeptide or a *single* layer of the alpha-helix form (10–15 Å).

Robertson's Unit Membrane

In the late 1950s, electron microscopy provided additional information about the structure of the plasma membrane. J. D. Robertson was a pioneer in this area, showing that membranes fixed with osmium tetroxide revealed a characteristic *trilaminar* appearance consisting of two parallel outer dark (osmiophilic) layers and a central light (osmiophobic) layer (Fig. 15–7). The osmiophilic layers typically measured 20–25 Å in thickness and osmiophobic layers measured 25–35 Å, yielding a *total* thickness of 65–85 Å. This value compared favorably with the thickness predicted on the basis of chemical studies. However, the thickness of the osmiophilic outer layer was much greater than that predicted for the outer protein coat, while the thickness of the central osmiophobic layer was too small to be the bimolecular lipid leaflet (see Fig. 15–6). To account for these apparent discrepancies, Robertson suggested that the dark layers might be produced by osmium ions binding to *both* the polar amino acid side chains of the protein and the polar ends of the phospholipid molecules, while the light central layer represented only the nonpolar fatty acid chains of each phospholipid that would not bind osmium. In some cells, the outer dark line was thicker than the internal dark line, and was presumed to be due to the binding of additional osmium ions by adsorbed glycoproteins or other osmiophilic molecules.

Robertson and others demonstrated that the trilaminar pattern was characteristic of many other cellular membranes, including the endoplasmic reticulum. In view of the underlying unity in the appearance of the cell membranes studied, Robertson proposed his now famous **unit membrane** model. According to Robertson, the unit membrane consisted of a bimolecular lipid leaflet sandwiched between outer and inner layers of protein organized in the pleated sheet configuration. Such an arrangement was presumed to be basically the same in all cell membranes. While acknowledging specific chemical differences between membranes (i.e., the particular molecular species that make up each membrane differ), Robertson proposed that the pattern of molecular organization was fundamentally the same.

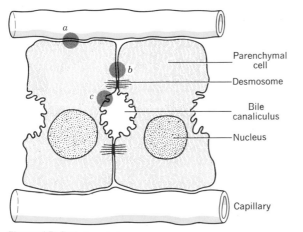

Figure 15–8

Three faces of the liver cell. *(a)* Juxtaposition of plasma membrane and sinusoid capillary membrane. *(b)* Juxtaposition with neighboring cell. *(c)* Juxtaposition with bile canaliculus. Each face contains specific membrane constituents and properties.

Robertson extended his unit membrane model to include the notion that continuity exists between the membranes of the nuclear envelope and the plasma membrane via the endoplasmic reticulum. The occurrences of such continuities have been confirmed in electron microscopic studies of many different cells and tissues. Furthermore, Robertson suggested that vesicular organelles might arise from this continuous membrane system and that they are subsequently pinched off to form separate structures. There is some evidence in support of this notion in the case of microbodies (Chapter 19).

From its inception, Robertson's unit membrane model was highly controversial and almost continuously argued and questioned. Nonetheless, the model held a cornerstone position in membrane biology until the late 1960s, when a number of new findings made the unit membrane no longer tenable. Although there can be no doubt about the similar electron microscopic appearance of nearly all membranes, so strict a chemical interpretation to account for the uniformity is no longer supportable. Among the more important observations and viewpoints that argue against a generalized unit membrane model are the following:

1. Cellular membranes vary in biological functions, in kinds and relative amounts of phospholipids, and in quantitative lipid-to-protein ratios. Consider, for example, the plasma membrane of a parenchymal cell in the liver. The membrane forms junctions with at least three different neighboring structures (Fig. 15–8): (*a*) One face of the liver cell plasma membrane forms a junction with the membranes of the *capillary sinusoids,* and across this face pass various substances exchanged between the bloodstream and the liver cell cytosol (e.g., sugars, amino acids, insulin). The membrane in this region would be expected to contain specific transferases and hormone receptors; (*b*) a second face (called the *contiguous face*) is formed with the plasma membranes of neighboring cells, and it is across this face that intercellular exchange takes place; (*c*) a third face of the liver cell is in contact with the *bile canaliculi* into which bile salts and bile pigments are continuously discharged; here too the organization and composition of the plasma membrane would be expected to be rather specific. W. H. Evans and others have shown that the different faces of the liver cell plasma membrane are clearly distinguishable in enzyme content and molecular composition. The epithelial cells that line the small intestine also reveal a differential distribution of membrane proteins. The portion of the membrane facing the intestinal lumen is rich in glycoproteins, while the opposite membrane face contains *sodium pumps* (discussed later). The membrane uniformity that is apparent during electron microscopy of cells belies the chemical heterogeneity that actually exists.

2. There is no rigorous evidence unequivocally relating the trilaminar electron microscopic appearance of membranes to a specific arrangement of protein and lipid. Indeed, the chemistry of the interaction between protein and lipid and electron-dense stains such as osmium tetroxide is unclear. For example, the trilaminar appearance of some cell membranes is not altered by the preliminary extraction of the membrane lipids. Moreover, it has been shown chemically that osmium can react with the unsaturated fatty acid side chains of phospholipids.

3. There is now ample evidence to substantiate that the protein and lipid composition of the plasma membrane does not remain constant. Instead, the composition changes as protein and lipid molecules are added to the membrane, removed from the membrane, and redistributed through the membrane.

For all intents and purposes, the only aspect of Robertson's unit membrane model that remains undisputed is the existence in the membrane of a bimolecular leaflet of lipids.

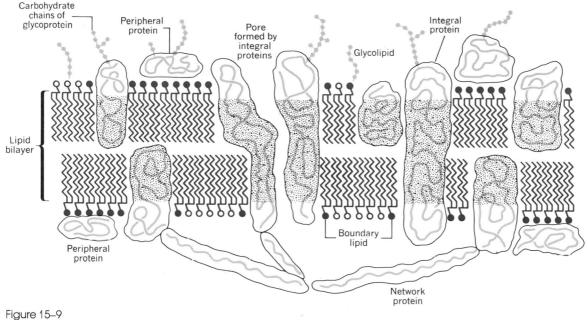

Figure 15–9

The *fluid-mosaic* model of membrane structure.

Yet even this should more properly be credited to the earlier studies of Gorter and Grendel. In the time since their original studies were conducted, the quantitative distributions of membrane lipids in erythrocytes have been reevaluated using newer and improved techniques that guarantee total lipid extraction from the membranes and more accurate measurements of cell surface area. Some of the new methods account also for the specific physical dimensions of various lipid molecules known to be present in the membrane (fatty acids, phospholipids, neutral lipids, etc.) in making calculations of the lipid surface area. For the most part, these studies indicate that indeed there is enough membrane lipid present to form a layer two molecules thick.

Other sources of support for the presence of the lipid bilayer include results of X-ray diffraction analysis of dispersions of isolated cell membranes and electron spin resonance (ESR) studies that indicate that the lipid molecules are oriented with their long axes perpendicular to the plane of the membrane. Attempts have been made to reconstruct lipid bilayers using molecular models for the various membrane lipids and by considering the attractive and repulsive forces resulting from van der Waals interactions, hydrogen bonds, and electrostatic interactions. These studies indicate

that a stable bimolecular lipid leaflet can be formed by a tail-to-tail interdigitation of the lipids.

The dimensions, electron microscopic appearance, permeability, surface tension, and electrical capacitance of artificially created lipid bilayers (i.e., **liposomes,** see Chapter 6) are very similar to those of natural biological membranes. Finally, freeze-fracture techniques (see Chapter 1) used in transmission electron microscopy produce images of plasma membranes and ER membranes, suggesting a natural plane of cleavage at the center of the membrane. A plane susceptible to such cleavage would be provided by the space at the center of the bimolecular lipid leaflet. This will be pursued in the next section.

The Fluid-Mosaic Model of Membrane Structure

At the present time, the most widely accepted model of membrane structure is the **fluid-mosaic model** (an expression introduced by S. J. Singer and G. Nicolson to describe both the properties and organization of the membrane). According to this model (Fig. 15–9), the membrane contains a bimolecular lipid layer, the surface of which is interrupted by proteins. Some proteins are attached at the polar surface of the lipid (i.e., the **peripheral, or extrinsic, proteins**), while others penetrate the bilayer or span the membrane entirely (i.e., the **integral, or intrinsic, proteins**). The peripheral proteins and those parts of the integral proteins that occur on the outer membrane surface frequently contain chains of sugars (i.e., they are glycoproteins). The sugar chains are believed to be involved in a variety of physiological phenomena including the adhesion of cells to their neighbors. Membrane lipid is primarily phospholipid, although quantities of neutral lipids may also be present. Some of the lipid at the outer surface is complexed with carbohydrate to form glycolipid.

Freeze-Fractured Membranes

The fluid mosaic model of membrane structure is beautifully supported by visual evidence provided when freeze-fractured membranes are examined with the transmission electron microscope. D. Branton, who pioneered this field, showed that membranes rapidly frozen at the temperature of liquid nitrogen and cut or chipped with a microtome blade readily fracture along specific planes. When the plane of the fracture intersects the plane of the membrane, the membrane is split along the center of the lipid bilayer, producing two "half-membranes" called the *E* half and the *P* half. The *E* half is that portion of the membrane that faced the cell *exterior,* while the *P* half corresponds to the portion that faced the protoplasm (cytosol). One side of each half-membrane is the original membrane surface, called the *E* and *P* faces, while the other side is the newly exposed **fracture face,** called the *E* fracture face (EF) and *P* fracture face (PF). The fracture faces are extremely delicate and are not examined directly. Instead, a thin film of platinum and carbon is evaporated onto the surface of the fracture faces to produce a **replica** (see Chapter 1), which is then examined by transmission electron microscopy.

In many instances, before the replica is made, water (as well as other volatile materials) on or near the fracture surfaces is eliminated by sublimation (i.e., by carefully raising the temperature of the sample). This step, which used to be called "freeze etching," exposes additional surface features of the fracture face (see Fig. 15–10 and especially Fig. 1–11).

Electron micrographs of freeze-fractured cells show the membranes to be covered by numerous small particles (Fig. 15–11). There is convincing evidence that the particles are membrane proteins (e.g., they disappear when the membranes are first treated with proteolytic enzymes). This suggests that the plane of fracture passes around the protein molecules rather than through them. This relationship is depicted in Figure 15–12. The relatively uniform background apparent in the fracture face in Figure 15–11 corresponds to the surface of one-half of the lipid bilayer.

Membrane Proteins

Peripheral (Extrinsic) Proteins

Peripheral or extrinsic membrane proteins are generally loosely attached to the membrane and are more readily removed than are the integral proteins. Peripheral proteins are rich in amino acids with hydrophilic side chains that permit interaction with the surrounding water and with the polar surface of the lipid bilayer. Peripheral proteins on the cell's exterior membrane surface often contain chains of sugars.

Integral (Intrinsic) Proteins

Integral or intrinsic membrane proteins contain both hydrophilic and hydrophobic regions. Those portions of the protein that are buried in the lipid bilayer are rich in amino acids with hydrophobic side chains. The latter are believed to form hydrophobic bonds with the hydrocarbon tails of the membrane phospholipids. Portions of integral proteins that project outward from the lipid bilayer are rich in hydrophilic amino acids; those projecting from the outer membrane surface may contain carbohydrate chains.

Integral Proteins That Span the Membrane. It was M. Bretscher who first demonstrated the existence of integral proteins that span the entire membrane. In a series of elegant experiments, Bretscher showed that radioactive ligands specific for membrane proteins of the erythrocyte were bound in smaller quantities to intact cells than to disrupted cells. Disruption of the cells was shown to expose portions of the membrane proteins previously facing the cell interior, thereby allowing additional radioactive ligand to associate with the protein.

T. L. Steck developed a technique for converting fragments of disrupted erythrocyte membranes into small vesicles that were either "right-side-out" (i.e., the external

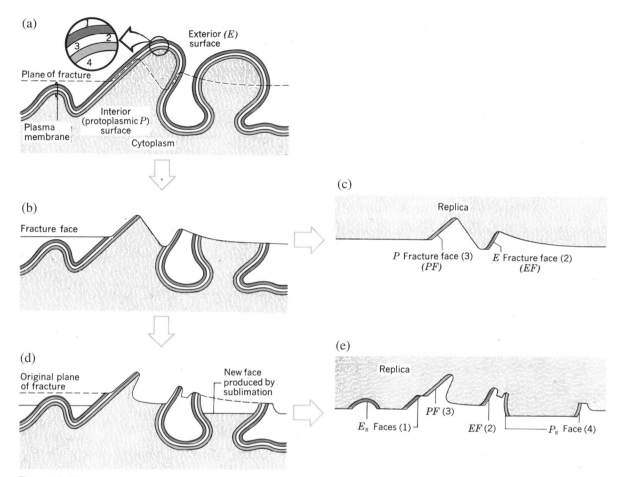

Figure 15–10

Membrane faces exposed by freeze-fracturing alone and by freeze-fracturing followed by sublimation. *(a)* The fracture passes along the lipid bilayers for some distance following its intersection with the membrane. *(b)* This exposes the *PF* and *EF* fracture faces seen in the replica *(c)*. Alternatively, the surface of the specimen is lowered by sublimation *(d)*, exposing the outer and inner surfaces (i.e., the E_s and P_s faces) as well as the fracture faces *(e)*.

Figure 15–11
Freeze-fractured and sublimated plasma membrane of an erythrocyte. The fracture face of the protoplasmic half of the membrane (i.e., *PF*) is covered with particles believed to be integral proteins. Sublimation reveals the outer surface of the exterior half of the membrane (i.e., E_S). The fracture face is seen to be separated by a step (St) from the outer surface of the membrane. Circled arrow indicates direction of shadowing. Magnification 90,000×. (Photomicrograph courtesy of Dr. C. Stolinski.)

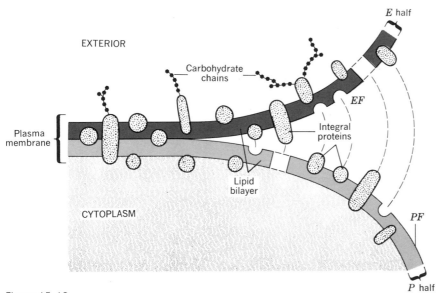

Figure 15–12
Freeze-fracturing of the plasma membrane. The fracture plane occurs at the center of the lipid bilayer and passes over (or under) the integral membrane proteins.

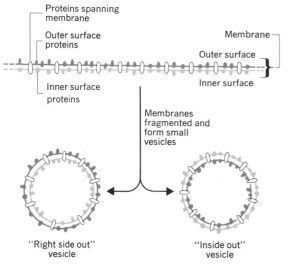

Figure 15–13

Experiments of T. L. Steck showing that proteins of the erythrocyte membrane may be confined to one or the other surface or may span the membrane. Membranes were fragmented and allowed to form suspensions of vesicles. When the vesicles were "right-side-out," proteolytic enzymes added to the suspension destroyed those proteins normally present at the outer surface of intact membranes but had no effect on proteins normally present at the inner surface. In contrast, when "inside-out" vesicles were similarly treated, proteins normally present at the inner surface were destroyed. Proteins spanning the membrane were destroyed in both cases.

face of the membrane also formed the external face of the vesicle) or "inside out" (Fig. 15–13). When proteolytic enzymes were added to separate suspensions of each type of vesicle, certain of their membrane proteins were found to be equally susceptible to digestion and could therefore be enzymatically attacked from either membrane surface. These proteins clearly spanned the membrane. Other proteins were susceptible to enzymatic digestion only when present in right-side-out or inside-out vesicles, indicating their differential distribution in the membrane's outer and inner surfaces.

Integral proteins that span the entire membrane contain outer regions that are hydrophilic and a central region that is hydrophobic. Carbohydrate associated with the hydrophilic region facing the cell's surroundings is believed to play a role in maintaining the orientation of the protein within the membrane. The hydrophilic sugars, together with the hydrophilic side chains of amino acids in the outer

region of the protein, effectively prevent reorientation of the protein in the direction of the hydrocarbon core of the lipid bilayer.

Asymmetric Distribution of Membrane Proteins

The outer and inner regions of the cell membrane do not contain either the same types or equal amounts of the various peripheral and integral proteins. For example, the outer half of the erythrocyte membrane contains far less protein than does the inner half. In addition, various membrane proteins may be present in significantly different quantities; the membranes of some cells contain a hundred times as many molecules of one protein species as another. Moreover, regardless of absolute quantity, all copies of a given membrane protein species have exactly the same orientation in the membrane. The differential distribution of proteins in the various regions of the plasma membrane within a single cell was described earlier in connection with liver parenchymal cells and intestinal epithelium. This irregular distribution of membrane proteins is known as *membrane asymmetry*. Not only plasma membranes but also membranes of the endoplasmic reticulum and vesicular organelles (e.g., mitochondria) are asymmetric.

Mobility of Membrane Proteins

When cells are grown in culture, there is an occasional fusion of one cell with another to form a larger cell. The frequency of cell fusion can be greatly increased by adding *Sendai virus* to the cell culture. In the presence of this virus, even different strains of cells can be induced to fuse, producing *hybrid* cells or **heterokaryons.** D. Frye and M. Edidin utilized this phenomenon to demonstrate that membrane proteins may not maintain fixed positions in the membrane but may move about laterally through the bilayer. Frye and Edidin induced the fusion of human and mouse cells to form heterokaryons and, using fluorescent antibody labels, followed the distribution of human and mouse membrane proteins in the heterokaryon during the time interval that followed fusion. At the onset of fusion, human and mouse membrane proteins were respectively restricted to their "halves" of the hybrid cell, but in less than an hour both protein types became uniformly distributed through the membrane (Fig. 15–14). The distribution

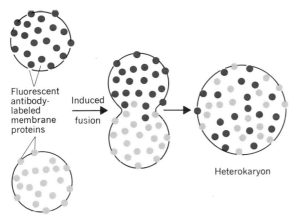

Figure 15–14
Movement of proteins in the plasma membrane. When fluorescent antibody labeling of the membrane proteins of different cells is followed by *Sendai virus*–induced cell fusion, the proteins are soon distributed throughout the membrane of the heterokaryon.

Table 15–1
Enzymes Present in Plasma Membrane

Acetylphosphatase
Acetylcholinesterase
Acid phosphatase
Adenosine triphosphatase
Adenylate cyclase
Alkaline phosphatase
Alkaline phosphodiesterase
Aminopeptidase (several forms)
Cellobiase
Cholesterol esterase
Guanylate cyclase
Lactase
Maltase
Monoglyceride lipase
NADH oxidizing enzymes
NAD glycohydrolase
5′-Nucleotidase
Phospholipase A
Sialidase
Sphingomyelinase
Sucrase
UDP glycosidase

of the membrane proteins was not dependent on the availability of ATP and was not prevented by metabolic inhibitors, indicating that lateral movement of proteins in the membrane occurred by diffusion.

Not all membrane proteins are capable of lateral diffusion. G. Nicolson and others have obtained evidence suggesting that some integral proteins are restrained within the membrane by a protein *network* lying just under the membrane's inner surface (Fig. 15–9). In many cells, this network is associated with a system of cytoplasmic filaments and microtubules that radiate through the cytosol forming a **cytoskeleton** (see below).

Enzymatic Properties of Membrane Proteins

Membrane proteins have been shown to possess enzymatic activity. Table 15–1 lists *some* of the enzymes that are now recognized as constituents of the plasma membrane of various cells. To this list of proteins must be added *receptor* proteins (such as the insulin-binding sites of the liver cell membrane) and *structural* proteins.

Ectoenzymes and Endoenzymes. Enzymes disposed in the plasma membrane may be characterized according to the membrane face containing the enzymic activity. Accordingly, **ectoenzymes** are those enzymes whose catalytic activity is associated with the exterior surface of the plasma membrane; the activity of plasma membrane **endoenzymes** is associated with the interior of the cell. Many (perhaps all) plasma membrane ectoenzymes are glycoproteins.

Isolation and Characterization of Membrane Proteins

Because of the relative ease with which they may be purified, the plasma membranes of erythrocytes provided much of the early information on the chemistry of proteins (and lipids) present in membranes. Now, however, plasma membranes can be obtained from many cell types in a reasonably uncontaminated state using various forms of density gradient centrifugation. Nonetheless, the individual protein constituents of the membrane are not so easily extricated for individual study because of their high degree of insolubility. Varying degrees of success in extracting proteins from the plasma membrane have been achieved using organic detergents (especially sodium dodecyl sulfate, SDS) and concentrated solutions of urea, *n*-butanol, and ethylene diamine tetraacetic acid (EDTA). These chemicals have a disaggregating effect on membranes, causing the release of many of the membrane proteins by dissociating the bonds that link the proteins together or to other membrane constituents. Often, the removal of these agents from a preparation of solubilized membrane proteins is quickly followed by the reassociation or reaggregation of the proteins to form an intractable matrix.

Once solubilized, the membrane proteins can be separated into discrete classes using electrophoresis, chromatography, or other procedures (see Fig. 13–10). This generally demands that the dissociating agents be present in the separating medium (the electrophoresis gel, the column eluent, etc.); otherwise, application of the membrane extract to the medium is followed by membrane protein reaggregation into insoluble complexes that will not separate into distinct fractions. For example, the separation of liver plasma membrane proteins shown in Figure 13–10 is achieved only if the electrophoresis gel contains SDS. The solubility problem has been one of the greatest barriers to progress in isolating and fully characterizing the proteins of membranes.

Some of the plasma membrane enzymes listed in Table 15–1 have not actually been isolated from the membrane, since removal and isolation of the enzyme is *not* a prerequisite for establishing its presence. Instead, the enzyme activity can be measured directly in the (unsolubilized) membrane preparation.

Membrane Lipids

Much more is known about the specific lipid composition of cell membranes, because the lipids are more readily extracted from the membranes using a variety of organic solvents. Once extracted from isolated membranes, the lipids may be separated and identified using chromatographic or other procedures. Nearly all the membranes studied so far appear to contain the same types of lipid molecules. Phospholipids such as phosphatidyl ethanolamine, phosphatidyl serine, phosphatidyl inositol, phosphatidyl choline (lecithin), and sphingomyelin are the most common constituents, but cholesterol may also be present. The chemical structures of these lipids may be found in Chapter 6. Table 15–2 lists the most common lipids found in a variety of cell membranes and also shows their protein-to-lipid weight ratios; the latter vary considerably.

Table 15–2
Lipids Present in the Plasma Membrane

Plasma Membrane	Major Lipids Present	Protein/Lipid (wt/wt)
Liver cell	Cholesterol, phosphatidyl choline, phosphatidyl ethanolamine, phosphatidyl serine, sphingomyelin	1.0–1.4
Intestinal epithelial cell	Cholesterol, phosphatidyl choline, phosphatidyl ethanolamine, phospfhatidyl serine, sphingomyelin	4.6
Erythrocyte	Phosphatidyl inositol, cholesterol, phosphatidyl choline, phosphatidyl ethanolamine, phosphatidyl serine, sphingomyelin	1.6–1.8
Myelin	Cholesterol, cerebrosides, phosphatidyl ethanolamine, phosphatidyl choline,	0.25
Gram-positive bacteria	Diphosphatidyl glycerol, phosphatidyl glycerol, phosphatidyl ethanolamine	2.0–4.0

Mobility of Membrane Lipids

Lipids exhibit a higher degree of mobility in membranes than do proteins, although lateral mobility is very much greater than transverse ("flip-flop") mobility. A single lipid molecule may move several microns laterally through the membrane in just 1 or 2 seconds! Lipid molecules that are in direct contact with membrane proteins are not as mobile as lipid molecules that are solely in contact with one another; such immobilized lipid is called **boundary lipid** (Fig. 15–9).

The mobility of lipid and protein molecules in the plasma membrane attests to the membrane's fluidity. C. F. Fox and H. M. McConnell have shown that the degree of fluidity is dependent, in turn, on the types and the balance of fatty acid side chains of phospholipids in the membrane.

Fatty acid side chains of membrane phospholipids can be either **saturated** or **unsaturated.** In saturated side chains, all the carbon—carbon bonds are single, with the remaining carbon bonds carrying hydrogen atoms; in unsaturated side chains, one or more pairs of neighboring carbon atoms are linked by double bonds (see Chapter 6). In phospholipid layers consisting exclusively of saturated fatty acids, the side chains are aligned next to one another in an ordered, crystalline array; the result is a relatively rigid structure (Fig. 15–15). In phospholipid layers consisting of a mixture of saturated and unsaturated fatty acid side chains, the packing of neighboring molecules is less orderly (and therefore more fluid). The double bonds of the unsaturated side chains produce bends in the hydrocarbon chains, and these give rise to structural deformations that prevent formation of the more rigid crystalline structure. The greater the number of double bonds, the more disordered (and fluid) is the lipid bilayer (Fig. 15–15).

The rigidity of lipid layers is also affected by temperature. Almost everyone is familiar with the "melting" of fats and waxes at elevated temperatures. In order to maintain membrane fluidity, cells living at low temperatures have higher proportions of unsaturated fatty acids in their membranes than do cells at higher temperatures. Evidence also exists suggesting that cells can alter the balance of saturated and unsaturated fatty acids in their membranes as an adjustment to changing temperature or the demands for altered membrane fluidity.

In recent years, the degree of membrane fluidity has been linked to the ability of various metabolites and hormones

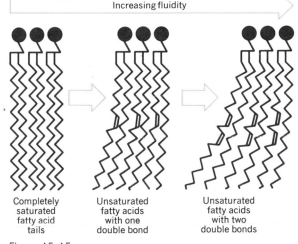

Increasing fluidity

Completely saturated fatty acid tails

Unsaturated fatty acids with one double bond

Unsaturated fatty acids with two double bonds

Figure 15–15

Disruption of the orderly stacking of phospholipids having saturated fatty acid tails by unsaturated fatty acids having one or more double bonds.

Table 15–3

Distribution of Lipids in the Erythrocyte Membrane

	Interior Lipid Monolayer (%)	Exterior Lipid Monolayer (%)
Total	50	50
Sphingomyelin	6	20
Phosphatidyl choline	9	23
Phosphatidyl ethanolamine	25	6
Phosphatidyl serine	10	0
Phosphatidyl inositol	0	0

to bind to surface receptors. An increase in membrane fluidity may be accompanied by the withdrawal of exposed receptors (i.e., they are drawn deeper into the lipid bilayer), while a decrease in membrane fluidity is accompanied by greater accessibility of the receptor through increased exposure above the bilayer.

Lipid Asymmetry

The various membrane lipids are not equally distributed in both monolayers, although the asymmetry is not nearly as marked as in the case of protein. The distribution of lipids in the erythrocyte membrane is shown in Table 15–3 and reveals that the choline phosphatides are primarily in the

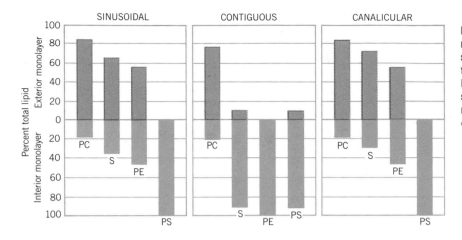

Figure 15–16

Relative distributions of lipids in the internal and external monolayers of the sinusoidal, contiguous, and canalicular faces of the liver cell plasma membrane. *PC*, phosphatidyl choline; *S*, sphingomyelin; *PE*, phosphatidyl ethanolamine; *PS*, phosphatidyl serine. See also Figure 15–8.

outer monolayer and the aminophosphatides are in the inner monolayer. Cholesterol is believed to be present in both surfaces in large amounts. Although lipid asymmetry is a general property of membranes, the type of asymmetry varies considerably from one membrane to another. Asymmetry, once established, is most likely maintained because of the high activation energy that would be required to move the polar groups through the hydrophobic center of the bilayer. Just as proteins are differentially distributed in the plasma membrane areas comprising the various functional faces of a tissue cell, so are the lipids. This is vividly seen in Figure 15–16, which shows the distributions of phosphatidyl choline, sphingomyelin, phosphatidyl ethanolamine, and phosphatidyl serine in the three major plasma membrane regions of the liver parenchymal cell.

Membrane Carbohydrate

It has already been noted that carbohydrate is present in the plasma membrane as short, sometimes branched chains of sugars attached either to exterior peripheral proteins (forming glycoproteins), to the exterior portions of intrinsic proteins, or to the polar ends of phospholipid molecules in the outer lipid layer (forming glycolipid). No membrane carbohydrate is located at the interior surface.

The oligosaccharide chains of membrane glycoproteins and glycolipids are formed by various combinations of six principal sugars: D-*galactose*, D-*mannose*, L-*fucose*, *N-acetylneuraminic acid* (also called *sialic acid*), *N-acetyl-D-glucosamine* and *N-acetyl-D-galactosamine* (see Chapters 4 and 5 for chemical structures). All of these may be derived from glucose.

Possible Functions of Membrane Carbohydrate

Several roles have been suggested for the carbohydrate present on the outer surface of the plasma membrane. One possibility is that because they are highly hydrophilic, the sugars help to orient the glycoproteins (and glycolipids) in the membrane so that they are kept in contact with the external aqueous environment and are unlikely either to rotate toward the interior or to diffuse transversely. Certain plasma transport proteins, hormones, and enzymes are glycoproteins, and in these molecules, carbohydrate is important to physiological activity. It would therefore not be inappropriate to expect that in certain glycoproteins of the plasma membrane the carbohydrate moeity is basic to either enzymatic or some other activity.

Surface carbohydrate is clearly responsible for the various human blood types (e.g., ABO types, MN types) and other tissue types. That is, the sugar sequence and the arrangement of the sugar chains in the membranes of blood cells of an individual with type A blood differ from those of an individual with type B blood, and so on. The carbohydrate is responsible for cell type specificity and is therefore fundamental to the specific antigenic properties of cell membranes. These antigenic properties are linked in some manner to the body's *immune system* and the ability of that system to distinguish between cells that should be present in the organism (i.e., native cells) and foreign cells. Foreign cells (such as bacteria, other microorganisms, transplanted tissue, or transfused blood) may be recognized as foreign because their membrane glycoproteins contain different carbohydrate markers than those present in the individual's own tissues. Such a situation triggers the immune response. In contrast, an individual's own plasma membrane carbohydrate organization is recognized as being native (referred to as "recognition of self") and does not normally trigger an immunological response. Of course, neither does blood transfusion or tissue transplantation if the carbohydrate organization in the membranes of the "donor's" and "recipient's" cells is the same. Cell-specific membrane carbohydrate organization is considered further below in connection with the actions of **lectins** and **antibodies.**

Oppenheimer, Roseman, Roth, and others have clearly implicated surface carbohydrate in the adhesion of a cell to its neighbors in a tissue; presumably, the carbohydrate acts as an adhesive maintaining the integrity of the tissue by linking neighboring cells together. **Contact inhibition,** the phenomenon in which cells grown in culture stop dividing when they touch one another (thereby limiting the growth of the population), may possibly be attributable to a mechanism triggered by interaction of carbohydrates on the surfaces of neighboring cells.

Lectins, Antibodies, Antigens, and the Plasma, Membrane

Lectins

Lectins are a special class of proteins (found principally in plants, especially legumes, and also in some invertebrates) that have a high affinity for sugars and combine with them in much the same manner as an enzyme combines with its substrate or an antibody combines with an antigen. Because the interaction of the lectin with sugar is specific (see Table 15–4), lectins can be used to map the distribution of sugars on the cell surface.

Following their discovery in plants some 90 years ago by H. Stillmark, the lectins were for some time called **phytohemagglutinins** because of their ability to cause the agglutination of red blood cells. However, lectins will agglutinate many kinds of cells, including bacteria. Lectin molecules contain two or more sugar-binding sites, and when large numbers of lectins bind *simultaneously* to sugars on the surfaces of separate cells (thereby cross-linking the cells), the result is agglutination (Fig. 15–17). It should be noted that binding of the lectin to the cell surface sugars can occur *without* ensuing agglutination *if no cross-linking takes place.* The presence and extent of cross-linking is dependent on the balance of lectin concentration and the

Table 15–4
Some Lectins and Their Sugar Specificities

Lectin	Sugar Specificity
Concanavalin A (Con A) (from jack beans)	D-Mannose
Wheat germ agglutinin (WGA)	N-Acetyl-D-glucosamine
Ricinis communis agglutinin (RCA)	D-Galactose
Soybean agglutinin (SBA)	N-Acetyl-D-galactosamine D-Galactose
Lima bean lectin	N-Acetyl-D-galactosamine
Limulus lectin	N-Acetylneuraminic acid

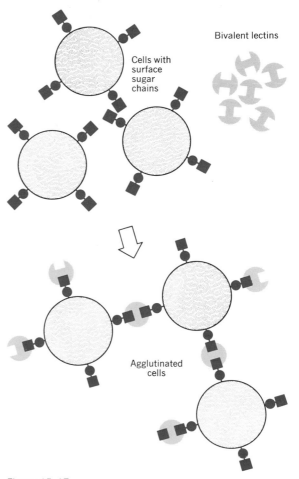

Figure 15–17
Agglutination of cells by *lectins.*

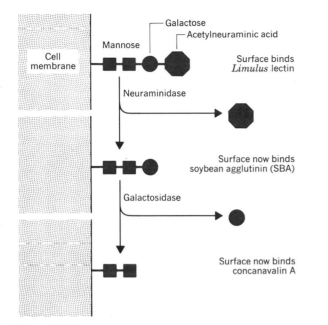

Figure 15–18

Structure of surface carbohydrate is established by successive treatments with lectins and terminal sugar-cleaving enzymes. In this illustration, binding of *Limulus* lectin by the cells indicates the presence of terminal acetylneuraminic acid groups. Their removal using *neuraminidase* is followed by binding of soybean agglutinin, indicating that the acetylneuraminic acid was linked either to galactose or to acetylgalactosamine. Which of these two alternatives is correct is revealed by sensitivity to the specific enzyme added in the next round. In the illustration, it is *galactosidase* that now allows concanavalin A binding, thereby indicating that the terminus was galactose. Removal of galactose and binding of Con A indicates that the next sugar is mannose.

malignant cells can be caused to agglutinate at much lower lectin concentrations than are required to agglutinate normal cells. It has been found that the increased agglutinability of the malignant cells results from increased glycoprotein mobility in the lipid bilayer of the plasma membrane. Since the malignant cell membrane is more fluid, lectins are able to cluster the glycoproteins in the membrane (i.e., draw them together) and thereby make it possible to form greater numbers of cross-bridges.

How lectins can be used to determine the composition of the sugar chains of surface carbohydrate is illustrated in Figure 15–18. The lectins that are bound by unmodified cell membranes establish the nature of the terminal sugars; these may then be cleaved from the remaining carbohydrate using a specific enzyme. The newly exposed terminal sugars are now examined for their lectin-binding characteristics, following which another round of enzymatic sugar removal is carried out. Repetition of this sequence of treatments progressively reveals the order of sugars.

Antigens and Antibodies

An antigen may be loosely defined as any molecule that has the capacity to stimulate antibody production by the immune system of higher animals. Typically, antigens are glycoproteins in the membranes of cells or in other particles foreign to the animal. For example, the antigens present in the membranes of bacterial cells or in the coats of viruses act to stimulate antibody production by the immune system of the infected animal. The antibodies or immunoglobulins (see Chapter 4) produced in response to the presence of the antigen combine with the antigen to form a complex, and this is followed by a series of reactions in which the antigen-bearing agents (e.g., the bacteria) are destroyed.

Antibodies are synthesized by lymphocytes, a subpopulation of white blood cells produced either in the bone marrow (B-lymphocytes) or in the thymus gland (T-lymphocytes). The reaction between antibody and antigen is very specific, a particular antibody combining with only one type of antigen. An enormous variety of B- and T-lymphocytes are present in the body's tissues, each capable of manufacturing only a single antibody type (and therefore capable of reacting with a single type of antigen or foreign cell). The nature of this antibody-synthesizing specificity is also considered in Chapter 24 in connection with the clinical production of **monoclonal antibodies.**

numbers of surface sugars. In this respect, the lectin–sugar interaction is much like that of an antibody–antigen reaction (see below). However, unlike antibodies, which chemically are very similar proteins (Chapter 4), lectins are of diverse structure, organization, and size.

Lectins will bind free sugars as well as sugars attached to cell membranes. Consequently, lectin-induced cell agglutination can be blocked by preliminary addition of the appropriate free sugar to a suspension of cells.

Lectins have been used to verify that the plasma membranes of malignant cells and normal cells differ. Malignant cells are much more readily agglutinated by lectins than the normal cells from which they are derived; that is, the

Some of the antibodies manufactured by a lymphocyte are maintained in its plasma membrane. In the presence of the corresponding antigen, a reaction takes place on the surface of the lymphocytes that acts as a stimulus to the production and secretion of additional quantities of antibody. The surface reactions also trigger lymphocyte proliferation, so that even larger quantities of antibody become available. Following the initial reaction on the lymphocyte cell surface, subsequent antigen–antibody reactions may not involve lymphocytes directly. Instead, the secreted antibodies react with either free antigens or more likely with antigens in the invading cells' membranes. The involvement of either lymphocyte membranes or foreign cell membranes (or both) in the antigen–antibody reaction makes these molecules especially valuable tools for studying the properties of the cell surface.

It should be clear that the surface antibodies of the lymphocytes of one animal can serve as antigens if these lymphocytes are transferred to the bloodstream of another animal. The transferred lymphocytes will be treated much like any other foreign cell and serve to stimulate the production of **anti-immunoglobulin antibodies** (AIA). AIA has been especially useful in probing the distribution of glycoproteins in the cell membrane.

M. C. Raff and S. dePetris prepared AIA, which they then coupled with *ferritin*. Ferritin is a liver protein rich in iron and readily discernible as dark spots by transmission electron microscopy (i.e., the iron renders the ferritin electron-dense). When ferritin-coupled AIA is added to suspensions of lymphocytes, it combines with their surface immunoglobulins, thereby assisting in their identification. Raff and dePetris showed that when lymphocytes are incubated at 4°C with ferritin-coupled AIA, the electron-dense spots are distributed all over the membrane surface, whereas at 20°C, the ferritin-coupled AIA is clustered together at one pole of the cell. As in the case with lectins described earlier, the AIA is able to *cluster* the surface immunoglobulin at 20°C but not at 4°C because at the higher temperature the plasma membranes of the cell are more fluid.

Origin of the Plasma Membrane and Its Protein and Lipid Asymmetry

All of the cell's proteins are synthesized by ribosomes including, of course, those proteins that are destined for inclusion in the plasma membrane. Cytoplasmic ribosomes in eucaryotic cells occur in two states: (1) "attached"—ribosomes associated with the membranes of the endoplasmic reticulum; and (2) "free"—ribosomes freely dispersed in the cytosol. The differences between these ribosomes are considered in Chapter 22 together with the mechanics of protein synthesis. Both attached and free ribosomes are believed to contribute proteins to the plasma membrane.

Synthesis of Membrane Proteins and the "Signal Hypothesis"

Principally as a result of the work of G. Blobel, D. D. Sabatini, C. M. Redman, C. Milstein, J. E. Rothman, and H. F. Lodish, the mechanism that routes newly synthesized proteins to their proper destinations in the cell is gradually unfolding. A major contribution to this end has been the confirmation of the **signal hypothesis** proposed in the early 1970s by Blobel and Sabatini. According to this hypothesis (Fig. 15–19a), mRNA molecules encoding proteins to be secreted from the cell contain a special segment, called the "signal." The signal segment encodes a chain of hydrophobic amino acids 15 to 30 residues long that appears at the beginning of the polypeptide chain. When a ribosome attaches to the mRNA in the cytosol and begins to translate the message, the hydrophobic signal sequence or **signal peptide** emerging from the ribosome is recognized and bound by a receptor in the membranes of the endoplasmic reticulum. The hydrophobic character of signal sequences makes them well-suited for entry into the lipid bilayer of the ER membranes. Ribosomes synthesizing polypeptides that lack a signal sequence are not bound by the ER. Attachment of the ribosome and signal peptide to the ER is followed by the formation of a pore in the membrane. The signal sequence followed by the remainder of the elongating polypeptide passes through this pore into the lumenal phase of the ER. An enzyme (**signal peptidase**) attached to the lumenal surface of the ER cleaves the signal sequence from the growing polypeptide. The completed protein is eventually released from the ribosome into the intracisternal space and the ribosome detaches from the membrane and the mRNA. Proteins discharged into the ER cisternae in this manner are ultimately conveyed to the Golgi apparatus for chemical modification prior to secretion (Chapter 18).

Confirmation of the signal hypothesis was quickly fol-

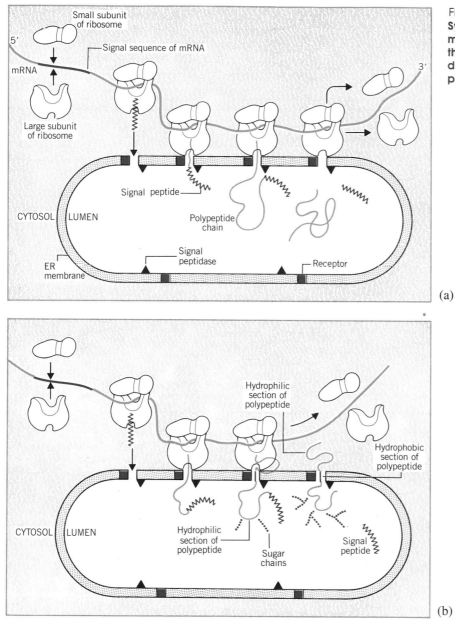

Figure 15–19

Synthesis of secretory proteins (a) and membrane proteins (b) by ribosomes that transiently associate with the endoplasmic reticulum (see text for explanation).

lowed by the proposal of a related mechanism that accounts for the disposition of proteins *within* the membranes of the ER and the plasma membrane itself. The mechanism is illustrated in Figure 15–19b. Synthesis of a signal peptide triggers ribosome attachment to the ER and peptide entry into the membrane. For hydrophilic peripheral proteins and proteins that possess regions facing the lumenal phase of the ER or the polar exterior of the cell, the signal sequence is followed by the synthesis of the hydrophilic segment of the polypeptide. This is threaded through a transient hydrophilic pore formed in the membrane, If the membrane protein being synthesized is an integral protein, the hydrophobic segment that is anchored in the lipid bilayer is synthesized next, accompanied by the loss of the hydrophilic pore. For proteins that span the membrane, synthesis is completed with the production of the internal hydrophilic

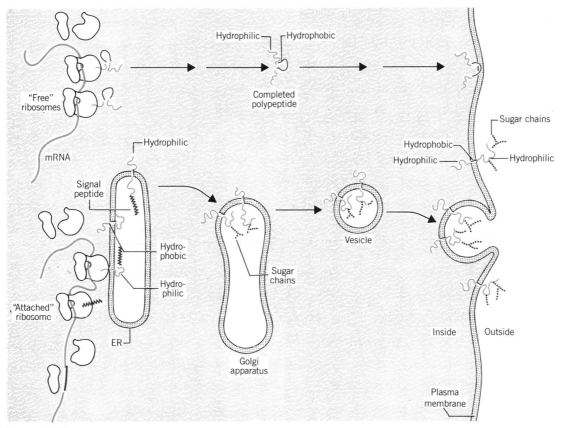

Figure 15–20
Mechanisms proposed for the origin of plasma membrane proteins. Proteins synthesized by free ribosomes may be inserted into the plasma membrane following the completion of the protein's synthesis and its release from the ribosome (top). Other proteins may be inserted into ER membranes during their synthesis by attached ribosomes. These proteins ultimately reach the plasma membrane in the form of small vesicles dispatched from the Golgi apparatus (see text for additional explanation).

segment. The ribosome detaches from the ER following completion of the protein whose signal sequence is removed by signal peptidase.

By comparing parts *a* and *b* of Figure 15–19, it is seen that the major distinction between the synthesis of secretory proteins and membrane proteins is that secretory proteins are released into the lumenal phase of the ER, whereas membrane proteins remain anchored in the ER. Addition of sugars to presumptive membrane glycoproteins may occur soon after the hydrophilic portions of the molecules enter the ER cisternae. Presumptive plasma membrane glycoprotein migrates from the ER to the Golgi apparatus. Although the mechanism for this transfer is still uncertain,

it is believed to take place either by dispatchment of small vesicles from the ER which then migrate to and fuse with the Golgi membranes (see Chapter 18) or by lateral "flow" along the ER membranes to the Golgi. Glycosylation of the membrane proteins is completed in the Golgi apparatus. The Golgi apparatus dispatches completed plasma membrane glycoproteins as small vesicles that migrate to and fuse with the plasma membrane. The overall process is summarized in Figure 15–20, which also shows that the intracellular/extracellular orientation of the membrane protein is maintained throughout its passage from the ER to the plasma membrane.

Although a similar mechanism is not precluded for the

synthesis of integral and peripheral membrane proteins that face the *interior* of the cell, these proteins could be synthesized by free ribosomes. Following release of these proteins in the cytosol they may diffuse to the plasma membrane. Integral proteins would be spontaneously inserted into the membrane by a hydrophobic segment (Fig. 15–20), whereas peripheral proteins would attach to the membrane through polar interactions. Peripheral proteins reaching the plasma membrane in this manner could not pass through the membrane to the exterior surface because they could not traverse the hydrophobic membrane core.

As already noted, procaryotic cells do not contain an endoplasmic reticulum. Secretory proteins and new plasma membrane proteins are synthesized by ribosomes that attach to the inner surface of the plasma membrane. As in eucaryotic cells, ribosome attachment is specified through the synthesis of a signal peptide encoded in the protein's mRNA. The protein is dispatched through the membrane and into the extracellular space.

Synthesis of Membrane Lipids

In eucaryotic cells, phospholipid synthesis is associated with the endoplasmic reticulum, whereas in procaryotic cells lipid synthesis is a property of the cytoplasmic half of the plasma membrane. It is therefore likely that lipid synthesis in eucaryotic cells takes place in the cytoplasmic half of the ER membranes. Following synthesis, the phospholipid becomes, at least temporarily, part of the interior monolayer but may be either enzymatically translocated to the outer layer or flip-flopped between the two layers. In view of the fact that outer monolayer lipids are derived from the inner monolayer, an absolute asymmetry is precluded.

Since in procaryotic cells, lipid synthesis occurs in the plasma membrane, incorporation into the membrane's structure is direct. In eucaryotic cells, however, presumptive plasma membrane lipid must make its way from the ER to the plasma membrane. This is believed to occur by one or both of two processes. Newly synthesized lipids inserted into ER membranes may make their way to the plasma membrane by the same mechanism that translocates ER membrane proteins (Fig. 15–20). That is, *both* membrane proteins and lipids pass from the ER to the Golgi apparatus and are later dispatched to the plasma membrane via small vesicles. The cytosol of eucaryotic cells contains a number of **phospholipid transport proteins** that function to transfer phospholipid molecules from one cellular membrane to another. These transport proteins might also play a major role in mediating the passage of phospholipids from the membranes of the ER to the plasma membrane.

Special Cell Surface Properties Revealed by Erythrocytes

In view of the fact that the mature erythrocyte lacks organelles, this cell has always been a popular source of plasma membranes. Indeed, the pioneering studies of Gorter and Grendel, which were the first to indicate the existence of the lipid bilayer, were carried out using erythrocytes. It is now clear that the erythrocyte membrane may possess a number of unusual properties in addition to others that are believed to be widespread. Notwithstanding its apparent chemical and structural specializations, the erythrocyte is being studied more extensively today than at any time previously.

Like the plasma membranes of other cells, the red blood cell membrane is asymmetric. The lipid asymmetry of the erythrocyte membrane has already been described (Table 15–3). With regard to protein asymmetry, the peripheral proteins account for about 40% of all membrane proteins *but are restricted to the interior surface*. The most abundant of these proteins and the first to be isolated is a molecule called *spectrin*. Spectrin is believed to be an important component of a weblike network of proteins on the interior membrane surface.

There are two major integral proteins, and both apparently span the lipid bilayer. One of these, called *glycophorin-A*, has been fully sequenced and reveals several very interesting properties. Glycophorin-A consists of a single chain of 131 amino acids; 16 short carbohydrate chains are linked to residues near the N-terminus of the polypeptide (primarily to serine and threonine side chains), the carbohydrate accounting for about 60% of the total mass of the glycoprotein. The N-terminal region of glycophorin-A is thought to project beyond the exterior membrane surface. The C-terminal end of glycophorin-A is rich in acidic amino acids, especially glutamic acid, and is believed to project into the cell interior. A segment of about 20 amino acids in the middle of the polypeptide consists exclusively of nonpolar and hydrophobic amino acids and apparently is that portion of glycophorin-A that spans the lipid bilayer. The C-terminal ends of the glycophorin-A molecules are thought to interact with the spectrin molecules on the inner surface of the membrane.

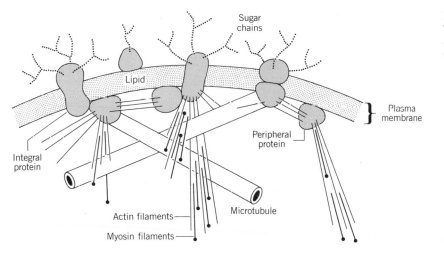

Figure 15–21

The *cytoskeleton* is a network of filaments and microtubules anchored at its margins to the undersurface of the plasma membrane.

The Cytoskeleton

Antibodies against spectrin cause aggregation of spectrin molecules on the inner surface of the erythrocyte membrane. Clustering of spectrin is accompanied by a corresponding rearrangement of glycophorin-A in the membrane, confirming the existence of connections between these two membrane proteins. Spectrin molecules are in turn associated with actin filaments that form a network just below the cell surface. The existence of such a **cytoskeleton** explains the unexpectedly low rate of lateral movement of glycophorin-A in the red cell membrane. The cytoskeleton serves to maintain a specific spatial relationship among the proteins in the membrane.

In other cells, the cytoskeleton is more extensive and includes myosin filaments and microtubules as well as actin (Fig. 15–21). The network of filaments and microtubules radiates through much of the cytosol (see Figs. 1–18 and 23–2) and provides points of attachment for many of the cellular organelles. While the rearrangement of cytoskeletal components just below the cell surface manifests itself in the redistribution of integral membrane proteins, major movements of the cytoskeleton may be fundamental to such gross activities as cellular motion and endocytosis and exocytosis. The cytoskeleton is considered further in Chapter 23.

The differentiation of the erythrocyte in the bone marrow is accompanied by a major reorganization of the cell in which organelles like the nucleus, mitochondria, intracellular membranes, ribosomes, and so forth are progressively lost. Despite its seeming simplicity, the erythrocyte retains a characteristic shape. In humans (and in most other mammals), the cell takes the form of a *biconcave disk* having a diameter of about 8 μm (Fig. 15–22), although changes in shape are readily induced by variations in osmotic pres-

Figure 15–22

(a) **Erythrocyte.** *(b)* **Cross section showing biconcavity.** *(c)* **Effect of lateral force on erythrocyte shape.**

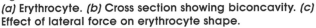

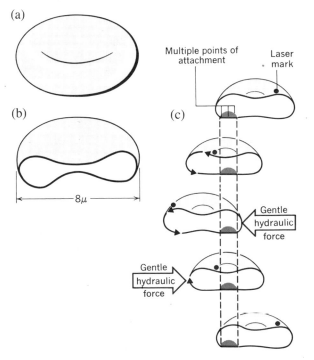

sure (see later in this chapter). The biconcave shape of the erythrocyte is important in its biological function, since such a shape maximizes oxygen diffusion from the cell into the tissues and promotes efficient stacking of cells (*rouleaux* formation) as they circulate through the narrow capillary passageways.

The characteristic shape of the erythrocyte was for some time believed to be due to the differential lateral distribution of lipid (and perhaps also protein) in the membrane. However, it is now known that the erythrocyte membrane is uniformly organized laterally throughout its surface. Despite the existence of a positive hydrostatic pressure internally and the absence of internal structure, the cell does not normally assume the spherical shape expected. The biconcave shape must be related in some manner to the properties and lateral arrangement of neighboring lipid and protein molecules in the cell's membrane or to constraints resulting from the organization of the paracrystalline hemoglobin content.

In some rather startling experiments bearing on this problem, B. Bull and J. D. Brailsford have shown that when an erythrocyte is attached by a portion of its undersurface to a glass slide and a laser used to make a visible mark on the membrane's surface so that membrane movement can be followed, slight lateral displacement of the cell using hydraulic force is accompanied by the membrane *rolling* in the direction of the force, much as a tractor track does (Fig. 15–22). The laser mark travels over the cell surface, following the contours of the biconcave shape. In other words, though rolling laterally, the biconcave shape of the cell is maintained with the vertical biconcavity moving parallel to the glass surface.

Erythrocytes traveling through capillaries and other blood vessels are often arranged as stacks or rouleaux, with their biconcave faces juxtaposed. Such an orderly procession makes it possible for so large a number of cells to pass through the body tissues in short periods of time. P. B. Canham has shown that when a cell in a rouleaux is struck by a laser beam of sufficient energy, the cell membrane is disrupted and the cell lyses (bursts). However, several cells in the rouleaux on either side of the target cell (and not directly affected by the laser) also are seen to undergo gradual lysis. This "contagious lysis" apparently results from the fact that the membranes of erythrocytes in rouleaux transiently adhere to one another. Sudden movements

of the membrane of one cell (as during lysis) create sufficient shear forces at contact points with neighboring cells that their membranes are also affected. The nature of membrane interaction between neighboring erythrocytes in a rouleaux is unknown, but it is clear that much remains to be learned about the "simple" red blood cell.

Cell–Cell Junctions and Other Specializations of the Plasma Membrane

The plasma membranes of neighboring cells in a tissue frequently exhibit specialized junctional regions that play a role in cell–cell adhesion and in intercellular transport. The most common of these junctions are (1) **tight junctions (zonula occludens)**, (2) **intermediate junctions** or **belt desmosomes** (also referred to as **terminal bars** or **zonula adherens**), (3) **spot desmosomes (macula adherens)**, (4) **gap junctions** or **connexons** (also called **nexuses**), and (5) **plasmodesmata** (Fig. 15–23 to 15–26, and Fig. 15–31).

Tight Junctions

In tight junctions, the plasma membranes of neighboring cells fuse with one another at one or more points. At these points of fusion, the external half-membrane of one cell may form a continuous leaflet with the external half-membrane of the adjacent cell. Tight junctions occur at the same circumferential level of the cell, so that they give rise to *belts* of fusion points with neighboring cells. The belt obliterates the intercellular space and acts as a barrier to the flow of materials between the cell surfaces. Intracellularly, the belts of tight junctions are reinforced by a network of fine filaments that radiate into the cytoplasm. Tight junctions between cells are formed by two interdigitating rows of membrane particles (probably integral membrane proteins), one row contributed by each cell. These rows are called *sealing strands* (Fig. 15–24*d*) and act in much the same manner as the two halves of a zipper. The number of rows of sealing strands linking neighboring cells and the extent to which they interconnect to form a network vary from one type of tissue to another. It is also thought that sealing strands act to deter the movements of other proteins within the plasma membrane. In this way, the functional specialization of different faces of the membrane can be maintained (see above).

Figure 15–23
Specializations of the plasma membrane.

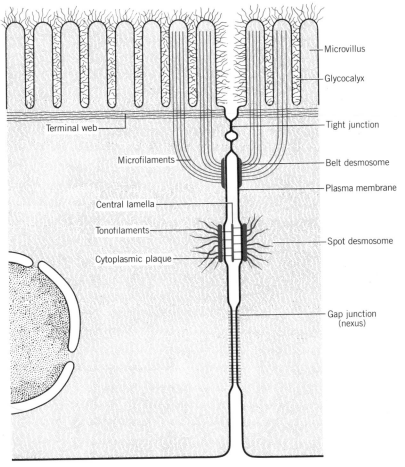

Intermediate Junctions

Intermediate junctions, or belt desmosomes, are girdles of contractile filaments appressed to the cytoplasmic surfaces of juxtaposed membranes. The girdles of contractile filaments have been shown to contain actin and appear also to interweave with another web of filaments that extends into microvilli (Fig. 15–23).

Spot Desmosomes

Unlike tight junctions and belt desmosomes, spot desmosomes (or macula adherens) do not form a belt around the cell. Instead, they are discrete buttonlike attachment points scattered over the opposing membrane surfaces (Figs. 15–23, 15–24 and 15–25). In the region of a spot desmosome, the intercellular space between the adjacent plasma membranes narrows to a distance of about 300 Å. The intercellular space contains a central dense band (the **central lamella**). The cytoplasm adjacent to this region of the plasma membrane is divided into two regions: a lucid region that lies immediately adjacent to the membrane and a neighboring dense band called the **cytoplasmic plaque.** Microfilaments called **tonofilaments** arise from this region and radiate into the cytoplasm, where they may also form links with other spot desmosomes. Tiny rodlike connectives appear to link the central lamella with the cytoplasmic plaque. Spot desmosomes are believed to be the strongest points of attachment between neighboring cells.

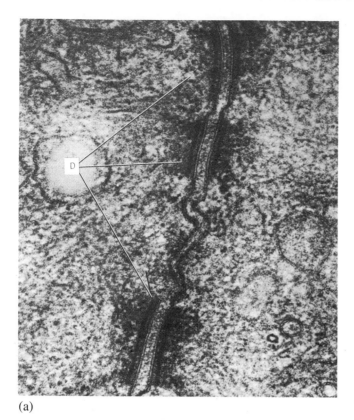

(a)

(c)

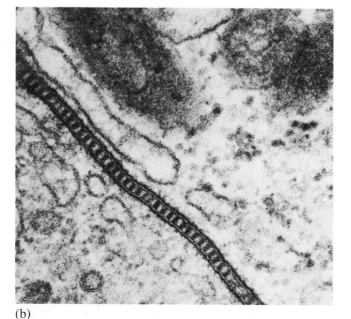

(b)

Figure 15–24

Types of cell junctions. *(a* and *b)* Typical appearances of *desmosomes* (D) and *gap* junctions when seen in thin-sections. *(c)* Freeze-fracture replica through several gap junctions (G$_j$); each junction is seen to be comprised of a number of closely packed particles. *(d)* Freeze-fracture view of a *tight junction* (T$_j$) formed by a number of *sealing strands* of membrane proteins; in this view, the sealing strands (narrow ridges [right] and grooves [left]) create an occlusive zone that acts as a permeability barrier to vertical flow of macromolecules between the two cell surfaces. (Photomicrographs courtesy of Dr. N. B. Gilula; *a, b,* and *c* from Gilula, N. B., *Cell Cummunications* (R. P. Cox, Ed.), John Wiley & Sons, Inc., 1974; *d* from Friend, D. S., and Gilula, N. B., *J. Cell Biol. 53,* 758 (1972). Reprinted with permission from The Rockefeller University Press.)

(d)

Epithelium is a good example of a tissue in which neighboring cells contain the various junctions just described. Beginning at the lumenal surface of the epithelium (i.e., the *apical* surface of the cells) and "descending" through toward the basement membrane, the junctions occur in a characteristic order: a belt of tight junctions is followed by a belt of intermediate junctions; spot desmosomes are more abundant near the basal ends of the cells.

Gap Junctions

The **gap junction** or **nexus** is probably the most complex modification of adjacent plasma membranes. In the gap junction, the opposing plasma membranes are separated by a distance of only 30 to 40 Å and the space penetrated by a number of quasi-cylindrical structures that run from the cytoplasmic surface of one plasma membrane to the other (Fig. 15–24b and c; Fig. 15–26). Each cylinder is called a **connexon** and is composed of six protein subunits having molecular weights of about 30,000 daltons. Each subunit is approximately rod-shaped, about 75 Å long and 25 Å in diameter. The rods protrude slightly further from the cytoplasmic face of each membrane than from the extracellular face. Of special interest is the finding by P. N. Unwin and G. Zampighi that the long axis of each rod is *not* parallel to the long axis of the cylinder; instead, there is a slight angle of inclination (Fig. 15–26). It has been suggested that this angle is varied in order to alter the size of the channel formed in the cylinder. Accordingly, when the connexon is open it forms a channel with a diameter up to about 20 Å. The channel is hydrophilic and allows intercellular passage of a wide variety of molecules having molecular weights up to about 800 daltons (i.e., ions, amino acids, sugars, nucleotides, vitamins, and certain hormones). Rotation of the ends of the subunits acts to close the channel and interrupt the passage of materials between cells.

In freeze-fracture studies, gap junctions are revealed as patches of closely packed connexons (Fig. 15–24). In some tissues, the sealing strands characteristic of tight junctions occur at the margins of the connexon patches.

Microvilli and Surface Ruffles

In some tissues, much of the plasma membrane at the apical surface of the cells is modified to form numerous fingerlike outfoldings called **microvilli** (Fig. 15–27). The microvilli, which usually are about 1-2 μm long, greatly increase the surface area of the membrane and are especially abundant but not restricted to cells having a transport activity such as absorptive epithelium or glandular epithelium. In absorptive epithelium, the microvilli are often referred to as the *brush border*. Individual cells in culture may also display microvilli.

Microvilli contain bundles of about 20 to 30 *actin* filaments (see Fig. 15–29) that are extensions of the cytoskeleton; at one end, the filaments are anchored to the undersurface of the plasma membrane, and at the other end they enter the **terminal web** that forms a plane just below the base of the microvilli (Fig. 15–23). Filaments of the terminal web insert into the desmosomes that encircle the cell.

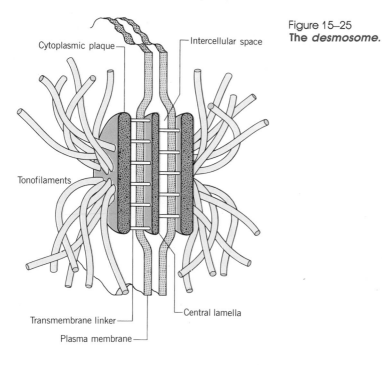

Cytoplasmic plaque

Intercellular space

Tonofilaments

Transmembrane linker

Plasma membrane

Central lamella

Figure 15–25
The *desmosome*.

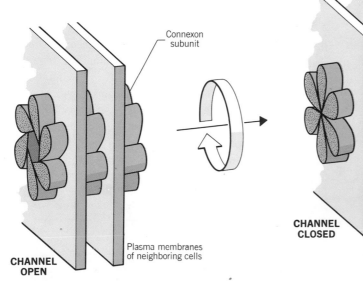

Connexon
subunit

Plasma membranes
of neighboring cells

**CHANNEL
OPEN**

**CHANNEL
CLOSED**

Figure 15–26
The *gap junction*.

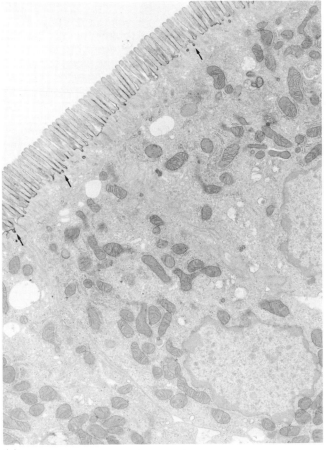

(a)

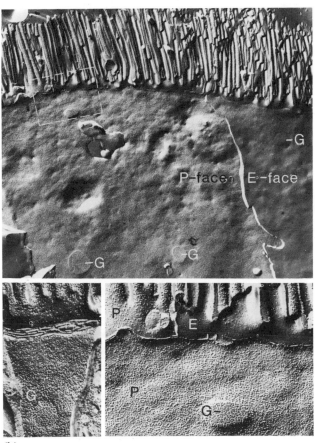

(b)

Figure 15–27

Microvilli. (a) Thin section; arrows show invaginations at the bases of certain microvilli. *(b)* Free-fracture; *G,* gap junction. The terminal web at the base of the microvilli and the membrane particles that cover the *P*-fracture face are seen more clearly in the enlarged inserts. (*a* courtesy fo Dr. R. Rodewald; from *J. Cell Biol. 85,* 18, 1980; copyright The Rockefeller University Press. *b* courtesy of Dr. A. Schiller; from *Cell and Tissue Res. 221,* 431, 1981; copyright Springer-Verlag.)

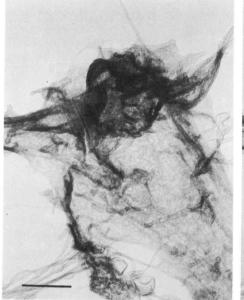

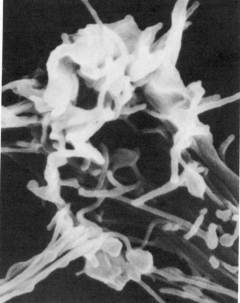

Figure 15–28
High-voltage TEM (top) and SEM (bottom) stereo views of cell surface *ruffles* and *lamellipodia*. As seen in the TEM view, the ruffles are supported by extensions of the microtrabecular lattice and cytoskeleton. Note how the supportive elements seen in the top stereo pair conform in shape and distribution to the cell surface shape seen in the bottom SEM stereo pair. The reference bar is 1 μm. (Courtesy of Drs. K. R. Porter and M. E. Stearns.)

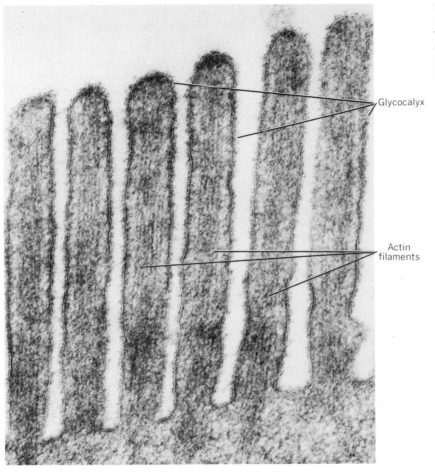

Figure 15–29
***Glycocalyx* covering the surface of micro-villi. The network-like glycocalyx or "fuzzy coat" consists of carbohydrate chains of membrane glycoproteins and glycolipids. (Courtesy of S. Shultz.)**

Glycocalyx

Actin filaments

Depending upon physiological conditions, microvilli may be drawn inward toward the cell or reextended. On the basis of their extensive studies of brush borders, M. S. Mooseker and L. G. Tilney have suggested that movements of the microvilli may be brought about by an interaction between certain actin filaments of each microvillus and *myosin* filaments in the terminal web. The interaction is reminiscent of that which occurs between actin and myosin in contracting muscle cells (see also Chapters 23 and 24).

The plasma membranes of many cells give rise to thin, undulating folds called **surface ruffles** or **lamellipodia** (see Fig. 1–18). The intracellular **microtrabecular lattice** and the cytoskeleton that forms and supports this structure is readily discerned in high-voltage electron photomicrographs of whole cells (Fig. 15–28). Ruffles are especially characteristic of cells carrying out endocytosis (discussed later in the chapter).

The Glycocalyx.   Carbohydrate chains that are part of glycoprotein and glycolipid molecules of the plasma membrane form a network-like coat on the outer surface of the membrane. This coat is called the **glycocalyx** or "fuzzy coat." Some tissues such as the absorptive epithelium of the intestine have an especially thick glycocalyx that surrounds the microvilli and is readily discerned during electron microscopy (Fig. 15–23 and 15–29).

Plasmodesmata

In plant tissues, the cytoplasm of neighboring cells may be connected through numerous narrow channels that penetrate the fibrous cell wall separating the cells. The channels, called **plasmodesmata,** are formed by outward extensions of the plasma membranes of adjacent cells (Fig. 15–30) and are much larger than the channels of gap junctions.

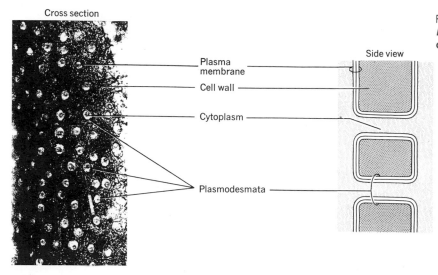

Cross section

Side view

Plasma membrane

Cell wall

Cytoplasm

Plasmodesmata

Figure 15–30
Plasmodesmata (photomicrograph courtesy of Dr. E. G. Pollock).

Plasmodesmata provide for the direct exchange of materials between the cytosols of neighboring cells in the plant tissue.

Passive Movements of Materials Through Cell Membranes

In this section, we consider some of the fundamental principles that govern the passive movements of water, ions, and various other molecules through cell membranes. The term *passive* is intended to denote that the movement of the substance through the membrane is not associated with any chemical or metabolic activities in the membrane. Passage through the membrane in such instances is regulated by such factors as the *concentration gradient* across the membrane and the chemical and physical relationships between the membrane and substances inside and outside the cell. Later, we will direct our attention to movements through the membrane that are accompanied by chemical changes, metabolic activity, or gross molecular rearrangements within the membrane itself.

Osmosis and Diffusion Across Membranes

Membranes that allow substances to pass through them are said to be *permeable* to the substance. Nearly all plasma membranes are permeable to water. If water (or some other solvent) is the only substance that can pass through the membrane, the membrane is said to be **semipermeable.**

Membranes that display a gradation of permeability to water and dissolved solutes (i.e., membranes that permit water to pass through more readily than salts, sugars, etc.) are said to be **selectively permeable.**

Water molecules are continuously moving into and out of cells through the plasma membrane. Such movements are generally not discernible as changes in cell size or shape because the flux in each direction is the same. When the concentrations of solutes inside and outside the cell differ, the water flux in one direction may be greater than in the other direction, and the cell may swell or shrink. Water moves from a region of *low* solute concentration to a region of *higher* solute concentration in order to establish a concentration equilibrium. The movement of water (or some other solvent) in response to such a solute **concentration gradient** is known as **osmosis.**

Osmosis may readily be demonstrated using *artificial* membranes such as *cellophane,* which is permeable to small molecules such as salts, sugars, and amino acids, but is impermeable to large molecules such as proteins. If a cellophane bag filled with a concentrated salt solution is connected to a vertical length of glass tubing and is then immersed in a container of distilled water, water will pass into the bag by osmosis. The entry of water into the cellophane bag will cause the level of water in the glass tubing to rise (Fig. 15–31). Salts permeate cellophane membranes more slowly than water, so that some time elapses before the salt molecules pass out of the bag into the surrounding

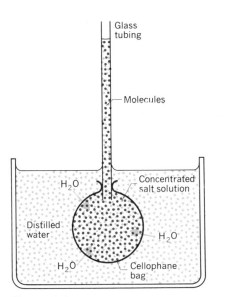

Figure 15–31
Osmosis of water into a cellophane bag containing a concentrated salt solution. Because the solute concentration inside the bag is greater than outside the bag (i.e., the solute concentration outside the bag is zero), water enters through the cellophane membrane by osmosis. The influx of water causes the water level in the attached glass tubing to rise.

water. The movement of solute molecules (in this case, salt molecules) from a region of *high* concentration (inside the bag) to one of *lower* concentration (outside the bag) occurs by the process of **diffusion.** In the case illustrated in Figure 15–31, the initial movement of water into the cellophane bag is followed by the outward diffusion of salt from the bag. As the solute concentration inside the bag decreases, the liquid level in the glass tubing falls as water molecules leave the bag by osmosis. The movements of salt and water molecules continue until the salt concentration inside and outside the bag are equal.

Consider a case in which the cellophane bag is filled with a solution containing an impermeable solute. As in the previous instance, water will enter the bag by osmosis, causing the liquid level in the glass tubing to rise. Since the solute is impermeable, it cannot diffuse from the bag, and a concentration equilibrium across the membrane cannot be achieved. Consequently, water will continue to enter the bag and rise in the glass tubing until a height is reached at which the pressure at the base of the water column is just great enough to prevent any further water influx. In theory, the water will remain at this level indefinitely. The

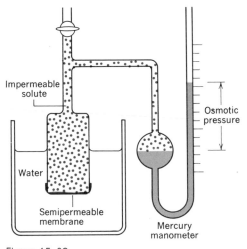

Figure 15–32
An *osmometer.*

pressure that is created inside the bag by the impermeable solute and that supports the column of water is called **osmotic pressure.** Its numerical value can be approximated by measuring the height of the water column. Usually osmotic pressure is expressed in *millimeters of mercury* (i.e., mm Hg) rather than in inches of water. Devices that are used to measure osmotic pressure are called *osmometers* (Fig. 15–32).

Osmosis and Diffusion Across Cell Membranes

Cellular phenomena associated with osmosis and diffusion across the plasma membrane are readily demonstrated using red blood cells, sea urchin eggs, or certain plant cells. The *plasma* in which the red blood cells are normally suspended contains the same concentration of *impermeable* salt (0.15 M NaCl) as the erythrocyte cytoplasm (0.15 M KCl). Normal plasma is said to be *isotonic* to the red cell. If the plasma is diluted with water, its salt concentration will decrease, and the plasma will become *hypotonic* to the red blood cell. Any suspending medium containing an impermeable solute concentration that is *lower* than the corresponding solute concentration in the cells suspended in that medium is considered hypotonic. In the case of hypotonic plasma, water will enter the red cells by osmosis, causing the cells to swell. The same effect can be produced by placing red blood cells in any hypotonic solution. Just

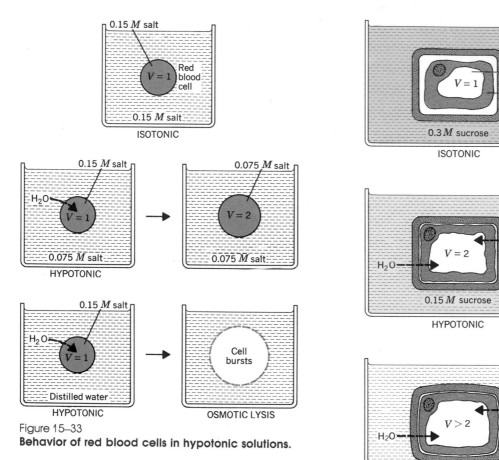

Figure 15–33
Behavior of red blood cells in hypotonic solutions.

how much water will enter the cell depends upon how hypotonic the suspending medium is. For example, if the red cells are suspended in plasma containing *one-half* the normal salt concentration, water will enter the cells until they swell to twice their original volume. This will reduce the internal salt concentration to one-half its former value, bringing the internal and external salt concentrations into equilibrium (Fig. 15–33).

If the cells are suspended in a solution of even greater hypotonicity, then proportionately more water will have to enter the cell to reduce the internal salt concentration to that outside the cell. Obviously, cells can tolerate only a certain amount of swelling before the membrane ruptures, spilling the cell contents into the surrounding medium; this is called **osmotic lysis.** Red blood cells lyse when suspended in very dilute salt solutions or in distilled water (Fig. 15–33). In the specific case of the red blood cell, this

Figure 15–34
Behavior of plant cells in hypotonic solutions.

phenomenon is called *hemolysis,* since hemoglobin from the red blood cell is released into the suspending medium. Other animal cells behave in a similar manner in appropriately hypotonic media.

Plant cells generally do not lyse even when placed in distilled water because cell swelling is limited by the rather inflexible cellulose cell wall. In hypotonic solutions, plant cells swell as water enters the cytoplasmic vacuoles by osmosis. This forces the cytoplasm to the margins of the cell wall. Under these conditions, the plant tissue becomes **turgid** (Fig. 15–34).

Equal concentrations of impermeable salts and nondissociating (nonionizing) molecules (such as sucrose and

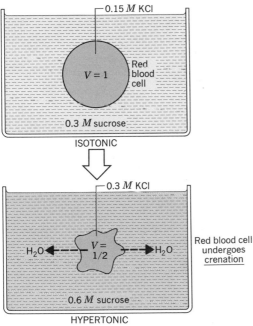

Figure 15–35
Crenation of red blood cells in hypertonic solution.

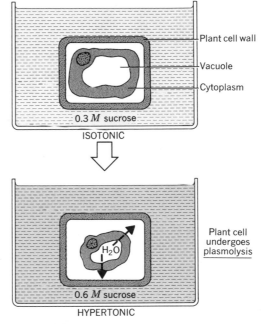

Figure 15–36
Plasmolysis of cells in hypertonic solution.

other sugars) do *not* exhibit the same osmotic effects. For example, 0.15 *M* NaCl exerts twice the osmotic pressure as does 0.15 *M* sucrose. This is because 0.15 *M* NaCl undergoes dissociation in water to produce twice the number of particles (i.e., ions) per cubic centimeter as 0.15 *M* sucrose. Thus, 0.15 *M* NaCl, 0.15 *M* KCl, 0.10 *M* $CaCl_2$, and 0.3 *M* sucrose would all exert the *same* osmotic pressure, since they produce the same particle concentrations in water.

Most cell membranes, including the membrane of the red blood cell, are impermeable to sucrose. Therefore, both 0.15 *M* NaCl and 0.3 *M* sucrose are isotonic to red blood cells. Solutions that contain higher concentrations of impermeable solute than are found inside cells are said to be *hypertonic*. What happens to red blood cells when they are placed in hypertonic sucrose is depicted in Figure 15–35. Water moves by osmosis from the cells into the medium until the concentration of solute inside and outside the cell are the same. The shrinkage associated with such water loss is called **crenation.** Other animal cells behave in a similar manner.

When plant cells are placed in hypertonic media, they undergo **plasmolysis,** that is, water passes from the cytoplasmic vacuoles into the space between the cell wall and cell membrane (Fig. 15–36).

So far, we have considered only the movement of water into and out of the cell. When the concentration of a permeable solute is greater outside a cell than inside, the solute molecules diffuse into the cell until the external and internal concentrations are in equilibrium. In most instances, the combined volume of all the cells present in a cell suspension is only a small fraction of the total volume of the suspending medium. Therefore, the quantity of solute entering (or leaving) the cells has a negligible effect on the solute concentration of the medium.

Consider what happens when red blood cells are placed in an isotonic saline solution (e.g., 0.15 *M* NaCl) containing 0.1 *M* urea. The red blood cell membrane is permeable to urea, and since red blood cells normally contain little or no urea, urea molecules will diffuse into the cell until the intracellular urea concentration reaches 0.1 *M*. Even after a concentration equilibrium is achieved, urea molecules will continue to pass through the cell membrane into and out of the cell. However, the migration in each direction will be the same, so that no *net* change in the urea concentration of the cell occurs. If the cells are transferred to an isotonic saline solution lacking urea, urea molecules will diffuse back out of the cell. Many substances enter and leave cells by diffusion through the plasma membrane.

The Permeability Constant.   The quantity of solute diffusing from one region to another depends on the concentration difference between the two regions—that is, the **concentration gradient.** Concentration gradients often exist between the internal and external environments of a cell. If the solute can permeate the cell membrane, diffusion in the direction of the concentration gradient into (or out of) the cell ensues.

The **diffusion rate** for a solute is given by **Fick's equation:**

$$\frac{dS}{dt} = D\,A\,\frac{C_1 - C_2}{x} \tag{15-1}$$

where dS/dt =   the number of moles of solute, S, diffusing from region 1 to region 2 in the time interval dt.

D =   the **diffusion coefficient** of the solute. Each solute has a specific diffusion coefficient; the units of the diffusion coefficient are moles per unit cross-sectional area per unit concentration gradient per unit time. This reduces to the dimensions square centimeters per second. The diffusion coefficient of a solute is determined in part by the solute molecule's size and shape.

A =   the cross-sectional area through which diffusion occurs.

C_1 and C_2 =   the concentrations of S in regions 1 and 2, respectively.

x =   the distance between regions 1 and 2

$\dfrac{C_1 - C_2}{x}$ =   the concentration gradient

In the case of diffusion into or out of the cell, we may let

x =   the membrane thickness,

$C_1 = C_{out}$ =   the concentration of S outside of the cell.

$C_2 = C_{in}$ =   the concentration of S inside of the cell.

A new term, the **permeability constant K,** which describes the diffusion of a particular substance across a particular membrane, may be substituted for D/x in equation 15–1 to yield

$$\frac{dS}{dt} = K\,A\,(C_{out} - C_{in}) \tag{15-2}$$

The dimensions of K are centimeter per second. Since the total amount of substance S in a cell of volume V is VC_{in}, substitution of VdC_{in} for dS in equation 15–2 followed by a transposition yields

$$\frac{dC_{in}}{dt} = K\,\frac{A}{V}\,(C_{out} - C_{in}) \tag{15-3}$$

Thus, it can be seen that other things being equal, the rate of change of the internal solute concentration (i.e., dC_{in}/dt) depends upon the surface area to volume ratio of the cell (i.e., A/V). Since the ratio A/V is at a minimum for spherical objects, the rate at which solute is exchanged by diffusion between a cell and its environment is increased by deviation from spherical shape.

If the external volume is sufficiently large in comparison with the cell volume (this generally is the case), then C_{out} will not change significantly with time as a result of diffusion. That is, during the time interval between t_0 and t_1, C_{out} will remain constant. Therefore, equation 15–3 may be solved by integration as follows:

$$\int_{C_{out} - C_{in}\,\text{at}\,t1}^{C_{out} - C_{in}\,\text{at}\,t_0} \frac{dC_{in}}{C_{out} - C_{in}} = \int_{t_1}^{t_0} K\frac{A}{V}\,dt \tag{15-4}$$

$$\ln(C_{out} - C_{in}\,\text{at}\,t_0) - \ln(C_{out} - C_{in}\,\text{at}\,t_1) = K\,\frac{A}{V}\,\Delta t \tag{15-5}$$

where Δt equals the time interval between t_0 and t_1. Solving equation 15–5 for K, we obtain

$$K = \frac{V}{A\Delta t}\ln\frac{C_{out} - C_{in}\,\text{at}\,t_0}{C_{out} - C_{in}\,\text{at}\,t_1} \tag{15-6}$$

C_{out}, C_{in}, A, V, and Δt may be determined experimentally, so that equation 15–6 can be employed to find the permeability constant. Evaluation of the permeability constants for a variety of substances entering a cell permits much more to be learned about the chemical nature and behavior of the cell membrane. The permeability constants of glycol, urea, glycerol, and sucrose for the membranes of the red blood cell, the bacterium *Beggiatoa*, and the alga *Chara* are compared in Table 15–5. The marked differences that can be seen in these values reflect the variation that must exist in the composition and organization of the plasma membranes of these cells.

Table 15–5
Some Permeability Constants

Substance	Permeability Constant (cm/sec × 10⁵)		
	Red Blood Cell	Beggiatoa	Chara
Glycol	0.21	1.39	1.2
Urea	7.8	1.58	0.11
Glycerol	0.0017	1.06	0.021
Sucrose	ᵃ	0.14	0.0008

ᵃSucrose does not permeate the red blood cell membrane.

Factors Influencing Permeability. A number of different molecular parameters influence the ability of a substance to permeate cell membranes. Important among these are the **distribution coefficient, molecular size,** and **charge.** The distribution coefficient of a substance relates its solubility in oil to its solubility in water. In general, the more *nonpolar* a substance, the more soluble it is in oil and the less soluble it is in water. In contrast, the more polar a substance is, the less soluble it is in oil and the more soluble it is in water.

One of the most extensive studies of the relationship between distribution coefficient and membrane permeability was that carried out in the 1930s by R. Collander and H. Barlund using the unicellular alga, *Chara.* They found that, in general, the higher the distribution coefficient of a compound, the greater its membrane permeability. Overton had made similar but less quantitative observations many years earlier, and this led him to propose that the cell membrane was composed of lipid. The relationship between lipid solubility and membrane permeability is the basis for the proposal that lipid-soluble substances readily pass into or out of the cell by dissolving through lipid regions of the cell membrane.

For chemically related substances, permeability increases with increasing distribution coefficient *regardless of molecular size.* However, in general, where two molecules have similar distribution coefficients, the smaller molecule is more permeable than the larger molecule. The fact that water has a very high permeability constant in spite of its low lipid solubility led to the suggestion that there are *hydrophilic* pores in the cell membrane through which small polar molecules may more readily diffuse. The permeability constant for water is about 1×10^{-4} cm/sec. This value

is only 0.001% of the rate at which water molecules diffuse across a water layer the thickness of the cell membrane. This implies that if they are present, hydrophilic pores must cover only a small percentage of the surface area of the cell. In the case of the red blood cell, A. K. Solomon has estimated that 0.06% of the surface area is occupied by such pores.

In addition to the distribution coefficient, the *effective size* of a molecule influences its membrane permeability. In general, small molecules are more permeable than large molecules. Substances of very high molecular weight (such as starch, glycogen, and many proteins) are usually unable to permeate the membrane at all. This is not to imply that cells cannot incorporate or eliminate very large molecules. Indeed, they do; however, the process involves mechanisms other than diffusion (see below).

Electrolytes enter cells more slowly than do **nonelectrolytes** of similar effective molecular size, and *strong* electrolytes enter more slowly than *weak* electrolytes. Since pH influences the degree of ionization of many electrolytes, permeability to electrolytes varies with the pH. In general, monovalent ions (e.g., Na^+, K^+, I^-, Cl^-) permeate membranes more readily than divalent ions (e.g., Ca^{++}, Mg^{++}, $SO_4^=$), which in turn are more permeable than trivalent ions (e.g., Fe^{+++}). The relative permeability of the membrane to ions of the same valency number but different sign (i.e., K^+ vs. Cl^-) depends on the particular cell considered.

Not all ions of the same valency and sign are equally permeable. Most differences can be explained in terms of the relative effective sizes of the ions in an aqueous medium. The charge on the ion attracts neighboring water molecules, causing them to align themselves around the ion to form **spheres of hydration.** The effective size of the ion is therefore determined by the number of these hydration spheres (Fig. 15–37). Ions of low molecular or atomic weight attract more water molecules than do ions of higher molecular weight because there are fewer electron shells around the atomic nuclei to neutralize the ionic charge. Accordingly, the effective size of Li^+ (atomic number = 3) is greater than Na^+ (atomic number = 19) and so on. Therefore, K^+ is more permeable than Na^+, and Na^+ is more permeable than Li^+.

Ions may enter or leave a cell through pores in the plasma membrane or the endoplasmic reticulum. Cations are at-

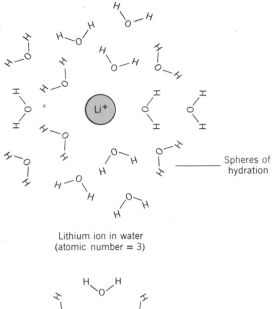

Lithium ion in water
(atomic number = 3)

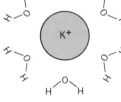

Potassium ion in water
(atomic number = 19)

Figure 15–37
Sphere of hydration formed around lithium and potassium ions in water. Ions of lower atomic or molecular weight attract and orient more water molecules than ions of higher atomic or molecular weight. Therefore, their *effective size* is greater.

tracted to and pass more readily through pores lined by negative charges (e.g., negatively charged R groups of amino acids of membrane proteins). Anions enter or leave the cell through positively charged pores.

The Gibbs-Donnan Effect

Proteins behave like ions because the R groups of their amino acids may bear positive or negative charges. The net ionic charge of the protein molecule depends on the *relative* numbers of positive and negative side groups. Most soluble proteins behave like *anions* because they possess more negative than positive sites. Unlike many other ions, proteins are generally too large to permeate cell membranes. The **Gibbs-Donnan Effect** (after J. W. Gibbs and F. G. Donnan) describes the effect that proteins have on the equilibrium distributions of small ions across a selectively permeable membrane. When a selectively permeable membrane separates an electrolyte solution containing proteins from one that lacks proteins, the concentrations reached at equilibrium for each permeable ionic species will *not* be the same on both sides of the membrane. Instead, the concentration of small ions having the *same* sign as the protein will be lower on the side of the membrane containing the protein (i.e., inside the cell), while the concentration of small ions of *opposite* sign will be higher. The Gibbs-Donnan effect does not influence the equilibrium distributions of nonionized substances such as glucose, urea, and the like. The Gibbs-Donnan phenomenon is responsible for the rupture and lysis of bacteria and other organisms when antibodies attach to their surface. The reason is that the binding of antibody sets off the ''complement'' reactions in which a series of plasma proteins (in a cascade type activation) form enzymes that puncture a hole in the invading cell's membrane. This is followed by an inrush of ions and water that leads to cell lysis.

The Gibbs-Donnan effect may be illustrated as follows. Suppose that two chambers of *equal* and *fixed* volumes are separated by a membrane permeable to water and small ions but impermeable to proteins. Into chamber 1 is placed a sodium proteinate (NaPr) solution of concentration C_1, and into chamber 2, a NaCl solution of concentration C_2. Since the chloride ions are permeable and present at a greater concentration in chamber 2 than in chamber 1 (i.e., chamber 1 initially has no Cl^-), Cl^- will diffuse into chamber 1. In order to preserve the electrical neutrality of each chamber, the diffusion of Cl^- into chamber 1 must be accompanied by the diffusion of an equivalent amount of Na^+. Therefore, chamber 2 will be left with equal but reduced concentrations of Na^+ and Cl^-. According to the law of mass action, **at equilibrium the products of the diffusable ion concentrations in each chamber will be equal.** Therefore, if we let X equal the concentration of Cl^- diffusing from chamber 2 into chamber 1 (X will of course also equal the concentration of Na^+ accompanying Cl^-), then at equilibrium

$$(C_1 + X)(X) = (C_2 - X)(C_2 - X)$$
$$C_1 X + X^2 = (C_2)^2 - 2C_2 X + X^2$$
$$C_1 X = (C_2)^2 - 2C_2 X$$

(15–7)

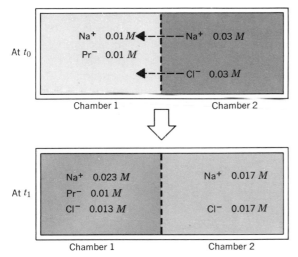

Figure 15–38

Gibbs-Donnan effect: Initial (t_0) and equilibrium (t_1) distributions of ions across a selectively permeable membrane separating two chambers of equal and fixed volume, when one ion (Pr⁻) is impermeable. At equilibrium the sum of anions equals the sum of cations on each side of membrane.

and

$$X = \frac{(C_2)^2}{C_1 + 2C_2} \qquad (15\text{–}8)$$

Equation 15–8 is called the Gibbs-Donnan equilibrium equation and may be used to determine the equilibrium distribution of ions between the protein-containing and protein-free portions of the two-chamber system.

Let us consider a specific case in which chamber 1 initially contains 0.01 M sodium proteinate (i.e., 0.01 M Na⁺ and 0.01 M Pr⁻) and chamber 2 contains 0.03 M NaCl (i.e., 0.03 M Na⁺ and 0.03 M Cl⁻). The concentration of Cl⁻ (and Na⁺) diffusing across the membrane and into chamber 1 from chamber 2 may be determined using equation 15–8 as follows:

$$X = \frac{(0.03)^2}{0.01 + 2(0.03)}$$

$$= 0.013$$

Therefore, the equilibrium distribution of ions would be (see also Fig. 15–38):

Chamber 1	*Chamber 2*
Na⁺ 0.023 M	Na⁺ 0.017 M
Cl⁻ 0.013 M	Cl⁻ 0.017 M
Pr⁻ 0.010 M	Pr⁻ 0.000 M

Note that (1) an electrical balance exists *within* each chamber, (2) the osmotic pressure is greater in chamber 1 than chamber 2 as a result of the higher total concentration of particles (no osmosis occurs because the chamber volumes are fixed), and (3) there are more sodium ions but fewer chloride ions in chamber 1 than in chamber 2.

By extrapolating the observations made for the artificial two-chamber system just described to cells (i.e., chamber 1) and their surrounding milieu (i.e., chamber 2), it may be seen that the Gibbs-Donnan effect would result in an unequal distribution of permeable ions across the cell membrane. That is, a *stable* ionic concentration gradient would be established. Just such a phenomenon accounts in part for the resting electrochemical membrane potentials possessed by many cells (see Chapter 24).

In the cases we have considered so far, the chamber volumes have been equal and fixed. This, of course, is not necessarily the case for a cell and its surroundings. The flexibility of the plasma membrane permits notable fluctuations in cell shape and volume. The greater osmotic pressure created inside the cell by the Gibbs-Donnan effect causes water to pass into the cell by osmosis, increasing the cell's volume and lowering its ion concentration. Since this disturbs the Gibbs-Donnan equilibrium previously established, a new equilibrium is achieved by the passage of additional permeable ions into the cell. Osmosis halts when the hydrostatic pressure created by water inside the cell balances the osmotic pressure generated by the dissolved cell solute.

Facilitated (Mediated) Diffusion Through the Cell Membrane

A variety of compounds including sugars and amino acids pass through the plasma membrane and into the cell at a much higher rate than would be expected on the basis of their size, charge, distribution coefficient, or magnitude of the concentration gradient. The increased rate of transport through the membrane is believed to be facilitated by specific membrane *carrier* substances and is called **facilitated** (or **mediated**) **diffusion.**

During facilitated diffusion, the rate at which the solute permeates the membrane increases with increasing solute concentration *up to a limit*. Above this limiting concentration, no increase in the rate of transport across the membrane is observed. In other words, facilitated diffusion exhibits **saturation kinetics** (Fig. 15–39) and is therefore similar to the relationship between reaction rate and substrate concentration in enzyme-catalyzed reactions (see Chapter 8). Other characteristics of facilitated diffusion are also similar to enzyme catalysis. Transport is *specific*; for

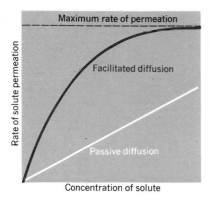

Figure 15–39
Kinetics of facilitated diffusion. Note the similarity to the kinetics of enzyme-catalyzed reactions.

example, in the erythrocyte, the inward diffusion of glucose, but not fructose or lactose, is facilitated. The rate of solute permeation can also be affected by the presence of structurally similar chemical compounds, much as in *competitive enzyme inhibition*. Facilitated diffusion exhibits *pH dependency*. Although facilitated diffusion results in a more rapid attainment of a concentration equilibrium across the membrane than passive diffusion, *the normal equilibrium concentrations are not altered*. Substances are *not* transported through the membrane *against a concentration gradient*. Although facilitated diffusion is not affected by chemicals that act as metabolic inhibitors, it is affected by *enzyme inhibitors* such as sulfhydryl blocking agents.

Facilitated diffusion is believed to result from the interaction of solute with specific membrane molecules, presumably proteins, thereby forming a **carrier–solute complex.**

The complex undergoes a positional change within the membrane in such a way that the solute now faces the other membrane surface and is released from the carrier (Fig. 15–40). An alternative suggestion is that the carrier may be a small molecule with the solute–carrier complex formed by an enzyme-catalyzed reaction within the membrane. Once formed, the solute–carrier complex diffuses to the other side of the membrane, where the solute is released in a second reaction.

A good and well-defined example of facilitated diffusion occurs in the bacterium *E. coli*. The sugar *lactose* does not readily permeate *E. coli* cells and cannot be hydrolyzed by cytoplasmic extracts of these cells. However, when *E. coli* is in a medium containing lactose, the enzyme *beta-galactosidase*, which hydrolyzes lactose, soon appears in the cell cytoplasm. The presence of the substrate in the growth medium is said to *induce* the formation of the enzyme by the cells. Such cells are able to take up galactose from the medium as well as metabolize it. Certain mutants of *E. coli* can be induced in this way to form beta-galactosidase even though lactose remains impermeable and cannot be incorporated by the cells. Finally, other *E. coli* mutants can be found that are able to incorporate lactose but cannot metabolize it once inside the cell.

It is now clear from these studies that in wild-type cells the presence of lactose in the growth medium induces *both* the formation of the hydrolytic enzyme *and* a carrier system—called a **permease** or **translocase.** The sets of genes controlling the formation of the permease and the hydrolytic enzyme are coordinately induced (see Chapter 11) in the presence of lactose. The loss or alteration of either set of genes (as in the *E. coli* mutants) results in a corresponding

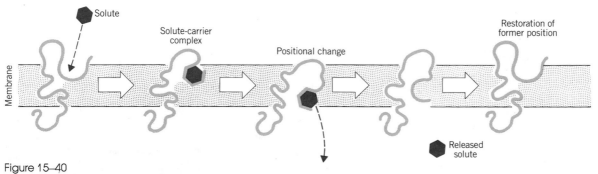

Figure 15–40
Stages of facilitated diffusion (see text for details).

inability to induce the formation of the enzyme or the permease.

Active Transport

During diffusion (passive or facilitated), substances pass through the plasma membrane until some sort of equilibrium is achieved. The equilibrium may be of the Gibbs-Donnan variety or may be a simple concentration equilibrium. Both involve an interplay between the concentrations of soluble solute inside and outside the cell. Cells can also accumulate solutes in quantities far in excess of that expected by any of the above mechanisms if the solute is rendered insoluble once it has entered the cell, since insoluble materials do not contribute to concentration gradients. Alternatively, once inside the cell, a solute may enter a metabolic pathway and be chemically altered, thereby reducing the concentration of that particular solute and allowing additional solute permeation. In all the cases we have so far considered, solute permeation of the membrane hinges on the presence of a concentration gradient, with the solute moving in the direction of the gradient.

Substances can also move through the plasma membrane into or out of the cell *against a concentration gradient*. This requires the expenditure of energy on the part of the cell and is called **active transport.** Active transport ceases when cells are (1) cooled to very low temperatures (such as 2–4°C), (2) treated with metabolic poisons such as cyanide or iodoacetic acid, or (3) deprived of a source of energy. The best understood and most exhaustively studied instances of active transport are those that involve the movements of sodium and potassium ions across the plasma membrane of erythrocytes, nerve cells, and *Nitella* cells and that result in an ionic concentration gradient across the cell membrane. The mechanism that establishes and maintains these gradients appears to be basically similar in all of these cells and can be illustrated with the erythrocyte.

The Na$^+$/K$^+$ Exchange Pump

The cytoplasm of the mammalian erythrocyte contains 0.150 M K$^+$, whereas the surrounding blood plasma contains only 0.005 M K$^+$. In contrast, the erythrocyte contains only 0.030 M Na$^+$ while the plasma contains 0.144 M Na$^+$. Hence, marked K$^+$ and Na$^+$ concentration gradients exist across the cell membrane. In Chapter 14 we

noted that tracer studies utilizing radioactive isotopes of Na and K clearly established that these ions are permeable to the erythrocyte membrane and are constantly diffusing through it. Yet, in spite of this permeability, Na$^+$ and K$^+$ concentration gradients across the membrane are maintained. The gradients are maintained because sodium ions diffusing into the cell from the plasma under the influence of the concentration gradient are transported outward again, and potassium ions diffusing out of the cell are replaced by the inward transport of K$^+$ from the plasma. That these movements involve active metabolic processes is clearly demonstrated when the temperature of a blood sample is reduced from 37°C (normal mammalian blood temperature) to 4°C, when cyanide is added to the blood, or when plasma glucose consumed during erythrocyte metabolism is not resupplied to the blood sample. Under these conditions, cell metabolism is interrupted and is followed by the inward diffusion of Na$^+$ and the outward diffusion of K$^+$ until the ionic concentrations on both sides of the erythrocyte membrane are in a passive equilibrium.

In the case of red blood cells and nerve cells, the active transport of Na$^+$ and K$^+$ appears to be linked. That is, the mechanism responsible for the outward transport of Na$^+$ simultaneously transports K$^+$ inward. An enzyme isolated from nerve cell membranes and believed to be involved in Na$^+$ and K$^+$ transport has been shown to have two sites that bind one or more of each of these cations. The enzyme is believed to be an integral protein spanning the lipid bilayer.

Active transport of Na$^+$ and K$^+$ through the membranes of nerve cells and erythrocytes requires ATP, and ATP cannot be replaced by other nucleoside triphosphates such as GTP, UTP, and ITP. ATP is converted to ADP during active transport by a membrane-bound Na$^+$- and K$^+$-stimulated ATPase. This enzyme and that involved in the transport of Na$^+$ and K$^+$ may be one and the same. The membranes of cells from many other mammalian tissues seem to possess a similar ATPase activity.

Two K$^+$ and three Na$^+$ are transported through the membrane for each molecule of ATP dephosphorylated. Transport of Na$^+$ and K$^+$ through the plasma membrane is believed to occur in the following stages (see Fig. 15–41). Three sodium ions and one molecule of ATP inside the cell are bound to specific sites on the enzyme carrier, while two potassium ions are bound to a site on the same enzyme facing the exterior of the cell. Binding of the

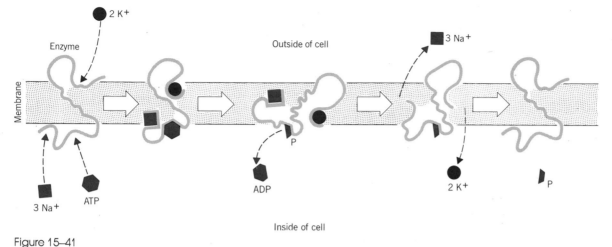

Figure 15–41
The Na⁺/K⁺ exchange pump (see text for details).

substrates results in and is followed by a change in the tertiary structure of the carrier molecule such that the bound sodium and potassium ions are ''translocated'' across the membrane. It is presumed that at some stage during this process, ATP is split, releasing ADP. Translocation is followed by an alteration of the binding sites such that the sodium ions are ''released'' outside the cell, while the potassium ions are released inside the cell. Once the ions are released, the carrier undergoes another change in structure, priming it for another round of the transport cycle. This stage, called ''recovery,'' is accompanied by the release of inorganic phosphate. Although this model is widely accepted, it has also been suggested that the enzyme site that binds Na⁺ on the inside of the cell binds K⁺ on the outside following translocation, while the site that initially binds K⁺ on the outside binds Na⁺ on the inside following translocation. In this manner, the recovery phase would result in an additional movement of ions through the membrane and would be more efficient.

Cotransport

Amino acids, sugars, and other metabolites are also actively transported through the plasma membrane into the cell. In many cells, the transport of these metabolites is coupled to the movements of sodium ions, as shown in Figure 15–42. The Na⁺/K⁺ exchange pump creates a steep concentration gradient across the plasma membrane favoring the inward diffusion of Na⁺. Indeed for every *two* K⁺ pumped into the cell, *three* Na⁺ are pumped out. Carrier proteins in the

membranes bind both Na⁺ and the metabolite, following which a change in the carrier's structure brings both substrates to the cell interior, where they are released. Release of the Na⁺ internally is followed by its active extrusion back through the membrane. The latter event is coupled to ATP hydrolysis and results in the maintenance of the steep Na⁺ gradient. In a sense, the steep Na⁺ gradient acts as the driving force for the inward transport of metabolites, and the simultaneous movements of Na⁺ together with metabolites into the cell constitute **cotransport.** Since the ATP-dependent Na⁺/K⁺ pump pumps 3 Na⁺ for every 2 K⁺, an electrical gradient is created across the membrane. For this reason, the Na⁺/K⁺ exchange pump is called an **electrogenic** pump. As will be seen in Chapter 16, the potential energy of an electrogenic pump is coupled to ATP synthesis in mitochondria.

"Simple" Active Transport

The passage of some substances through membranes against a concentration gradient is unidirectional but not coupled to ionic movement even though ATP is consumed in the process. Such movement is called **simple active transport.** The carrier enzyme cyclically binds the solute at one membrane surface and releases it at the other. The cycle is accompanied at some point by the hydrolysis of ATP.

Bulk Transport Into and Out of Cells

Before we consider several mechanisms responsible for bulk movement of materials into and out of cells, it is

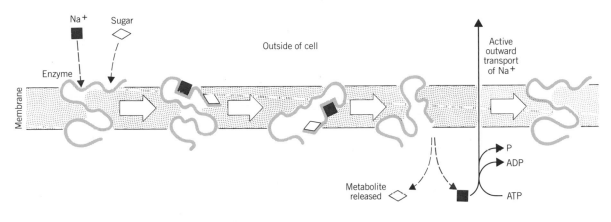

Figure 15–42

Cotransport. The coupled transport of Na+ plus metabolite (in this case, sugar).

important to recall what is meant by "inside the cell" and "outside the cell" (see also Chapter 1). For convenience, we have so far depicted the plasma membrane as a more-or-less continuous smooth sheet enclosing the cell. In reality, this is an oversimplification, for in most cells the plasma membrane exhibits numerous outfoldings and infoldings. Outfoldings of the plasma membrane cover microvilli, cilia, flagella, and other cytoplasmic extensions. Infoldings of the plasma membrane form small pockets and narrow channels that descend into the cytoplasm. Some of these infoldings may join the network of cisternae that form the endoplasmic reticulum and that furrows the cytoplasm. This implies that the intracisternal space of the endoplasmic reticulum may be in direct continuity with the surrounding cell environment and that the movement of materials between the lumenal phase and the cell surroundings does not require passage across any membranes. Consequently, any materials that are in the cisternae of the endoplasmic reticulum (i.e., in the lumenal phase) may be regarded as "outside" the cell. In order to get "inside" the cell, substances in the lumenal phase must pass through the membranous walls of the endoplasmic reticulum and into the cytosol (or hyaloplasm).

Many intracellular vesicles appear to be derived from the endoplasmic reticulum by being "pinched off" from the latter. For example, **transitional vesicles** that merge with the forming face of the cell's Golgi bodies (Chapter 18) and **peroxisomes** and other **microbodies** (Chapter 19) may be formed in this way. Technically then, the contents of

these vesicles are "outside" the cell and are separated from the cytosol by membranes. Some vesicles are derived from invaginations of the plasma membrane; the contents of these vesicles are also to be considered as "outside" the

Figure 15–43

Relationship of external and internal halves of the plasma membrane to the inner and outer faces of vesicles in exocytosis and endocytosis.

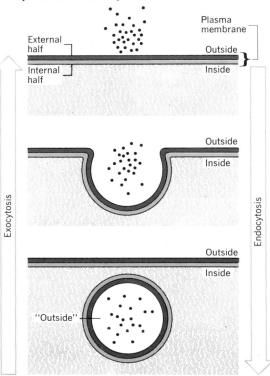

cell. These relationships are depicted in Figure 15–43, in which the external (or cisternal) and internal (or cytosol) halves of the membrane are distinguished. Particles within such a vesicle are in contact with the external face of the membrane.

From the preceding discussion, it should be clear that while substances may be directly exchanged between the cell surroundings and the cytosol through the plasma membrane, many exchanges also occur across membranes of cytoplasmic vesicles and the endoplasmic reticulum. These exchanges are mediated by the same mechanisms described above in connection with movements directly through the plasma membrane (diffusion, facilitated diffusion, active transport, etc.).

The formation of cytoplasmic vesicles from the plasma membrane and the consequent entrapment within these vesicles of substances formerly in the cell surroundings is called **endocytosis.** Several different kinds of endocytosis have been described, including **pinocytosis, rhopheocytosis,** and **phagocytosis** and these are discussed below. Movements of materials from the cell into the surroundings by the fusion of cytoplasmic vesicles with the plasma membrane constitute **exocytosis.** Endocytosis continuously removes small portions of the plasma membrane, while exocytosis continuously adds to the membrane. From a quantitative standpoint, the turnover is quite impressive, for in some cells (e.g., cultured fibroblasts and macrophages) a membrane area equal to the total surface area of the cell is turned over every few hours. Since tracer studies indicate that many membrane constituents have a half-life of several days, it has been suggested that some of the membranous material internalized during endocytosis is returned to the plasma membrane during exocytosis. Endocytosis and exocytosis have many characteristics in common and are the subject of the next section.

Endocytosis

Pinocytosis.   Using time-lapse photography to study tissue culture cells, W. H. Lewis in 1931 described what seemed to be a curious phenomenon in which small amounts of culture medium were trapped in invaginations of the plasma membrane and then pinched off to form small cytoplasmic vesicles. Since the entire process appeared much like some form of organized cell drinking, Lewis termed the phenomenon **pinocytosis** (*pinos* means ''I drink'' in Greek). Lewis' observations with tissue culture cells were confirmed in 1934 by S. O. Mast and W. L. Doyle studying amoebae in which pinocytosis is readily observed with the light microscope. Using electron microscopy, it became clear in the 1950s that pinocytosis is a common phenomenon occurring at different times in many kinds of cells and tissues including leucocytes, kidney cells, intestinal epithelium, liver macrophages, and plant root cells.

Pinocytosis is induced by the presence of appropriate concentrations of proteins, amino acids, or certain ions in the medium surrounding the cell. The first step in the process (Fig. 15–44) involves the binding of the inducer substance to specific receptor sites on the plasma membrane. This is followed by invagination of the membrane to form either **pinocytic vesicles** or narrow channels (Fig. 15–44). Although binding of the inducer is not inhibited by cyanide or low temperature, the formation of pinocytic vesicles is, and it is therefore dependent on cell metabolism. Actin filaments are associated with the margins of pinocytic vesicles and are believed to play some role in the vesicle's invagination. Pinocytic vesicles detach from the plasma membrane and migrate toward the interior of the cell where they may fragment into smaller vesicles or coalesce to form larger ones. Unless the vesicles are ''tagged'' by inducing pinocytosis in the presence of radioactive tracers, they soon become indistinguishable from other vacuoles in the cell.

Two levels of pinocytosis can be distinguished. In one of these, called **micropinocytosis,** the vesicles formed have a diameter of about $0.1 \mu m$ and are derived from small depressions in the cell surface **(caveolae).** In **macropinocytosis,** the vesicles are considerably larger, having diameters of 1 to 2 μm. The larger vesicles are formed either from larger invaginations of the plasma membrane or from surface ruffles (Figs. 15–29 and 15–30). Although pinocytosis is induced by the presence of specific substances in the cell surroundings, other materials are also enclosed by the pinocytic vesicles, including water, salts, and so on. These substances, together with the inducer molecules, may enter the cytosol from the vesicle by diffusion, active transport, or related transport mechanisms.

Coated Pits and Receptosomes.   In recent years, it has become apparent that another form of pinocytosis is carried out by many kinds of cells. It is called ''concentrative receptor-mediated endocytosis.'' In this form of pinocyto-

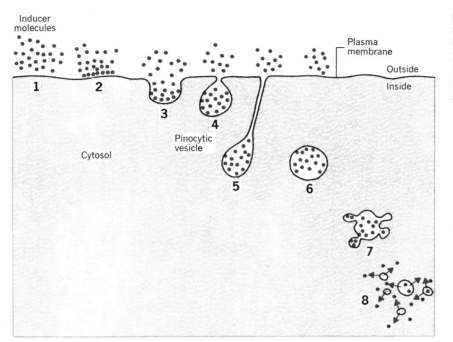

Figure 15–44
Stages of pinocytosis: 1–2, binding of inducer molecules to plasma membrane; 3–5, invagination of the membrane; 6–8, detachment from plasma membrane and fragmentation into smaller vesicles.

sis, binding of the metabolite (usually referred to in this instance as **ligand**) to the plasma membrane receptor is followed by lateral movement of the ligand–receptor complex through the membrane toward a **clathrin-coated pit** (Fig. 15–45). These pits are scattered through the plasma membrane, accounting for about 2 to 4% of its total surface area. Since lateral movement through the membrane by the ligand-receptor complex is quite rapid, a coated pit is encountered within a few seconds. Having entered one of these pits, movement of the ligand–receptor complex is terminated, and as a result, many such complexes can be concentrated in a small area of the plasma membrane. Since different kinds of receptors can be clustered in the same coated pit, the receptors probably share a property in common. The clathrin that "coats" the pit is a peripheral protein whose subunits (molecular weight about 185,000)

Figure 15–45
Involvement of *clathrin*-coated pits in endocytosis. Ligand molecules bind to membrane receptors, which move laterally in the membrane to clathrin-coated pits. Small vesicles called *receptosomes* detach from the lower margins of the pits and carry the internalized ligand molecules deeper into the cell.

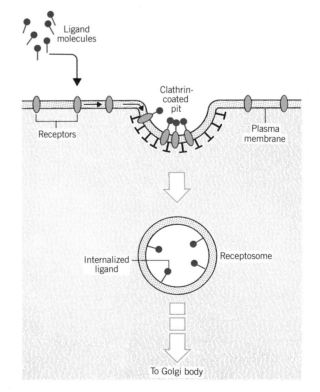

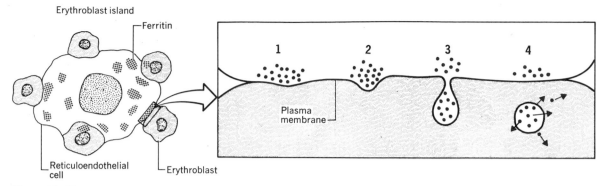

Figure 15–46

Rhopheocytosis **in the bone marrow. Stages 1–4 show successive evagination and invagination of adjacent plasma membranes, transferring ferritin and some cytoplasm from the reticuloendothelial cell to the erythroblast.**

form a basketlike net on the cytosol side of the pit. The clathrin is believed to interact with integral membrane proteins.

Once a clathrin-coated pit contains a large number of ligand–receptor complexes, it invaginates further into the cytosol. Eventually, a small vesicle called a **receptosome** is dispatched from the pit into the cytosol. Receptosomes are devoid of clathrin, leading to the suggestion that the vesicle is comprised of new membranous material. This preserves the number of clathrin-coated pits in the membrane. Once the ligand has been "internalized" in this fashion, it is released from the receptor. Receptors may be returned to the plasma membrane intact and may not accompany the receptosome during its entire journey. If the receptosome formed contains proteins, other macromolecules, or particles, it may fuse with a lysosome so that enzymatic digestion of the entrained material can ensue.

Rhopheocytosis. **Rhopheocytosis** is a special bulk transport mechanism in which quantities of cytoplasm, together with their inclusions, are transferred from one cell to another. Rhopheocytosis was first demonstrated in bone marrow tissue by M. Bessis. In the marrow, maturing red blood cells (erythroblasts) are attached to reticuloendothelial cells to form large numbers of *erythroblast islands* (Fig. 15–46). The reticuloendothelial cells of the marrow contain large quantities of iron derived during the breakdown of hemoglobin from old red blood cells. In the reticuloendothelial cells, the iron is converted to *ferritin*—a high-molecular-

weight protein containing up to 23% iron by weight. Because of their high density, clusters of ferritin molecules in reticuloendothelial cells can be visualized by transmission electron microscopy and their fate can therefore be studied. During maturation of red blood cells, ferritin granules, together with small amounts of cytoplasm, are transferred from reticuloendothelial cells to erythroblasts by the simultaneous evagination and invagination of their adjacent plasma membranes (Fig. 15–46). Once inside the erythroblast, iron derived from ferritin (along with iron that is obtained directly from the circulating plasma) passes into the cytosol and is used in the synthesis of new hemoglobin molecules.

Phagocytosis. Phagocytosis, which was first described by E. Metchnikoff in the late nineteenth century, is similar to pinocytosis but involves the engulfment of much larger quantities of particulate material. For example, entire ciliates, rotifers, or other microscopic organisms may be phagocytosed by an amoeba and enclosed within one or more vacuoles called **phagosomes, food vacuoles,** or **food cups** (see Fig. 15–47). During phagocytosis, the "prey" may be temporarily immobilized by secretions from the phagocytic cell. The phagocytosis of ciliates is characterized by the flowing of the amoeba's cytoplasm into footlike projections (**pseudopodia**) that gradually encircle and fully encapsulate the ciliate. Using a similar mechanism, certain white blood cells phagocytose hundreds of bacteria. The removal and destruction of old red blood cells in the liver,

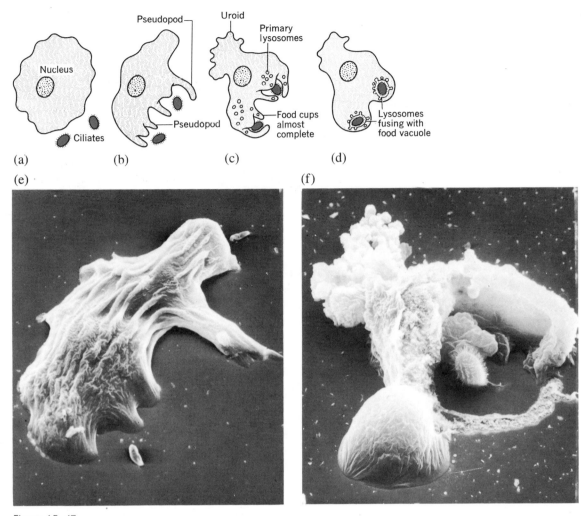

Figure 15–47

Phagocytosis. In the presence of phagocytosable material (a), the cell forms pseudopodia (b) that entrap the prey in vacuoles (c and d). The fusion of primary lysosomes with these vacuoles is followed by digestion of the prey. Scanning electron micrographs are an amoeba prior to (e) and during (f) phagocytosis and correspond to stages b and d of the diagrams. (Photomicrographs kindly provided by Dr. K. W. Jeon. Copyright © 1976, J. Photozoology 23, 83.)

spleen, and bone marrow by reticuloendothelial cells in these organs also occurs by phagocytosis. Following phagocytosis, the phagosomes fuse with *primary lysosomes* in the cell. The hydrolytic enzymes from these lysosomes digest the engulfed material, converting it to a form that may be transported across the vacuolar membranes and into the cytosol (see Chapter 19).

Exocytosis

Exocytosis is the mechanism by which large quantities of material enclosed within a cell vacuole are transferred to the cell surroundings by fusion of the vacuole with the plasma membrane. In a sense, the process is the reverse of pinocytosis or phagocytosis, the contents of the vacuole

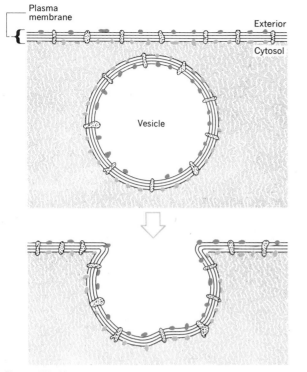

Figure 15–48
Fusion of an exocytic vesicle with the plasma membrane.

being emptied into the extracellular space. The best understood form of exocytosis is secretion. When the secretory vesicle touches the plasma membrane, lipids in both membranes merge to form a common bimolecular leaflet. In effect, the vesicle's membrane is incorporated into the plasma membrane, while at the same time the vesicle's contents are discharged to the exterior (Fig. 15–48). Exocytosis returns to the plasma membrane portions that are lost by endocytosis. In the many cells that are continuously carrying out both processes some sort of an equilibrium exists so that no major changes in the total surface area of the cell take place. Among the intracellular structures actively involved in the processing of cytoplasmic vesicles are Golgi bodies and lysosomes; these organelles will be considered in this regard in Chapters 18 and 19.

Summary

The cell is delimited at its surface by the **plasma membrane** and the cytoplasm is furrowed by the membranes of

the **endoplasmic reticulum** so that the cell volume is divided into two phases: the **cytosol** or **hyaloplasm** and the **lumenal** or **intracisternal phase.** The cell's membranes actively participate in the exchanges of material between the cytosol and the surroundings. Early studies on the chemistry of the plasma membrane focused on the content and organization of its lipids resulting in the **bimolecular lipid leaflet** model. This concept was subsequently modified to account for the membrane proteins, resulting first in the **unit membrane** model and more recently in the **fluid mosaic** model. The latter model appears to be more consistent with biochemical studies and electron microscopic analyses of plasma membranes.

Both the proteins and lipids of the membrane are asymmetrically distributed across the inner and outer halves. In many cells, the proteins in the inner half of the membrane are anchored in position by a **cytoskeletal network** consisting of filaments and microtubules. Carbohydrate chains associated with protein (and lipid) in the outer membrane surface play roles in maintaining membrane organization, in transmembrane transport, in cell-to-cell adhesion, and in providing membrane antigenic properties. In some cells, the carbohydrate forms an outer covering called the **glycocalyx.**

In tissues, the plasma membranes of neighboring cells exhibit specialized junctional regions including **tight junctions, desmosomes, gap junctions,** and **plasmodesmata.** Such junctions are important in maintaining tissue integrity and in regulating the passage of a substance across a tissue and from cell to cell.

The movements of materials across cell membranes may be achieved by passive or active mechanisms. The passive movements of solutes (by **diffusion**) and water (by **osmosis**) across the membrane occur principally as the result of concentration gradients. In some cases, diffusion through the membrane is **facilitated** by carrier molecules possessing the properties of enzymes. Substances can also move through the plasma membrane against a concentration gradient; this consumes cellular energy (i.e., ATP is hydrolyzed in the process) and is termed **active transport.** For individual ions and small molecules, active transport is effected by membrane-associated enzymes acting as pumps. However, transport into and out of cells can be in *bulk* and is then called **endocytosis** and **exocytosis.** Bulk transport involves gross movements of the plasma membrane, the three most common forms of bulk transport being **pinocytosis, rhopheocytosis,** and **phagocytosis.**

References and Suggested Reading

Articles and Reviews

Branton, D., Membrane structure. *Annu. Rev. Plant Physiol.*, Vol. 20 (L. Machlis, W. R. Briggs, and R. B. Park, Eds.), Annual Reviews, Palo Alto, Calif. (1969).

Capaldi, R. A. A dynamic model of cell membranes. *Sci. Am. 230*(3), 26 (March 1974).

Collander, R., The permeability of plant protoplasts to small molecules. *Physiol. Plantarum 2,* 300 (1949).

da Silva, P. P., and Kachar, B., On tight junctions. *Cell 28,* 441 (1982).

Davis, B. D., and Tai, P. C., The mechanism of protein secretion across membranes. *Nature 283,* 433 (1980).

Evans, W. H., Communication between cells. *Nature 283,* 521 (1980).

Evans, W. H., A biochemical dissection of the functional polarity of the plasma membrane of the hepatocyte. *Biochim. Biophys. Acta 604,* 27 (1980).

Fox, C. F., The structure of cell membranes. *Sci. Am. 226*(2), 30 (Feb. 1972).

Gorter, E., and Grendel, F., On bimolecular layers of lipoids in the chromocytes of the blood. *J. Exp. Med. 41,* 349, (1925).

Hendler, R. W., Biological membrane ultrastructure. *Physiol. Rev. 51,* 66 (1971).

Kaplan, D. M., and Criddle, R. S., Membrane structural proteins. *Physiol. Rev. 51,* 249 (1971).

Keynes, R. D., Ion channels in the nerve-cell membrane. *Sci. Am. 240*(3), 126 (Mar. 1979).

Knutton, S., The mechanism of virus-induced cell fusion. *Micron 9,* 133 (1978).

Korn, E. D., Structure of biological membranes. *Science 153,* 1491 (1966).

Lodish, H. F., and Rothman, J. E., The assembly of cell membranes. *Sci. Am. 240*(1), 48 (Jan. 1979).

Lodish, H. F., Braell, W. A., Schwartz, A. L., Strous, G. J. A. M., and Zilberstein A., Synthesis and assembly of membrane and organelle proteins. *Int. Rev. Cytology* (A. L. Muggleton-Harris, Ed.) Suppl. 12, Academic Press, New York (1981).

Luria, S. E., Colicins and the energetics of cell membranes. *Sci. Am. 233*(6), 30 (Dec. 1975).

Marchesi, V., Furthmayr, H., and Tomita, M., The red cell membrane. *Annu. Rev. Biochemistry,* Vol. 45 (E. E. Snell et al., Eds.), Annual Reviews, Palo Alto, Calif. (1976).

Marx, J. L., Newly made proteins zip through the cell. *Science 207,* 164 (1980).

Pastan, I. H., and Willingham, M. C., Journey to the center of the cell: role of the receptosome. *Science 214,* 504 (1981).

Paul, S. M., and Skolnick, P., The red cell as fluid droplet: tank treadlike motion of the human erythrocyte membranes in shear flow. *Science 202,* 894 (1978).

Raff, M. C., Cell surface immunology. *Sci. Am. 234*(5), 30 (May 1976).

Ravazzola, M., and Orci, L., Intercellular junctions in the rat parathyroid gland: A freeze-fracture study. *Rev. Biol. Cellulaire 28,* 137 (1977).

Robertson, J. D., The membrane of the living cell. *Sci. Am. 206*(4), 65 (Apr. 1962).

Robertson, J. D., Membrane structure. *J. Cell Biol. 91,* 189 (1981).

Rothman, J. E., and Lenard, J., Membrane asymmetry. *Science 195,* 743 (1977).

Sabatini, D. D., Kreibich, G., Morimoto, T., and Adesnik, M., Mechanism for the incorporation of proteins in membranes and organelles. *J. Cell Biol.* 92, 1 (1982).

Satir, B., The final steps in secretion. *Sci. Am. 233*(4), 28 (Oct. 1975).

Scott, D. E., Hall, M. N., and Silhavy, T. S., A mechanism of protein localization: the signal hypothesis and bacteria. *J. Cell Biol. 86,* 701 (1980).

Sharon, N., Lectins. *Sci. Am. 236*(6), 108 (June 1977).

Staehelin, L. A., and Hull, B. E., Junctions between living cells. *Sci. Am. 238*(5), 141 (May 1978).

Stolinski, C., Freeze-fracture replication in biological research: development, current practice and future prospects. *Micron 8,* 87 (1977).

Unwin, P. N. T., and Zampighi, G. Z., Structure of the junction between communicating cells. *Nature 283,* 545 (1980).

Weatherbee, J. A., Membranes and cell movement. *Int. Rev. Cytol.* (A. L. Muggleton-Harris, Ed.). Suppl. 12, Academic Press, New York (1981).

Books, Monographs, and Symposia

Bessis, M., Weed, R. I., and Leblond, P. F. (Eds.), *Red Cell Shape,* Springer-Verlag, New York, 1973.

Bonting, S. L., and de Pont, J. J. (Eds.), *Membrane Transport,* Elsevier/North-Holland, Amsterdam, 1981.

Branton, D., and Park, R. B. (Eds.), *Papers on Biological Membrane Structure,* Little, Brown, Boston, 1968.

Evans, W. H., *Preparation and Characterisation of Mammalian Plasma Membranes,* North-Holland, Amsterdam, 1979.

Finean, J. B., and Michell, R. H. (Eds.), *Membrane Structure,* Elsevier/North-Holland, Amsterdam, 1981.

Gomperts, B. D., *The Plasma Membrane,* Academic Press, London, 1977.

Harrison, R., and Lunt, G. G., *Biological Membranes* (2nd ed.), Halsted Press, New York, 1980.

Hendler, R. W., *Protein Biosynthesis and Membrane Biochemistry,* Wiley, New York, 1968.

Kates. M., and Kuksis, A., *Membrane Fluidity,* Human Press, Clifton, N. J., 1980.

Lewis, J. A., Synthesis of proteins by ribosomes attached to membranes, in *Biochemistry of Cellular Regulation, Vol. 1, Gene Expression* (M. J. Clemens, Ed.), CRC Press, Boca Raton, Florida, 1980.

Schwartz, L. M., and Azar, M. M. (Eds.), *Advanced Cell Biology,* Van Nostrand Reinhold, New York, 1981.

Schweiger, H. G. (Ed.) *International Cell Biology 1980–1981,* Springer-Verlag, Berlin, 1981.

Stein, W. D., *The Movement of Molecules Across Cell Membranes,* Academic Press, New York, 1967.

Weiss, L., *The Cell Periphery, Metastasis and Other Contact Phenomena,* North-Holland, Amsterdam, 1967.

Chapter 16

THE MITOCHONDRION

Most of the energy-requiring, or *endergonic,* reactions carried out by cells either directly or indirectly consume **adenosine triphosphate** (ATP). This "energy-rich" substance is converted to adenosine diphosphate (ADP) and occasionally to adenosine monophosphate (AMP) (Fig. 16–1). Cells have evolved three major ways of producing this vital source of chemical energy: (1) ATP is produced in the cytosol during the chain of exergonic reactions called **glycolysis** in which sugars are catabolized; (2) ATP may be produced within the **chloroplasts** of certain plant cells, utilizing the energy of sunlight; and (3) ATP may be produced within the **mitochondria** present in virtually all plant and animal cells by the **oxidation** of a variety of elementary substrates. It is for this reason that mitochondria are often referred to as the cell's "powerhouses."

ATP production in chloroplasts is considered in detail in Chapter 17, which also deals with photosynthesis, and further discussion of that subject is deferred until then. ATP synthesis during glycolysis occurs by **substrate-level phosphorylations** in which the phosphate is enzymatically transferred directly from the substrate to ADP to form ATP (see Chapters 9 and 10). Some ATP is also generated in mitochondria by substrate-level phosphorylations, but the greater amount is formed by special **electron transport system** (ETS) oxidation-reduction reactions. Although much of this chapter is devoted to a description of the manner in which mitochondria function in ATP production,

it is to be noted that mitochondria also play several other important roles; for example, in eucaryotic cells the mitochondria are responsible for the oxidation of fatty acids and other lipids and are one of the sites for lengthening the carbon chain of fatty acids. Some subunits of cytochromes are also synthesized in mitochondria, and final assembly occurs there.

In recent years, it has been clearly established that mitochondria contain their own genetic apparatus, as well as the ribosomes and enzymes for the synthesis of an array of enzymatic and structural mitochondrial proteins. Mitochondria are capable of semiautonomous proliferation within the cell. These subjects are dealt with here and also in Chapters 20 and 21.

Discovery of Mitochondria

Mitochondria were first observed and isolated from cells about 130 years ago when Kölliker mechanically teased these organelles from insect striated muscle tissue and studied their osmotic behavior in various salt solutions. Kölliker, whose work extended over several decades beginning around 1850, concluded that these "granules" were independent structures not directly connected to the interior structure of the cell. In 1890, Altmann identified stains specific for these granules which he named "bioblasts"; this term was superseded when Benda introduced

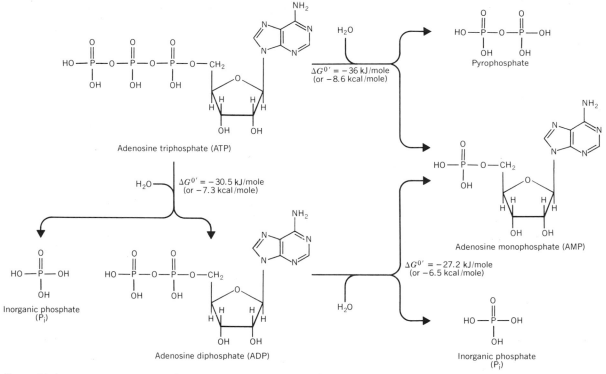

Figure 16–1
Hydrolysis of ATP to form ADP and AMP.

the expression "mitochondrion" (Greek; *mito-* = "thread" + *chondrion* = "granule") because of the threadlike appearance of these granules when examined with the light microscope. In 1900, Michaelis introduced the use of the supravital dye Janus green B to specifically stain mitochondria to the exclusion of other cellular components and showed that oxidative reactions in the mitochondria caused color changes in the dye. Janus green B is still frequently employed as a histochemical marker of mitochondria.

In 1910 Warburg showed that the "large granule" fraction isolated by low-speed centrifugation of tissues disrupted by grinding contained enzymes catalyzing oxidative cellular reactions, and Kingsbury in 1912 suggested that the mitochondria were the specific loci of the oxidation. Warburg's findings were confirmed in the 1930s by Claude, Bensley, and Hoerr, who employed more sophisticated methods to isolate a mitochondrial preparation essentially free of contaminating cell structures. They disrupted liver tissue using procedures almost identical to those currently used to prepare a conventional "homogenate" (see Chapter 12) and isolated the mitochondria by repeated differential centrifugation; the final mitochondrial preparation was then examined biochemically. At about the same time, Sir Hans Krebs elucidated the various reactions of the tricarboxylic acid (TCA) cycle (or Krebs cycle; see below), and by 1950 Lehninger, Green, Kennedy, Hogeboom, and others had clearly shown that these reactions, as well as those of fatty acid oxidation and oxidative phosphorylation (i.e., ATP generation), were properties of mitochondria.

The Origin and Evolution of Mitochondria

About 30 years ago, it was generally believed that mitochondria arose de novo from other cellular structures such as microbodies, the outer nuclear membrane, or the endoplasmic reticulum. As recently as 12 years ago, the prospect of de novo formation of mitochondria was still considered. However, the wealth of modern biochemical, physiological, and microscopic evidence indicates that new mitochondria

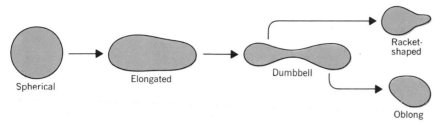

Figure 16–2
Division of mitochondria by a fissionlike process produces organelles of diverse shapes.

Spherical Elongated Dumbbell Racket-shaped Oblong

arise by the growth and division of mitochondria already present in the cell. Using time-lapse microcinematography mitochondria are observed to divide by a fissionlike process. The mitochondrion appears to stretch out in a tubelike manner, constrict near the center assuming a dumbbell-shape, and then pinch off to form two halves that once again round up (Fig. 16–2). Direct observations of mitochondrial division are also supported by biochemical studies using radioactive tracers and autoradiography (see Chapter 14). D. J. L. Luck labeled the mitochondria of *Neurospora* cells and followed the change in mitochondrial radioactivity through several generations of cells. With each cell division the amount of tracer associated with the mitochondria of the daughter cells was reduced to about one-half its previous level.

There does not appear to be a specific method for apportioning mitochondria between the two daughter cells during cell division. In most cells the cytoplasm becomes very turbulent during cell division, and therefore a random distribution of the organelles between the daughter cells is enhanced. In some instances, the mitochondria appear to adhere to the spindle fibers and may be drawn into the two daughter cells. During cell division in certain yeasts, the daughter cell forms as a small *bud* emerging from the parent cell. Rather than an equal division of the cytoplasm between the daughter cells, little parent cytoplasm enters the bud along with the divided nucleus. Occasionally these buds fail to obtain any mitochondria and the growth of these cells is much slower than that of normal cells, since the cells rely for the most part on glycolysis in the cytosol for ATP production. Fletcher and Sanadi in 1961 and Wilson and Dove in 1965 studied the turnover rate for mitochondria by using radioactively labeled compounds. The former workers labeled mitochondrial proteins and lipids and followed the disappearance of the label from the cells with time. The latter workers utilized a technique employing tritiated water, (i.e., ^3H_2O). Both studies indicated that

there is a complete intracellular turnover of mitochondria every 20 days.

L. Margulis has made some interesting speculations concerning the evolutionary origin of mitochondria in eucaryotic cells. Mitochondria contain DNA and ribosomes of their own and these have a number of functions that are independent of the nuclear and cytoplasmic ribosomal processes of the cell. (Protein synthesis and ribosome formation in mitochondria are discussed in Chapter 22.) The discovery of the presence of DNA and ribosomes in mitochondria served as the initial inspiration for the idea that mitochondria may at one time have been independent organisms; that is, mitochondria may have evolved from procaryote-like cells. This notion is supported by the fact that mitochondrial DNA, like procaryotic DNA, is a single, circular molecule (i.e., no free ends) and mitochondrial ribosomes are made up of subunits whose sedimentation coefficients and protein and RNA composition are more like those of procaryotic ribosomes than eucaryotic ribosomes. It has been proposed that one type of aerobic procaryote invaded or infected another and that through time, the infective procaryote evolved morphologically and physiologically into the mitochondrion. This fascinating theory of the evolution of mitochondria must be evaluated in the light of evidence that indicates that most mitochondrial proteins are encoded by nuclear DNA and are synthesized by ribosomes in the cytosol or on the endoplasmic reticulum. It would therefore appear that there is a greater transfer of information from nuclear DNA to the mitochondrion than arises from DNA found within the mitochondrion itself.

The genetic code of DNA isolated from the mitochondria of yeast, *Drosophila,* and human tissues has been shown to be slightly different from that of nuclear DNA. The noncoding sequences also differ. For example, DNA from human mitochondria contains 16,569 nucleotide pairs that encode 2 rRNA molecules, 22 or 23 tRNAs and 10 to 13 proteins. Very few noncoding nucleotide sequences are present. Eucaryotic nuclear DNA usually contains large noncoding nucleotide sequences and encodes a greater number of tRNAs (at least 32) and rRNAs (4). Frederick Sanger, who received a Nobel Prize in 1980 for his analyses of DNA nucleotide sequences, contends that mammalian mitochondrial genetic systems cannot be classified either as procaryotic or eucaryotic.

Structure of the Mitochondrion

The size, shape, and structural organization of mitochondria, as well as the number of these organelles per cell and their intracellular location, vary considerably depending on the organism, tissue, and physiological state of the cell examined.

Some cells, usually unicellular organisms, contain a single mitochondrion. Figure 16–3 contains a photomicrograph of the single mitochondrion in the motile swarm spore of *Blastocladiella emersonia,* a fungus, and a model of the single mitochondrion of *Chlorella fusca,* an alga. At the other extreme are cells such as *Chaos chaos,* an amoeba, which contains several hundred thousand mitochondria. Cells of higher animals also contain various numbers of mitochondria. Sperm cells have fewer than 100 mitochondria. Kidney cells generally contain less than 1000, while liver cells can contain several thousand. As described in Chapter 1, procaryotic cells such as bacteria and blue-green algae do not contain mitochondria. The functions associated with mitochondria are carried out in the cytosol or are associated with the plasma membrane.

The distribution of the mitochondria in the cell can change with time. In *Blastocladiella* (Fig. 16–3), the single mitochondrion is at the base of the flagellum. Mitochondria also appear to be concentrated in metabolically active areas of cells. In epithelial cells lining the lumen of the small intestine, the mitochondria occur in greater numbers near the cell surface that is adjacent to the lumen (where active absorption of digestive products is occurring). In general, when mitochondria are present in greater numbers in one part of the cell than another, it is usually near a site where significant ATP utilization is occurring. For example, in muscle tissue the mitochondria are aligned in rows parallel to the contractile fibrils (Fig. 16–4). In many plant cells, cyclosis, the active streaming of the cytoplasm about the cells, tends to distribute the mitochondria uniformly.

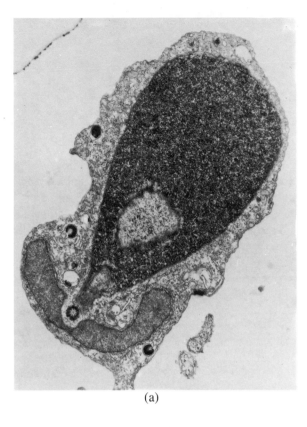

(a)

The number and distribution of mitochondria in a cell are closely related to the activity of the cell and its organelles. Cells that are actively growing, producing especially large amounts of some product such as digestive enzymes, actively transporting materials into the cell, or undergoing movement may display increased numbers of mitochondria during periods of activity and reduced numbers during periods of quiescence. In yeast cells grown anaerobically, successive generations contain fewer and fewer mitochondria. However, cells that have been cultured in the absence of oxygen rapidly produce greater numbers of mitochondria if oxygen and appropriate nutrients are added to the culture. An increased rate of cell growth and division also occurs since greater numbers of mitochondria produce more ATP in order to facilitate absorption of nutrients and the synthesis of cell components.

The size and shape of mitochondria, like the number in a cell, vary from one tissue to another and with the physiological state of the cells. Most mitochondria are ovoid bodies having a diameter between 0.5 and 1.0 μm and a

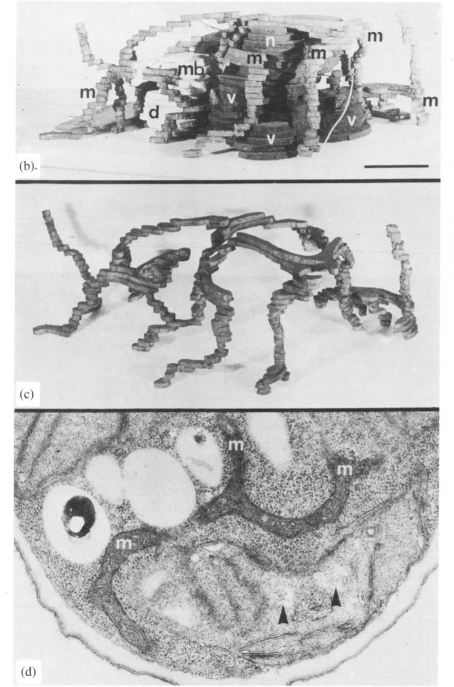

(b).

(c)

(d)

Figure 16–3

(a) The single mitochondrion present in the fungus *Blastocladiella emersonia;* the mitochondrion is located at the base of the cell's flagellum. (Courtesy of Dr. W. E. Barstow.) *(b), (c),* and *(d)* The mitochondrion of the alga *Chlorella fusca.* In *(b)* the mitochondrion *(m),* cell nucleus *(n),* vacuoles *(v),* Golgi body or dictyosome *(d),* and microbody *(mb)* are depicted as three-dimensional models produced by reconstruction of serial sections through the cell. In *(c)* the mitochondrion is seen alone. The torturous character of the mitochondrion is also apparent in the photomicrograph *(d)* of the cell; arrows point to DNA-containing areas of the cell's chloroplast. (Courtesy of Dr. B. E. S. Gunning; from Atkinson, A., John, P. C. L., and Gunning, B. E. S., *Protoplasma 81,* p. 77, 1974, Springer-Verlag, Berlin.)

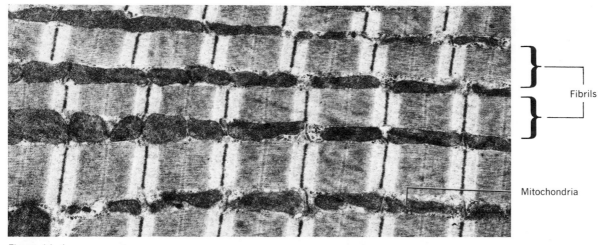

Fibrils

Mitochondria

Figure 16–4

The distribution of mitochondria within cells often reflects localized cellular activity. In this electron photomicrograph of muscle tissue, the mitochondria are arranged in rows between clusters of contractile fibrils. (Photomicrograph courtesy of R. Chao.)

length up to 7 μm. Usually, the lower the number of mitochondria per cell, the larger are the individual organelles. In many electron photomicrographs, mitochondria appear to be dumbbell-shaped or racket-shaped. These odd shapes may be a reflection of the fission process by which mitochondria are believed to proliferate. Dumbbell shapes are seen just prior to separation and the ''handles'' (or ''tails'') of racket-shaped mitochondria may be the bridges that connect the separating mitochondrial halves (Fig. 16–5).

From the point of view of fine structure, mitochondria are especially interesting and very intricate organelles. Because mitochondria are so small, light microscopy reveals little about their structure. The contemporary model of mitochondrial architecture is therefore based on decades of study using the transmission electron microscope. Figure 16–6 illustrates the typical organization revealed in thin sections of these organelles. Recently, some rather spectacular and highly informative photomicrographs of mitochondrial structure have been obtained by scanning electron microscopy of organelles that have been cracked open (Fig. 16–7; also see the book cover). In the past, SEM has not offered the degree of resolution needed to reveal organelle substructure. However, using a special fracturing technique and a novel high-resolution SEM, K. Tanaka, Y. Masun-

aga, and T. Naguro have obtained detailed photomicrographs of mitochondria and a number of other organelles. After tissue fixation in OsO_4, the samples are frozen, cracked open, and then treated with dimethylsulfoxide (DMSO). After a second treatment with osmium, followed by dehydration with ethanol and critical point drying, the fine structure is revealed in bold relief. The three-dimensional effect is striking and amazingly consistent with three dimensional models formulated on the basis of the earlier TEM studies (see Fig. 1–23). Because mitochondria from diverse sources exhibit certain features in common, a ''generalized'' organelle may be described (Fig. 16–8). The mitochondrion is enclosed by two distinct membranes called the **outer** and **inner membranes** (Figs. 16–6, 16–7). The inner membrane separates the organelle's volume into two phases: the *matrix,* which is a gel-like fluid enclosed by the inner membrane, and the fluid-filled **intermembrane space** between the inner and outer membranes. The matrix and intermembrane spaces as well as outer and inner membranes themselves contain a variety of enzymes (Table 16–1). The matrix contains a number of the enzymes of the Krebs cycle (tricarboxylic acid cycle, or TCA cycle) as well as salts and water. Suspended in the matrix are strands of circular DNA (Fig. 16–9) and ribosomes. A number of other inclusions have been observed in the mi-

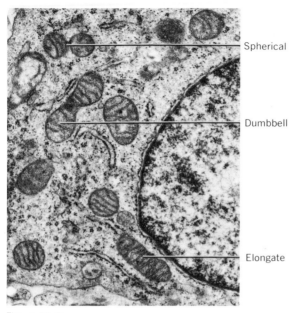

Figure 16–5
**Various shapes of mitochondria seen in thin sections.
(Courtesy of R. Chao.)**

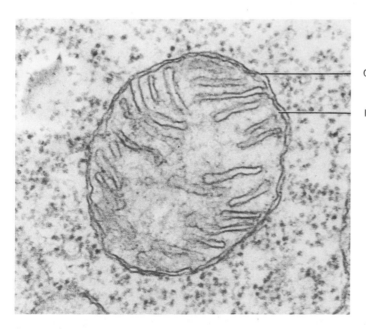

Figure 16–6
**Outer and inner membranes of the mito-
chondrion.**

Outer membrane

Inner membrane

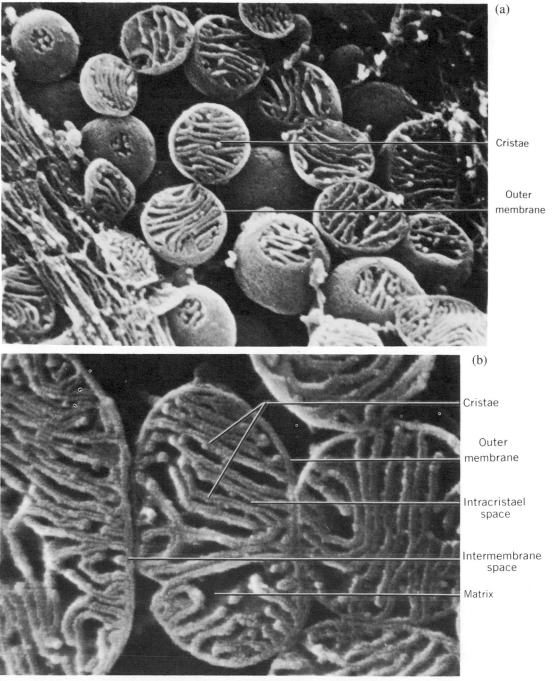

(a)

Cristae

Outer membrane

(b)

Cristae

Outer membrane

Intracristael space

Intermembrane space

Matrix

Figure 16–7

Scanning electron photomicrographs of mitochondria of rabbit cardiac tissue. *(a)* The outer membranes of the mitochondria have been cracked away by a freeze-fracture technique revealing the cristael membranes within the organelles. *(b)* In this higher magnification view, the inter- membrane and intracristael space may be discerned; the small bumps along the matrical surface of the cristae are believed to be the *inner membrane spheres* (see text). (Photomicrographs courtesy of Dr. Y. Masunaga and Dr. K Tanaka, *J. Yonago Med. Assn. 30,* 519, 1979).

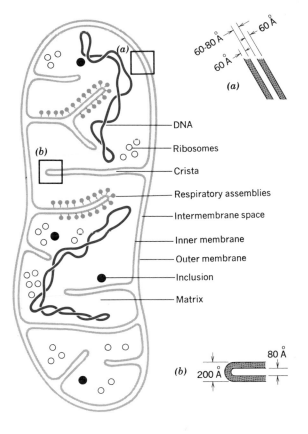

Figure 16–8
The generalized mitochondrion.

DNA

Ribosomes

Crista

Respiratory assemblies

Intermembrane space

Inner membrane

Outer membrane

Inclusion

Matrix

Figure 16–9
DNA fibrils in mitochondria. M, mitochondrion; N, cell nucleus. (Photomicrograph courtesy of Dr. E. G. Pollock.)

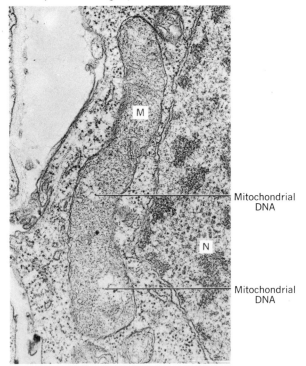

Mitochondrial DNA

Mitochondrial DNA

Table 16–1
Location of Some Mitochondrial Enzymes

Outer membrane:
  Monoamine oxidase
  Fatty acid thiokinases
  Kynurenine hydroxylase
  Rotenone-insensitive cytochrome c reductase
Space between the membranes:
  Adenylate kinase
  Nucleoside diphosphokinase
Inner membrane:
  Respiratory chain enzymes
  ATP-synthesizing enzymes
  α-Keto acid dehydrogenases
  Succinate dehydrogenase
  D-β-Hydroxybutyrate dehydrogenase
  Carnitine fatty acyl transferase
Matrix:
  Pyruvate dehydrogenase complex
  Citrate synthase
  Isocitrate dehydrogenase
  Fumarase
  Malate dehydrogenase
  Aconitase
  Glutamate dehydrogenase
  Fatty acid oxidation enzymes

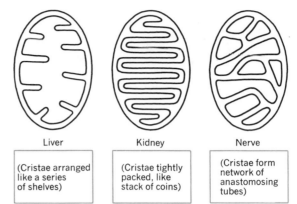

Liver
(Cristae arranged like a series of shelves)

Kidney
(Cristae tightly packed, like stack of coins)

Nerve
(Cristae form network of anastomosing tubes)

Figure 16–10
Various arrangements of cristae in mitochondria.

tochondrial matrix of diverse kinds of cells. These include filaments and tubules, what appear to be crystalline protein inclusions, and a number of small granules. The intermembrane space contains some enzymes (Table 16–1), but generally is devoid of particulate inclusions.

The Mitochondrial Membranes

The inner and outer membranes are distinctly different in structure and function. Although accurate measurement of the thicknesses of the membranes by electron microscopy is difficult because various fixatives promote different degrees of swelling, the inner membrane appears to be somewhat thicker (6.0–8.0 nm) than the outer membrane (about 6.0 nm). The inner membrane has a greater surface area because it possesses folds that extend into the matrix. These projections, called **cristae**, vary in number and shape. With distinct exceptions, the cristae of mitochondria in higher animal cells may almost bridge the matrix. Usually the cristae lie parallel to one another across the long axis of

the mitochondrion, but in some cells they run longitudinally or form a branching network (Fig. 16–10). In protozoa and many plants, the cristae form a set of tubes that project into the matrix from all sides, sometimes twisting in different directions. The number of cristae may increase or decrease depending upon the level of aerobic activity. Active aerobic tissue cells producing large amounts of ATP generally contain mitochondria with extensive cristae.

The organization of protein and lipid in the outer and inner mitochondrial membranes has been the subject of intense study for many years. Chemically, the two membranes are qualitatively and quantitatively distinct, differing from one another and also from other intracellular membranes. The inner membrane is much richer in protein than the outer membrane and the proteins themselves are more deeply embedded in the membrane. The outer membrane contains two to three times more phospholipid than the inner membrane and contains most of the membrane cholesterol. In contrast, the inner membrane is rich in cardiolipin. Differences in the two membranes are also apparent in freeze-fracture views of the E and P faces (Fig. 16–11). The P fracture face (PF) of the outer membrane contains more than three times as many particles as the E fracture face (EF). In contrast, the EF of the inner membrane contains almost as many particles as the PF of the outer membrane, while the inner membrane's PF contains twice as many particles as its EF. This is not surprising in view of the manifold enzymatic activities of the matrical surface of the inner membrane (see below). The organization of protein and lipid in the inner membrane is depicted in

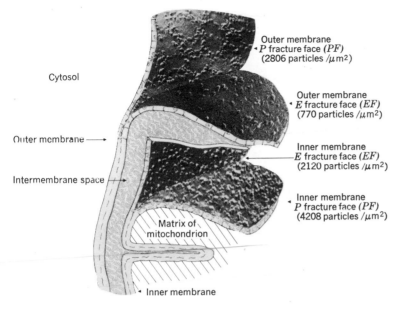

Cytosol

Outer membrane —

Intermembrane space

Matrix of mitochondrion

Inner membrane —

Outer membrane
◄ *P* fracture face *(PF)*
(2806 particles /μm^2)

Outer membrane
◄ *E* fracture face *(EF)*
(770 particles /μm^2)

Inner membrane
E fracture face *(EF)*
(2120 particles /μm^2)

◄ Inner membrane
P fracture face *(PF)*
(4208 particles /μm^2)

Figure 16–11
Fracture faces of and particle distributions in the outer and inner membranes of the mitochondrion. (Courtesy of Dr. L. Packer, copyright © New York Academy of Science, *Ann. N. Y. Acad. Sci. 227,* **167, 1974.)**

Figure 16–12
Sjöstrand and Barajas model of the structure of mitochondrial membranes. In this model, the enzymes and other proteins of the membrane are depicted as globular bodies of various shapes. Lipid molecules in areas consisting of a lipid bilayer are shown as "T"-shaped structures. (Courtesy of Dr. F. Sjöstrand, copyright © Academic Press, *J. Ultrastr. Res. 32,* **298, 1970.)**

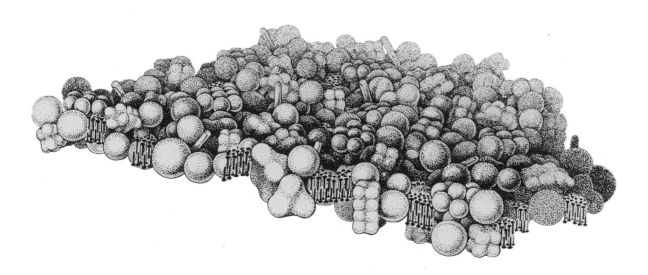

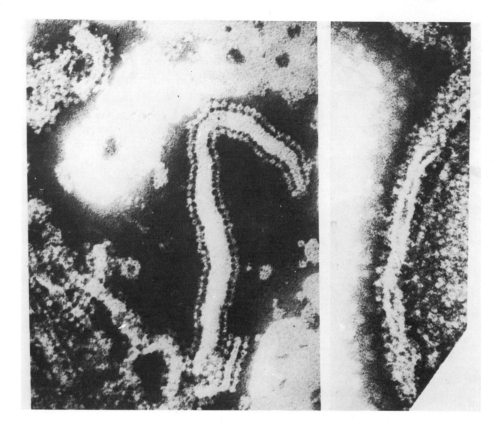

Figure 16–13
Electron photomicrographs of negatively stained fragments of mitochondria showing the inner membrane and the *inner membrane spheres*. The spheres are attached by stalks to the matrical surface of the membrane. (Courtesy of Dr. B. Chance.)

Figure 16–12. Many of the proteins of the mitochondrial membranes are enzymes that catalyze the reaction sequences discussed later in the chapter. Attached to the matrix side of the inner membrane are many spherical particles 8.0 to 9.0 nm in diameter. These **inner membrane spheres,** first described by Fernandez-Moran in 1962, are seen in the scanning electron photomicrographs of Figure 16–7 and the negatively stained preparation shown in Figure 16–13. They appear to be spaced regularly along the membrane and are borne on stalks. These inner membrane spheres have been identified as the primary sites of oxidative phosphorylation and electron transport (discussed later).

Each crista is formed by an infolding of the inner mitochondrial membrane. Recent fine structural studies by F. Sjöstrand using a freeze-fracture technique suggest that little or no space exists between these two layers (Fig. 16–14). Instead they appear to form a single structure containing large globular proteins about 15 nm in diameter and

small quantities of lipid at the matrical surfaces. Sjöstrand also contends that the inwardly projecting cristael membranes are anchored to the inner membrane by one or more *stalks* or **peduncles** rather than arising as shelflike folds (Fig. 16–15). The scanning electron photomicrographs of Figure 16–7 reveal little space between the two cristael membranes and also suggest pointlike attachments of the cristae to the inner membrane.

In addition to their structural differences, the outer and inner membranes differ significantly in permeability. The outer membrane is permeable to a wide variety of substances having a molecular weight up to about 5,000 daltons. When the fluid from the intermembrane space is isolated, it reflects the water-soluble small-molecular-weight components of the cytosol. In contrast, the inner membrane has a limited permeability, especially to substances with molecular weights above 100 to 150 daltons.

The difference in permeability of the two membranes can be used to advantage to separate the outer membrane from

(a)

Figure 16–14

(*a*) Cristael membranes of a mitochondrion revealed by freeze-fracturing and etching technique; the small particles in the membranes are believed to be proteins (e.g., the enzymes associated with the membrane); (*b*) composite diagram and photomicrograph showing the angle of fracture through the cristael membranes. According to F. Sjöstrand, the intracristael space is reduced and bridged at various points by large globular proteins, which are probably enzyme complexes. (Photomicrographs courtesy of Dr. F. Sjöstrand, *J. Ultrastr. Res. 69,* 378, 1979.)

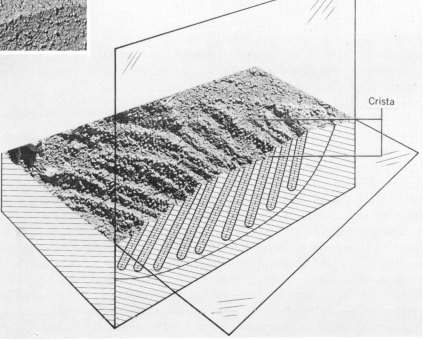

Crista

(b)

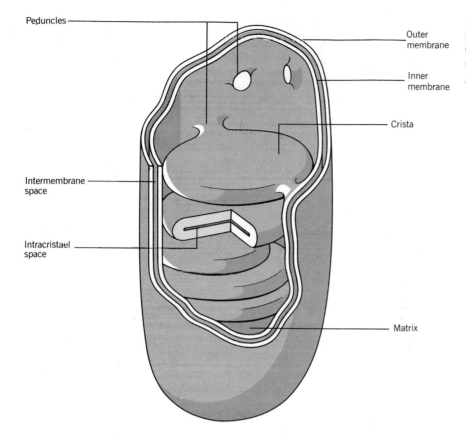

Peduncles

Outer membrane

Inner membrane

Crista

Intermembrane space

Intracristael space

Matrix

Figure 16–15
Model of the mitochondrion showing that the cristae are attached to the inner membrane through one or more stalks or *peduncles*.

the inner. Although several variations of the method exist, the basic procedure is to disrupt the cells and isolate intact mitochondria by differential or density gradient centrifugation. The mitochondria are then placed in a hypotonic solution, causing them to swell. Usually, a phosphate buffer solution is used for this purpose. As the mitochondria swell, their outer membranes rupture and fragment; the inner membranes also swell, causing a loss of organized cristael structure, but the membranes do not break. The mitochondria are then transferred to a hypertonic solution, causing the matrix to shrink, pulling the inner membrane away from the outer membrane. The hypertonic solution is frequently a sucrose solution containing ATP and Mg^{++}. Since the inner membrane may be attached to the outer membrane at several points (probably by proteinaceous connecting strands), some workers use digitonin, EDTA, or sonication to aid the separation. Resuspension in an isotonic solution allows the matrix to reassume its normal size and typical morphology. Isolation of the separated membrane

fractions can be achieved by centrifugation. The results of a typical isolation are shown in Figure 16–16.

Conformational States of Mitochondria

Changes in the level or type of physiological activity occurring in mitochondria can be related to differences in the morphological appearance of the organelles. Accordingly, C. R. Hackenbrock has shown that the **orthodox conformational state** (Fig. 16–17) of mitochondria (the appearance most commonly seen in photomicrographs) is typical of inactive organelles. The **condensed conformational state** (Fig. 16–17) corresponds to periods in which the phosphorylation of ADP to form ATP and electron transport are at high metabolic levels. In the condensed conformational state, the cristae are more randomly distributed and the intermembrane space is greatly enlarged. The transition from the orthodox to the condensed conformational state is

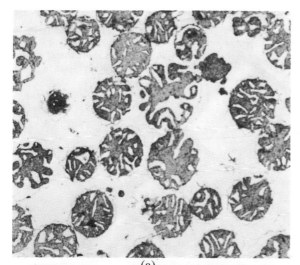
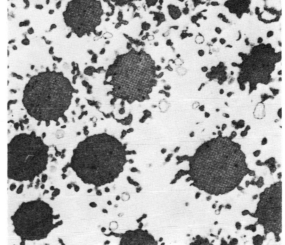

(a)

(b)

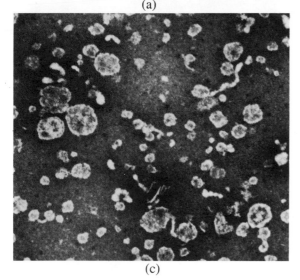

(c)

Figure 16–16

Mitochondrial preparations isolated from rat liver tissue. *(a)* Freshly isolated whole mitochondria. *(b)* Inner membrane and matrix isolated after digitonin treatment and differential centrifugation. *(c)* Outer membrane preparation isolated after digitonin treatment and differential centrifugation. (Courtesy of Dr. C. A. Schnaitman, copyright © The Rockefeller Press, *J. Cell. Biol. 38,* 170, 1968.)

triggered by the binding of ADP to *ADP-ATP translocase* molecules in the inner membrane.

Over the years, numerous investigators have studied the relationship between the structural organization of the mitochondria and its specialized metabolic functions. As seen in Table 16–1, the location of many enzymes has been determined, and it is generally possible to assign specific functions to the outer membrane, intermembrane space, the inner membrane, and the matrix. In some cases, the enzyme activity can be assigned to a specific surface of the membrane.

Among the most thoroughly studied processes unique to mitochondria are **substrate oxidation, respiratory chain oxidation-reductions,** and **oxidative phosphorylation** (Figure 16–18). Products of metabolic reactions in the cytosol (such as pyruvate formation during glycolysis) enter the mitochondrion to be oxidized by the Krebs or tricarboxylic acid (TCA) cycle enzymes. The enzymes that catalyze these reactions (except *succinic dehydrogenase;* see below) are believed to be localized in the matrix or on the surface of the inner membrane that faces the matrix.

As a result of the Krebs cycle oxidations, CO_2 and water

Orthodox conformation

Condensed conformation

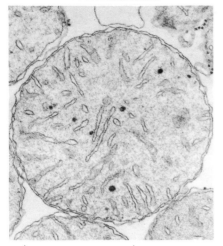

0.5 μm

0.5 μm

Figure 16–17

Orthodox and *condensed* conformational states of mitochondria. (Courtesy of Dr. C. R. Hackenbrock; copyright © The Rockefeller Press, *J. Cell. Biol. 37,* 345, 1968.)

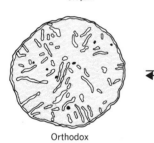

Orthodox

Condensed

Swollen

Contracted

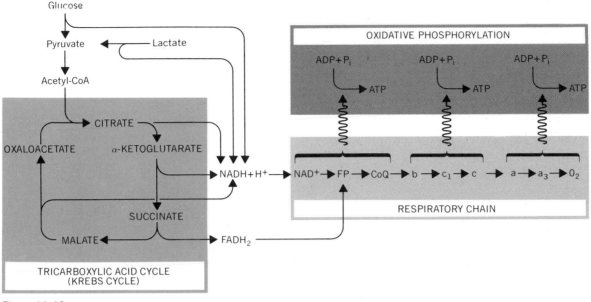

Figure 16–18
Relationship between the tricarboxylic acid (Krebs) cycle, respiratory chain oxidation-reductions, and oxidative phosphorylation reactions.

are formed as end products, and a number of special oxidation-reduction compounds are reduced. These compounds (e.g., NADH; see Chapters 3 and 10) subsequently participate in the initial step of a sequence of oxidation and reduction reactions, called the **respiratory** or **electron transport chain,** that is specifically associated with the inner membrane of the mitochondrion. The result of this chain of reactions is the reduction of O_2 to form H_2O. A third major process, called **oxidative phosphorylation,** is coupled to the respiratory chain and brings about the conversion of ADP to ATP. The respiratory chain reactions and oxidative phosphorylation are intimately associated with the inner membrane spheres.

In effect, pyruvate and other small molecules produced during cytosol metabolism diffuse through the permeable outer mitochondrial membrane and across the intermembrane space. It is upon entering the inner membrane that the three major reaction sequences (Krebs or TCA cycle, respiratory chain oxidation-reductions, and oxidative phosphorylation) begin. Pyridine nucleotides that are reduced during reactions in the cytosol (e.g., NADH produced in the glycolytic pathway and NADPH produced in the pentose phosphate pathway; see Chapter 10) may also permeate

the outer mitochondrial membrane, ultimately transferring their reductive capacity to the respiratory chain (see below).

Tricarboxylic Acid (Krebs) Cycle

The **tricarboxylic acid cycle** is frequently also called the **Krebs cycle** because the major steps of the cycle were worked out in 1937 by Sir Hans Krebs. At that time, radioactively labeled compounds were not available for biological studies and the cellular site of the reactions was not known with certainty. The experiments performed by Krebs and the logic and reasoning with which his findings were interpreted in order to formulate the metabolic pathway are recognized as milestones in our progressive unraveling of cellular metabolism. In 1948 E. P. Kennedy and A. L. Lehninger showed that mitochondria isolated from liver tissue homogenates by differential centrifugation contained the enzyme activities of the Krebs cycle reactions. Krebs received the Nobel Prize in 1953 for his elucidation of this most important metabolic pathway.

The Krebs cycle reactions are utilized by the cell to further metabolize a number of products of other reactions in the cytosol. These can be products as diverse as amino

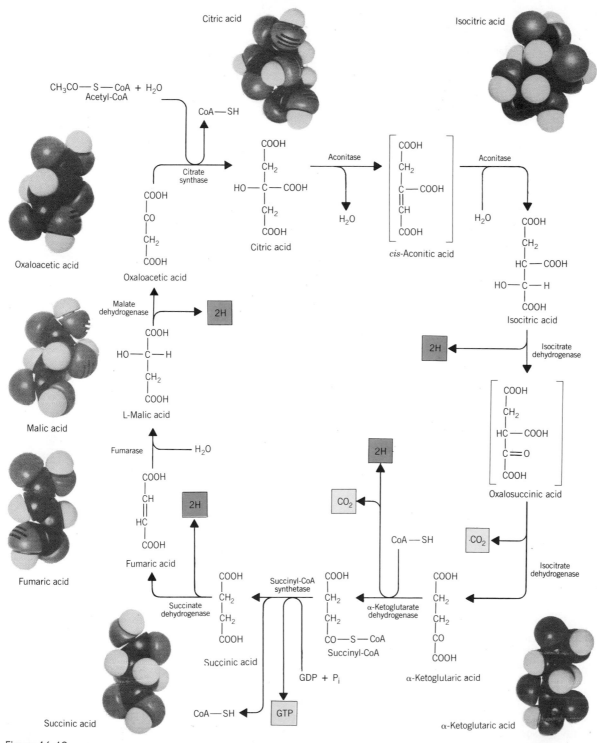

Figure 16–19
The tricarboxylic acid or Krebs cycle. (Photos by Kathy Bendo.)

acids, fatty acids, and pyruvic acid. However, the greatest metabolic merit of the cycle is its oxidation of pyruvate—the product of carbohydrate metabolism in the cytosol. The oxidation of this compound in the TCA cycle provides the reducing power to make significant amounts of ATP. The reactions are summarized in Figure 16–19 and are described below.

Oxidation of Pyruvate

The oxidation of pyruvate is brought about by a set of reactions catalyzed by the **pyruvate dehydrogenase complex.** This complex contains three enzymes (E_1, *pyruvate dehydrogenase*; E_2, *dihydrolipoyl transacetylase*; and E_3, *dihydrolipoyl dehydrogenase*) and five coenzymes and is located in the matrix of the mitochondrion. The first step is the decarboxylation of the pyruvate to yield CO_2 and an α-hydroxyethyl unit attached through *thiamine pyrophosphate* (TPP, a coenzyme) to enzyme E_1. That is,

$$
\begin{array}{c}
E_1\text{—TPP} \\
+ \\
CH_3COCOOH \\
\text{\textbackslash} CO_2 \\
\downarrow \\
E_1\text{—TPP—CHOH–CH}_3
\end{array}
\qquad (16\text{–}1)
$$

The hydroxyethyl group is then dehydrogenated (oxidized) to form an acetyl group, which is transferred to one of the sulfur atoms of *lipoic acid,* a coenzyme of E_2.

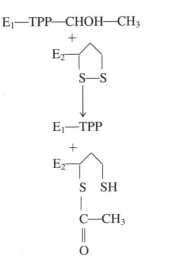

$$(16\text{–}2)$$

The acetyl group is then transferred to coenzyme A (Fig. 3–11) forming *acetyl CoA,* which separates from the enzyme complex. One hydrogen from the lipoyl group and one from coenzyme A are transferred to flavin adenine dinucleotide (FAD)(Fig. 3–11), a coenzyme of E_3.

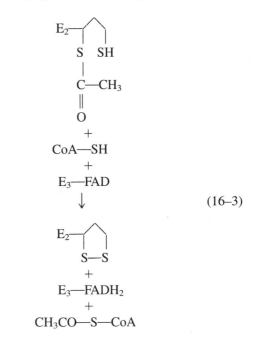

$$(16\text{–}3)$$

The resulting $FADH_2$ is then reoxidized by the transfer of hydrogens to NAD^+.

$$FADH_2 + NAD^+ \rightarrow FAD + NADH + H^+ \qquad (16\text{–}4)$$

The NADH and H^+ are then released from the enzyme complex. Thus, the overall reaction is

$$
\begin{array}{c}
\text{pyruvate} + NAD^+ + \text{CoA} \\
\downarrow \\
\text{acetyl CoA} + NADH + H^+ + CO_2
\end{array}
\qquad (16\text{–}5)
$$

The $\Delta G^{o\prime}$ (see Chapter 9) is -33.5 kJ mole^{-1} (-8.0 kcal mole^{-1}), indicating that the reaction is highly exergonic, with the equilibrium lying well to the right. In most animal tissues, the reaction is considered irreversible.

The Krebs Cycle Reactions

The first step of the Krebs cycle is a **condensation** reaction between acetyl CoA and oxaloacetate, a product of the cycle itself. The reaction is catalyzed by *citrate synthetase* (also called *citrate synthase*), with citrate produced and CoA released for reutilization in the prior reaction.

$$\text{acetyl CoA } + \text{ oxaloacetate } + \text{ H}_2\text{O}$$
$$\downarrow \qquad\qquad (16\text{–}6)$$
$$\text{citrate } + \text{ CoA}$$

The $\Delta G^{\circ\prime}$ is -32.2 kJ mole^{-1} (-7.7 kcal mole^{-1}). This reaction is the "pacemaker" or primary rate-limiting reaction of the cycle. The rate is controlled by the availability of acetyl CoA and oxaloacetate. Succinyl CoA, the product of a later step in the cycle, is a competitive inhibitor of this reaction, since it competes with acetyl CoA for the active site on the enzyme.

The enzymatic conversion of citrate to isocitrate is a two-step process in which the intermediate, *cis-aconitate* remains attached to the enzyme *aconitase* and therefore is frequently not shown in the Krebs cycle. The equilibrium of this reaction is toward citrate, $\Delta G^{\circ\prime}$ is 6.7 kJ mole^{-1} (1.59 kcal mole^{-1}), but the isocitrate is rapidly oxidized in the next step, thus shifting the direction of the reaction by removal of the product.

$$\text{citrate} \xrightarrow{\text{H}_2\text{O}} \left[\begin{array}{c} \textit{cis-}\text{aconitic} \\ \text{acid} \end{array} \right] \xrightarrow{\text{H}_2\text{O}} \text{isocitrate} \quad (16\text{–}7)$$

The oxidation of isocitrate to α-ketoglutarate is also a two-step process, with the intermediate remaining attached to the enzyme *isocitrate dehydrogenase*. The first step is an oxidation, two hydrogens being transferred to NAD$^+$. Actually, there are two isocitrate dehydrogenases in the mitochondrial matrix. One is linked to NAD$^+$ and the other to NADP$^+$. An NADP$^+$-linked isocitrate dehydrogenase is also found in the cytosol. However, it appears that the NAD$^+$-isocitrate dehydrogenase is the most active form in Krebs cycle reactions. This is an allosteric enzyme and specifically stimulated by ADP. When large amounts of

ATP are consumed in the cell and corresponding quantities of ADP produced, the rising ADP level acts as a positive effector for the allosteric enzyme. Alternatively, when the level of ATP rises, the accompanying fall in the level of ADP diminishes the positive effect on the allosteric enzyme. The overall effect of ADP-dependent NAD$^+$-isocitrate dehydrogenase in controlling the level of Krebs cycle activity is considered secondary, however, to the citrate synthetase step.

The second step of the reaction catalyzed by isocitrate dehydrogenase is the decarboxylation of the β-carboxyl group. The $\Delta G^{\circ\prime}$ for the entire reaction shown below is -20.9 kJ mole^{-1} (-5.0 kcal mole^{-1}).

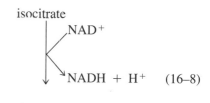

$$(16\text{–}8)$$

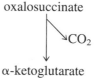

The oxidation of α-ketoglutarate to succinyl CoA involves an enzyme complex called the α-**ketoglutarate dehydrogenase complex.** In the sequence of reactions catalyzed by this complex, CO_2 is removed by first complexing with thiamine pyrophosphate. During addition of CoA, two hydrogen atoms removed from lipoic acid and coenzyme A reduce NAD$^+$ to NADH $+$ H$^+$. The $\Delta G^{\circ\prime}$ is -33.5 kJ mole^{-1} (-8.0 kcal mole^{-1}).

$$\alpha\text{-ketoglutarate } + \text{ NAD}^+ + \text{ CoA}$$
$$\downarrow \qquad\qquad (16\text{–}9)$$
$$\text{succinyl CoA } + \text{ CO}_2 + \text{ NADH } + \text{ H}^+$$

The removal of CoA from succinyl CoA is coupled to substrate level phosphorylation of GDP and is catalyzed by the enzyme *succinyl CoA synthetase.*

succinyl CoA | P_i + GDP

$$\downarrow \qquad\qquad (16\text{-}10)$$

succinate + GTP + CoA

The $\Delta G^{o\prime}$ is -2.9 kJ mole^{-1} (-0.7 kcal mole^{-1}). Phosphate is attached to the enzyme–succinyl CoA complex before it is transferred to GDP. In *E. coli*, the phosphate is transferred to ADP; in animal and most plant tissues GTP is formed and then the GTP donates the phosphate to ADP to form ATP.

$$GTP + ADP \rightleftharpoons ATP + GDP \qquad (16\text{-}11)$$

Succinate dehydrogenase catalyzes the oxidation of succinate to fumarate. This is the one enzyme of the TCA cycle reactions that has been shown to be firmly bound to the inner surface of the inner membrane and not associated with the matrix as are the other enzymes. This enzyme contains a flavin adenine dinucleotide (FAD) as a coenzyme, and it is this coenzyme that accepts the two hydrogens removed from the succinate during oxidation.

$$succinate + FAD \longrightarrow fumarate + FADH_2 \qquad (16\text{-}12)$$

The $\Delta G^{o\prime}$ is 0.

Fumarate is converted to malate by *fumarase*. The reaction has a $\Delta G^{o\prime}$ of -3.7 kJ mole^{-1} (-0.88 kcal mole^{-1}).

$$fumarate + H_2O \longrightarrow malate \qquad (16\text{-}13)$$

Malate is oxidized by *malate dehydrogenase,* which is an NAD$^+$-containing enzyme. Although the $\Delta G^{o\prime}$ is 29.7 kJ mole^{-1} (7.1 kcal mole^{-1}) and indicates that the isolated reaction

$$malate + NAD^+ \rightarrow oxaloacetate + NADH + H^+$$
$$(16\text{-}14)$$

is endergonic, the products of the reaction are readily removed in vivo and the reaction is therefore driven in the direction as written. Malate dehydrogenase is found not only in the matrix of mitochondria but also in the cytosol. The oxaloacetate produced by the reaction can serve to initiate the cycle again by combining with acetyl CoA to form citrate.

Summary of the TCA Cycle

It is possible to account for all the atoms that pass through the TCA cycle. There are two carbons in acetyl CoA, and during the cycle, one is converted to CO_2 at the isocitrate dehydrogenase step (eq. 16–8) and the other is converted to CO_2 at the α-ketoglutarate dehydrogenase step (eq. 16–9). Although these are not the same carbon atoms that entered the cycle, a balance is achieved—two carbons enter the cycle and two leave the cycle.

The same accounting of carbons can be made when starting with glucose.

$$C_6H_{12}O_6 + 6\, O_2 \rightarrow 6\, CO_2 + 6\, H_2O \qquad (16\text{-}15)$$

This monosaccharide is broken down during glycolysis in the cytosol (Fig. 16–20; see also Chapter 10 for a detailed description of glycolysis) to form two molecules of pyruvate. Each pyruvate molecule loses one carbon to CO_2 as it enters the Krebs cycle at the pyruvate dehydrogenase step (eqs. 16–1, 16–2) and another at each of the two steps described above in the TCA cycle:

$$\begin{array}{ccc} \text{glucose} & \longrightarrow \ 2 \text{ pyruvate} \longrightarrow & 6\ CO_2 \\ \text{(6 carbons)} & \text{(3 carbons each)} & \end{array} \quad (16\text{-}16)$$

An accounting can also be made for the hydrogen atoms during glycolysis and the Krebs cycle reactions. One glucose ($C_6H_{12}O_6$) and 6 H_2O contribute a total of 24 hydrogens. During glycolysis, four hydrogens are removed at the glyceraldehyde-3-phosphate dehydrogenase step (reaction 10–14, Fig. 10-2) to form 2 NADH + 2 H$^+$ in the cytosol and the two pyruvates from glucose each yield two more hydrogens (2 NADH + 2 H$^+$) at the pyruvate dehydrogenase step (eqs. 16–1, 16–2). Four more are transferred to NAD$^+$ at the isocitrate dehydrogenase step (eq. 16–8), four are transferred to NAD$^+$ at the α-ketoglutarate dehydrogenase step (eq. 16–8), four more are transferred to FAD at the succinate dehydrogenase step (eq. 16–12), and finally another four are transferred to NAD$^+$ at the malate dehydrogenase step (eq. 16–14). This accounts for all 24

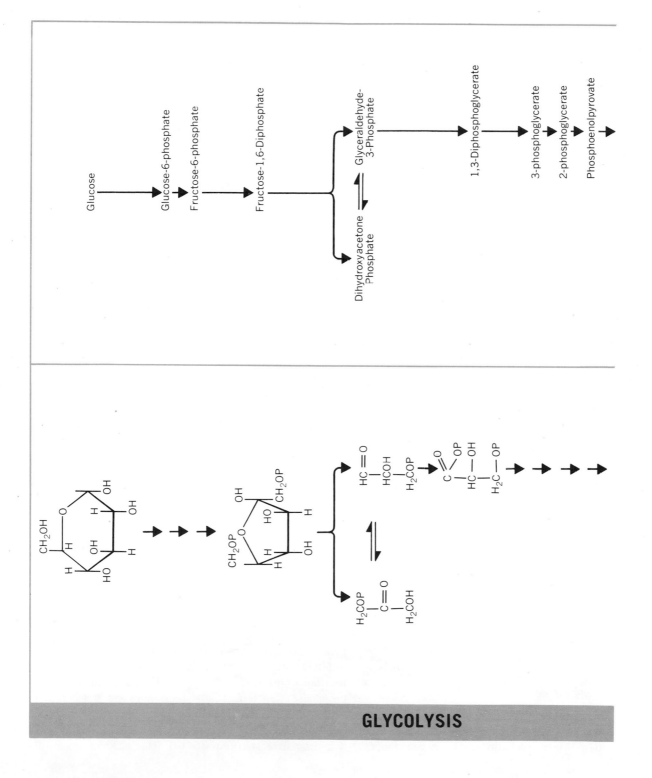

GLYCOLYSIS

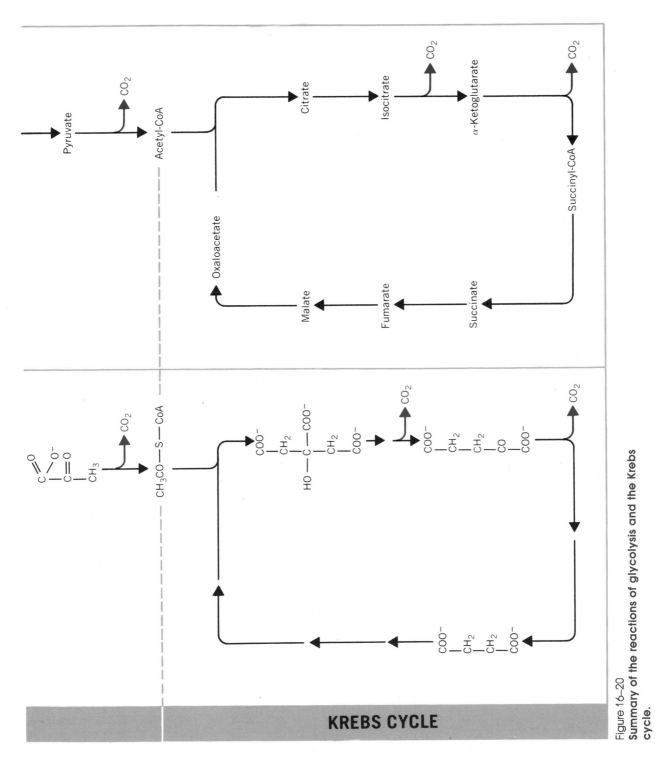

Figure 16–20
Summary of the reactions of glycolysis and the Krebs cycle.

hydrogens. Although these are not the *identical* hydrogen atoms that were present in the original glucose and water molecules, a numerical balance is nonetheless achieved.

In a similar way, the oxygens also balance. Glucose contributes six oxygens, and the six water molecules added mean that 12 oxygens enter the system. Twelve oxygen molecules are present in the 6 CO_2 molecules produced. Taken altogether, the overall equation for the oxidation of glucose through the Krebs cycle reactions is:

$$C_6H_{12}O_6 + 6\ H_2O + \begin{Bmatrix} 10\ NAD^+ \\ 2\ FAD \end{Bmatrix}$$
$$\downarrow \qquad\qquad (16\text{-}17)$$
$$6\ CO_2 + \begin{Bmatrix} 10\ NADH + 10\ H^+ \\ 2\ FADH_2 \end{Bmatrix}$$

The physiological importance of the Krebs cycle reactions is the production of tremendous chemical reducing power in the matrix of the mitochondrion by the accumulation of NADH, H^+, and $FADH_2$. These compounds enter the electron transport system reactions, and their reducing potential serves to drive the oxidative phosphorylation reactions that produce ATP.

Electron Transport System

The NADH, H^+, and $FADH_2$ that accumulate as the TCA cycle operates form an enormous potential energy pool. The reduced coenzymes are reoxidized and transfer their reducing capabilities through a sequence of compounds to oxygen, which accepts electrons and H^+ to form water. The sequence of compounds through which the electrons and H^+ pass is called the **electron transport system.** Each transfer of electrons (or H^+) from one compond to the next results in the oxidation (electron loss) of the donor molecule and the reduction (electron gain) of the acceptor molecule.

Such a transfer occurred in reaction 16–3 in the matrix during pyruvate dehydrogenase action. The enzyme-linked FAD (E_3-FAD) accepted the electrons from the substrate and was reduced, thereby forming E_3-$FADH_2$. Subsequently, the E_3-$FADH_2$ transferred hydrogens and electrons to NAD^+, which therefore was reduced while the $FADH_2$ was reoxidized (Fig. 16–21).

In the electron transport system the electrons are shuttled from one to another of about 10 intermediate compounds before they eventually reach oxygen to form water. Although it is possible to directly reoxidize NADH with oxygen, in the mitochondria this is prevented. The enzymes of the electron transport system appear to be arranged in the inner membrane in such a manner that the transfer of the electrons *must* proceed through the specified series of intermediates. With each electron transfer, there is an energy change and the energy released from the electron transfer sequence is utilized to phosphorylate ADP—a process termed **oxidative phosphorylation** and described in the next section.

Oxidation-Reduction Reactions

In Chapter 3 the tendency of acids and bases to donate or accept protons was described. In acid–base systems, one compound acts as the proton (H^+) donor and the other as the proton acceptor, the donor being the acid, and the acceptor the base. The two form a conjugate acid–base pair. In a similar manner, oxidizing and reducing agents function in pairs. In this case, they are called **redox pairs** or **redox couples.** The member of the pair that donates the electron is called the **reducing agent** or **reductant,** and the electron acceptor is called the **oxidizing agent,** or **oxidant.** The terms *donating* and *accepting* fail to convey the nature of the amounts of energy change involved in the reaction. In effect, the reducing agent has a certain ability (power) to retain electrons, as does the oxidizing agent. In a redox couple, one member attracts electrons more strongly than

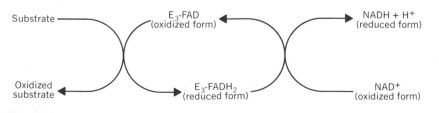

Figure 16–21
Sequential reduction and oxidation of FAD linked to E_3 (dihydrolipoyl dehydrogenase).

Table 16–2
Oxidation-Reduction Potentials of the Electron Transport System (Values are based on two-electron transfers at pH = 7.0 and 25–30°C.)

Electrode equation	E'_0 V
$2H + 2e^- \rightleftharpoons H_2$	-0.421
$NAD^+ + 2H^+ + 2e^- \rightleftharpoons NADH + H^+$	-0.320
$NADP^+ + 2H^+ + 2e^- \rightleftharpoons NADPH + H^+$	-0.324
Ubiquinone $+ 2H^+ + 2e^- \rightleftharpoons$ ubiquinol	$+0.10$
2 cytochrome $b_{(ox)} + 2e^- \rightleftharpoons 2$ cytochrome $b_{(red)}$	$+0.030$
2 cytochrome $c_{ox} + 2e^- \rightleftharpoons 2$ cytochrome c_{red}	$+0.254$
2 cytochrome $a_{3(ox)} + 2e^- \rightleftharpoons 2$ cytochrome $a_{3(red)}$	$+0.385$
$\frac{1}{2}O_2 + 2H^+ + 2e^- \rightleftharpoons H_2O$	$+0.816$

the other, and in effect the oxidizing agent can pull the electrons away from the reducing agent. This power to gain (or lose) electrons can be measured and is called the **oxidation-reduction potential** or **redox potential** and is expressed in volts. (Redox potentials are described in Chapter 9.)

For convenience, the potential of most redox couples is measured against a reference standard, which usually is a hydrogen electrode. The measurements are made under standard conditions using 1.0 M concentrations of oxidant and reductant at 25°C and at pH 7.0. The hydrogen electrode is equilibrated with H_2 gas at 1 atmosphere and the $[H^+]$ is 1.0 M at 25°C. When the pH is adjusted to 7.0 (i.e., $[H^+] = 10^{-7}M$), the redox potential of the reference hydrogen electrode is -0.42 volts. The redox potentials of the electron transport system intermediates are given in Table 16–2. The greater the E'_0 value, the more strongly the oxidant binds electrons. However, an oxidant having a lower E'_0 than another compound becomes the reductant of or electron donor to that compound.

The Henderson-Hasselbach equation given in Chapter 3 (eq. 3–5) describes the relationship between pH and the dissociation constant. In a similar way, the *Nernst equation* shows for a redox pair the relationship between the standard redox potential (E'_0), the observed potential at any concentration, and the concentration ratio of the oxidant and reductant:

$$E_h = E'_0 + \frac{2.3\ RT}{n\mathscr{F}} \log \frac{[oxidant]}{[reductant]}$$

where E'_0 is the standard redox potential, E_h is the observed electrode potential, R is the *gas constant* (8.31 J deg^{-1} mole^{-1}), T is the absolute temperature, n is the number of electrons transferred, and $\mathscr{F}$ is the faraday (96.4 kJ volt^{-1}). When two electrons are transferred at a time (which is usual in biological systems) and the constants are combined, the Nernst equation becomes

$$E_h = E'_0 - 0.03 \log \frac{[oxidant]}{[reductant]}$$

Classes of Electron-Transport-System (ETS) Compounds

There are five groups of compounds associated with the electron transport system. Three of these groups consist of enzymes whose coenzymes or prosthetic groups are known to be directly responsible for the transfer. They are (1) *pyridine-linked dehydrogenases,* which have either NAD^+ or $NADP^+$ as coenzymes; (2) *flavin-linked dehydrogenases,* which are linked to flavin adenine dinucleotides (FAD) or flavin mononucleotides (FMN); (3) *coenzyme Q* or *ubiquinone,* a lipid-soluble coenzyme functioning in electron transport; (4) the *cytochromes,* which contain iron-porphyrin prosthetic groups, and (5) the iron-sulfur proteins.

Pyridine-linked dehydrogenases require as their coenzyme either NAD^+ or $NADP^+$ (see Fig. 3–10 for structure of $NADP^+$ and the chemistry of its reduction by hydrogen);

both NAD^+ and $NADP^+$ can accept two electrons at a time. There are about 200 dehydrogenases for which NAD^+ or $NADP^+$ serve as coenzymes. Although the NAD^+ and $NADP^+$ dehydrogenases are found both in the cytosol and in the mitochondria and are known to transfer electrons between compounds in both places, it appears that only the NAD^+-linked compounds are involved in the electron transport system.

Flavin-linked dehydrogenases (often called flavoproteins, *FP*) require either FAD (Fig. 3–11) or FMN. Both are prosthetic groups whose isoalloxazine ring can accept two hydrogen atoms. (Prosthetic groups are firmly bound to the protein, whereas coenzymes are not.) Flavin-linked enzymes are involved in a number of enzyme systems, the more common of which are associated with fatty acid oxidation, amino acid oxidation, and Krebs cycle activity (e.g., pyruvate dehydrogenase and succinate dehydrogenase). It is not uncommon for flavin prosthetic groups and NAD^+ coenzymes to be linked to the same protein in dehydrogenases.

Ubiquinones were so named because of their occurrence in so many different organisms (*ubiquitous* means "present everywhere") and their chemical resemblance to quinone (Fig. 16–22). They are found in several different forms and are related to the *plastoquinones* of chloroplasts. The form present in mitochondria is often called coenzyme Q (CoQ) and accepts two hydrogen atoms at a time.

The *cytochromes* are proteins containing iron-porphyrin (or *heme*) groups (Fig. 16–23). There is a large number of cytochromes in cells; most are found in mitochondria, although some function in the endoplasmic reticulum and in chloroplasts. In mitochondria, five cytochromes appear to be associated with the inner membrane and are identified as cytochrome b, c_1, c, a, and a_3. Some occur in two or more forms, but all transfer electrons by reversible valence changes of the iron atom ($Fe^{+++} \rightleftharpoons Fe^{++}$). In cytochromes b, c_1, c, and a, the manner of binding of iron in the porphyrin ring and its association with the protein prevents the iron ligands from forming with oxygen, and therefore these reduced cytochromes cannot be directly oxidized

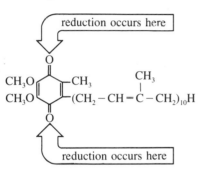

Figure 16–22
Chemical structure and sites of reduction of *ubiquinone*.

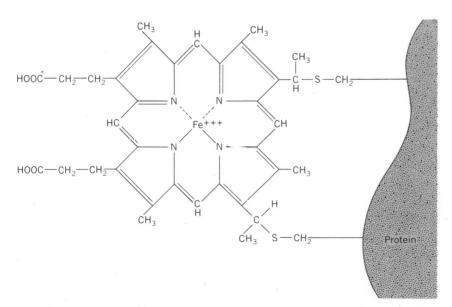

Figure 16–23
Cytochrome c. In cytochrome c, the heme prosthetic group is covalently bonded to the protein. In the other cytochromes (as in hemoglobin), the heme group is anchored to the protein by non-covalent bonds.

by molecular oxygen. Cytochrome a_3, which is the terminal carrier in the electron transport system and which together with cytochrome a forms the *cytochrome oxidase* complex, is an exception and can be directly oxidized by oxygen. In addition to cytochromes a and a_3, cytochrome oxidase contains copper. Electrons received from cytochrome c are picked up by cytochrome a and then transferred from a_3 to oxygen by $Cu^{++} \rightleftharpoons Cu^{+}$ intermediation.

Iron sulfur proteins of mitochondria are electron carriers containing iron and sulfur in equal amounts. The iron is reversibly oxidized during the electron transfer. Iron-sulfur groups occur in certain flavoproteins and are termed *iron-sulfur centers*. Four are associated with the NADH dehydrogenase complex, two with cytochrome b, and one with cytochrome c_1.

Electron Transport Pathway

It has taken almost half a century to work out the chain of mitochondrial electron transfers shown in Figure 16–24. During the early 1900s, several investigators (notably T. Thunberg) discovered a number of dehydrogenase enzymes. In 1913, O. Warburg discovered that cyanide inhibits oxygen consumption but does not interfere with dehydrogenases. He proposed the existence of iron-containing "respiratory enzymes," now recognized as the cytochromes. The flavoproteins were identified by A. Szent-Gyorgyi as the intermediates between dehydrogenases and the respiratory enzymes. R. A. Morton discovered ubiqui-none, and a number of other investigators, notably Keilin, Kuhn, Green, Chance, Racker, and Lehninger added details about the chemistry of the intermediates in electron transfer and the sequence of the reactions. Among the Nobel Prize recipients who contributed to our early understanding of electron transport are Warburg, Szent-Gyorgyi, and Kuhn.

Today we visualize the electron transport chain as an orderly sequence of compounds through which the hydrogens and electrons are passed, two at a time. The fact that the electrons do not jump or "short-circuit" to stronger oxidizing agents is probably related to the specific physical positioning of the coenzymes in the inner membrane. However, hydrogen or electrons may enter the chain at various points.

Most electrons are removed from the substrates in the cytosol or the matrix of the mitochondria by NAD^{+}- (or $NADP^{+}$-) linked dehydrogenases. The NADH produced in the matrix conveys the electrons to the NAD^{+}-flavoprotein-linked dehydrogenase in the inner membrane of the mitochondrion. As shown in Figure 16–24, the NADH is then oxidized by FAD, which now being reduced is in turn reoxidized by CoQ. Reduced CoQ is subsequently reoxidized by cytochrome b, and the reduced iron in cytochrome b is then reoxidized by cytochrome c_1. Cytochrome c_1 is reoxidized by cytochrome c, and finally, cytochrome oxidase accepts the electrons from cytochrome c, passing them from coenzyme a to a_3 and subsequently to oxygen. The acceptance of the electrons by oxygen occurs at the same time as the incorporation of $2H^{+}$, thereby forming water.

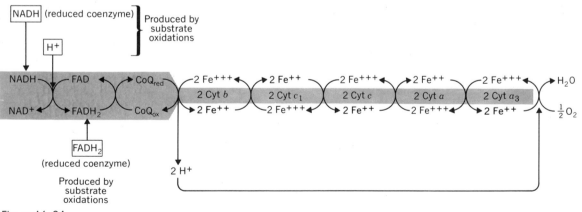

Figure 16–24

Oxidation-reduction reactions that conduct electrons and H⁺ to O₂, thereby forming water.

During one complete transfer, each of the coenzymes in the chain is reoxidized and is available to be reduced again. Pyruvate dehydrogenase, malate dehydrogenase, isocitrate dehydrogenase, and α-ketoglutarate dehydrogenase are examples of enzymes initiating the electron transfer chain through NAD^+. Succinate dehydrogenase oxidizes succinate, and the electrons removed are bound to FAD. The FAD_{suc} is reoxidized by giving up its electrons to CoQ, thereby bypassing the NAD^+ step. Fatty acyl-CoA and glycerol phosphate are oxidized in a similar manner with entry of electrons directly from the flavoprotein to CoQ.

Balance of Electrons from Glycolysis and TCA Cycle Metabolism

As was pointed out in the earlier section on TCA metabolism, 10 molecules of NAD^+ and 2 molecules of FAD are reduced when 1 molecule of glucose is metabolized via glycolysis and the TCA cycle. Each of these 12 compounds passes pairs of electrons through the electron transport system to oxygen. For each pair of electrons, $\frac{1}{2}$ O_2 is consumed and 1 H_2O is formed. Therefore, for 12 electron pairs, 6 O_2 are consumed and 12 H_2O are produced. You will recall that 6 of the 12 waters are consumed in the TCA cycle, so that in the overall oxidation of glucose there is a net production of only 6 H_2O.

$$C_6H_{12}O_6 + 6\,O_2 \longrightarrow 6\,CO_2 + 6\,H_2O$$

The Energetics of Electron Transport

The standard free energy change, $\Delta G^{\circ\prime}$ (see Chapter 9), can be calculated from the redox potentials using the following formula

$$\Delta G^{\circ\prime} = n\mathscr{F}\,\Delta E_0^\prime$$

where n = the number of electrons transferred at a time (usually 2), $\mathscr{F}$ = the number of faradays (96.4 kJ volt^{-1}), and $\Delta E_0^\prime$ is the change in standard redox potential (the $E_0^\prime$ of the accepting redox couple less the $E_0^\prime$ of the donating couple). The $\Delta G^{\circ\prime}$ for the entire chain from NAD to oxygen would be -220.5 kJ mole^{-1} (or -52.7 kcal mole^{-1}). The stepwise sequence is shown in Figure 16–25. The fate of the energy released by the transfer of electrons is discussed later in the chapter under oxidative phosphorylation. It is interesting to note at this point that enough energy is produced to phosphorylate several moles of ADP

at a standard free energy of formation of 30.5 kJ mole^{-1} (or 7.3 kcal mole^{-1}). Generally, only three molecules of ATP are formed, corresponding to the three steps producing sufficient energy for the phosphorylation (Fig. 16–25).

Transport of Protons

Transport of H^+ closely parallels electron transport. Some of the electron carriers accept and transfer electrons in the form of complete hydrogen atoms. These are NAD^+, $NADP^+$, FAD, FMN, and ubiquinone. FAD and FMN carry pairs of hydrogen atoms, while NAD^+ and $NADP^+$ carry a hydrogen that bears one extra electron (a hydride ion); a second H^+, however, associates closely with the NADH or NADPH. The cytochromes, on the other hand, only pick up and release electrons, each cytochrome carrying only one electron at a time. Therefore, there is speculation that the cytochromes work in pairs.

Electron Transport Inhibitors

Inhibitors of the electron transport system may be assigned to three main groups. One group includes those inhibitors that act on NAD^+ (or $NADP^+$) dehydrogenase and ubiquinone. Examples are *rotenone* (a plant substance once used by some South American Indians as a fish poison and occasionally used today in insecticides), *amytal,* and *piericidin* (an antibiotic similar in structure to and competitive with ubiquinone). The second group includes those that block electron transport between cytochrome *b* and *c;* an example is *antimycin A* an antibiotic from *Streptomyces griseus.* The third group includes those inhibitors that block transport of electrons from cytochromes *a* and a_3 to oxygen by cytochrome oxidase; included in this group are cyanide, azide, hydrogen sulfide, and carbon monoxide.

NAD [P]⁺-Transhydrogenases

The electron transport chain usually accepts NADH as the initial donor of hydrogen. In many animal tissues and in some eucaryotic microorganisms, hydrogens from NADPH are also accepted. The entry of these hydrogens into mitochondria is apparently brought about by the enzyme *NAD[P]⁺-transhydrogenase,* which is located in the outer mitochondrial membrane and which catalyzes the transfer of hydrogen to NAD^+ located inside the mitochondrion; i.e.,

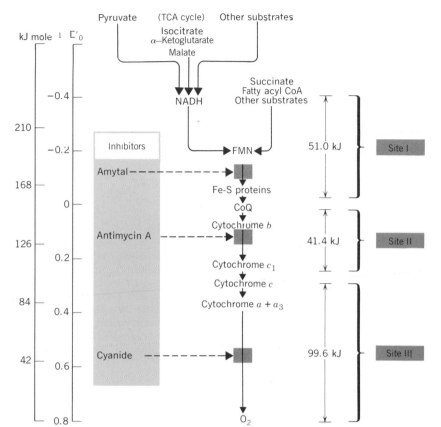

Figure 16–25
Changes in free energy as electrons pass through the electron transport system. The three changes that are believed to generate sufficient energy to bring about the phosphorylation of ADP to form ATP are identified as *Sites I, II,* and *III.*

$$(NADPH + H^+) + NAD^+ \longrightarrow (NADH + H^+) + NADP^+$$

Oxidative Phosphorylation

That phosphorylation is linked to oxidative metabolism was first proposed in the 1930s. The first evidence was the observation that phosphate was removed from the medium when TCA cycle intermediates were metabolized by cells. Subsequently, it was shown that the disappearance of inorganic phosphate was accompanied by the accumulation of organic phosphates such as glucose-6-phosphate and fructose-6-phosphate. Shortly thereafter, ATP was recognized as the primary product of oxidative phosphorylation. In 1948 Kennedy and Lehninger established that these oxidations and phosphorylations occurred exclusively in the mitochondria. By 1951 Lehninger had shown that the Krebs cycle oxidations could be bypassed entirely by adding

NADH and H^+. For each NADH and H^+ added, three phosphates and one oxygen were consumed, and it was therefore concluded that the electron transport chain was important for ATP synthesis.

The exergonic nature of the reactions of electron transfer beginning with NAD^+-dehydrogenases and ending with cytochrome oxidase provides the energy to drive the endergonic oxidative phosphorylation. The exergonic reactions can be summarized as

$$NADH + H^+ + \tfrac{1}{2} O_2 \longrightarrow NAD^+ + H_2O$$

which has a $\Delta G^{o\prime} = -220.5$ kJ mole^{-1} (52.7 kcal mole^{-1}). The endergonic reactions summarized as

$$3\ ADP + 3\ P_i \longrightarrow 3\ ATP$$

are associated with a standard free energy change of 91.6

kJ (i.e., $\Delta G^{o'} = 30.5$ kJ per mole or 7.3 kcal per mole). The efficiency of the coupling of the two systems is about 42% (i.e., 91.6/220.5).

The sites of the coupling are known. Figure 16–25 shows the major energy changes along the electron transfer chain. The step from NADH to CoQ represents the reactions of the dehydrogenases that link NAD$^+$ and FAD to CoQ. This reaction sequence (**site I**) provides enough energy (51.0 kJ or 12.2 kcal mole^{-1}) for the formation of the first ATP. The step from CoQ to cytochrome b does not provide enough energy for the second phosphorylation, but the following step (from cytochrome b to cytochrome c) does yield sufficient energy and is identified as the second coupling site (**site II**). The step from cytochrome c to cytochrome a is only moderately exergonic, but the final step from cytochrome a to oxygen is highly exergonic and is easily identified as a site of phosphorylation (**site III**). Thus, a pair of electrons transferred from one carrier to the next by these oxidation-reduction reactions generates the energy for the formation of ATP at sites I, II, and III. However, not all electrons from oxidized substrates enter the electron transport chain at an initial NAD$^+$ position. Some electrons, like those from succinate, pass from the coenzyme FAD of succinate dehydrogenase directly to CoQ. When this happens, site I is skipped and, therefore, only two ATPs are formed per pair of electrons transferred.

Molecular Events in Oxidative Phosphorylation

Despite numerous experiments, the mechanism by which energy from the electron transport system is used for oxidative phosphorylation is far from clear. It would appear that there are yet to be discovered coupling factors and enzymes involved in the process. Among the more significant observations that have been made are the identification of (1) mechanisms for controlling the *rate* of electron transport and oxidative phosphorylation, (2) inhibitors that uncouple electron transport from oxidative phosphorylation, (3) partial oxidative phosphorylation reactions, and (4) coupling factors. Maximal electron transport in mitochondria can occur only if there is ample ADP available to act as an acceptor of inorganic phosphate. When ADP is lacking, the rate of respiration is low and is called **state 4 respiration** (Fig. 16–26). When ADP is added, the rate of oxygen consumption rises and the ADP is phosphorylated.

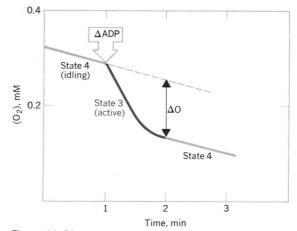

Figure 16–26

Changes in oxygen consumption and physiological state with the addition to ADP to mitochondria. Initially, the rate of oxygen consumption is low, due to the lack of ADP (most of that available has been converted to ATP). This "idling" state is called *state 4 respiration*. Addition of ADP induces the more active *state 3 respiration*, which continues until all of the added ADP is converted to ATP.

This elevated level of respiration is called **state 3 respiration.** Once the ADP is consumed *state 4 respiration* is reestablished. The influence of ADP on the respiratory rate is called *acceptor control*, the "acceptor" being ADP. Interestingly, the configuration of mitochondria changes between state 4 and state 3 respiration. During state 4 respiration, the mitochondrion assumes the orthodox conformation (Fig. 16–17) but changes to the condensed conformation when added ADP induces state 3.

A number of chemical agents uncouple oxidative phosphorylation from the ETS (e.g., 2,4-dinitrophenol, dicumarol, and the salicylanilides). The addition of these compounds has two interesting effects. First, they speed up electron transport and oxygen consumption even in the absence of ADP, but there is no synthesis of ATP. In other words, ADP acceptor control is uncoupled and so is energy transfer. Second, in the presence of the uncoupling agent, the hydrolysis of ATP occurs—the *opposite* of the normal goals of mitochondrial activity.

Oligomycin is an inhibitor of oxidative phosphorylation and oxygen consumption, but it does not prevent electron transfer. A group of agents called ionophores also prevent oxidative phosphorylation by preventing transfer of energy,

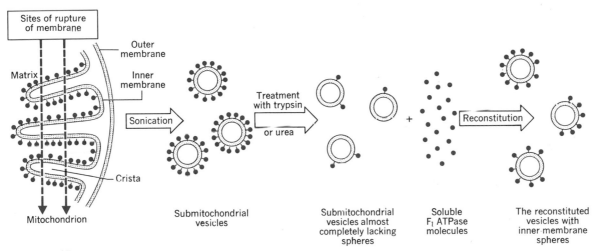

Figure 16–27

Under appropriate conditions the disruption of mitochondria by sonification is followed by the formation of small vesicles from fragments of the inner membrane. These vesicles function normally in electron transfer and oxidative phosphorylation. The inner membrane spheres can be detached from the outside surfaces of these vesicles using trypsin or urea. When this is done, the capacity for oxidative phosphorylation is lost. Reconstitution of vesicles and spheres is accompanied by the restoration of these activities.

but they are effective only in the presence of certain monovalent cations such as K^+ and Na^+.

Some of the reactions of oxidative phosphorylation are known. An *ATPase* is present in the inner membrane and is stimulated by 2,4-dinitrophenol but inhibited by oligomycin. Another reaction is the exchange of inorganic phosphate with the terminal phosphate of ATP, a reaction called *phosphate-ATP exchange*. This reaction is inhibited by both 2,4-dinitrophenol and oligomycin. An exchange of atoms also occurs between the oxygens of water and those of inorganic phosphate, called *phosphate–water exchange*. The terminal phosphate can also be exchanged between ATP and ADP, a process inhibited by 2,4-dinitrophenol and oligomycin and called the *ADP-ATP exchange* reaction.

Some interesting studies that relate structure and function in mitochondria have been carried out using **submitochondrial vesicles** formed spontaneously from fragments of the inner membrane. In these studies, the inner membranes of mitochondria isolated from cells are first separated from outer membranes and then subjected to sonification or detergent action. The resulting fragments of the cristael membranes then round up to form vesicles with the inner membrane spheres on the outside surface (Fig. 16–27). If *closed* vesicles are formed, they exhibit functioning ETS and oxidative phosphorylation reactions. No functional reactions are associated with open vesicles or nonvesicular fragments.

If the submitochondrial vesicles are treated with urea or trypsin, the inner membrane spheres are removed and can be separated from the vesicles, which lose their capacity to carry out oxidative phosphorylation. However, if the spheres are added back to the vesicles, the capacity to perform oxidative phosphorylation is regained. The spheres are called *coupling factor one* (F_1). It has also been shown that the isolated coupling factor contains an enzyme that hydrolyzes ATP. This enzyme has been purified and found to have a molecular weight of about 360,000, a diameter of 9 nm, and a requirement for Mg^{++}. In intact mitochondria the enzyme probably catalyzes the synthesis of ATP from ADP and inorganic phosphate, rather than the reverse, and as such could be termed *ATP-synthetase*. The enzyme's function is not inhibited by oligomycin. However, another factor has been isolated that when present with F_1 renders the ATPase sensitive to oligomycin. This latter factor is called F_o or *oligomycin-sensitivity-conferring factor (OSCF)*. F_o is also a large protein; it has been speculated that F_o could constitute the stalk that attaches the sphere to the inner membrane.

The Chemiosmotic-Coupling Hypothesis

The exact mechanism that couples the energy of electron transfer to the phosphorylation of ADP remains somewhat speculative. However, at the present time the **chemiosmotic-coupling hypothesis** proposed by P. Mitchell is the most widely supported model. Since its initial proposal in 1961, the hypothesis has undergone modification by Mitchell and by others. For his contributions to the development of this model, Mitchell received the Nobel Prize in 1978. According to Mitchell, a gradient of hydrogen ions is created across the inner membrane as membrane carriers pump H^+ from the matrix into the intermembrane area of intact mitochondria or into the submitochondrial vesicles formed from fragmented inner membranes. This process creates an **electrochemical gradient** (designated $\Delta\mu_{H^+}$) across the membrane consisting of a negative-inside membrane potential ($\Delta\psi$) and an acid-outside gradient of H^+ (ΔpH). An intact membrane is essential for the establishment of this electrochemical gradient. The gradient, in turn, provides the energy to drive the synthesis of ATP.

It is proposed that the molecules involved in electron transport are not randomly arranged in or on the inner membrane but are spatially oriented in a specific manner. As electrons are transferred through the sequence of membrane carriers, there is an accompanying expulsion of H^+ from the matrix. This causes the matrix to become especially alkaline (low H^+ and high OH^- content). In Figure 16–28, these events are depicted with the electron transport intermediates arranged in order in the membrane (including an as yet unidentified carrier, X). In some models, the order (FMN, CoQ, cyt. *b*) is altered (FMN, cyt. *b*, CoQ) but still accounts for the same overall observations. The vectorial flow of protons from matrix to intermembrane space shown in Figure 16–28 creates the electrochemical gradient, which is the source of energy for phosphorylation. As oxidative phosphorylation proceeds, the alkaline sink within the matrix and the external acid pool are both neutralized. Continued electron transport is necessary in order to maintain the pH gradient that drives phosphorylation. For many years, it was believed that two H^+ were expelled from the

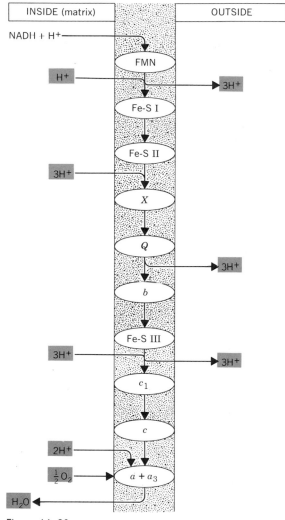

Figure 16–28

Transfer of H^+ through the inner mitochondrial membrane from the matrix during electron transfer. According to Mitchell's *chemiosmotic-coupling hypothesis,* **the electrochemical gradient established by this movement is the source of energy for oxidative phosphorylation (see also Fig. 16–29).**

matrix at each of the three electron transfer sites coupled to phosphorylation (i.e., Sites I, II, and III in Fig. 16–25). It is now apparent that at least three H^+ (and perhaps as many as four) are ejected from the matrix per two electrons passing through these coupling sites.

There is good evidence, both direct and indirect, to

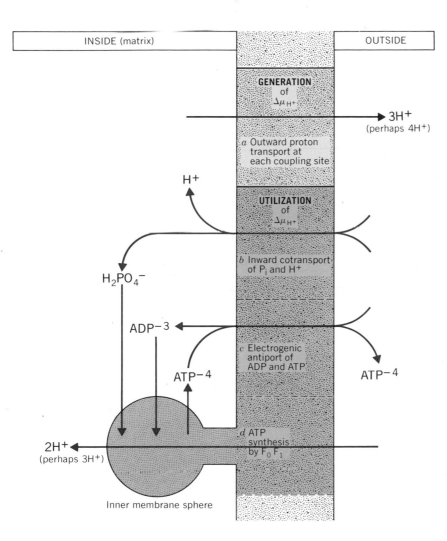

Figure 16–29
Chemiosomotic-coupling. *(a)* Passage of
two e⁻ through one coupling site pumps
at least three protons (perhaps as many
as four) out of the matrix; this generates
the electrochemical gradient. *(b)* and
(c) Inward transport of P_i and ADP and
outward transport of ATP are coupled to
the return of one proton to the matrix
along the proton gradient. *(d)* Reentry of
the other two (or three) protons into the
matrix through the inner membrane
spheres causes a change in F_O or F_1 that
promotes ADP and P_i binding at one catalytic site and ATP release at another.

support the chemiosmotic-coupling hypothesis. As will be seen in Chapter 17, phosphorylation of ADP also occurs in chloroplasts and can be explained by a similar chemiosmotic-coupling mechanism that creates a proton gradient across the organelle's thylakoid membranes. Indeed, A. Jagendorf has shown that a burst of ADP phosphorylation occurs when chloroplasts isolated from cells at neutral pH are placed in a buffer solution at low pH.

While electron transport is accompanied by the outward transport of H⁺, ATP synthesis is linked to the inward movement of H⁺. Since ATP synthesis occurs on the matrical surface of the inner membrane, ADP and inorganic phosphate must be transported from outside the mitochondrion across the membrane and into the matrix. The inward

transport of P_i and ADP and the outward transport of ATP are also driven by the electrochemical gradient (see Fig. 16–29). One H⁺ reenters the matrix along the pH gradient during the inward movement of inorganic phosphate, and at least two more H⁺ (perhaps as many as three more) reenter the matrix for each molecule of ATP synthesized.

Although the chemiosmotic-coupling model is widely accepted, the precise mechanism resulting in ATP synthesis remains uncertain. Also remaining to be explained is how the protons are transported chemically. The similarities in structure, properties, and function of ATP synthetases and hydrogen ion pumps that have recently been isolated from various animal, plant, and microbial cells suggests that some answers will be forthcoming soon.

Other Functions of Mitochondria

Mitochondria are generally described as the powerhouses of the cell, and, therefore, most interest is directed to the processes that evolve energy, namely, the TCA cycle, electron transport, and oxidative phosphorylation. However, a great number of other processes also take place in mitochondria.

The Glyoxylate Cycle

This metabolic pathway was described in Chapter 10. The enzymes are localized in the matrix, and some of them are the same as those of the TCA cycle. In effect, the glyoxylate cycle is a modified form of the Krebs cycle, but its function appears to be primarily associated with conversion of acetate produced by the catabolism of fatty acids into oxaloacetate. Oxaloacetate is also an important intermediate in the conversion of fatty acids to carbohydrates. While animals lack certain enzymes of the glyoxylate cycle and are incapable of converting fatty acids to carbohydrates using this pathway, plants and microorganisms have functional glyoxylate systems. In higher plants, some of the enzymes of the system (e.g., *isocitrate lyase* and *malate synthetase*) are localized in specific organelles called *glyoxysomes* (Chapter 9) as well as in mitochondria.

Fatty Acid Oxidation

Free fatty acids are rarely found in more than trace quantities in cells because they are highly toxic. The fatty acids associated with mono-, di-, and triglycerides and in phospholipids are generally hydrolyzed from the glycerol in the cytosol and immediately activated for transport into the matrix of the mitochondrion, where they are then oxidized. These reactions were also described in Chapter 10.

Fatty Acid Chain Elongation

Fatty acids are generally synthesized by the smooth endoplasmic reticulum. However, there are a number of enzymes in mitochondria that catalyze the elongation of palmitic and other saturated fatty acids by successive additions of acetyl CoA to the carboxyl end. In smooth ER, both unsaturated and saturated fatty acids are elongated but by the addition of malonyl CoA rather than acetyl CoA. Synthesis of fatty acids is discussed in Chapter 10.

Superoxide Dismutase and Catalase

During electron transport, a number of toxic reductive products of oxygen are formed. The most common are the superoxide radical (O_2^-) and hydrogen peroxide (H_2O_2). A protective enzyme has been identified in mitochondria called *superoxide dismutase,* which converts O_2^- and H^+ into hydrogen peroxide. The hydrogen peroxide, in turn, is decomposed by *catalase* in peroxisomes (Chapter 19).

$$2\ O_2^- + 2H^+ \xrightarrow{\text{dismutase}} H_2O_2 + O_2$$

$$H_2O_2 \xrightarrow{\text{catalase}} H_2O + \tfrac{1}{2}\ O_2$$

Amphibolic and Anaplerotic Reactions

The intermediates of the TCA cycle can act as precursors of a variety of products of anabolic pathways (Table 16–3). Thus the TCA cycle is **amphibolic** in that it can act in both a catabolic and anabolic manner. These anabolic reactions can drain intermediates away from the TCA cycle and thus deplete the oxaloacetate supply necessary to keep the cycle functioning. Reactions that supply the TCA cycle with intermediates that replace those lost are called **anaplerotic reactions.** Several anaplerotic reactions of the Krebs cycle are shown in Table 16–4.

Permeability of the Inner Membrane

As noted earlier most metabolites can readily pass through the outer mitochondrial membrane. In contrast, the inner membrane is not freely permeable. Except in special cases, sugars cannot pass through the membrane. Ions such as Na^+, K^+, and Cl^- are impermeable as are NAD^+, NADH, $NADP^+$, NADPH, AMP, CDP, GDP, CTP, GTP, CoA, and acetyl CoA. These compounds collect in the matrix and are unable to mix or exchange with pools of these molecules in the cytosol. There are, however, special transport systems in the inner membrane that are specific for select metabolites, and via these systems, ADP, ATP, P_i, pyruvate, and a number of TCA cycle intermediates are transported. The transport systems are species-specific and in most cases are associated with membrane proteins called

Table 16–3
TCA Precursors of Anabolic Pathways

TCA Cycle Intermediate	Pathway or Anabolic Reaction
Citrate	Citrate + ATP + CoA → oxaloacetate + ADP + P_1 + acetyl CoA └──→ fatty acid biosynthesis
Isocitrate	Isocitrate → glyoxylate → malate → carbohydrate synthesis
α-Ketoglutarate	α-Ketoglutarate + alanine → pyruvate + glutamate ┐
Succinyl CoA	Heme biosynthesis
Malate	Carbohydrate synthesis
Oxaloacetate	Oxaloacetate + alanine → pyruvate + aspartate → protein synthesis

Table 16–4
Anaplerotic Reactions of the TCA Cycle

TCA Cycle Intermediate Produced	Non-TCA Cycle Reaction Generating the TCA Cycle Intermediate
α-Ketoglutarate	Glutamate + pyruvate → α-ketoglutarate + alanine (from protein breakdown)
Succinate	From glyoxylate cycle
Malate	Malic enzyme: pyruvate + CO_2 + NADPH + H^+ → malate + $NADP^+$
Oxaloacetate	Pyruvate carboxylase: pyruvate + CO_2 + ATP + H_2O → oxalacetate + ADP + P_i
Oxaloacetate	Aspartate + pyruvate → oxaloacetate + alanine (from protein breakdown)

carriers, translocases, or *porters.* Table 16–5 lists a number of these systems and indicates their functions.

These systems can function passively, facilitating the exchange of metabolites when there is a favorable concentration gradient, or they can function in an active manner if coupled to the energy-producing electron transport system. During electron transport, there is an accumulation of hydroxyl ions in the matrix, but these hydroxyl ions can be transported externally by a *phosphate carrier,* which si-

multaneously carries inorganic phosphate into the matrix on an exchange basis (Fig. 16–30). The phosphate accumulated can be partly consumed in ATP synthesis, with the ADP-ATP carrier functioning to bring ADP into the matrix. The phosphate can also be exchanged with external dicarboxylate or tricarboxylate required in TCA cycle reactions using the appropriate carboxylate carrier.

One of the clearest connections between active transport across the inner mitochondrial membrane and the electron

Table 16–5
Mitochondrial Membrane Transport Systems

System	Exchange
Dicarboxylate carrier	Exchange on mole-for-mole basis of malate, succinate, fumarate, and phosphate between matrix and cytosol.
Tricarboxylate carrier	Exchange on mole-for-mole basis citrate and isocitrate between matrix and cytosol.
	Exchange citrate or isocitrate for dicarboxylate.
Aspartate-glutamate carrier	Exchange aspartate for glutamate across membrane.
α-Ketoglutarate-malate carrier	Specifically exchange α-ketoglutarate for malate across membrane.
ADP-ATP carrier	Exchange of ADP for ATP.

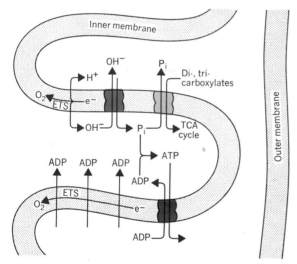

Figure 16–30
Carriers in the inner mitochondrial membrane transfer phosphate, di- and tricarboxylic acids, and other metabolites between the matrix and intermembrane space.

transport system involves the movement of calcium ions into the matrix. The movement of Ca^{++} into the matrix is accompanied by an equivalent uptake of phosphate. It has been shown that for each pair of electrons transported through the electron transport system from NADH to oxygen, six Ca^{++} are transported across the inner membrane into the matrix. However, when the Ca^{++} is transported, there is no phosphorylation of ADP. Studies using specific inhibitors indicate that the same sites responsible for phosphorylation along the electron transport chain are the energy-providing sites for the active transport of Ca^{++} that must function in connection with the phosphate carrier.

Cytosol-Matrix Exchange of NADH and NADPH

Reduced coenzymes such as NADH and NADPH do not permeate the inner membrane of the mitochondrion to any significant extent. However, reduced pyridine nucleotides are known to be produced in a number of reactions in the cytosol (the reduction of NAD^+ at the glyceraldehyde step in glycolysis is an important example) and the reoxidation of NADH occurs via the mitochondrion. The mechanism involves a set of reactions called a *shuttle* (Fig. 16–31).

Glycerol Phosphate Shuttle

The **glycerol phosphate shuttle** (Fig. 16 31) involves (1) *glycerol phosphate dehydrogenase* in the cytosol, (2) glycerol phosphate dehydrogenase on the outer surface of the inner mitochondrial membrane, and (3) the reduction of CoQ in the electron transport chain. Dihydroxyacetone phosphate, NADH, and H^+ react in the cytosol to form glycerol phosphate, which diffuses through the outer mitochondrial membrane to the outer surface of the inner membrane. There the glycerol phosphate reacts with the

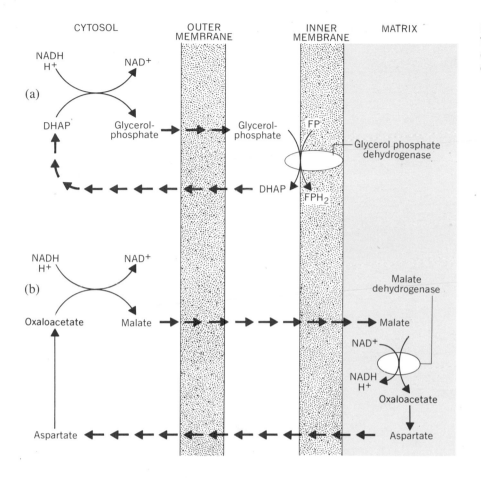

Figure 16–31
The glycerol phosphate and malate-aspartate *shuttles*. DHAP = dihydroxy-acetone phosphate.

membrane dehydrogenase to form dihydroxyacetone phosphate, which returns to the cytosol. The membrane-bound dehydrogenase employs a flavoprotein (FP) as a coenzyme, and the FP becomes reduced during the reaction. Subsequently, the electrons from the FPH$_2$ are passed directly into the electron transport system at the CoQ step. Because the NAD$^+$ → FP step of electron transport is skipped, only two ATP molecules are generated for each pair of electrons that enters in this fashion from the cytosol.

Malate-Aspartate Shuttle

Another shuttle, the **malate-aspartate shuttle,** can also transport the hydrogens accepted during the reduction of NAD$^+$ in the cytosol across the inner membrane (Fig. 16–31). H$^+$ transported in this manner into the matrix enters

the electron transport chain via NADH, and as a result three ATP are generated for each pair of electrons.

Total ATP Production from Catabolism of Glucose

The complete oxidation of a molecule of glucose by glycolysis, the TCA cycle, and electron transport is accompanied by the *gross* production of 38 or 40 ATP or *net* production of 36 or 38 ATP. The ATPs result from substrate-level phosphorylations and electron transport coupled to oxidative phosphorylation. An accounting of ATP production is described in Figure 16–32. The different totals for ATP production (i.e., gross production either of 38 or 40) depend upon which shuttle is used to transport hydrogen from NADH into the mitochondrion. Hydrogen from the

ATP Consumed	Metabolic Pathway	Substrate-Level Phosphorylation	ETS-Level Phosphorylation
	1 glucose		
1 ATP			
1 ATP			
	2 phospho-glyceraldehydes		
	. . . .		4 or 6 ATP
	2 diphospho-glyceric acids		
	. . . .	2 ATP	
	. . . .	2 ATP	
	2 pyruvates		
	. . . .		6 ATP
	2 acetyl CoA		
	2 citrate		
	. . . .		6 ATP
	2 α-ketoglutarate		
	. . . .		6 ATP
	2 succinyl CoA		
	. . . .	2 GTP(ATP)	
	2 succinate		
	. . . .		4 ATP
	2 fumarate		
	. . . .		6 ATP
	2 oxalacetate		
2 ATP		6 ATP	32 or 34 ATP

Figure 16–32
Balance sheet for ATP production and consumption during glycolysis, the tricarboxylic acid cycle, and electron transport.

glycerol phosphate shuttle is accepted by FP, and as a result, one coupling step is bypassed. Hydrogen from the malate-aspartate shuttle enters the electron transport chain earlier, so that all three coupling sites are utilized.

Summary

Mitochondria are the ''powerhouses'' of the cell and the primary sites of cell **oxidations.** Within these organelles, elementary substrates produced by the breakdown of carbohydrates, lipids, or nitrogenous macromolecules in other cell locations are oxidized to CO_2 and water. Some of the energy of these exergonic oxidations is conserved in the phosphorylation of ADP to form ATP.

Mitochondria occur in almost every type of aerobic eucaryotic cell. Although their shape may vary, all mitochondria contain two structurally and functionally different membranes—an **inner** and an **outer membrane.** Between the two membranes is the fluid-filled **intermembrane space.** The inner membrane surrounds the mitochondrial **matrix.** Projections of the inner membrane into the matrix are called **cristae.** New mitochondria form by the fission of other mitochondria.

The oxidative and phosphorylation reactions occur in the inner membrane or in the matrix. The tricarboxylic acid or **Krebs cycle** reactions constitute the first phase of the oxidation of substrates such as acetate. In these reactions, the molecules are enzymatically degraded to CO_2 and water. Hydrogen and electrons released from the metabolic intermediates reduce NAD^+ and FAD to NADH and $FADH_2$. Some ATP is formed at the *substrate level* by these reactions. Most of the ATP generated in mitochondria is via the **electron transport system** (ETS), which reoxidizes the NADH and $FADH_2$ formed in the Krebs cycle. The ETS functions as a multistep series of oxidation-reduction reactions, transferring electrons through a set of intermediates associated with the *inner membrane* and ultimately reducing molecular oxygen to water. These intermediates include *pyridine* or *pyridine-linked dehydrogenases, flavin-linked dehydrogenases, ubiquinones, iron-sulfur proteins,* and *cytochromes.*

Associated with the ETS is a mechanism for the formation of ATP called **oxidative phosphorylation.** For each pair of electrons transferred from NADH, three molecules of inorganic phosphate are added to three ADP to form three ATP (two ATP for each $FADH_2$ that is reoxidized). Oxidative phosphorylation is a function of the inner membrane. Mitchell's **chemiosmotic-coupling hypothesis** is the most widely accepted model explaining the source of the energy that drives phosphorylation of ADP to form ATP.

The mitochondrion is also the site of the **glyoxylate cycle** reactions, **fatty acid oxidation, fatty acid chain elongation,** *superoxide dismutase* and *catalase* reactions, and a number of **amphibolic** and **anaplerotic** reactions.

References and Suggested Reading

Articles and Reviews

Cohen, S. S., Mitochondria and chloroplasts revisited. *Am. Sci. 61,* 437 (1973).

Cross, R. L., The mechanism and regulation of ATP synthesis by F_1-ATPases. *Annu. Rev. Biochem. 50,* 681 (1981).

Ernster, L., and Schatz, G., Mitochondria: a historical review. *J. Cell Biol. 91,* 227 (1981).

Fernandez-Moran, H., Cell membrane ultrastructure. *Circulation 26,* 1039 (1962).

Fillingame, R. H., The proton-translocating pumps of oxidative phosphorylation. *Annu. Rev. Biochem. 49,* 1079 (1980).

Fletcher, M. J., and Sanadi, D. R., Turnover of rat liver mitochondria. *Biochim. Biophys. Acta 51,* 356 (1961).

Guppy, M., and Ballantyne, J., The importance of water and oxygen in the evolution of hydrogen shuttle mechanisms. *Comp. Biochem. Physiol. 69B,* 1 (1981).

Hinkle, P. C., and McCarty, R. E., How cells make ATP. *Sci. Am. 238*(3) 104 (Mar. 1978).

Mitchell, P., Keilin's respiratory chain concept and its chemiosmotic consequences. *Science 206,* 1148 (1979).

Raven, P. H., A multiple origin for plastids and mitochondria. *Science 169,* 641 (1970).

Schwartz, R. M., and Dayhoff, M. O., Origins of procaryotes, eucaryotes, mitochondria, and chloroplasts. *Science 199,* 395 (1978).

Sjöstrand, F. S., The structure of mitochondrial membranes: a new concept. *J. Ultr. Res. 64,* 217 (1978).

Sjöstrand, F. S., The interpretation of pictures of freeze-fractured biological material. *J. Ultr. Res. 69,* 378 (1979).

Tanaka, K., Scanning electron microscopy of intracellular structures. *Int. Rev. Cytol. 68,* 97 (1980).

Tedeschi, H., The mitochondrial membrane potential. *Biol. Rev. 55,* 171 (1980).

Wilson, J. E., and Dove, J. L., Turnover of mitochondria in rat liver, kidney and heart. *J. Elisha Mitchell Sci. Soc. 81,* 21 (1965).

Books, Monographs, and Symposia

Kozlov, I. A., Mitochondrial transhydrogenase: general principles of functioning, in *Current Topics in Membranes and Transport* (A. Kleinzeller and F. Bonner, Eds.) Academic Press, New York, 1982.

Lehninger, A., *Biochemistry* (2nd ed.), Worth, New York, 1975.

Lehninger, A. L., and Reynafarje, B., Cycles in the function of mitochondrial membrane transport systems, in *Current Topics in Cellular*

Regulation (R. W. Estabrook and P. Srere, Eds.), Vol. 18, Academic Press, New York, 1981.

Lipmann, F., The ATP-phosphate cycle, in *Current Topics in Cellular Regulation* (R. W. Estabrook and P. Srere, Eds.), Vol. 18, Academic Press, New York, 1981.

Margulis, L., *Symbiosis in Cell Evolution.* W. H. Freeman, San Francisco, 1981.

Meijer, A. J., and van Dam, K., Mitochondrial ion transport, in *Membrane Transport* (S. L. Bonting and J. J. de Pont, Eds.) Elsevier/North-Holland, Amsterdam, 1981.

Racker, E., *Membranes of Mitochondria and Chloroplasts.* Van Nostrand-Reinhold, New York, 1970.

Roodyn, D. B., and Wilkie, D., *The Biosynthesis of Mitochondria,* Methuen, London, 1968.

Stryer, L., *Biochemistry* (2nd ed.), W. H. Freeman, San Francisco, 1981.

Tedeschi, H., *Mitochondria: Structure, Biogenesis and Transducing Functions,* Springer-Verlag, New York, 1976.

Wehrle, J. P., The role of electrogenic proton translocation in mitochondrial oxidative phosphorylation, in *Current Topics in Membranes and Transport* (A. Kleinzeller and F. Bonner, Eds.) Academic Press, New York, 1982.

Chloroplasts are organelles found in the green tissues of plants and are responsible for the absorption of light energy, the synthesis of carbohydrates, and the evolution of molecular oxygen. The sum of these three processes is called **photosynthesis.** Light energy captured by the chloroplast is converted into potential chemical energy in the form of carbohydrates and starts the "energy chain" in nature. The oxygen evolved during the capture of light energy becomes the ultimate oxidizing agent for cellular metabolic reactions, as described in Chapters 10 and 16. Mitochondrial oxidations, as well as other oxidations in plant, animal, and many microbial cells, depend on this primary source of oxygen. It is currently believed that the entire supply of oxygen in the atmosphere today was derived from and is presently maintained by photosynthesis.

A **chloroplast** is any membrane-encased organelle containing **chlorophyll** that belongs to a group of related organelles in plants called *plastids*. The plastids have a variety of morphological forms, carry out diverse functions, and store many different compounds. For example, the **amyloplast** is the starch-storing plastid of potato tubers; the **chromoplast** is the lycopene-containing plastid that gives the fruit of tomatoes its red color. Each of the diverse plastids is believed to arise from a common **proplastid** precursor. All of the major groups of plants, with the exception of the fungi, contain chloroplasts. There are two fundamental types of chloroplasts each associated with one of two types of photosynthesis: C_3-photosynthesis or C_4-photosynthesis. C_3-photosynthesis converts (or "fixes") carbon dioxide into 3-carbon acids, whereas C_4-photosynthesis fixes CO_2 into 4-carbon acids. Plants that carry out C_4-photosynthesis (called C_4-plants) are generally grasses or plants that are endemic to environments with little water. C_3-plants (those that carry out C_3-photosynthesis) are commonly the broad-leafed flowering plants, cone-bearing plants, or those living in areas of adequate moisture. The characteristics of these two types of photosynthesis and their associated chloroplast types are discussed in detail later in this chapter. There may be a single chloroplast or dozens of chloroplasts in a cell. Most frequently, the simpler plants, such as the algae, contain a single chloroplast, while the higher plants, such as the cone-bearing and flowering plants, have many chloroplasts in each cell.

Photosynthesis occurs in many procaryotic organisms such as the cyanobacteria and photosynthetic bacteria. These procaryotes do not have true chloroplasts; instead, they have lamellated structures called **chromatophores** that carry out only the light-absorbing reactions of photosynthesis but not the carbohydrate-synthesizing reactions. For many laboratory studies of photosynthesis, the single-celled plant *Chlorella* is employed. This alga has one cup-shaped chloroplast that practically fills the cell. The organism can easily and conveniently be cultured with artificial lighting in the laboratory in solutions of inorganic salts. Because of

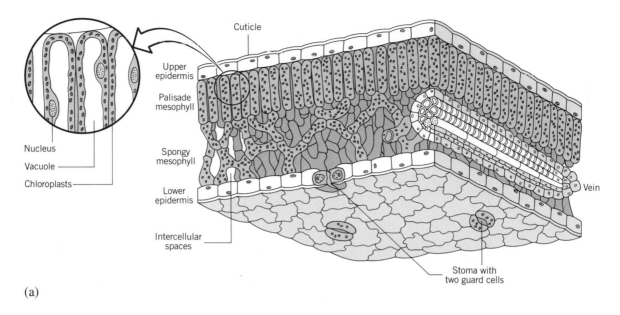

(a)

Figure 17–1

Arrangements of chloroplast-containing cells in leaves of plants carrying out C$_3$-photosynthesis *(a)* and C$_4$-photosynthesis *(b)*. In plants employing C$_3$-photosynthesis (e.g., spinach, parsley, and other broad-leafed plants), chloroplasts are present in the *palisade layer* and in the *spongy mesophyll* layers sandwiched between the upper and lower *epidermis*. In plants that carry out C$_4$-photosynthesis (e.g., corn, sugar cane, and certain grasses), there is no palisade layer, but the veins of the leaf are encased by chloroplast-containing *bundle sheath cells*.

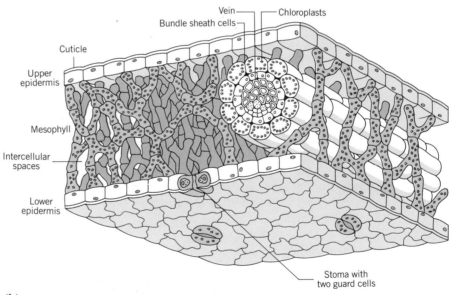

(b)

the one-to-one relationship between cell and chloroplast and the ease with which the cells can be grown and enumerated, quantitative studies using chloroplast preparations of known content are possible.

For studies with chloroplasts of higher plants, leaves are generally used, and spinach and parsley leaves are probably the most popular source. In these leaves, the chloroplasts are found in greatest numbers in two internal tissues, the **palisade mesophyll** and the **spongy mesophyll** (Fig. 17–1*a*). Both tissues lie between an upper and lower epidermis and have thin cell walls that are easily broken. The number of chloroplasts in each cell varies within an organism with changing environmental conditions and varies greatly from one species to another. In spinach, there are between 20 and 40 chloroplasts in each palisade parenchyma cell. In the palisade parenchyma, the chloroplasts lie along the side walls of the cell, the center of the cell being filled with large vacuoles. In the spongy parenchyma, the chloroplasts are more randomly distributed throughout the cytoplasm of the cell. In many genera, cytoplasmic streaming (i.e., *cyclosis*) moves the chloroplasts about the cell, and in a few instances, an active amoeboid-type of movement of the chloroplast has been observed. Their positions in the cell and movement about the cell maximize the exposure of the chloroplasts to light.

Chloroplasts are routinely isolated from plant tissues by differential centrifugation following the disruption of the cells. Leaves are homogenized in an ice-cold isotonic saline solution such as 0.35 M NaCl buffered at pH 8.0. The disruption is generally carried out with short spins in a Waring blender. After a preliminary filtration through nylon gauze (20 μm pore size) to remove the larger particles of debris (cell nuclei, tissue fragments, and unbroken cells), the chloroplasts are separated by centrifugation at 200 g for 1 minute. The chloroplast-rich pellet is then resuspended and centrifuged again at 2000 g for 45 seconds to resediment the chloroplasts. Chloroplast preparations obtained by this procedure are generally mixtures of intact and broken organelles. Since the chemical composition, rate of photosynthetic activity, and other properties of intact chloroplasts differ significantly from those of damaged organelles, it is often desirable to separate the two populations. This may be accomplished by rate or isopycnic density gradient centrifugation of the chloroplast preparation using sucrose, Ficoll, or Ludox gradients.

Chloroplast size is quite variable. Although the average diameter of a chloroplast in higher plant cells is between 4 and 6 μm, the size may fluctuate according to the amount of available illumination. In sunlight, chlorophyll is more readily synthesized by the plant, and the chloroplasts increase in size; in the shade, chlorophyll synthesis declines, and there is a corresponding reduction in chloroplast size. Changes in chloroplast shape are also observed after short-term exposure of plants to light. Short-term light exposure produces a small but measurable decrease in chloroplast volume. Presumably, this is due to a light-induced production of ATP, for the addition of ATP to chloroplasts in the dark causes a reduction in volume. Polyploid cells contain larger chloroplasts than comparable diploid cells.

The shape of most chloroplasts in higher plants is spheroid, ovoid, or discoid (Fig. 17–2). Other irregular shapes sometimes occur but are more common in lower plants. For example, in algae, cup-shaped chloroplasts as well as spiral bands, star shapes, and digitate forms are observed. The shape and structure of chloroplasts can also be altered by the presence of starch granules. During periods of active photosynthesis, the sugars formed in the chloroplasts are polymerized into starches that precipitate as small granules. The starch granules are usually ellipsoidal and may be up to 1.5 μm long.

Fine Structure of the Chloroplast

The chloroplasts of C_3-plants are composed of two membrane layers similar to those of mitochondria. Each membrane is about 70 to 80 Å thick and the two membranes are separated by a space of about 70 to 100 Å. The **outer membrane,** which lacks folds or projections, serves to delimit the organelle and regulate the transport of materials between the cytoplasm and the interior of the organelle. The **inner membrane** parallels the outer membrane, but inward folds of this membrane are extensive. The inner membrane gives rise to a series of internal parallel membranes called **lamellae** (Fig. 17–3). The lamellae are suspended in a granular fluid or matrix that appears somewhat electron-dense in electron photomicrographs. This matrix is referred to as the **stroma.**

Most of the lamellae in the chloroplasts of higher plants are organized to form disk-shaped sacs called **small thylakoids.** The small thylakoids are often arranged in stacks called **grana** (one stack is a *granum*) having a diameter of about 300 to 600 nm. Since these thylakoids are round, the grana appear much like a stack of coins (Figs. 17–3a, b). A typical chloroplast has between 40 and 60 grana, and each may be composed of 2 to 100 small flattened thylakoids. Frequently, a small portion of the thylakoid extends radially into the stroma forming a branching tube, or **large thylakoid,** that communicates with other small thylakoids and grana. Collectively, the branching and anastomosing network is called the **stroma lamellae.**

Structure of the Thylakoid

The adjacent membranes of neighboring thylakoids within each of the grana form thick layers called **grana lamellae.**

(a)

Figure 17–2

TEM photomicrographs of chloroplasts of sugar beet *(a),* **grass** *(b),* **and** *Chlorella (c) (chl,* **chloroplast;** *py,* **pyrenoid;** *st,* **starch).** *(d)* **SEM photomicrograph of** *Callisia* **chloroplasts. (***a* **and** *b* **courtesy of Dr. W. Laetsch;** *c* **courtesy of Drs. G. Dempsey, D. Lawrence, and V. Cassie; copyright © 1980 by Blackwell Scientific Publications,** *Phycologia 19,* **13;** *d* **courtesy of Drs. R. Wise and J. Harris; copyright © 1980,** *Cytologia 45,* **113.)**

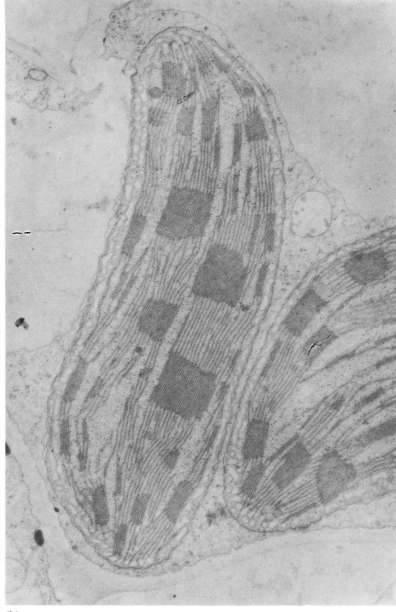

(b)

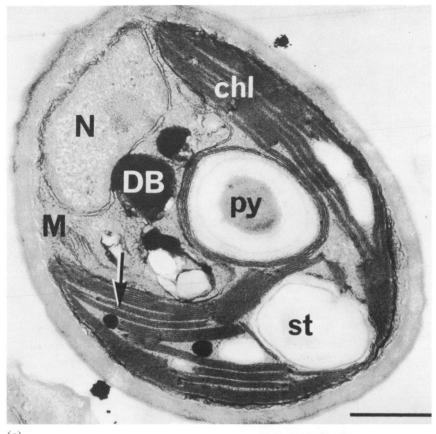

(c)

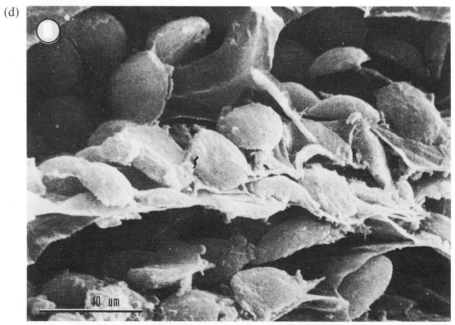

(d)

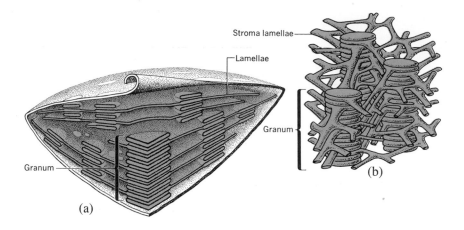

(a)

(b)

Figure 17–3
(a) Cross section through a chloroplast showing the arrangement of lamellae and grana. *(b)* Grana and stroma thylakoids formed by the lamellae. *(c)* Chloroplast of a corn cell. (Courtesy of Dr. N. Rascio; copyright © 1981 by Springer-Verlag, *Protoplasma 105,* 241).

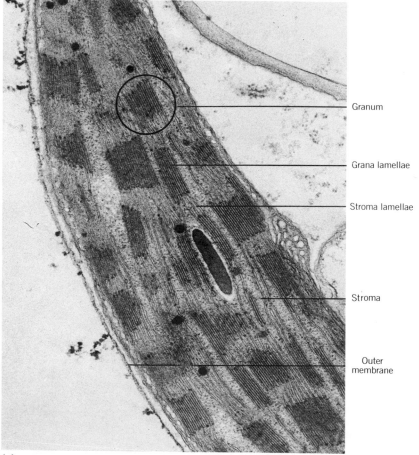

(c)

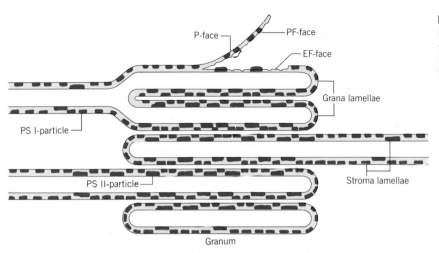

Figure 17–4
Diagram of the stroma thylakoid lamellae and grana lamellae showing the arrangement of particles on the membrane fracture faces.

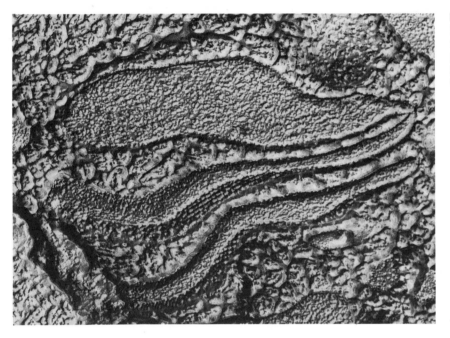

FIGURE 17–5
Freeze-fractured and sublimed grana membrane showing large and small membrane particles (Courtesy of Dr. K. Muhlethaler; copyright © Springer-Verlag, 1965, *Planta 67*, 305).

Electron photomicrographs of grana lamellae fixed with glutaraldehyde and stained with osmium reveal a five-layered arrangement consisting of three dark 40 Å thick osmiophilic layers enclosing two 17 Å thick osmiophobic spaces. Freeze-fracture techniques indicate that the grana membranes contain numerous particles. The particles, which are primarily protein in composition, appear to be of two basic sizes, 105 Å and 140 Å nm in diameter (Fig. 17–4). The stroma lamellae contain mainly the smaller-size particles. The larger particles are present on the EF face of the grana membranes (see Chapter 15 for an explanation of the membrane fracture faces) and are associated with photosystem II of photosynthesis (described later in the chapter). The smaller particles are associated with photosystem I and are present in about equal numbers on the PF face of both the grana and stroma membranes (Fig. 17–5).

Stroma Structures

The granular stroma contains a variety of particles. The presence of starch granules was noted earlier. Electron micrographs also reveal a number of osmiophilic granules and groups of ellipsoidal structures called *stromacenters*. Of particular interest are the strands of DNA scattered through the stroma and ribosomelike particles. Some of these structures are shown in Figure 17–2, which also clearly depicts the relationship between the stroma and grana lamellae.

The chloroplasts of C_4-plants have the same general structures as those of C_3-plants (an outer enclosing membrane, grana and stroma thylakoids, a stroma with DNA, ribosomes, and stromacenters), but there are some notable differences. The C_4-plant chloroplasts all have a *peripheral ticulum,* which is a group of anastomosing tubules about the periphery. In many C_4-plants such as corn there are two types of chloroplasts. Those found in the mesophyll cells (see Fig. 17–1b) are similar to the chloroplasts of C_3-plants but lack starch grains. The chloroplasts found in the cells surrounding the leaf veins (called bundle sheath cells) contain starch granules but have elongated grana.

Chemical Composition of Chloroplasts

The organic constituent present in greatest quantity in the chloroplast is protein, which may represent up to 69% of the dry weight (Table 17–1). In leaf cells, 75% of the total cell nitrogen is found within the chloroplasts. Both structural and soluble proteins have been identified, but only a few of these have been extracted and purified. A peptide analysis by SDS gel electrophoresis shows compositional differences between stroma and grana lamellae, but the differences are primarily quantitative rather than qualitative. The relative compositions of stroma lamellae and grana lamellae are shown in Table 17–2.

Essentially all the pigments and cytochromes are located in the lamellae. The stroma lacks these compounds but contains DNA and RNA, which are not present in the lamellae. Most of the RNA is associated with the ribosomes of the stroma. The amount of DNA is low; estimates are 10^{-15} to 10^{-14} grams per chloroplast or about 0.03% of its dry weight. However, this is enough information to account for the synthesis of some chloroplast proteins, including some of the enzymes of photosynthesis. The disposition of chloroplast DNA during chloroplast division is unclear.

Table 17–1
Chemical Composition of Spinach Chloroplasts

Component	Percentage of Chloroplast Dry Weight	
	Chloroplasts Isolated in Water	Values Corrected for Loss of Soluble Protein
Total protein	50	69
Water-insoluble protein	50	31
Water-soluble protein	0	38
Total lipid	34	21
Chlorophyll	8	5
Carotenoids	1.1	0.7
Ribonucleic acids	—	1.0–7.5
Deoxyribonucleic acids	—	0.02–0.1
Carbohydrate (starch, etc.)	Variable	

Source: Modified from J. T. O. Kirk and R. A. E. Tilney-Bassett, *The Plastids,* W. H. Freeman and Co., San Francisco, 1967.

Table 17–2
Major Components of Stroma and Grana Lamellae

	Stroma Lamellae	Grana Lamellae
Total chlorophyll	278[a]	401
Chlorophyll *a*	238	281
Chlorophyll *b*	40	130
P_{700}	2.5	0.6
β-Carotene	21	17
Lutein	10	29
Violaxanthin	15	20
Neoxanthin	8	16
Phospholipid	76	66
Monogalactosyl diglyceride	231	214
Digalactosyl diglyceride	172	185
Sulfolipid	65	59
Cyt *b* (total)	1.0	3.4
Cyt *f*	0.5	0.7
Manganese	0.3	3.2

[a]Values in micromoles of component per gram of membrane protein.

Lipid and lipid-soluble pigments account for about 34% of the dry weight of the spinach chloroplast. An exceedingly large number of different lipid compounds have been identified. The more common lipids are the galactosyl diglycerides, phospholipids, quinones (including vitamin K), and sterols.

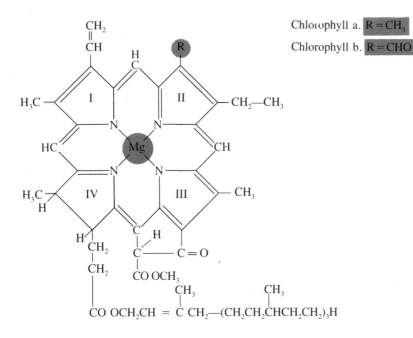

Chlorophyll a. R = CH₃
Chlorophyll b. R = CHO

Figure 17–6
Chemical structure of chlorophylls *a* and *b*.

The Chlorophylls

The green pigments of the chloroplast and the main sources of the color of green plants are the chlorophylls. Although a large number of chemically distinct chlorophylls have been identified in a variety of different plants, the structures of these chlorophylls are basically the same. (The structures of chlorophylls *a* and *b* are given in Fig. 17–6.) It is customary to identify each chlorophyll by a different letter. All photosynthetic plants have been found to contain chlorophyll *a*, but the presence of the secondary chlorophylls *b*, *c*, or *d* depends on the type of plant. Higher plants usually have chlorophyll *b*. In the photosynthetic bacteria, a chlorophyll called **bacteriochlorophyll** occurs in place of chlorophyll *a*. Together, chlorophylls *a* and *b* represent about 8% of the dry weight of the spinach chloroplast, with an *a:b* weight ratio of 2.1 to 3.5. In most plants, the *a:b* ratio varies according to the light intensity. For example, alpine plants, which receive light of high intensity, have an average ratio of 5.5. The ratio is 2.3 in shade plants.

Each chlorophyll has a characteristic light absorption spectrum. The in vitro light absorption spectra of chlorophylls *a* and *b* are shown in Figure 17–7. Extracted chlorophyll *a* has absorption maxima at 430 and 670 nm, whereas the absorption maxima of chlorophyll *b* occur at 455 and 640 nm. In vitro absorption maxima of other plant and bacterial pigments are indicated in Table 17–3. The absorption spectrum and maxima of plant pigments vary according to the solvent used for extraction. Therefore, it is not surprising that values obtained during in vivo measurements differ from those yielded by extracts. For example, in vivo studies of chlorophyll *a* indicate that its native absorption maximum occurs at 677 nm. One very

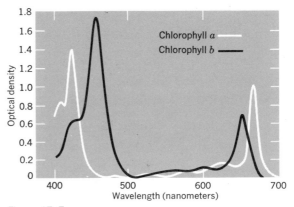

Figure 17–7
In vitro absorption spectra of chlorophylls *a* and *b* in an ether solvent.

Table 17–3

Absorbtion Maxima[a] of Plant and Bacterial Pigments

Pigment	Wavelength (nm)	Occurrence
Chlorophyll a	430, 670	All green plants
Chlorophyll b	455, 640	Higher plants; green algae
Chlorophyll c	445, 625	Diatoms; brown algae
Bacteriochlorophyll	365, 605, 770	Purple and green bacteria
α-Carotene	420, 440, 470	Leaves; some algae
β-Carotene	425, 450, 480	Some plants
γ-Carotene	440, 460, 495	Some plants
Luteol	425, 445, 475	Green leaves; red and brown algae
Violaxanthol	425, 450, 475	Some leaves
Fucoxanthol	425, 450, 475	Diatoms; brown algae
Phycoerythrins	490, 546, 576	Red and blue-green algae
Phycocyanins	618	Red and blue-green algae
Allophycoxanthin	654	Red and blue-green algae

[a]Absorption maxima vary according to the solvent in which the pigment is dissolved.

Source. From E. Rabinowitch and Govindjee, *Photosynthesis*, Wiley, New York, 1969.

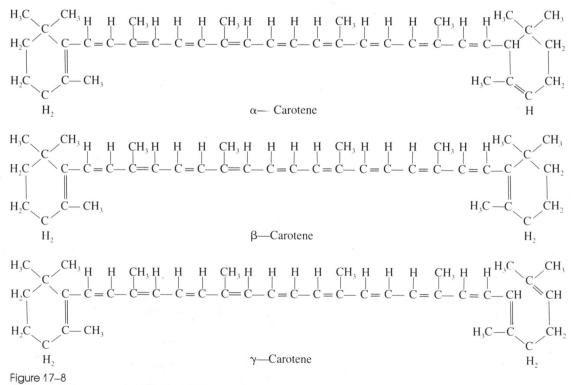

Figure 17–8

Chemical structure of alpha (α), beta (β), and gamma (γ) *carotene*.

important form of chlorophyll *a* that is readily bleached by light has an absorption maximum at 700 nm. This form, which represents only about 0.1% of the total chlorophyll *a* molecules present in a sample, is called P_{700} or chlorophyll a_1. The roles of the various chlorophylls in photosynthesis are discussed later in the chapter.

The Carotenoids

The carotenoids are all long-chain isoprenoid compounds having an alternating series of single and double bonds. Although these compounds are synthesized only in plant tissue and participate in photosynthesis, they also serve as precursors of vitamin A in animal tissues. Most carotenoids are yellow, orange, or red. The formulas of α, β, and γ carotene are shown in Figure 17–8, and their absorption maxima are given in Table 17–3. Most of these pigments are located in the chloroplast lamellae and are believed to function as accessory pigments for light absorption during photosynthesis.

Location and Arrangement of the Pigments

Both the chlorophylls and the carotenoids are located almost exclusively in the chloroplast lamellae. The lamellae are about 52% protein and 48% lipid, and the two pigments reside primarily in the lipid component. Some lipid is also represented by the osmiophilic granules of the stroma, but these are not believed to contain chlorophyll. Because each chlorophyll molecule has a hydrophilic portion (the tetrapyrrole) and a lipophilic portion (the phytyl chain), the chlorophyll molecules are thought to be aligned in a specific manner within the lamellae. The pyrrole groups form weak bonds with the lamellar protein, while the carotenoids are dissolved in the lipid adjacent to the chlorophyll molecules.

The chloroplast stroma contains many of the enzymes associated with photosynthesis. Chloroplast protein synthesis also takes place in the stroma. Circular DNA strands about 40 μm long have been isolated from the chloroplast along with ribosomes and polyribosomes. The DNA strands have 115 to 200 kbp (kilobase pairs) and contain genes for about 180 to 280 polypeptides. Chloroplast ribosomes belong to the 70 *S* class and contain 23 *S* and 16 *S* RNA (see Chapter 22); thus, they are smaller than those found in the cytoplasm of plant (and animal) cells. Chloroplast ribosomes are numerous and account for about one-half of the cellular protein synthesized.

Development of Chloroplasts

New chloroplasts are produced by the division of mature chloroplasts (Fig. 17–9) or by development from proplastids in the cell. The cells of young shoots of higher plants may contain 20 to 40 proplastids (Fig. 17–10). These small bodies, which are ovoid to round in shape and are surrounded by two membranes, can develop into a number of different plastids in addition to chloroplasts. As the proplastid develops into a chloroplast, the inner membrane gives rise to internal membranes, which then form the lamellae and thylakoids. Proplastids increase in numbers by a form of division (Fig. 17–10*b*).

The control of chloroplast development is not fully understood. The differentiation of new organelles appears to rely upon an interaction between genetic information present in the cell nucleus and information present in the chloroplast itself. Chloroplasts contain DNA that is transcribed and translated within the organelle to form some of the chloroplast proteins; however, other chloroplast proteins are derived from transcription and translation of nuclear DNA.

The intriguing but controversial hypothesis that chloroplasts are the evolutionary products of bacteria-like organisms that "invaded" eucaryotic cells and established a symbiotic relationship with the host cell is similar to the proposal for the origin of mitochondria discussed in Chapter 16. In the case of the chloroplast, the host could have been a heterotrophic cell. Interestingly, chloroplasts are found in the cells of a subgroup of *Nudibranchs* (marine snails lacking a shell) that feed on algae. As the algae cell cytoplasm passes through the intestine of the Nudibranch, whole chloroplasts are absorbed into the tissue and may persist there for the life of the animal. When the animal is in light, oxygen evolution by the absorbed chloroplasts can be detected.

Photosynthesis—Historical Background

The overall reactions of photosynthesis may be summarized by the equation

$$6\,CO_2 + 12\,H_2O \xrightarrow[\text{chlorophyll}]{\text{light}} C_6H_{12}O_6 + 6\,O_2 + 6\,H_2O$$

(a)

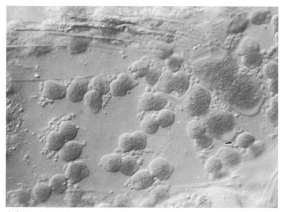

(b)

(c)

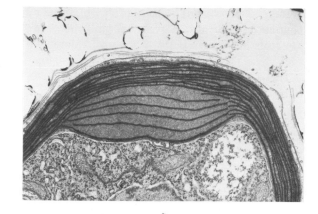

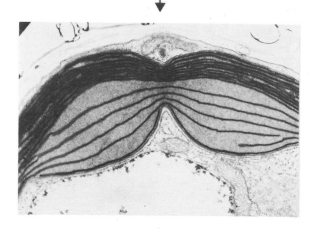

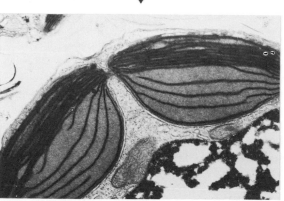

(d)

Figure 17–9
Electron photomicrographs of dividing chloroplasts. *(a)* In this SEM view of spinach cells, the chloroplasts are arranged in a single layer close to the cell wall; their constricted (i.e., dumbbell) shape is believed to represent a terminal stage in the division of the organelles by a fissionlike process. The small bodies clinging to the surface of the chloroplasts are probably mitochondria. Magnification, 6000×. *(b)* Freeze-fracture TEM view of a constricted chloroplast; magnification, 16000×. *(c)* Chloroplasts revealing different degrees of constriction are seen in this light photomicrograph of a living spinach cell; magnification, 900×. *(d)* Sequence of chloroplast division stages in *Cricosphaera. (a, b,* and *c* courtesy of Dr. J. V. Possingham; copyright © 1980 by Company of Biologists, Ltd.; *J. Cell Science 46,* 87. *d* courtesy of Drs. T. Hori and I. Inouye; copyright © 1981 by Springer-Verlag, *Protoplasma 106,* 121.)

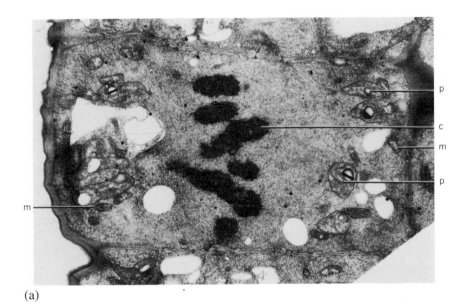

(a)

Figure 17–10
Proplastids. (*a*) In this TEM view of a dividing spinach cell, several proplastids (p) and mitochondria (m) are seen on either side of the metaphase chromosomes (c); magnification, 9000×. *(b)* Dividing proplastid in a spinach root tip cell; magnification, 44000×. (Courtesy of Dr. J. V. Possingham; copyright © 1981, Societe Francaise de Microscopie Electronique; *Biologie Cellulaire 41,* 203.)

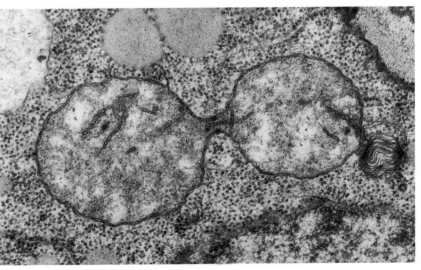

(b)

Studies by many individuals over the past 300 years have led to our present understanding of this process. One of the first studies by *Jan Baptista van Helmont* (Dutch, 1577–1640) was published in 1648 after his death. Van Helmont planted a 5-lb willow shoot in a large pot containing 200 lb of soil. The plant was regularly watered over a 5-year period and was then carefully removed and weighed. While the willow tree was found to weigh almost 170 lb, the original soil weighed only a few ounces less than the original 200 lb. Van Helmont concluded that the increase in weight of the willow tree was due primarily to the addition of water, for he was not aware of the role played by gases in the air in the growth of the plant. In 1727 *Stephen Hales* proposed that part of a plant's nourishment came from the air and that light was involved in the process. Today, we also recognize that the small quantity of material removed from the soil included minerals, nitrogen, phosphate, and calcium salts.

Joseph Priestley (English, 1733–1804) was the first to show that plants exchange gases with the atmosphere. In particular, Priestley found that oxygen was evolved by plants during photosynthesis. Late in the eighteenth century, *Jan Ingenhousz* (Dutch, 1730–1799) showed that oxygen was produced by the green parts of plants when exposed to light. Furthermore, he observed that the amount of oxygen produced varied according to the amount of light to which the plant was exposed. In 1804 *Nicholas de Saussure* (Swiss, 1767–1845) found that the increase in the carbon content of plants resulted from the accumulation of carbon from carbon dioxide in the air. De Saussure also showed that leaves respire in darkness—that is, they take in oxygen and release carbon dioxide.

The studies of *Julius Sachs* in the mid-nineteenth century showed that chlorophyll was confined to the chloroplasts and was not distributed throughout the plant cell. He also showed that sunlight caused chloroplasts to absorb carbon dioxide and that chlorophyll is formed in chloroplasts only in the presence of light. Sachs also noted that one of the products of photosynthesis was starch. It was not until 1918 that Wilstätter and Stoll isolated and characterized the green pigments chlorophyll *a* and *b*.

On the basis of these early findings, the reactions of photosynthesis were described by the equation

$$6\ CO_2 + 6\ H_2O \xrightarrow[\text{chlorophyll}]{\text{sunlight}} (C_6H_{12}O_6)_n + 6\ O_2$$

However, photosynthesis is not simply the fusing of carbon dioxide and water molecules through the use of light energy. Instead, two complex series of chemical reactions are involved. One set of reactions, called the **photochemical** or **light reactions,** occurs in the lamellae of the chloroplasts. In these reactions, light energy is absorbed and used to form ATP, and water molecules are split, releasing oxygen and hydrogen. The hydrogen is used in the reduction of $NADP^+$ and is not evolved as a gas. The second set of reactions, called the **synthetic** or **dark reactions,** occurs in the stroma of the chloroplast, and although these reactions do not require light, they do depend on the availability of ATP and NADPH from the photochemical reactions in order to reduce the carbon dioxide to form sugars.

Photosynthesis—The Photochemical (Light) Reactions

The Absorption of Light by Chlorophyll

The absorption of electromagnetic energy by any atom or molecule often involves a shift of electrons from one atomic orbital to another. Each electron possesses energy, and the amount is determined by the location of the electron orbital in space and the speed at which the electron moves. When an atom absorbs light energy, an electron is either raised to an orbital of higher energy level or accelerated in its orbit. In either case, certain discrete quantities of energy are required, for when light photons have either too much or too little energy, they are not absorbed. Electrons may orbit in pairs within an orbital, the members of the same orbital spinning in *opposite* directions. Most atoms at their lowest energy level (i.e., **ground state**) have all their electrons paired in this fashion and are said to be in the **singlet** (i.e., *S*) state. When a photon of light is absorbed and an electron is thereby raised to a higher, unoccupied orbital, it may continue to spin in the direction opposite its former partner (in which case, the atom is still in the *S* state), or it may spin in the same direction as its former partner (in which case the atom is said to be in the **triplet** or *T* state).

In a molecule, some electrons orbit exclusively about specific atomic nuclei; others may be shared between two nuclei forming a bond (called localized or π^* electrons), or may orbit about several nuclei (called delocalized or π electrons). The absorption of a light photon may move a π electron to a π^* position.

Chlorophyll is normally a singlet in its ground state.

Although the absorption of light causes some π electrons to be raised to a π* orbital, chlorophyll remains in a singlet state. When red (680 nm) light is absorbed, an electron is raised to a higher, $S\pi^*$, orbital. Blue (430 nm) light possesses more energy per photon than red light, and its absorption raises an electron to a higher (but still $S\pi^*$) energy state. These transitions are summarized in Figure 17–11.

Once a molecule has absorbed light energy and is in an excited state, the ground state may be reestablished in three different ways: (1) the energy may be reemitted in the form of radiation of longer wavelength (i.e., fluorescence); (2) the energy may be converted to heat; and (3) the excited molecule may transfer its excess energy to another molecule. The transfer of energy from one molecule to another often involves the exchange of a high-energy electron for one of lower energy. Referring to Figure 17–11, the blue light absorbed by chlorophyll raises an electron to the $S^b\pi^*$ state. By losing energy in the form of heat, that electron could "drop" back to the $S^a\pi^*$ state. The electron could then drop back from the $S^a\pi^*$ state to the $S\pi$ state (i.e., the ground state) by the immediate loss of energy as heat or fluorescence. Another possibility also exists, for the electron could drop from the $S^a\pi^*$ state to a triplet (i.e., $T\pi^*$) state by heat loss and then to the ground state by either phosphorescence (delayed fluorescence) or heat loss. Concentrated solutions of extracted chlorophyll will strongly fluoresce red when placed in a beam of sunlight or UVL. Finally, it should also be noted that if the chlorophyll molecule is sufficiently excited, it may not return to the ground state by heat loss or reradiation; instead, the excited electron may be transferred to another molecule, temporarily leaving the chlorophyll in an oxidized state.

Primary Photochemical Events in Photosynthesis

As shown in Figure 17–11, chlorophyll absorbs both blue and red light, and this raises electrons to $S^b\pi^*$ or $S^a\pi^*$ states. The return of an electron from the $S^b\pi^*$ state to the $S^a\pi^*$ state is extremely fast (about 10^{-12} seconds) and does not afford an opportunity for the energy to be lost by fluorescence or by transfer to another molecule. Consequently, the energy is lost as heat. The decay of an electron from the $S^a\pi^*$ state, *does* permit the transfer of energy to another molecule, and this is the event that initiates photosynthesis. Consequently, a photon of red light is just as

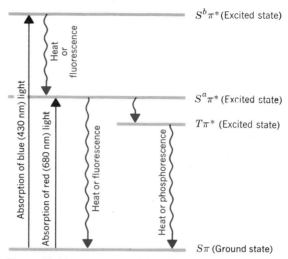

Figure 17–11
Main electronic states in the excitement of chlorophyll by red and blue light.

effective in initiating photosynthesis as a photon of blue light, even though the former is much less energetic.

As we have already noted, the chloroplast contains many different pigment molecules (e.g., chlorophylls, carotenoids, phycobilins, etc.) and the electrons of these molecules may be excited to various energy states by the absorption of light. As these excited accessory pigment molecules return to the ground state, the resulting energy is transferred to chlorophyll *a* molecules, causing their excitation. Since the chlorophyll *a* molecules present in a thylakoid vary in their absorption maxima, varying quantities of energy are required to raise their electrons to the $S^a\pi^*$ state. The molecule that requires the least energy is postulated to be a pigment that absorbs long, red wavelengths—namely the P_{700} molecule. It seems reasonable that light energy captured by the accessory pigments and transferred to chlorophyll *a* is in turn transferred from the latter to P_{700}. Each accessory pigment or chlorophyll *a* molecule can pass its energy on only to a pigment having an absorption maximum of longer wavelength because these require less energy to be activated to the $S^a\pi^*$ state. Since P_{700} has its absorption maximum at the longest wavelength, it serves as the final energy trap in a part of the photosynthetic unit called the *reaction center*.

Two Photosystems

When some plants are exposed to light containing only wavelengths of 690 nm or longer, photosynthetic efficiency decreases. The effect is called the *red drop* (Fig. 17–12). Since the absorbed energy is funneled to P_{700}, it would be expected that the absorption of light by the accessory pigments, chlorophyll *a* or even P_{700}, should be equally efficient. The efficiency can be increased through the addition of shorter wavelength radiations. This *enhancement* phenomenon can increase the photosynthetic rate 30 to 40% above the rate obtained by either the short wavelength or long wavelength alone. The synergistic effect of the two different wavelengths led early investigators to conclude that two distinct photochemical reaction systems exist. In one of these (called *photosystem I*, PS-I), P_{700} serves as the ultimate energy trap; in the other (called *photosystem II*, PS-II), the ultimate energy trap is a chlorophyll that has its absorption maximum at 680 nm and is called P_{680}.

In higher plants, photosystem I is a unit containing several hundred molecules of chlorophyll (mostly type *a*), about 50 carotenoids, one cytochrome *f*, one plastocyanin, two cytochrome b_{564}, one or two ferredoxin molecules, and one molecule or a dimer of chlorophyll P_{700}. Photosystem II has about 200 molecules of chlorophyll of both *a* and *b* types absorbing at less than 680 nm, 50 carotenoids, one P_{680} molecule or dimer (the primary electron donor), a primary electron acceptor that is believed to be a quinone, four plastoquinones, six Mn atoms, and two cytochrome b_{559} molecules.

Sequence of Energy (Electron) Flow

The absorption of light energy by chlorophyll P_{700} or P_{680} alters the state of the orbiting electrons, and if the energy is not lost by reradiation or heat, the excited electrons can be transferred to another compound. Such an electron loss "bleaches" the chlorophyll and leaves it in an oxidized state. Oxidized P_{700} may be reduced by accepting an electron from photosystem II, while the reduction of the oxidized chlorophyll of photosystem II is brought about by the oxidation of water.

In 1938 R. Hill demonstrated that isolated chloroplasts exposed to light could evolve oxygen and reduce a variety of compounds without consuming carbon dioxide. Three years later S. Ruben and M. Kamen, using the isotope ^{18}O,

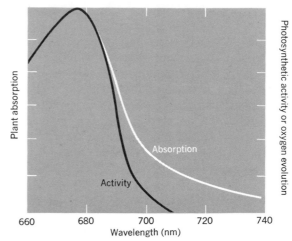

Figure 17–12
The photosynthetic activity or *action spectrum* (i.e., oxygen evolution spectrum) and *absorption spectrum* of the green alga *Ulva*. The difference between the two curves at wavelengths above 590 nm is called the *red drop*.

were able to show that the oxygen liberated during whole plant photosynthesis was derived from water molecules. That is,

$$6\ CO_2 + 12\ H_2{}^{18}O \xrightarrow{\text{light}} C_6H_{12}O_6 + 6\ {}^{18}O_2 + 6\ H_2O$$

The method by which the water molecules are split is unknown, although four water molecules are required for the evolution of one oxygen molecule and four quanta of light are necessary:

$$4\ H_2O \xrightarrow{4\ h\nu} O_2 + 4\ (H^+ + e^-) + 2\ H_2O$$

Protein-bound Mn^{++} and Cl^- may also be required. Some of these reactions are summarized in Figure 17–13.

Redox Reactions

Electrons that are released from chlorophyll P_{700} and P_{680} pass from one to another of a series of electron transfer molecules in the lamellae. Each one of these molecules has a specific electrical potential called a **redox potential** (see Chapter 9). The relative values of the redox potentials of the molecules involved in the transfer of electrons deter-

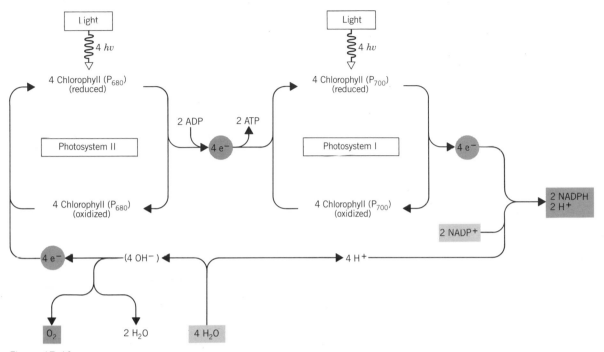

Figure 17–13

Absorption of light energy by photosystems I and II and the resulting movements of electrons. Electrons transferred from photosystem II to photosystem I are regained by splitting water molecules. Electrons from photosystem I are used together with H⁺ from water to reduce NADP⁺ (see text for additional details.)

mine the order in which each molecule participates in the transfer. During electron transfer, the molecules accept electrons from less positive (more negative) molecules and donate electrons to more positive (less negative) molecules. When a molecule accepts electrons, it is said to be *reduced,* and when electrons are given up, the molecule is said to be *oxidized.* The transfer of electrons following the absorption of light energy by chlorophyll therefore involves a sequence of oxidation-reduction reactions. This sequence is shown in Figure 17–14, which also identifies the intermediate electron acceptors and their redox potentials.

Photosystem I is located in the membranes of both the grana thylakoids and stroma thylakoids, whereas photosystem II is found only in the membranes of the grana thylakoids. Therefore, photosystem I functions independently in the stroma thylakoids but functions in conjunction with photosystem II in the grana thylakoids. Absorption of light

energy by P_{700} of photosystem I causes the loss of electrons to an intermediate iron-sulfur protein and then to ferredoxin itself (Fig. 17–14). Ferredoxin is reoxidized by the transfer of electrons to either of two compounds: *ferredoxin-NADP reductase* or cytochrome *b* (see below). Ferredoxin-NADP reductase transfers electrons to FAD, which also accepts H⁺ from water and reduces the NADP; that is,

$$NADP^+ + 2\,(H^+ + e^-) \longrightarrow NADPH + H^+$$

The NADPH and H⁺ spill into the stroma, where the NADPH is reoxidized in the dark reactions (described later).

In the alternative mechanism of ferredoxin reoxidation, cytochrome *b* accepts electrons from ferredoxin and initiates a sequence of redox reactions passing the electrons on to cytochrome *f,* plastocyanin, and ultimately back to P_{700},

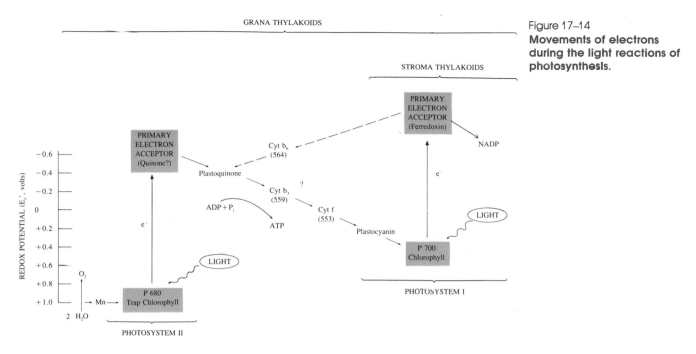

Figure 17–14
Movements of electrons during the light reactions of photosynthesis.

completing a cycle. The function of this cyclic set of redox reactions is coupled to the phosphorylation of ADP (*cyclic photophosphorylation,* described below).

In the grana lamellae, the P_{700} oxidized by the absorption of light can be reduced by the cyclic return of the electrons from ferredoxin or by the transfer of electrons from photosystem II. In photosystem II (Fig. 17–14), light energy is ultimately transferred to chlorophyll P_{680} by the accessory pigments or other chlorophyll molecules. The ejected electrons are absorbed by an electron acceptor, which is currently unidentified but could be a quinone. This photoevent initiates another set of redox reactions in which the electrons successively pass from the unidentified electron acceptor to plastoquinone, cytochrome b_{559}, the cytochrome *f,* plastocyanin, and ultimately P_{700}. The electrons in photosystem II do not cycle back to chlorophyll P_{680}, but are transferred by redox reactions to P_{700} and this is coupled to phosphorylation of ADP (*noncyclic photophosphorylation,* described below).

The reduction of the oxidized P_{680} chlorophyll is brought about by the oxidation of water via an enzyme system that is closely associated with the structure of the thylakoid and that also produces molecular oxygen. The enzyme system

presumably processes four protons simultaneously to produce O_2.

$$2\ H_2O \longrightarrow O_2 + 4\ H^+ + 4\ e^-$$

Mn^{++} is known to be a cofactor in the system. The electrons produced act to reduce the light-oxidized P_{680}, and the H^+ forms a pool available for the reduction of $NADP^+$.

Cyclic and Noncyclic Photophosphorylation

Phosphorylation of ADP occurs in chloroplasts during the light reactions and is called **photophosphorylation.** The photophosphorylation occurs in the lamellae of the stroma and grana thylakoids as a part of both photosystems I and II. There is a clear similarity between the mechanism of photophosphorylation in chloroplasts and electron transport system phosphorylation in mitochondria (Chapter 16). In both organelles, the mechanism is closely associated with a membrane and with a compartment separated from the rest of the organelle. In the chloroplast, the compartment is the space inside the thylakoids; in the mitochondrion, it is the matrix enclosed by the inner membrane. In both organelles, the phosphorylation that occurs is coupled to electron transport through a sequence of redox reactions. Several of the molecules that participate in the redox reactions are similar, including the quinones and cytochromes.

Cyclic photophosphorylation occurs when electrons released by P_{700} are shunted back to P_{700} through ferredoxin, cytochrome b, cytochrome f, and plastocyanin. The energy from these exergonic redox reactions is coupled to phosphorylation. **Noncyclic photophosphorylation** occurs when electrons released by P_{680} of photosystem II are shuttled via plastoquinone, cytochrome b, cytochrome f, and plastocyanin to P_{700} of photosystem I. The energy from these noncyclic exergonic redox reactions is coupled to phosphorylation.

At the present time, the mechanism of coupling is unknown, although there is evidence to support the concept that a proton gradient generated across the thylakoid membrane supports chemiosmotic coupling like that described in Chapter 16 for mitochondria and diagrammed in Figure 17–15. Chloroplasts illuminated in an unbuffered medium quickly cause the medium to become alkaline, implying that protons are transported from the medium into the thylakoid. In darkness the system reequilibrates. If the medium is made alkaline, phosphorylation of ADP occurs in the dark. Compounds that degrade the thylakoid membrane (such as detergents) also cause leakage of protons from the thylakoid and prevent phosphorylation.

The standard free energy change that takes place when ATP is formed from ADP and P_i is about 30.5 kJ/mole (or 7.3 kcal/mole); this is equivalent to a redox potential of about 0.45 to 0.61 volts. Most of the measurements that have been made indicate that one molecule of ATP is produced for each pair of electrons moved from photosystem II to photosystem I (i.e., noncyclic photophosphorylation). Several recent studies strongly suggest that more than one molecule of ATP is produced by the reactions that begin with the splitting of water (i.e., **photolysis**) and end with the reduction of NADP$^+$. The second ADP phosphorylation is believed to be coupled to the photolysis. If the number of ATP molecules produced during the light reactions is to balance the number consumed by the dark reactions, then three ATP must be produced for each pair of electrons transported from H_2O to NADP$^+$. The third ATP may be generated by cyclic photophosphorylation.

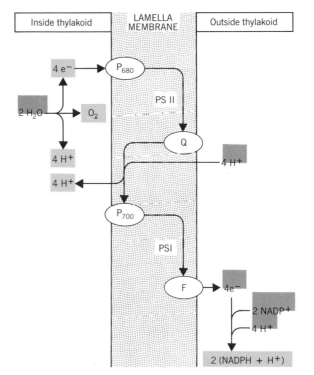

Figure 17–15
Chemiosmotic coupling model for the accumulation of protons inside thylakoids in the presence of light. *Q*, electron acceptor of photosystem II (*PS II*); *F*, electron acceptor of *PS I*.

Another light-induced phosphorylation of ADP to form ATP called *pseudocyclic photophosphorylation* occurs when ferredoxin reduced in photosystem I transfers electrons through a series of reactions to O_2 (not to NADP$^+$ or P_{700}), with the result that water and ATP are formed.

Summary of the Light Reactions

Two photosystems function during the light reactions of photosynthesis. As each system absorbs four quanta of light energy, chlorophyll P_{680} (system II) and P_{700} (system I) are activated, passing two pairs of electrons to acceptor molecules. Chlorophyll P_{680} is then returned to its reduced state by the oxidation of two water molecules, and in the process one molecule of oxygen is evolved. P_{700} is returned to its reduced state by the flow of electrons from photosystem II through intermediate oxidation-reduction compounds. The transfer of two pairs of electrons from water through photosystem II to photosystem I causes the noncyclic photophosphorylation of two ADP molecules producing two ATP and two H_2O. The two pairs of electrons released from photosystem I reduce two $NADP^+$ to two NADPH ($+ H^+$). Therefore, the net result is

$$2 H_2O + 2 NADP^+ + (2 ADP + 2 P_i)$$

$$\downarrow 8 \text{ quanta}$$

$$O_2 + 2 NADPH + H^+ + (2 ATP + 2 H_2O)$$

The two ATP and two NADPH molecules produced by the light reactions occurring in the grana lamellae are used in the synthetic (dark) reactions that take place in the stroma. The latter reactions fix CO_2 into sugars. For each CO_2 fixed, two NADPH (and two H^+) and three ATP molecules are required (see below). Where the third ATP molecule is formed is not yet clear; it may be obtained from cyclic photophosphorylation or pseudocyclic photophosphorylation.

Photosynthesis—Synthetic (Dark) Reactions

The Calvin Cycle

The elucidation of the sequence of chemical reactions that result in the incorporation of carbon dioxide into sugars and starches relied heavily on the use of radioactive isotopes. Using ^{14}C-labeled carbon dioxide, it was possible to add $^{14}CO_2$ at known times to an actively photosynthesizing system, halt the process a short time later, and then identify the compounds into which the labeled carbon dioxide had become incorporated. Identification of the ^{14}C-containing intermediates was carried out using combined paper chromatography (Chapter 13) and autoradiography (Chapter 14).

Ruben and co-workers first showed that the active form of CO_2 in the chloroplast was carbonic acid. M. Calvin and co-workers established the sequence of reactions that follow the formation of carbonic acid and its entry into the chloroplast. They added $^{14}CO_2$ to cultures of the alga *Chlorella* and allowed the cells to photosynthesize for given periods

of time (usually between 2 and 60 seconds). The *Chlorella* cells were then killed and the soluble cell components extracted and concentrated. The extracts containing radioactive carbon were chromatographed on paper, and the spots containing radioactivity were identified. In 1961, Calvin received the Nobel Prize for this most important series of experiments.

When photosynthesis in the presence of $^{14}CO_2$ was allowed to proceed for only 2 seconds, the major labeled compound identified was 3-phosphoglyceric acid (PGA). After 7 seconds, sugar phosphates and diphosphates were found in addition to PGA. A 60-second exposure to $^{14}CO_2$ produced labeled phosphoenolpyruvic acid (PEP), carboxylic acids, and amino acids. Using many different time intervals, the entire sequence of reactions was uncovered, and it was found that many of the steps were the reverse of those in the glycolytic pathway (Fig. 17–16).

In the chloroplast stroma, CO_2 in the form of carbonic acid reacts with the sugar ribulose diphosphate (RuDP) to form an unstable 6-carbon compound that immediately splits to form two molecules of 3-phosphoglyceric acid (PGA). The enzyme catalyzing this reaction is *ribulose-1,5-diphosphate carboxylase* and the radioactive carbon of $^{14}CO_2$ is incorporated into the carboxyl group of PGA.

PGA is then reduced to 3-phosphoglyceraldehyde (PGAL) in two steps. First, each PGA is phosphorylated by ATP and then reduced by NADPH. The ATP and NADPH were produced by photochemical reactions in the grana lamellae. Thus, for each molecule of CO_2 fixed and converted to PGAL, two ATP and two NADPH molecules are required. These reactions are catalyzed by a kinase and dehydrogenase.

Some of the PGAL is isomerized by *triose phosphate isomerase* to form dihydroxyacetone phosphate (DHAP). The enzyme *aldolase* then condenses PGAL and DHAP to produce fructose-1,6-diphosphate (FDP). *Fructose diphosphatase* splits off the phosphate group of the first carbon atom, producing fructose-6-phosphate (F6P). F6P may then be converted to fructose, glucose, or starch.

F6P and PGAL are also used for the resynthesis of RuDP (Fig. 17–17). F6P and PGAL are converted to erythrose-4-phosphate (E4P) and xylulose-5-phosphate (X5P). An aldolase then catalyzes the condensation E4P and DHAP to form sedoheptulose-1,7-diphosphate (SDP), which is then converted to sedoheptulose-7-phosphate (S7P). S7P and PGAL also react to form ribose-5-phosphate (R5P) and

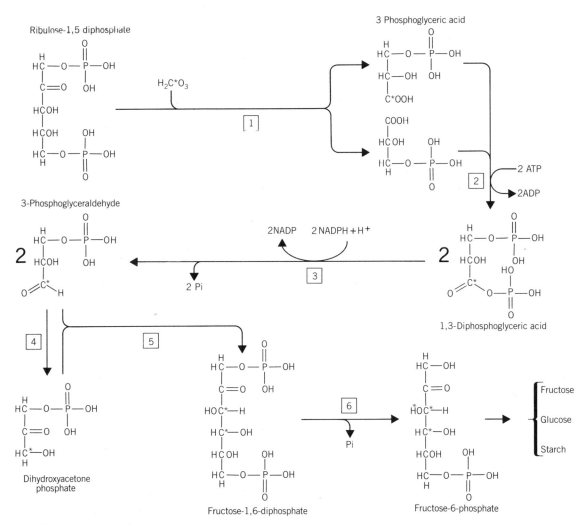

Figure 17–16
**Initial steps in the fixation of CO₂ during photosynthesis. C* = ¹⁴C, which
traces the path of carbon through the reaction sequence. Enzymes: (1) RDP-
carboxylase or carboxydismutase; (2) phosphoglyceric acid kinase; (3)
triose phosphate dehydrogenase; (4) isomerase; (5) fructose diphosphate
aldolase; (6) fructose diphosphatase.**

X5P. The R5P is then isomerized to form ribulose-5-phosphate (Ru5P). Ru5P is also formed from X5P. Finally, ATP phosphorylates Ru5P to form RuDP.

For each CO_2 fixed, one RuDP, two ATP, and two NADPH are consumed and one phosphate sugar is produced. In actuality, all the RuDP is resynthesized. Fixation of every three molecules of CO_2 results in the formation of six PGAL. Five of these are recycled to replenish the pool

of RuDP. The extra PGAL is the photosynthetic product and is used for sugar and starch synthesis. Figure 17–18 summarizes all the steps and the requisite numbers of molecules participating in the dark reactions; the pathway is known as the *Calvin cycle*.

Since additional ATP is consumed in the formation of RuDP, a total of nine ATP and six NADPH are required for the fixation of three CO_2 molecules. Since six CO_2

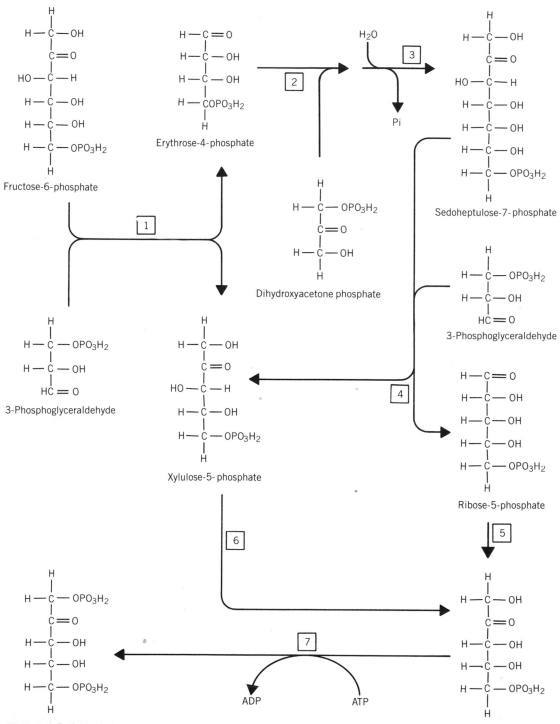

Figure 17–17

Resynthesis of ribulose-1, 5-diphosphate. Enzymes: (1) transketolase; (2) aldolase; (3) sedoheptulose-1,7-diphosphatase; (4) transketolase; (5) isomerase; (6) epimerase; (7) kinase.

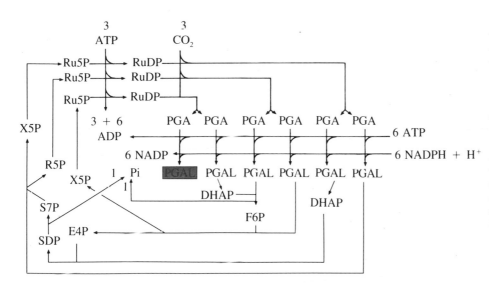

Figure 17–18
The path of carbon in photosynthesis. The molecule of PGAL in the box is the photosynthetic product and is employed for sugar and starch synthesis. Ru5P, ribulose-5-phosphate; RuDP, ribulose-1,5-diphosphate; PGA, 3-phosphoglyceric acid; PGAL, 3-phosphoglyceraldehyde; DHAP, dihydroxyacetone phosphate; F6P, fructose-6-phosphate; SDP, sedoheptulose-1,7-diphosphate; S7P, sedoheptulose-7-phosphate; E4P, erythrose-4-phosphate; X5P, xylulose-5-phosphate; R5P, ribose-5-phosphate; P_i, inorganic phosphate.

molecules are required to produce one 6-carbon molecule, 18 ATP molecules are consumed in the fixation (or three ATP/CO_2).

It has been estimated that about 5×10^{16} g of carbon are fixed annually by photosynthesis and this corresponds to a storage of 20.1×10^{17} kJ or 4.8×10^{17} kcal of energy. Since about 28.0×10^{21} kJ or 6.7×10^{21} kcal of light energy fall on the earth each year, photosynthesis traps a mere 0.0072%.

C₄-Photosynthesis (Hatch-Slack Pathway)

The C_3 or Calvin cycle is not the only metabolic pathway found in plants for fixing CO_2. In 1967, M. D. Hatch and C. R. Slack showed that in some plants CO_2 is also fixed into 4-carbon compounds. This mechanism (which operates *in conjunction* with the Calvin cycle in the leaves of these plants) has since come to be known as the **Hatch-Slack** or **C₄-pathway.** Plants that employ the Hatch-Slack pathway (e.g., corn, sugar cane, and certain grasses) are characteristically found in arid environments and possess an unusual leaf anatomy (Fig. 17–1*b*). In the leaves of plants employing only the Calvin cycle, the C_4-reactions occur in both the palisade mesophyll and spongy mesophyll layers. However, in plants that can also carry out the C_4-reactions, the Calvin cycle occurs only in the layer of **bundle sheath cells** that surrounds the veins of the leaf. The C_4-reactions take place in the **mesophyll** layers that lie immediately

under the upper and lower epidermis. In addition to their unusual leaf anatomy, plants using the Hatch-Slack pathway are characterized by high photosynthetic and growth rates, low photorespiration rates, and low rates of water loss through the stomates of the leaves.

Like the Calvin cycle, the sequence of reactions that constitutes the C_4-pathway (and the relationship of these reactions to the Calvin cycle) was worked out using ^{14}C-labeled carbon dioxide. The reactions are summarized in Figure 17–19. Carbon dioxide passes from the air surrounding the leaves through the epidermis and into the mesophyll cells. Here the CO_2 condenses with phosphoenolpyruvate (PEP, a 3-carbon compound) to form oxaloacetate (a 4-carbon compound). The reaction is catalyzed by the enzyme *phosphoenolpyruvate carboxylase*. In some plant species, the oxaloacetate is then converted to malate, while in others it is converted to aspartate. The malate (or aspartate) then passes into the bundle sheath cells where it is decarboxylated within the cells' chloroplasts. As seen in Fig. 17–19, the CO_2 that is released contains the same carbon atom as the CO_2 that entered the mesophyll cells. This carbon dioxide then enters the Calvin cycle reactions of the bundle sheath cells. The 3-carbon compounds formed after decarboxylating malate or aspartate are transported back into the mesophyll, where they are reconverted to PEP. The sugars that accumulate from the Calvin cycle reactions are temporarily converted to starch under active photosynthetic conditions. At night or during darkness or dim light, the

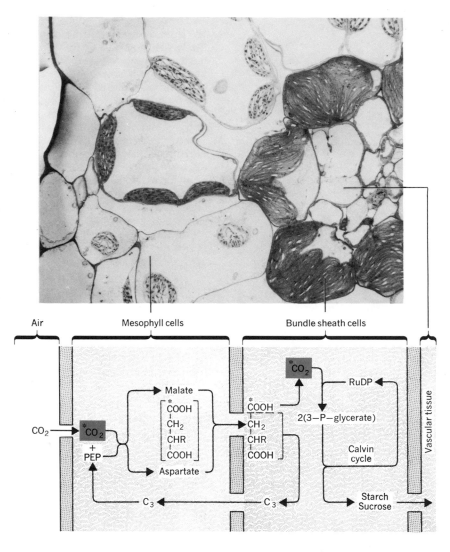

Figure 17–19
Electron photomicrograph of bundle sheath cells and adjacent mesophyll cells. (Courtesy of Dr. C. C. Black, *Plant Physiol. 47*, 15 (1971)). The diagram shows the C_4 photosynthetic reactions associated with each type of cell.

starches are converted back to sugars and transported out of the leaf by the vascular tissue of the veins (vascular bundles).

The presence of the C_4 pathway may seem like a needless addition to the Calvin cycle system. However, there is evidence indicating that the mesophyll is able to build up a high concentration of fixed CO_2 by this method, which could provide evolutionary advantages. In addition, the rapid and efficient fixing and storing of CO_2 as 4-carbon acids decreases the leaf's need to have a large number of stomates (openings in the leaf epidermis that allow CO_2 and other gases to diffuse into the leaf). Open stomates, while allowing the passage of CO_2 into the plant, also allow water to escape from the plant—a disadvantage to plants in arid climates!

Crassulacean Acid Metabolism

Crassulacean acid metabolism (CAM) is a special form of metabolism associated with photosynthesis that is carried out by members of the plant family *Crassulaceae* (succulent herbs such as *Sedum*). Plants carrying out this form of metabolism have closed stomates during the daylight hours and therefore cannot absorb sufficient CO_2 for photosynthesis. But during the night (dark) hours, the stomates open and the leaf cells can fix CO_2 in the dark by combining it with PEP to form oxaloacetate. The oxaloacetate is converted into malate for storage. During the following daylight hours, the malate is decarboxylated, and the CO_2 is utilized by the Calvin cycle reactions. The 3-carbon pyruvate remaining after decarboxylation is converted first into PEP and then into phosphoglyceric acid and is utilized by the Calvin cycle as well. The PEP required for the dark fixation of CO_2 is derived from some of the starch produced from the Calvin cycle products (Fig. 17–20).

Bacterial Photosynthesis

Photosynthesis in procaryotic organisms occurs in lamellar membrane systems called **chromatophores.** The chromatophores contain the pigments for the photochemical reactions but none of the subsequent biosynthetic enzymes. The pigment system includes the chlorophylls, carotenoids, and in some cases phycobilins. However, in bacteria, **bacteriochlorophyll** is the ultimate light-trapping molecule (not chlorophyll *a*).

The most important distinction between plant and bacterial photosynthesis is that water is not used as the reducing agent and oxygen is not an end product. The power to reduce CO_2 may come from molecular hydrogen, H_2S, or organic compounds. Two major groups of bacteria that carry out photosynthesis are the green and purple sulfur bacteria; these organisms utilize H_2S and produce sulfur and sulfate; that is

$$6\ CO_2\ +\ 12\ H_2S \xrightarrow[\text{bacteriochlorophyll}]{\text{light energy}}$$

$$C_6H_{12}O_6\ +\ 6\ H_2O\ +\ 12S$$

During photosynthesis, sulfur accumulates as granules of elemental sulfur and may be further metabolized later.

Nonsulfur purple bacteria use organic compounds such as acetic acid as electron donors. The acetic acid is anaerobically oxidized via the Krebs cycle reactions (Chapter 10). Acetic acid can also be reduced to hydroxybutyric acid. Certain members of the sulfur and nonsulfur purple

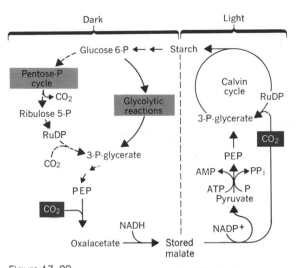

Figure 17–20

The major reaction steps of the light and dark phases of crassulacean acid metabolism.

bacteria can use molecular hydrogen to reduce either CO_2 or acetic acid; that is,

$$6\ CO_2\ +\ 12\ H_2 \longrightarrow C_6H_{12}O_6\ +\ 6\ H_2O$$

and

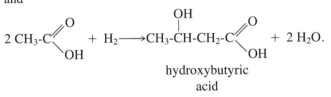

hydroxybutyric acid

Other Plastids

The chloroplast is only one of several different plastids found in plant cells. Other plastids such as **etioplasts, amyloplasts,** and **chromoplasts** have different structures (Fig. 17–21) and functions. They are all called plastids because they appear to develop from a common structure or from one another.

Proplastids are small, generally colorless structures found in young or dividing cells. They have little internal structure but are delimited by a double membrane. Proplastids give rise to other types of plastids.

Etioplasts are prevalent in the leaves of plants grown in

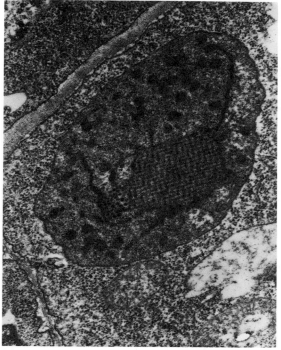
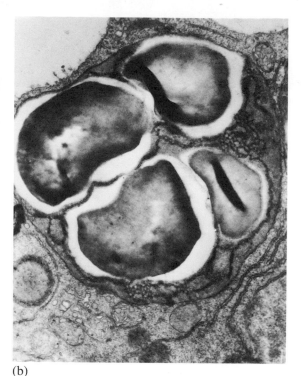

(a)   (b)

Figure 17–21

(a) Etioplast **in a cell of a bean plant grown in darkness. Note the prominent** *prolamellar body* **inside the organelle. Magnification, 29000×.** *(b) Amyloplast* **in cotyledon cell of a spinach plant. The organelle contains prolamellar bodies at the margins of starch granules. Magnification, 16000×. (Photomicrographs courtesy of Dr. J. V. Possingham; copyright © 1976,** *Plant Physiol. 57,* **41.)**

the dark. Their ellipsoidal and sometimes irregular structure is also bounded by a double membrane. Internally, etioplasts contain one or more paracrystalline *prolamellar bodies* (Fig. 17–21*a*) and a number of flattened vesicles called *primary thylakoids*. Etioplasts develop into functional chloroplasts upon exposure to light.

The outer membrane of the **amyloplast** encloses the stroma, containing one to eight starch granules. In certain plant tissues such as the potato tuber, the starch granules within the amyloplasts may become so large that they rupture the encasing membrane. Starch granules of amyloplasts are typically composed of concentric layers of starch.

Chromoplasts contain carotenoids and are responsible for imparting color (yellow, orange, red) to certain portions of plants such as flower petals, fruit, and some roots. The chromoplasts of carrots contain large quantities of lipid that reduces their overall density to less than 1 g/ml; consequently, during centrifugation of root homogenates, the amyloplasts rise to the surface of the tube. Chromoplast structure is quite diverse; they may be round, ellipsoidal, or even needle-shaped, and the carotenoids that they contain may be localized in droplets or in crystalline structures. The function of the chromoplast is not clear, but in many cases (e.g., flowers and fruit), the color that they produce probably plays a role in attracting insects and other animals for pollination or seed dispersal.

A number of other, less frequently occurring plastids have been described, such as the oil-filled **elaioplasts** including the sterol-rich **sterinochloroplasts,** and the protein-containing **proteoplasts.**

Summary

In eucaryotes, **photosynthesis,** like the mitochondrial reactions, is concerned with the formation of ATP and involves hydrogen and electron transport in compounds like NADPH and cytochromes. The two processes differ in that photosynthesis uses light rather than chemical substrates as the source of energy, CO_2 and water are consumed rather than produced, and O_2 and carbohydrate are produced rather than consumed. The overall reaction

$$6\ CO_2\ +\ 12\ H_2O\ \xrightarrow{\text{light}}(C_6H_{12}O_6)_n\ +\ 6\ O_2\ +\ 6\ H_2O$$

can be broken down into a light phase (in which **photolysis** of water occurs) and a dark phase (in which CO_2 fixation occurs). In the light phase, visible light is absorbed by chlorophyll or a variety of other pigments located in the membranous **thylakoids** of the chloroplast. The light energy excites the molecules, inducing them to reemit light or heat or transfer the energy to **chlorophyll** P_{700} or P_{680} molecules. The activation of chlorophyll P_{680} triggers the reactions of **photosystem II,** which generate ATP by the process called **noncyclic photophosphorylation** and terminate with the reduction of P_{700}. Absorption of light energy directly activates **photosystem I.** This photosystem can also produce ATP (by the process of **cyclic photophosphorylation**) or the reduction of $NADP^+$ to NADPH. The ATP and NADPH transported into the **stroma** of the chloroplast are consumed in the dark reactions.

During the dark reactions of C_3-plants, CO_2 is fixed by binding to ribulose diphosphate and is subsequently reduced by NADPH. ATP acts as the source of energy for these endergonic reactions. The final product is carbohydrate, usually in the form of a sugar, which may be stored as starch. In C_4-plants, CO_2 is first fixed as a 4-carbon acid in the mesophyll cell chloroplasts and then transferred to the bundle sheath cells where the **Calvin cycle** reactions occur and starch is stored.

Chloroplasts are found only in eucaryotic plant cells. They form from proplastids, which may be responsible for the formation of other plastids, such as **chromoplasts** and **etioplasts,** or they form by a division of preexisting chloroplasts. Plastids associated with C_3 photosynthesis have inner membranes arranged in layers and organized into **grana.** Plastids associated with C_4 photosynthesis may have a similar structure but frequently lack grana.

References and Suggested Reading

Articles and Reviews

Arnon, D. I., The role of light in photosynthesis. *Sci. Am. 203*(5), 104 (Nov. 1960).

Bassham, J. A., The path of carbon in photosynthesis. *Sci. Am. 206*(6), 88 (June 1962).

Bogorad, L., Chloroplasts. *J. Cell Biology, 91,* 256s (1981).

Calvin, M., The path of carbon in photosynthesis. *Science 135,* 879 (1962).

Calvin, M., and Androes, G. M., Primary quantum conversion in photosynthesis. *Science 138,* 867 (1962).

Cheniae, G. M., Photosystem II and O_2 evolution. *Annu. Rev. Plant Physiol. 21,* 467 (1970).

Ellis, R. J., Chloroplast proteins. *Annu. Rev. Plant Physiol. 32,* 111 (1981).

Hatch, M. D., and Slack, C. R., Photosynthetic CO_2-fixation pathways. *Annu. Rev. Plant Physiol. 21,* 141 (1970).

Heber, U., and Heldt, H. W., The chloroplast envelope. *Annu. Rev. Plant Physiol. 32,* 139 (1981).

Jope, C. A., Atchison, B. A., and Pringle, R. C., A computer analysis of a spiral, string-of-grana model of the three-dimensional structure of chloroplasts. *Bot. Gaz. 141,* 37 (1980).

Oquist, G., Samuelsson, G., and Bishop, N. I., On the role of β carotene in photosystem II. *Physiol. Plant 50,* 63 (1980).

Rabinowitch, E. I., and Govindjee, The role of chlorophyll in photosynthesis. *Sci. Am. 213*(1), 74 (July 1965).

Shavit, N., Energy transduction in chloroplasts. *Annu. Rev. Biochem. 49,* 111 (1980).

Stacy-French, C., Fifty years of photosynthesis. *Annu. Rev. Plant Physiol. 30,* 1 (1979).

Stainer, R. Y., Photosynthetic mechanisms in bacteria and plants; development of a unitary concept. *Bacteriol. Rev. 25,* 1 (1961).

von Wettstein, D., Genetics and submicroscopic cytology of plastids. *Hereditas 43,* 303 (1957).

Whatley, J. M., and Whatley, F. R., Chloroplast evolution. *New Phytologist 87,* 233 (1981).

Wildman, S. G., Jope, C. A., and Atchison, B. A., Light microscopic analysis of the three dimensional structure of higher plant chloroplasts. *Bot. Gaz. 141,* 24 (1980).

Books, Monographs, and Symposia

Bonner, J., and Varner, J. E., *Plant Biochemistry* (3rd ed.), Academic Press, New York, 1976.

Clayton, R. K., *Molecular Physics in Photosynthesis,* Blaisdell, New York, 1965.

Graber, P. Phosphorylation in chloroplasts, in *Current Topics in Membranes and Transport* (A. Kleinzeller and F. Bonner, Eds.) Academic Press, New York, 1982.

Hall, D. O., and Whatley, F. R., The chloroplast, in *Enzyme Cytology* (D. B. Roodyn, Ed.), Academic Press, London, 1967.

Hatch, M. D., Osmond, C. B., and Slatyer, R. O., *Photosynthesis and Photorespiration,* Wiley-Interscience, New York, 1971.

Junge, W., Electrogenic reactions and proton pumping in green plant photosynthesis, in *Current Topics in Membranes and Transport* (A. Kleinzeller and F. Bonner, Eds.) Academic Press, New York, 1982.

Kok, B., Photosynthesis: The path of energy, in *Plant Biochemistry* (J. Bonner and J. E. Varner, Eds.), Academic Press, New York, 1965.

Mahler, H. R., and Cordes, E. H., *Biological Chemistry,* Harper & Row, New York, 1966.

Margulis, L., *Symbiosis in Cell Evolution,* W. H. Freeman, San Francisco, 1981.

Nobel, P. S., *Plant Cell Physiology,* W. H. Freeman, San Francisco, 1970.

Nobel, P. S., *Biophysical Plant Physiology,* W. H. Freeman, San Francisco, 1974.

Novikoff, A. B., and Holtzman, E., *Cells and Organelles,* Holt, Rinehart and Winston, New York, 1970.

Park, R. B., The chloroplast, in *Plant Biochemistry* (J. Bonner and J. E. Varner, Eds.), Academic Press, New York, 1970.

Price, C. A., *Molecular Approaches to Plant Physiology,* McGraw-Hill, New York, 1970.

Rabinowitch, E., and Govindjee, *Photosynthesis,* Wiley, New York, 1969.

Reinert, J. (Ed.), *Chloroplasts,* Springer-Verlag, Berlin, 1980.

Chapter 18
THE GOLGI APPARATUS

The **Golgi apparatus** or **Golgi bodies** of eucaryotic cells are organelles that play a variety of functions including (1) the packaging of secretory materials that are to be discharged from the cell, (2) the **processing** of proteins (e.g., **glycosylation, phosphorylation, sulfation,** and **selective proteolysis**), (3) the synthesis of certain polysaccharides and glycolipids, (4) the sorting of proteins destined for various locations in the cell, and (5) the proliferation of membranous elements for the plasma membrane.

The Golgi apparatus was first described in 1898 by C. Golgi, a pioneer of cytology and cytochemistry (Fig. 18–1). Although Golgi referred to the structure as the "internal reticular apparatus" of the cell, the organelle was later renamed in his honor. Golgi received the Nobel Prize in 1906 for his many cytological findings and his cytochemical and histochemical innovations. For many decades after its original description, the Golgi apparatus remained a controversial cell structure because it was not readily identifiable in all cells. To visualize these organelles, Golgi employed stains containing silver, osmium, and other heavy metals, and it was believed by many other cytologists that the organelle was an *artifact* produced by precipitation of the metal within the cell. The controversy was not put to rest until the 1950s when the existence of the Golgi apparatus was confirmed by electron microscopic studies; these studies also shed light on the details of the apparatus' organization (Fig. 18–2).

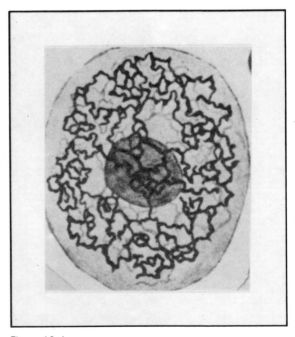

Figure 18–1
The "internal reticular apparatus" of the cell as first depicted by Camillo Golgi in 1898.

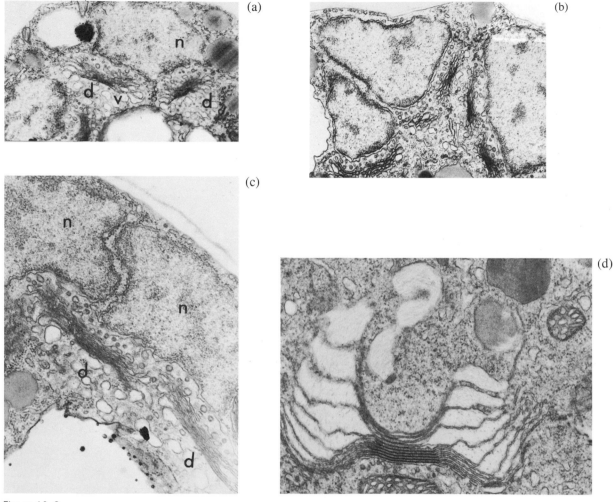

Figure 18–2

Appearance of the Golgi apparatus in transmission electron photomicrographs of thin sections. *d*, dictyosome (Golgi body); *n*, nucleus; *v*, vesicles. (Micrographs *a*, *b*, and *c* courtesy of Drs. J. DiOrio and W. F. Millington, copyright © 1978 by Springer-Verlag, *Protoplasma 97*, 329; *d* courtesy of Dr. R. Wetherbee, copyright © 1978 by Springer-Verlag, *Protoplasma 95*, 347.)

Structure of the Golgi Apparatus

The contemporary model of the structure of the Golgi apparatus is based upon some 25 years of transmission electron microscopic study; the photomicrographs reproduced in Fig. 18–2 show the typical appearance of the organelle. Although Golgi bodies differ somewhat in organization from one type of tissue or cell to another, they character-istically take the form of a stack of flattened, oval cisternae surrounded at the circumference and above and below by vesicles and tubular structures (Fig. 18–3). Golgi bodies are also referred to as **dictyosomes,** which means "stack-like bodies." In its strictest sense, the dictyosome does not include the vesicles that fuse with or are discharged from the cisternae. In most Golgi bodies, the number of cisternae is less than 10, the lumen of each cisterna varying in width

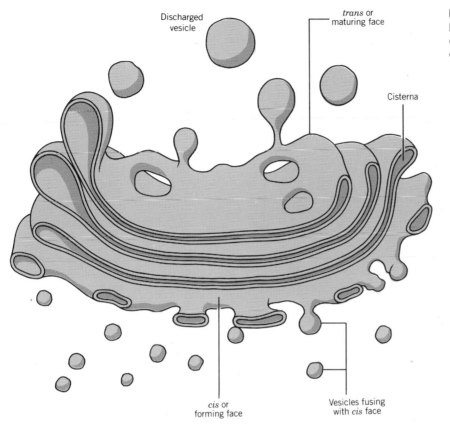

Discharged
vesicle

trans or
maturing face

Cisterna

cis or
forming face

Vesicles fusing
with *cis* face

Figure 18–3
Diagrammatic reconstruction of Golgi apparatus based upon electron microscopic studies.

from about 500 to 1000 nm. The margins of each cisterna are gently curved so that the entire Golgi body takes on a cuplike or bowl-like appearance. The cisterna at the *convex* end of the dictyosome comprises what is called the **forming face** or **cis face** while the cisterna at the concave end comprises the **maturing face** or **trans face.** The small vesicles that are adjacent to the *cis* face are believed to fuse with and contribute additional structure to the Golgi body. The vesicles near the *trans* face are larger and are believed to be formed from the uppermost cisterna. Small vesicles may also be discharged from the margins of the cisternae between the *cis* and *trans* faces. Very recently, a number of scanning electron photomicrographs of Golgi bodies have been obtained by Y. Kinose and K. Tanaka (Fig. 18–4); these are remarkably consistent with the models of the organelle that are based on transmission studies.

In many cells, particularly those in which the Golgi bodies' main function is related to secretion, the faces of the Golgi bodies are arranged in a specific manner. The forming face is located next to either the nucleus or a specialized portion of the endoplasmic reticulum that lacks bound ribosomes and is called "transitional" ER. The maturing face is usually directed toward the plasma membrane. It is believed that the nuclear membrane and smooth endoplasmic reticulum are the source of the small vesicles that fuse with the *cis* face, while some of the larger vesicles arising from the maturing face are **secretory vesicles** and subsequently fuse with the plasma membrane. When vesicles released from the maturing face of the dictyosome form an internal cell structure, such as the developing **acrosome** of sperm cells (Fig. 18–5), the maturing face is directed toward the site of deposition. Thus, in the case of the sperm cell acrosome, the maturing face is proximal to the nucleus. Because of their intimate morphological and physiological associations with the endoplasmic reticulum and intracellular vesicles such as lysosomes and secretory granules, Golgi bodies are sometimes referred to as elements of the **endomembrane** or **vacuolar system** of cells.

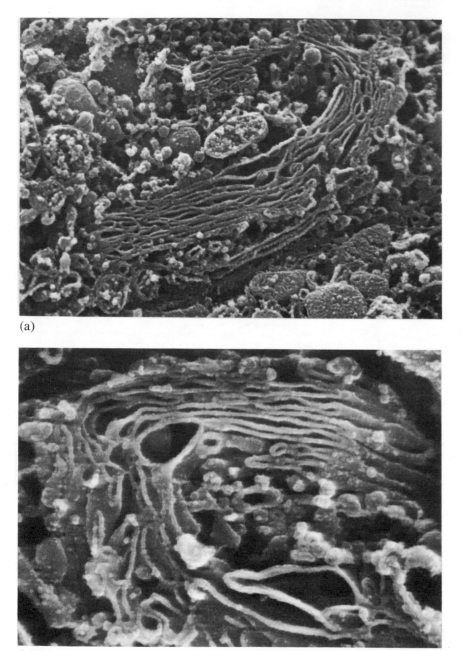

(a)

(b)

Figure 18–4
Scanning electron photomicrographs of Golgi bodies from rat epididymal tissue. (Courtesy of Dr. Y. Kinose and K. Tanaka, *J. Yonago Med. Assn. 30,* 527, 1979).

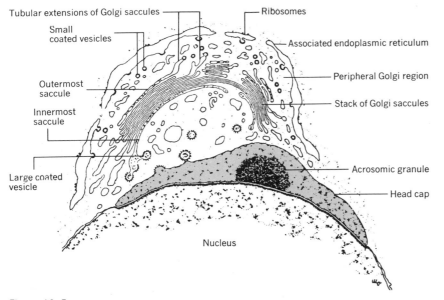

Tubular extensions of Golgi saccules
Small coated vesicles
Ribosomes
Associated endoplasmic reticulum
Outermost saccule
Peripheral Golgi region
Innermost saccule
Stack of Golgi saccules
Large coated vesicle
Acrosomic granule
Head cap
Nucleus

Figure 18–5

Contributions of the Golgi apparatus to the forming *acrosome* during sper-matid development. The uppermost Golgi elements are part of the *cis* or forming face; the vesicles contributing to the acrosomal membrane are de-rived primarily from the *trans* or maturing face. (Copyright © Wistar Institute; F. Susi et al., *Am. J. Anat. 130*, 262, 1971.)

The cisternae membranes of Golgi bodies are *smooth;* that is, they are not lined by ribosomes or other particles. At their margins, the cisternae may be perforated, creating **fenestrae,** or they may give rise to tubular extensions that branch and anastomose with each other (Fig. 18–3). The size and number of Golgi bodies vary from one type of cell to another and according to the cell's metabolic activity. The organelle is therefore to be thought of as being almost continuously in a state of flux, as new cisternae are formed and others dispatch vesicles. Some cells are reported to have only one Golgi body, while others may have hundreds. Since one of the major functions of the Golgi apparatus is secretion, it is not surprising that the size and number of Golgi bodies increase during periods of active cellular se-cretion. In plant cells, the number of Golgi bodies increases during cell division, when these organelles secrete materials that form the **cell plate,** which then develops into the cell wall that separates the two new cells. The **goblet cells** found in the intestinal epithelium contain only a single, large Golgi body located in the region of the cell where **mucigen granules** are stored prior to their secretion (Fig.

18–6). In these cells, the size of the Golgi body increases dramatically during periods of digestive activity.

Origin of Golgi Bodies

There are three proposed sources of new Golgi bodies in cells: (1) vesicles derived from the outer membrane of the nuclear envelope or the endoplasmic reticulum, (2) other cytoplasmic vesicles or structures, and (3) division of Golgi bodies already present in the cell. That cisternae of the Golgi apparatus may be formed from vesicles arising from the outer nuclear membrane or the endoplasmic reticulum is supported by electron microscopic evidence (Fig. 18–6b). These vesicles, called *transition vesicles,* migrate to the forming face of the Golgi body, fuse there with existing cisterna membranes, and in so doing contribute to the or-ganelle's growth. However, evidence is less than sufficient to indicate that a *complete* Golgi body can be formed in this manner.

Aggregations of transition vesicles occur in areas of the cytoplasm referred to as **zones of exclusion,** which are free

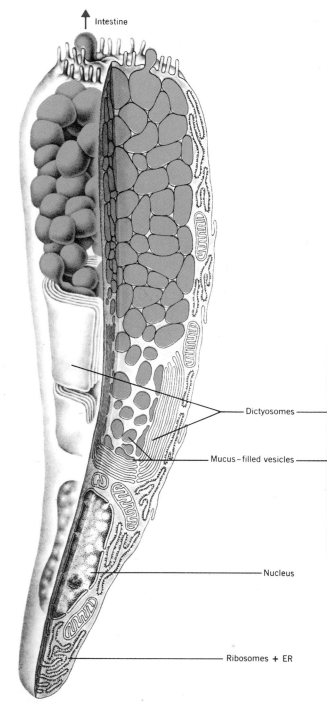

Intestine

Dictyosomes

Mucus–filled vesicles

Nucleus

Ribosomes + ER

Figure 18–6
Goblet cell of the intestinal epithelium showing the location of the dictyosomes (Golgi bodies) and mucigen granules. (Electron photomicrograph courtesy of Dr. E. G. Pollock.)

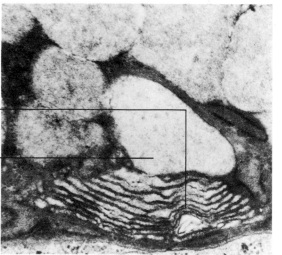

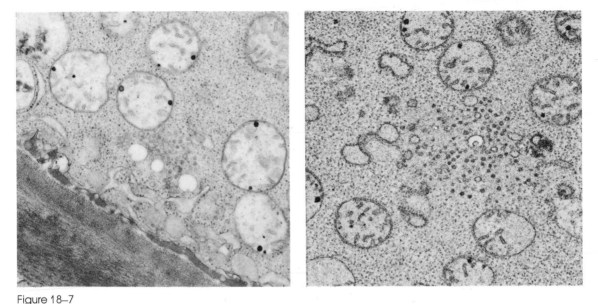

Figure 18–7
Zones of exclusion (clusters of tiny vesicles) believed to give rise to Golgi bodies. (Courtesy of Drs. H. H. Mollenhauer and J. Morre, in *Origin and Continuity of Cell Organelles*, J. Reinert and H. Ursprung, Eds., copyright © 1971, Springer-Verlag.)

of ribosomes. These zones are often surrounded by membranes of the endoplasmic reticulum or the nuclear envelope. Small Golgi bodies, which are presumed to represent early stages in the development of the organelle, are found in the zones of exclusion. The cells of dormant seeds of higher plants generally lack Golgi bodies but they do display zones of exclusion containing aggregations of small vesicles. Photomicrographs of cells in early stages of germination suggest progressive development of Golgi bodies in these zones of exclusion (Fig. 18–7) and the development of Golgi bodies coincides with the disappearance of the aggregations of vesicles.

In frog oocytes, Golgi bodies appear to develop from clusters of vesicles in zones of exclusion (Fig. 18–8). R. Ward and E. Ward suggest that the vesicles arise de novo from fine fibers in the zones and not from vesicles dispatched from the endoplasmic reticulum. That these early stages of Golgi body development are dependent on protein synthesis was shown by G. Werz, who found that actinomycin *D* (an inhibitor of protein synthesis) prevents the formation of Golgi body prestages in *Acetabularia*.

During cell division in both animals and plants the number of Golgi bodies increases, for the number of organelles in each daughter cell just after division is about the same as the number in the parent cell prior to division. Photomicrographs obtained at successive stages of cell division indicate that more Golgi bodies are present in cells that are in the metaphase or anaphase of mitosis than in earlier stages, but direct evidence for the division of the Golgi bodies is still lacking. In the multinucleate alga *Botrydium granulatum*, a single Golgi body is seen at each pole of the dividing cell just prior to the formation of the spindle. By late metaphase, two Golgi bodies are present at each end of the spindle and are separated by the centriole. Observations of this sort support the notion that new Golgi bodies may be formed by the division of existing organelles.

Development of the Golgi Apparatus

Because it is not possible to observe living Golgi bodies clearly, it has not been possible to directly follow the developmental sequence of a complete organelle. However, developmental sequences have been worked out by observing dictyosomes of cells at different stages of growth and

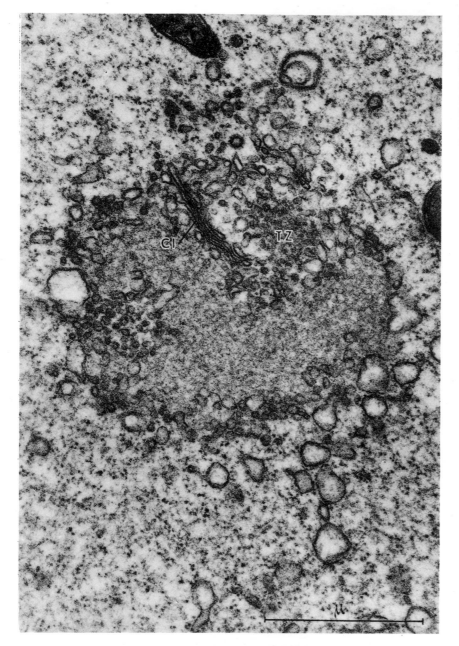

Figure 18–8
Development of the Golgi apparatus in the frog oocyte. In advance of the appearance of a definitive Golgi apparatus, transition zones (TZ) containing numerous small vesicles are seen in the zone of exclusion along with the first cisternae (CI). (Courtesy of Drs. R. Ward and E. Ward; copyright © 1968. *J. Microscopie 7*, 1007.)

then correlating differences in dictyosome appearance with other developmental changes in the cell.

For example, when cultures of the ciliate *Tetrahymena pyriformis* growing by fission are in the logarithmic phase, no typical Golgi structures are seen in the cells. However, individual, smooth, flattened **saccules** can be seen in the oral region. These saccules are of non-uniform shape and are not associated with one another. *Tetrahymena* can be induced to conjugate if the cells are starved for 12 to 48 hours. During starvation, the saccules in the oral region

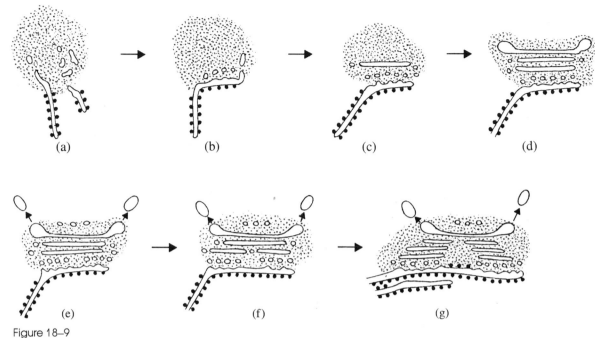

Figure 18–9
Proposed formation of a Golgi body from endolasmic reticulum *(a–c)* and subsequent developmental stages: layered cisternae *(c, d)*, formation of secretory vesicles *(e)*, division *(f, g)*. (Diagram courtesy of Dr. J. Morre et al., in *Origin and Continuity of Cell Organelles*, J. Reinert and H. Ursprung, Eds., copyright © 1971 Springer-Verlag.)

become aligned to form stacks. No vesicles are seen around these stacks, suggesting that the stacks are formed from the preexisting saccules and do not arise de novo. This also suggests that the stacks are not processing proteins or dispatching products. When starved cells of opposite mating types are mixed together, conjugation occurs within three hours. Five minutes after mixing, and well before conjugation begins, the stacked saccules in the oral region enlarge and numerous new vesicles can be seen. The structure now has the appearance of a Golgi apparatus and is processing material. By the time the mating cells separate, the Golgi bodies are well developed. If the cells are re-fed, the ordered arrangement of membranous sacs is lost.

The aggregation of small Golgi components to form a completed organelle is also seen in such diverse cells as developing zoospores and sperm. In embryonic liver cells, the Golgi apparatus appears to form from a localization of tubular cisternae. As the cells mature and differentiate, the tubular cisternae become platelike, and still later secretory

vesicles form at the outer edges of the cisternae. This sequence is illustrated in Figure 18–9.

Functions of the Golgi Apparatus

Secretion

Although the Golgi apparatus is involved in many different cellular processes, its principal role in most cells is in **secretion.** Two sets of experiments bear on the role of the Golgi apparatus in secretion. In 1964 L. Caro and G. Palade showed that the Golgi apparatus in the *acinar cells* of the pancreas is involved in the packaging of enzyme precursors into **zymogen** granules prior to secretion. Caro and Palade injected radioactive amino acids into rats and followed the movements of the "label" using autoradiography (Chapter 14). This type of experiment is called a "pulse-chase" because the initial short-term application of labeled amino acids (i.e., the "pulse") is immediately followed by the more prolonged application of unlabeled forms (i.e., the "chase"). Although amino acid metabolism and protein synthesis are not interrupted, the metabolic fate of the labeled amino acids can be traced through the cell with time. As might be expected, after a three-minute pulse, the label appeared almost exclusively in the rough endoplasmic reticulum, since this is the region of protein synthesis. Following the three-minute pulse, nonlabeled amino acids were added for 17 minutes (e.g., a total of 20 minutes from the beginning of the pulse). Although some label was found in the rough endoplasmic reticulum as before, most of the label had shifted to the Golgi apparatus. When the chase was continued for an additional 100 minutes (120 total), almost all the label had left the endoplasmic reticulum and the Golgi apparatus and was now found in the zymogen granules and in the lumen outside of the cells (as a result of the contents of the vesicles being discharged at the plasma membrane). These experiments showed that the path of the amino acids is first into proteins in the rough ER and that these proteins are then transferred into the cisternae of the Golgi apparatus and then into the zymogen granules.

In 1966 using similar autoradiographic techniques, M. Neutra and C. P. Leblond studied the secretion of mucous by the *goblet* cells of intestinal epithelium. Mucous is a glycoprotein in which glucose and glucose derivatives are linked together forming polysaccharide side chains of the protein molecules (Fig. 18–10). Glucose labeled with tritium (^{3}H) was used to follow the assembly and fate of the glycoproteins. Fifteen minutes after injection of the radioactive sugar, the label was most concentrated in the cisternae of the Golgi apparatus. This label did not enter or associate with the rough endoplasmic reticulum first. After a 20-minute chase, the label appeared in the mucous vesicles, and after four hours, most had been released through the plasma membrane into the intestinal lumen. This experiment not only revealed the path of glucose through the cell but also demonstrated that the final stages of assembly of the glycoprotein occur *in the Golgi apparatus*. Using the goblet cell as an example, Figure 18–11 depicts the central role of the Golgi apparatus in the packaging of newly synthesized proteins into vesicles for secretion. The assembly and processing of large molecules in the Golgi apparatus is not unique to goblet cells. Cartilage cells assemble glycoproteins in the cisternae of their Golgi bodies, and sulfate groups have been shown to be added as well. Pectins and cellulose are assembled in the Golgi bodies of plant cells prior to deposition onto the forming cell plate or cell wall.

Proliferation of Cellular Membranes

In addition to its role in secretion, the Golgi apparatus also plays a role in the preparation of proteins for organelles such as lysosomes and the plasma membrane (Fig. 18–12). Proteins destined to be incorporated into lysosomes or the plasma membrane are synthesized by ribosomes attached to the endoplasmic reticulum (i.e., rough ER). Some of these proteins are released into the lumenal phase of the ER while others remain anchored to the ER membranes. Within minutes of their synthesis, these proteins appear in the *cis* face of the Golgi apparatus. The mechanism by which lumenal phase and membrane-associated ER proteins reach the Golgi apparatus is still uncertain but is believed to involve one or a combination of the following processes. Transfer to the Golgi apparatus may be mediated by dispatchment of small vesicles from the ER that migrate to and fuse with the cisterna that comprises the *cis* face. Newly synthesized proteins discharged into the lumenal phase of the ER would be enclosed within these vesicles, while proteins that were left anchored in the ER membrane would be constituents of the vesicle wall. Alternatively, proteins in the lumenal phase of the ER reach the cisterna of the *cis* face by diffusion along transient connections between the channels, while ER membrane-anchored proteins reach the Golgi apparatus by lateral flow within the membranes that form these connections. Glycosylation of proteins is initiated en route to the Golgi apparatus but is completed within the Golgi. As processing of these proteins progresses, they are successively transferred from one cisterna to the next, ultimately reaching the *trans* face. The transfer may take the form of small intermediary vesicles, diffusion, or lateral membrane flow through continuities between adjacent cisternae.

Proteins that are destined to be components of lysosomal membranes or the plasma membrane and that are anchored to the ER membranes at the time of synthesis are presumed to move from the ER to the *cis* face and from the *cis* face to the *trans* face as membrane components. The membrane of the vesicles discharged from the *trans* face contain these proteins (see Fig. 18–12). Soluble lysosomal proteins (as well as proteins destined for secretion) move from the lumenal phase of the ER through the cisternael space and

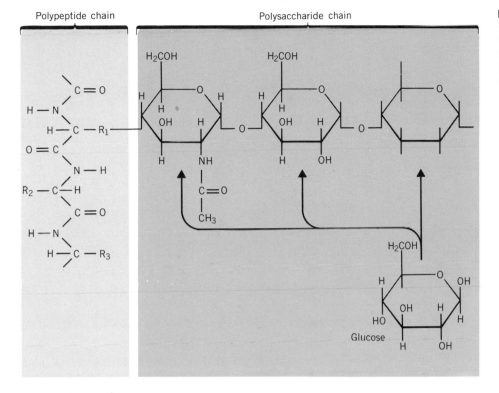

Figure 18–10
Mucous glycoprotein showing attachment of polysaccharide side chain to the polypeptide chain; the chain is composed of various derivatives of glucose.

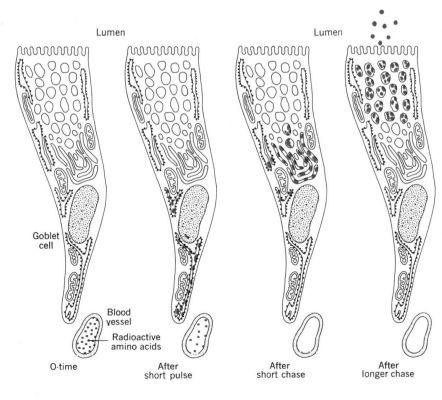

Figure 18–11
Incorporation of amino acids into secretory proteins by the goblet cells of the intestine. Amino acids removed from the bloodstream are used in protein synthesis by rough endoplasmic reticulum. The proteins are conveyed to the Golgi apparatus for incorporation into secretory vesicles. Glycosylation of the proteins (to form glycoproteins) occurs within the Golgi apparatus. Vesicles detach from the maturing face of the Golgi apparatus and migrate to the plasma membrane, where they are discharged into the intestinal lumen.

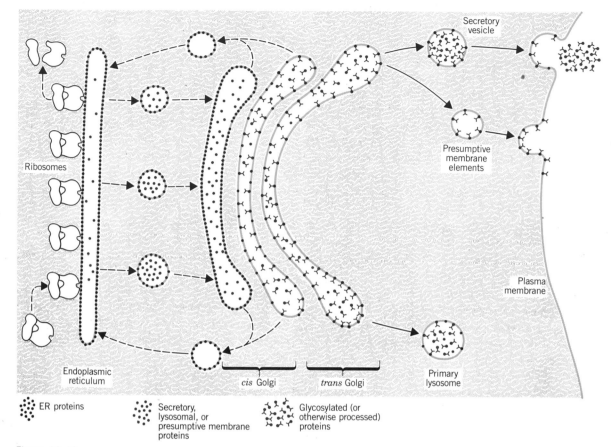

Ribosomes

Endoplasmic
reticulum

cis Golgi

trans Golgi

Primary
lysosome

Secretory
vesicle

Presumptive
membrane
elements

Plasma
membrane

⠿ ER proteins

⠿ Secretory,
lysosomal, or
presumptive membrane
proteins

Glycosylated (or
otherwise processed)
proteins

Figure 18–12

Functions of the Golgi apparatus. Proteins from the endoplasmic reticulum (black and colored circles) enter the *cis* face of the Golgi. As these are passed from one cisterna to the next, ER proteins (black circles) are removed from the margins of the cisterna by budding vesicles and are returned to the ER. Proteins destined for secretion or incorporation into organelles (e.g., lysosomes, plasma membrane; colored circles) are successively purified and are processed. These proteins are discharged as vesicles from the *trans* face (see text for additional details.)

are enclosed by the membranes of discharged vesicles.

Vesicles containing membrane-bound and soluble lysosomal enzymes are called *primary* lysosomes and are discussed at length in connection with lysosome function in the next chapter. Vesicles containing secretory proteins fuse with the plasma membrane and empty their contents outside the cell. This process was discussed earlier in the chapter in connection with the secretory functions of the small intestine and pancreas. Vesicle membranes that are studded with presumptive plasma membrane proteins also fuse with the plasma membrane. As seen in Figure 18–12, those

portions of the protein that faced the lumenal phase of the ER and that were glycosylated (or sulfated, etc.) in the Golgi cisternae face the exterior of the cell once they are incorporated into the plasma membrane.

As noted in Chapter 1, one of the characteristic features of many plant tissues is the presence of one or more large **vacuoles.** Some of these vacuoles have been shown to contain hydrolytic enzymes comparable to those present in lysosomes. For this reason, it has been suggested that the Golgi bodies of plant cells may give rise to some, if not all, of these vacuoles.

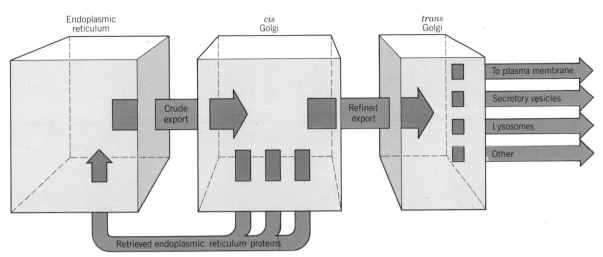

Figure 18–13

Division of labor among the *cis* and *trans* portions of the Golgi apparatus and the endoplasmic reticulum (see also Figure 18–12).

Sorting Problems of the Golgi Apparatus

It is clear from the preceding discussion that proteins destined for secretion granules, lysosomes, and the plasma membrane enter the *cis* face of a Golgi body from a common source—the rough endoplasmic reticulum. These proteins are delivered to the *cis* face of the Golgi along with a vast excess of ER membrane proteins. Yet the vesicles dispatched from the *trans* face are essentially free of ER proteins. Thus, the Golgi apparatus also acts to progressively sort these proteins. ER proteins are believed to be returned to the endoplasmic reticulum by some of the small vesicles released from the margins of the *cis* cisternae (Fig. 18–13). J. E. Rothman has recently suggested that a Golgi body may actually be two organelles in tandem: the *cis* Golgi and the *trans* Golgi. Beginning at the forming face, the *cis* Golgi consists of all but the last one or two cisternae, and its role is to sort and process proteins reaching these cisternae from the ER. The *trans* Golgi consists of the last one or two cisternae of the maturing face, which act to receive the refined proteins and distribute them through vesicles to their specific locations throughout the cell.

Cell-Specific Functions of the Golgi Apparatus

Although a discussion of each of the functions of the Golgi apparatus in different kinds of cells and tissues is beyond the scope of this book, two important examples are discussed below. A list of other well-studied functions is presented in Table 18–1.

Formation of the Cell Plate and Cell Wall in Plant Tissues

In plants, the cell plate and cell wall form during anaphase and telophase of mitosis and meiosis II (Chapter 20). During these final stages of nuclear division, the chromosomes are already separated into two masses in the cell that will become nuclei. Between these two nuclear masses, pectin and hemicellulose are deposited slowly, forming a plate in the center of the cell, which ultimately grows to the side walls, cutting and separating the protoplasts in two, and thereby producing the two daughter cells. Prior to anaphase, the cell's Golgi bodies are found outside the spindle. During anaphase, vesicles that appear to be released from the Golgi apparatus invade the center of the spindle (Fig. 18–14) and aggregate about the spindle fibers. These vesicles are the source of the carbohydrate that forms the cell plate and eventually the cell wall. The nature of the carbohydrate secreted by the vesicles is controversial. Some investigators believe that complete cellulose fibers are secreted, while others believe that the final stages of fiber assembly occur after secretion. In either case, the Golgi apparatus is clearly involved in the secretion of the carbohydrate that forms the wall between the two cell halves.

Table 18–1
Specific Functions of Golgi Structures

Cell	Tissue or Organ	Golgi Function
Exocrine	Pancreas	Secretion of zymogen (proteases, lipases, carbohydrases and nucleases)
Gland cell	Parotid gland	Secretion of zymogen
Goblet cell	Intestinal epithelium	Secretion of mucous and zymogens
Follicle cells	Thyroid gland	Prethyroglobulin
Plasma cells	Blood	Immunoglobulins
Myelocytes, sympathetic ganglia, Schwann cells	Nervous tissue	Sulfation reactions
Endothelial cells	Blood vessels	Sulfation reactions
Liver cells	Liver	Lipid secretion (lipid transformation?)
Alveolar epithelium	Mammary gland	Secretion of milk proteins (and lactose?)
Paneth cells	Intestines	Secretion of proteins (chitinase?)
Brunner's gland cell	Intestines	Synthesis and secretion of mucopolysaccharides, enzymes, hormones
Connective tissue	Amblystoma limb	Synthesis (?) and secretion of collagen
Cornea	Avian eye	Secretion of collagen
Plant cells	Most	Secretion of pectin and cellulose

The plasma membrane of plant cells does not pinch inward or grow inward during cell division as occurs in animal cells. Instead, the membrane forms on both sides of the developing cell plate and grows outward with it. Formation of the membrane results from fusion of the vesicle discharged from the Golgi apparatus.

Acrosome Development in Sperm

The development of the acrosome of sperm cells is an excellent example of the involvement of the Golgi apparatus in the formation of another cellular organelle. The acrosome is a membrane-bound structure at the anterior end of the sperm cells of most animals. A part of the acrosome membrane appears to be involved in recognition and binding of the sperm cell to the surface of the egg cell during fertilization. The acrosome contains hydrolytic enzymes of which **hyaluronidase** is the most abundant; it causes the breakdown of the protective surfaces of the egg. As shown in Figure 18–5, the singular large Golgi body of the sperm cell discharges coated vesicles that migrate to the forming acrosome. At the surface of the acrosome, these coated vesicles fuse with the acrosome membrane contributing to the acrosome's growth.

Since the acrosome is made up of hydrolytic enzymes, it has been suggested that the acrosome is nothing more than a giant lysosome. As the acrosome grows, the Golgi body becomes reduced in size, and in many mature sperm cells disappears entirely. The outer membrane of the acrosome fuses with the plasma membrane. In mouse sperm

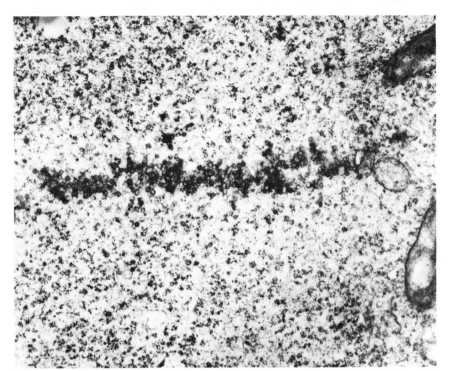

Figure 18–14
Clusters of small vesicles released from the Golgi apparatus during anaphase in *Zea mays* root apex cell. The vesicles migrate toward the equatorial region, where they contribute polysaccharides to the forming cell plate. (Photomicrograph courtesy of Dr. G. Whaley et al., copyright © Academic Press, *J. Ultrastruc. Res. 15,* 173, 1966.)

cells, it has been shown that the area of the plasma membrane that fuses with the acrosomal membrane contains a large number of *concanavalin A* binding sites. The increased number of glycoproteins in the membrane is attributed to its origin in the Golgi apparatus.

Summary

The **cisternae** of the **Golgi apparatus** develop either from the outer portion of the nuclear envelope or from the endoplasmic reticulum. The cisternae form a cup-shaped stack of flattened vesicles (sometimes called a **dictyosome**), with smaller vesicles joining the stack on one face, and leaving the stack at the other face.

Proteins destined for secretion, insertion in the plasma membrane, or deposition in intracellular organelles such as lysosomes and vacuoles reach the **forming face** (or **cis face**) of the Golgi apparatus from neighboring rough endoplasmic reticulum. In the cisternae comprising the *cis face,* these proteins are glycosylated, sulfated, or *processed* in other ways. Lipid and carbohydrate synthesis also occur in the *cis face*. Vesicles are discharged from the **maturing face** (or **trans face**) of the Golgi apparatus. These vesicles contain Golgi products destined for other cellular compartments.

Golgi bodies also play specific roles in different types of cells. For example, in plant cells Golgi bodies are associated with the formation of the **cell plate.** In sperm cells, growth of the acrosome results from the continuous influx of new materials from the Golgi apparatus.

References and Suggested Reading

Articles and Reviews

Beams, H., and Kessel, R., The Golgi apparatus: Structure and function. *Int. Rev. Cytol. 23,* 209 (1968).

Brodie, D. A., Bead rings at the endoplasmic reticulum–Golgi complex boundary. *Cell Biol. 90,* 92 (1981).

Farquhar, M. G., and Palade, G. E., The Golgi apparatus (complex)— (1954-1981)—from artifact to center stage. *J. Cell Biol. 91,* 77 (1981).

Hino, Y., Minakami, S., and Muratami, H., The comparison of Golgi subfractions isolated from mitochondrial fraction with those from microsomal fraction. *Exp. Cell Res. 133,* 171 (1981).

Lodish, H. F., and Rothman, J. E., The assembly of cell membranes. *Sci. Am. 240*(1), 48 (Jan. 1979).

Lodish, H. F., Braell, W. A., Schwatrz, A. L., Strous, G. J., and Zilberstein, A., Synthesis and assembly of membrane and organelle proteins. *Int. Rev. Cytol. Suppl.* 12, 247 (1981).

Morre, D. J., and Ovtracht, L., Dynamics of the Golgi apparatus: membrane differentiation and membrane flow. *Int. Rev. Cytol.* (Suppl. *5*), 61 (1977).

Morre, D. J., and Ovtracht, L., Structure of rat liver Golgi apparatus. *J. Ultrastruc. Res. 74,* 28 (1981).

Neutra, M., and Leblond, C., The Golgi apparatus. *Sci. Am. 220*(2), 100 (Feb. 1969).

Northcote, D., Chemistry of the plant cell wall. *Annu. Rev. Plant Physiol. 23,* 113 (1972).

Rothman, J. E., The Golgi apparatus: two organelles in tandem. *Science 213,* 1212 (1981).

Susi, R., Leblond, C., and Clermont, Y., Changes in the Golgi apparatus during spermiogenesis in the rat. *Am. J. Anat. 130,* 251 (1971).

Triemer, R. E., Role of Golgi apparatus in mucilage production and cyst formation in *Euglena gracilis. J. Phycol. 16,* 46 (1980).

Books, Monographs, and Symposia

Fleischer, B., and Fleischer, S., The Golgi apparatus, in *Advanced Cell Biology* (L. M. Schwartz and M. M. Azar, Eds.), Van Nostrand-Reinhold, New York, 1981.

Moore, D., Mollenhauer, H., and Bracker, C., Origin and continuity of Golgi apparatus, in *Origin and Continuity of Cell Organelles* (J. Reinhert and H. Ursprung, Eds.), Springer-Verlag, Berlin, 1971.

Waley, W. G., *The Golgi Apparatus,* Springer-Verlag, New York, 1975.

Chapter 19
Lysosomes and Microbodies

Our knowledge of the structure, composition, and function of **lysosomes** and **microbodies** is considerably more recent than that of most other cell organelles. Although a variety of small bodies seen in plant and animal cells (including what are now termed lysosomes) had been variously called "microbodies" or "cytosomes" for many years, the diversity of their composition and action was not recognized until the 1950s. The "discovery" of lysosomes and microbodies during the 1950s may be attributed to the growing sophistication of electron microscopy, the application of gentler procedures for dispersing tissues and cells, and the development of improved methods for separating, fractionating, and chemically characterizing the subcellular complexes released from disrupted cells. At the present time, lysosomes are recognized as a separate category of organelles, whereas **peroxisomes** and **glyoxysomes** are collectively called microbodies.

Lysosomes

The existence of lysosomes was subtly suggested for the first time in 1949 in the results of a series of experiments by Nobel Prize laureate Christian de Duve. These experiments were designed to identify the cellular locus of the two enzymes *glucose-6-phosphatase* and *acid phosphatase*. Liver tissue homogenates were separated into nuclear, mitochondrial, microsomal, and cytosol fractions by differ-

ential centrifugation (see Chapter 12), and enzyme assays were performed on each of the collected fractions. Although results with glucose-6-phosphatase clearly indicated that this enzyme was bound to particles sedimenting with the microsome fraction, observations on the distribution of acid phosphatase were at first rather confusing. The confusion centered around three seemingly peculiar but nonetheless reproducible findings: (1) the acid phosphatase activities of tissue homogenates prepared for centrifugation using a glass tube and close-fitting plunger (Dounce homogenizer) were about one-tenth the value observed when tissue was more vigorously dispersed using a Waring blender; (2) the total (i.e., combined) enzyme activity of the isolated centrifugal fractions analyzed following differential centrifugation was about twice the activity of the original homogenate; and (3) after storage for several days in a freezer, both the enzyme activity of the homogenate and the collected fractions, especially the mitochondrial fraction, increased dramatically (Table 19–1).

These observations were explained when de Duve showed that acid phosphatase activity was confined to sedimentable particles, the surrounding membranes of which limited the accessibility of the substrate (beta-glycerophosphate) used in the enzyme assay. Only when these membranes were disrupted and the acid phosphatase released from the particles was the enzyme activity demonstrable. This occurred during vigorous dispersion in the Waring

Table 19–1
Distribution of Acid Phosphatase Enzyme Activity in Liver Tissue Fractions Prepared by Differential Centrifugation

Fraction	Acid Phosphatase Activity (μg phosphate released/20 min)	
	Before Freezing and Storage	After Freezing and Storage for 5 Days
Whole homogenate	10	89
Nuclear fraction	2	10
Mitochondrial fraction	7	46
Microsomal fraction	6	10
Soluble fraction	6	9

Source. Based on the published results of J. Berthet and C. de Duve, *Biochem. J. 50,* 174 (1951), and C. de Duve, The lysosome in retrospect, in *Lysosomes in Biology and Pathology* (J. T. Dingle and H. B. Fell, Eds.), North-Holland Publishing Co., Amsterdam, 1969, p. 6.

Table 19–2
Some Enzymes Present in Lysosomes

Enzyme	Substrate
Proteases and peptidases	
Cathepsin A, B, C, D and E	Various proteins and peptides
Collagenase	Collagen
Arylamidase	Amino acid arylamides
Peptidase	Peptides
Nucleases	
Acid ribonuclease	RNA
Acid deoxyribonuclease	DNA
Phosphatases	
Acid phosphatase	Phosphate monoesters
Phosphodiesterase	Oligonucleotides, phosphodiesters
Phosphatidic acid phosphatase	Phosphatidic acids
Enzymes acting on carbohydrate chains of glycoproteins and glycolipids	
Beta-galactosidase	Beta-galactosides
Acetylhexosaminidase	Acetylhexosaminides, heparin sulfate
Beta-glucosidase	Beta-glucosides
Alpha-glucosidase	Glycogen
Alpha-mannosidase	Alpha-mannosides
Sialidase	Sialic acid derivatives
Enzymes acting on glycosaminoglycans	
Lysozyme	Mucopolysaccharides, bacterial cell walls
Hyaluronidase	Hyaluronic acid, chondroitin sulfates
Beta-glucuronidase	Polysaccharides, mucopolysaccharides
Arylsulfatase, A, B	Arylsulfates, cerebroside sulfates, chondroitin sulfate
Enzymes acting on lipids	
Phospholipase	Lecithin, phosphatidyl ethanolamine
Esterase	Fatty acid esters
Sphingomyelinase	Sphingomyelin

blender, which disrupted virtually all particles present. In contrast, only about 10% of the acid phosphatase–containing particles were disrupted by homogenization with Dounce homogenizer, thus accounting for the low activity of the enzyme in the homogenate (Table 19–1). Some additional enzyme activity was released during and following centrifugal fractionation, but much larger quantities of enzyme were released by the membrane disruption that occurred during freezing and thawing.

At first, de Duve and his co-workers did not recognize that the acid phosphatase activity was associated with a distinct population of cellular particles. Instead, on the basis of the observed "latent" activity of the mitochondrial fraction obtained by differential centrifugation (compare the values before and after freezing in Table 19–1), de Duve believed that acid phosphatase resided within the mitochondria. Continued studies during the early 1950s in which the mitochondrial fraction was further divided centrifugally into a number of subfractions revealed that acid phosphatase was absent from fractions containing rapidly sedimenting mitochondria but was present in high concentrations in fractions containing slowly sedimenting mitochondria. This observation, together with a newly developed appreciation of the potential contamination of sediments occurring during differential centrifugation (see Chapter 12), led de Duve to suspect that the acid phospha-

tase might, in fact, be associated with a special class of particles distinct from the mitochondria. Added credence was given to this idea by the finding that four other acid hydrolases, namely, *beta-glucuronidase, cathepsin, acid ribonuclease,* and *acid deoxyribonuclease,* were distributed through the centrifugal fractions in an identical manner. Thus, five hydrolytic enzymes, each having an acid pH optimum and acting on completely different substrates, appeared to be present in the same cell particle. On the basis of the *lytic* effects of all of these enzymes, de Duve named the particles "lysosomes." A number of additional enzymes have subsequently been identified in lysosomes (Table 19–2). Most of the chemical substances present in cells including proteins, polysaccharides, nucleic acids, and lipids are degraded by these enzymes.

It is interesting to note that the initial postulation of the existence of lysosomes was made by de Duve purely on biochemical grounds. However, in 1955, A. Novikoff, working with de Duve, examined centrifugal fractions rich in acid phosphatase activity using the electron microscope and provided the first morphological evidence supporting the existence of these particles. The lysosomes were identified as small, dense membrane-enclosed particles distinct from the mitochondria.

In recent years, sophisticated centrifugal methods have been devised for obtaining preparations that are rich in lysosomes. Nearly all preparations obtained by differential centrifugation are contaminated with quantities of mitochondria. Although the *average* sedimentation coefficient of mitochondria is greater than that of lysosomes, mitochondria are polydisperse with respect to size, so that the smaller mitochondria invariably sediment with the lysosomes. Moreover, in tissues containing peroxisomes (such as liver and kidney), the range of sedimentation coefficients for these organelles is almost identical to that of the lysosomes. Consequently, it is virtually impossible to obtain lysosome preparations that do not also contain large numbers of peroxisomes. Somewhat greater success is obtained when *isopycnic* density gradient centrifugation is used in the last stages of the isolation procedure, for the equilibrium densities of lysosomes (1.22 g/cm^3), mitochondria (1.19 g/cm^3), and peroxisomes (1.23–1.25 g/cm^3) in sucrose density gradients are slightly different. Most density gradient procedures used to prepare lysosomes are modifications of the technique developed by W. C. Schneider (Fig. 19–1). Using this technique, most of the mitochondria are banded isopycnically at a density of about 1.19 g/cm^3, while most of the lysosomes form a separate zone at about 1.22 g/cm^3 and can be recovered independently from the density gradient.

By far the greatest purity of lysosomes is obtained from tissues of animals previously treated with Triton WR-1339 (a polyethylene glycol derivative of polymerized *p-tert-octyl phenol*), dextran (a polymer of glucose), and Thorotrast (colloidal thorium hydroxide). These compounds are rapidly incorporated in large quantities by the cell's lysosomes, significantly altering their density. For example, the incorporation of Triton WR-1339 reduces the average density of the lysosome from 1.22 to about 1.10 g/cm^3. It is interesting to note that although the density of lysosomes incorporating Triton WR-1339 is significantly reduced, their size is increased; the result is that Triton WR-1339-loaded lysosomes have the same sedimentation coefficient as do normal lysosomes but have a lower density.

The latent enzymic effect originally noted by de Duve and his co-workers is still employed as a major criterion in evaluating the effectiveness of any lysosome isolation. Accordingly, the lysosome preparation is incubated under the appropriate conditions with the hydrolase substrate before and after treatments known to disrupt the lysosome membranes. If the original preparation contains intact lysosomes, then no substrate is hydrolyzed before treatment (most substrates of the lysosomal hydrolases are unable to permeate the lysosome's membrane); however, disruption of the membrane (by sonification, repeated freezing and thawing, addition of lytic agents such as bile salts, digitonin, Triton X-100, etc.), and release of the lysosomal enzymes are quickly followed by hydrolysis of the added substrates.

Structure and Forms of Lysosomes

Lysosomes are a structurally heterogeneous group of organelles varying dramatically in size and morphology. As a result, it is difficult to identify lysosomes strictly on the basis of morphological criteria. When lysosome-rich fractions were initially isolated centrifugally by de Duve and Novikoff and examined with the electron microscope, it was found that the suspected lysosomes were generally smaller than mitochondria. Typically, they varied in diameter from about 0.1 to 0.8 μm, were bounded by a single membrane, and were usually somewhat electron-dense. Identification of lysosomes in sections of whole cells is considerably more difficult because other small, dense organelles are also bounded by a single membrane. The application of cytochemical procedures at the level of the electron microscope in which the lysosomes are identified

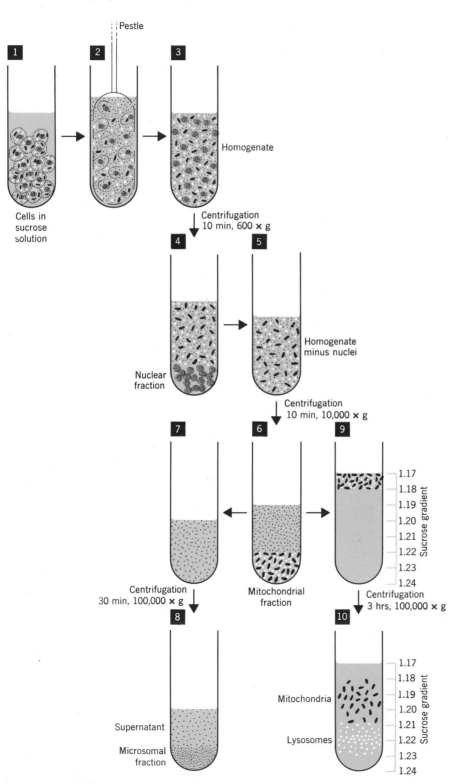

Figure 19–1
Steps in the isolation of lysosomes by combined differential and sucrose density gradient centrifugation.

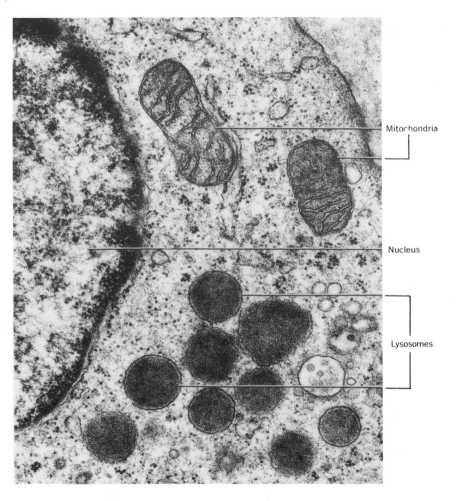

Mitochondria

Nucleus

Lysosomes

Figure 19–2
**Electron photomicrograph of a cluster
of lysosomes; these clusters are often
seen next to mitochondria. (Photomicrograph courtesy of R. Chao.)**

on the basis of their enzyme content is much more reliable. Notable among such procedures is the one introduced in 1952 by G. Gomori that is routinely employed in variously modified forms for the identification of lysosomes on the basis of their high acid phosphatase content. In the Gomori method, the tissue to be examined is incubated at pH 5.0 in a medium containing beta-glycerophosphate (a substrate for acid phosphatase) and a lead salt (such as lead nitrate). Phosphate enzymatically cleaved from the substrate during incubation combines with the lead ions to form insoluble lead phosphate, which precipitates at the locus of enzyme activity. Since the lead phosphate is electron-dense, lysosomes appear as dark, granular organelles in the electron microscope (Fig. 19–2). For identification with the light microscope, ammonium sulfide may be used to convert the lead phosphate to lead sulfide, which appears black. The

Gomori reaction may be carried out with fixed and sectioned material, as well as with fresh tissue, albeit with reduced efficiency as a result of some enzyme inactivation during and following fixation.

Several different lysosomal forms have been identified within individual cells, including (1) **primary lysosomes,** (2) **secondary lysosomes,** and (3) **residual bodies.**

Primary Lysosomes. **Primary lysosomes,** or **protolysosomes,** are newly produced organelles bounded by a single membrane and varying greatly in size. The primary lysosome is a virgin particle in that its digestive enzymes have not yet taken part in hydrolysis.

Secondary Lysosomes. Two different kinds of **secondary lysosomes** can be identified: **heterophagic vacuoles** (also

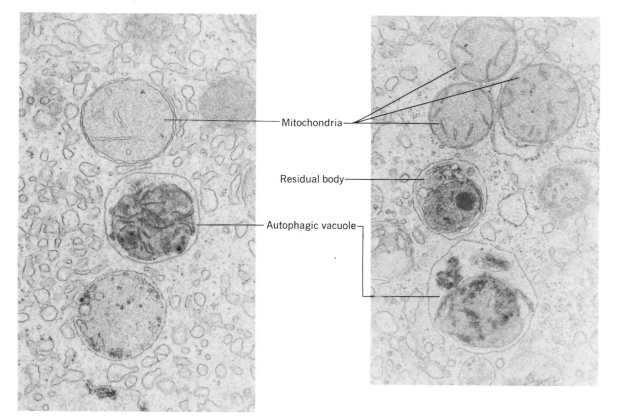

Figure 19–3

Autophagic vacuoles containing partially degraded mitochondria. (Electron photomicrographs courtesy of Dr. Z. Hruban.)

called **heterolysosomes** or **phagolysosomes**) and **autophagic vacuoles** (also called **autolysosomes**). Heterophagic vacuoles are formed by the *fusion* (see below) of primary lysosomes with cytoplasmic vacuoles containing *extracellular* substances brought into the cell by any of a variety of endocytic processes (see Chapter 15). Following fusion, the hydrolases of the primary lysosome are released into the vacuole (called a **phagosome**). Autophagic vacuoles contain particles isolated from the cell's own cytoplasm, including mitochondria, microbodies, and smooth and rough fragments of the endoplasmic reticulum. The autodigestion of cellular organelles is a normal event during cell growth and repair and is especially prevalent in differentiating and dedifferentiating tissues and tissues under stress. Autophagic vacuoles containing partially degraded mitochondria are shown in Figure 19–3. The formation of

heterophagic and autophagic vacuoles is soon followed by enzymatic digestion of the vacuolar contents. As digestion proceeds, it becomes increasingly difficult to identify the nature of the original secondary lysosome; the more general term **digestive vacuole** is used to describe the organelle at this stage.

Residual Bodies. Endocytosed substances and parts of autophagocytosed organelles that are not digested within the secondary lysosomes and transferred to the cytoplasm are retained (usually temporarily) within the vacuoles as residues. Lysosomes containing such residues are called **residual bodies** (sometimes also called **telolysosomes** or **dense bodies**). The undigested residues often take the form of *whorls* of membranes, grains, amorphous masses, ferritin-like particles, or myelin figures (Fig. 19–4). Residual bod-

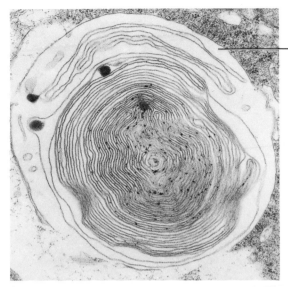

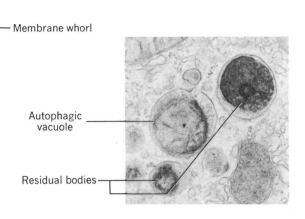

Membrane whorl

Autophagic
vacuole

Residual bodies

Figure 19–4
Left: a membrane whorl. (Electron photomicrograph courtesy of
Dr. E. G. Pollock.) *Right:* residual bodies. (Electron photomicro-
graph courtesy of Dr. Z. Hruban.)

ies often fail to display the degree of hydrolytic activity associated with the primary and secondary lysosomes.

Formation and Function of Lysosomes (The "Vacuolar System")

Lysosomal enzymes are concerned with the degradation of metabolites and not with cellular synthetic or transfer reactions. The specific cellular origin of the lysosomal acid hydrolases and the mechanism by which they are incorporated into primary lysosomes are still uncertain. However, on the basis of numerous cytochemical observations, two theories have gained widespread acceptance. According to Novikoff and others, the acid hydrolases that are destined for lysosomes are synthesized by ribosomes of the rough ER in the vicinity of the Golgi bodies (Chapter 18). Some of these hydrolases are discharged into the lumenal phase of the ER, while others remain anchored in the ER membranes. Through either the dispatchment of tiny vesicles from the ER or via direct communication through cisternae, the hydrolases make their way to the *cis* face of the Golgi body. After purification and processing in successive Golgi cisternae, the hydrolases are released from the *trans* face of the Golgi apparatus in the form of primary lysosomes.

This concept, depicted diagrammatically in Figure 19–5, is supported by a large number of observations made with a variety of tissues. The process is reminiscent of the formation of zymogen granules for secretion by Golgi bodies. Because of the intimate association between Golgi bodies, ER, and primary lysosomes, regions of cells containing these organelles are sometimes called GERL complexes.

An alternative proposal based on more limited observations suggests that primary lysosomes may be produced directly by dilations of rough endoplasmic reticulum. To reconcile these observations, it has been suggested that the Golgi apparatus may be involved in the formation of primary lysosomes in cells rich in *smooth* endoplasmic reticulum, whereas direct formation by *rough* endoplasmic reticulum occurs in cells in which that form of cytomembrane predominates.

Heterophagy. Extracellular materials brought into the cell by endocytosis are enclosed within vacuoles called **phagosomes.** These materials may later be rejected unaltered by exocytosis, or the phagosomes may fuse with one or more primary lysosomes that empty their digestive hydrolases into the newly formed particle (now called a *second-*

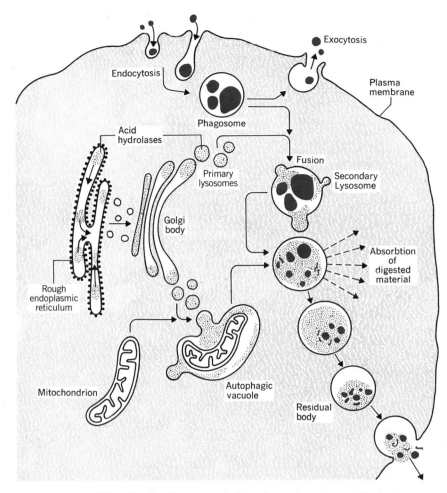

Figure 19–5
Formation and function of lysosomes in cellular *heterophagy* and *autophagy*.

ary lysosome, Fig. 19–5). Lysosomal digestion of endocytosed material is termed **heterophagy.**

The fusion of primary lysosomes with phagosomes has been demonstrated in vivo in a number of tissues using various exogenous **markers** introduced into the organism. These markers, which include *horseradish peroxidase, ferritin,* and *hemoglobin,* are engulfed by the tissue cells and are later detected within secondary lysosomes along with lysosomal hydrolases.

Z. A. Cohn and B. Benson employed ^{3}H-labeled leucine to trace the fate of newly synthesized hydrolases in peritoneal phagocytes of mice. They found that pinocytic activity was greatly increased when these cells were incubated in blood serum. Autoradiographic analysis revealed that the labeled hydrolases appeared first in the Golgi region of the cells and later within pinocytic vesicles (i.e., phagosomes).

Their observations support the proposal that secondary lysosomes are formed by the fusion of phagosomes and primary lysosomes. Moreover, Cohn and Benson also found that the rate at which hydrolases were produced by the cells was related to the level of pinocytic activity, suggesting that the production of primary lysosomes may somehow be regulated by endocytosis (see below).

In some cells, several small primary lysosomes may fuse with a single large phagosome; in other cells, large primary lysosomes sequentially fuse with a number of small phagosomes. The contents of the secondary lysosome change dramatically with time as (1) the contents of the lysosome are enzymatically degraded, (2) new materials are introduced through fusion of additional phagosomes, and (3) additional hydrolases are added by the fusion of new primary lysosomes. The hydrolases in the secondary lysosome break down the endocytosed materials, producing a variety of useful substances (e.g., amino acids, sugars) as well as some useless waste products. It is generally agreed that usable materials make their way across the membrane of the secondary lysosome and enter the cell cytoplasm, where they participate in cellular metabolism. This transfer probably takes the form of passive diffusion or facilitated or active transport. Eventually, digestion and absorption are terminated, leaving only residues and denatured enzymes within the vacuole, which is now referred to as a **residual body.** In many cells, residual bodies fuse with the plasma membrane, and this is followed by exocytosis (Fig. 19–5). In some cells (especially those of higher organisms), residual bodies accumulate within the cytoplasm or continue to increase in size, eventually interfering with the normal activities of the cell and resulting in cell death. Progressive lysosome engorgement is believed to be involved in the aging process.

Autophagy. The isolation and digestion of portions of a cell's own cytoplasmic constituents by its lysosomes occurs in normal cells and is termed **autophagy** (Fig. 19–5). The phenomenon is most dramatic in the tissues of organs undergoing regression (changes in the uterus following delivery, during metamorphosis in insects, etc.). Autophagic vacuoles containing partially degraded mitochondria, smooth and rough endoplasmic reticulum, microbodies, glycogen particles, or other cytoplasmic structures are frequently observed in tissue sections examined with the electron microscope. Cellular autophagy results in a continuous turnover of mitochondria in liver tissue. The half-life of the liver mitochondrion is about 10 days and corresponds to the destruction of one mitochondrion per liver cell every 15 minutes.

Distribution of Lysosomes

Since their initial discovery in mammalian liver, lysosomes have been identified in many different cells and tissues; some of these are listed in Table 19–3. The greatest variety of tissues found to contain lysosomes occurs in animals. Although most studies have been carried out using mammalian tissues, lysosomes have been identified in insects, marine invertebrates, fish, amphibians, reptiles, and birds. Lysosomes are particularly numerous in epithelial cells of absorptive, secretory, and excretory organs (liver, kidneys,

Table 19–3
Cells and Tissues Containing Lysosomes

Protozoa	Nerve cells
Amoeba	Brain
Campanella	Intestinal epithelium
Tetrahymena	Lung epithelium
Paramecium	Uterine epithelium
Euglena	Macrophages of spleen,
Plants	bone marrow, liver,
Onion seeds	and connective tissue
Corn seedlings	Thyroid gland
Tobacco seedlings	Adrenal gland
Tissue culture cells	Bone
HeLa cells	Urinary bladder
Fibroblasts	Uterus
Monocytes	Ovaries
Macrophages	Blood
Chick cells	(leucocytes and
Lymphocytes	platelets)
Animal tissues	
Liver	
Kidney	

etc.). They are also present in large numbers in the epithelial cells of the intestines, lungs, and uterus. Phagocytic cells and cells of the reticuloendothelial system (e.g., bone marrow, spleen, and liver) have also been found to contain large numbers of lysosomes. Few lysosomes occur in muscle cells or in acinar cells of the pancreas. Lysosomes are produced by certain cells in tissue culture (HeLa cells, monocytes, lymphocytes, etc.). Although it has a number of functions not shared by lysosomes of animal cells, the large vacuole of many plant cells is a modified lysosome. Some of the various roles played by the lysosomes are summarized in Table 19–4.

Leucocytes, especially granulocytes, are a particularly rich source of lysosomes, and this is related to their physiological role as scavengers of microorganisms or other foreign particles in the blood. Following phagocytosis of a bacterium by a leucocyte, numerous lysosomes fuse with the endocytic vacuole containing the microorganism and initiate its digestion. The lysosomes of granular leucocytes are especially large and readily visible by light microscopy. Once the lysosome content of the leucocyte is exhausted, the white blood cell dies.

Table 19–4
Some Functions of Lysosomes

1. Nutrition via a digestive role in protozoa and many metazoan cells
2. Nutrition via cellular autophagy during unfavorable environmental conditions
3. Lysis of organelles during cellular differentiation and metamorphosis
4. Scavenging of worn-out cell parts and denatured proteins
5. Destruction of aged red blood cells and dead cells
6. Defense against invading bacteria and viruses by circulating macrophages
7. Dissolution of blood clots and thrombi
8. Keratinization of skin
9. Secretion of hydrolases by sperm for egg penetration during fertilization
10. Yolk digestion during embryonic development
11. Bone resorption
12. Reabsorption in kidney and urinary bladder

Plant Vacuoles

Many plant cells contain one or more **vacuoles** (see Fig. 1–33), which possess some of the properties of lysosomes. In immature and actively dividing plant cells the vacuoles are quite small. As the cells mature the vacuoles coalesce to form larger compartments. Mature cells of higher plants usually have a larger, central vacuole that may occupy more than 80% of the total cell volume.

The membrane enclosing the plant cell vacuole is called the **tonoplast.** Like lysosomes, plant cell vacuoles contain hydrolytic enzymes. In addition, they usually contain sugars, salts, acids, nitrogenous compounds such as alkaloids, and anthocyanin pigments. The pH of the plant vacuole may be as high as 9 or 10 due to large quantities of alkaline substances or as low as 3 due to the accumulation of quantities of acids (e. g., citric, oxalic, and tartaric acids).

The plant vacuole is the major contributor to cell **turgor** (Chapter 15), giving support to the individual plant cell and contributing to the rigidity of the leaves and younger parts of the plant. Water accumulation in the vacuole as a result of the osmotic effects of the dissolved substances causes the vacuole to expand, pushing outward against the cytoplasm and cell wall. When there is a lack of water the turgor diminishes, and this results in wilting.

Lysosome Precursors in Bacteria

Although bacterial cells do not possess lysosomes, they do contain a variety of hydrolases that are believed to be localized in the space between the plasma membrane and the cell wall. These hydrolases may be synthesized by ribosomes attached to the plasma membrane and then dispatched through it. The bacterial hydrolases play a digestive role, breaking down complex substrates in the cell's environment and providing smaller molecules required for cell growth. Bacterial hydrolases also participate in **sporulation** and **autolysis.** Although the latter process destroys the individual cells involved, it is highly beneficial to the bacterial population as a whole, for it provides for the survival of a small number of cells under unfavorable environmental conditions. Infolding of the bacterial membrane to form internalized extracellular pockets containing both hydrolases and their substrates would provide the ''evolutionary link'' with lysosomes of animal and plant cells.

Regulation of Lysosome Production

As noted earlier, the mechanism proposed for primary lysosome formation is strikingly similar to that proposed for zymogen granule formation in pancreatic cells and other instances of secretory protein synthesis. This similarity does not seem so unusual when one considers the following. The enzymatic contents of primary lysosomes are discharged into vacuoles that are derived from the plasma membrane and contain extracellular materials. Consequently, the mechanism is similar to secretion except that the extracellular space into which the secretory products pass is internalized as a vacuole (i.e., phagosome). In secretory cells, the production of new secretory products is regulated by a feedback mechanism in which secretion itself acts as a stimulus for the production of additional secretory materials. The experiments of Cohn and Benson described above demonstrate the relationship that exists between endocytic activity and lysosomal enzyme synthesis. It has therefore been suggested that the passage of phagosomes into the Golgi regions of the cell is followed by the discharge of some primary lysosomes and that this triggers the synthesis of new acid hydrolases.

Disposition and Action of the Lysosomal Hydrolases

Many of the lysosome's enzymes are released into the surrounding environment when these organelles are physically or chemically disrupted. Those enzymes that are so readily solubilized are believed to be located in the interior of the organelle. Other lysosomal hydrolases cannot be solubilized or are extracted with great difficulty and are thought to be an integral part of the lysosome membrane together with other proteins and lipids. Some of the enzymes known to be present in lysosomes are listed in Table 19–2; it is to be noted that while this list is extensive, it is by no means complete.

All the substrates of lysosomal enzymes are either polymers or complex compounds and include proteins, DNA, RNA, polysaccharides, carbohydrate side chains of glycoproteins and glycolipids, lipids, and phosphates. The lysosomal breakdown of proteins into amino acids illustrates how these enzymes act in concert. The initial hydrolysis of protein is effected by *cathepsins D* and *E* and also by *collagenase*. These enzymes cleave peptide bonds and produce peptide fragments of varying length. The peptides, together with previously undigested proteins, are further hydrolyzed to individual amino acids by *cathepsins A* and *B. Cathepsin C, arylamidase,* and the lysosomal *dipeptidases* act on specific peptides, producing additional amino acids.

The breakdown of DNA and RNA is initiated by the enzymes *acid deoxyribonuclease* and *acid ribonuclease.* The resulting oligonucleotides are then degraded first by *phosphodiesterase* and then by *acid phosphatase,* producing nucleosides and inorganic phosphate. Lysosomes also possess all the enzymes necessary for hydrolysis of lipids and polysaccharides.

As noted earlier, some lysosomal enzymes are part of the membrane encasing the organelle. Among the enzymes found to be integral parts of the lysosome membrane are *acetylglucosaminidase, glucosidase,* and *sialidase. Arylsulfatase, acid phosphatase, ribonuclease,* and *glucuronidase* may also be bound to the membrane under certain conditions.

Enzymes freed from disrupted lysosomes exhibit a wide variation in stability. Some are particularly resistant to au-

tolysis and retain their activity for months when appropriately refrigerated; others lose their activity only a few hours following tissue disruption. Of the lysosomal enzymes isolated and characterized to date, a number have been shown to be glycoproteins, including *cathepsin C, acid deoxyribonuclease, glucuronidase,* and *acetylglucosaminidase.*

Microbodies

As noted at the beginning of the chapter, the term *microbody* was used by cell biologists and cytologists for many years to describe a variety of different small cellular components. More recently, it has usually been restricted to organelles possessing *flavin oxidases* and *catalase.* Organelles possessing these activities are typically spherical or ovoid structures having a single bounding membrane, a diameter of about 0.5 to 1.5 μm, and an amorphous granular matrix, occasionally with crystalloid inclusions (Fig. 19–6). The organelles vary somewhat in structure, appearance, and function from one tissue to another and from species to species. Certain microbodies exhibit specific biochemical characteristics as well as specific distributions among animal, plant, and microbial cells. Included here are the **peroxisomes** and **glyoxysomes.**

Peroxisomes

The modern usage of the term *microbody* dates back to 1954 and the work of J. Rhodin, who described the structure and properties of these organelles in mouse kidney tissue. Since then, organelles of similar organization have been reported in many other animal tissues and also in plants (Fig. 19–6). In 1965, de Duve showed that microbodies of rat liver contained a number of *oxidases* that transfer hydrogen atoms to molecular oxygen, thereby forming hydrogen peroxide (Fig. 19–7). de Duve coined the term **peroxisomes** for these organelles, although a true peroxidatic activity is generally demonstrable only in vitro. In vivo, conditions favor the removal (or degradation) of hydrogen peroxide by *catalase* rather than by a *peroxidase.* However, since hydrogen peroxide is an intermediate in the reaction, the term *peroxisome* may be appropriate. The chemical and enzymatic relationships between an oxidase, peroxidase, and catalase are shown in Figure 19–7.

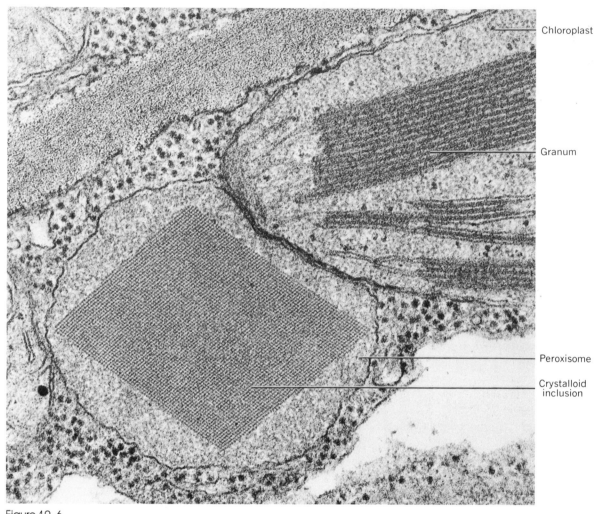

Figure 19–6
Electron photomicrograph of a peroxisome in a tobacco leaf cell; note how closely the peroxisome is appressed to a neighboring chloroplast. (Photomicrograph courtesy of Dr. E. H. Newcomb; from Frederick et al., *Protoplasma 84*, 1 (1975).)

A number of enzymes are characteristically present in peroxisomes including *uric acid oxidase, D-amino acid oxidase, acyl-CoA oxidase, polyamine oxidase, β-hydroxy-acid oxidase, NADH-glyoxylate reductase, NADP-isocitrate dehydrogenase,* and *catalase*. When *uric acid oxidase* is present in large amounts, it frequently takes the form of a paracrystalline "nucleoid" at the center of the organelle.

The functions of peroxisomes in animal cells are diverse.

Peroxisomal *catalase* is thought to be involved in the degradation of H_2O_2, which is extremely toxic, the source of the peroxide being other peroxisomal reactions (e.g., those catalyzed by the *flavin oxidases*). Uric acid oxidase is important in the catabolic pathway that degrades purines. The early observation that there is an abundance of peroxisomes in cells engaged in lipid metabolism suggested that these organelles may be involved in lipid metabolism. Recently,

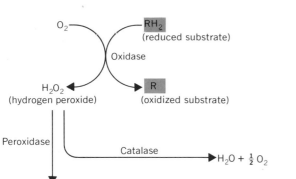

Figure 19–7
Chemical interrelationship between *oxidase*, *peroxidase*, and *catalase* enzymes.

it has been shown that liver peroxisomes contain a major system for the beta-oxidation of fatty acids (see Chapter 10); however, the enzymes are different from those of mitochondria (Chapter 16), although they produce the same end product, acetyl CoA.

Electron microscopic studies of tissue sections often reveal a close proximity between peroxisomes and mitochondria. This is not surprising since the products of peroxisomal activity may serve as substrates for mitochondrial activity. For example, glyoxylate produced in peroxisomes may be converted there to glycine by transamination. After passage to neighboring mitochondria, the glycine may be further metabolized in a variety of ways including conversion to other amino acids or incorporation into heme.

We saw earlier that one of the characteristic features of lysosomal enzyme activity is its *latency*. No latency is exhibited by the peroxisomal enzymes, as relatively large molecules (including the peroxisomal enzyme substrates) readily permeate the peroxisome membrane.

Isolation of Peroxisomes. The sedimentation coefficients and densities of peroxisomes in sucrose gradients are close to those of lysosomes and account for the fact that for some time peroxisomal enzyme activities were ascribed to lysosomes. Density gradient centrifugation of the "mitochondrial fraction" prepared by preliminary differential centrifugation is the method of choice for isolating peroxisomes. The greatest success in peroxisome purification is obtained if the lysosomes are first allowed to accumulate Triton WR-

1339 (see earlier discussion). Triton-loaded lysosomes are considerably less dense than normal lysosomes and so they are easily "floated" away from the peroxisomes during density gradient centrifugation.

Formation of Peroxisomes. For many years, it has been held that peroxisomes of both plant and animal tissues arise as outgrowths of the endoplasmic reticulum and that the peroxisomal enzymes are dispatched into the cisternae of the endoplasmic reticulum by attached ribosomes. The enzymes make their way into the organelle prior to its physical separation from the ER. Certainly, this idea is supported by ultrastructural evidence (especially for plants). However, recent findings using liver cells suggest instead that at least in this tissue many peroxisomal enzymes are synthesized by unattached ribosomes (i.e., ribosomes that are not bound to the ER) and are released into the cytosol. From there, the enzymes are slowly taken up by preexisting peroxisomes. Studies with peroxisomal catalase indicate that the subunits of the enzyme and the heme enter the microbody and are then assembled to form the functional enzyme. Peroxisomes often appear in clusters when examined by electron microscopy and occasional dumbbell-shaped images are observed. This suggests that the "tails" seen on peroxisomes are connections to other peroxisomes. Indeed, either permanent or transient connections between peroxisomes might form a "peroxisome reticulum." Such interconnections would explain the remarkable biochemical homogeneity of peroxisomes and the synchronous turnover of peroxisomal proteins. Some new peroxisomes may be formed either by the fission of preexisting peroxisomes or by budding from the peroxisome reticulum.

Glyoxysomes

In 1967 R. W. Breidenbach and H. Beevers discovered that microbodies of the fat-storing cells of germinating fatty seeds contain enzymes of the **glyoxylate cycle** (see Chapter 10) in addition to peroxisomal enzymes. They used the term **glyoxysomes** for these particles. Glyoxysomes not only contain the glyoxylate cycle enzymes *isocitrate lyase* and *malate synthetase* but also contain several of the essential enzymes of the Krebs cycle (Chapter 16); these enzymes therefore function simultaneously in both mitochondria and glyoxysomes.

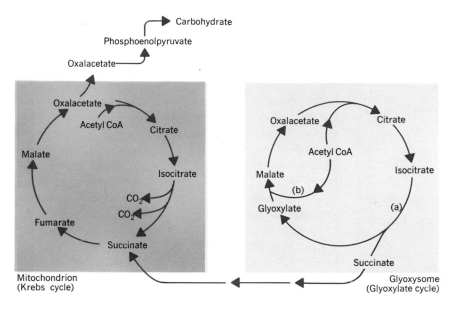

Figure 19–8
Comparison of the Krebs cycle in mitochondria and the glyoxylate cycle in glyoxysomes. Enzymes [a] and [b] are *isocitrate lyase* **and** *malate synthetase,* **respectively.**

The relationship between the Krebs cycle and the glyoxylate cycle is shown in Figure 19–8. Both cycles employ the same reactions to produce isocitrate from acetyl CoA and oxaloacetate, but beyond this point the pathways differ. In the Krebs cycle, isocitrate is successively decarboxylated to produce succinate and two molecules of CO_2. In the glyoxylate cycle, isocitrate is converted to succinate and glyoxylate. Instead of being lost as two molecules of CO_2, the 2-carbon glyoxylate condenses with another acetyl CoA to form the 4-carbon dicarboxylic acid, malate. The four carbon atoms of the two acetyl CoA molecules are thus conserved as one 4-carbon compound, which, after conversion to succinate and migration to the mitochondrion, may be converted to oxaloacetate. The oxaloacetate may then be utilized in gluconeogenesis.

Oxaloacetate formed in mitochondria from glyoxysomal succinate is presumed to serve as a direct precursor of phosphoenolpyruvate (PEP). The conversion of PEP to carbohydrate occurs essentially by the reversal of the steps of glycolysis (see Chapter 10). Glyoxysome-containing tissues are thus able to convert simple 2-carbon sources such as acetate into carbohydrate. In some tissues, such as fat-storing cells in seeds, the acetate is obtained through the degradation of fatty acids (Chapter 10); therefore, glyoxysomes participate in the conversion of fat to carbohydrate.

Distribution and Origin of Glyoxysomes. The glyoxylate cycle is especially significant for cells growing exclusively on acetate or fatty acids (e.g., a number of microorganisms), where the cycle acts as a source of 4-carbon dicarboxylic acids. Certain microorganisms including *Euglena, Chlorella, Neurospora,* and *Polytomella* contain "glyoxysome-like" particles, but the term *glyoxysome* is normally reserved for the organelles of fat-storing endosperm or cotyledons of germinating fatty seeds; glyoxysomes are not found elsewhere in plants.

Reports have appeared in the literature from time to time indicating that DNA is present in glyoxysomes, and this raises the possibility that these organelles possess some degree of autonomy. However, this notion is not generally accepted. Glyoxysomal enzymes appear to be synthesized by free ribosomes in the cytosol and are often taken up by preexisting glyoxysomes. The origin of new glyoxysomes remains uncertain. Glyoxysomes may arise as outgrowths of the ER or may be produced by the fission or budding of organelles already present in the cell.

In concluding this discussion of microbodies, it is important to note that some organelles fitting the general microscopic description of microbodies do not clearly fit into either the peroxisome or glyoxysome category when evaluated in terms of their enzymatic activities. It is entirely possible that microbodies may be associated with varying activities depending upon the specialization of the cell and that microbodies exist whose actions and cellular functions remain to be determined. It is clear that one cannot name particles solely on the basis of microscopic characteristics and expect that all will function identically.

Summary

The realization that certain hydrolase activities are associated with a distinct class of organelles called **lysosomes** is quite recent. Previously, these activities were believed to be localized in mitochondria. Lysosomes function in the intracellular digestion of poorly functioning or superfluous organelles, as well as endocytosed materials. Several forms of lysosomes may be identified, including **primary lysosomes, secondary lysosomes** (e.g., **heterophagic** and **autophagic** vacuoles), and **residual bodies.** Primary lysosomes are released from the maturing or *trans* face of Golgi bodies, their hydrolase content initially derived from rough endoplasmic reticulum. Fusion of primary lysosomes with **phagosomes** forms the secondary lysosomes, in which digestion occurs. Usable products of this digestive activity are transferred to the cytoplasm. Undigested or unabsorbed materials remain in the residual bodies, which may accumulate in the cell or fuse with the plasma membrane during **exocytosis.**

Two distinct but related classes of microbodies occur in cells; these are **peroxisomes** and **glyoxysomes.** Peroxisomes are found in both animal and plant tissues and contain a number of *flavin oxidases* that produce hydrogen peroxide during their degradative activity. The potentially harmful peroxides are further degraded by peroxisomal catalase. Glyoxysomes are found in certain plant tissues and contain enzymes of the **glyoxylate cycle** in addition to peroxisomal enzymes.

References and Suggested Reading

Articles and Reviews

Bainton, D. F., The discovery of lysosomes. *J. Cell Biol. 91,* 66 (1981).

Beevers, H., Microbodies in higher plants. *Annu. Rev. Plant Physiol. 30,* 159 (1979).

de Duve, C., The lysosome. *Sci. Am. 208*(5), 64 (May 1963).

de Duve, C., The peroxisome: A new cytoplasmic organelle. *Proc. R. Soc. Lond. 173,* 71 (1969).

Fahimi, H. D., and Yokota, S., Ultrastructural and cytochemical aspects of animal peroxisomes—some recent observations, in *International Cell Biology* (H. G. Schweiger, Ed.), Springer-Verlag, Berlin, 1981.

Farquhar, M. G., and Palade, G. E., The Golgi apparatus (complex)—(1954-1981)—from artifact to center stage. *J. Cell Biol. 91,* 77s (1981).

Lazarow, P. B., Functions and biogenesis of peroxisomes, in *International Cell Biology* (H. G. Schweiger, Ed.), Springer-Verlag, Berlin, 1981.

Leighton, F., Poole, B., Beaufay, H., Baudhuin, P., Coffey, J. W., Fowler, S., and de Duve, C., The large-scale separation of peroxisomes, mitochondria, and lysosomes from the livers of rats injected with Triton WR-1339. *J. Cell. Biol 37,* 482 (1968).

Lodish, H. F., Braell, W. A., Schwartz, A. L., Strous, G. J., and Zilberstein, A., Synthesis and assembly of membrane and organelle proteins. *Int. Rev. Cytol. Suppl.* 12, 247 (1981).

Newcomb, E. H., and Tandon, S. R., Microbodies (peroxisomes) in soybean root nodules, in *International Cell Biology* (H. G. Schweiger, Ed.), Springer-Verlag, Berlin, 1981.

Tolbert, N. E., Metabolic pathways in peroxisomes and glyoxysomes. *Annu. Rev. Biochem. 50,* 133 (1981).

Tolbert, N. E., and Essner, E., Microbodies: peroxisomes and glyoxysomes. *J. Cell Biol. 91,* 271 (1981).

Willingham, M. C., Haigler, H. T., Dickson, R. B., and Postan, T. H., Receptor-mediated endocytosis in cultured cells: coated pits, receptosomes, and lysosomes, in *International Cell Biology* (H. G. Schweiger, Ed.), Springer-Verlag, Berlin, 1981.

Books, Monographs, and Symposia

Breidenbach, R. W., Microbodies, in *Plant Biochemistry* (3rd ed.) (J. Bonner and J. E. Varner, Eds.), Academic Press, New York, 1976.

Dean, R. T., *Lysosomes,* Camelot Press, Southampton, U. K., 1977.

de Duve, C., and Wattiaux, R., Function of lysosomes, in *Annual Review of Physiology* (V. E. Hall, A. C. Giese, and R. R. Sonnenschein, Eds.), Annual Reviews, Palo Alto, Calif., 1966.

de Reuck, A. V. S., and Cameron, M. P. (Eds.), *CIBA Foundation Symposium on Lysosomes,* Little, Brown, Boston, 1963.

Dingle, J. T., and Fell, H. B. (Eds.), *Lysosomes in Biology and Pathology* (Parts I and II), North-Holland, Amsterdam, 1969.

Dingle, J. T., *Lysosomes,* North-Holland, Amsterdam, 1977.

Fleischer, B., and Fleischer, S., The Golgi apparatus, in *Advanced Cell Biology* (L. M. Schwartz and M. M. Azar, Eds.), Van Nostrand Reinhold, New York, 1981.

Hruban, Z., and Rechcigl, M., *Microbodies and Related Particles,* Academic Press, New York, 1969.

Schweiger, H. G. (Ed.) *International Cell Biology (1980–1981),* Springer-Verlag, Berlin, 1981.

Strauss, W., Lysosomes, phagosomes and related particles, in *Enzyme Cytology* (D. B. Roodyn, Ed.), Academic Press, London, 1967.

Chapter 20
THE NUCLEUS AND CELL DIVISION

This is the first of two successive chapters dealing with the biology of the cell nucleus. In this chapter, we will examine the structural and chemical organization of the major nuclear components and the gross nuclear changes that characterize cell division. In Chapter 21, we will be concerned with the operation of the genetic apparatus, focusing on the molecular basis of gene expression.

Most of the functions of the nucleus are intimately related to events that occur in the cytoplasm; for example, translation of mRNA by cytoplasmic ribosomes (a mechanism described in Chapter 22) is the end result of the transcription of DNA in the nucleus. Structurally too, the nucleus and cytoplasm are interrelated. The outer membrane of the nuclear envelope is continuous with membranes of the endoplasmic reticulum. Events in the cytoplasm also affect nuclear function. Therefore, while the nucleus is perhaps the single dominant structural feature of most eucaryotic cells, it is not to be viewed as an isolated organelle.

Structure of the Nucleus

The nucleus of a eucaryotic cell (Fig. 20–1) is delimited by a pair of membranes called the **nuclear envelope.** The outer and inner membranes of this envelope are separated by a narrow space called the **perinuclear space** but fuse with each other at the margins of **pores.** The fluid of the cytoplasm (the cytosol) is continuous through the nuclear

pores with the fluid of the nucleus, called the **nucleoplasm.** Although ribosomes may be attached to the cytoplasmic surface of the outer nuclear membrane, these ribosomes are not considered nuclear structures. The nucleoplasm contains a number of discrete structures including one or more **nucleoli, chromosomes** (as many as several hundred in some eucaryotic cells), and other structures or regions that appear at different times depending upon nuclear activity.

Procaryotic cells do not have a "true" nucleus; indeed, identification of a cell as **procaryotic** ("pre-nucleus") or **eucaryotic** ("true-nucleus") is based upon the presence or absence of this membrane-enclosed organelle. In procaryotic cells, most of the genetic material is confined to an area of the cell called the **nucleoid** (Fig. 1–37), but the nucleoid is not enclosed by a membrane.

Chromosomes

The first descriptions of the chromosomes of eucaryotic cells appeared in the period between 1840 and 1880, but it was not until 1888 that Waldeyer introduced the term **chromosome** ("colored body") for these structures. Chromosomes are composed of **chromatin,** which readily binds basic stains. Because the chromatin is highly condensed during cell division, the chromosomes are readily seen and described by light microscopy. The number of chromosomes in the cell nucleus varies considerably among dif-

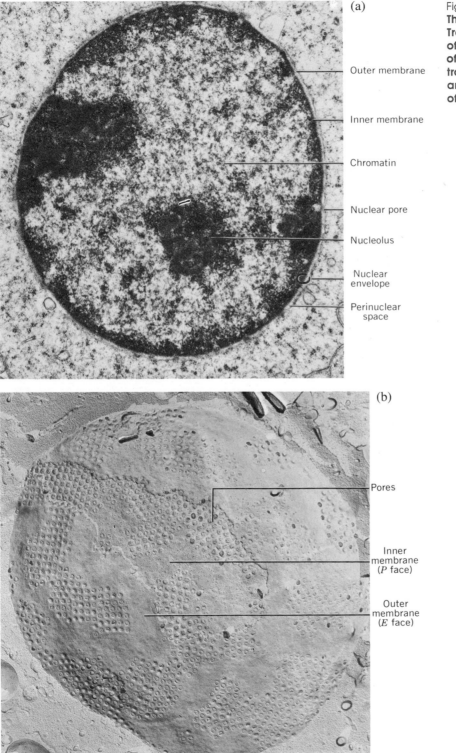

(a)

Outer membrane

Inner membrane

Chromatin

Nuclear pore

Nucleolus

Nuclear envelope

Perinuclear space

(b)

Pores

Inner membrane (*P* face)

Outer membrane (*E* face)

Figure 20–1

The nucleus of the eucaryotic cell. *(a)* Transmission electron photomicrograph of a thin-sectioned nucleus. (Courtesy of R. Chao.) *(b)* Transmission electron photomicrograph of freeze-etched and fractured cell nucleus. (Courtesy of Dr. E. G. Pollock).

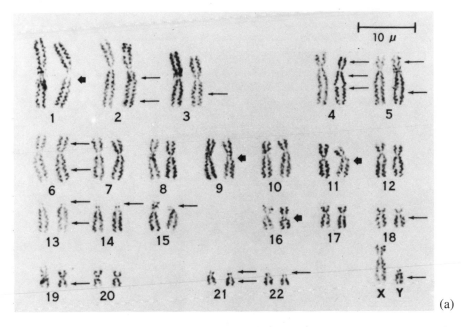

(a)

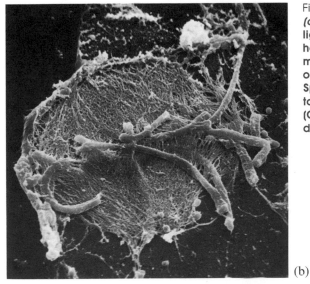

(b)

Figure 20–2

(a) The 23 pairs of homologous human chromsomes as seen by light microscopy. Each chromosome consists of paired chromatids held together at the centromere. These chromosomes are from a male cell since the pair of sex chromosomes consists of one *X* and one *Y* chromosome. (Courtesy of Dr. Y. Ohnuki; copyright © 1968 by Springer-Verlag, *Chromosoma 25,* 416.) *(b)* Scanning electronphotomicrograph of metaphase chromosomes on the mitotic spindle. (Courtesy of Dr. W. K. Heneen; copyright © 1980, Société Française de Microscopie Électronique; *Biol. Cellulaire 37,* 13.)

ferent animal and plant species; however, each species has a specific *chromosome number.* For example, cells of the nematode *Ascaris megalocephala* have only 2 chromosomes, human cells have 46, and certain protozoa have over 300. Unrelated organisms may have the *same* chromosome number. The potato plant has 48 chromosomes, but so do plum trees and chimpanzees. Within a species, each chromosome exhibits a specific and characteristic

shape during the metaphase (see below) of cell division. The unique appearance of the metaphase chromosomes is retained from one generation of cells and organisms to the next (Fig. 20–2).

Chromosome shape and size change during the stages of nuclear division. Most chromosomes have two **arms,** one on each side of the **primary constriction** or **centromere** (also called **kinetochore**). Metaphase chromosomes have

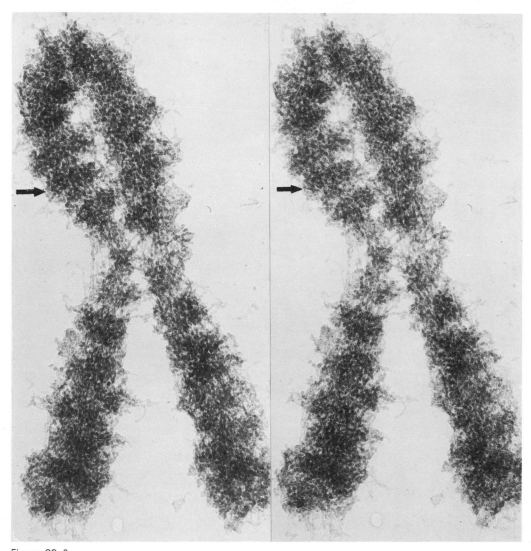

Figure 20–3
Stereophotomicrographs of a metaphase chromosome. Coiling of the chromatin threads to form a helical structure is apparent. Arrow identifies one gyre of the helix. (Courtesy of Dr. Hans Ris.)

already undergone replication so that each chromosome consists of **sister chromatids** and therefore appears to have *two* sets of arms (Fig. 20–3). The centromere is the site of attachment of the chromosome to the microtubules of the **spindle** and acts as the focus of chromosome (or chromatid) movement during the anaphase of division. Chromosomes that lack a centromere are said to be **acentric** and fail to segregate normally during division. *Secondary* and *tertiary constrictions* may also be identified. Secondary constrictions are associated with nucleoli and are called **nucleolar**

organizer regions (**NOR**). The significance of the tertiary constrictions is unclear, but their characteristic disposition in each chromosome helps to distinguish one chromosome from another.

Fine Structure of Chromosomes

The condensed chromosomes visible during mitosis are composed of an organized array of **chromatin fibers** (Fig. 20–4), but in their decondensed state they give rise to a

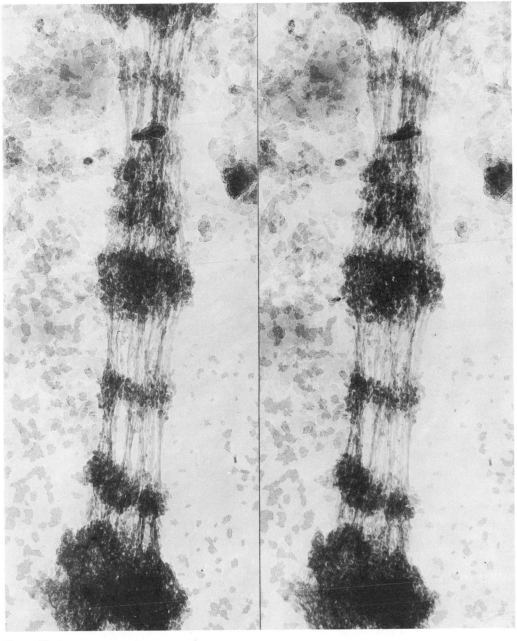

Figure 20–4
Stereophotomicrographs of a *polytene* (i.e., multistranded) chromosome in a fat cell of the larva of *Drosophila melanogaster*. The chromosome consists of 24 nm chromatin fibers, which are slightly stretched in the regions between successive bands. (Courtesy of Dr. Hans Ris; copyright © 1976 by Intertal, International Publications Co.)

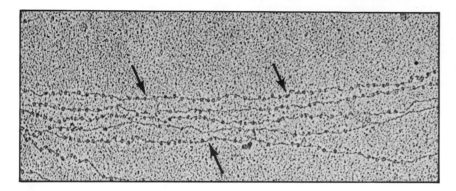

Figure 20–5
Electronphotomicrograph of chromatin fibers revealing a succession of *nucleosomes* (arrows). (Courtesy of Dr. F. Puvion-Dutilleul.)

highly dispersed network. Each chromatin fiber is believed to contain one molecule of DNA. The average diameter of *type A* chromatin fibers is about 10 nm, whereas the diameter of the DNA double helix is only 2 nm (see Fig. 7–8). The difference is due to the coiling of the DNA molecule and the presence of large quantities of proteins (primarily histones and enzymes) and some RNA. *Type B* fibers are even thicker (20–30 nm), the increased thickness resulting from additional levels of coiling. Chromatin of varying thickness exists in the interphase nucleus and is presumed to consist of alternating areas of type A and B fibers.

An indication of the extent of DNA coiling that occurs in a chromosome is obtained from the **packing ratio** (i.e., the length of DNA divided by the length of the chromatin fiber or whole chromosome). For decondensed interphase chromatin this ratio is about 10:1 but for metaphase chromosomes is more than 1000:1. Several levels of coiling occur during the condensation of chromatin that precedes mitosis.

The Nucleosome

In recent years it has been shown that chromatin consists of a repeating pattern of bodies, called **nucleosomes** (Fig. 20–5), formed by the association of DNA and histone molecules. Nucleosomes are quasi-cylindrical in shape and have a diameter of about 10 nm and a length of 5.5 nm. The histones form a core around which the DNA is wound to form a superhelix (Fig. 20–6). Nucleosomes can be released from chromosomes by partially digesting the chromatin with a nuclease. Digestion by the nuclease is progressive and initially releases nucleosomes containing 160

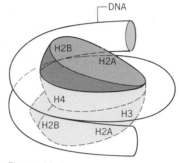

Figure 20–6
Diagram of structure proposed for the *nucleosome*. The nucleosome consists of an octamer of histones (two each of histones H2A, H2B, H3, and H4) around which the DNA makes nearly two full turns. Histone H1 (not shown in the diagram) is associated with the lengths of DNA between nucleosomes.

to 240 base pairs of DNA and includes the **spacer** (or **linker**) DNA that joins successive nucleosomes (Fig. 20–7). Continued digestion produces an intermediate product containing about 160 to 170 base pairs, the core of histones, and a separate histone termed H1. Finally, a stable unit is formed that consists of the core of histones around which is wound a length of DNA containing about 146 base pairs.

The histone core is an **octamer** consisting of two each of four histones identified as H2A, H2B, H3, and H4 (Fig. 20–6). The DNA forms nearly two complete turns around the core. The H3 and H4 histones are believed to form a subnucleosome particle that holds the main DNA loop, while the two H2A and H2B histones are associated with the DNA nucleotide sequences at either end of the loop. Histones H2A and H2B have been found to be rich in the

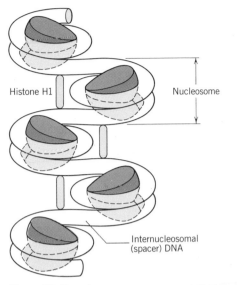

Figure 20–7
Diagram of a chain of nucleosomes. The position of histone H1 is not known with certainty.

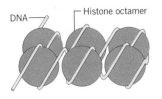

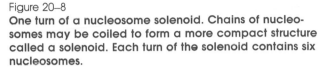

Figure 20–8
One turn of a nucleosome solenoid. Chains of nucleosomes may be coiled to form a more compact structure called a solenoid. Each turn of the solenoid contains six nucleosomes.

basic amino acid lysine; this would account for their ability to bind these stretches of the DNA.

Histone H1, also rich in lysine, is associated with the spacer DNA. The core particle, the spacer DNA, and H1 comprise a unit called a **mononucleosome** or **chromatosome.** In the complete chromatosome, the two points at which the DNA molecule enters and leaves the core particle are located near one another on the particle's surface. When H1 is digested away, the strands tend to be displaced further from one another. Thus it seems that histone H1 locks the entering and exiting stretches of DNA into position and is closely associated with the core.

The frequency of occurrence of nucleosomes along chromatin fibers varies in different tissues. Although the number of base pairs in the DNA wound around the core particle remains the same in all cells, spacer length varies among species and among tissues within a species. Spacer DNA may contain as few as zero bases (e.g., in yeast cells) or as many as 80 bases (e.g., in sea urchin sperm cells). The significance of the variation in spacer length is not yet clear.

As noted earlier, two classes of chromatin fibers exist having diameters of 10 and 30 nm. It appears that the 10 nm fibers are formed from a loose, linear array of nucleosomes, whereas 30 nm fibers are formed by the helical coiling of 10 nm fibers. Each turn of the helix in a 30 nm fiber consists of about six nucleosomes, and successive turns form a structure called a *solenoid* (Fig. 20–8). Whether or not chains of nucleosomes in isolated chromatin exist in the linear or solenoid configuration is determined by the ionic strength of the medium. The loose, linear array exists at low ionic strength, but the solenoid structure will form as the ionic strength is raised. In the linear form, the axis of each nucleosome is parallel to the main axis of the fiber, but as the fiber twists to assume the more compact solenoid configuration, the nucleosome axes are nearly perpendicular to the main axis of the solenoid. These configurational changes are believed to have physiological significance, for they are believed to underly the initial stages of chromatin condensation that characterizes the onset of mitosis and the decondensation that follows mitosis. Histone H1 is essential to the transition from the loose, linear array to the solenoid structure. In the absence of H1, no ordered structures are formed.

Bacterial and Viral Chromosomes

The chromosomes of bacteria and viruses are single nucleic acid molecules. In most cases, they consist of double helical DNA, but in certain viruses they consist of only a single strand of DNA or RNA. The DNA may be linear or circular, depending on the cell or virus.

Bacteria. The chromosomes of *E. coli* and *B. subtilis* have been studied more extensively than those of any other bacteria. In both species the chromosome contains a single,

Figure 20–9
Chromosome released from an *E. coli* cell. The chromosome contains many supercoiled domains. The reference bar is 1 μm long. (Courtesy of Dr. A. Worcel, copyright © 1974 by Academic Press, *J. Mol. Biol. 82,* 108.)

circular DNA molecule more than a millimeter long. The chromosome of *E. coli* is about 1100 μm long and has a diameter of 2 to 4 nm and a molecular weight of 2.5 × 10^9 daltons. It is condensed into an area less than 1 μm² in the cell. When mild procedures are used to extract *E.*

coli DNA into solutions rich in cations, the chromosome remains in a highly condensed state (Fig. 20–9). In this form, the chromosome consists of several dozen loops. In each loop the DNA is highly twisted and forms a **supercoil** (see below). Treatment of the chromosome with nucleases

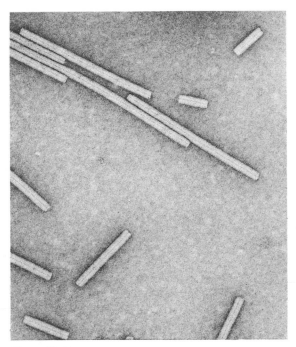

Figure 20–10
The tobacco mosaic virus. (Photomicrograph courtesy of Dr. F. Eiserling.)

that degrade RNA frees some of the loops so that the chromosome takes on a more open appearance. The loops are believed to be stabilized by RNA and protein.

Viruses. The nucleic acid of a virus may be either RNA or DNA. Among the RNA viruses, the nucleic acid occurs in four different structural states: (1) single-stranded, linear forms (e.g., tobacco mosaic virus); (2) single-stranded circular forms (e.g., encephalomyocarditis and potato spindle tuber viruses); (3) double-stranded, linear forms (e.g., mammalian reoviruses); and (4) double-stranded, circular forms (e.g., mycoviruses). One of the best studied RNA viruses and the first to be prepared in crystallized form is the tobacco mosaic virus (TMV). This virus has the shape of a cylinder 300 nm long and 18 nm in diameter (Fig. 20–10). Within the protein coat is the viral chromosome, which consists of a single 3300 nm long RNA molecule comprised of 6400 nucleotides. In order to fit within the virus, the RNA molecule is twisted to form a right-handed helix containing 49 nucleotides per turn and a pitch of 2.3 nm. The diameter of the helix is about 8 nm. Running through the center of the helix is an aqueous core about 3 nm in diameter. The protein sheath or **capsid** associated with the outer surface of the RNA helix consists of 2130 identical protein subunits.

The nucleic acids of the DNA viruses also occur in four different structural states: (1) single-stranded, linear forms (e.g., the parvoviruses); (2) single-stranded, circular forms (e.g., ϕX174 and M13); (3) double-stranded, linear forms (e.g., Herpes virus, the T-even bacteriophages, and lambda phage); and (4) double-stranded circular forms (e.g., SV 40 and Polyoma viruses). The double-stranded DNAs of the T_4 and lambda bacteriophages are shown in (Fig. 20–11).

Supercoiling

The organization of eucaryotic DNA into nucleosomes demonstrates that DNA is capable of several different levels of coiling and folding. The first level is the coiling of the two polynucleotide chains around one another to form the conventional double-helix structure (i.e., that shown in Fig. 7–8). In the second level of coiling, the double helix is wound around the histone core of the nucleosome. Finally, nucleosomes may be arranged helically to form the so-called solenoid structure. In bacteria, the length of the chromosome is many times greater than the diameter of the nucleoid area in which the DNA is concentrated; thus, it is clear that a great deal of folding of procaryotic DNA does occur. DNA folding in procaryotic cells and the formation of solenoids in eucaryotic cells is called **supercoiling.** Viral nucleic acid is also supercoiled.

If *one* of the two strands that form a circular, double-helical DNA molecule is broken to produce *two* free ends and one of the free ends is then rotated around the unbroken strand while the other free end remains fixed, the resulting structure is a *supercoil*. If the rotation of the free end is in the same direction as the double-helix, a *positive* supercoil is produced; if the rotation of the free end is left-handed, a *negative* supercoil results. Left-handed rotation serves to ''unwind'' the DNA double-helix, while right-handed rotation ''overwinds'' the double-helix. Circular DNAs of procaryotic cells are characteristically positively supercoiled (Fig. 20–12).

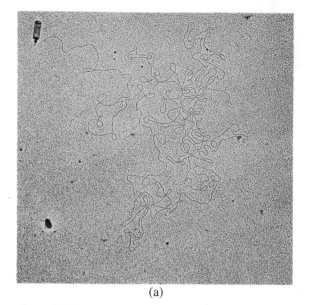

(a)

Figure 20–11

(a) DNA molecule released from a T$_4$ bacterio-phage (top left corner of the photomicrograph). *(b)* DNA from the *lambda* phage; the reference bar is 1 μm long. (Courtesy of Dr. Hans Ris, copyright © 1963 by Cold Spring Harbor Laboratory, *C. S. H. Symp. Quant. Biol. 28, 2.)*

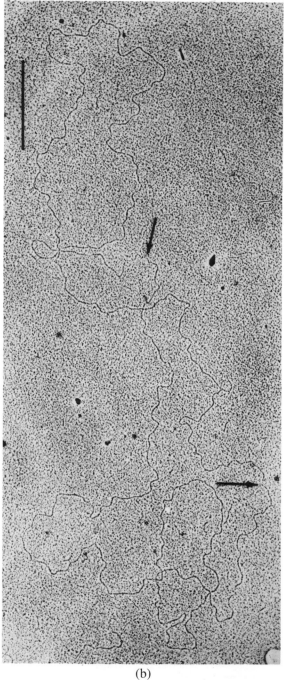

(b)

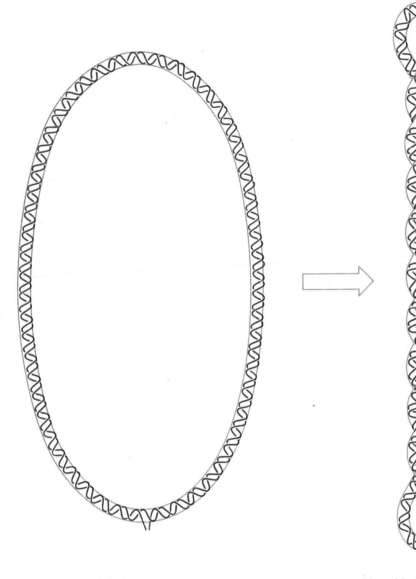

Figure 20–12

Supercoiling of DNA. When one of the two strands of a circular DNA molecule is nicked *(left)* and one free end rotated in a clockwise direction around the unnicked strand, a positive supercoil is formed *(right).*

Plasmids and Episomes

Plasmids are *extrachromosomal* genetic elements found in bacteria that take the form of relatively small (i.e., they contain only a small number of genes), circular, double-stranded DNA molecules (Fig. 20–13). Plasmid DNA is usually positively supercoiled. There may be several dozen plasmids in one cell, and although they are capable of autonomous replication, the number of plasmids remains fairly constant from one generation of cells to the next.

Plasmids by definition are extrachromosomal, but under certain conditions plasmid DNA can be incorporated into the cell's main chromosome. A plasmid that is capable of integration into the main chromosome is referred to as an *episome.* Some plasmids are transferred from one cell to another by viruses.

In *E. coli* cells, there are three classes of plasmids: (1) *F* (i.e., *fertility*) factors, (2) *R* (i.e., *resistance*) factors, and (3) *col* (*colicinergic*) factors. The *F* factors promote sexual *conjugation* in *E. coli,* the *R* factors offer the bacteria

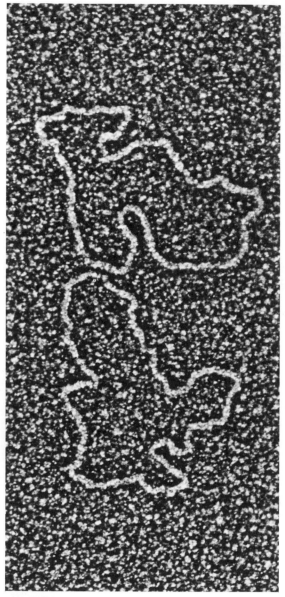

Figure 20–13
Plasmids isolated from *E. coli* cells. (Photomicrograph courtesy of Dr. E. Namork.)

The *F* Factor and Conjugation. The *F* factor of *E. coli* has a circumference of about 20 μm and contains several families of genes. Not all *E. coli* cells contain this plasmid; those that do are called *F*⁺ cells, while those that lack the *F* factor are called *F*⁻. Among the genes of this plasmid are those that control the development of *pili* or tubelike extensions from the surface of the *E. coli* cells. When these pili contact the surfaces of neighboring cells, they form a channel through which plasmid DNA can pass from one cell to another. Before transfer between two conjugating *E. coli* cells begins, a nick is produced in one of the two strands of the circular DNA molecule; it is at this point that replication of the DNA is initiated. The unbroken strand serves as a template for the production of a new complementary strand. As this synthesis proceeds, the nicked strand is displaced through the pili into the neighboring cell. The transferred strand then acts as template in the neighboring cell for the synthesis of a complementary DNA strand, and upon completion of synthesis, the ends of both strands are closed to form a circular molecule. Thus, the *F*⁺ cell and the *F*⁻ cell with which conjugation occurred have complete copies of the plasmid.

Mitochondria and Chloroplast DNA

Mitochondria and chloroplasts contain DNA. Mitochondrial DNA (called *mtDNA*) from animal cells is double-stranded and circular, whereas mtDNA from plant cells is double-stranded and linear. In multicellular animals, the mtDNA is about 4.4 to 5.6 μm in length and has a molecular weight of about 9 to 11×10^6 daltons. In protozoa and yeast mtDNA ranges in length from 16 to 25 μm. Plant mtDNA is the longest, having lengths up to 62 μm. The mtDNA is found in the matrix of the mitochondria (see Fig. 16–9) and at a specific point along its length is attached to the inner mitochondrial membrane.

Chloroplast DNA (*chlDNA*) is circular and typically has a length of 35 to 45 μm, although some chlDNA may be as long as 200 μm. Linear chlDNAs have been isolated from some plants, but these are generally considered to be fragments of larger circular molecules.

Forms of Chromatin

For convenience, the cell cycle (see Chapter 2) may be divided into two major phases—the **interphase** (in which

resistance to certain drugs, and the *col* factors give the bacteria the capacity to secrete *colicins,* which have an antibiotic function (i.e., they kill bacteria that lack the *col* factor).

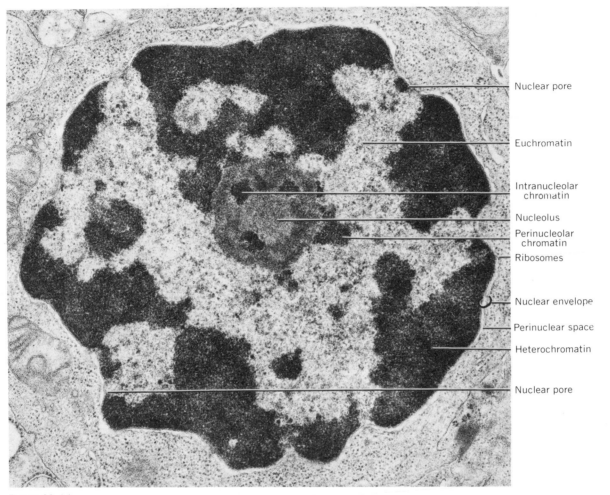

Nuclear pore

Euchromatin

Intranucleolar
chromatin

Nucleolus

Perinucleolar
chromatin

Ribosomes

Nuclear envelope

Perinuclear space

Heterochromatin

Nuclear pore

Figure 20–14

Types and distribution of chromatin in the interphase nucleus (see also Fig. 20–15). (Photomicrograph courtesy of R. Chao.)

the cell is engaged in its cell-specific or tissue-specific activities) and the **mitotic phase** in which the cell undergoes division. During **mitosis** (discussed later in the chapter), chromosomes of eucaryotic cells *condense,* and when stained with basic dyes, the chromatin is easily studied by microscopy. During the *interphase* of the cell cycle, most of the chromatin does not exist in the condensed state, and it is difficult, if not impossible, to distinguish individual chromosomes. However, there are areas of the nucleus called **chromocenters** that do stain deeply even during interphase. In 1928, E. Heitz identified these chromocenters

as portions of chromosomes that remain in the condensed state throughout the cell cycle.

Depending upon their staining properties, two different types of chromatin may be distinguished in the interphase nucleus (Figs. 20–14 and 20–15). Portions of chromosomes that stain lightly are only partially condensed; this chromatin is termed **euchromatin** and usually represents most of the chromatin that disperses after mitosis is completed. In the darkly staining regions, the chromatin remains in the condensed state and is called **heterochromatin.** Usually there is some condensed chromatin around the nucleolus,

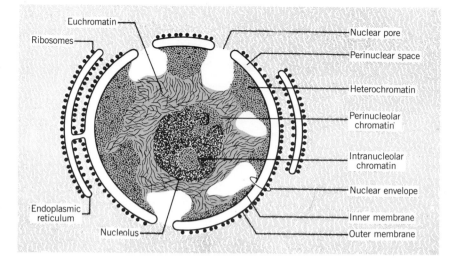

Figure 20–15
The interphase nucleus (compare with Fig. 20–14).

called **perinucleolar chromatin,** and some inside the nucleolus, called **intranucleolar chromatin.** The perinucleolar and intranucleolar chromatin appear to be connected; together they are referred to as **nucleolar chromatin.**

Dense clumps of deeply staining chromatin often occur in close contact with the inner membrane of the nuclear envelope and are referred to as **condensed peripheral chromatin** (Fig. 20–14). Between the peripheral chromatin and the nucleolar chromatin are regions of lightly staining chromatin called **dispersed chromatin.**

Heterochromatin can be further divided into two types: **constitutive heterochromatin** and **facultative heterochromatin.** In constitutive chromatin, the DNA is permanently inactive and remains in the condensed state throughout the cell cycle. Facultative heterochromatin is not permanently maintained in the condensed state but does undergo periodic dispersal and during these times is transcriptionally active.

Heterochromatin is characterized by its especially high content of repetitive DNA sequences and contains very few, if any, structural genes (i.e., genes that encode proteins). However, heterochromatin may be involved in the regulation of gene expression, for when euchromatic genes of known functions are relocated adjacent to heterochromatin, their expression is modified. Euchromatin is believed to contain the structural genes and is expressed when decondensed in the interphase cell.

Chemically, chromatin consists of DNA, proteins (primarily histones, but also "nonhistone" proteins), and some

RNA. The histone:DNA weight ratio varies from about 0.8 to 1.3 and averages about 1.1. The ratio varies not only with the species but also among the tissues of a single organism. Histones are constituents of the chromatin of all eucaryotic organisms except fungi, which therefore resemble procaryotic organisms in this respect. RNA rarely accounts for more than 5% of the chromatin.

Chromosome–Chromosome Associations

It has been known for some time that certain associations occur among the chromosomes arranged on the metaphase plate during nuclear division. There may be size assortment on the spindle such that long chromosomes are on the outside and short chromosomes are on the inside. Certain chromosomes are known to group together or have arms directed toward one another. G. Hoskins has shown that human chromosomes are linked together by interchromosomal "connectives." The connectives are composed of DNA and protein and hold the chromosomes together even when teased from a cell by microsurgical methods. Fibers linking separate chromosomes have been visualized using electron microscopy.

Polytene Chromosomes

In the salivary glands of dipteran flies, of which the genus *Drosophila* has been most extensively studied, and in certain other tissues as well, the interphase nucleus is characterized by extremely large chromosomes. These so-called giant chromosomes are actually about 1000 parallel and tightly packed copies of the *same* chromosome (Fig. 20–4). Because of their multiple structure, they are called **polytene** chromosomes. Along the length of the polytene chromosome, disks or bands of variably staining intensities can be distinguished. The identification of the genetic content of these bands (and interband regions) has played a major part in the genetic analysis of dipteran organisms.

Some of the bands of giant chromosomes appear to be swollen or "puffed," the specific bands exhibiting puffing varying from one tissue to another even within the same organism. It is now well established that the puffed bands of these interphase chromosomes correspond to regions in which the DNA is being actively transcribed into mRNA (Chapter 21).

The Nucleolus

The nucleolus is the most dominant internal feature of the cell nucleus and was among the first subcellular structures described by microscopy (by Fontana in 1774). The nucleolus is not separated from the remaining nucleoplasm by a membrane, but in many cells its margins are associated with chromatin (i.e., the perinucleolar chromatin, see Fig. 20–14). Depending upon the physiological state of the cell, nucleoli vary in number, size, and appearance. Usually, the nucleolus is attached to a particular chromosome, which produces this structure at a certain site called the **nucleolar organizing region (NOR).**

The functions of the nucleolus are considered in detail in Chapter 22, which deals with protein synthesis. It is clear that one of the primary functions of the nucleolus is the synthesis of most of the rRNA species found in the small and large subunits of ribosomes and the packaging of these rRNAs with ribosomal proteins to form preribosomal particles.

The Nuclear Envelope

The presence of a membrane separating the nuclear material from the surrounding cytoplasm is one of the principal characteristics distinguishing eucaryotic organisms from procaryotic organisms. The existence of a membrane delimiting the nucleus was first demonstrated by O. Hertwig in 1893. However, little interest was shown in studying this membrane until electron microscopy revealed that it was not simply a single membrane, but rather a double membrane in which the outer membrane had features that clearly distinguished it from the inner membrane. The two membranes lie close together, one surrounding the other. The two membranes fuse together at the **nuclear pores** but elsewhere are separated by the **perinuclear space,** which varies in thickness from about 10 to 50 nm (Fig. 20–14). The folded appearance at the nuclear pores and the thin perinuclear space between the membranes aptly justifies the term **nuclear envelope** to describe the entire structure. A fibrous material called the **lamina densa** lies just below the inner nuclear membrane and is believed to give structural support to the envelope and also provide anchorage for the condensed peripheral heterochromatin.

The functions of the nuclear envelope are diverse. Clearly, it acts to compartmentalize the nucleoplasm. Mitochondria, Golgi bodies, lysosomes, chloroplasts, and other cytoplasmic organelles do not enter the nucleus, and the chromosomes and nucleoli rarely exit into the surrounding cytoplasm. However, ribosome precursors from the nucleolus and mRNA and tRNA molecules do leave the nucleus through the nuclear pores. Also, molecules enter the nucleus from the cytoplasm, triggering a number of nucleocytoplasmic interactions; some of these are described in Chapter 24.

The nuclear envelope is not just a physical barrier, but also functions in connection with cellular activities taking place on either side. Ribosomes may be attached to the cytosol face of the outer nuclear membrane, which thereby takes on a structural appearance like rough endoplasmic reticulum (Chapter 22); indeed, the outer membrane may have periodic continuities with the endoplasmic reticulum. The intimate relationship between the nuclear membranes and the endoplasmic reticulum is not surprising in view of the fact that (1) a new nuclear envelope forms at the end of mitosis by the coalescence of membranes of the endoplasmic reticulum and (2) new membranous elements for various cellular organelles are produced by the rough endoplasmic reticulum (Chapter 15).

The inner membrane of the nuclear envelope may have special functions associated with the chromosomes. As noted earlier, in the interphase nuclei of many cells, heterochromatin is closely appressed to the inner membrane. This association does not occur at the nuclear pores; instead, the chromosomes seem to be firmly attached to the

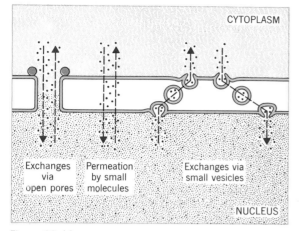

Figure 20–16
Various avenues for transport of material between the nucleus and the extranuclear cytoplasm.

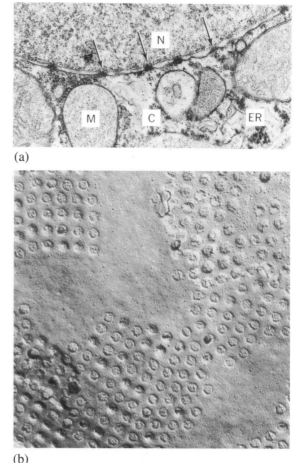

(b)

Figure 20–17
The nuclear envelope. *(a)* **Thin section through the envelope.** *C*, **cytoplasm;** *N*, **nucleus; arrows point to pores that are plugged with granular material;** *M*, **mitochondrion;** *ER*, **endoplasmic reticulum. (Photomicrograph courtesy of R. Chao.)** *(b)* **Freeze-fractured nuclear envelope and pores. (Courtesy of Dr. E. G. Pollock.)**

interpore areas. The role of the association between chromatin and the nuclear envelope is unclear but may be related to chromosome replication and orientation during interphase and the early prophase of mitosis and meiosis.

Some materials such as ribonucleoprotein complexes are exchanged between the nucleoplasm and the cytoplasm through open nuclear pores. However, the nuclear pores are not the only avenues for nucleocytoplasmic exchanges (Fig. 20–16). For example, small molecules and ions readily permeate both nuclear membranes. Larger molecules and particles may pass through the membrane by formation of small pockets and vesicles that traverse the envelope and empty on the other side. It is also possible that small sections of the entire envelope evaginate or break away and undergo dissolution (as is known to occur during mitosis and meiosis).

Although many cellular membranes contain pores, the pores of the nuclear envelope possess a special character since they are openings formed by the fusion of *two* membranes (Fig. 20–17). The number of nuclear pores varies widely among different cells but in general ranges from 1 to about 60 pores per square micrometer of membrane. Unlike pores in other cytoplasmic membranes, the nuclear pore is a complex of structures. The **orifice** is essentially circular, with occasional evidence of a polygonal shape. Seen in freeze-fractured preparations, the pores appear as rings and therefore are called *annuli*. The inside diameter of an annulus is markedly constant for a given cell type but

may yield variable values depending on the technique used to fix and examine the cell. The usual range is 60 to 100 nm, with most nuclear envelopes having pore orifice diameters of about 70 nm.

The rim of the pore contains eight granular *subunits* symmetrically placed about the circumference (Fig. 20–18). The center of the pore often contains distinct substructures such as conical projections from the sides into the center. More often than not, the lumen of the pore is filled

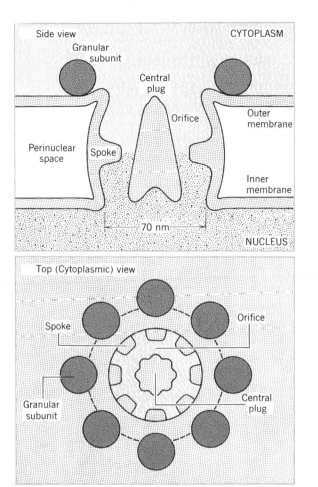

Figure 20–18
Model for the organization of the nuclear pore. Eight *granular subunits* **are symmetrically arranged on the outer cytoplasmic margin of the pore.** *Annular projections* **or** *spokes* **are directed into the** *orifice* **and toward the** *central plug.*

with "annular material" that projects into both the nucleoplasm and cytoplasm often flowing radially beyond the pore rim.

Mitosis

During nuclear division or **mitosis,** there is a progressive change in the structure and appearance of the chromosomes. Although mitosis is a continuous process, for convenience it is divided into four major stages (Figs. 20–19 and 20–

20). **Prophase,** the first stage of mitosis, is characterized by the condensation of the chromosomes, the disappearance of the nucleolus and nuclear envelope, and the formation of the microtubules of the **spindle.** If, prior to prophase, the cell contained a centriole, then the centriole is also replicated during this phase, with the two resulting centrioles migrating to opposite ends of the spindle. The chromosomes become distinguishable by light microscopy as a result of their progressive shortening and thickenning during prophase and eventually are seen to be composed of

Figure 20–19
The stages of mitosis (see also Fig. 20–20.) (From E. J. Gardner and D. P. Snustad, *Principles of Genetics,* **John Wiley & Sons, New York, 1981, p. 45.)**

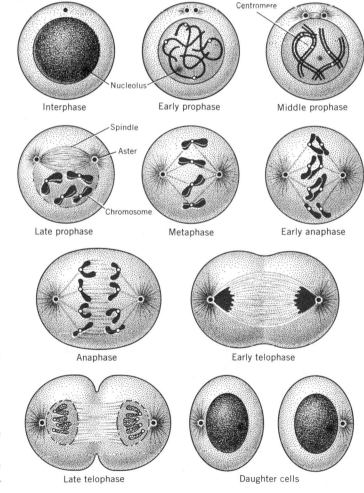

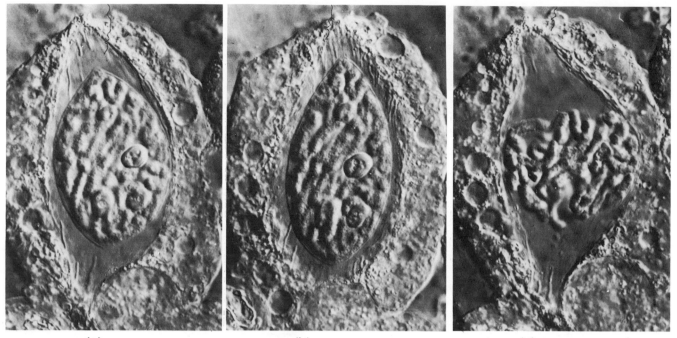

(a) Interphase   **(b)** Early prophase   **(c)** Late prophase

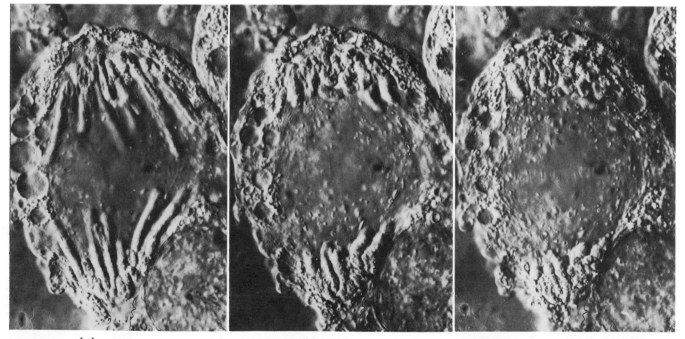

(g) Late anaphase   **(h)** Early telophase   **(i)** Telophase and beginning of cell plate formation

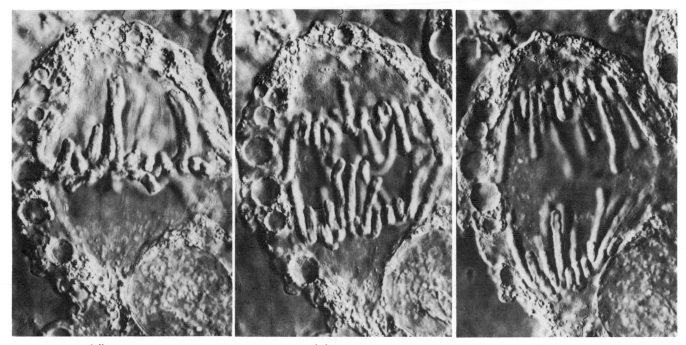

(d) Metaphase   **(e)** Anaphase   **(f)** Late anaphase

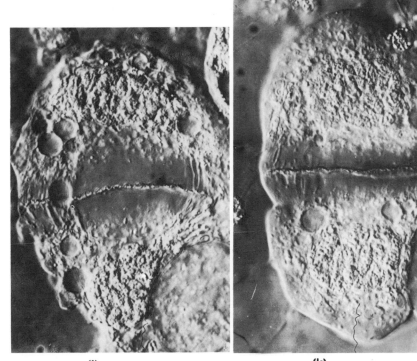

(j) Late telophase   **(k)** Interphase

Figure 20–20
Stages of mitosis as seen in a cell of the *blood lily (Haemanthus).* (Photo-micrographs courtesy of Dr. A. S. Bajer; see also Fig. 20–19.)

two *sister chromatids* held together at the centromere *or* **kinetochore.** The sister chromatids are the products of the replication of chromosomal DNA during the interphase of the cell cycle.

Toward the end of prophase (sometimes called **prometaphase,** since the second phase of mitosis is called the **metaphase**), the chromosomes migrate toward the center of the spindle. Despite decades of study, the cause of this movement remains unclear. During *metaphase,* the centromeres of each chromosome are aligned midway across the spindle on a plane called the **equatorial plate.** At this point, the centromeres are linked to separate spindle fibers such that each chromatid is attached to a *different* pole of the spindle. Some of the spindle fibers do not form associations with any chromosomes and extend directly from one pole to the other. The centromeres are duplicated so that each chromatid becomes an independent chromosome.

The onset of **anaphase** is characterized by the movement of the chromosomes toward opposite poles of the spindle. During anaphase, a process called **cytokinesis** begins and divides the cell into two halves, thereby physically separating the two complements of chromosomes. Cytokinesis is distinct from nuclear division but is usually synchronized so that it occurs during the later stages of mitosis. The final phase of mitosis, called the **telophase,** involves the decondensation of the chromosomes after they have reached the poles of the spindle. During the telophase, nucleoli are reformed together with a nuclear envelope enclosing the chromosomes.

Movement of Chromosomes During Anaphase.   During the anaphase of mitosis, the centromere of each chromosome advances toward one of the two poles of the spindle, with the arms of the chromosomes lagging behind (Fig. 20–20). This arrangement suggests that the chromosomes are being *pulled* toward the pole, and for a number of years many investigators have attempted to determine the nature of the mechanism that is responsible for this movement. Among the different models that have been proposed are (1) the *microtubule* model, (2) the *sliding cytoplasmic filament* model, and (3) the *dynamic equilibrium* model.

The microtubule model was first proposed in the late 1960s by J. R. McIntosh. According to McIntosh, the microtubules extend inward from the poles of the spindle and overlap at the center of the cell. During anaphase, the microtubules slide past one another in the region of overlap, thereby extending the cell and pushing the poles further apart. At the same time that the microtubules slide past each other, they are being disassembled at the poles so that the overall length of the microtubules is decreasing. The diminishing length of the microtubules is accompanied by the movement of the chromosome centromeres that are attached to the microtubules closer to the poles. Since microtubule sliding is known to be involved in other forms of movement in cells, such as the beating of cilia and flagella (Chapter 23), this model is entirely plausible. However, it appears not to be the complete story.

In 1974 A. Forer showed that actin microfilaments are present in the spindle, and later myosin filaments were also shown to be present. This prompted the suggestion that it is the sliding of the actin and myosin filaments past one another that accounts for chromosome movement. The role of the microtubules was relegated to that of serving as a structural framework on which the chromosomes are mounted. Movements of the cytoplasmic filaments are not reflected by corresponding movement of the chromosomes until anaphase begins and the microtubules begin to break down at the two poles of the spindle. In effect, the disassembly of the microtubules at the poles frees the chromosomes to move in response to the sliding of the actin and myosin.

According to the dynamic equilibrium model proposed in 1967 by S. Inoué, the microtubules that comprise the spindle fibers are in a dynamic equilibrium with a pool of microtubule subunits. The addition of new subunits to fibers that extend from one pole of the spindle to the other but do not have attached chromosomes serves to increase the interpolar distance, while at the same time the removal of subunits from either the polar ends or the centromere ends of other spindle fibers serves to shorten them and draw the chromosomes closer to the poles. In addition to biochemical evidence supporting this latter notion is the recognition that the movement of chromosomes toward the poles of the spindle during anaphase is a relatively slow process when compared to the beating motions of cilia and flagella and the contraction of muscle cells. Indeed, anaphase movement takes place at a rate that is more akin to the rates at which filaments grow or shorten by either an influx or loss of subunits.

From the preceding discussion it is apparent that several

different models attempting to account for chromosome movement during the anaphase of mitosis are supported by independent observations. This is not to suggest that the actual mechanism may not involve some combination of all of the models.

Meiosis

Meiosis is a form of nuclear division that is of fundamental importance among sexually reproducing organisms. An in-depth discussion of meiosis on a cellular as well as a genetic basis is beyond the scope of this book and is normally treated at length in textbooks on genetics (see the list of references at the end of this chapter). However, for the sake of completeness we will consider some of the major meiotic events and their implications.

Meiosis occurs in eucaryotic organisms whose cells contain the **diploid** number ($2n$) of chromosomes. Diploid infers ''double'' in the sense that the genetic information present in any one chromosome can also be found in an identical (or somewhat modified) form in a second chromosome in the nucleus. The two chromosomes forming such pairs are said to be homologous.

As noted earlier in this chapter, human cells contain 46 chromosomes or 23 homologous pairs (i.e., in humans $n = 23$). The 46 chromosomes of the *zygote* formed at fertilization are derived equally from the sperm cell and egg cell of the male and female parents. Each of these **gametes** contributes one member of each homologous pair of chromosomes. Once the zygote is formed, mitosis produces the billions of cells that ultimately make up the whole organism. Since sperm cells and egg cells contain only one member of each homologous pair of chromosomes, they are said to be **haploid**. It is meiosis that produces haploid cells, the process being restricted to the reproductive tissues (i.e., ovaries and testes).

During meiosis, the replicated chromosomes of the nucleus are apportioned among four daughter nuclei, each nucleus acquiring half the number of chromosomes of a diploid cell. Although the resulting cell nuclei contain only half the diploid number of chromosomes, the chromosome set is genetically complete, for each nucleus acquires one member of each pair of homologous chromosomes. The homologous chromosomes are assorted randomly at anaphase, and this accounts in part for the genetic variation that characterizes sexually reproducing organisms. Additional genetic variation occurs during the prophase of the first nuclear division by a process called **crossing over.** The genetic implications of random assortment and crossing over are principal subjects of genetics courses.

The various stages of meiosis (Fig. 20–21) may be summarized as follows:

Meiotic Division I

Prophase I

1. Leptotene stage (leptonema). The chromosomes become visible as threadlike structures as condensation of the chromatin begins, and each chromosome can be seen to consist of two chromatids.

2. Zygotene stage (zygonema). Homologous chromosomes are aligned side by side so that **allelic** genes (i.e., those encoding products of similar or identical function) are situated adjacent to one another. This phenomenon is called **synapsis.** The unit consisting of two synapsed and duplicated homologous chromosomes is called a **bivalent** or **tetrad.** As synapsis progresses, a protein framework joining adjacent, nonsister chromatids of each tetrad is formed at one or more points in the narrow space separating the homologues. It is in the region of these **synaptinemal complexes** that crossing over occurs. Crossing over or **chiasmata** results from the cleavage by **endonucleases** of the DNA in corresponding positions of two nonsister chromatids, followed by the transposition and rejoining of the free ends of homologous strands (see Fig. 20–22 for details). As a result of crossing over, new combinations of genes are formed in the homologous chromosomes.

3. Pachytene stage (pachynema). During this stage the chromatids become increasingly distinct as condensation continues.

4. Diplotene stage (diplonema). The diplotene stage is characterized by the separation of the paired homologous chromosomes except at points where chiasma were formed.

5. Diakinesis. Diakinesis brings the prophase of nuclear division to an end. During this stage chromosome condensation is completed.

Metaphase I

In this phase, the spindle apparatus forms, much as it does

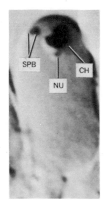

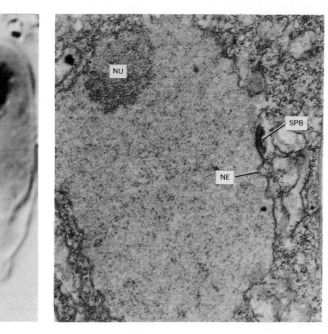

Figure 20–21
The stages of meiosis as seen in the basidium of *Pholiota*. Views obtained by light *(left)* and electron *(right)* microscopy are paired. See the text for description of the phases. *CH*, chromatin; *CM*, continuous microtubules; *ER*, endoplasmic reticulum; *KC*, kinetochore (centromere); *M*, membranes; *MT*, microtubules; *N*, nucleus; *NE*, nuclear envelope; *NU*, nucleolus; *PER*, perinuclear endoplasmic reticulum; *SPB*, spindle pole body. (Courtesy of Dr. K. Wells, copyright © 1978 by Springer-Verlag, *Protoplasma 94*, 85.)

(a) Prophase I

(b) Metaphase I

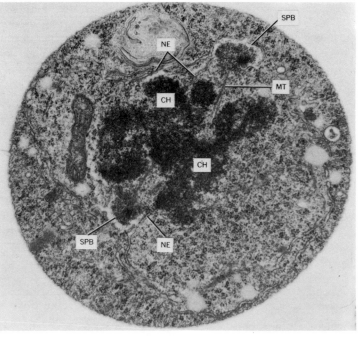

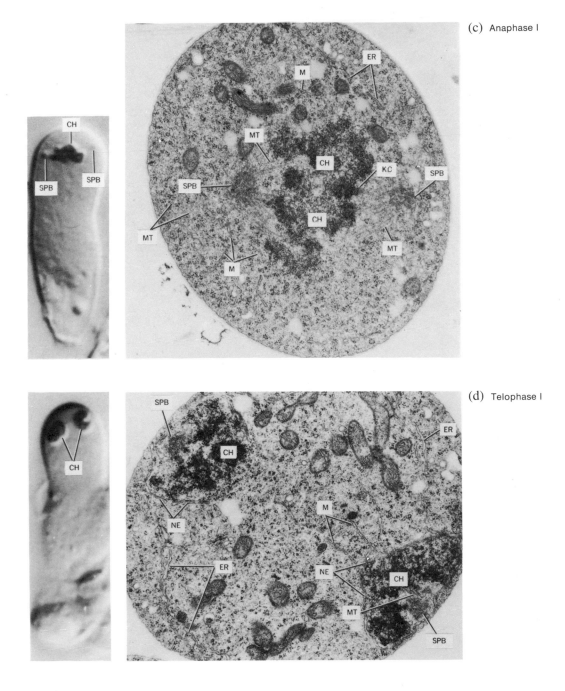

(c) Anaphase I

(d) Telophase I

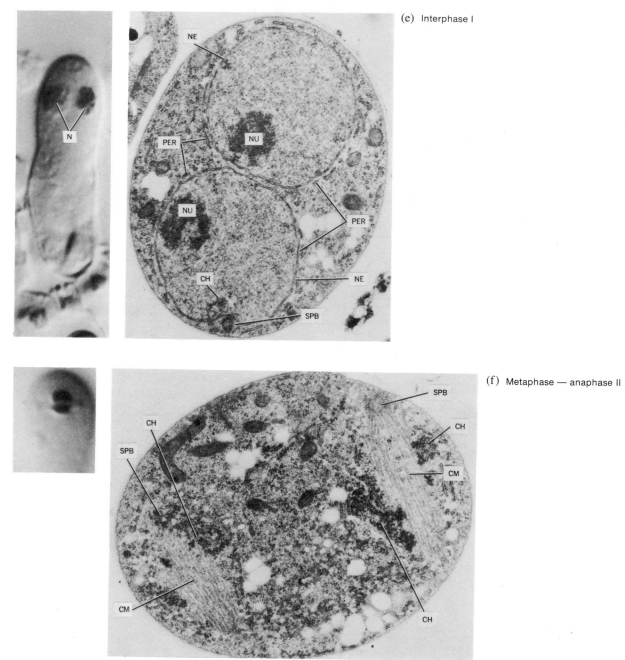

(e) Interphase I

(f) Metaphase — anaphase II

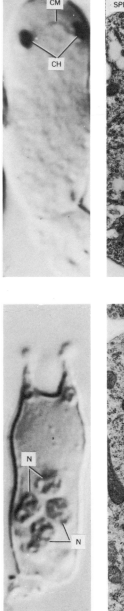

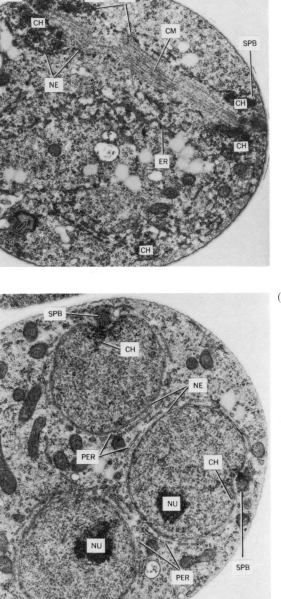

(g) Anaphase II

(h) Postmeiosis

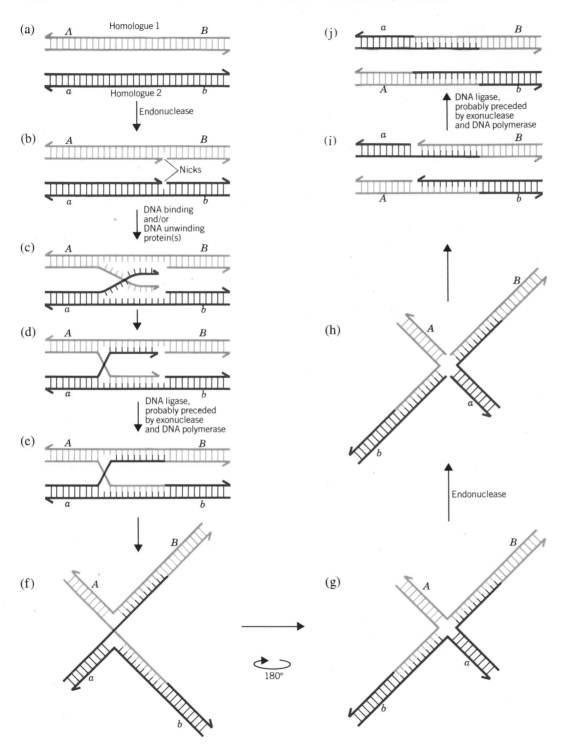

(a) Homologue 1 — A ... B

Homologue 2 — a ... b

Endonuclease

(b) Nicks

DNA binding and/or DNA unwinding protein(s)

(c)

(d)

DNA ligase, probably preceded by exonuclease and DNA polymerase

(e)

(f)

180°

(g)

Endonuclease

(h)

DNA ligase, probably preceded by exonuclease and DNA polymerase

(i)

(j)

Figure 20–22
Crossing over. *(a)* **Synapsis of homologous chromosomes (gray and color) carrying markers A, B and a, b. The antiparallel nature of the DNA strands of each chromosome is shown by the arrowheads at the ends of the strands.** *(b)* **An** *endonuclease* **nicks one strand in each chromosome.** *(c)* **Free ends are displaced and form base pairs** *(d)* **with the unnicked complementary strand of the homologous chromosome.** *(e)* **Free ends are rejoined by a** *DNA ligase*, **leaving the two chromosomes joined by single-stranded** *bridges (f)* **and** *(g)*. **Cleavage at the bridges by an endonuclease** *(h)* **frees the two chromosomes** *(i)*, **following which the remaining nicks are ligated** *(j)*. (From E. J. Gardner and D. P. Snustad, *Principles of Genetics*, John Wiley & Sons, New York, 1981, p. 176.)

in mitosis, and the tetrads align on the equatorial plate. The centromeres of homologous chromosomes attach to spindle fibers arising from opposite poles of the cell.

Anaphase I

Homologous chromosomes (but *not* sister chromatids) of each tetrad separate from each other and move to opposite poles of the spindle.

Telophase I

Telophase I brings the first meiotic division to a conclusion as the separated homologous chromosomes aggregate at their respective poles so that two nuclear areas are distinguishable. In most organisms, a new nuclear envelope is formed and some decondensation of the chromosomes occurs.

Interkinesis (or Interphase)

Interkinesis is the period between the end of telophase I and the onset of prophase II. This period is usually quite short. The DNA of the two nuclei produced by the first meiotic division does *not* engage in replication during interkinesis.

Meiotic Division II

Prophase II

The events characterizing this phase are similar to mitotic prophase, although each cell nucleus has only half the number of chromosomes as does a cell in prophase I; that is, the nucleus is already *haploid*. Each chromosome remains composed of the two sister chromatids formed prior to prophase I.

Metaphase II

The events occurring in this phase are similar to those in mitotic metaphase. The paired chromatids migrate to the center of the spindle and are attached there to the microtubules.

Anaphase II

The events occurring in this phase are similar to those in mitotic anaphase, but differ from those of anaphase I. In anaphase II, sister chromatids separate from one another and are drawn to opposite poles of the spindle. (Recall that sister chromatids do *not* separate in anaphase I.)

Telophase II

The events occurring in this phase are similar to those in mitotic telophase. The separated chromosome groups are re-enclosed in a newly developed nuclear envelope and begin decondensation.

Meiosis produces four cells, each with the haploid number of chromosomes. In many higher animals and some plants, meiosis in the female reproductive tissues is accompanied by an uneven division of the cytoplasm, in which case one of the two cells formed during telophase I is a nonfunctional **polar body** and may not enter prophase II (Fig. 20–23). In some organisms (such as humans) the polar body completes meiosis, but the two smaller polar bodies produced during telophase II are similarly nonfunctional. The second meiotic division of the larger cell produced during telophase I is also unequal and produces an additional polar body. During the production of spermatozoa in the male reproductive tissues, division of the cytoplasm is equal, but remarkable cytoplasmic differentiation of the four spherical haploid **spermatids** produced by meiosis is required (Fig. 20–23) before functional, flagellated spermatozoa are produced.

Cytokinesis

In most instances, eucaryotic cells contain only one nucleus and the 1:1 ratio is perpetuated by the synchronization of division of the cytoplasm with nuclear division. Mitotic divisions without cytoplasmic divisions occur in many species to produce multinucleate or *coenocytic* cells. *Striated muscle* cells, *mycelia* of molds, and some phloem cells of plants are examples of multinucleate cells. Few eucaryotic cells lack a nucleus. The *sieve cells* of plants have no

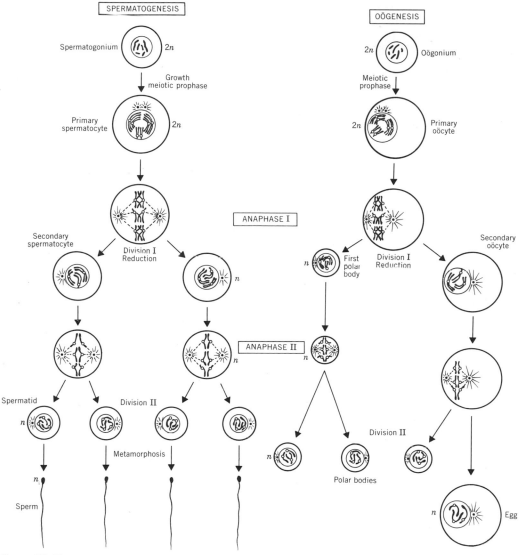

Figure 20–23

Stages of meiosis in the testes (left) and ovaries (right). Spermatogenesis produces four viable sperm cells, but oogenesis produces a single viable egg with three *polar bodies* (see text for details). (From E. J. Gardner and D. P. Snustad, *Principles of Genetics*, John Wiley & Sons, New York, 1981, p. 48.)

nucleus but are always in close proximity to a nucleated *companion* cell. Mammalian erythrocytes (red blood cells) lose their nuclei during the final stages of differentiation in the bone marrow (Chapter 24).

In general, animal cells and many lower plant cells divide into two cells by a pinching or furrowing action in the cytoplasm that lies between the two newly formed nuclei. In plant cells, which possess a cell wall, the division of the cell occurs by the formation of a **cell plate.** The plate begins forming at the center of the spindle near the end of anaphase. It grows outward until the cell is cleaved into two halves. While *furrowing* appears to be aided by the formation of **cytoplasmic filaments** (discussed in Chapter 23), the cell plate is formed by the progressive accumulation of carbohydrate and lipid materials that are derived from the Golgi apparatus (Chapter 18).

Summary

Eucaryotic cell nuclei are delimited by a pair of membranes that form the **nuclear envelope.** These membranes fuse with each other around the margins of **nuclear pores.** A variety of substances pass between the nucleus and cytoplasm through these pores. Much of a cell's DNA is within the nucleus and is organized into a number of **chromosomes.** During **mitosis** and **meiosis,** the chromosomes condense, thereby assuming a distinctive shape and size. Metaphase chromosomes are characterized by a **primary constriction** or **centromere** and varying numbers of **secondary constrictions.** In eucaryotic cells, the double helical DNA is associated with histones to form chains of small bodies called **nucleosomes.**

The chromosomes of bacteria and viruses are single nucleic acid molecules. The nucleic acid may be single-stranded DNA or RNA or may be double helix DNA. The chromosomes of procaryotic cells are not contained within a nuclear envelope, but in bacteria they may be attached to the plasma membrane. **Plasmids** are small double-stranded circular pieces of DNA that exist separately from the main chromosome of bacteria. Some plasmids are capable of being incorporated into the main chromosome and are called **episomes.**

During the interphase of the cell cycle of eucaryotic organisms, the chromatin shows varying degrees of **condensation.** The more condensed portions of interphase chromatin are called **heterochromatin** and the less condensed regions **euchromatin.** Euchromatin contains most of the cell's **structural genes** and is transcriptionally active. Some of the heterochromatin may be more concerned with the control of **gene expression.**

References and Suggested Reading

Articles and Reviews

Bauer, W. R., Crick, F. H. C., and White, J. H., Supercoiled DNA. *Sci. Am. 243*(1), 118 (July 1980).

Dickerson, R. E., Drew, H. R., Conner, B. N., Wing, R. M., Fratini, A. V., and Kopka, M. L. (1982) The anatomy of A-, B-, and Z-DNA. *Science 216,* 475.

Dubochet, J., and Noll, M., Nucleosome arcs and helices. *Science 202,* 280 (1978).

Gall, J. G., Chromosome structure and the C-value paradox. *J. Cell Biol. 91,* 3 (1981).

Gellert, M., DNA topoisomerases. *Annu. Rev. Biochem. 50,* 879 (1981).

Khorana, H. G., Total synthesis of a gene. *Science 203,* 614 (1979).

Kornberg, R. D., Structure of chromatin. *Annu. Rev. Biochem. 46,* 931 (1977).

Kornberg, R. D., and Klug, A., The nucleosome. *Sci. Am. 244*(2), 52 (Feb. 1981).

Landy, A., and Ross, W., Viral integration and excision: structure of the lambda ''att'' sites. *Science 197,* 1147 (1977).

McGhee, J. D., and Felsanfeld, G., Nucleosome structure. *Annu. Rev. Biochem. 49,* 1115–1156 (1980).

Miller, O. L., The nucleolus, chromosomes, and visualization of genetic activity. *J. Cell Biol. 91,* 15 (1981).

Unwin, P. N. T., and Milligan, R. A. (1982) A large particle associated with the perimeter of the nuclear pore complex. *J. Cell Biol. 93* 63.

Books, Monographs, and Symposia

Adams, R. L. P., Burdon, R. H., Campbell, A. M., and Smellie, R. M. S., *Davidson's Biochemistry of the Nucleic Acids* (8th ed.), Academic Press, New York, 1976.

Bonner, J., and Varner, J. E., *Plant Biochemistry,* Academic Press, New York, 1976.

Busch, H. (Ed.), *The Cell Nucleus,* Vols. I, II, III, Academic Press, New York, 1974.

DuPraw, E. J., *DNA and Chromosomes,* Holt, Rinehart and Winston, New York, 1970.

Gardner, E. J., and Snustad, D. P., *Principles of Genetics* (6th ed.), Wiley, New York, 1981.

Lehninger, A. L., *Biochemistry* (2nd ed.), Worth, New York, 1975.

Old, R. W., and Primrose, S. B., *Principles of Gene Manipulation,* University of California Press, Berkeley, Calif., 1980.

Swanson, C. P., Merz, T., and Young, W. J., *Cytogenetics,* Prentice-Hall, Englewood Cliffs, N.J., 1981.

Chapter 21

THE NUCLEUS AND MOLECULAR GENETICS

In Chapter 20, we were concerned primarily with the chemical composition and organization of the chromatin of the nucleus and with the behavior of the chromosomes during mitosis and meiosis. In this chapter, we will examine (1) the mechanisms by which DNA is replicated and transcribed into RNA, (2) the special properties that characterize eucaryotic, procaryotic, viral, and plasmid chromosomes, and (3) the emerging recombinant DNA technology and its implications in genetic engineering.

Replication of DNA

It has now been nearly 30 years since J. D. Watson, F. H. C. Crick, and M. H. F. Wilkins established the double-helical nature of the DNA molecule (Chapter 7) and suggested how each of the two strands of the double helix could act as a template for the duplication or *replication* of a new partner. Since then, the precise mechanism by which DNA replication is achieved has been intensively studied in many laboratories. In 1957, Nobel laureate A. Kornberg synthesized DNA in vitro and isolated *DNA polymerase* (now known as *DNA polymerase I*), an enzyme involved in the replication mechanism. In 1958, M. S. Meselson and F. W. Stahl showed that the succession of nucleotides in each of the two DNA strands remains intact during replication and that each original strand of the double helix becomes one of the two strands in the new double-helical

DNA molecules. They termed this process **semiconservative replication.** In 1963 J. Cairns was able to visualize replication of the chromosome of *E. coli* using a combination of microscopy and autoradiography.

Although semiconservative replication of DNA is now known to be universal, in eucaryotes replication originates simultaneously at many sites in the chromosome and proceeds along the DNA molecule in both directions from these sites. This is referred to as **bidirectional replication.** Bidirectional replication also occurs in procaryotes, but it originates at a single site on the chromosome. **Unidirectional replication,** in which replication occurs only in one direction along the chromosome, is much rarer and occurs in only a small number of the procaryotes that have been studied. In recent years, much has been learned about how replication is initiated, the nature of the enzymes that effect the process, and the manner in which "errors" introduced during replication are subsequently corrected. These and other recent findings are discussed in the sections that follow.

Replication as a "Semiconservative" Process

During replication of DNA, the two polynucleotide chains of the "parent" DNA double helix separate and each serves as a "template" for the synthesis of a new, *complementary* polynucleotide chain. During strand separation, the nitro-

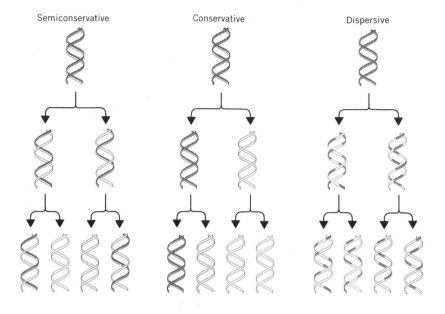

Semiconservative   Conservative   Dispersive

Figure 21–1
Three possible modes of DNA replication: *semiconservative* (left), *conservative* (center), and *dispersive* (right) (see text for explanation). (From E. J. Gardner and D. P. Snustad, *Principles of Genetics*, John Wiley & Sons, 1981, P. 84.)

gen bases of each original strand are exposed and establish sites for the association of free nucleotides. These nucleotides are then enzymatically linked together to form a new complementary strand (see Chapter 7, Fig. 7–10). Since deoxyadenylic acid (dAMP) can form hydrogen bonds only with thymine positions of the template strand (and dGMP can bond only to cytosine, dCMP only to guanine, and dTMP only to adenine), the newly synthesized strand will be identical to the original complementary strand. As a result, two new double helixes are formed, each consisting of one polynucleotide strand from the parent double helix (i.e., this is the strand that is conserved intact) and a newly synthesized polynucleotide strand. Since each of the two double helixes conserves only one of the parent polynucleotide strands, the process is said to be *semiconservative*.

Semiconservative replication of DNA was predicted by the original Watson–Crick model but was not verified until the classical studies of M. S. Meselson and F. W. Stahl were conducted. At the time of their experiments, two other modes of replication were deemed equally possible (Fig. 21–1): (1) *conservative replication,* in which *both* strands of the parent double helix would be conserved and the new DNA molecule would consist of two newly synthesized strands; and (2) *dispersive replication,* in which replication would involve fragmentation of the parent double helix and

the interspersing of pieces of the parent strands with newly synthesized pieces, thereby forming the two new double helixes.

Meselson and Stahl verified the semiconservative nature of DNA replication in a series of elegant experiments using isotopically labeled DNA and a form of isopycnic density gradient centrifugation (see Chapter 12). They cultured *E. coli* cells in a medium in which the nitrogen was ^{15}N (a "heavy" isotope of nitrogen, but not a radioisotope) instead of the commonly occurring and lighter ^{14}N. In time, the purines and pyrimidines of DNA in new cells contained ^{15}N (where ^{14}N normally occurs) and thus the DNA molecules were denser. DNA in which the nitrogen atoms are ^{15}N can be distinguished from DNA containing ^{14}N because during isopycnic centrifugation, the two different DNAs band at different density positions in the centrifuge tube (Fig. 21–2).

Meselson and Stahl centrifuged DNA isolated from the cells for two to three days at very high rotational speeds in centrifuge tubes initially containing a uniform solution of CsCl. During centrifugation, density gradients were automatically formed in the tubes as a result of the equilibrium that was established between the *sedimentation* of CsCl toward the bottom of the tube and *diffusion* of the salt toward the top of the tube. This form of centrifugation,

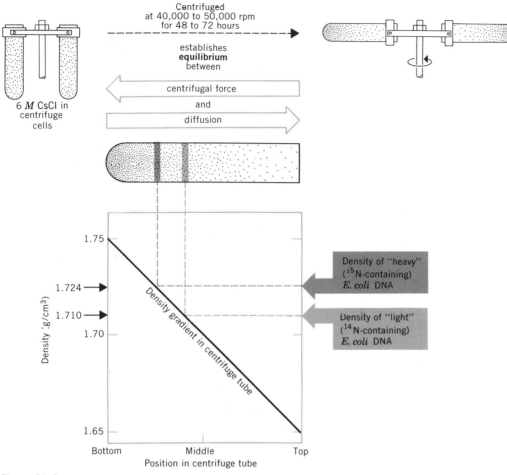

Centrifuged
at 40,000 to 50,000 rpm
for 48 to 72 hours

6 *M* CsCl in
centrifuge
cells

establishes
equilibrium
between

centrifugal force

and

diffusion

Density gradient in centrifuge tube

1.75

1.724 →

1.710 →

1.70

1.65

Density (g/cm^3)

Bottom   Middle   Top

Position in centrifuge tube

Density of "heavy"
(^{15}N-containing)
E. coli DNA

Density of "light"
(^{14}N-containing)
E. coli DNA

Figure 21–2
**Separation of ^{15}N-containing DNA from ^{14}N-containing DNA by equilibrium
isopycnic centrifugation (see text for details). (From E. J. Gardner and D. P.
Snustad, *Principles of Genetics*, John Wiley & Sons, 1981, p. 86.)**

called *equilibrium isopycnic centrifugation,* was discussed in Chapter 12. Depending upon its content of ^{15}N and ^{14}N the DNA bands at a specific position in the density gradient. Since the DNA synthesized by cells grown in ^{15}N would be denser than ^{14}N-containing DNA it would band further down the tube (Fig. 21–2).

Cells grown for some time in the presence of ^{15}N medium were washed free of the medium and transferred to ^{14}N-containing medium and allowed to continue to grow for specific lengths of time (i.e., for various numbers of generations). DNA isolated from cells grown for one genera-

tion of time in the ^{14}N medium had a density intermediate to the DNA from cells grown only in ^{15}N-containing medium (identified as *generation 0* in Fig. 21–3) or only in ^{14}N-containing medium (the *controls* of Fig. 21–3). Such a result immediately ruled out the possibility that DNA replication was conservative, since conservative replication would have yielded two DNA bands in the density gradient for *generation 1* (i.e., F$_1$) cells. The single band of intermediate density (identified as "hybrid" DNA in Fig. 21–3) consisted of DNA molecules in which one strand contained ^{15}N and the other contained ^{14}N. When the incuba-

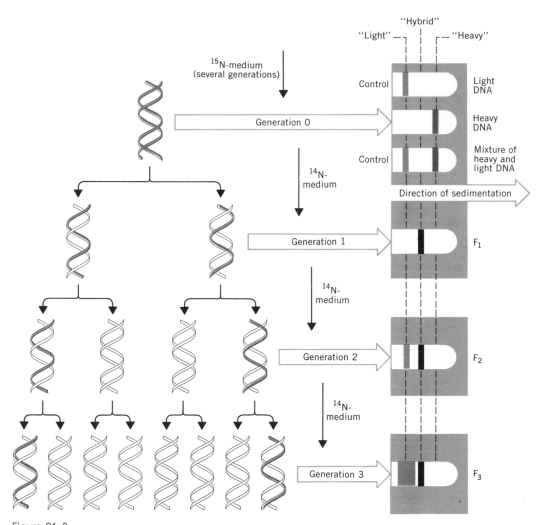

Figure 21–3
Results of the Meselson-Stahl experiments (right) and their interpretation (left) (see text for explanation). (From E. J. Gardner and D. P. Snustad, _Principles of Genetics,_ John Wiley & Sons, 1981, p. 87.)

tion in ^{14}N-medium was carried out for two generations of time (i.e., _generation 2_), _two_ DNA bands were formed—one at the same density position as the DNA from cells grown exclusively in ^{14}N medium (i.e., "light controls") and one of intermediate density. Subsequent generations produced greater numbers of DNA molecules that banded at the "light" (^{14}N-containing DNA) position in the density gradient. These results are consistent only with the model of semiconservative replication. Dispersive replication would have produced a single band for each generation and the band would have been found at successively lighter

density positions in the gradient. Studies using other procaryotes as well as eucaryotes indicate that semiconservative replication of DNA is probably the universal mechanism.

Replication by Addition of Nucleotides in the 5′ → 3′ Direction

Each nucleotide of a DNA strand is joined to the next nucleotide by a _phosphodiester_ bond that links the 3′ carbon of its deoxyribose to the 5′ carbon of the deoxyribose of

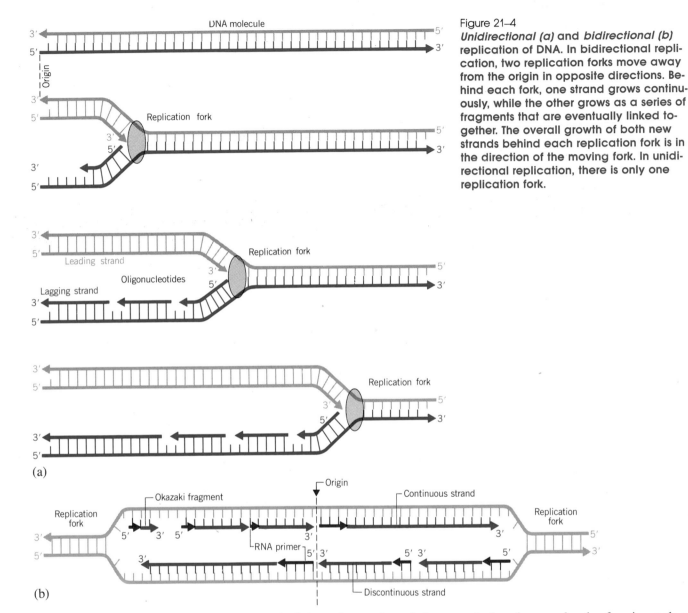

Figure 21–4

Unidirectional (a) and *bidirectional (b)* replication of DNA. In bidirectional replication, two replication forks move away from the origin in opposite directions. Behind each fork, one strand grows continuously, while the other grows as a series of fragments that are eventually linked together. The overall growth of both new strands behind each replication fork is in the direction of the moving fork. In unidirectional replication, there is only one replication fork.

the next nucleotide (Chapter 7). At one end of the polynucleotide chain, there is an hydroxyl group attached to the 3′ carbon of the last nucleotide, and at the other end, there is a phosphate group attached to the 5′ carbon. The two chains of a double helix have *opposite polarities* and are said to be **antiparallel;** that is, each end of the double helix contains the 5′ end of one strand and the 3′ end of the other. During replication, new nucleotides are added to the phosphodiester bonds. Attachment of a nucleotide to a growing strand always takes place at the terminal 3′ position of that strand. In other words, the forming polynucleotide chain "grows" from its 5′ end toward its 3′ end.

Unidirectional and Bidirectional Replication

Replication begins at a point on the chromosome where the two parental strands begin to separate; this point is called the **origin.** Addition of complementary nucleotides to form two new strands takes place along both parent strand templates starting from that point (Fig. 21–4). In *unidirectional* replication, growth proceeds along both strands in the *same* direction leading from the origin. For one of the parental template strands, synthesis of the new complementary strand takes place by the continuous addition of nucleotides to the available 3′ end of the forming strand. The growing strand is called the **leading strand** or **continuous strand.** The 5′ end of this strand is located at the origin and its 3′ end at the moving **replication fork** (i.e., the progressing point of separation of the parental strands).

The other polynucleotide strand being formed is called the **lagging strand** or **discontinuous strand.** The elongation of this strand takes place by a somewhat modified mechanism. In contrast to the leading strand, the lagging strand has its 3′ position at the origin and its 5′ position at the replication fork. If nucleotides were sequentially added to the end of the lagging strand at the replication fork, then this strand's growth would proceed in a 3′ → 5′ direction. *This does not occur.* Instead, growth takes place by the synthesis of a number of short polynucleotide chains between the replication fork and the origin. Each chain is laid down in the direction 5′ to 3′ and these are later linked together and to the 5′ end of the lagging strand. As a result, the *overall* direction of growth of the lagging strand is the same as that of the leading strand. The unusual growth pattern that characterizes the synthesis of the lagging strand explains why it is also referred to as the "discontinuous" strand.

In **bidirectional replication** (Fig. 21–4), *two* replication forks are formed at the origin and these move away from the origin in *both* directions as the double helix unwinds. The synthesis of the complementary strands also occurs in both directions. Behind *each* fork there is a set of leading and lagging strands. As in the case of unidirectional replication, elongation of the two leading strands is continuous, while elongation of the two lagging strands is discontinuous. It is to be noted that regardless of whether replication is unidirectional or bidirectional, the addition of nucleotides always occurs in the direction from 5′ to 3′, as new nucleotides are added to available 3′ ends of either the continuous strand or the discontinuous strand.

Discontinuous synthesis of lagging strands was first demonstrated by R. Okazaki. Okazaki incubated *E. coli* cells in a medium containing ^{3}H-thymidine for very short periods of time (a pulse of only 15 seconds) and then examined the distribution of the radioisotope in newly synthesized DNA. The radioisotope was found in a number of polynucleotides (1000–2000 nucleotides long), now referred to as **Okazaki fragments** (Fig. 21–4). When pulsed cells were transferred to unlabeled medium for varying lengths of time prior to analysis, the radioactive label was recovered in much longer stretches of DNA. This is because the Okazaki fragments produced during the short tritium pulse had been linked together and added to the end of the lagging strand. In eucaryotic cells, Okazaki fragments are usually smaller (about 100–200 nucleotides long).

Visualization of Replication in *E. coli*

Using autoradiography, J. Cairns was able to visualize replicating bacterial chromosomes. Cairns placed *E. coli* cells in a medium containing ^{3}H-thymidine for various periods of time so that the radioactive thymidine was incorporated into the DNA as the chromosome replicated in successive generations of cells. Cells were removed from the medium after various periods of incubation and *gently* lysed to release the chromosome from the cell (the shear forces created by harsh lysis break the chromosome into small pieces). The chromosomes were then transferred to glass slides and coated with a photographic emulsion sensitive to the low-energy beta particles emitted by the ^{3}H-thymidine. After exposing the emulsion to the beta rays, the emulsion was developed and examined by light microscopy (see Chapter 14 for a discussion of combined autoradiographic-microscopic techniques). Wherever labeled thymidine occurred in a chromosome, the emulsion was exposed and created a visible grain. A chromosome not engaged in replication appeared as a circular structure formed from a close succession of exposure spots. Chromosomes "caught in the act" of replication gave rise to what are called *theta* structures because they have the appearance of the Greek letter theta (i.e., θ) (Fig. 21–5). The theta structures reveal the positions of the replication forks in the circular chromosome and provide added credence for the semiconservative model of replication.

The Replicon and the Replication Sequence

The sequence of events that takes place during replication appears to be as follows (Fig. 21–6). Parental strand separation begins at a site called the **origin,** which contains a special nucleotide sequence and directs the association of a number of proteins. An ATP-dependent **unwinding enzyme** (sometimes called *helicase*) promotes separation of

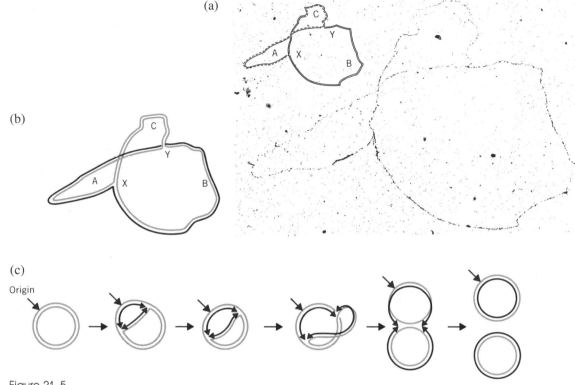

Figure 21–5

(a) Autoradiogram of a replicating *E. coli* chromosome. Loops A and B have completed replication, whereas loop C remains to be replicated. X and Y are the replication forks. In *(b)* the newly synthesized strand is shown in color and the original strand in gray. *(c)* Diagram illustrating successive stages during bidirectional replication. *(a* Courtesy of Dr. J. Cairns; copyright © 1963 by Cold Spring Harbor Laboratory. *b* and *c* modified from E. J. Gardner and D. P. Snustad, *Principles of Genetics,* John Wiley & Sons, 1981, p. 88.)

the two parental strands and establishes a **replication fork** that will progressively move away from the origin (Fig. 21–6a). Behind the replication fork, the single DNA strands are stabilized by a set of proteins called **helix-destabilizing proteins** (Fig. 21–6b). The action of the helix unwinding enzyme introduces a positive *supercoil* into the duplex DNA ahead of the replication fork. Enzymes called **topo-isomerases** relax the supercoil by attaching to the transiently supercoiled duplex, nicking one of the strands, and rotating it through the unbroken strand. The nick is then resealed.

Prior to DNA synthesis beginning at the origin, short RNA polynucleotides are formed that are complementary to the DNA template. These stretches of RNA are called *primers* and are also laid down in the 5′ to 3′ direction. DNA nucleotides are then added one at a time to the free 3′ ends of the RNA primers. Since growth of the lagging strand is discontinuous, several RNA primers and Okazaki fragments are formed. Note that an RNA primer must be formed for each Okazaki fragment to be laid down (Fig. 21–6c). The enzymes required for the synthesis of the RNA primers are called **primases.**

Elongation of the leading strand and synthesis of the Okazaki fragments are catalyzed by *DNA polymerase III*. The substrates of DNA polymerase III are the deoxynucleoside triphosphates (e.g., dATP, dGTP, dCTP, and

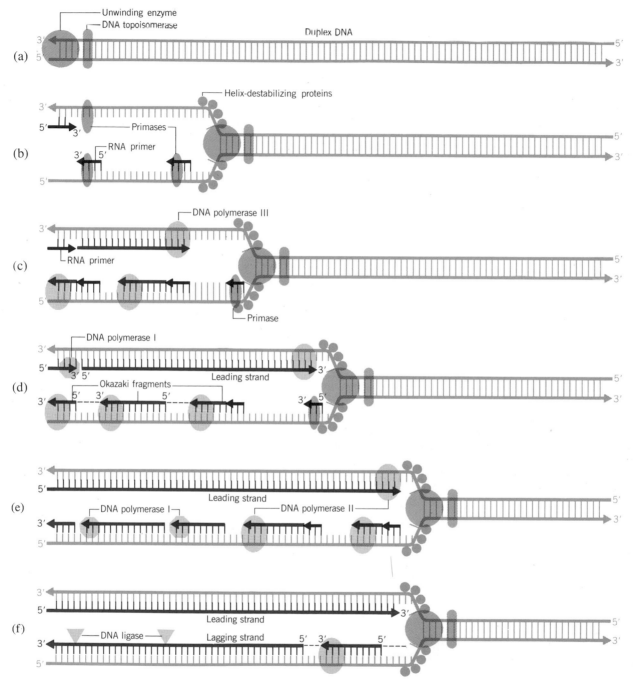

Figure 21–6
Stages of DNA replication.

dTTP). Addition of a nucleotide to the available 3′ position of the continuously growing leading strand or an Okazaki fragment of the lagging strand involves removal of pyrophosphate to yield a deoxynucleoside monophosphate (e.g., dAMP, dGMP, dCMP, or dTMP). Upon completion of the Okazaki fragments, the RNA primers are enzymatically excised and the Okazaki fragments are joined (Fig. 21–6d, e). The gaps left by the removal of the primers are filled by addition of nucleotides, a process that is mediated by *DNA polymerase I*. When the final nucleotide is added to fill the gap, the enzyme *DNA ligase I* forms the phosphodiester bond that links the free end (Fig. 21–6f).

During unidirectional replication, the replication fork fully circles the chromosome and the resulting DNA molecules are separated. For procaryotes in which replication is bidirectional, the replication forks proceed around the chromosome until they meet (Fig. 21–5). In eucaryotic chromosomes where there are many replicating units or **replicons,** all replicons are linked together before the chromatids can be separated. The replicon consists of that segment of a chromosome that includes an origin and two **termination points** (i.e., points where replication ends).

Much remains to be learned about the enzymes that catalyze the reactions of replication. So far, the major obstacle to such studies has been the difficulty of isolating and purifying the enzymes, either individually or in complexes. This is due in part to the fact that some of them may be associated with membranes. In procaryotic cells, the replication forks are bound to the plasma membrane.

At the present time, about two dozen different proteins have been shown to be involved with the replication of DNA in *E. coli* cells. Several of the *E. coli* proteins are involved with **prepriming reactions,** that is, the reactions that occur before the formation of RNA primers.

Chromosome Replication in Eucaryotes

Replication of the chromosomes of eucaryotic cells is a much more complex process than that in procaryotes and the DNA viruses. The added complexity is due in part to the greater length of the eucaryotic chromosome. Nonetheless, there are still a number of fundamental similarities between DNA replication in eucaryotes, procaryotes, and viruses. For example, (1) replication is semiconservative; (2) RNA primers are used; (3) replication occurs in both directions; (4) new DNA is synthesized along the lagging

strand as a series of Okazaki fragments; and (5) the overall direction of elongation is 5′ to 3′.

Semiconservative Replication in Eucaryotes

Verification of semiconservative replication of DNA in eucaryotic cells was provided in a series of experiments carried out by J. H. Taylor and P. Woods in 1957. Taylor and Woods grew *Vicia faba* bean root tip cells for one generation of time in a medium containing ^{3}H-thymidine and then transferred the root tips to ^{3}H-free medium, where they were allowed to enter another round of cell division. However, *colchicine* was added to the medium in order to prevent anaphase separation of sister chromatids (see Chapter 20). The cells were then examined by combined light microscopy and autoradiography and the sister chromatids of each chromosome found to contain equal amounts of radioactive label (i.e., equal numbers of exposure grains were seen in each chromatid; Fig. 21–7). When cells were allowed to duplicate the chromosomes and enter a second round of mitosis in unlabeled medium, the label was found in only one of the two chromatids comprising the metaphase chromosomes. The sister chromatid not distinguishable by autoradiography during this second metaphase consisted entirely of unlabeled nucleotides. These observations are consistent with a semiconservative model of chromosome replication.

Replicons of Eucaryotic Chromosomes

While the replicating procaryotic chromosome may consist of one replicon, the average chromosome of a eucaryotic cell contains thousands of replicons. Replication of DNA does not begin in all replicons at the same time of the *S* phase (Fig. 21–8). Instead, some sort of control mechanism turns each replicon on at a specific time. Using the pulse labeling technique combined with autoradiography, it is possible to show that certain regions of a chromosome always replicate early in the *S* phase while other regions are active at later times. Euchromatic regions are usually replicated first and heterochromatic regions later in the *S* phase.

In eucaryotes, three classes of DNA polymerase have been identified: these are called α, β, and γ forms. The enzymes in each class are characterized by their molecular size, chromatographic behavior, and primer-template spec-

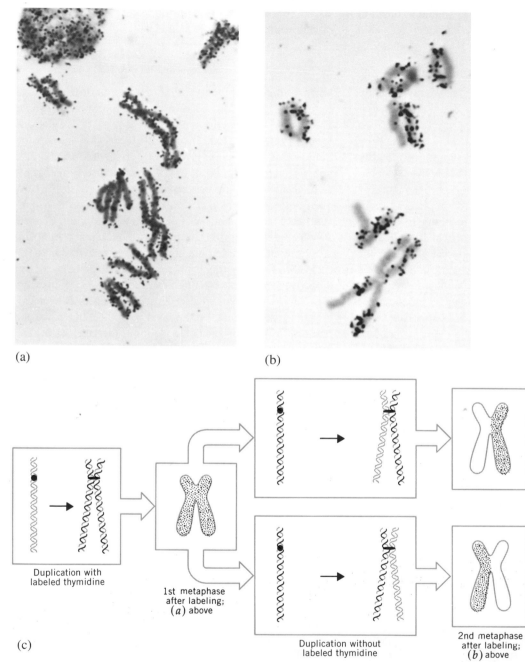

(a)

(b)

(c)

Duplication with
labeled thymidine

1st metaphase
after labeling;
(*a*) above

Duplication without
labeled thymidine

2nd metaphase
after labeling;
(*b*) above

Figure 21–7

Semiconservative replication of DNA in eucaryotic cells.
(a) and *(b)* Autoradiographs of *Vicia faba* root tip cells at
the first *(a)* and second *(b)* metaphases after an initial
replication in the presence of ³H-thymidine, *(c)* Interpretation of the results seen in *a* and *b*. ³H-thymidine-labeled
DNA strands and radioactive chromatids are shown in
color (*a* and *b* from J. H. Taylor, The replication and organization of DNA in chromosomes, *Molecular Genetics,* Part
I, J. H. Taylor (Ed.), Academic Press, New York, 1963; *c* from
E. J. Gardner and D. P. Snustad, *Principles of Genetics,*
John Wiley & Sons, 1981, p. 128.)

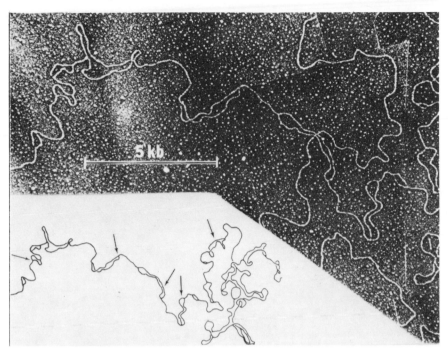

Figure 21–8
Electron photomicrograph of *Drosophila* DNA showing multiple sites of replication (arrows). (Courtesy of *Proc. Natl. Acad. Sci. 71*, 135. 1974.)

ificity. The αDNA polymerase, which has been reported to occur in multiple forms, fills in the gaps left between successive Okazaki fragments when the RNA primer is removed. Following insertion of the last nucleotide, an enzyme called *DNA ligase I* links the free ends together. The αDNA polymerase is responsible solely for mitochondrial DNA replication. The γDNA polymerase is responsible for DNA synthesis at the replication forks.

Disassembly and Reassembly of Nucleosomes

Replication of eucaryotic DNA necessitates a preliminary disassembly of nucleosomes. The nucleosomes are then reassembled after the replication fork has moved some distance along the chromosome (Fig. 21–9). On the **forward arm,** the last reassembled nucleosome is about 125 nucleotides from the 3′ end of the leading strand. On the **retrograde arm,** the last reassembled nucleosome is about 250 nucleotides from the 5′ end of the first Okazaki fragment. The histone molecules that are part of the nucleosome in front of the replication fork are probably transferred to one of the two DNA duplexes behind the fork, while the other DNA duplex combines with newly synthesized histone octamers to form a completely new nucleosome. The new and old octamers may be distributed at random between the forward and retrograde arms.

Repetitive DNA

The DNA of procaryotic chromosomes consists essentially of *unique sequences*. That is, except for the genes encoding ribosomal RNAs (see Chapter 22), base-pair sequences are not repeated. In contrast, the chromosomes of eucaryotic organisms contain *repetitive DNA*—that is, base sequences that are repeated many times. Indeed, it has been estimated that 20 to 50 percent of the genome of eucaryotes may be represented by repetitive DNA. The rest of the DNA consists of *unique sequences* (i.e., single copies). The amount of DNA represented by repeated sequences is determined by allowing fragmented and denatured DNA to reanneal by complementary base pairing. The rate of reannealing is directly related to the quantity of repetitive DNA present. The method is described in some detail in Chapter 7. It is generally acknowledged that the unique sequences represent structural genes for messenger RNA and proteins. There

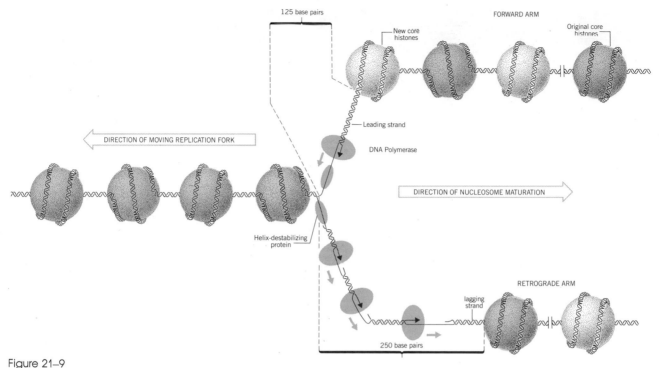

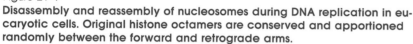

Figure 21–9

Disassembly and reassembly of nucleosomes during DNA replication in eucaryotic cells. Original histone octamers are conserved and apportioned randomly between the forward and retrograde arms.

are two broad classes of repetitive DNA: *moderately repetitive* (from as few as 2 to as many as 100,000 copies of a particular sequence) and *highly repetitive* (more than 100,000 copies). Moderately repetitive DNA includes the genes for certain proteins such as the histones and ribosomal proteins and the various ribosomal RNAs. The functions of the highly repetitive sequences are uncertain; they are located in the inactively expressed heterochromatic regions of the nucleus (Chapter 20). Apparently, some moderately repetitive DNA can migrate from one chromosome to another; such *nomadic* sequences may be analogous to the *transposable elements* in procaryotic cells (see below) and may have a role in the selective gene expression that characterizes an organism's development.

Replication of Viral DNA

DNA viruses are unable to independently replicate their DNA. Instead, they infect a cell (called the *host*) and utilize the metabolic machinery of the host to replicate their own DNA. The extent to which the virus relies on the host cell's enzymatic apparatus is related to the size of the viral genome (Table 21–1). Viruses like φX174, whose genome has only nine genes and 5375 nucleotides encoding the formation of coat proteins, rely entirely on the host cell to provide the enzymes for DNA replication. In viruses with larger genomes (such as the T4 bacteriophage that attacks *E. coli* and contains several dozen genes), the DNA encodes enzymes of the viruses' own replicative process as well as coat proteins.

When lysogenic phages, such as the λ phage that attacks *E. coli,* infect a metabolically active host cell, a virulent infection results and soon leads to lysis of the host and release of new phage. The infecting phage introduces its DNA into the host, redirecting the host's metabolism so that the phage DNA is replicated many times and enclosed in newly synthesized coat proteins. However, when λ phage infects a host cell that is metabolically *inactive* (i.e., the

Table 21–1
Sizes and other characteristics of certain viruses

Virus	Related Viruses	Particle Shape	DNA			Comments
			Mol. Wt. × 10^{-6}	Number of Base Pairs, in kb	Shape	
SV 40	Polyoma	Polyhedron	3.4	5.1	Duplex, circular	Animal cell host
φX174	S13	Polyhedron	1.8	5.4	Single strand, circular	Duplex replicative form
M13	fd, f1	Filament	1.9	5.7	Single strand, circular	Duplex replicative form
T7	T3	Head, short tail	23	35.4	Duplex, linear	Terminal redundancy
λ	φ80, 434, P2, 186	Head, tail	32	49	Duplex, linear	Cohesive ends form replicative circles, lysogenic
T5		Head, tail	76	115	Duplex, linear	Nicked
T4	T2, T6	Head, tail	120	180	Duplex, linear	Terminal redundancy; permuted
R17	MS2, f2 Qβ	Polyhedron	1.0	3.0	Single strand, linear	RNA, not DNA

Source. From A. Kornberg, *DNA Synthesis*, W. H. Freeman and Co., San Francisco, copyright © 1974.

glucose supply of the cell is exhausted and its cyclic AMP level is elevated), a **temperate** infection occurs producing what is called a **lysogenic response.** In this instance, the inserted viral DNA is integrated into the DNA of the host and is replicated along with the host's DNA during the growth and division of the host cell. The replication employs the native mechanism of the host cell. The inserted phage DNA (now called **prophage**) may be replicated in many successive generations of host cells. Eventually, the prophage DNA is expressed, producing new virus particles that are released upon lysis of the host cell.

The life cycle of a virus is characterized by the following stages:

1. Virus adsorption. In this stage, the virus attaches to the surface of the host cell. The specificity between the virus and the host is achieved through the interaction between specific viral adsorption proteins called *pilot proteins* and specific host cell-surface receptors. The pilot proteins not only attach the virus to the host but may also position the viral DNA (or RNA in the case of the RNA viruses) for insertion into the most appropriate part of the cell.

2. DNA (or RNA) insertion. In this stage, the viral nucleic acid is inserted through the host cell envelope and plasma membrane into the cytosol. This process may be aided by the pilot proteins, which may be carried into the cell along with the viral nucleic acid.

3. Expression of viral nucleic acid. Once in the host cell, the viral DNA is expressed in some manner. In the case of the small single-stranded viruses like φX174 and M13, a second strand complementary to the single strand is synthesized, yielding a duplex form. For the double-stranded DNA viruses, certain segments necessary for virus replication are transcribed and translated, while for the RNA viruses, certain portions are directly translated.

4. Replication of virus nucleic acid. The mechanism of DNA replication varies from one type of virus to another, but ultimately many complete double (or single) strands of the viral nucleic acid are produced in preparation for final ''packaging.''

5. Virus assembly. The mature strands of viral nucleic acid are enclosed within a shell or coat of proteins.

6. Release. Finally, the assembled viruses are released from the host cell. If release is accompanied by cell lysis, then the host cell is killed. However, in some cases viruses emerge from the host cell in the form of small *buds* and do not cause serious harm to the host cell. Occasionally, part of the host cell's plasma membrane is used to envelope the emerging virus particle.

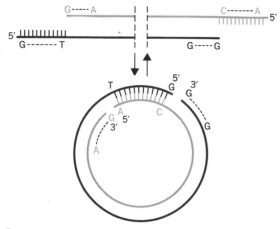

Figure 21–10
Structure of the linear form of λ phage DNA showing the overlapping ends. The linear form can be converted to the circular form by annealing at the ends. (Redrawn from A. Kornberg, *DNA Synthesis*, W. H. Freeman and Company, copyright © 1974.)

Lambda (λ) Phage

E. coli is the host cell for a number of different lysogenic phages. One of these, the **lambda (λ) phage,** is interesting because its double-stranded DNA has a single-stranded tail at each end of the molecule and these tails are complementary to each other (Fig. 21–10). When the λ phage DNA is inserted into *E. coli,* the complementary bases at the ends pair up (i.e., they anneal), thereby forming a circular DNA molecule. As noted above, when the DNA enters a metabolically active cell, replication of the DNA occurs immediately and is followed by synthesis of viral coat proteins, assembly of the viruses, cell lysis, and release of the viruses. On the other hand, in metabolically inactive cells, integration of the λDNA into the *E. coli* chromosome occurs. Integration into the *E. coli* genome requires an enzyme (called *integrase*) encoded in one of the virus' genes (called *int*); the viral DNA is inserted at a specific locus in the *E. coli* chromosome called *att* (Fig. 21–11). Accordingly, the *E. coli* chromosome is broken, the λDNA inserted, and the λ prophage formed. The prophage may be replicated along with the *E. coli* DNA for many thousands of generations.

Subsequent *excision* of the λDNA from the *E. coli* chromosome is also an unusual phenomenon and requires the products of the *int* and *xis* genes. On rare occasions during excision, a part of the host chromosome is removed along with the λDNA. The *E. coli* genes that are removed with the λDNA are those for galactose utilization (i.e., *gal*) and biotin synthesis (i.e., *bio*). These genes are located on each side of the *att* site. A most important implication of all of this is that when these viruses are released and infect new cells, they can transfer the genes of the former host to the new host. The transfer of genes in this way is called **transduction.** The cell receiving these genes is said to be *transduced,* and the result is a form of **genetic recombination** (see below and Chapter 20).

φX174 Phage and Rolling Circle Replication

φX174 is yet another bacterophage that infects *E. coli* cells. Its chromosome is circular and single stranded. The single-stranded DNA inserted into the *E. coli* cell is called the + strand (i.e., the positive strand) and serves as the template for the synthesis of a complementary − strand (negative strand). The synthesis of the − strand begins with the synthesis of a short RNA primer at the origin, and DNA polymerase III then adds deoxyribonucleotides to the 3′ end of the growing chain until the entire circular template has been used (Fig. 21–12). The RNA primer is then excised and the gap filled by the action of DNA polymerase I. DNA ligase covalently closes the free ends of the minus strand, with the result that a double-stranded, circular chromosome is formed.

The double-stranded circular chromosome (called the *parental replicative form* or simply RF-form) then participates in **rolling circle replication,** which produces additional double-stranded circular chromosomes (called *progeny RF*). Rolling circle replication begins when an enzyme called a *nickase* breaks the + strand at the origin—a site that is probably attached to the plasma membrane. The 5′ end of the nicked + strand rolls away from the − strand and DNA synthesis ensues using both exposed single strands as template. A new positive strand is formed along the negative strand beginning at its exposed 3′ end, and a new negative strand is formed discontinuously along the unfolded + strand beginning at a point near the 5′ end of the + strand. As synthesis proceeds, the + strands continue to roll away from the − strand. When a complete − strand has been formed along the + strand template, a nickase frees the double-stranded molecule and a ligase joins their free ends. Thus, a complete circular DNA mol-

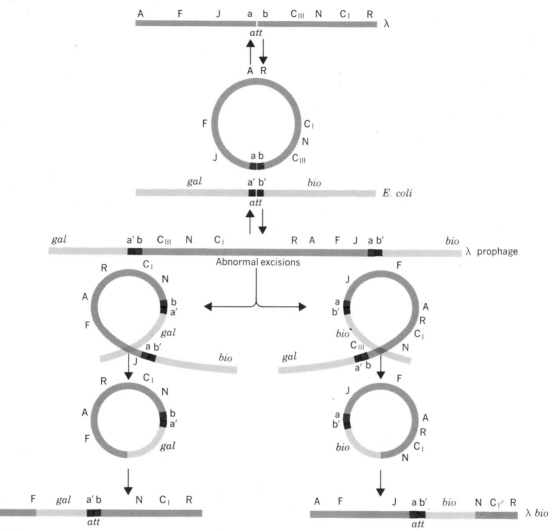

Figure 21–11

Diagram of mechanics of integration and excision of λ prophage DNA into chromosome of *E. coli.* (Redrawn from A. Kornberg. *DNA Synthesis*, W. H. Freeman and Company, copyright © 1974.)

ecule (or progeny RF) results. Rolling circle replication by parental RF produces additional progeny.

Progeny RF are converted to single-stranded molecules by nicking the + strand and unrolling it from the double helix (Fig. 21–12); these positive strands may then be converted to circular, single-stranded molecules through the action of a ligase that links the strands' free ends. Circular single-stranded molecules may then be encased in newly synthesized viral proteins to form complete φX174 viruses.

Rolling circle replication also occurs during the transfer of plasmid DNA between conjugating bacteria (see Chapter 20).

DNA Repair

The native structure of a DNA molecule may be altered in many ways, producing a defective form; for example, bases along one of the two strands may be chemically altered or

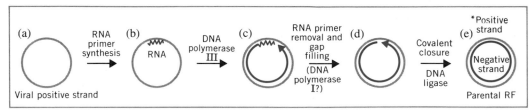

Stage I: Viral positive strand⟶parental RF

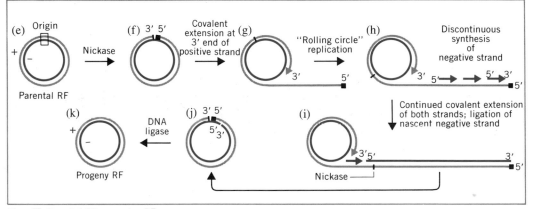

Stage II: Parental RF⟶progeny RF

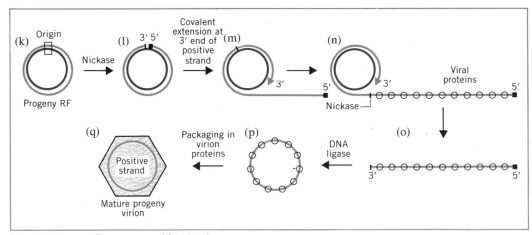

Stage III: Progeny RF⟶progeny positive strands

Figure 21–12
Rolling circle replication of the single-stranded DNA of the φX174 bacterio-phage (see text for details). (Modified from E. J. Gardner and D. P. Snustad, *Principles of Genetics,* **John Wiley & Sons, 1981, p. 102.)**

removed entirely; adjacent bases along one strand may become linked together to form a **dimer;** or breaks may be introduced into one or both chains. Such alterations interfere with proper replication, change normal recombination events, or produce useless transcription products.

Several repair mechanisms are known to exist in cells that can correct errors present in one or both strands of a molecule. Among the most common of these are **excision repair** and **post-replication repair.** Excision repair (Fig. 21-13) has been demonstrated in both procaryotes and eu-

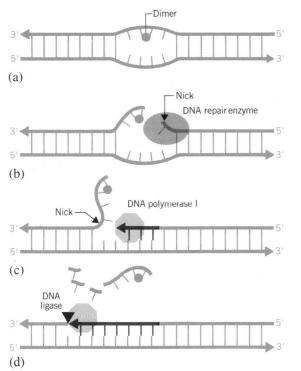

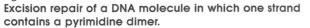

(a)

(b)

(c)

(d)

Figure 21–13

Excision repair of a DNA molecule in which one strand contains a pyrimidine dimer.

caryotes and is used to correct for the presence of pyrimidine dimers within a strand. Such dimers are produced, for example, by exposure to ultraviolet radiation. P. Howard-Flanders, D. Pettijohn, P. Hanawalt, and others have shown that the nucleotides forming the dimer are enzymatically excised and the resulting gap in the strand then filled by DNA polymerase I. The repair enzyme nicks the defective strand at the 5′ end of the damaged region, following which DNA polymerase I binds at the nick and adds the appropriate complementary nucleotides to the free 3′ end. After adding a succession of nucleotides, the DNA polymerase produces a second nick in the strand, releasing a short polynucleotide that contains the dimer. After this, DNA ligase closes the nick, thereby completing the repair process.

If a segment of one of the two strands of a double helix contains a dimer, then during replication the complementary strand that is synthesized will lack several nucleotides in the vicinity of the chain corresponding to the position of

the dimer (Fig. 21-14); thus, the new strand has a gap in it and the original (defective) strand is single stranded in the region of the dimer. Such an error cannot be corrected by excision repair, but is corrected by *post-replication* repair. In post-replication repair, the normal duplex is nicked and a repair enzyme realigns the single-stranded region with the complementary region of the sister duplex (i.e., the repair enzyme switches the free end of the nicked strand into the gap.) The result is a **cross-strand exchange.** Examination of the process, which is depicted in Figure 21-14, reveals that the upper heteroduplex can now be repaired by the action of DNA polymerase. However, we are still left with two double helixes that are cross-stranded. Another repair enzyme produces two nicks in the cross-stranded helices and recombines the free ends to produce two double helixes that are no longer cross-stranded. The original dimer, which is now part of the lower double helix, can be corrected by the excision repair enzymes.

Transcription of DNA

Transcription is the synthesis of RNA using DNA as template. In Chapter 7, we considered the chemistry of RNA and the basic principles of the transcription process and we examined the mechanisms that regulate what portions of the genome of procaryotic and eucaryotic cells are expressed.

Transcription of DNA produces three types of RNA: messenger RNA (mRNA), ribosomal RNA (rRNA), and transfer RNA (tRNA). The initial RNA transcripts contain many more bases than the final functional ribonucleic acid molecules; the extra nucleotides are removed by a mechanism called **RNA processing.** The net effect of RNA processing (described at length in Chapter 22) is that certain nucleotides are removed, others are added, and a number are chemically modified.

Unlike replication, transcription does not progress along the entire length of a chromosome. Instead, certain parts of the chromosome are transcribed. Moreover, only one of the two strands of a DNA duplex is transcribed; this strand is called the *sense strand* (Fig. 21–15). The ribonucleotides are added to the growing 3′-OH end of the RNA transcript by *DNA-dependent RNA polymerase*. RNA polymerases are *multimeric* proteins; *E. coli* RNA polymerase consists of six polypeptides and one of these (called the **sigma factor**) is responsible for initiation of transcription. The

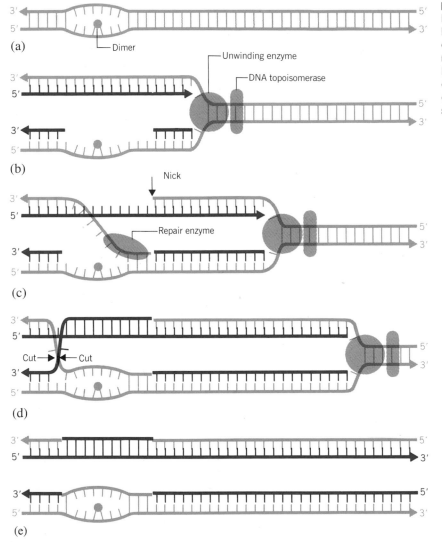

Figure 21–14
Post-replication repair of a DNA molecule in which the presence of a dimer in one original strand results in a gap in the newly synthesized complementary strand. Filling the gap requires a cross-strand exchange. After the gap is filled, the dimer can be eliminated by excision repair, as shown in Figure 21–13.

sigma factor ensures that the enzyme binds to the initiation site by recognizing the *promoter* (see Chapter 11) of the DNA. Like DNA, RNA is synthesized in the 5'→3' direction but proceeds without the need for primers. A transient *heteroduplex* consisting of the sense strand and a portion of the newly synthesized RNA is formed (Fig. 21–15*b*). The transcript soon separates from the sense strand, the heteroduplex structure being maintained only in the region of RNA elongation (Fig. 21–15*c*). As a result, a number of RNA polymerases can simultaneously transcribe the same region of the DNA (Fig. 21–15*d, e;* see also Fig. 22–10).

Bacterial Recombination

Recombination is the exchange of genetic material between chromosomes that results in new combinations of genes in successive generations of cells. In eucaryotic cells, recombination occurs during *crossing over* in *meiosis* (see Chapter 20). Although crossing over cannot occur in haploid organisms, genetic recombinations in procaryotic cells do result from transduction, conjugation, and cellular transformation.

In Chapter 20 we saw that bacterial *plasmids* can be transferred from one cell to another during conjugation and

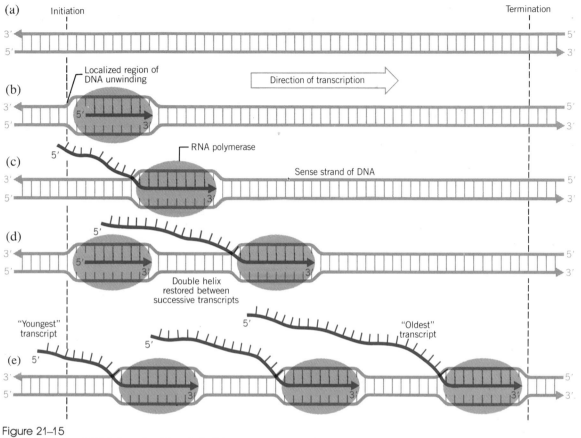

Figure 21–15
Transcription of DNA (see text for details).

may then be integrated into the cell's main chromosome. The integration of plasmids is dependent upon **IS elements**—stretches of DNA (750–1400 base pairs long) found in both plasmids and the main chromosomes of bacteria. These IS elements are *transposable* in that they can move from one position to another in a chromosome.

Genes, Recombinant DNA, and Gene Cloning

For many years, the gene has been thought of as a unit of *structure*, its definition based upon the results of recombination experiments and the analysis of mutant cells. According to this older view, a gene was the smallest unit of inheritance that was not subdivisible by recombination as well as the smallest unit capable of independent mutation.

Today, we recognize that a nucleotide pair is the smallest structural unit that fits the recombination and mutation definitions, although it certainly is not reasonable to call a nucleotide pair a gene. Therefore, a *functional* definition is more appropriate, and the gene is better defined as a unit of inheritance that encodes one polypeptide chain or one RNA molecule.

The complete analysis of the nucleotide sequence of a gene is a substantial undertaking. Nevertheless, the primary structures of a number of genes are now known, including most of the genes encoding the various globin chains of human hemoglobin (see Chapter 22). Even small chromosomes, such as the chromosome of the virus φX174, have been fully sequenced. This chromosome, which consists of 5375 nucleotides, contains nine genes whose starting and ending locations are known. One of the more remarkable

Table 21–2
Recognition Sequences and Cleavage Sites of Representative Restriction Endonucleases

Enzyme	Source	Recognition Sequence[a] and Cleavage Sites[b]	Number of Recognition Sequences per Chromosome of:	
			Phage λ	SV-40 Virus
EcoR1	*Escherichia coli*	↓ GAA TTC ● CTT AAG ↑	5	1
HindII	*Hemophilus influenzae*	↓ GTPy PuAC[c] ● CAPu PyTG ↑	34	5
HindIII	*Hemophilus influenzae*	↓ AAG CTT ● TTC GAA ↑	6	6
HpaI	*Hemophilus parainfluenzae*	↓ GTT AAC ● CAA TTG ↑	11	3
HpaII	*Hemophilus parainfluenzae*	↓ CC GG ● GG CC ↑	750	1

[a] The axis of symmetry in each palindromic recognition sequence is indicated by the colored dot.

[b] The position of each bond cleaved is indicated by a colored arrow. Note that the cuts are staggered (at different positions in the two complementary strands) with some restriction nucleases.

[c] Pu indicates that either purine (adenine or guanine) may be present at this position; Py indicates that either pyrimidine (thymine or cytosine) may be present.

Source. From E. J. Gardner and D. P. Snustad, *Principles of Genetics,* John Wiley & Sons, 1981.

findings of recent years is that the nine genes do not occupy nine separate segments of the chromosome. Instead, two of the genes are located entirely within the coding sequences of two other genes. As will be seen in Chapter 22, the chromosomes of eucaryotic cells contain noncoding nucleotide sequences *between* and also *within* encoding sequences.

In the last several years, methods have been developed that make it possible to experimentally carry out recombination and in so doing manipulate the gene content and genetic expression of cells. For example, it is now possible to isolate genes of one species of cell and transfer these genes to the genome of a cell of a different species, where they may then be expressed. The DNA that results from this experimental recombination is popularly known as **recombinant DNA.** The potential value of so dramatic a technological advance is staggering. Recombinant DNA methods are founded on the discovery of and ability to

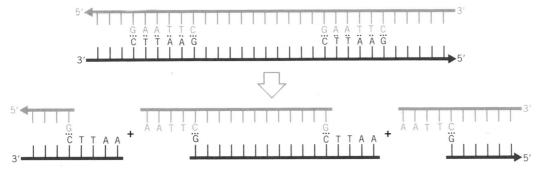

Figure 21–16
Cleavage of DNA by a restriction endonuclease that makes "staggered cuts." The newly exposed ends of the DNA fragments have unpaired bases. (Modified from E. J. Gardner and D. P. Snustad, *Principles of Genetics*, John Wiley & Sons, 1981, p. 381.)

utilize a variety of **restriction enzymes** that cleave DNA molecules at specific locations in their nucleotide sequences (Table 21–2).

Gene Cloning Using the "Shotgun" Method. The so-called *shotgun* approach to gene cloning methods is conducted as follows. First, cells are disrupted and their DNA isolated by centrifugation. The DNA is then treated with one or more of the *restriction endonucleases* (Table 21–2) that cleave the DNA into specific fragments. These endonucleases are site-specific, cleaving the DNA wherever certain nucleotide sequences, called **restriction sites,** occur. Some restriction endonucleases make "staggered cuts" in the DNA so that the exposed cut ends have a number of unpaired bases (Fig. 21–16). Because the nucleotide sequences that are recognized by the restriction endonucleases are **palindromes** (i.e., base sequences that read the same in the forward direction along one chain and the backward direction along the other), the single-stranded ends are complementary.

When treated with the same restriction enzymes, plasmids isolated from bacteria or the chromosomes of viruses such as the λ phage also produce linear molecules having complementary, single-stranded ends. The cellular DNA fragments and the plasmid or virus DNA are then mixed together in the presence of DNA ligase under conditions that permit them to be spliced together. The result is *recombinant DNA* of variable composition depending upon which DNA fragments combined with the plasmid or viral DNA. *E. coli* cells can then be made to take up the recom-

binant DNA by using calcium salt solutions to increase the permeability of the cell's plasma membrane to large molecules. Once incorporated by the bacteria, the recombinant DNA is replicated along with the remaining cellular DNA during growth (Fig. 21–17).

Identification of the cells containing the specific genes to be cloned demands specific assay procedures. For example, if the gene to be cloned is one that codes for the production of an enzyme involved in the synthesis of the amino acid histidine (i.e., the *his* gene), then the recombinant DNA would be introduced into a mutant variety of *E. coli* cells unable to synthesize histidine. After transfer of the recombinant DNA to the cells, they would be cultured using a medium that lacked histidine. In this way, the cells that had not incorporated recombinant DNA containing the *his* gene would be unable to produce the necessary enzymes for histidine synthesis and would not survive in the histidine-free medium. Only the cells acquiring the *his* gene would survive; thus, the appropriate cells are selected and the gene is cloned as these cells grow and divide. A different selection method is used for each recombinant gene to be cloned.

Gene Cloning Methods Using Reverse Transcription. In some instances it is possible to isolate from cells the mRNA for specific proteins. For example, globin chain mRNA can be isolated from reticulocytes; albumin mRNA can be isolated from chicken oviduct cells; proinsulin mRNA can be isolated from pancreas cells; and interferon mRNA can be isolated from white blood cells. The isolated mRNAs can

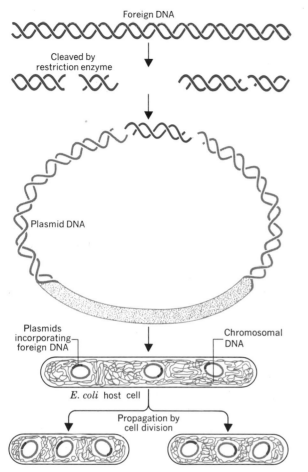

Foreign DNA

Cleaved by
restriction enzyme

Plasmid DNA

Plasmids
incorporating
foreign DNA

Chromosomal
DNA

E. coli host cell

Propagation by
cell division

Figure 21–17

Shotgun approach to gene cloning. Chromosomal DNA is cleaved by restriction endonucleases forming a number of fragments. These are then spliced into bacterial plasmids, thereby producing recombinant DNA. *E. coli* cells can be induced to take up the recombinant DNA by altering the permeability of the cell membrane. Once the recombinant DNA is incorporated, it is replicated along with the cell's native DNA during cell growth and proliferation.

the recombinant DNA used in this approach is specific (not random recombinant DNA molecules, as are produced in the shotgun method), the necessity of using special *E. coli* mutants for cell selection is precluded. The incorporation of DNA molecules produced by reverse transcription into plasmids requires the use of enzymes called **terminal transferases** instead of restriction endonucleases, because reverse transcription does not produce DNA molecules that have offset complementary nucleotide chains at their ends (i.e., the enzymes do not produce staggered cuts). Terminal transferases add single-stranded tails to the 3′ ends of the DNA strands.

Genetic Engineering of Mammalian Cells

Genes from a number of different eucaryotic cells have been excised and then cloned in bacterial cells using the restriction endonuclease and reverse transcription techniques. The bacteria containing the recombinant DNA are thus able to produce a number of eucaryotic proteins. The hormone *proinsulin* was the first such eucaryotic protein whose production was achieved in cultures of bacteria transformed using recombinant DNA. The gene encoding this protein was synthesized using reverse transcriptase and then transferred to the bacterial cells. Human *interferons* (proteins that act as antiviral agents) are also being made by applying the recombinant DNA technology to bacterial cells. In this instance, interferon mRNA is isolated from human white blood cells.

Interferon. The prospect of large-scale production of human **interferon** using recombinant DNA methods has attracted a great deal of public and scientific attention in recent years because interferon has also been found to attenuate the growth of transformed human cells. In a number of instances, cancerous tumors have been shown to go into remission after patients have been treated with interferon. Interferon was discovered in the 1950s by A. Isaacs and J. Lindenmann, who were studying a phenomenon called ''virus interference'' in which cells treated with one type of virus resisted infection by other types. Isaacs and Lindenmann found that virus interference was not due to one virus blocking the growth of another; rather, the infection by one virus was shown to induce a cell to produce and *secrete* a substance that conferred viral resistance to other cells. Because the substance *interfered* with the ca-

then be used as template for the synthesis of complementary DNA strands (called *c*DNA) using **reverse transcriptase,** an enzyme encoded in the genome of certain RNA viruses (Fig. 21–18; see also Chapter 7).

Single stranded DNA produced by reverse transcription can be used as template for the synthesis of a complementary DNA strand. The resulting double-stranded DNA can then be inserted into an opened plasmid or phage chromosome and cloned in bacteria as described above. Since

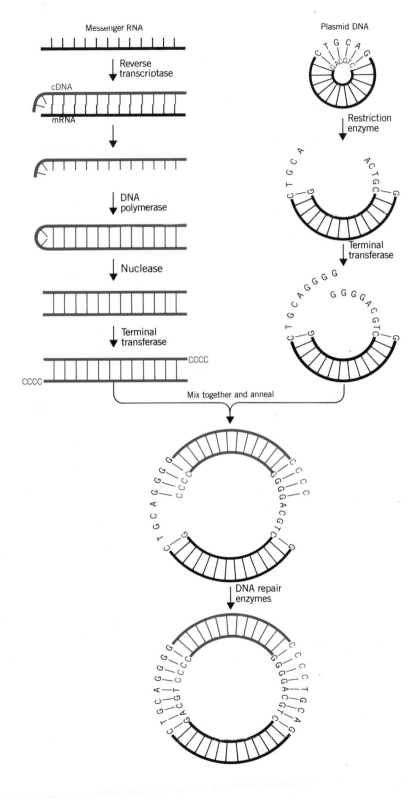

Figure 21–18
**Gene cloning employing reverse tran-
scription. Messenger RNA for a specific
protein is isolated and reverse transcriptase used to make a complementary DNA
strand (cDNA). DNA polymerase is then
used to convert the cDNA to the double-
stranded form. A terminal transferase then
adds specific nucleotides, in this case
CCCC, to each 3′ end of the molecule.
Plasmid DNA is opened using restriction
enzymes and the sequence GGGG added
to the free ends by a terminal transferase.
Finally, the plasmid and DNA are mixed
together and allowed to anneal, following which gaps in the recombinant DNA
are filled by repair enzymes. The recombinant DNA may now be incorporated into
bacteria and cloned.**

pacity for viruses to produce an infection, they called the substance ''interferon.''

Human interferons are now known to be a class of glycoproteins produced by a number of different cells including lymphocytes and fibroblasts. Synthesis of the interferons by these cells is triggered by the derepression of the interferon genes of the cell's genome (see Chapter 11). Once released from the cells, the interferons react with receptors in the plasma membranes of other cells and, in some yet uncertian way, block viral infection of these cells. Since interferon is produced and secreted in very small quantities, intensive study of the action of interferon has been hampered for decades because of the difficulty of isolating sufficient amounts of this material.

Since certain cancers are believed to be transmitted by viruses and interferons have been shown to diminish the growth of tumors, it is tempting to suggest that interferon administration could prevent or cure virus-mediated cancers. However, it has *not* been shown that the affects of interferon on tumor growth result directly from antiviral activity. Rather, the interferon may act indirectly by stimulating other body defense mechanisms. Moreover, most cancers are not produced by viruses. Application of the recombinant DNA technology for large-scale interferon production should soon provide the scientific community with enough of this most interesting material to make definitive tests of its clinical value possible. It is clear that we are at present seeing only the ''tip of the iceberg'' insofar as the benefits that may be accrued by mankind as the use of bacteria for cloning human genes and producing human proteins becomes more and more commonplace.

The genes of eucarotic cells can also be cloned in other eucaryotic cells, but this is a much more difficult task. This is because the membranes of mammalian cells are not so readily made permeable to recombinant DNA. One method for transferring genes to mammalian cells that has had some success is the use of a whole virus as a vector for such transfer. The gene to be transferred is first linked to the viral DNA and the recombinant DNA-containing virus used to infect the animal cells. The gene encoding the beta chain of rabbit hemoglobin has been successfully incorporated into the genome of the simian virus (SV 40). When these viruses are then allowed to infect mammalian cells, the beta globin chain gene is transferred along with the native viral genes and rabbit beta globin chains are subsequently synthesized in these cells. Some success has also been achieved using mechanical microinjection procedures to transfer recombinant DNA-containing plasmids directly into the nuclei of mammalian cells.

Summary

In all cells, DNA replication is **semiconservative.** In procaryotic cells, replication may be **unidirectional** or **bidirectional,** but in eucaryotes it is always bidirectional. The overall growth of the two new complementary DNA strands during replication proceeds in the $5' \rightarrow 3'$ direction. The elongation of one strand is continuous, while the other grows as a series of **Okazaki fragments,** which are eventually linked together. Many of the enzymes involved in the replication process have now been identified.

Viruses are incapable of directly replicating their nucleic acids. Instead, they employ the metabolic machinery of a *host* cell. Replication proceeds in any of a variety of ways depending upon whether the viral nucleic acid is double-stranded DNA, single-stranded DNA, or RNA.

Although **genetic recombination** is a natural phenomenon in procaryotes and eucaryotes, recombination can also be performed experimentally. Of special significance is the use of **recombinant DNA** methods for transferring human genes to and cloning them in bacteria.

References and Suggested Reading

Articles and Reviews

Anderson, W. F., and Diacumkos, E. G., Genetic engineering in mammalian cells, *Sci. Am. 245*(1), 106 (July 1981).

Brown, D. D., Gene expresesion in eucaryotes. *Science 211,* 667 (1981).

Chambon, P., Split genes. *Sci. Am 244*(5), 60 (May 1981).

Cohen, S. N., and Shapiro, J. A., Transposable genetic elements. *Sci Am. 242*(2), 40 (Feb. 1980).

DePamphilis, M. L., and Wassarman, P. M., Replication of eukaryotic chromosomes: a close-up of the replication fork. *Annu. Rev. Biochem. 49,* 627 (1980).

Dickerson, R. E., Drew, H. R., Conner, B. N., Wing, R. M., Fratini, A. V., and Kopka, M. L., The anatomy of A-, B-, and Z-DNA. *Science 216*, 475 (1982).

Gilbert, W., and Villa-Komoroff, L., Useful proteins from recombinant bacteria. *Sci. Am. 242*(4), 74 (April 1980).

Grobstein, C., The recombinant-DNA debate. *Sci. Am. 237*(1), 22 (July 1977).

Howard-Flanders, P., Inducible repair of DNA. *Sci. Am. 245*(5), 62 (Nov. 1981).

Khorana, H. G., Total synthesis of a gene. *Science 203*, 614 (1979).

Landy, A., and Ross, W., Viral integration and excision: structure of the lambda "att" site. *Science 197*, 1147 (1977).

Lehman, A. R., and Karran, P., DNA repair. *Intl. Rev. Cytol. 72*, 101 (1981).

Novick, R. P., Plasmids. *Sci. Am. 243*(6), 102 (Dec. 1980).

Schimke, R. T., Gene amplification and drug resistance. *Sci. Am 243*(5), 60 (Nov. 1980).

Sinsheimer, R. L., Recombinant DNA. *Annu. Rev. Biochem. 46*, 415 (1977).

Books, Monographs, and Symposia

Bonner, J., and Varner, J. E., *Plant Biochemistry,* Academic Press, New York, 1976.

Butterworth, P. H. W., and Beebee, T. J. C., Transcriptional regulation in eukaryotic cells, in *Biochemistry of Cellular Regulation, Vol. 1, Gene Expression* (M. J. Clemens, Ed.), CRC Press, Boca Raton, Florida, 1980.

Gardner, E. J., and Snustad, D. P., *Principles of Genetics* (6th ed.), Wiley, New York, 1980.

Green, C. D., The regulation of gene expression by steroid hormones in animal cells, in *Biochemistry of Cellular Regulation, Vol. 1, Gene Expression* (M. J. Clemens, Ed.), CRC Press, Boca Raton, Florida, 1980.

Kornberg, A., *DNA Replication,* W. H. Freeman, San Francisco, 1980.

Lehninger, A. L., *Biochemistry,* (2nd ed.), Worth, New York, 1975.

Old, R. W., and Primrose, S. B., *Principles of Gene Manipulation,* University of California Press, Berkeley, 1980.

Swanson, C. P., Merz, T., and Young W. J., *Cytogenetics,* Prentice-Hall, Englewood Cliffs. N.J., 1981.

Chapter 22

RIBOSOMES AND THE SYNTHESIS OF PROTEINS

Until the 1930s, it was the prevailing view that DNA was found only in animal cells and RNA only in plant cells. This view was dispelled by a number of findings in the 1930s that definitively established that both DNA and RNA are present in animal and plant cells. Moreover, J. Brachet and T. Caspersson showed that the bulk of the RNA was present in the cytoplasm and that cells actively engaged in protein synthesis (such as pancreas cells and the silk-gland cells of silk worms) contain greater amounts of RNA than cells that do not actively produce protein. Albert Claude showed in the 1940s that the cytoplasmic RNA was included in tiny particles of ribonucleoprotein later to be called "ribosomes."

Protein Turnover in Cells

The rate of breakdown and replacement of protein in cells was badly misunderstood prior to 1939. In growing animals generally and in secretory tissues in particular (e.g., the liver, pancreas, and endocrine glands), active synthesis of protein was known. However, the amount of protein synthesis taking place in other tissues of the adult was believed to be very low and confined to that necessary to replace protein lost in damaged or dying cells. These small protein losses, together with the catabolism of dietary amino acids, were believed to be responsible for the urea and ammonia measureable in urine. Proteins were thus regarded as highly

stable constituents lasting virtually the entire lifetime of the cell.

The first serious challenge to the "wear and tear" view of protein turnover came as a result of the work of R. Schoenheimer in 1938. Schoenheimer synthesized a number of amino acids in which the ^{15}N content of the alpha-amino nitrogen was considerably increased over the natural amount of this isotope. Schoenheimer then injected ^{15}N-containing glycine and leucine into rats and noted that these labeled amino acids were incorporated into the proteins of many tissues very rapidly. Although ^{15}N is not a radioactive isotope of nitrogen, it may nonetheless be distinguished chemically from the more common ^{14}N form and is called a "heavy" isotope of nitrogen. The results clearly indicated that protein synthesis in adult animals is not restricted to growing and secretory tissues but occurs in nearly all cells and that tissue proteins are in a continuous state of metabolic flux, being broken down and replaced by newly synthesized molecules.

Although the radioactive isotope of carbon, ^{14}C, was produced in the Berkeley cyclotron in 1940, it was not until 1947 that ^{14}C-labeled amino acids became available for use as biological tracers. The availability of radioactive amino acids was followed by a series of classical tracer experiments by H. Borsook, T. Hultin, P. Zamecnik, and P. Siekevitz which verified the findings of Schoenheimer that most tissues readily incorporate amino acids into protein and also added crucial details to the newly emerging view of protein synthesis and metabolic turnover.

The first attempts to determine the subcellular site of amino acid incorporation into protein were carried out in 1950 by Borsook. Minutes after injecting ^{14}C-labeled amino acids into the bloodstreams of guinea pigs, Borsook removed the animals' liver and, using the technique of differential centrifugation (see Chapter 12), prepared subcellular fractions of the tissue. Borsook showed that it was the **microsomal fraction** that contained the highest degree of radioactivity and suggested that the microsomes were the repository of the cell's protein-synthesizing apparatus. In the same year, Hultin demonstrated that it was the microsomal fraction of chick liver tissue homogenates that incorporated intravenously injected ^{15}N-glycine into protein.

By 1952, Siekevitz and Zamecnik had been able to demonstrate the in vitro incorporation of ^{14}C-labeled amino acids into liver cell proteins by both tissue slices and tissue homogenates. By measuring and comparing protein synthetic activity in cell-free whole homogenates, individual cell subfractions, and various combinations of subfractions, Siekevitz showed that the incorporation of amino acids into proteins by microsomes was dependent on an energy source provided by the mitochondrial subfraction and required enzymes and other factors present in the cytosol. The demonstration that amino acid incorporation into protein required metabolic energy laid to rest a view popular in the 1940s that polypeptide synthesis might be brought about by the reversal of protein hydrolysis. It is especially interesting to note that Siekevitz demonstrated the existence in the cytosol of a $MgCl_2$-precipitable factor required for protein synthesis. Since $MgCl_2$ was known to precipitate RNA, Siekevitz suggested that RNA might somehow be involved in protein synthesis, a fact not fully recognized until many years later.

The studies described above established the general cytological and chemical basis of protein biosynthesis. Exhaustive research since the 1950s by dozens of groups of investigators has revealed the step-by-step, reaction-by-reaction details of the process and has given us an astounding insight into the molecular organization of the cell's protein-synthesizing apparatus. Although most of this chapter is devoted to the examination of ribosome structure and to the chemical events that accompany protein synthesis, much of what is to be presented is better understood if *first* placed in perspective with a brief, preliminary overview of the subject; this will set the foundation for the more comprehensive study that follows. For simplicity, this synopsis will be concerned only with *cytoplasmic* protein synthesis in eucaryotic cells.

A Preliminary Overview of Protein Biosynthesis

The variety of proteins synthesized by a cell and the specific amino acid compositions of each protein are ultimately governed by the cellular DNA. This DNA is enzymatically **transcribed** in the cell nucleus to produce a host of RNAs (Chapter 21), including ribosomal RNA (rRNA), messenger RNA (mRNA), and transfer RNA (tRNA). The base sequences of these RNAs are *complementary* to the base sequences of the DNA molecules transcribed. rRNA is ultimately incorporated into the cytoplasmic ribosomes, which may be *free* in the cytosol or *attached* to the surface of the endoplasmic reticulum that faces the cytosol. Each ribosome consist of two parts or *subunits*–a *small* subunit and a *large* subunit. The small subunit binds mRNA entering the cytosol from the nucleus, and the functional complex is completed with the subsequent addition of the large subunit. Attached ribosomes are linked to the endoplasmic reticulum via the large subunit.

The nucleotides of mRNA are arranged as a linear sequence of *codons* (also known as triplets), each codon consisting of three successive nitrogenous bases. The codon sequence of each mRNA molecule contains all the information necessary to (1) properly *initiate* polypeptide synthesis on the ribosome, (2) designate the specific *sequence* of amino acids to be incorporated (i.e., the primary structure of the polypeptide), and (3) *terminate* polypeptide synthesis and *release* the completed polypeptide. Table 22–1 shows the various codons of mRNA and their meanings in protein synthesis. This is called the "genetic code." The code is said to be **degenerate** because in certain instances, a single amino acid may be coded for by more than one codon.

Molecules of tRNA entering the cytosol from the nucleus combine with amino acids; this is a molecule-specific association in that each amino acid species is enzymatically combined with a particular type (or species) of tRNA. The products, called **aminoacyl-tRNA,** represent the form in which amino acids are incorporated into newly synthesized protein. Each species of tRNA contains, among other func-

Table 22–1
The Genetic Code

First Base	Second Base				Third Base
	U	C	A	G	
U	phe	ser	tyr	cys	U
	phe	ser	tyr	cys	C
	leu	ser	"stop" (ochre)	"stop" (opal)	A
	leu	ser	"stop" (amber)	try	G
C	leu	pro	his	arg	U
	leu	pro	his	arg	C
	leu	pro	gln	arg	A
	leu	pro	gln	arg	G
A	ile	thr	asn	ser	U
	ile	thr	asn	ser	C
	ile	thr	lys	arg	A
	met ("start")	thr	lys	arg	G
G	val	ala	asp	gly	U
	val	ala	asp	gly	C
	val	ala	glu	gly	A
	val ("start")	ala	glu	gly	G

tional groups, an **anticodon** (sequence of three bases) that is recognized by a corresponding (probably complementary) codon of mRNA and ensures that the correct amino acid will be incorporated into its proper position in the primary structure of the polypeptide being synthesized.

Once the mRNA-ribosome complex has been formed, amino acids bound to their specific tRNA molecules are sequentially brought to the ribosome and incorporated into the growing polypeptide chain. This process, called **translation,** is believed to take place by an orderly and linear movement of the mRNA along the ribosome (or vice versa) so that each codon is translated in sequence. The elongation of the polypeptide chain takes place by a series of enzyme-catalyzed reactions occurring on two adjacent sites of the ribosome; these are the **amino acid** (or **acceptor) site** and the **peptide** (or **donor) site.** To understand the process of elongation, consider an intermediate stage in the synthesis of a polypeptide. At this time, the growing polypeptide chain is attached to the peptide site of the ribosome by a molecule of tRNA and is termed **peptidyl-tRNA.** The mRNA codon located near the vacant amino acid site specifies the particular aminoacyl-tRNA that can be bound there. With a new aminoacyl-tRNA in position in the amino acid site, the bond linking the growing polypeptide to its tRNA is broken and replaced by a peptide bond with the amino acid of aminoacyl-tRNA. This leaves the peptidyl-tRNA (which is now one amino acid longer) temporarily in the amino acid site. The tRNA molecule released in the process reenters the cytosol, where it may combine with another amino acid to be used in protein synthesis. Formation of the peptide bond is followed by a shift of the peptidyl-tRNA to the peptide site, once again leaving the amino acid site vacant. This shift is accompanied by the movement of the ribosome and/or mRNA so that the next codon is in position adjacent to the amino acid site and may now be translated. Thus, tRNA molecules employed in bringing amino acids to the ribosome are transiently bound first to the amino acid site and then to the peptide site before returning to the cytosol.

Amino acids are sequentially added to the growing polypeptide until its primary structure is complete. Once the end of the message encoded in the strand of mRNA is reached, the completed protein is released from the ribosome. The ribosome separates from the mRNA and dissociates into its two subunits; these may be used again in another round of protein synthesis.

Many mRNA molecules are large enough to be simultaneously translated by a number of ribosomes. These ribosomes move in series along the mRNA, or the mRNA is moved through the ribosomes, the net result being that its coded message is translated into a number of *identical* proteins. The release of one ribosome at the end of the message is accompanied by the attachment of a new ribosome at the beginning of the message. Such strings of ribosomes are called **polysomes,** and most cellular protein synthesis takes place on these structures. Although each mRNA molecule may be attached to several ribosomes, each ribosome synthesizes but a single protein chain before dissociating into its subunits. The mechanics of protein synthesis described briefly here for perspective only is treated in detail in later sections of this chapter.

Table 22–2

Properties and Composition of Eucaryotic and Procaryotic Ribosomes

	Eucaryotes	Procaryotes
Monomers		
Sedimentation coefficient	80 *S*	70 *S*
Molecular weight	4.5×10^6	2.7×10^6
Number of RNAs	4	3
Number of Proteins	70	55
Small subunit		
Sedimentation coefficient	40 *S*	30 *S*
Molecular weight	1.5×10^6	0.9×10^6
RNAs present	18 *S*	16 *S*
	(M.W., 0.7×10^6)	(M.W., 0.6×10^6)
	(2110 nucleotides)	(1600 nucleotides)
Number of proteins	30	21
	(total M.W., 0.78×10^6)	(total M.W., 0.3×10^6)
Large subunit		
Sedimentation coefficient	60 *S*	50 *S*
Molecular weight	3×10^6	1.7×10^6
RNAs present	5 *S*	5 *S*
	(M.W., 3.2×10^4)	(M.W., 3.2×10^4)
	(120 nucleotides)	(120 nucleotides)
	5.8 *S*	23 *S*
	(M.W., 5×10^4)	(M.W., 1.1×10^6)
	(150 nucleotides)	(3200 nucleotides)
	28 *S*	
	(M.W., 1.7×10^6)	
	(5000 nucleotides)	
Number of proteins	40	34
	(total M.W., 1.37×10^6)	(total M.W., 0.5×10^6)

Structure, Composition, and Assembly of Ribosomes

In this section, we are concerned with the organization, composition, and assembly of the cytoplasmic ribosomes of procaryotic and eucaryotic cells. Organellar ribosomes (e.g., chloroplast and mitochondrial ribosomes) will be considered separately later in the chapter. Although functionally analogous, many differences exist between the ribosomes of procaryotic and eucaryotic cells (Table 22–2). Considerably more is known about the structure and composition of bacterial ribosomes than ribosomes of eucaryotic cells, as will become evident during the discussion that

follows. Most of the work on procaryotic ribosomes has been carried out using *E. coli*. Although some variations are observed among the procaryotes, findings using *E. coli* are generally representative.

Ribosomes in the cytoplasm of eucaryotic cells have a sedimentation coefficient of about 80 *S* (M. W., about 4.5 $\times 10^6$) and are composed of 40 *S* and 60 *S* subunits. In procaryotic cells, ribosomes are typically about 70 *S* (M. W., about 2.7×10^6) and are formed from 30 *S* and 50 *S* subunits. The complete ribosome formed by combination of the subunits is also referred to as a **monomer.** Although ribosomes from both procaryotic and eucaryotic sources are about 30 to 45% protein (by weight), with the remainder

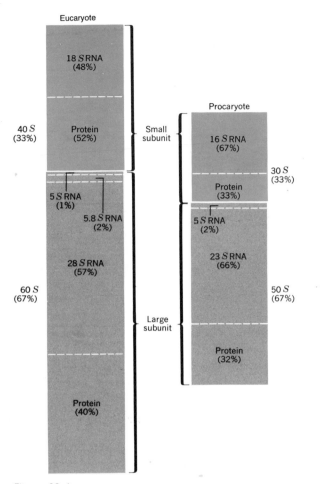

Figure 22–1
Geometric representation of the relative sizes and RNA and protein complements of eucaryotic and procaryotic ribosomes. Each area corresponds to the total molecular weight of that component. (Numbers in parentheses are the weight percentages of the components.)

being ribonucleic acid, the specific protein and RNA components of these two major classes of ribosomes differ (Table 22–1 and Fig. 22–1); carbohydrate and lipid are virtually absent. Magnesium ions (and perhaps other cations) play an important role in maintaining the structure of the ribosome. Dissociation into subunits occurs when Mg^{++} is removed. The precise role (or roles) of Mg^{++} remains uncertain, although interaction with ionized phosphate of subunit RNA is presumed.

Procaryotic Ribosomes

RNA Content. The small subunit of procaryote ribosomes contains one molecule of an RNA called 16 S RNA (M. W., 0.6×10^6), while the large subunit contains two RNA molecules, a 23 S RNA (M. W., 1.1×10^6) and a 5 S RNA (M. W., 3.2×10^4) (see Table 22–2). All three rRNAs are products of closely linked genes transcribed in the sequence 16 $S \rightarrow$ 23 $S \rightarrow$ 5S. This assures an equal proportion of each unit. A polynucleotide containing the 16 S *transcript* is enzymatically cleaved (by an *endoribonuclease*) from the growing RNA strand once transcription has entered the 23 S region of the DNA (referred to as (rDNA). A second polynucleotide containing the 23 S transcript is similarly released once the 5 S region is reached. A final product contains the 5 S transcript. The initial transcription products are successively "trimmed" to form the 16 S, 23 S, and 5 S RNAs finally incorporated into the ribosomal subunits. Figure 22–2 presents the scheme of maturation of the procaryotic rRNAs. For clarity, the incorporation of the ribosomal proteins is not shown. Ribosomal proteins combine with the rRNAs at various stages of subunit assembly: some are incorporated during "pre-rRNA" transcription, others following pre-rRNA cleavage from the growing polynucleotide and during trimming, and still others once the mature rRNA products are formed. Certain proteins bind to the rRNAs only transiently and are not found in the fully assembled subunits.

Multiple copies of the rRNA genes occur in the genomes of procaryotic (and eucaryotic) cells (Table 22–3); this is known as **reiteration.** In *E. coli* the number of rRNA genes is estimated to be between 5 and 10 and accounts for about 0.4% of the cell's total DNA. The primary structures of the three procaryotic rRNAs have been extensively studied; 5 S RNA was identified in 1963 and, being the smallest of the three rRNAs (about 120 nucleotides), was sequenced first (in 1967).

Analyses of the nucleotide sequence of the *E. coli* 16 S RNA (1542 nucleotides) and 23 S RNA (2904 nucleotides) have recently been completed. The secondary structure of the 16 S RNA is shown in Figure 22–3. Methylation of certain bases in the sequence of 16 S RNA (and also in the sequence of 23 S RNA) occurs while transcription is taking place. No methylation of 5 S RNA nucleotides occurs. Unlike 5 S RNA, in which duplication of certain sequences occurs, no repeated sequences are found in 16 S and 23 S

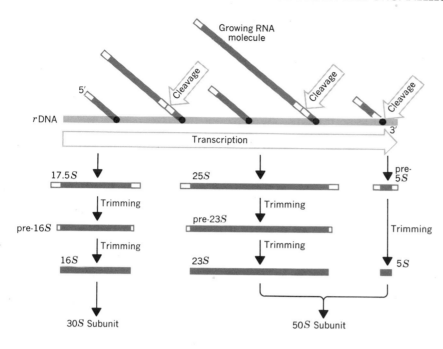

Figure 22–2
"Processing" of procaryotic rRNAs (see text for explanation). Each rectangular block represents an rRNA molecule. Colored portions correspond to regions of the rRNA that become final products and are incorporated into ribosomal subunits. Open portions are "spacer" segments transcribed from rDNA but eliminated during processing; these spacers represent about 20% of the original RNA transcript.

Table 22–3
Reiteration of rRNA in Various Cells

Cell or Tissue	Number of Genes per Genome
Liver	750
HeLa cells	1100
Xenopus (toad)	900
Drosophila melanogaster	260
Tobacco leaves	1500
Saccharomyces cerevisiae	140
E. coli	5–10
Bacillus subtilis	9–10
B. megaterium	35–45

Table 22–4
Average Molecular Weights of Procaryotic and Eucaryotic Ribosome Proteins

	Procaryote	Eucaryote
Small subunit	18,900	25,300
Large subunit	16,400	28,100

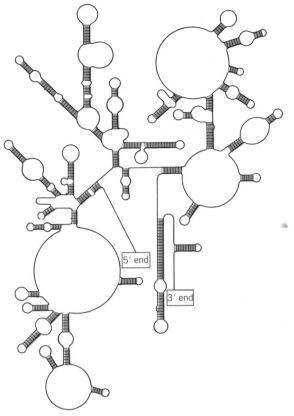

Figure 22–3

Secondary structure of the 16 *S* RNA component of the small subunit of *E. coli* ribosomes. The double-helical regions (shaded) are stabilized by conventional base pairing.

RNA. The rRNAs contain a number of double-helical regions that are stabilized by conventional, complementary base-pairing (Chapter 7). In *E. coli* 16 *S* rRNA, about half of all nucleotides present are involved in base pairing. Several **palindromes** (base pair sequences reading the same from either the 5′ or 3′ ends) exist in 16 *S* RNA, and these may play a role in restricting the formation of the double-helical regions. In 16 *S* RNA, a 7-nucleotide segment of the chain at the 3′ end is believed to interact with mRNA, leading to its binding during the initiation of translation. The 5 *S* and 23 *S* RNAs interact with one another in the large subunit, and both appear to be involved in aminoacyl-tRNA and peptidyl-tRNA binding during polypeptide chain

elongation. Since the 16 *S* RNA as well as several proteins of the small subunit interact with 23 *S* RNA, the latter may also have a role in subunit association.

Ribosomal RNA transcripts are not translated into proteins (i.e., rRNAs cannot serve as messengers); however, ribosomal proteins are the products of a typical transcription-translation process.

Protein Content. Nomura, Kurland, and others have established that the small procaryotic ribosomal subunit contains 21 protein molecules (identified as *S*1, *S*2, *S*3, . . . *S*21) and the large subunit 34 proteins (*L*1, *L*2, *L*3, . . . *L*34). Each ribosomal subunit contains *no more than one copy* of each of the *S* and *L* proteins. All the ribosomal proteins have been isolated and characterized. The small subunit proteins range in molecular weight from 10,900 to 65,000, the large subunit proteins vary in molecular weight from 9600 to 31,500 (Table 22–4). Most of the ribosomal proteins are basic in nature, being rich in basic amino acids and having isoelectric points around pH 10 or higher. About 40 of the 55 ribosomal proteins have been fully sequenced. This, together with the RNA observations described earlier and the discussion of protein synthesis presented later in the chapter, suggests that the procaryotic ribosome may well be the first organelle completely understood in terms of molecular structure and function. An exhaustive analysis of the primary structures of procaryotic ribosomal proteins in order to evaluate their degree of **homology** (see Chapter 4) indicates that these proteins did *not* have a common evolutionary ancestor. Homologies among them do not occur more often than would be expected on a random basis.

rRNA–Protein Interaction. Wittmann, Traut, Stoffler, Kurland, Nomura, and others have studied the relationships between the three rRNAs and the ribosomal proteins and have shown that about 30 proteins bind specifically and *directly* to the rRNAs (i.e., the *primary* binding proteins). Those proteins that do not bind directly to rRNA (i.e., the *secondary* binding proteins) presumably interact with the primary binding proteins in the assembled ribosome.

Best understood is the RNA–protein interaction in the 30 *S* subunit. Figure 22–4 shows the approximate regions of 16 *S* RNA with which the proteins associate. In addition to RNA–protein interaction, there is considerable protein–protein interaction, including interactions among the primary binding proteins and with the secondary binders. It

S6, S8, S15, S18   S7, S9, S13, S19

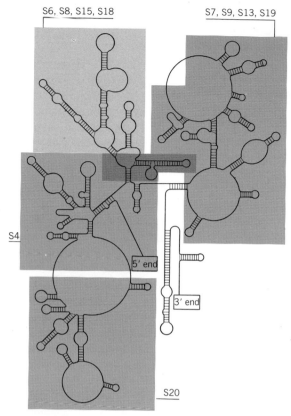

S4

5' end

3' end

S20

Figure 22–4
Areas of *E. coli* 16 *S* RNA with which various small subunit proteins associate.

is believed that the backbone of the 16 *S* RNA polynucleotide winds its way among the proteins, with interactions occurring between hairpin turns of the RNA and surface residues of the protein molecules.

The association of the RNA and protein complements of the 50 *S* ribosomal subunit is not as well understood as the 30 *S* subunits; however, the situation appears to be analogous in that certain proteins bind to specific regions of the 23 *S* and 5 *S* RNAs and protein–protein interactions are numerous.

Assembly of Procaryotic Ribosomes

Since all of the proteins and RNAs of the procaryotic ribosome subunits may be isolated, it is possible through recombination studies to examine the assembly process.

Nomura and others have shown that the assembly of individual subunits and their association to form functional ribosomes (i.e., ribosomes capable of translating mRNA into protein) occurs *spontaneously* in vitro when all the individual rRNAs and protein components are available. Thus the ribosome is capable of *self-assembly,* and this is believed to be the mechanism in situ. The assembly is promoted by the unique and complementary structures of the ribosomal protein and RNA molecules and proceeds through the formation of hydrogen bonds and hydrophobic interactions. There is *order* to the assembly in that certain proteins combine with the rRNAs prior to the addition of others. Cooperativity also exists, since addition of certain proteins to the growing subunit facilitates addition and binding of others.

No self-assembly takes place when *L* proteins are added to 16 *S* RNA or when *S* proteins are added to 5 *S* and 23 *S* RNA. However, it is interesting to note that RNA from the 30 *S* subunit of one procaryotic species will combine with the *S* proteins of another procaryote to form functional subunits. The same is true for 50 *S* subunit proteins and RNAs from different procaryotes. Assembly of *hybrid subunits* and formation of functional monomers from these occur in spite of the fact that ribosomal proteins and RNAs from different procaryotes have different primary structures. It is clear that their secondary and tertiary structures, which are very similar, are the more important in guiding rRNA-protein interactions. Although some proteins from yeast, reticulocyte, and rat liver cell ribosomes can be replaced by *E. coli* ribosomal proteins, *hybrid monomers* formed from these procaryotic/eucaryotic subunits will not function in protein synthesis.

Model of the Procaryotic Ribosome

Based upon the available electron microscopic data, results of small-angle X-ray analysis, and of course, chemical studies, several proposals can be made about the structure of the ribosome monomer and its subunits. The 30 *S* subunit approximates a *prolate* ellipsoid of revolution (Fig. 22–5*a*). A transverse partition or groove encircles the long axis of the subunit, dividing it into segments of one-third (i.e., the *head*) and two-thirds (i.e., the *base*). A small protuberance called a **platform** extends from the base. The 50 *S* subunit is somewhat more spherical and possesses a flattened region on one surface (Fig. 22–5*a*). Extending from the main body

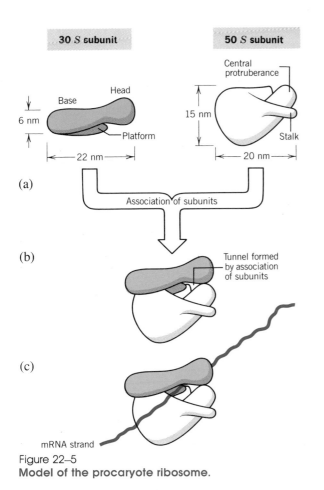

30 S subunit   **50 S subunit**

Central
protruberance

Head

Base

6 nm

15 nm

Platform

Stalk

22 nm

20 nm

(a)

Association of subunits

(b)

Tunnel formed
by association
of subunits

(c)

mRNA strand

Figure 22–5
Model of the procaryote ribosome.

protected from cleavage by nuclease when associated with the ribosome. The observation that nascent (i.e., growing) polypeptides are also protected from proteolysis suggest that they, too, are located within the ribosome—in either the same or perhaps a separate channel (see below).

Genes for Ribosomal RNA and Protein

The genome of *E. coli* and other procaryotes consists of a single, long, circular DNA molecule supercoiled and packed into the nuclear region of the cell. The *E. coli* chromosome is about 1100 nm long and appears to contain at least three separate regions coding for rRNA. Each region contains closely linked 5 *S*, 23 *S*, and 16 *S* rDNA genes. Since some 5 to 10 copies of each gene occur in the genome, more than one copy of each gene is likely present in each rDNA region. Genes coding for ribosomal proteins are present in at least two separate regions of the *E. coli* chromosome. The same regions appear also to contain genes for *RNA polymerase,* some *transfer RNAs,* and the *elongation factors* required for protein biosynthesis (see below). The genes are distributed among at least four operons (Chapter 11), each operon containing genes for a dozen or more proteins (Fig. 22–6).

Eucaryotic Ribosomes

The cytoplasmic ribosomes of eucaryotic cells differ from those of procaryotes in both size and chemical composition (Table 22–2, Fig. 22–1). The monomer has a sedimentation coefficient of 80 *S* and is formed from 40 *S* and 60 *S* subunits. In addition, ribosomes occur in two states in the cytoplasm. They may be associated with cellular membranes such as those of the endoplasmic reticulum (i.e., "attached" ribosomes) and engaged in the synthesis of secretory or membrane proteins, or they may be freely distributed in the cytosol. The functional differences be-

of the large subunit are a **stalk** and **central protuberance.** Association of the subunits to form the 70 *S* monomer is accompanied by a deformation of the 30 *S* subunit at its transverse partition (Fig. 22–5b). The subunits are thereby joined on either side of a **tunnel.** There is considerable morphological and biochemical evidence supporting the idea that the tunnel in the monomer accommodates messenger RNA and the aminoacyl-tRNAs during protein synthesis (Fig. 22–5c). For example, (1) in many electron photomicrographs of polyribosomes, the thin mRNA stand seems to "disappear" into the ribosomes; (2) in vitro experiments have shown that when the synthetic messenger polyU is associated with the 70 *S* monomer, the polynucleotide is protected from ribonuclease attack over a length of about 70 to 120 nucleotides; and (3) transfer RNA is

Figure 22–6
One of the ribosomal protein operons of the *E. coli* genome. Note that genes for both *S* and *L* proteins occur in the same operon. *P*, promoter gene.

---(L15)-(L30)-(S5)-(L18)-(L6)-(S8)-(S14)-(L5)-(L24)-(L14)-(P)---

tween attached and free ribosomes will be pursued later, but let us turn to a consideration of the chemical and morphological characteristics of eucaryotic ribosomes.

RNA Content. The small subunit of eucaryotic ribosomes contains one molecule of 18 S RNA (M. W., 7×10^6), while the large subunit contains 28 S (M. W., 1.7×10^6), 5 S (M. W., 3.2×10^4), and 5.8 S (M. W., 5×10^4) RNAs. Hence, in addition to molecular weight or size differences, a major distinction between the RNA complements of procaryotic and eucaryotic ribosomes is the presence of an additional molecule of RNA in the large subunit of eucaryotes. Of the four rRNAs, the 5.8 S molecule has only recently been discovered and characterized (5.8 S RNA had variously been referred to previously as lRNA, 7 S RNA and 5.5 S RNA). The 5.8 S RNA eluded earlier identification because of its intimate association with 28 S RNA in the ribosome.

18 S, 5.8 S, and 28 S rRNAs are the transcription products of closely linked genes in the chromosomes of the **nucleolar organizing region** (**NOR**) of the cell nucleus. Considerable redundancy exists since hundreds, perhaps even thousands, of copies of these rRNA genes are believed to be present (see Table 22–3). The genes for 5 S RNA are *not* present in the NOR but occur elsewhere in the nucleus. Consequently, unlike procaryotes in which the 5 S RNA genes are linked to the genes for other rRNAs, the 5 S RNA genes of eucaryotes occur separately in the nucleus. This difference, together with other observations to be noted later, supports a contention that the 5 S rRNAs of procaryotic and eucaryotic ribosomes are not analogous; instead, it is the eucaryotic 5.8 S RNA that is the ''counterpart'' of procaryotic 5 S RNA.

Figure 22–7 depicts the transcription and post-transcriptional processing of eucaryotic rRNAs. It should be noted that 5 S RNA is a *primary* transcription product and is not the product of post-transcriptional trimming (another distinction from procaryotic 5 S RNA; see Fig. 22–2). Whereas the precursors of the procaryotic rRNAs are sequentially cleaved from the growing transcript, a *single, high-molecular-weight transcript, 45 S RNA*, containing the precursors of 18 S, 5.8 S, and 28 S rRNAs, is produced in eucaryotes. About half of the 45 S RNA molecule is represented by *spacer* sequences that are trimmed during final processing. The first processing step (Fig. 22–7) divides the 45 S RNA into two parts; the larger of these (41

S RNA) eventually gives rise to 5.8 S and 28 S RNA, while 18 S RNA is derived from the smaller product. Not shown in Figure 22–7 but discussed later is the incorporation of the ribosomal proteins.

It is natural when comparing procaryotic and eucaryotic cells to look for structures or molecules of similar or even identical function. With regard to ribosome structure and composition, the analogy of 16 S RNA (of procaryotes) and 18 S RNA (of eucaryotes) is obvious, since both are parts of the small subunits of ribosomes and also have other features in common. Similarly, an analogy exists between 23 S RNA (of procaryotes) and 28 S RNA (of eucaryotes). But what about procaryotic 5 S RNA and eucaryotic 5 S and 5.8 S RNA? Eucaryotic 5 S RNA is similar in size (about 120 nucleotides) to procaryotic 5 S RNA and also lacks modified nucleotides. In contrast, eucaryotic 5.8 S RNA is larger (about 150 nucleotides) and contains small numbers of modified nucleotides. Notwithstanding these differences, there is significant albeit not yet conclusive evidence for the contention that eucaryotic 5.8 S (not 5 S) RNA is analogous to procaryotic 5 S RNA. For example, (1) the additional nucleotides of 5.8 S RNA occur for the most part in two sections at the 5′ and 3′ ends of the polynucleotide chain; the central portion reveals primary nucleotide sequences more closely related to procaryotic 5 S RNA than to eucaryotic 5 S RNA; (2) as noted earlier, procaryotic 5 S RNA and eucaryotic 5.8 S RNAs are transcription products of closely linked rRNA genes and undergo post-transcriptional processing; and (3) there is evidence to support the proposal that 5.8 S RNA, like procaryotic 5 S RNA, interacts at the A site of the ribosome, whereas eucaryotic 5 S RNA interacts with tRNA during the initiation phase of protein synthesis.

Protein Content. Various studies have established that the small subunits of eucaryotic ribosomes contain 30 proteins ($S1$, $S2$, $S3$, etc.), and the large subunits 40 proteins ($L1$, $L2$, $L3$, etc.). The proteins of eucaryotic ribosomes are not only more numerous but also have greater average molecular weights (Table 22–4). From a chemical standpoint, eucaryotic ribosomal proteins have similar general properties as those in procaryotes (rich in basic amino acids, high isoelectric point, etc). Certain eucaryotic and procaryotic ribosomal proteins reveal homologous regions, and these homologous proteins appear also to be functionally similar.

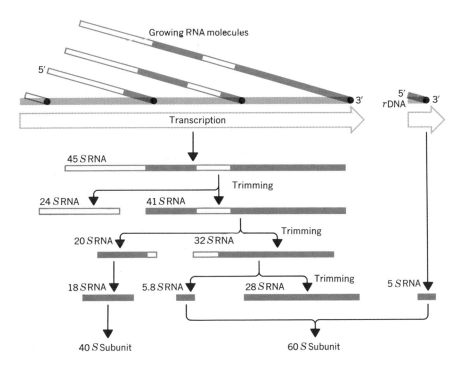

Growing RNA molecules

5′

3′

Transcription

5′ *r*DNA 3′

45 *S* RNA

Trimming

24 *S* RNA
41 *S* RNA

Trimming

20 *S* RNA
32 *S* RNA

18 *S* RNA
5.8 *S* RNA
28 *S* RNA
Trimming
5 *S* RNA

40 *S* Subunit
60 *S* Subunit

Figure 22–7
Processing of eucaryotic rRNAs (see text for explanation). Each rectangular block represents an rRNA molecule. Colored portions correspond to regions of the rRNA that become final products and are incorporated into the ribosomal subunits. Open portions are "spacer" segments transcribed from rDNA but eliminated during processing; these spacers represent about 50% of the original RNA transcript.

Nucleolar Organizing Region. Eucaryotic cells contain several hundred copies of the genes coding for rRNA. These genes are arranged in a tandem fashion on one or more chromosomes of the nucleus. The DNA sequences between successive rDNA regions are not transcribed and represent *spacer* DNA. The rRNA genes and the spacer segments are usually looped off the main axis of the chromosome and are referred to as the *nucleolar organizing region* (NOR). It is here that most of the rRNA is synthesized. The NOR coalesces with nuclear proteins and forms visible bodies known as *nucleoli*. Most eucaryotic cells contain one or a few nucleoli, but certain egg cells are a striking exception. The oocytes of amphibians (e.g., the clawed toad, *Xenopus laevis*) are extremely large cells and are engaged in the synthesis of especially large quantities of cellular protein. These cells produce large numbers of ribosomes in order to provide the means to sustain such quantitative protein synthesis. Accordingly, it is not unusual to find hundreds or thousands of nucleoli (and NORs) in the nuclei of these cells. Such large numbers of nucleoli are the result of gene amplification—the differential replication of the rRNA genes of the genome. The ribosomes produced in the oocyte serve its needs for protein synthesis for the period prior to fertilization through the first few weeks of embryonic development.

By gently dispersing nuclear fractions isolated from oocytes of the amphibian *Triturus viridescens* and "spreading" the material on grids, O. L. Miller and B. R. Beatty in 1969 were able to obtain photomicrographs of transcription *in progress*. Since then, the same approach has been extended by a number of other investigators to mammalian oocytes and to spermatocytes and embryo cells from various organisms. The visualization of transcriptional activity is achieved most easily with spread *nucleoli* because of the high degree of rDNA gene amplification (Fig. 22–8). The tandem rDNA genes are serially transcribed by RNA polymerases to produce 45 *S* rRNA. The rRNA (apparently complexed with protein) appears as a series of fibrils of varying length extending radially from an axial, linear DNA fiber (Fig. 22–9). These feather-shaped or "Christmas tree" regions are called *matrix units*. The spaces between successive matrix units are nontranscribed *spacer* segments.

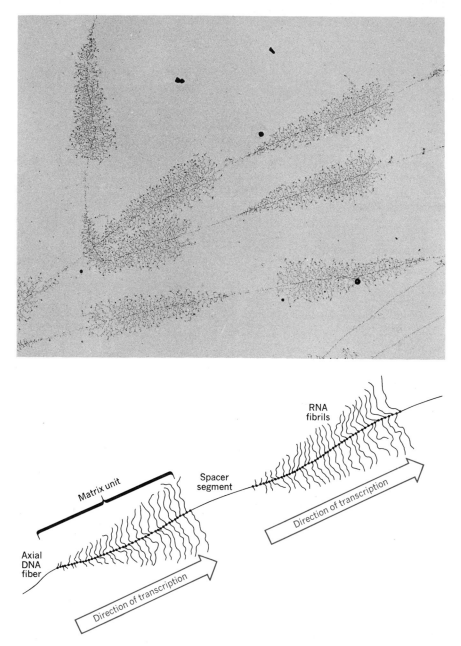

Figure 22–8
Visualization of transcription of rDNA genes. The thin axial DNA fiber is being transcribed simultaneously by a number of RNA polymerase enzymes (small black dots; see also Fig. 22–9). The transcripts (ribonucleoprotein complexes) appear as fine fibrils extending radially away from the DNA axis. Magnification, 18,000×. (Electron photomicrograph courtesy of Dr. O. L. Miller, from O. L. Miller and B. R. Beatty, *Science 164*, 956, 1969. Copyright © 1969 by the American Association for the Advancement of Science.)

The ribonucleoprotein (RNP) fibrils are seen to be in various stages of completion. The short fibrils near the tip of each feather are RNA molecules whose synthesis has only just begun, while the longest fibrils represent RNA molecules whose synthesis is almost complete. Hence, *the direction of rDNA transcription is apparent in the photomicrograph*. In high-magnification views (Fig. 22–9), even the *RNA polymerase* enzyme molecules carrying out the

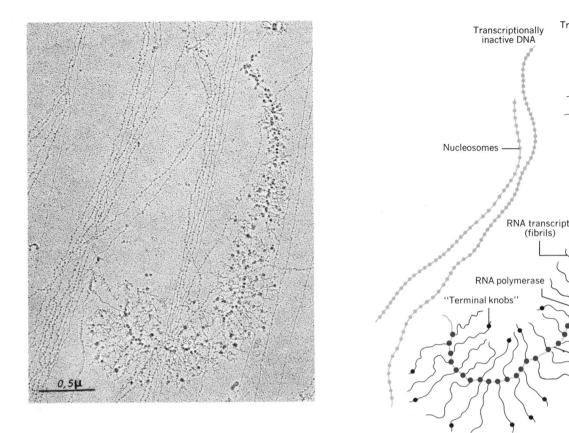

Figure 22–9
High-magnification electron photomicrograph of matrix unit and neighboring nontranscribed DNA fibers (see legend of Figure 22–8 and text for explanation). (Photo courtesy of Dr. F. Puvion-Dutilleul.)

transcription of the DNA are visible along the axial DNA fiber.

Success in visualizing transcription has not been restricted to nucleolar genes. Almost identical results have been obtained with nonnucleolar chromatin. Here, however, the RNA transcripts represent messenger RNA.

Dispersed and spread nuclear fractions contain nontranscribing DNA as well as matrix units (Fig. 22–9). The succession of *nucleosomes* (Chapter 20) reveals itself as a series of beadlike structures along the DNA fiber. Regions in which DNA is undergoing replication (called **replicons**) can also be seen (Fig. 22–10). S. L. McKnight and O. L. Miller have shown that DNA of homologous "daughter"

fibers of the replicon also occurs as chains of nucleosomes, suggesting that replication may not require dissociation of nucleosomes or that nucleosomes are almost immediately reformed (see Chapter 21). Transcriptional activity can be identified within a replicon (Fig. 22–11), indicating that the newly synthesized DNA is almost immediately available for transcription. The growing RNA fibrils are seen in homologous regions of *both* chromatid arms of the replicon.

Assembly of Eucaryotic Ribosomes. The assembly of eucaryotic ribosomes is more complex than that of procaryotes; the principal stages of the process are outlined in

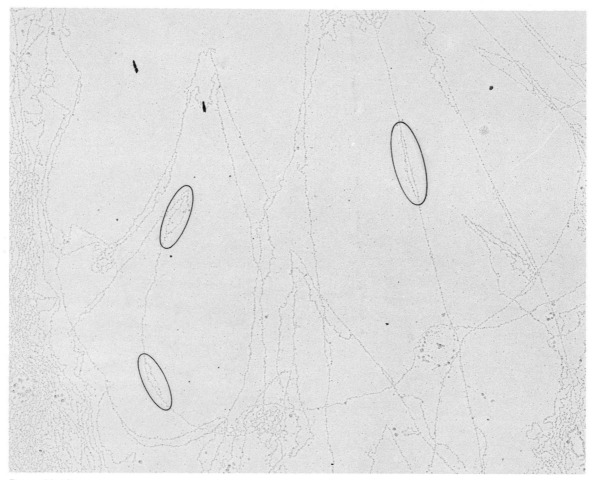

Figure 22–10
**DNA fibers from dispersed nuclei exhibiting regions of replication (i.e.,
replicons). Three replicons are specifically identified in the figure. Magnification, 26,000×. (Electron photomicrograph courtesy of Dr. S. L. McKnight.
Copyright © 1977 by MIT Press, *Cell 12*, 795.)**

Figure 22–12. Transcribed 45 *S* RNA combines with proteins in the nucleolus to form ribonucleoprotein complexes
(RNP). However, not all the protein molecules of the complex become a part of the completed ribosomal subunit.
Instead, certain proteins are released as RNA processing
ensues; these ''nucleolar proteins'' return to a nucleolar
pool and are reutilized. Those proteins that are retained
during processing and become part of the completed subunits are, of course, legitimately called ''ribosomal proteins.'' Enzymatic cleavage of the RNP complex during
processing produces three classes of fragments. One fragment contains spacer RNA and nucleolar proteins. (It
should be noted that the spacer RNA is produced by transcription of rDNA and *not* the spacer DNA between genes.)
The spacer RNA is hydrolyzed, and the free nucleolar
proteins return to the pool. A second RNP fragment contains a complex of 18 *S* RNA and certain ribosomal proteins
that give rise to 40 *S* ribosome subunits in the cytoplasm.
The third RNP fragment, which contains 28 *S*, 5.8 *S* RNA,
and ribosomal proteins, combines with 5 *S* RNA transcribed

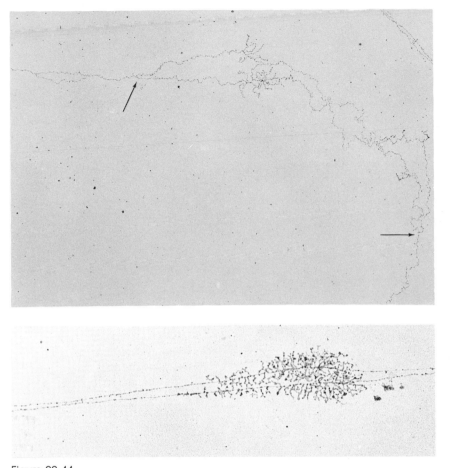

Figure 22–11
Visualization of homologous transcriptional activity in the two daughter chromatids of a replicon. Magnification: upper photomicrograph, 15,000 ×; lower photomicrograph, 27,000×. (Courtesy of Dr. S. L. McKnight. Copyright © 1977 by MIT Press, *Cell 12*, 795.)

from extranuclcolar rRNA genes, and the complex exits the nucleus to give rise to 60 *S* subunits in the cytoplasm. Like the genes for 45 *S* RNA, the extranucleolar 5 *S* RNA genes occur in multiple tandem copies. Among the various proteins synthesized in the cytoplasm using ribosome subunits derived from the nucleus are the ribosomal proteins themselves. These apparently reenter the nucleus for incorporation into new RNP complexes.

Model of Eucaryotic Ribosomes

In spite of the differences in overall sizes (as manifested in the greater molecular weights, sedimentation coefficients, and numbers of RNAs and proteins), the cytoplasmic ribosomes of eucaryotes are remarkably similar in morphology to those of procaryotes. As in 30 *S* subunits of procaryote ribosomes, the 40 *S* eucaryote subunit is divided

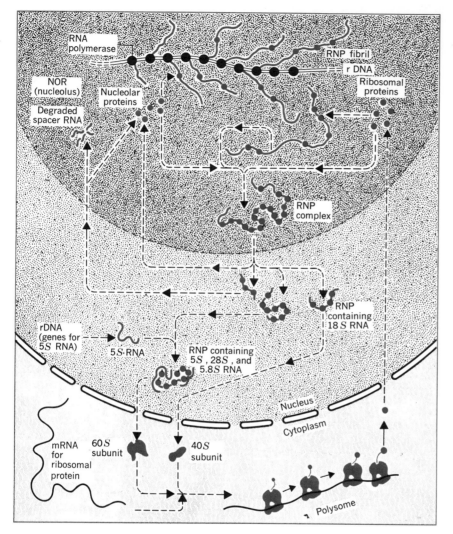

Figure 22–12
Synthesis and assembly of the components of eucaryotic ribosomes.

into head and base segments by a transverse groove (Fig. 22–13). What differences do exist between procaryote and eucaryote small subunits appear to be restricted to the upper third (i.e., the head segment). The 60 S subunit is generally rounder in shape than the small subunit. One side of the large subunit is flattened, and it is this side that becomes confluent with the transverse groove of the small subunit during the formation of the monomer (Fig. 22–13). The resulting tunnel formed in the ribosome is believed to accommodate the mRNA strand during translation.

Free and Attached Ribosomes

The cytoplasmic ribosomes of eucaryotic cells can be divided into two classes: (1) *attached* ribosomes and (2) *free* ribosomes (Fig. 22–14). Attached ribosomes are ribosomes associated with intracellular membranes, primarily the endoplasmic reticulum, whereas free ribosomes are distributed through the hyaloplasm or cytosol. Although all animal and plant cells contain both attached and free ribosomes, the proportion of each varies from one tissue to another and

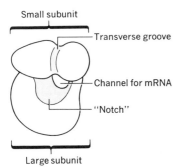

Small subunit

Transverse groove

Channel for mRNA

"Notch"

Large subunit

Figure 22–13

Model of the eucaryotic cytoplasmic ribosome. Although the detailed structure is not as well understood as that of the procaryotic ribosome (see Fig. 22–5), the overall contours appear to be similar.

can be caused to shift within a tissue in response to the administration of certain substances, notably hormones and growth factors.

Membranes of the endoplasmic reticulum (ER) that contain attached ribosomes constitute what is called "rough" ER (or RER), while membranes that are devoid of ribosomes are called "smooth" ER (SER) (see Chapter 1). The ribosomes of RER are attached to the hyaloplasmic surface of membranes (as opposed to the lumenal or cisternal surface). Attachment to the membrane occurs through the large (60 S) subunit.

For many years, there has been considerable controversy about the functions of attached and free ribosomes. The currently accepted view suggests that proteins destined to be secreted from the cell or to be incorporated into such intracellular structures as lysosomes and peroxisomes (which may or may not release their contents to the cell exterior) are synthesized on attached ribosomes. For example, many of the proteins circulating in the blood plasma are derived via secretion by the liver, and these plasma proteins are known to be synthesized exclusively by the attached ribosome of the liver cells. Most proteins destined to become constituents of the ER membranes or the plasma membrane are also synthesized by attached ribosomes (see also Chapter 15). Most, *but not all,* proteins destined for use in the cytosol are synthesized by free ribosomes. Exceptions include certain hormones like thyroglobulin, which is secreted by the thyroid gland and is synthesized by free ribosomes. Milk proteins produced by mammary gland cells are also synthesized by free ribosomes.

When cells are disrupted, the sheets of endoplasmic reticulum are broken into small vesicular fragments (Fig. 22–15a), which may be isolated with the *microsomal phase* by centrifugation. As noted in Chapter 12, this phase is quite heterogeneous and contains a variety of small particles in addition to fragmented endoplasmic reticulum. Fragmentation of the endoplasmic reticulum produces two kinds of vesicles; fragments of RER form vesicles whose outer surface is studded with ribosomes, while SER vesicles are free of ribosomes. The volume within the vesicle corresponds to the lumenal or cisternal phase of the cell.

In a series of elegant experiments, D. Sabatini and C. M. Redman examined protein synthesis in vitro by ribosome-laden vesicles. They were able to demonstrate that radioactive amino acids incorporated into protein by the ribosomes did not appear in the suspending medium but were recovered instead within the vesicles (Fig. 22–15b). These results argue strongly in favor of the proposal that attached ribosomes synthesize proteins for secretion, since the interior of the vesicles corresponds to the cisternal phase of the cell. Sabatini and Redman called this directional synthesis of protein **vectorial synthesis.** Once the protein is released into the cisternae, it is transported to the Golgi apparatus for packaging (Chapter 19).

Although it is widely accepted that proteins destined to be secreted from the cell are synthesized on membrane-bound ribosomes, there are several pieces of evidence that indicate that membrane-bound ribosomes may have other functions as well. For example, J. R. Tata and others, working with muscle and nerve tissue, have observed the synthesis by RER ribosomes of small quantities of proteins for intracellular utilization. The rapid proliferation of RER during periods of active cell growth also suggests a nonsecretory function for attached ribosomes. The enzyme *serine dehydrogenase* has been shown to be specifically synthesized by attached ribosomes in liver cells, and yet this is an intracellular enzyme. Preliminary evidence suggests that some mitochondrial enzymes may be differentially synthesized by RER ribosomes.

Working with liver tissue, Sabatini and his colleagues have shown that ER membranes to which ribosomes are attached contain certain proteins absent in ribosome-free ER membranes. Biochemical studies coupled with freeze-fracture electron microscopy suggest that these RER membrane proteins are binding sites for ribosomes. Figure 22–16 depicts a contemporary model for the synthesis of

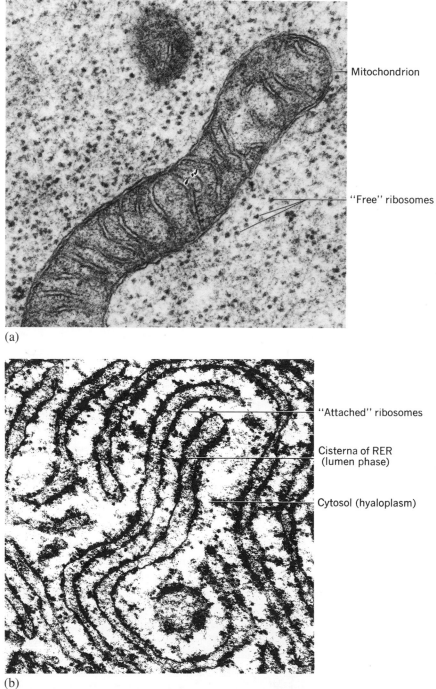

(a)

Mitochondrion

"Free" ribosomes

(b)

"Attached" ribosomes

Cisterna of RER
(lumen phase)

Cytosol (hyaloplasm)

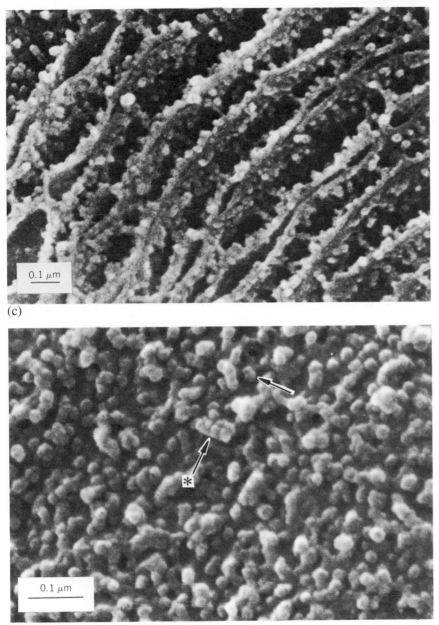

(c)

(d)

Figure 22–14
Free (a) and *attached (b), (c), (d)* ribosomes. The SEM views seen in parts *(c)* and *(d)* are particularly unusual. In *(c)*, several neighboring membranous layers of the rough endoplasmic reticulum are seen. In *(d)*, the attached ribosomes of one layer are seen from the hyaloplasmic side; for some of these ribosomes, the subunits may be discerned. (*a* and *b* courtesy of R. Chao; *c* and *d* courtesy of Dr. K. Tanaka. *Intern. Rev. Cytol. 68*, 102.)

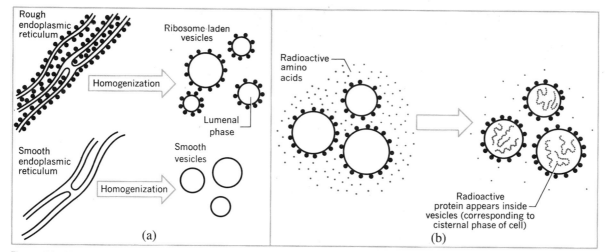

Figure 22–15

(a) **Production of microsomal vesicles during homogenization of endo-plasmic reticulum.** *(b)* **Vectorial synthesis of protein by ribosome-laden vesicles.**

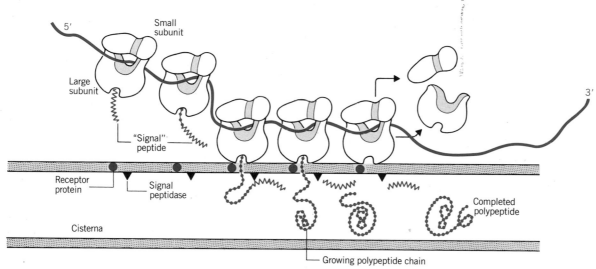

Figure 22–16
Synthesis of secretory proteins by attached ribosomes.

secretory proteins by attached ribosomes. Synthesis of a "signal" region of the polypeptide precedes ribosome attachment to the membrane. The **signal peptide** interacts with a **signal receptor protein** in the membrane, directing the association of the large subunit with a ribosome receptor site. The elongating polypeptide is presumed to exit the ribosome through a channel in the large subunit. Prior to the discharge of the completed proteins into the ER cisternae, the signal peptide is removed by **signal peptidase** (an enzyme in the lumenal face of the membrane).

Ribosomes of Organelles

The mitochondria and chloroplasts of eucaryotic cells contain their own DNA and protein-synthesizing apparatus. Although reports of the presence of DNA in mitochondria and chloroplasts have appeared periodically in the scientific literature since the 1920s, such an astounding proposition was not generally accepted until the 1960s, when H. Ris and M. Nass independently verified the existence of DNA fibrils in these organelles in animal and plant cells using electron microscopic methods originally developed to visualize DNA in procaryotic cells (Fig. 22–17). Little attention was given to the possibility that mitochondria and chloroplasts might contain ribosomes and other elements involved in protein synthesis until the presence of DNA in these organelles was established. With such an impetus provided, it did not take long before ribosomes were indeed indentified in and isolated from chloroplasts and mitochondria.

Chloroplast Ribosomes

Chloroplast ribosomes have a sedimentation coefficient of 70 *S* and consist of 50 *S* and 33 *S* subunits. In this respect, chloroplast ribosomes are similar to those of procaryotic cells but distinct from eucaryotic cytosol ribosomes. The large subunit of the chloroplast ribosome contains 5 *S* and 23 *S* RNAs and the small subunit of 16 *S* RNA (Table 22–5).

Mitochondrial Ribosomes

Unlike chloroplast ribosomes, which are similar in all groups of organisms studied, mitochondrial ribosomes are quite heterogeneous. With respect to their disposition within the mitochondrion, the ribosomes occur either free in the *matrix* of the organelle or are associated with the cristael membranes. (Ribosomes attached to the cytosol side of the outer mitochondrial membrane are cytoplasmic ribosomes possibly engaged in vectorial synthesis of certain intramitochondrial proteins; see below.)

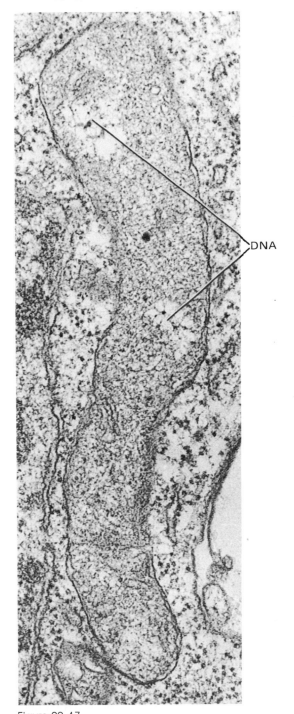

Figure 22–17
DNA fibrils in mitochondria. (Photomicrograph courtesy of Dr. E. G. Pollock.)

Table 22–5
Properties of Chloroplast and Mitochondrial Ribosomes

	Sedimentation Coefficient	
	Particle	RNAs
Chloroplast ribosomes		
Monomer	70 S	
Large subunit	50 S	5 S and 23 S
Small subunit	33 S	16 S
Mitochondrial ribosomes		
Animal cells		
Monomer	50–60 S	
Large subunit	40–45 S	16–18 S
Small subunit	30–35 S	12–13 S
Yeast, fungi, and protists		
Monomer	70–80 S	
Large subunit	50–55 S	21–24 S
Small subunit	32–38 S	14–16 S
Higher plant cells		
Monomer	70–80 S	
Large subunit	50–60 S	>23 S
Small subunit	40–44 S	>16 S

Although mitochondrial ribosomes from all sources studied consist of two subunits, there is considerable variation in the sizes of the subunits and the monomers formed from them (Table 22–5). Mitochondrial ribosomes of yeast, fungi, protists, and higher plants are characterized by a sedimentation coefficient of 70 to 80 S, whereas those of animal cells have a sedimentation coefficient of only 50 to 60 S. Unlike the ribosomes of chloroplasts, procaryotic cells, and eucaryotic cytoplasm, mitochondrial ribosomes contain only *two* species of rRNA.

Protein Synthesis in Chloroplasts and Mitochondria

The amounts of DNA present in chloroplasts and mitochondria is only about 10 to 15% of that necessary to encode for the hundreds of different proteins present in these organelles. Therefore, most of the proteins of chloroplasts and mitochondria are the products of genetic information in the cell nucleus. Experimentally, ribosomal RNAs of chloroplasts and mitochondria can be shown to hybridize with their organellar DNA, and it is generally

agreed that these RNAs are synthesized in the organelles. The origins of only a small number of chloroplast and mitochondrial proteins have been established to date, but the results are interesting and perhaps also surprising. Most, if not all, of the proteins that make up the organelles' ribosomes are synthesized in the cytoplasm of the cell and apparently make their way to the organelle for assembly (along with rRNA) into ribosomes. Entry into the organelle may be via vectorial discharge. The synthesis of several organelle enzymes, including *ribulosediphosphate carboxylase* (of chloroplasts) and *cytochrome oxidase,* and an *ATPase* (of mitochondria) appears to be a "joint operation" of the organelle and the cytoplasm. For example, of the seven polypeptides that make up cytochrome oxidase, four are synthesized on cytosol ribosomes and three on mitochondrial ribosomes. There is some evidence that a few of the chloroplast membrane proteins are synthesized on chloroplast ribosomes.

Much remains to be learned about the total function of organelle ribosomes; however, it is clear that while most chloroplast and mitochondrial proteins are the products of the nucleocytoplasmic protein-synthesizing machinery of the cell, a small number of proteins and portions of certain multisubunit enzymes are the products of genetic material intrinsic to the organelle and the result of translation on organelle ribosomes.

Mechanism of Protein Synthesis

Thousands of experiments carried out during the past 20 years involving hundreds of scientists have slowly and painstakingly revealed the intricate details of the mechanism by which proteins are synthesized in procaryotic and eucaryotic cells. The overwhelming majority of these studies were conducted using two particular kinds of cells, namely, the bacterium *E. coli* and the mammalian **reticulocyte.** Bacterial cells represent highly desirable sources for studying protein synthesis, since the cells themselves are readily obtained and conveniently cultured in the laboratory. Moreover, since the ribosomes of bacteria are not attached to intracellular membranes, they are readily isolated from disrupted cells.

The mammalian immature red blood cell or *reticulocyte* has been the overwhelming favorite among scientists studying protein synthesis in eucaryotic cells for a number of very important reasons; these are best appreciated by briefly considering the origin and features of this unusual cell.

In mammals, red blood cells are produced in the bone marrow and pass through a number of characteristic developmental stages before the mature red cell or *erythrocyte* enters the circulating blood (see Chapter 24). In its early stages of development, the red blood cell possesses most of the structural elements that characterize typical animal cells (nucleus, mitochondria, lysosomes, endoplasmic reticulum, etc.), but nearly all of these structures are lost by the reticulocyte stage. The reticulocyte has only a small number of ribosomes, soluble enzymes, and other soluble constituents with which it completes the synthesis of hemoglobin begun at earlier stages. Indeed, hemoglobin accounts for nearly all the protein being synthesized by the cell, and in this respect the reticulocyte is a more desirable source for studying protein synthesis than bacteria, which synthesize many different proteins. Because the reticulocyte contains no organelles other than ribosomes, the latter are readily isolated following lysis of the cell. The cell is called a reticulocyte because its cytoplasm displays a fine reticulum when stained with certain basic dyes (such as methylene blue), the reticulum being formed in part by precipitation of the residual cytoplasmic RNA and ribosomes.

In the bone marrow, the reticulocyte is transformed to the mature erythrocyte by the loss of its remaining ribosomes and the termination or completion of hemoglobin synthesis. The erythrocyte is then released from the bone marrow and enters the circulating blood.

The separation of reticulocytes from other bone marrow cells in order to follow protein (i.e., hemoglobin) synthesis is no simple matter, and for this reason, marrow tissue is rarely used as the source of reticulocytes. Instead, reticulocytes are obtained using the following procedure. The experimental animal (usually a rabbit or a rat) is rendered severely anemic either by removing a large portion of its blood or by introducing a hemolytic agent (typically, phenylhydrazine) into its bloodstream. The hemolytic agent quickly produces an extensive intravascular hemolysis. In either instance, the resulting anemia is followed within several days by a marked increase in red blood cell production in the bone marrow and the premature release of reticulocytes into the circulating blood. The bloodstream becomes literally flooded with reticulocytes, which may account for nearly all circulating red blood cells in severely anemic animals. Thus, large numbers of reticulocytes can easily be obtained by removing a blood sample from these anemic animals, and it is therefore unnecessary to employ bone marrow tissue itself.

Protein synthesis can be studied using either intact cells or a "cell-free" system. That is, under appropriate experimental conditions, not only do *whole* cells incorporate amino acids into new protein but disrupted cells or simply isolated ribosomes supplemented with all the requisite soluble components (amino acids, tRNA, mRNA, enzymes, cofactors, etc.) also carry out protein synthesis. In recent years, other cells including HeLa cells, liver cells, and yeast cells have been employed to study protein synthesis, and these studies have confirmed most of the observations originally made using *E. coli* and reticulocytes. Some important differences do exist in the mechanism of protein synthesis in procaryotic and eucaryotic cells, but the overall process is fundamentally the same. Those differences that do exist will be noted below as we consider the details of this process.

Protein synthesis involves a number of distinct and sequential steps including (1) **activation** of amino acids, (2) formation of an **initiation complex** between messenger RNA and the ribosome subunits, (3) polypeptide chain **initiation,** (4) chain **elongation,** (5) chain **termination** and **release** of the completed polypeptide, and (6) **dissociation** of the messenger RNA-ribosome complex. Before we consider each of these stages individually, it is worthwhile for perspective to discuss first the pioneering studies of H. M. Dintzis, who in the early 1960s established the *linearity* of chain elongation and determined the *direction* in which polypeptide chain assembly takes place. His work not only provided an insight into the complexity of the process yet to be revealed, but his brilliant selection of methodology served as a model and guide for many of the studies subsequently carried out by other scientists investigating the mechanics of protein synthesis.

Linearity and Direction of Polypeptide Chain Assembly—the Experiments of H. M. Dintzis

Dintzis' experiments were carried out to determine whether assembly of a polypeptide chain (1) began at one terminus and proceeded in order toward the other (and if so, which end was synthesized first), (2) began near the middle and then grew toward both termini simultaneously, or (3) occurred simultaneously at several (or many) points along the polypeptide chain with the eventual linking up of all segments. For his studies, Dintzis used reticulocytes isolated from the blood of rabbits made anemic by the injection of phenylhydrazine.

To follow hemoglobin synthesis in these cells, Dintzis used radioisotopically labeled leucine. Leucine was selected for the following reason: this particular amino acid is more or less uniformly distributed through the primary structure of human alpha and beta globin chains (see Chapter 4), and although the primary structure of rabbit hemoglobin had not been worked out at the time, there was good reason to suspect that it would be quite similar to that of human hemoglobin. This supposition was subsequently verified.

As a starting point, Dintzis proposed a model for protein synthesis according to which the polypeptide chains are assembled in sequence beginning at one end. Therefore, at any chosen instant in time, say t_0, we would expect to find polypeptide chains of various lengths (i.e., varying degrees of completion) attached to the mRNA–ribosome complexes in the cell. Such partially completed polypeptide chains are called *nascent* chains. If at t_0 cells are briefly incubated in a medium that permits continued nascent chain growth, and if radioactive amino acids are included in that medium, then we would expect that at time t_1, each nascent chain would have increased in length by adding a section containing the radioactive label. (This type of experiment is called a "pulse label" experiment.) If the time period in which the cells are incubated in labeled medium is sufficiently short in comparison to the time required for synthesis of a whole polypeptide, then only a few chains will be completed in this interval and released from the ribosome (i.e., those nascent polypeptide chains that were already near completion at the time the incubation was begun.) The remaining chains will still be attached to their respective mRNA–ribosome complexes. For short incubations, the radioactivity present in completed and released chains should therefore be confined to those regions of the chain synthesized *last*.

It should be clear from the preceding discussion that the longer the time interval of incubation in labeled medium (i.e., from t_0 to t_2 or t_3, etc.), the more chains will be completed and released and therefore the further "back" along the polypeptide's primary structure will radioactive label be found. That is, for any period of incubation equal to or shorter than the time required for complete assembly of whole polypeptide chains, one should find a "gradient of radioactivity" among those chains that were completed

in that time interval, such that the highest radioactivity is toward the end of the chain synthesized last and the lowest radioactivity is toward the end synthesized first. Figure 22–18 depicts this concept in diagrammatic terms.

If incubation in the presence of radioactive amino acids is carried out for a period of time greater than that required for the assembly of a whole chain, then not only will all of those chains already begun at t_0 be completed using the label but *new* chains containing the label throughout will also be synthesized and released. Therefore, after prolonged incubation in the presence of labeled amino acids, we would expect to find *no* "gradient of radioactivity" in completed chains. Instead, the radioactive amino acids would be more or less distributed uniformly through the entire length of the polypeptide.

To test this model, it is necessary to isolate the polypeptide chains synthesized during the periods of incubation of varying length and to determine and compare the amounts of radioactive amino acids in various segments along their total lengths.

Dintzis incubated the rabbit reticulocytes at 15°C in a medium that included ^{3}H-labeled leucine, as well as all the other materials necessary for continued hemoglobin synthesis. He selected 15°C rather than 37°C (the normal environmental temperature of these mammalian cells) because at this lower temperature, protein synthesis is sufficiently slowed down to permit these experiments to be carried out more easily. After incubation for varying periods of short duration, the reticulocytes were removed, washed, and lysed and the lysate separated into three fractions by centrifugation. A low-speed centrifugation removed plasma membranes and other large particulate debris from the lysate, while a high-speed centrifugation of this supernatant provided a pellet containing ribosomes with nascent polypeptides and a second supernatant containing hemoglobin—including those molecules nascent at t_0 but completed and released during the period of incubation.

The hemoglobin from these experiments was dissociated into its constituent heme and globin parts and the globin separated into alpha and beta chains by ion-exchange column chromatography. The alpha and beta chains were then treated with the proteolytic enzyme *trypsin,* which cleaves peptide bonds on the alpha-carboxyl side of lysine and arginine residues. Therefore, each chain is split into a specific number of peptide segments. For clarity, only five such segments are depicted in Figure 22–18, although rabbit

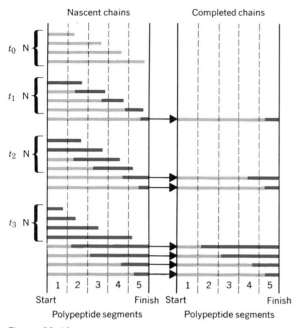

Nascent chains   Completed chains

t_0 N

t_1 N

t_2 N

t_3 N

1 2 3 4 5   1 2 3 4 5
Start   Finish Start   Finish
Polypeptide segments   Polypeptide segments

Figure 22–18
**Model according to Dintzis for the linear growth of na-
scent polypeptide chains. Nascent (N), unlabeled chains
at time t_0 are shown in black. Radioactive segments
added in *separate* experiments by times t_1, t_2, etc., are
shown in color. Polypeptide chains completed and released from ribosomes are shown to the right, while those
remaining nascent are indicated by N. In this hypothetical polypeptide, trypsin digestion produces five polypeptide segments of various lengths (see text discussion for
further explanation).**

alpha and beta globin chains regularly provided 35 segments. These peptides were then separated from one another by a technique called "fingerprinting," which combines paper electrophoresis with paper chromatography to
produce a two-dimensional distribution of separate peptides
across the sheet of filter paper. Having separated the peptide
fragments from one another, the next task was to determine
their specific ^{3}H-labeled leucine contents and to compare
the results after varying periods of incubation in labeled
medium. Not all peptide fragments produced by trypsin
digestion of globin chains would be expected to contain
leucine residues, and in fact, only nine fragments in each
chain did. Obtaining quantitative data on the specific radioactivity in each peptide fragment, comparing the results
for several experiments, and determining whether these
results support or contradict the proposed model for chain

assembly required that two major problems be solved. One
problem is that the total yield of peptide fragments un-
avoidably varied from one experiment to the other as a
result of differential losses at each stage in the isolation
procedures. The other problem is that if the model is correct, the radioactivity of an ^{3}H-leucine-containing peptide
fragment would not only vary as a function of its position
along the primary structure of the globin chain but would
depend also on the number of leucine residues in that
fragment. That is, it is necessary to compensate in some
manner for the differential numbers of leucine residues in
each peptide.

Dintzis solved these problems by using what he called
an "internal standard." At the end of each pulse incubation
experiment, he added hemoglobin that was uniformly labeled with ^{14}C-leucine to the ^{3}H-leucine-containing samples, and the *mixture* was then carried through the stages
of digestion and fingerprinting. The uniformly labeled
hemoglobin was prepared by long-term incubation of reticulocytes in medium containing ^{14}C-leucine; this yielded a
preparation in which, for each hemoglobin molecule, either
all the leucine positions were occupied by the radioactive
form of the amino acid or *none* of them was labeled. (This
would depend upon whether the hemoglobin molecule was
synthesized *before* or *after* the reticulocytes were placed in
the labeled medium.) By expressing the radioactivity of
each resulting peptide fragment as the ratio ^{3}H-leucine:[14]C-
leucine, Dintzis simultaneously circumvented both the
problems of *differential losses* of peptides during the course
of the experiment and the *differential numbers* of leucine
residues in the peptide fragments. Internal standards of this
kind are now almost routinely used in experiments of this
nature.

The kinds of results obtained by Dintzis are shown in
Figure 22–19. The numbers assigned to each peptide in
order to identify and compare them in different "finger-
prints" are more or less arbitrary. Dintzis arranged the data
for each of the labeled peptides from very short (e.g., four-
minute) incubations to yield a curve showing increasing
radioactivity as a function of a selected peptide sequence.
This in itself is not significant, since any collection of
numerical data can be arranged in increasing order. What
was important was that if the selected peptide sequence was
then held constant when plotting the data from a series of
incubations of longer duration, the resulting curves gave
rise to a family of decreasing slopes (Fig. 22–19). Exam-

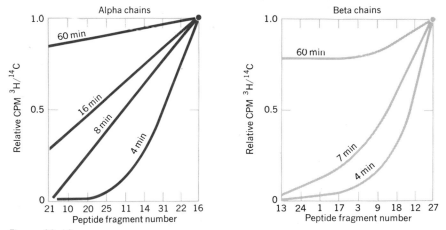

Figure 22–19

Distributions of radioactivity among the peptides produced by tryptic digestion of alpha and beta chains of rabbit hemoglobin released from reticulocyte ribosomes after various periods of incubation at 15°C (see text for explanation).

ining the graph for alpha chains in Figure 22–19, it can be seen that after a four-minute incubation, appreciable radioactivity was recovered in only four peptide fragments (i.e., numbers 14, 31, 22, and 16). As a result of longer incubations, proportionately more radioactivity was found in these peptides and in others that were not labeled at four minutes. Even after 60 minutes of incubation, a gradient of radioactivity still persisted for the peptide sequence, although by this time the slope was approaching zero. These data are consistent with the model of linear growth proposed by Dintzis and are not consistent with the other alternatives. Accordingly, peptide fragment 16 would be near the end of the polypeptide chain synthesized last and peptide fragment 21 would be near the end of the chain synthesized first. It can also be seen from the data in Figure 22–19 that after a seven-minute incubation in radioactive medium, some labeled leucine was found in all the peptides, indicating that only seven minutes are required for the synthesis of a complete alpha or beta globin chain at 15°C. An extrapolation of this to the normal environmental temperature for these cells would suggest that about 1.5

minutes are required for complete chain assembly at 37°C. Since the alpha and beta globin chains of rabbit hemoglobin each contain about 150 amino acids, this corresponds to the elongation of the chain at an average rate of about two amino acids per second.

It was subsequently found that peptide fragment 16 from the alpha globin chain digest contained the *C-terminal* amino acid. This indicated that it is the C-terminus that is synthesized last and that synthesis must therefore begin with the *N-terminal* amino acid of the polypeptide.

At about the same time that Dintzis was carrying out his experiments, J. Bishop, J. Leahy, and R. Schweet, also working with rabbit reticulocytes, reached similar conclusions about the direction of polypeptide chain assembly. The N-terminal amino acid of the globin chains is valine. Bishop, Leahy, and Schweet incubated reticulocytes in ^{14}C-labeled valine and then isolated the reticulocyte ribosomes together with their nascent globin chains. The ribosomes were then incubated in an in vitro system providing for continued growth of the nascent chains but containing ^{12}C-valine (i.e., ordinary valine). After incubation, the amount of N-terminal ^{14}C-valine was compared with that in other regions of the globin chains completed and released from the ribosomes and was found to be significantly higher. These results supported the concept that chain growth began at the N-terminus. A. Yoshida and T. Tobita, studying the synthesis of an amylase from bacteria, and R. E. Canfield and C. B. Anfinsen, studying egg white lysozyme synthesis, also reached similar conclusions about the direction of protein synthesis.

In addition to showing the regular progression of polypeptide addition to growing globin chains, the data of Dintzis (and later Naughton and Dintzis, and others) also reveal differences in the *instantaneous* rates of chain elongation along the polypeptide. This was first suggested by S. W. Englander and L. A. Page, who proposed that curves such as those obtained in the pulse label experiments of Dintzis were also *profiles* of nascent chain lengths at t_0. They noted that the increment of radioactivity between one leucine-containing peptide and another reflected the number of nascent chains having their *growing ends between these two leucines at t_0* and that the slopes at various points along the curve are *inversely proportional* to the rates of chain growth through these points. Consider as an example the seven-minute curve for beta chains shown in Figure 22–19. According to Englander and Page, the rate of chain elongation through the region of the chain containing peptide 1 would be considerably faster than the rate of chain growth through the region containing peptide 12. (Compare the slopes of the curve in these two regions.) If a uniform (i.e., constant) rate of growth occurred over the entire length of the polypeptide, the data resulting from pulse label experiments would yield a family of straight lines. Using bone marrow cells and carrying out pulse label experiments similar to those of Dintzis, R. M. Winslow and V. M. Ingram obtained similar findings for the synthesis of alpha and beta chains of human hemoglobin A, namely, that the rate of chain growth is greater during the synthesis of the first half of the polypeptide than during the synthesis of the second half. Some hypothetical curves based on pulse label experiments of the type conducted by Dintzis, Winslow and Ingram, and others are shown in Figure 22–20, along with an explanation of what these curves imply about the rates of chain growth and the instantaneous distribution of nascent chain lengths on the cell's ribosomes. Recently, A. J. Morris showed that the decreases in the rate of assembly of globin chains occur specifically in the vicinity of amino

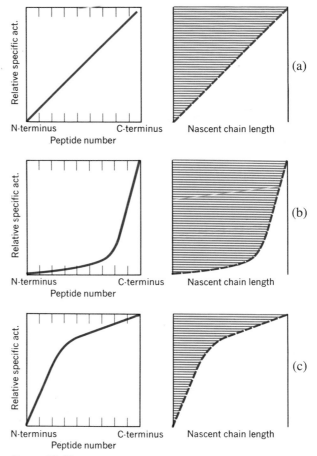

Figure 22–20

Hypothetical curves from pulse label experiments similar to those of Dintzis showing implications of these curves regarding the rates of chain growth and nascent peptide chain lengths at t_0. Curve a would result from a situation in which there were equal numbers of nascent chains of all lengths at t_0, the growth rate being constant. Curve b (which is similar to Dintzis' results with hemoglobin) would result from a case in which there were more long nascent chains than short nascent chains, the growth rate being rapid at first but slowing down toward the end of the chain. Curve c would result from a case in which there were more short chains than long chains, the growth rate being slow at first but increasing toward the end of the chain.

acids, 40, 57, 89, and 120–145 of the primary structure. Since it is clear from the hemoglobin data that polypeptide chain growth does not necessarily proceed at a constant rate, we would expect that in polysomes, the distances between successive ribosomes along the strand of messenger RNA would not be equal but would reflect the differential rates of translation of the mRNA code; electronmicroscopic studies indicated that this is, in fact, the case.

There are several possible explanations for these findings. For example, it could be that not all mRNA codons are translated at the same rate and that the rate of chain elongation over a given region of the polypeptide's primary structure depends on the types and amounts of amino acids and tRNAs present in the cell. It is also believed that for many proteins the assumption of tertiary structure begins during the course of chain growth, and this, too, might influence the rate of chain elongation.

Processing and Structure of Transfer RNA

The first stage in the incorporation of an amino acid into a growing polypeptide chain involves the "activation" of the amino acid, that is, the enzymatic attachment of the amino acid to a specific transfer RNA molecule capable of inserting that amino acid into its appropriate position in the polypeptide chain being assembled on the mRNA–ribosome complex. Each tRNA molecule is specific for a particular amino acid. In a given tissue or cell, each amino acid–specific tRNA can exist in multiple forms called **isoaccepting species.** For example, *E. coli* contains five different (i.e., "isoaccepting") tRNAs capable of combining with leucine. Altogether there may be as many as 50 different tRNAs in a cell or tissue.

tRNA Processing. Transfer RNA is produced in precursor form by *RNA polymerase* transcription of DNA. The precursor, which may contain the base sequence of more than one tRNA, has extra nucleotides at the 3′ and 5′ ends and also internally; these are subsequently cleaved by *endonucleases* and *exonucleases*. All mature tRNAs contain the sequence C-C-A at the 3′ end of the molecule, and this segment may be added following transcription by an appropriate *nucleotidyl transferase*. Nucleosides at various positions in the primary structure may be modified enzymatically (see Table 22–6) to produce the final tRNA product capable of aminoacylation. In some species of tRNA, as many as 16% of the bases may be modified.

Structure of tRNA. The first successful complete purification of tRNA was achieved by R. W. Holley using countercurrent distribution; in 1965 Holley reported the primary structure of yeast alanine tRNA. The primary structure was

Table 22–6
Nucleosides of tRNA

Nucleoside	Symbol
Unmodified	
Adenosine	A
Uridine	U
Guanosine	G
Cytidine	C
Modified	
Inosine	I
Methylinosine	I^{Me}
Methylguanosine	G^{Me}
Dimethylguanosine	G^{DiMe}
2′-*O*-methylguanosine	$G^{2'-OMe}$
Thymine riboside	T
Pseudouridine	ψ
Dihydrouridine	D
Acetylcytidine	C^{Ac}
Methylcytidine	C^{Me}
Methyladenosine	A^{Me}

determined using small polynucleotide fragments produced by enzymatic digestion of the isolated tRNA by pancreatic *ribonuclease* and *phosphodiesterase*. Since Holley's pioneering work, more than 75 tRNAs have been fully sequenced, and all exhibit similar primary, secondary, and tertiary structures. Holley was awarded the Nobel Prize in 1968 for his pioneering studies.

The tRNAs contain a linear sequence of 75 to 85 nucleotides that can be arranged to form the classical "cloverleaf" pattern shown in Figure 22–21*a* and originally proposed by Holley. There are five folded regions: the **amino acid arm,** the **dihydrouridine (DHU) arm,** the

Figure 22–21
**Structure of tRNA. (*a*) Cloverleaf pattern showing secondary structure resulting from hydrogen bonding (i.e.,
●——●) in the helical stems of each arm. Invariant and semi-invariant positions are indicated with the nucleoside symbol (see Table 22–10). Pu, purine; Py, pyrimidine; G*, guanosine or 2′-*O*-methylguanosine; A*, adenosine or 1-methyladenosine; α and β are variable regions containing up to four nucleosides. (*b*) Tertiary structure of tRNA proposed by Kim; (*c*) rearrangement of the cloverleaf secondary structure to more clearly show the L-shape tertiary structure; ribbonlike regions form helical segments through hydrogen bonding. (*d*) Stereo pair of tRNA. (By permission of Dr. S. H. Kim.)**

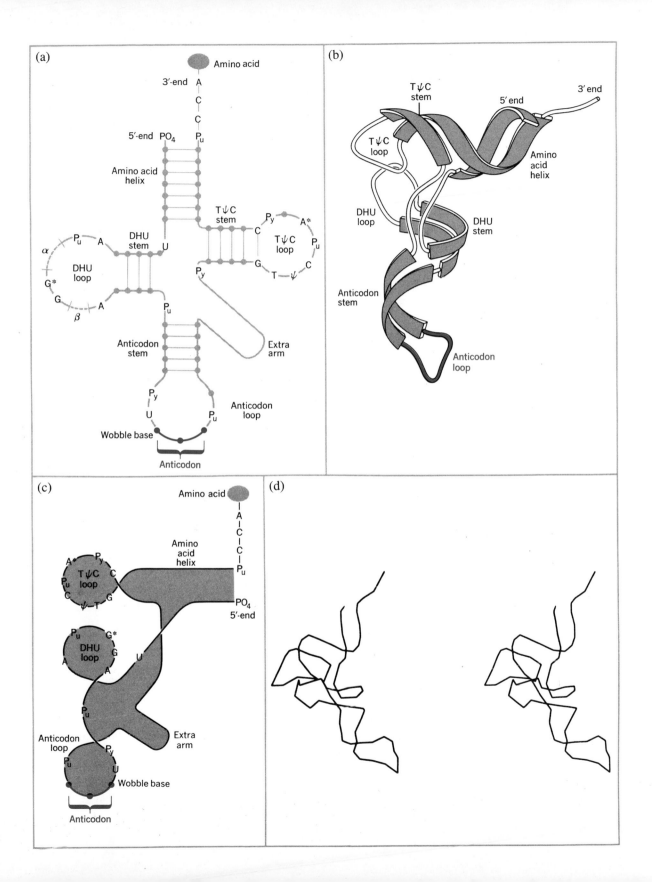

anticodon arm, the **TψC arm,** and the **extra** (or **variable**) **arm.** Each arm consists of a double-helical *stem* stabilized by base pairing. All except the amino acid arm possess a *loop* region containing unpaired bases. With only a few exceptions, the base pairing that creates the secondary structure of the helical regions is of the conventional Watson-Crick type involving hydrogen bonds between A and U and between G and C. To distinguish these bonds from other hydrogen bonds important in maintaining the tertiary structure of the tRNA, the hydrogen bonds of tRNA are denoted as *secondary* hydrogen bonds and as *tertiary* hydrogen bonds accordingly.

As seen in Figure 22–21a, certain positions are invariant or semi-invariant among all tRNAs so far sequenced. (The term *semi-invariant* is used to denote a position invariably occupied by the same *type* of base: purine or pyrimidine.) For example, the four unpaired bases that terminate the sequence at the 3′ end of the molecule are always purine-C-C-A; the last of these bases (i.e., adenine) forms the bond with the amino acid (see below). Most of the invariant and semi-invariant positions are found in the DHU loop and in the TψC loop. The invariant residues form hydrogen bonds with one another that are crucial to the maintenance of the characteristic tertiary structure of the tRNAs and also provide recognition sites for interactions with enzymes and with the ribosome. The loop of the anticodon arm contains seven bases, three of which form the *anticodon*.

One of the characteristic features of tRNA is that a large proportion of the nucleosides are modified. Table 22–6 lists some of the more than 40 modified nucleosides regularly occurring in the tRNAs. Most modifications involved methylation of the regular bases (i.e., A, U, G, and C) or methylation of the 2′ hydroxyl oxygen of the riboses. The role (or roles) of the modified bases are not known with certainty, but suggested roles include the *prevention* of base pairing (1) with the tRNA molecule in order to provide a characteristic tertiary structure and (2) between tRNA and mRNA during translation. It is interesting that the base at the 3′ side of the anticodon is nearly always a modified purine when the first base of the corresponding codon is either A or U.

During translation, the three bases of the anticodon form hydrogen bonds with the corresponding codon bases of mRNA. An examination of Table 22–1 reveals that a single amino acid may be coded for by two or more codon sequences. For example, the codon for alanine may be GCU, GCC, GCA, or GCG—the third base being seemingly un-

Table 22–7

Degeneracy in Codon-Anticodon Recognition

Base Occupying First Position of tRNA Anticodon	Base Occupying Third Position of mRNA Codon
I	U, C, or A
A	U only
G	U or C
C	G only
U	A or G

important. Keeping in mind that during translation tRNA and mRNA molecules are antiparallel, it would appear that the base occupying the *first* position of the anticodon may recognize one or more different bases occupying the *third* position in the codon (Table 22–7). (Remember that the numbering of the polynucleotide chain begins at the 5′ end and finishes at the 3′ end.) Francis Crick refers to this as the "wobble" base, implying that it may orient in different ways in order to accommodate the appropriate base pairing.

The most widely accepted model for the tertiary structure of tRNA is that proposed by S. H. Kim and is based principally on X-ray crystallographic studies of phenylalanine tRNA from yeast cells (Fig. 22–21b). The molecule has an L shape, with all double-helical regions being right-handed and antiparallel. The amino acid and TψC stems form one continuous double helix, and the DHU and anticodon stems form another. The two helices are perpendicular to each other, thereby forming the L, with the anticodon and C-C-A termini at opposite ends. The molecule is about 20 Å thick, which corresponds to the diameter of the RNA double helix, Figure 22–21c relates the tertiary structure of tRNA to the cloverleaf secondary structure. The three-dimensional appearance of the molecules is presented in Figure 22–21d.

Different regions of a tRNA molecule appear to serve as recognition and binding sites for various enzymes, ribosomal proteins, and other RNAs that interact with tRNA during the various stages of protein synthesis (Fig. 22–22). Although the amino acid is bound to the adenosine nucleoside at the 3′ end of the tRNA molecule, this region appears to have little if anything to do with codon–anticodon recognition or binding. This was elegantly demonstrated in a classic experiment by F. Lipman and F. Chapeville, who prepared a tRNA specific for cysteine (abbreviated

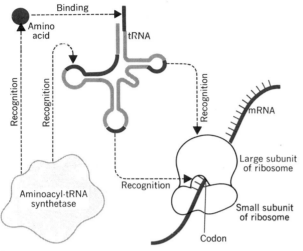

Figure 22–22
Relationship between various regions of tRNA and the enzymes, ribosomal proteins, and RNAs with which tRNA interacts during protein synthesis.

tRNA^{Cys}) that they then enzymatically combined with ^{14}C-cysteine to form cysteinyl-tRNA^{Cys} (or cys-tRNA^{Cys}). They treated the cys-tRNA^{Cys} with Raney nickel, which removes the sulfhydryl group from the cysteine residue, leaving ala-tRNA^{Cys} (i.e., a tRNA molecule specific for cysteine but containing bound ^{14}C-labeled alanine instead.) When the radioactive ala-tRNA^{Cys} was employed in an in vitro protein synthesizing system using a synthetic messenger RNA that coded for the production of a polypeptide rich in cysteine, the radioactive alanine residues were incorporated in place of cysteine. This showed that the specificity for codon recognition resided with the tRNA molecule and not with the amino acid.

Activation of Amino Acids

Amino acid activation involves two major steps (Fig. 22–23). In the first, the alpha-carboxyl group of the amino acid reacts with ATP to form an *aminoacyl-adenylate* and pyrophosphate. For each species of amino acid, there is at least one specific enzyme called an *aminoacyl-tRNA synthetase* (or *ligase*) that catalyzes the reaction, and a number of these amino-acid-activating enzymes have been isolated and studied. The aminoacyl-adenylate formed is not released from the enzyme but remains complexed to it, presumably by a linkage between the enzyme and the R group of the amino acid. In the second step, the aminoacyl-AMP

complex recognizes and reacts with a molecule of tRNA specific for that amino acid to form *aminoacyl-tRNA*, and the enzyme and AMP are released. The reaction between tRNA and the amino acid involves esterification to the 2′ or 3′ hydroxyl group of ribose in the terminal adenosine unit of tRNA by the alpha-carboxyl carbon atom of the amino acid. These reactions occur in the cytosol. The aminoacyl-tRNA thus formed can now participate in protein synthesis on the mRNA-ribosome complexes. A number of investigators contributed to the elucidation of the above reaction sequences, but most notable among them are P. Zamecnik, M. B. Hoagland, P. Berg, and E. J. Ofengand.

Formation of the Initiation Complex

Amino Acid and Peptide Sites of the Ribosome   In both procaryotes and eucaryotes, two sites on the intact ribosome are involved in protein synthesis; these are called the **amino acid site** and the **peptide site.** The peptide site is the region of the ribosome to which the growing polypeptide chain is bound by tRNA, while the amino acid site receives the tRNA bearing the next amino acid to be added to the chain. The two sites are believed to be shared by each of the two subunits making up the intact ribosome.

At first, it was believed that no special mechanism was required in order to initiate chain growth. It was proposed that a ribosome would simply attach to the 5′ end of the mRNA and proceed to translate successive codons. It is clear now that this is not the case and that a specific mechanism exists for initiating translation and preventing *out-of-phase* translation of the mRNA code.

The problem of out-of-phase translation warrants further consideration. Suppose that a section of mRNA contains the following codon sequence:

$$...\underline{U\,G\,U}\,\underline{A\,A\,G}\,\underline{G\,C\,U}\,\underline{A\,G\,A}...$$

This section of mRNA would therefore code for the addition of the amino acid sequence cys—lys—ala—arg to the growing polypeptide chain (see Table 22–1 for the genetic code). However, suppose that this section of mRNA was translated out-of-phase as follows:

$$..\underline{U}\,\underline{G\,U\,A}\,\underline{A\,G\,G}\,\underline{C\,U\,A}\,\underline{G\,A}...$$

In this case, the sequence val—arg—leu would be incor-

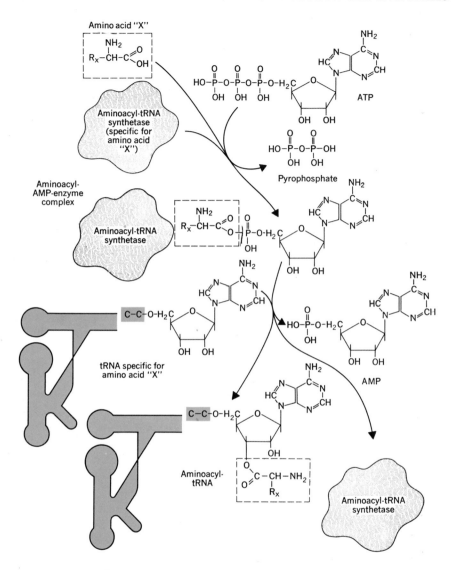

Figure 22-23
The reaction sequence of amino acid activation. Although in the final product shown the amino acid is esterified to the 3' hydroxyl position of the terminal ribose of tRNA, an equilibrium mixture of 2' and 3' esters may be formed.

rectly incorporated into the primary structure of the protein. The need for a mechanism that ensures that translation is begun in-phase and carried out in-phase is apparent.

Characteristics of mRNA. The mRNA molecules isolated from the cytosol of eucaryotic cells are typically about 1500 nucleotides long and consist of *both* translated and untranslated regions. Since mRNA of this size could at most encode a polypeptide about 500 amino acids long, it is most likely that eucaryotic mRNA is **monocistronic** (or **monogenic**); that is, the mRNA contains the codon sequence for no more than one polypeptide chain. By way of contrast, the mRNAs of procaryotic cells are quite variable in length since some mRNAs are transcribed from more than one gene; these are termed **polycistronic** (or **polygenic**) mRNAs. For example, five of the enzymes involved in the metabolic pathway leading to the synthesis of tryptophan in *E. coli* cells are encoded in a single polycistronic mRNA produced by the transcription of five closely linked genes (the mRNA contains more than 7000 nucleotides).

Interestingly, the various enzymes that are encoded in a polycistronic message are part of the *same* metabolic pathway.

The 5′-phosphate and 3′-OH ends of all eucaryotic mRNAs are similar. The 5′ end contains the sequence m^{7}GpppN′$_m$N″$_m$p . . ., referred to as the "cap" (Fig. 22–24) and the 3′ end of mRNA contains a "poly-A" sequence that is 20 to 250 nucleotides long and is called the "tail." Not all of the remainder of the mRNA molecule may be translated. For example, in addition to the cap and poly-A segments, the mRNA for the globin chains of hemoglobin contains a 150-nucleotide segment near the poly-A tail that is not translated into globin. Hemoglobin alpha and beta chain mRNAs have an estimated molecular weight of 200,000 to 220,000 daltons and contain 650 to 670 nucleotides. The alpha and beta globin chains are encoded by 423 to 444 nucleotides, respectively; the remainder of the mRNA contains a poly-A tail 50 to 75 nucleotides long, a nontranslated sequence of 150 to 175 nucleotides, and, of course, a cap segment. The special chemical nature of the cap region of mRNA is apparently involved in ensuring the proper initiation of translation. It is now fairly well established that a "leader sequence" also exists just following the cap region and just to the 5′ side of the start codon.

Processing of mRNA. Procaryotic cells such as *E. coli* contain a single chromosome about 3,000,000 base pairs long. This is sufficiently long to encode the primary structures of about 3000 proteins, and this number agrees fairly well with the number of different proteins believed to be present in this bacterium. Eucaryotic cells contain far greater quantities of DNA. For example, the human genome is believed to contain as many as 4,000,000,000 base pairs apportioned among each cell's 46 chromosomes. This is sufficient DNA to encode as many as ten million polypeptide chains—a number far greater than is believed to be present in human cells. The actual number of different polypeptides is more likely to be from about 30,000 to 150,000. The apparent discrepancy between the amount of structural gene DNA that "should" be present in human

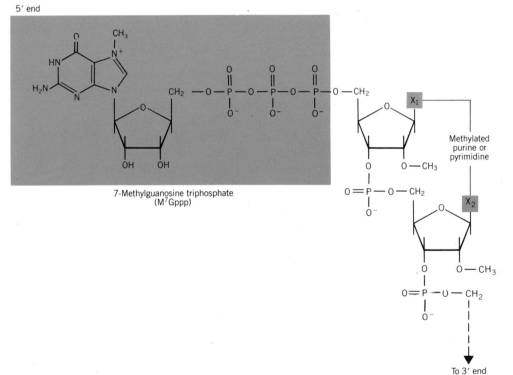

5′ end

7-Methylguanosine triphosphate
(M^{7}Gppp)

Methylated purine or pyrimidine

To 3′ end

Figure 22–24
Cap region of mRNA. The 5′ end of the molecule contains 7-methylguanosine triphosphate, followed by one or two nucleosides in which the sugar and/or base are methylated.

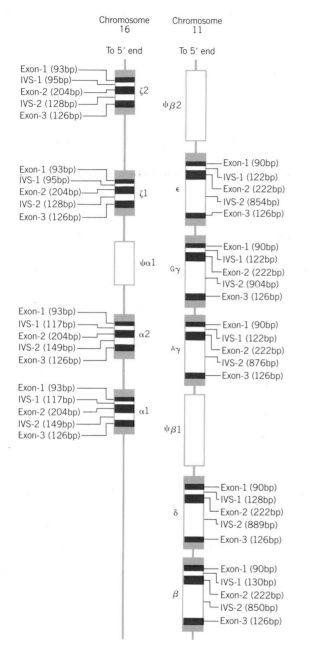

Chromosome 16 — To 5' end

Exon-1 (93bp)
IVS-1 (95bp)
Exon-2 (204bp)   ζ2
IVS-2 (128bp)
Exon-3 (126bp)

Exon-1 (93bp)
IVS-1 (95bp)
Exon-2 (204bp)   ζ1
IVS-2 (128bp)
Exon-3 (126bp)

ψα1

Exon-1 (93bp)
IVS-1 (117bp)
Exon-2 (204bp)   α2
IVS-2 (149bp)
Exon-3 (126bp)

Exon-1 (93bp)
IVS-1 (117bp)
Exon-2 (204bp)   α1
IVS-2 (149bp)
Exon-3 (126bp)

Chromosome 11 — To 5' end

ψβ2

Exon-1 (90bp)
IVS-1 (122bp)
ε   Exon-2 (222bp)
IVS-2 (854bp)
Exon-3 (126bp)

Exon-1 (90bp)
IVS-1 (122bp)
Gγ   Exon-2 (222bp)
IVS-2 (904bp)
Exon-3 (126bp)

Exon-1 (90bp)
IVS-1 (122bp)
Aγ   Exon-2 (222bp)
IVS-2 (876bp)
Exon-3 (126bp)

ψβ1

Exon-1 (90bp)
IVS-1 (128bp)
δ   Exon-2 (222bp)
IVS-2 (889bp)
Exon-3 (126bp)

Exon-1 (90bp)
IVS-1 (130bp)
β   Exon-2 (222bp)
IVS-2 (850bp)
Exon-3 (126bp)

Figure 22–25
Introns and exons of the globin chain genes. The α and ζ chain genes are located on chromosome 16 and are present as duplicates. A "pseudoalpha chain gene" (ψα1) is also present on this chromosome but no transcription or translation products of this gene have been found (i.e., the genes are "silent"). β, δ, Aγ, Gγ, and ε chain genes are on chromosome 11. Also present on this chromosome are silent beta genes ψβ1 and ψβ2. All of the coding sequences or *exons* of the globin chain genes are interrupted by noncoding *intervening sequences* or *introns*. The numbers of base pairs comprising each of the introns and exons are shown.

cells and the amount that *is* in fact present can in part be accounted for by one of the most astonishing findings of recent years, namely that *not all of the RNA produced by the transcription of a structural gene ends up in the mRNA to be translated.* For example, the genes that encode the beta chains of human hemoglobin have more than four times the number of base pairs as are needed to specify the primary structure of this polypeptide. The extra DNA is represented by two *intervening sequences* or **introns** consisting of 130 to 850 base pairs. The coding sequences, called **exons,** are not only interrupted by introns, but are also *flanked* by base sequences that do not encode amino acids for the globin chain. Thus the total coding sequence represents only a small portion of the total gene. Similar findings have been made for the alpha, gamma, delta, epsilon, eta, and zeta globin chains (Fig. 22–25) of human hemoglobin and the various globin chains of other mammalian hemoglobins. Intervening sequences are transcribed into RNA but the transcripts do not end up in the message because they are removed during *processing* (see below).

Intervening sequences in the structural genes of eucaryotic cells are not at all uncommon. Introns have been identified in the genes for ovalbumin, conalbumin, ovomucoid protein, lysozyme, thyroglobulin, tubulin, albumin, and various immunoglobulins. The *ovalbumin* genes of chickens are formed from eight exons, each separated by an intron. Beginning at the position of the gene encoding the 5' end of the mRNA molecule and ending with the region encoding the 3' end, the gene contains 7700 base pairs. This length is four times greater than the final mRNA, which is only 1,872 base pairs long. If we exclude the cap, the untranslated 3' segment, and the poly-A tail, then the gene is 7 times longer than the part of the message that is translated into polypeptide.

While the genes of vertebrates have been found to contain as many as 50 introns, some genes lack introns altogether. For example, the genes for the histones have no introns and neither does the gene for interferon. Apparently, introns are not essential constituents of eucaryotic genes. Introns often

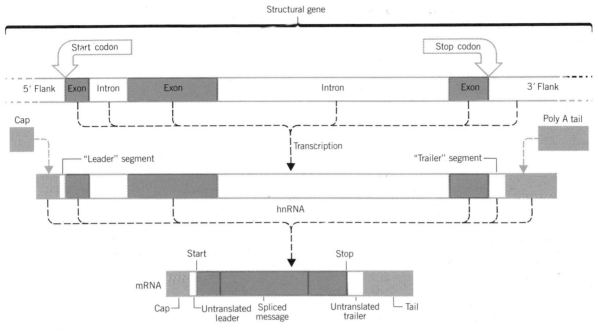

Figure 22–26
Transcription and processing of mRNA (see text for details).

separate exons that encode specific functional segments of a protein. The hydrophobic leader segment of secretory proteins (Chapter 15) is encoded by a separate exon, and the central exon of the globin chain genes encodes almost all of the amino acids that interact with the heme group. It has been suggested that by moving the exons apart, introns increase recombination frequency and thereby hasten evolution. Indeed, as you move up the evolutionary ladder, structural genes have greater numbers of introns.

At the present time, the synthesis and processing of mRNA is believed to take the following form (Fig. 22–26). The enzyme *RNA polymerase* produces continuous *primary transcripts* of all structural genes to be expressed in the cell. Each structural gene is transcribed in its entirety; that is, both exons and introns are transcribed. After the primary transcript is modified by addition of the cap segment and the poly-A tail, the introns are excised and the remaining exons are ligated to produce the mature messenger RNA. Each intron transcript begins with the sequence GU and ends with AG, suggesting that these base sequences have something to do with specifying the loci at which transcript cleavage is to occur. All intron base sequences appear to be unique; that is to say, they are not repeated elsewhere

in the cell's genome, although the intron sequences of related genes (e.g., the globin chain genes) are quite similar.

The nucleus of a cell contains a broad spectrum of RNAs, including unprocessed primary transcripts, partially processed transcripts, and completely processed transcripts. The spectrum of mRNAs found in the nucleus is often referred to as *heterogeneous nuclear RNA* or *hnRNA*. In Figure 22–26, the processed mRNA molecule contains an untranslated region between the stop codon and the poly-A tail; at the present time it is uncertain whether this region is derived from the 3′ flanking region of the gene or whether it is added during processing. In the case of the abnormal hemoglobin called "Hb constant spring," the alpha chains contain an extra amino acid segment at the C-terminal end of the chain. This is apparently due to a mutation that alters the stop codon with the result that during translation, the ribosome continues into the normally untranslated region of the mRNA until it eventually encounters another stop codon. In this instance, it is known that the untranslated 3′ region is derived from the flanking region of the gene. All of these mRNA processing events appear to take place within the nucleus of the cell, since neither the primary transcripts nor the processing intermediates are found in the cytosol.

Role of Formylmethionine.

In 1963 J. P. Waller reported the amazing finding that nearly one-half of all proteins in *E. coli* cells have the amino acid methionine in the N-terminal position. (Remember that protein synthesis begins at the N-terminus!) Then, in 1964, K. A. Marcker and F. Sanger discovered an unusual species of aminoacyl-tRNA in *E. coli—N-formylmethionyl-tRNA*—and suggested that this molecule may play a role in the special mechanism of chain elongation because the presence of the *N*-formyl group in the amino acid (leaving only the alpha-carboxyl group available for peptide bond formation) would restrict this residue to the N-terminus. The structural formulae of *N*-formylmethionyl-tRNA and methionyl-tRNA are shown in Figure 22-27.

Two transfer RNA molecules specific for methionine are present in *E. coli,* but only one of these can participate in the subsequent enzymatic formylation of the methionine residue. These tRNAs may be denoted as $tRNA^{Met}$ and $tRNA^{Met}_f$. The formation of *N*-formylmethionyl-tRNA occurs as follows:

(1) met + $tRNA^{Met}_f$ $\xrightarrow{\text{specific aminoacyl-tRNA synthetase}}$ met-$tRNA^{Met}_f$

(2) met-$tRNA^{Met}_f$ + formate $\xrightarrow{\text{formylating enzyme}}$ N-formylmet-$tRNA^{Met}_f$

The codon for methionine is A U G (Table 22–1), and when this codon occurs anywhere except at the beginning of mRNA it codes for met-$tRNA^{Met}$. However, when A U G occurs at the beginning of the message, it codes for *N*-formylmet-$tRNA^{Met}_f$ and chain initiation. For this reason, the A U G codon is also called the *initiator* or *start* codon. The picture is somewhat complicated by the fact that the codon G U G, which is a codon for valine, also codes for *N*-formylmet-$tRNA^{Met}_f$ when G U G occurs at the *beginning* of the message; anywhere else in the mRNA molecule, G U G codes for valine. The interaction that takes place between the aminoacyl-tRNA anticodon, and the mRNA codon thus depends on both base-pairing and the location of the codon in mRNA. It is now clear that *N*-formylmet's role as the initiating amino acid in protein synthesis is not

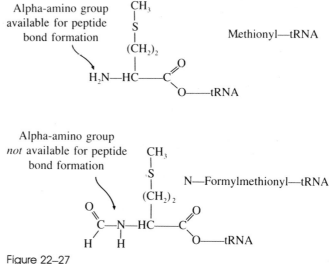

Figure 22–27
Methionyl-tRNA and formylmethionyl-tRNA.

restricted to *E. coli* but is a characteristic of procaryotes in general.

The process of initiation in eucaryotes is fundamentally similar to that in procaryotes. As in procaryotes, there are at least two methionine tRNAs that recognize the A U G codon of mRNA; however, only one these tRNAs can participate in chain initiation. The initiator methionyl-tRNA (met-$tRNA^{Met}_i$) can be enzymatically formylated in vitro, although this does *not* appear to take place under native circumstances, and there are no formylating enzymes present in the eucaryotic cytosol. The other methionyl-tRNA (met-$tRNA^{Met}$) recognizes A U G codons located internally in mRNA. Of special interest is the observation that the initiation of protein synthesis in mitochondria and chloroplasts takes place in much the same manner as in procaryotes. The initiating aminoacyl-tRNA is formylated using *formylase,* which is present in the organelles but absent from the cytosol.

Initiation Factors.

A number of factors present in the *soluble* phase of the cell are required in order to *initiate* protein synthesis; others, to be discussed more fully later, are needed for polypeptide *elongation* and for *termination* (Table 22–8). Procaryote initiation factors were discovered by the observation that washed *E. coli* ribosomes could not translate natural messengers unless supplemented with the

Table 22–8
Initiation, Elongation, and Termination Factors

	Procaryotes	Eucaryotes
Initiation factors	IF-1	eIF-1
	IF-2	eIF-2
	IF-3	eIF-3
		eIF-4A
		eIF-4B
		eIF-4C
		eIF-5
Elongation factors	EF-Tu	EF-1
	EF-Ts	
	EF-G	EF-2
Termination (release) factors	RF-1	RF
	RF-2	
	RF-3	

wash. Eucaryote initiation factors were discovered by the similar observation that washed reticulocyte ribosomes would not initiate globin synthesis unless the wash was added back.

In eucaryotes, polypeptide synthesis is initiated by a series of steps beginning with the formation of a complex between met-tRNA$_i^{Met}$, eIF-2, and GTP, which then attaches to the peptide site (*P* site) of the small ribosomal subunit to which eIF-3 has *already* been bound. This is followed by attachment of mRNA to the small subunit, with the initiator A U G codon near the 5′ end of the mRNA molecule aligned at the peptide site (Fig. 22–28). The next mRNA codon aligns at the amino acid site (*A* site). Association of mRNA with the small subunit requires eIF-1, eIF-4A, eIF-4B, and eIF-4C and is accompanied by the hydrolysis of one molecule of ATP. Addition of the large subunit follows release of eIF-2 and eIF-3, and this is followed by the hydrolysis of GTP and the release of GDP and P$_i$. The final product, called the 80 *S initiation complex,* may now proceed to translate the remainder of the message. Figure 22–28 serves also to describe initiation in procaryotic cells if (1) IF-1, IF-2, and IF-3 are substituted for eIF-1, eIF-2, and eIF-3, and (2) fmet-tRNA$_f^{Met}$ is substituted for met-tRNA$_i^{Met}$.

In certain cells, additional factors may be required for the formation of an initiation complex. For example, the initiation of globin synthesis in reticulocytes depends on the availability of heme; consequently, the synthesis of globin chains and heme are tightly coordinated.

As noted earlier, nearly half of all proteins in *E. coli* have a methionine residue in the N-terminal position. Since it is not a formylated residue, the formyl group must be removed from the methionine either during or immediately following polypeptide synthesis. Indeed, enzymes that are able to remove formate from formylmethionine residues of polypeptides are present in *E. coli* and other procaryotic cells. For those proteins that have amino acids other than methionine in the N-terminus (i.e., the majority of cell proteins), a mechanism must exist for removing methionine from the end of the polypeptide, and accordingly, aminopeptidases, which specifically cleave the peptide bond between methionine and the second amino acid of the polypeptide chain, have been identified. Consequently, for most *E. coli* proteins, formylmethionine is removed from the end of each polypeptide chain by enzymatically cleaving first the formyl group and subsequently the methionine residue itself.

Only a small percentage of eucaryotic proteins have methionine as the N-terminal amino acid. Like procaryotes, eucaryotes possess an aminopeptidase that removes methionine from the N-terminus of growing polypeptides. Studies with hemoglobin in which valine is the N-terminal residue of both the alpha and beta chains have revealed that the methionine that initially occupies the N-terminal position is removed only after the polypeptide is about 30 residues long. Until that point, the peptide bond linking methionine to valine is apparently protected in some fashion from cleavage by the enzyme. Indeed, other proteolytic enzymes, including papain, trypsin, chymotrypsin, and pronase, are unable to hydrolyze the peptide bonds of short nascent chains but can act on the outer segments of longer chains. This observation (and similar observations made for other proteins) is in accord with the current model of ribosomal structure in which the growing polypeptide chain is protected over that portion of its length residing in the interior of the organelle.

Influence of Magnesium Ions. Before we proceed to a discussion of chain elongation, some brief comments are warranted regarding the roles that magnesium ions (Mg^{++}) play in protein synthesis, for it has long been known that a critical, low level of Mg^{++} is necessary in order for protein synthesis to proceed normally. If the level of Mg^{++} falls below this, ribosomes dissociate into their subunits and protein synthesis ceases. On the other hand,

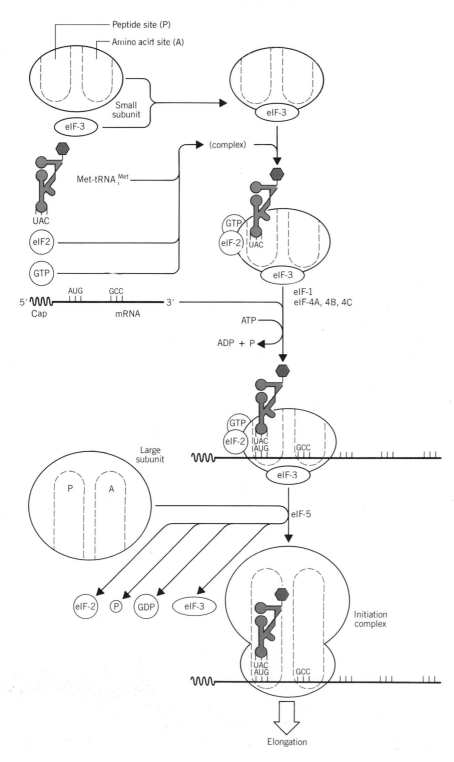

Figure 22–28
Stages in the formation of the initiation complex in eucaryotic cells (see text for explanation). For diagrammatic simplicity, the small and large subunits of the ribosome are shown as disk-like surfaces. The reader should take note that most of these events (as well as those depicted in Figs. 22–29, 22–30, and 22–31) occur *within* the channel(s) formed by the association of the subunits. The second codon, G C C (which codes for ala), is shown for illustrative purposes only.

an experimental increase in the Mg^{++} level above that which normally exists in cells is accompanied by all kinds of aberrations of the normal requirements for chain initiation and elongation. For example, protein synthesis can be carried out in vitro using synthetic messengers such as poly-U (the resulting polypeptide being polyphenylalanine). Indeed, most of the early studies of protein synthesis were carried out using synthetic messengers. Such syntheses occurred in spite of the absence of caps, initiator codons, tails, and so on, but *only* if the in vitro system was supplemented with a magnesium ion concentration far in excess of that which occurs normally in the cell.

Magnesium ions are thought to play at least two specific roles: (1) Mg^{++} seems to be a *cofactor* for several of the enzymes that mediate initiation and chain growth, and (2) Mg^{++} probably forms **salt bridges** (ionic bonds) with RNA and, in this manner, links the two ribosomal subunits together through their respective rRNAs; in some way Mg^{++} may also assist in binding mRNA to the ribosome (Fig. 22–29).

Chain Elongation

Once the initiator aminoacyl-tRNA is located in the peptide site of the ribosome, chain *elongation* ensues. Addition of the second and subsequent aminoacyl-tRNAs follows a similar pattern. GTP reacts with a soluble phase elongation factor, EF-1, to form a complex that then combines with aminoacyl-tRNA. The resulting EF-1-GTP aminoacyl-tRNA combination interacts with the ribosome so that the aminoacyl-tRNA becomes bound to the vacant amino acid site. This step is accompanied by hydrolysis of GTP and the release of inorganic phosphate and an EF-1-GDP complex (the latter can be recycled to EF-1-GTP using additional GTP) (Fig. 22–30). Occupation of both the *P* and *A* sites of the ribosome is followed by the formation of a peptide bond between the amino acid bound to tRNA in the *P* site and the amino acid that just entered the *A* site. In forming this bond, the alpha-carboxyl group of the amino acid attached to the terminal adenosine unit of tRNA in the *P* site is transferred to the free alpha-amino group of the amino acid held by its tRNA in the *A* site (Fig. 22–31). The formation of the peptide bond is catalyzed by the enzyme *peptide synthetase,* one of the proteins of the large subunit, and temporarily leaves polypeptidyl-tRNA in the

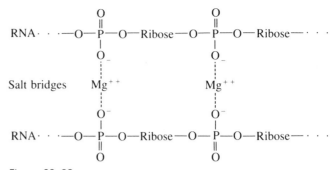

Figure 22–29
Salt bridges formed between RNA molecules by magnesium ions.

A site. A second elongation factor, EF-2 (also known as *translocase*), catalyzes a complex rearrangement of the ribosome in which the free tRNA at the *P* site is released and the peptidyl-tRNA and mRNA codon are shifted to the vacated *P* site. The ribosome is thus moved one codon further along the message, with peptidyl-tRNA of the *P* site elongated by one amino acid. The translocation step is accompanied by the hydrolysis of another molecule of GTP. With translocation completed, the ribosome is once again ready to accept aminoacyl-tRNA in the free *A* site. These steps of the elongation reactions are repeated for each new codon of the message entering the *A* site.

Chain elongation in procaryotic cells involved *three* soluble elongation factors: EF-Ts, EF-Tu, and EF-G. Binding of aminoacyl-tRNA to the *A* site requires an EF-Tu-GTP complex, and binding is followed by release of EF-Tu-GDP and inorganic phosphate. EF-Tu-GTP is replenished by EF-Ts-catalyzed transphosphorylation using GTP as substrate (Fig. 22–32). EF-G of procaryotes functions in the same manner as EF-2 of eucaryotes.

The enzymatic cleavage of met and/or formate from the N-terminus of procaryote proteins and met from the N-terminus of eucaryotic proteins takes place after a number of rounds of elongation have already been completed (Fig. 22–33). This leaves the amino acid coded for by the *second* mRNA codon in the "new" N-terminus.

Chain Termination

Chain *termination,* like initiation, involves a specific mechanism and does not occur automatically once the ribosome

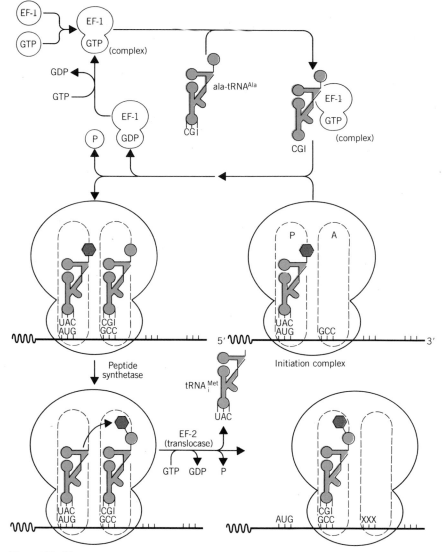

Figure 22–30
Chain elongation in eucaryotic cells. Elongation begins with addition to the initiation complex, but the reactions are the same whenever peptidyl-tRNA occupies the *P* site. For illustrative purposes only, G C C is shown as the second codon of the message. (See text and legend of Fig. 22–28 for other pertinent information.)

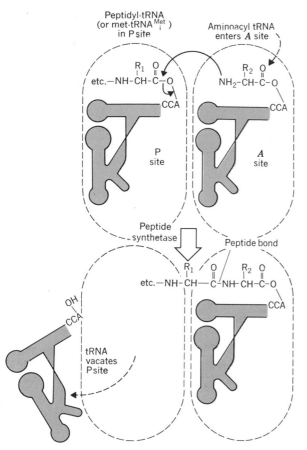

Figure 22–31
Action of *peptide synthetase*.

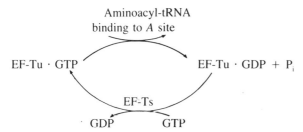

Figure 22–32
Utilization (and restoration) of EF-Tu·GTP complex during (and following) aminoacyl tRNA binding to the *A* site in procaryotic cells.

reaches the end of the message. An examination of Table 22–1 reveals that there are three triplets (sometimes referred to as the "nonsense" triplets) that do not code for any amino acid; these are the "amber" codon, U A G; the "ochre" codon, U A A; and the "opal" codon, U G A. Studies in both procaryotic and eucaryotic cells have implicated these codons in the process of chain termination.

The RNA of certain viruses that infect *E. coli* undergoes a mutation, with the result that virus coat protein is synthesized as incomplete polypeptide chains in an *E. coli* in vitro cell-free system. The incomplete polypeptides contain the N-terminus but not the C-terminus, suggesting that normal synthesis was interrupted and the partially completed chains released from the ribosomes. These mutations are suppressible; that is, *E. coli* mutants can be found that support the continued elongation of these polypeptides. However, in the resulting polypeptides, serine (code word U C G) or tryptophan (code word U G G) replace glutamine (code word C A G). This has been interpreted to mean that the phage mutation involved the change of codon C A G to U A G (i.e., C was mutated to U), which in normal *E. coli* cells resulted in incomplete chains, while in the mutant *E. coli* strain the U A G was being read as the codon for serine or tryptophan. Observations of this sort indicated that the normal role for the U A G codon is chain termination. Other studies with *E. coli* suggest similar roles for the U A A and U G A codons.

In humans, a point mutation in which the first position of the UAA codon at the end of the translated region of alpha globin chain mRNA is altered results in the translation of a major segment of the normally untranslated region (i.e., the ribosome continues past the terminator into the region near the 3′ end of the message that normally remains untranslated). The result is the production of alpha chains containing 31 extra amino acids at the C-terminal end of the polypeptide. Hemoglobins formed using these mutant alpha chains function abnormally (the abnormal hemoglobin is known as Hb "constant spring"). Similar point mutations have recently been identified for beta globin chains.

Once the C-terminal amino acid has been added to the end of the polypeptide, the polypeptidyl-tRNA is translocated from the *A* site to the *P* site of the ribosome, as described earlier. This moves one of the nonsense or terminator codons into position in the *A* site (Fig. 22–34).

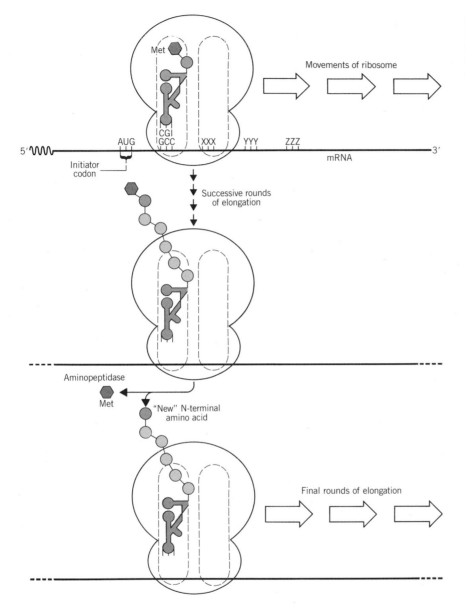

Figure 22–33
Removal of met from the N-terminal position of a growing polypeptide chain during elongation rounds in eucaryotic cells.

The terminator is not recognized by a particular tRNA or other RNA species. Instead, release of the completed polypeptide from its tRNA requires participation of a soluble protein called **release factor (RF).** One release factor has been identified in eucaryotic cells and three in procaryotes.

Binding of release factor and GTP to the free *A* site is followed by the activation of the peptidyl synthetase and translocase systems. The bond linking the completed polypeptide to tRNA is hydrolyzed, the polypeptide and tRNA released from the ribosome, and RF and GTP moved into the *P* site. Hydrolysis of GTP is followed by release of RF, GDP, and inorganic phosphate. At this time, the ribosome dissociates into its subunits, freeing mRNA.

For many proteins, the release of the completed polypeptide chain is followed by the spontaneous assumption of its functional secondary and tertiary structure. For others,

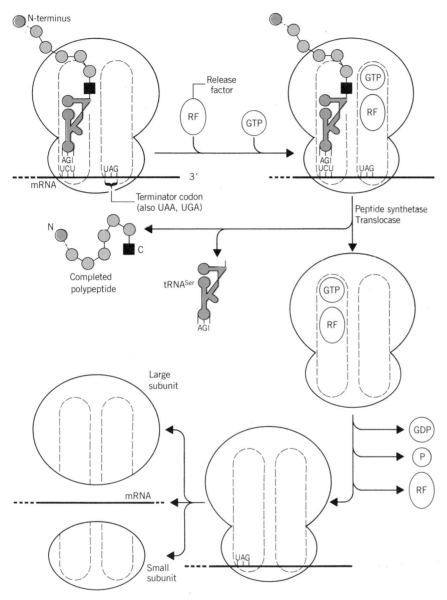

Figure 22–34
Stages of chain termination. (See text and legend of Fig. 22–28 for additional pertinent information.) Serine is shown as the C-terminal amino acid for illustrative purposes only.

all or part of the final secondary and tertiary structure is assumed *as the primary structure is being laid down.* In some cases, it is conceivable that the tertiary structure that is most favored thermodynamically is *not* the functional structure. Therefore, the *progressive assumption* of tertiary structure *during* elongation effectively reduces the number of possible alternative shapes that could be assumed by the polypeptide following release.

Polyribosomes (Polysomes)

Since the globin chains of hemoglobin each contain about 150 amino acids, their messenger RNAs must contain at least 450 nucleotides. Each nucleotide yields a linear translation of 3.4 Å (Chapter 7), so that the mRNA would be at least 1500 Å long. In contrast the diameter of a ribosome is only about 240 to 400 Å. These observations led Rich,

Warner, Knopf, and Hall in 1962 to propose that the translation of a single mRNA might be carried out simultaneously by several ribosomes (i.e., **polyribosomes**) attached to and moving in succession along the message. For example, in the case of globin chain synthesis, four or more ribosomes could be attached to the mRNA.

Rich and his co-workers incubated rabbit reticulocytes for short periods in a medium containing ^{14}C-labeled amino acids. During this brief incubation, radioactive segments were added to each nascent chain. Following this, the cells were lysed and the lysate fractionated by centrifugation through a sucrose density gradient. Fractions collected from the gradient at the conclusion of centrifugation were examined in two ways: (1) the distribution of ribosomes through the gradient was determined by measuring the ultraviolet light absorption of the ribosomal RNA, and (2) the distribution of nascent polypeptides was determined from the radioactivity of the collected fractions. Typical results are shown in Figure 22–35. Two UVL-absorbing regions were identified in the density gradient. The first (i.e., least rapidly sedimenting) peak (fractions 24 to 29), which corresponded to particles of about 80 *S* and represented *single* ribosomes, had no radioactivity associated with it. Instead, the radioactivity was distributed over a region of the gradient containing more rapidly sedimenting (i.e., larger) particles (i.e., fraction 10 to 20). This indicated that protein synthesis in reticulocytes took place on structures that were larger than individual ribosomes, and Rich suggested that these were groups of ribosomes held together by mRNA.

When the enzyme ribonuclease was added to the lysate prior to centrifugation, the rapidly sedimenting peak disappeared while the first peak increased in size and was now associated with the radioactivity. This result, of course, supported Rich's proposal. Further confirmation came from electron microscopic examination of the fractions, which revealed that the first peak contained single ribosomes, while the rapidly sedimenting peak contained clusters of ribosomes—the further down the gradient the sample was withdrawn for microscopic examination, the larger was the observed cluster size. The predominant size cluster contained five ribosomes (called a *pentamer*), with smaller numbers of clusters containing six ribosomes (*hexamers*) and four ribosomes (*tetramers*).

Subsequently, electron microscopic studies using negative staining techniques showed that the ribosomes were

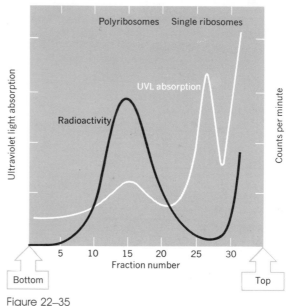

Figure 22–35

Distribution of ^{14}C-labeled amino acids and UVL-absorbing material in reticulocyte lysates subjected to density gradient centrifugation (see text for explanation).

connected by a thin thread about 10 to 15 Å thick and about 1500 Å long and were separated by gaps varying from 50 to 150 Å. This corresponded to the diameter of an RNA molecule and the approximate length predicted for the globin messenger. Some polysomes are shown in Figure 22–36.

In the model for polysome function originally proposed by Rich, the several ribosomes move along the mRNA strand, each synthesizing a polypeptide chain. When a ribosome reaches the end of the message, it detaches, while at the other end, another ribosome attaches to the mRNA.

Although it is now clear that polysome function in globin chain synthesis is not precisely as Rich predicted, the fundamentals of his model remain valid. The size of a polysome depends on both the length of the mRNA being translated and the amount of time required for initiation, elongation, and termination. For example, recent studies by S. H. Boyer have established that polysomes engaged in the synthesis of alpha globin chains contain an average of four ribosomes, whereas beta globin chain polysomes contain an average of six ribosomes. This occurs despite the fact that both globin chains are about the same size. The difference is due to the lower frequency with which

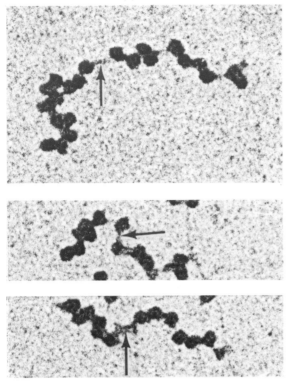

Figure 22–36
Polysomes. **The messenger RNA strands linking neighboring ribosomes of each polysome are indicated by arrows. Magnification, 200,000×. (Courtesy of Dr. O. L. Miller, from O. L. Miller, B. A. Hamkalo, and C. A. Thomas,** *Science 169,* **394, 1970. Copyright © 1970 by the American Association for the Advancement of Science.)**

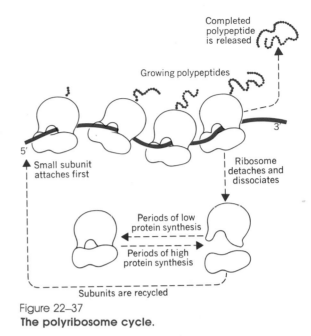

Figure 22–37
The polyribosome cycle.

initiation occurs on alpha globin chain mRNA (αmRNA), which in the case of globin chain synthesis is the *rate-limiting* step. Elongation and termination occur at about the same rates for both globin chains. It was noted earlier that equal amounts of alpha and beta chains are produced in normal erythrocytes. If this is so, one might ask how such a balance is maintained in view of the polysomal differences just noted. The balance results in part from the presence in these cells of larger quantities of αmRNA than βmRNA, which thus compensates for differences in the frequency of initiation. The larger quantities of αmRNA result, in turn, from the presence of *two pairs* of alpha chain structural genes and only one pair of beta chain structural genes (see Fig. 22–25 and Chapter 4).

The current model for polysome function in eucaryotic cells, shown in Figure 22–37, does not differ dramatically from that originally proposed by Rich. The detachment of the ribosome from mRNA is accompanied by the release of the completed (and probably folded) polypeptide chain. The ribosome immediately dissociates into its small and large subunits, which then enter a common cell subunit pool. During periods of reduced protein synthesis, subunits combine to form a pool of inactive but intact ribosomes (monosomes), but during periods of active cellular protein synthesis, these ribosomes again dissociate into subunits. As noted earlier, the small subunit attaches to the 5′ end of the mRNA before the large subunit There is some evidence that during periods of active protein synthesis, subunits recently released from the messenger are *preferentially* reused for translation because of their closer proximity to the 5′ end of mRNA than other subunits randomly spread through the cell.

Kinetics of Transcription and Translation in Maturing Erythrocytes. Because so much is known about hemoglobin and the development of erythrocytes, it is possible to carry out a number of calculations that give us some insight into the frequency and rate of transcription and translation. Since there are about 5.4×10^9 erythrocytes and 0.16 g of hemoglobin in each 1 cm^3 of human blood (these values are readily determined in the laboratory), this implies that a single red blood cell contains

0.16 g of hemoglobin per 5.4×10^9 cells =
$$3 \times 10^{-11} \text{ g of hemoglobin} \qquad (22\text{--}1)$$

The molecular weight of human hemoglobin A is about 64,500, and one gram-molecular weight (gMW) would contain 6.02×10^{23} molecules of hemoglobin (i.e., Avogadro's number); hence a single red blood cell contains

$$\frac{(3 \times 10^{-11} \text{ g Hb/cell}) (6.02 \times 10^{23} \text{ Hb molecules /gMW})}{(6.45 \times 10^4 \text{ g Hb/gMW})}$$

$$= 2.8 \times 10^8 \text{ molecules of hemoglobin} \qquad (22\text{--}2)$$

It has already been noted that the synthesis of hemoglobin in the maturing red blood cell occurs over a period of 3 days. Knowing this, and using the result of equation 22–2 above, it is possible to calculate the *average* number of hemoglobin molecules whose synthesis is completed *each second* during this period of differentiation. It would be

$$\frac{2.8 \times 10^8 \text{ molecules of Hb/cell}}{(3 \text{ days}) (24 \text{ hr/day}) (60 \text{ min/hr}) (60 \text{ sec/min})} =$$

$$\begin{array}{l} 1.1 \times 10^3 \text{ molecules of} \\ \text{hemoglobin completed} \\ \text{per second per cell} \end{array} \qquad (22\text{--}3)$$

Therefore, an average of 1100 molecules of hemoglobin are completed in each second of the 3-day maturation period. This, of course, assumes that synthesis is continuous and also uniform over the entire 3 days, which is not actually the case. Therefore, the real value would be greater than this during the period of peak synthesis and lower at other times. However, for purposes of this discussion we can assume continuous and uniform synthesis.

Since there are four globin chains in each hemoglobin molecule, there would be $4(1.1 \times 10^3)$ or 4400 globin chains completed per second in a single red blood cell. The experiments of Dintzis and others indicate that between 60 and 90 seconds are required for the synthesis of a single globin chain. If for convenience we use the larger value, then during any given second there would be

$$(90 \text{ sec}) (4400 \text{ chains/sec}) = 396,000 \text{ chains} \quad (22\text{--}4)$$

of globin in production in the cell. Since the most common form of alpha chain polysome in the cell is the tetramer, this means that there must be 198,000/4 or 49,500 molecules of αmRNA in the cell. The most common beta chain polysome is the hexamer, implying that there are 198,000/6 or 33,000 molecules of βmRNA. This assumes that each molecule of

mRNA is stable for the whole 3-day maturation period and is available at the outset of hemoglobin synthesis. Since neither assumption is entirely valid, the actual amounts of the globin mRNAs in the cell are most likely considerably higher during peak hemoglobin synthesis. Each erythroblast contains four structural genes for alpha chains and two structural genes for beta chains; therefore, each alpha chain structural gene would have to be transcribed into αmRNA 12,375 times (i.e., 49,500/4), and each beta chain structural gene would have to be transcribed into βmRNA 16,500 times (i.e., 33,000/2).

If the developing red blood cell contains an average of 49,500 molecules of αmRNA, and these are used to produce sufficient alpha globin chains for 2.8×10^8 molecules of hemoglobin, then the average αmRNA is translated 11,313 times. That is,

$$\frac{(2.8 \times 10^8 \text{ molecules Hb}) (2\alpha \text{ globin chains/Hb})}{(49,500 \text{ molecules } \alpha\text{mRNA})} \qquad (22\text{--}5)$$

$= 11,313$ alpha globin chains per αmRNA. The corresponding translation frequency for βmRNA would be

$$\frac{(2.8 \times 10^8 \text{ molecules Hb}) (2\beta \text{ globin chains / Hb})}{(33,000 \text{ molecules } \beta\text{mRNA})} \qquad (22\text{--}6)$$

$= 16,969$ times

These figures indicate that the globin mRNAs are extremely stable (recent experimental evidence points to a half-life of at least several hours), especially in comparison with procaryotic mRNAs, which may be translated only once.

In all of the above calculations, we have been considering *averages* only. The synthesis of hemoglobin in the maturing erythrocyte is *not* uniform or continuous throughout development. Instead, hemoglobin synthesis reaches a maximum in early development. Therefore, during the period of maximum synthesis more molecules of mRNA would be required, and the number of globin chains produced from a single mRNA molecule would be lower. Despite this, the high frequency of transcription and translation in this cell is clearly apparent.

In eucaryotic cells, transcription occurs in the cell nucleus and translation occurs later in the extranuclear hyaloplasm. In procaryotes, which have no nucleus, not only do transcription and translation occur in the same region of the cell but they also occur at the same time.

In bacteria such as *E. coli*, the transcription of DNA is accompanied by the translation of the *nascent messenger*

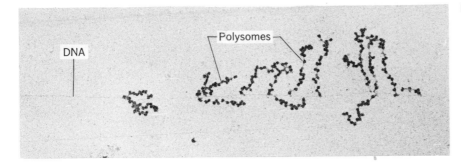

Figure 22–38
Visualization of the simultaneous transcription and translation of *E. coli* chromosomal DNA. Magnification 43,000 ×. (Courtesy of Drs. O. L. Miller and B. A. Hamkalo, from O. L. Miller, B. A. Hamkalo, and C. A. Thomas, Science *169*, 394, 1970. Copyright © 1970 by the American Association for the Advancement of Science.)

RNA. The mRNA is synthesized beginning with its 5′ end, and as soon as the mRNA strand is long enough, a ribosome attaches to the messenger and begins translation. As mRNA synthesis proceeds, more ribosomes attach to the elongating strand to form a polysome. O. L. Miller and others have provided elegant proof of this in the form of electron micrographs such as the one shown in Figure 22–38. In this figure, a portion of the *E. coli* chromosomal DNA appears as a thin filament being actively transcribed by a number of RNA polymerase molecules into mRNA strands. (RNA polymerase is the enzyme that transcribes DNA into RNA.) To each of these mRNA strands, ribosomes have attached to form polysomes. In Figure 22–38, it appears that transcription proceeds from left to right along the DNA, since the mRNA length (and therefore polysome size) exhibits a general increase in that direction. Unfortunately, electron micrographs do not reveal the growing nascent polypeptide chains on the ribosomes, since the amino acids that make up the polypeptides are too small to be resolved.

Cotranslational and Post-translational Protein Modification

Cotranslational Modification. In many instances, a number of changes are made in the structure and organization of polypeptide chains *during* their synthesis; these are called **cotranslational modifications** and include (1) **deformylation,** (2) **amino acid cleavage,** (3) **side-chain alteration,** (4) **disulfide bridge formation,** (5) **sugar addition,** and (6) **tertiary folding.**

Deformylation. In procaryotes and in eucaryotic mitochondria and chloroplasts, the formyl group of the N-terminal methionine of the growing polypeptide chain is enzymatically cleaved (see earlier).

Amino Acid Cleavage. In both procaryotes and eucaryotes, N-terminal methionine and occasionally other amino acids as well are enzymatically cleaved from the free N-terminus by an *aminopeptidase.*

Side-Chain Alteration. The R groups of certain amino acids are often altered following inclusion of the amino acid into the growing polypeptide chain. For example, during the synthesis of *collagen,* certain proline and lysine residues are hydroxylated (to form hydroxyproline and hydroxylysine, respectively). Other amino acid side chains may be phosphorylated (e.g., serine).

Disulfide Bridge Formation. Juxtaposed sulfhydryl groups of cysteine residues may be oxidized to form disulfide bridges. This normally occurs after tertiary folding (see below) orients these R groups into the necessary steric positions.

Sugar Addition. Sugars may be enzymatically attached to certain amino acids during the synthesis and completion of various *glycoproteins.*

Tertiary Folding. Although some proteins may spontaneously fold to form their biologically active tertiary structure following completion and release of the polypeptide from the ribosome (e.g., *ribonuclease,* see Chapter 4), others undergo tertiary folding during translation. In *E. coli,* tertiary folding of enzymatic polypeptides during their synthesis endows these nascent proteins with catalytic properties prior to termination and release.

Post-translational Modifications. **Post-translational modifications** are changes that occur in protein structure *after* completion and release of the polypeptide have taken place.

Some of the modifications already described as cotranslational may also occur following translation. For example, enzymatic hydroxylation and phosphorylation of amino acid side chains, the formation of disulfide bridges, and the addition of sugars to certain residues may occur following release of the completed polypeptide. Moreover, tertiary folding, although begun during translation, is completed following polypeptide release. Some modifications, however, are characteristically post-translational; included in this category are (1) **peptide cleavage,** (2) **quaternary association,** and (3) **addition of prosthetic groups.**

Peptide Cleavage. For some proteins, major changes in structure in the form of cleavage of specific bonds and removal of sections of the polypeptide occur following translation. For example, the A and B polypeptide chains that comprise the insulin molecule (Chapter 4) are produced by post-translational cleavage of a single translation product (Fig. 22–39) called *proinsulin,* which has no hormonal activity.

The activation of the zymogen *chymotrypsinogen* to form the digestive enzyme *chymotrypsin* (see Chapter 8) serves to illustrate the level of complexity that post-translational peptide cleavage can assume. Chymotrypsinogen is broken into two polypeptides by the enzyme *trypsin,* the product (which is still linked by disulfide bridges) being π-chymotrypsin (Fig. 22–40). π-chymotrypsin acts to catalyze its own conversion to the active digestive enzyme α-chymotrypsin. This activation involves the removal of two dipeptides of π-chymotrypsin, producing a product consisting of three interconnected polypeptide chains. Activation of chymotrypsinogen occurs in the small intestine, and this is where protein digestion takes place, whereas the pancreas, which is the site of production of the enzyme, secretes it in the zymogen form.

The post-translational modifications that produce insulin and chymotrypsin also demonstrate that *a protein consisting of more than one polypeptide chain may not be encoded by a corresponding number of mRNAs (or genes!) but may be encoded by a single mRNA (or gene).* Post-translational peptide cleavage may produce a series of separate polypeptide chains that ultimately make up the final protein product. Indeed, evidence is at hand that indicates that in some procaryotes and in the case of certain virsues, a single polypeptide chain may be cleaved to produce several individual proteins.

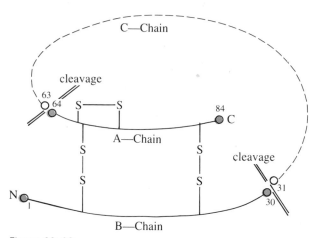

Figure 22–39
Post-translational cleavage of *proinsulin* to form *insulin.*

Quaternary Association. As noted in Chapter 4 and also earlier in this chapter, some proteins that possess quaternary structure are assembled by the spontaneous interaction of individual polypeptide chains. In the case of hemoglobin, for example, separate alpha and beta chains spontaneously combine to form asymmetric dimers, and these combine to form the functional tetramer. Assumption of quaternary structure is accompanied by the formation of stabilizing bonds between neighboring protein subunits and modification of the individual tertiary structures they previously possessed.

Addition of Prosthetic Groups. The prosthetic groups of enzymes and other proteins are attached following release of the completed polypeptide chains, and attachment may be spontaneous or catalyzed enzymatically. In the case of hemoglobin, the insertion of the heme groups occurs *after quaternary association is complete* and begins with alpha globin chains.

The completion of the hemoglobin molecule by the successive attachment of its four heme groups brings to a conclusion a synthetic process about which more is known than for any other complex protein. In this chapter, we

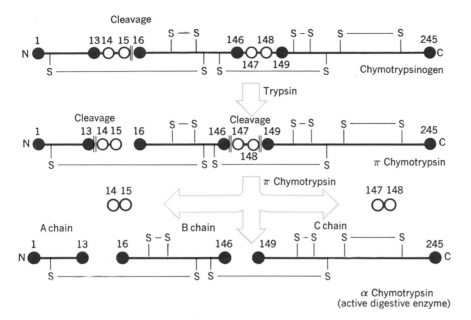

Figure 22–40
Post-translational cleavage of *chymotrypsinogen* to form *chymotrypsin*.

have seen that (1) heme regulates the initiation of globin chain synthesis; (2) globin chain elongation does not proceed at a uniform rate but that alpha and beta chains are produced in equal amounts; (3) asymmetric dimers spontaneously associate to form tetramers; and (4) heme insertion is sequential. To this must be added the long-known fact that heme acts as a negative effector of its own synthesis through feedback inhibition of an enzyme catalyzing an initial step in the heme biosynthetic pathway. In this fashion, the production of more heme than can be utilized for assembly of hemoglobin is avoided. Acting in concert, all of these mechanisms provide for the 1:1:1 ratio of alpha chain, beta chain, and heme group synthesis in the maturing red blood cell. Figure 22–41 summarizes our existing knowledge of the regulation of the synthesis of the hemoglobin protein.

Transfer RNA Specialization

When W. F. Anderson and J. M. Gilbert reported in 1969 that addition of isolated tRNA fractions to a reticulocyte cell-free system synthesizing globin chains altered the balance of alpha and beta globin chain production, the question was raised as to whether cells might have a rather specialized complement of tRNA. That this is indeed the case was borne out by the extensive studies of D. W. E. Smith on

the amounts and types of tRNAs present in the reticulocyte. As Figure 22–42 shows, a direct relationship exists between the frequency with which each type of amino acid occurs in the hemoglobin molecule and the abundance in the cell of the tRNAs specific for that amino acid species. This implies that cellular mechanisms exist that coordinate the production of various tRNAs according to the types and amounts of different amino acids present in the proteins of that cell—a most striking implication! This notion is supported by additional evidence of tRNA specialization in other cells, including silk-gland cells, lymphocytes, and cells of the pancreas and liver.

In Figure 22–42, the coordinates for met and leu appear to be exceptions to the linear distribution. For met, there appears to be an excess of $tRNA^{Met}$ (and $tRNA^{Met}_i$) (the additional met residues involved in chain initiation are already taken into account in the data), while for leu, there is a shortage of $tRNA^{Leu}$. The shortage of leucine tRNA in reticulocytes is believed to be yet another rate-limiting control factor for the production of globin chains in the maturing cell.

Inhibitors of Protein Synthesis

Many substances are known to act as inhibitors of various stages of protein synthesis. Included among these are a

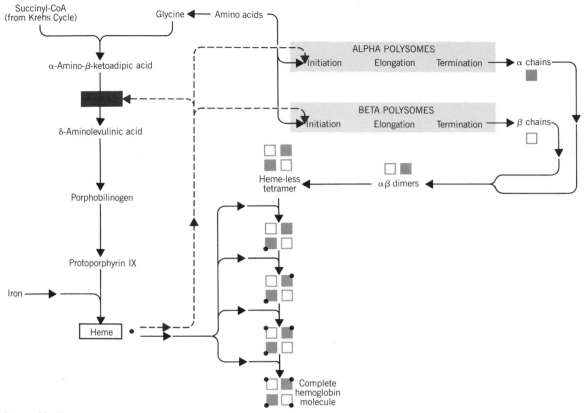

Figure 22–41
Synthesis of the hemoglobin molecule (solid arrows) and its regulation (dashed arrows).

number of **antibiotics** produced by one strain of microorganism and lethal to other strains of the same or a different species. Some of the best understood inhibitors of protein synthesis are listed in Table 22–9. Because the actions of many of these inhibitors are quite specific, they have proved extremely useful tools in the step-by-step elucidation of the mechanism of protein synthesis.

Inhibitors of Both Procaryotic and Eucaryotic Protein Synthesis

Aurintricarboxylic acid inhibits formation of the initiation complex by preventing the association of mRNA with the small ribosomal subunit. Inhibitors of initiation are readily distinguished from inhibitors blocking other stages of protein synthesis because of the *delay effect* that follows their administration. That is, protein synthesis continues for a

short time after administration of the inhibitor, since peptide chains whose growth had already begun are unaffected and grow to completion. *Edeine,* a polypeptide isolated from *Bacillus brevis,* inhibits the binding of aminoacyl-tRNA and N-formylmet-tRNA$_f^{Met}$ (in procaryotes) to the small subunit. *Fusidic acid* is a steroidal antibiotic; in procaryotes, it inhibits the binding of aminoacyl-tRNA to the ribosome, whereas in eucaryotes it inhibits translocation by reacting with elongation factor.

Puromycin, an inhibitor of protein synthesis, was one of the first of such inhibitors to have its specific effect determined. This antibiotic *mimics* aminoacyl-tRNA and binds to the free A site of ribosomes engaged in protein synthesis. Catalytic formation of a bond between the nascent polypeptide and puromycin is followed by the release of the *peptidyl-puromycin* from the ribosome, since no further elongation is possible. The specific effects of pur-

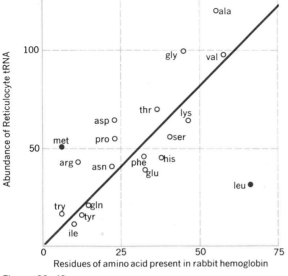

Figure 22–42
Transfer RNA specialization in the reticulocyte.

Table 22–9
Inhibitors of Protein Synthesis in Procaryotic and Eucaryotic Cells

Inhibitor	Effective in	
	Procaryotes	Eucaryotes
Anisomycin	−	+
Aurintricarboxylic acid	+	+
Chloramphenicol	+	−
Colicin E3	+	−
Cycloheximide	−	+
Diphtheria toxin	−	+
Edeine	+	+
Erythromycin	+	−
Fusidic acid	+	+
Pactamycin	−	+
Puromycin	+	+
Ricin	−	+
Sodium fluoride	−	+
Sparsomycin	−	+
Streptomycin	+	−
Tetracycline	+	+
Trichodermin	−	+

omycin have been used to advantage for studies of nascent chain length, the kinetics of chain elongation, and the identification of the effects of other antibiotics. **Tetracycline** inhibits protein synthesis by blocking aminoacyl-tRNA binding to the small subunit.

Inhibitors Specific for Procaryotes

Chloramphenicol (chloromycetin) binds to the large subunit of procaryotic ribosomes and interferes with the functioning of *peptide synthetase,* thereby inhibiting chain elongation. **Colicin E3** inhibits protein synthesis in procaryotes by interfering in some manner with the functioning of the small subunit. **Erythromycin** binds to ribosomes that are *not* engaged in protein synthesis, preventing their potential participation, but does not bind to ribosomes containing nascent chains (i.e., ribosomes that are part of a functioning polysome). **Streptomycin** was one of the earliest discovered antibiotics and was employed as an agent against bacterial infection for many years before its specific chem-

ical actions were known. Streptomycin binds to protein S12 of the small ribosome subunit, causing release of *N*-formylmet-tRNA$_f^{Met}$ from initiation complexes (thereby preventing initiation of chain growth) and also causing misreading of the codons of mRNA by ribosomes already involved in chain elongation.

Inhibitors Specific for Eucaryotes

Anisomycin is an antibiotic produced by *Streptomyces* that inhibits peptide bond formation when bound to the small ribosomal subunit. **Cycloheximide** binds to the large subunit, preventing the translocation of tRNA in the *A* site to the *P* site. **Diphtheria toxin** (produced by a strain of *Corynebacterium diphtheriae*) inhibits protein synthesis through its action on EF-2 (*translocase*). EF-2 exists in cells in two forms—ribosome-bound and free. Diphtheria toxin acts enzymatically to alter free EF-2, rendering the factor inactive. Ribosome-bound EF-2 is not susceptible to inactivation by the toxin.

Pactamycin (produced by a strain of *Streptomyces*) binds to free small subunits (not to small subunits already part of polysomes), where it prevents initiation by inhibiting binding of met-tRNA$_i^{Met}$ and formation of the initiation complex. The toxic effects of **ricin** (a protein present in the castor bean) have been known for nearly a century. Ricin consist of two polypepide chains (linked by disulfide bridges), one of which acts as the inhibitor once incorporated into the cell. Ricin acts on the large subunit, preventing formation of the 80 *S* initiation complex. Like ricin, **sodium fluoride** acts as an inhibitor of initiation; NaF blocks addition of the large subunit to mRNA.

Sparsomycin, another antibiotic produced by *Streptomyces,* inhibits the binding of the amino acid moiety of aminoacyl-tRNA to the large subunit and, in so doing, blocks peptide synthetase. *Trichodermin* is the only chemical compound so far identified as a specific inhibitor of the termination stage of polypeptide synthesis.

Inhibitors of Organellar Protein Synthesis

Protein synthesis by mitochondrial and chloroplast ribosomes is also subject to inhibition by certain antibiotics and other chemicals. Shortly after the initial demonstration of organellar protein synthesis, it was found that chloramphenicol, a strong inhibitor of procaryote protein synthesis, blocks synthesis in mitochondria and chloroplasts; while cycloheximide, which blocks eucaryote cytoplasmic ribosomal protein synthesis, is without effect on mitochondrial and chloroplast synthesis. These observations provided added credence for the notion that procaryotic cells, mitochondria, and chloroplasts have a common evolutionary origin. It is now clear, however, that the picture is considerably more complex. For example, streptomycin, which inhibits procaryotic protein synthesis, fails to inhibit mitochondrial protein synthesis in yeast cells. Other antibiotics inhibit mitochondrial protein synthesis but have no effect on procaryotes. Erythromycin inhibits the synthesis of proteins in procaryotes, yeast mitochondria, and chloroplasts but fails to block protein synthesis in mammalian mitochondria. The latter observation suggests that the nature of mitochondrial protein synthesis varies among different groups of eucaryotes. In general, mitochondria from higher eucaryotes are more resistant to inhibitors of procaryotic protein synthesis than are mitochondria from lower eucaryotes. In chloroplasts, protein synthesis is inhibited by the same agents that block this process in procaryotic cells.

The differential sensitivity of eucaryotic cytoplasmic and mitochondrial ribosomes to specific inhibitors provides a means for examining the sources of certain mitochondrial proteins. The synthesis of a mitochondrial protein in the presence of cycloheximide (a cytoplasmic inhibitor) indicates that the mitochondria are the source of the protein, whereas synthesis of the protein in the presence of chloramphenicol indicates that the mitochondrial protein is produced in the cytoplasm and then moves to the mitochondria. Experimentally, determinations of this sort are carried out by incubating cells in media containing both radioactively labeled amino acids and inhibitors. The synthesis of the mitochondrial protein is manifested by the appearance of radioactivity in the protein later isolated from the mitochondria.

Using this approach, it has been possible to show that perhaps 85 to 90% of all mitochondrial ribosomal proteins are synthesized in the cytoplasm and then make their way into the mitochondria where, together with mitochondrial rRNA, they are assembled into ribosomes. Of the seven polypeptides that make up the enzyme *cytochrome oxidase,* four are synthesized in the cytoplasm and three are synthesized in the mitochondria.

Summary

Using isotopes of nitrogen and carbon as tracers of amino acid metabolism, a number of investigators working in the 1940s and early 1950s showed that protein turnover in tissues was not simply a function of "wear and tear." Instead, cell protein undergoes continuous breakdown and synthesis, the latter being a property of the **microsomal fraction** of tissue homogenates. It is the **ribosomes** in the microsomal phase that carry out the assembly of the polypeptide chains that make up proteins, and these tiny organelles are currently the object of intensive biochemical and electron microscopic analyses. The ribosomes of procaryotic and eucaryotic cells are made up of two **subunits,** each subunit containing a specific combination of *proteins* and *ribosomal* ribonucleic acids (rRNA). Association of these subunits with **messenger RNA (mRNA)** is followed by *translation* of the mRNA base sequence (i.e., its **codons**) into the primary structure of polypeptides. **Transfer RNAs (tRNA)** combine with specific amino acids and enter the ribosome–mRNA complex. The succession of **aminoacyl-tRNA** species entering the complex and donating amino acid residues to the growing polypeptide is prescribed by the mRNA codon sequence. All the RNAs are **transcribed** from DNA and **processed (cleaved, trimmed,** or otherwise modified) prior to becoming functionally active.

Ribosomes may be *free* in the cytoplasm or *attached* to the endoplasmic reticulum membranes, each variety functioning in the assembly of different proteins. Individual polypeptides are assembled beginning at the N-terminus and proceeding toward the C-terminus. The rate of chain elongation may vary in different regions of the polypeptide. In most instances, a single message is translated by several ribosomes traveling in close succession along the mRNA molecule; such complexes are called **polysomes.** Initiation and termination of chain growth, as well as the elongation

cycle, require specific factors and are catalyzed by specific enzymes that are either constituents of the ribosomes or dissolved in the cytoplasm. Modifications of a protein may occur during or following translation. **Cotranslational** modifications include **deformylation** at the N-terminus, amino acid **cleavage, side-chain alteration,** formation of **disulfide bridges, addition of sugars,** and **tertiary folding.** Modifications that are characteristically **post-translational** include **peptide cleavage, quaternary association,** and **addition of prosthetic groups.**

Many substances, including a number of **antibiotics,** act as inhibitors of protein synthesis. Some inhibitors are effective only in procaryotes and some only in eucaryotes. Still others inhibit protein synthesis in both procaryotic and eucaryotic cells.

References and Suggested Reading

Articles and Reviews

Boyer, S. H., Smith, K. D., and Noyes, A. N., Immunological purification and characterization of hemoglobin chain-synthesizing polysomes. *Hemoglobin: Comparative Molecular Biology Models for the Study of Disease, Ann. N.Y. Acad. Sci. 241,* 204 (1974).

Brawerman, G., Characteristics and significance of the polyadenylate sequence in mammalian messenger RNA, in *Progress in Nucleic Acid Research and Molecular Biology,* Vol. 17 (W. E. Cohn, Ed.), Academic Press, New York, 1976, p. 118.

Brimacombe, R., Stoffler, G., and Wittmann, H. G., Ribosome structure, *Annu. Rev. Biochem. 47* (E. E. Snell et al., Eds), Annual Reviews, Inc., Palo Alto, 1978, p. 217.

Caskey, C. T., The universal RNA genetic code. *Q. Rev. Biophys. 3,* 295 (1970).

Clark, B. F. C., and Marcker, K. A., How proteins start. *Sci. Am. 218*(1), 36 (Jan., 1968).

Crick, F., Split genes and RNA splicing. *Science 204,* 264 (1979).

Darnell, J. E., Implications of RNA-RNA splicing in evolution of eukaryotic cells. *Science 202,* 1257 (1978).

Darnell, J. E., mRNA structure and function, in *Progress in Nucleic Acid Research and Molecular Biology,* Vol. 19 (W. E. Cohn and E. Volkin, Eds.), Academic Press, New York, 1976, p. 493.

Davis, B. D. and Tai, P. C., The mechanism of protein secretion across membranes. *Nature 283,* 433 (1980).

Dintzis, H. M., Assembly of the peptide chains of hemoglobin. *Proc. Natl. Acad. Sci. 47,* 247 (1961).

Efstratiadis, A., Posakony, J. W., Maniatis, T., Lawn, R. M., O'Connell, C., Spritz., R. A., DeRiel, J. K., Forget, B. G., Weissman, S. M., Slightom, J. L., Blechl, A. E., Smithies, O., Baralle, F. E., Shoulders, C. C., and Proudfoot, J. J., The structure and evolution of the human beta-globin gene family. *Cell 21,* 653 (1980).

Emr, S. D., Hall, M. N., and Silhavy, T. J., A mechanism of protein localization: the signal hypothesis in bacteria. *J. Cell Biol. 86,* 701 (1980).

Englander, S. W., and Page, L. A. Interpretation of data on sequential labeling of growing polypeptides. *Biochem. Biophys. Res. Commun. 19,* 565 (1965).

Erdmann, V. A., Structure and function of 5 S and 58 S RNA, in *Progress in Nucleic Acid Research and Molecular Biology,* Vol. 18 (W. E. Cohn, Ed.), Academic Press, New York, 1976, p. 45.

Furuichi, Y., Muthukrishnan, J. T., and Shatkin, A. J., Caps in eukaryotic mRNAs, in *Progress in Nucleic Acid Research and Molecular Biology,* Vo. 19 (W. E. Cohn and E. Volkin, Eds.), Academic Press, New York, 1976, p. 3.

Goodenough, U. W., and Levin, R. P., The genetic activity of mitochondria and chloroplasts, *Sci. Am. 223,* 22 (May, 1970).

Holley, R. W., The nucleotide sequence of a nucleic acid. *Sci. Am. 214*(2), 30 (Feb., 1966).

Hunt, T., The control of globin synthesis in rabbit reticulocytes, in *Hemoglobin: Comparative Molecular Biology Models for the Study of Disease, Ann. N. Y. Acad. Sci. 241,* 223 (1974).

Hunt, T., Control of globin synthesis. Haemoglobin: structure, function and synthesis, *Br. Med. Bull. 32,* 257 (1976).

Itano, H. A., Genetic regulation of peptide synthesis in hemoglobin, *J. Cell. Physiol. 67* (Suppl. 1), 65, (1966).

Kearns, D. R., High-resolution nuclear magnetic resonance investigations of the structure of tRNA in solution, in *Progress in Nucleic Acid Research and Molecular Biology,* Vol. 18 (W. E., Cohn, Ed.), Academic Press, New York, 1976, p. 91.

Kim, S. H., Three-dimensional structure of transfer RNA, in *Progress in Nucleic Acid Research and Molecular Biology,* Vol. 17 (W. E. Cohn, Ed.), Academic Press, New York, 1976, p. 182.

Lake, J. A., The ribosome. *Sci. Am. 245*(2), 84 (Aug. 1981).

Lawn, R. M., Efstratiadis, A., O'Connell, C., and Maniatis, T., The nucleotide sequence of the human beta-globin gene. *Cell 21* 647 (1980).

Lodish, H. F., Translational control of protein synthesis, in *Annual Review of Biochemistry,* Vol. 45 (E. E. Snell et al., Eds.), Annual Reviews Inc., Palo Alto, 1976, p. 39.

Marcker, K., and Sanger, F., *N*-formyl-methionyl-s-RNA. *J. Mol. Biol. 8,* 835 (1964).

McKnight, S. L., and Miller, O. L., Electron microscopic analysis of chromatic regulation in the cellular blastoderm *Drosophila melanogaster* embryo. *Cell 12,* 795 (1977).

Miller, O. L., The visualization of genes in action. *Sci. Am 228* 83, 34 (Mar. 1973).

Miller, O. L., and Beatty,, B. R., Visualization of nucleolar genes. *Science 164,* 955 (1969).

Miller, O. L., Hamkalo, B. A., and Thomas, C. A., Visualization of bacterial genes in action. *Science 169,* 392 (1970).

Morris, A. J., Slabaugh, R. C., and Protzel, A., Size characteristics of nascent globin chains (peptidyl tRNA) in the reticulocyte. Hemoglobin: comparative molecular biology models for the study of disease. *Ann. N. Y. Acad. Sci. 241,* 310 (1974).

Nikolaev, N., and Hadjiolov, A. A., Maturation of ribosomal ribonucleic acids and the biogenesis of ribosomes. *Prog. Biophys. Mol. Biol. 31,* 95 (1976).

Noller, H. F., and Woese, C. R., Secondary structure of 16 *S* ribosomal RNA. *Science 212,* 403 (1981).

Nomura, M., Ribosomes. *Sci. Am. 221*(4), 28 (Oct. 1969).

Ochoa, S., Regulation of protein synthesis. *Crit. Rev. Biochem. 7,* 7 (1979).

Perry, R. P., Processing of RNA, in *Annual Review of Biochemistry,* Vol. 45 (E. E. Snell et al., Eds.), Annual Reviews Inc., Palo Alto, 1976, p. 630.

Pestka, S., Insights into protein biosynthesis and ribosome function through inhibitors, in *Progress in Nucleic Acid Research and Molecular Biology.* Vol. 17 (W. E. Cohn, Ed.), Academic Press, New York, 1976, p. 217.

Poyton, D. O., Cooperative interaction between mitochondrial and nuclear genomes: cytochrome *c* oxidase assembly as a model, in *Current Topics in Cellular Regulation,* Vol. 17 (B. L. Horecker and E. R. Stadtman, Eds.), Academic Press, New York, 1980.

Proudfoot, N. J., and Brownlee, G. G., Nucleotide sequences of globin messenger RNA. Haemoglobin: structure, functions and synthesis. *Br. Med. Bull. 32,* 251 (1976).

Puvion-Dutilleul, F., Bachellerie, J-P., Zalta, J-P., and Bernhard, W., Morphology of ribosomal transcription units in isolated subnuclear fractions of mammalian cells. *Rev. Biol. Cell. 30,* 183 (1977).

Rich, A., Polyribosomes. *Sci. Am. 209*(6), 44 (Dec. 1963).

Rich, A., and Kim, S. H., The three dimensional structure of tRNA. *Sci. Am. 238*(1), 52 (Jan. 1978).

Rich, A., and RajBhandary, U. L., Transfer RNA: Molecular structure, sequence and properties, in *Annual Review of Biochemistry,* Vol. 45 (E. E. Snell et al., Eds.), Annual Reviews Inc., Palo Alto, 1976, p. 805.

Rich, A., Warner, J. R., and Goodman, H. M., The structure and function of polyribosomes. *Cold Spring Harbor Symp. Quant. Biol. 28,* 269 (1963).

Safer, B., and Anderson, W. F., The molecular mechanism of hemoglobin synthesis and its regulation in the reticulocyte. *Crit. Rev. Biochem. 5,* 261 (1978).

Satir, B., The final steps in secretion. *Sci. Am. 233*(3), 28 (Oct., 1975).

Shore, G. C., and Tata, J. R., Functions for polyribosome-membrane interactions in protein synthesis. *Biochim. Biophys. Acta 472,* 197 (1977).

Siekevitz, P., and Zamecnik, P. C., Ribosomes and protein synthesis. *J. Cell Biol. 91,* 53 (1981).

Smith, D. W. E., Reticulocyte transfer RNA and hemoglobin synthesis. *Science 190,* 529 (1975).

Spitnik-Elson, P., and Elson, D., Studies on the ribosome and its components, in *Progress in Nucleic Acid Research and Molecular Biology,* Vol. 17 (W. E. Cohn, Ed.), Academic Press, New York, 1976, p. 77.

Weissbach, H., and Ochoa, S., Soluble factors required for eukaryotic protein synthesis, in *Annual Review of Biochemistry,* Vol. 45 (E. E. Snell et al., Eds.), Annual Reviews Inc., Palo Alto, 1976, p. 191.

Winslow, R. M., and Ingram, V. M., Peptide chain synthesis of human hemoglobins A and A_2. *J. Biol. Chem. 241,* 1144 (1966).

Williamson, R., The processing of hnRNA and its relation to mRNA, in *Cell Biology: A Comprehensive Treatise,* Vol. 3 (L. Goldstein and D. M. Prescott, Eds.), Academic Press, New York, 1980.

Books, Monographs, and Symposia

Clemens, M. J., Translational control of protein synthesis in erythroid cells, in *Biochemistry of Cellular Regulation, Vol. 1, Gene Expression* (M. J. Clemens, Ed.), CRC Press, Boca Raton, Florida, 1980.

Lewin, B. M., *The Molecular Basis of Gene Expression,* Vol. 2., Wiley, New York, 1980.

Lewis, J. A., Synthesis of proteins by ribosomes attached to membranes, in *Biochemistry of Cellular Regulation, Vol. 1, Gene Expression* (M. J. Clemens, Ed.), CRC Press, Boca Raton, Florida, 1980.

Nomura, M., Tissieres, A., and Lengyel, P. (Eds.), *Ribosomes,* Cold Spring Harbor Monograph Series, Cold Spring Harbor Laboratory, N. Y., 1974.

Masters, C. J., and Holmes, R. S., *Haemoglobin, Isoenzymes and Tissue Differentiation. Frontiers of Biology,* Vol. 42, North-Holland, Amsterdam, 1975.

McConkey, E. H. (Ed.), *Protein Synthesis,* Vol. 1, Marcel Dekker, New York, 1971.

McConkey, E. H. (Ed.), *Protein Synthesis,* Vol. 2, Marcel Dekker, New York, 1976.

Tedeschi, H., *Mitochondria: Structure, Biogenesis and Transducing Function. Cell Biology Monographs,* Vol. 4, Springer-Verlag, Vienna, 1976.

Chapter 23
THE CYTOPLASMIC GROUND SUBSTANCE, FILAMENTS, AND MICROTUBULES

Although the cytoplasm of cells has been known since the 1950s to contain a variety of fiberlike structures called **filaments** and **microtubules,** the cytoplasm or **ground substance** itself had until recently been thought to be a structureless fluid. This view prevailed for many years despite the fact that the cytoplasm exhibits viscoelastic properties that are more like a gel than a liquid. However, the development of **high–voltage electron microscopy** in the 1970s provided the breakthrough that revealed that the ground substance is pervaded by a three-dimensional network of fine, gossamer threads, called the **microtrabecular lattice.** This lattice serves to anchor many of the cell's organelles in position and appears also to interconnect many cytoplasmic filaments and microtubules. In this chapter, we will examine the structure and functions of the microtrabecular lattice, the relationship of the lattice to the cytoplasmic filaments and microtubules, and the special roles played in cells by cytoplasmic filaments and microtubules.

The Ground Substance and the Microtrabecular Lattice

Structure of the Lattice

Conventional transmission electron microscopes have accelerating potentials of several thousand volts and produce a beam of electrons that penetrates tissue and cell slices

(i.e., sections) having thicknesses of about 0.2 μm. The development in the 1970s of *high-voltage electron microscopes* capable of accelerating electrons over a potential of *one million* volts has made it possible to examine specimens having the thickness of whole cells. Such microscopes are quite imposing, usually occupying two floors of a laboratory (they are about 30 feet tall) and weighing about 20 tons. Since the electron beam traverses the thickness of an entire cell, the interior of the cell is revealed *in depth*.

Principally as a result of the high-voltage electron microscopic studies of K. R. Porter, M. Schliwa, J. J. Wolosewick, and J. B. Tucker, the cytoplasm or ground substance of eucaryotic cells has been shown to be divided into two major phases: the **microtrabecular lattice** and the **intertrabecular spaces.** A portion of the lattice is seen in the high-voltage photomicrograph of Figure 23–1 and is depicted diagrammatically in Figure 23-2. The lattice is not apparent in electron photomicrographs of conventional thin sections because the sections lack the depth necessary to reveal the network. The microtrabeculae are rich in protein while the intertrabecular space is aqueous and serves to dissolve or suspend the great variety of small molecules concerned with cellular metabolism (e.g., glucose, amino acids, oxygen, and inorganic salts). At its margins, the lattice is attached to the plasma membrane of the cell, but it also interconnects many of the cytoplasmic organelles

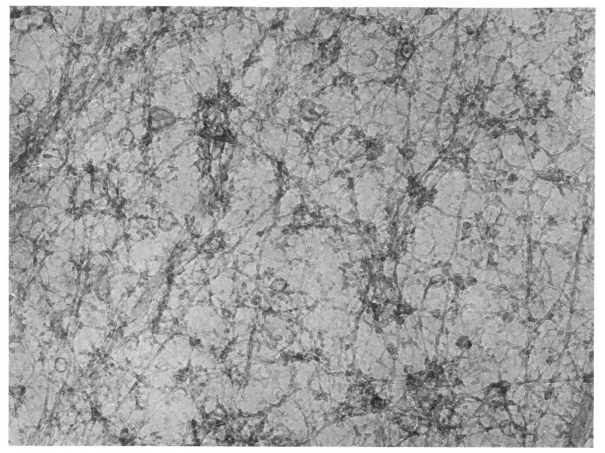

Figure 23–1

The *microtrabecular lattice* at the periphery of a kidney cell. The three-dimensional network of filaments that forms the lattice is revealed by high-voltage transmission electron microscopy of whole cells. The round bodies seen at a few of the intersections of the lattice are ribosomes. (Courtesy of Dr. M. Schliwa.)

(mitochondria, endoplasmic reticulum, ribosomes, polysomes, etc.) as well as the cytoplasmic filaments and microtubules that comprise the cytoskeleton (Fig. 23–2).

The microtrabecular lattice is not a rigid or static structure. Rather, it varies in response to changes in cell shape. Its disposition in a cell also varies according to the cellular environment. For example, when cells are grown in culture at low temperature (e.g., 4°C), they become spherical. The change in cell shape is due to a change in the cytoskeleton. First, the microtubules disassemble; then the filaments disappear; and finally, the microtrabeculae are modified, although they do not decompose entirely. Most notable is the separation of some of the microtrabeculae from the lattice and their aggregation into **gobbets** (Fig. 23–2). The loss of parts of the lattice produces gaps that allow the cell organelles to move about more freely and exhibit Brownian motion. If the cells are returned to their normal (higher) temperature, the microtrabecular lattice and the cytoskeleton reform, and the organelles are again confined in their movement.

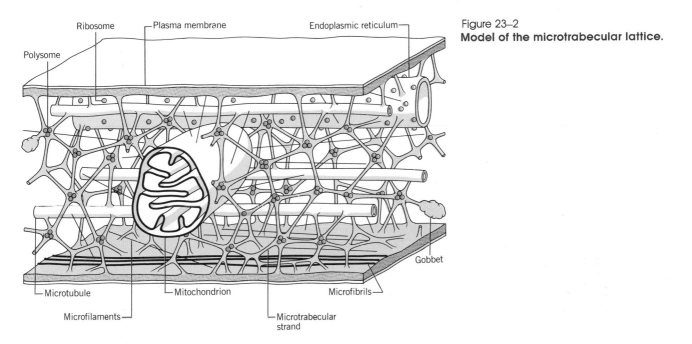

Figure 23–2
Model of the microtrabecular lattice.

Polysome
Ribosome
Plasma membrane
Endoplasmic reticulum
Gobbet
Microtubule
Microfilaments
Mitochondrion
Microtrabecular strand
Microfibrils

Changes in the chemical composition of the environment of a cell also produce reversible changes in the lattice. Cytochalasin B (Fig. 23–3), a drug obtained from the mold *Helminthosporium dematoiderum,* causes the microtrabeculae to thicken. High or low osmotic pressure, changes in the concentration of ions (e.g., Mg^{++} and Ca^{++}), and the presence of certain metabolic inhibitors also causes reversible changes in the lattice.

Chemistry of the Lattice

Chemically, the most important constituents of the lattice are proteins. Two-dimensional electrophoresis of extracts of the lattice suggests that more than 100 different proteins may comprise this polymeric network. Among the specific proteins that have been shown to be present are actin, myosin, and tubulin. These proteins are also major constituents of cytoplasmic filaments and microtubules (see below). A basic difference in the composition of microtrabeculae and the cytoskeletal elements (i.e., microtubules and cytoplasmic filaments) is revealed when both are treated with organic detergents such as Triton X-100. Cells treated with Triton X-100 lose the microtrabecular lattice along

with membranous structures such as mitochondria, endoplasmic reticulum, plasma membrane, and nuclear envelope, but they retain their cytoskeleton (Fig. 23–4). This behavior suggests that microtrabeculae have certain physical properties (and therefore a chemical composition) that

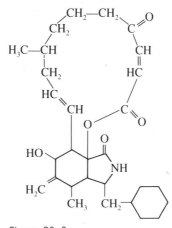

Figure 23–3
Structure of cytochalasin B.

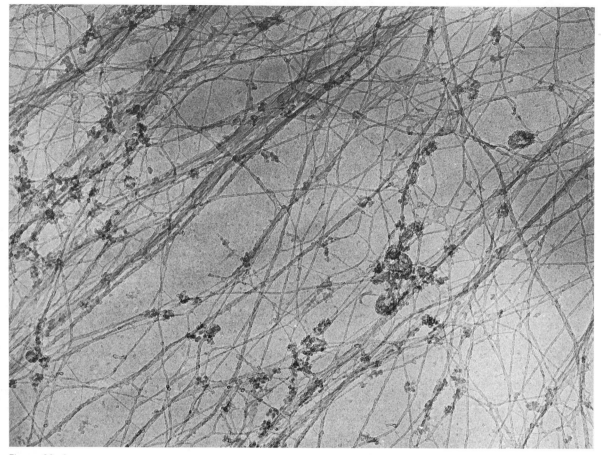

Figure 23–4

High-voltage transmission electron photomicrograph of a kidney cell cytoskeleton left behind as a residue after the microtrabecular lattice was extracted with the detergent Triton X-100. This detergent removes not only the lattice but also membranous organelles. (Courtesy of Dr. M. Schliwa.)

is similar to membranes but is unlike the cytoskeletal elements.

The observation that nonionic detergent extraction of cells leaves behind a residue consisting of microtubule and cytoplasmic filament proteins of the cytoskeleton has made it possible to study the cytoskeleton using conventional transmission electron microscopes. When the cytoskeletons of extracted cells are freeze-dried and replicas are produced using platinum, the cytoskeleton is revealed in considerable detail (Fig. 23–5).

Functions of the Microtrabecular Lattice

The microtrabecular lattice apparently serves as an intracellular scaffolding that suspends and organizes the diverse structural components of the cytoplasm, including the major cellular organelles. Acting in concert with the filaments and microtubules, the lattice plays an important role in maintaining cell shape and in cellular movements. Evidence also exists that suggests that a number of the enzymes of intermediary metabolism are bound to the lattice. It is possible

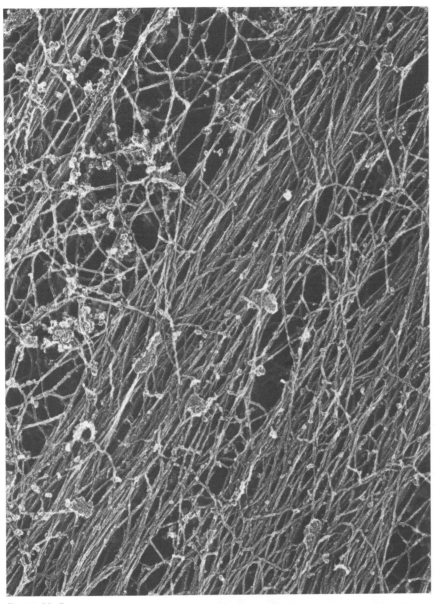

Figure 23–5
Platinum replica of a freeze-dried cytoskeleton of a fibroblast cell extracted with Triton X-100. Most of the filaments seen in this view are clustered together to form bundles running diagonally across the photo. Clusters of ribosomes are seen in areas where the filaments form a more diffuse network. (Courtesy of Drs. J. E. Heuser and M. W. Kirschner. Copyright © 1980 by The Rockefeller University Press; *J. Cell Biol. 86*, 215.)

that these enzymes are associated with the lattice in an *ordered* fashion; for example, the enzymes of a particular metabolic pathway may be bound to the lattice in such a manner that successive reactions of the pathway are spatially coordinated.

Cytoplasmic Filaments

Composition, Distribution, and Functions of Cytoplasmic Filaments

Cytoplasmic filaments are elongate, unbranched, proteinaceous strands that consist of bundles or groups of protein molecules sometimes wound into a helical shape. Although organized arrays of cytoplasmic filaments were first described for muscle cells and have been most extensively studied in this tissue, it is now apparent that essentially all eucaryotic cells contain them. Cytoplasmic filaments can be subdivided into three major classes based upon filament size; these are (1) **microfilaments** or **thin filaments,** which have a diameter of about 6 nm and are comprised primarily of the globular protein *actin;* (2) **intermediate filaments,** which have a diameter of about 10 nm and may contain the proteins *desmin, keratin,* or *vimentin;* and (3) **myosin filaments,** or **thick filaments,** which have a diameter of 13 to 22 nm and are rich in the fibrous protein *myosin.* Muscle cells, which are especially differentiated for contraction (discussed at length in Chpater 24), are rich in microfilaments and myosin filaments, but all cells contain cytoplasmic filaments, and it has been shown that actin can account for as much as 20% of the total cell protein in some non-muscle cells.

Although actin molecules are globular proteins (i.e., G-actin), in microfilaments the actin molecules are arranged as two helical chains (each of which is called a **protofilament**) that are twisted around each other to form a double helix (Fig. 23–6); in this form, the protein is called F-actin. Microfilaments are intimately associated with all cellular activities that involve movement. This is vividly demonstrated when cells are treated with cytochalasin B. In the presence of this substance, microfilaments dissociate, and the loss of the microfilaments is accompanied by a loss of certain cellular functions. Some of the more common processes sensitive to cytochalasin B are *phagocytosis, pinocytosis,* and *exocytosis* (see Chapter 15), *cytokinesis* (discussed later in this chapter), *cytoplasmic streaming* (in plant

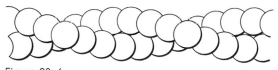

Figure 23–6

Arrangement of polymerized G-actin protofilaments in a microfilament. The two protofilaments form a right-handed double-helix.

cells), movements of *cilia* and *flagella,* movements of the *cytoskeleton,* and of course *muscle contraction* (Chapter 24).

Cytoplasmic Filaments of Muscle Cells

Striated, smooth, and *cardiac* muscle cells contain vast numbers of cytoplasmic filaments that function during the contraction of these cells. Nearly all of the cytoplasmic filaments or **myofilaments** in muscle cells are thin filaments (formed primarily from F-actin but containing tropomyosin and troponin) and thick filaments (formed from myosin). The two types of filaments are arranged in parallel rows and interact with each other through *cross-bridges* that enable the filaments to slide past one another and effect a shortening (i.e., contraction) of the cell. The contraction of striated muscle cells causes the muscle to shorten, thereby moving the skeleton. In striated muscle cells, the number of filaments is greatest and their geometric distribution is most highly ordered. Equally spaced about each thick filament are six thin filaments, and equally spaced about each thin filament are three thick filaments (Fig. 23–7). Units of several hundred thick and thin filaments are grouped together to form *myofibrils,* and each cell contains many hundreds of myofibrils. Further discussion of the structure of muscle and the biochemistry and mechanics of contraction is deferred until Chapter 24.

Cytokinesis

Cell division in animal cells involves two separate mechanisms: **mitosis** and **cytoplasmic cleavage,** or **cytokinesis.** In animal cells, cytokinesis begins toward the end of anaphase and is characterized by two events: (1) the cell begins to constrict about the midline of the spindle, and (2) dense

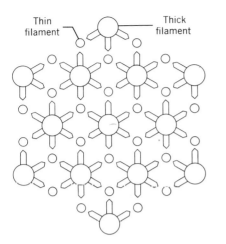

Thin filament   Thick filament

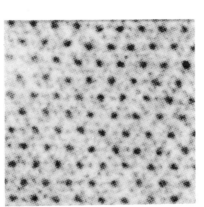

Figure 23–7

Cross section of a striated muscle myofibril showing geometric arrangement of thick and thin filaments. (Diagram courtesy of H. E. Huxley. Copyright © 1967 by Academic Press, Ltd; J. Molec. Biol. 30, 383.)

material is collected about the peripheral spindle fibers also along the midline. Both processes continue as the plasma membrane moves inward at the cleavage furrow. The material collecting at the midline of the spindle becomes quite dense, forming a structure known as the **midbody** (Fig. 23–8). Just before the infolding edges of the plasma membrane meet the fuse, the midbody disappears.

The furrowing or pinching-in of the plasma membrane at the cleavage furrow is reminiscent of the action of a purse string or of a rubber band tightening about a soft object. The process clearly involves the action of microfilaments seen in abundance in the area of the cleavage furrow. This conclusion is also supported by the observation that cytochalasin B inhibits the process. The presence of a band of microfilaments, the contractile ring, just underneath the plasma membrane in the area of constriction can be seen in Figure 23–9. These electron photomicrographs by T. E. Schroeder show an *Arbacia* sea urchin egg at various stages of cleavage. Prior to the onset of cleavage, no microfilaments are seen in the area that is about to constrict inwards (Fig. 23-9*a, d*). Once cleavage begins, microfilaments appear about the area of constriction, forming the *contractile ring (cr)*. The long axis of the microfilaments runs perpendicular to the plane of the sections seen in Figure 23–9*b* and *e*. Actually, the microfilaments are arranged in a circle about the furrow. As cleavage is completed, the spindle fibers and midbody fade and the contractile ring and microfilaments disappear (Fig. 23–29*c*). The molecular basis of the constriction that characterizes cytokinesis in animal cells is not yet clear, but it is speculated that the mechanism involves sliding actin filaments

(much like that known to occur during muscle cell contraction).

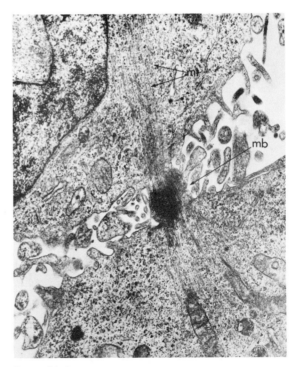

Figure 23–8

Final phase of cytokinesis in an animal cell. Only a few microtubules (m) of the spindle remain. Note the midbody (mb). (Courtesy of B. Byers. Copyright © 1968 by Springer-Verlag. Protoplasma 66, 423.)

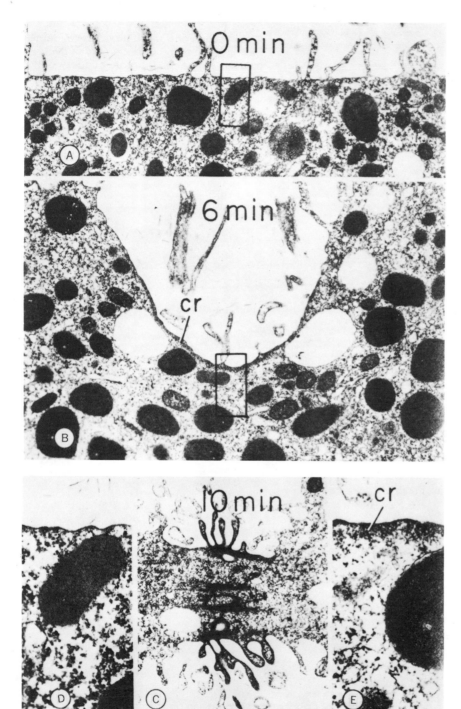

Figure 23–9
Cytokinesis in *Arbacia* sea urchin egg (see text for details). (Courtesy of T. E. Schroeder. Copyright © 1976 by Cold Spring Harbor Laboratory, *Cell Motility*, Book A, p. 268.)

Plasma Membrane Movement

Intestinal epithelial cells have many small projections called **microvilli** (see Chapter 15, especially Figs. 15–27 and 15–28) that cyclically shorten and extend into the lumen of the intestine. This action facilitates food absorption by increasing the cell surface area that is in contact with the intestinal lumen. Microfilaments of the intestinal microvilli are bundled into a core by the interaction of F-actin with the two proteins *villin* and *fimbrin*. The core is bound to the plasma membrane through a protein having a molecular weight of 110,000 daltons. There is no myosin in microvilli. The microfilaments are aligned parallel to the long axis of the microvilli with the actin filaments pointing toward the cell, very much like the way the thin filaments of a muscle cell sarcomere point toward a central plane (the M-line). Beneath the microvilli is another microfilament network called the **terminal web.** Myosin has been localized in the terminal web so that an actin-myosin contractile system could operate to pull the microvilli into the cell. A second possibility is that in the presence of high Ca^{++}, villin can act to sever F-actin into short fragments. This could cause the microvilli to collapse into the cell in a mechanism that is independent of myosin.

The Formation of Pseudopodia and Amoeboid Movement

The locomotion of amoebae, slime molds, leucocytes, and a number of other cells involves the formation of **pseudopodia**—large cytoplasmic extensions from the main body of the cell into which the remaining cytoplasm subsequently streams. Cells can form more than one pseudopod at a time, but continued locomotion in one direction requires the reversal of the process in the nondominant pseudopodia. In pseudopod-forming cells, the outer margin of the cytoplasm (called **ectoplasm**) is thick or *gel-like* and is relatively free of granules and other inclusions; the remaining cytoplasm (i.e., the **endoplasm**) is more fluid or *sol-like*. During locomotion, the fluid endoplasm flows forward into the advancing pseudopod, and as it reaches the anterior end of the pseudopod, it is transformed into gel, thereby forming new ectoplasm. At the rear of the moving cell, in a region called the **uroid** (see Fig. 15–47), the ectoplasm *solates,* streams deeper into the cell, and becomes endoplasm (Fig. 23–10).

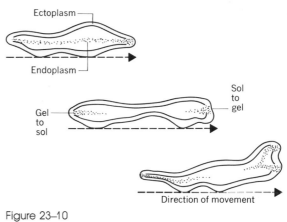

Figure 23–10
Amoeboid movement.

Two different views have dominated attempts to explain amoeboid movement. According to the hypothesis of R. D. Allen and others, the endoplasm in the anterior region of a pseudopod undergoes contraction as it is transformed into ectoplasm. In effect, these contractions *pull* the "viscoelastic" endoplasm forward and into the pseudopod. An older hypothesis advanced in the 1920s by S. O. Mast suggests that the ectoplasm toward the rear of the cell contracts, *pushing* the endoplasm forward.

Regardless of the site of origin of the forces that cause the cytoplasm to stream forward, overwhelming evidence implicates microfilaments in the process. For example, cytochalasin B inhibits amoeboid movement. Actin and myosin filaments have been identified in amoeboid cells, as has an ATPase similar to one known to be important in the chemistry of muscle contraction. Contractions of the ectoplasm of amoeboid cells can be induced by adding ATP and Ca^{++} to the cells. Although not fully understood, amoeboid movement appears to use an actin-myosin contractile system similar to that of muscle cells. Microtubules are also involved in the amoeboid movements of cells.

Microtubules

Microtubules have many features that distinguish them from microfilaments (Table 23–1). To begin with, the outside diameter of a microtubule (usually about 24 nm) is much greater than that of most microfilaments; moreover, microtubules are *hollow,* containing a central lumen about 15 nm in diameter. Microtubule length is quite variable. Some

Table 23–1

Comparison of the Properties of Most Microfilaments and Microtubules

	Microfilament	Microtubule
Diameter	30–60 Å	100–250 Å
Structure	Double-helical protofilament	Hollow tube of 13 protofilaments
Protein	Actin	Tubulin
Disassociating or inhibiting agent	Cytochalasin B	Colchicine, vinblastine, or vincristine
Subunit binding agent	ATP	GTP

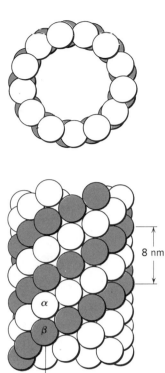

Heterodimer

Figure 23–11

Microtubule model seen in cross section and laterally. The circumference of the tubule usually has 13 subunits (see text for variations). Although subunits appear to be linear chains that are parallel to the axis of the tubule, the paired subunits (α + β heterodimers) join to form a helix about the central, hollow core. (Courtesy of J. Bryan. Copyright © 1974 by *Fed. Proc. 33*, 156.)

microtubules are less than 200 nm long, but in the long processes of nerve cells their lengths may be as great as 25 μm (i.e., 25,000 nm). Microtubules can also be distinguished from microfilaments chemically. Microtubules contain two major proteins called α *tubulin* and β *tubulin*. Each protein consists of a single polypeptide about 500 amino acids long (M.W., 55,000 daltons); both are similar in primary structure, indicating that they are probably derived from a common ancestral protein. Not only are the α and β tubulins nearly identical but tubulins from diverse species of cells are very similar, suggesting that they have hardly changed since they first appeared in eucaryotic organisms, or that tubulin is a highly conserved protein.

Alpha and beta tubulin molecules combine to form **heterodimers** and these serve as the basic building blocks of microtubules. The model of heterodimer organization shown in Figure 23–11 is based upon both chemical studies of microtubules and transmission electron microscopy. The microtubule is formed from a helical array of heterodimers with 13 subunits per turn of the helix. Neighboring heterodimers are linked to one another not only longitudinally but laterally as well.

Assembly of Microtubules

The current model for the manner in which tubulin subunits are assembled into a microtubule is based on in vitro studies. Under carefully controlled conditions (e.g., the appropriate concentration of tubulin and the absence of calcium),

alpha and beta subunits spontaneously form dimers that when present in high concentrations assemble into chains (Fig. 23–12); the chains then form a variety of intermediate structures including single and double rings, spirals, and stacked rings. The rings eventually open up to form linear chains or *protofilaments* that associate side by side to form sheets. When a sheet is sufficiently wide, it is curled to form a tube. The end result is the formation of short cylinders of dimers (Fig. 23–11). After a short cylinder is formed, continued growth occurs by the direct addition of more dimers. Growth occurs primarily by addition of dimers at one end of the tubule. It is believed that during certain microtubular functions (such as the operation of the anaphase spindle) the addition of dimers to one end of a

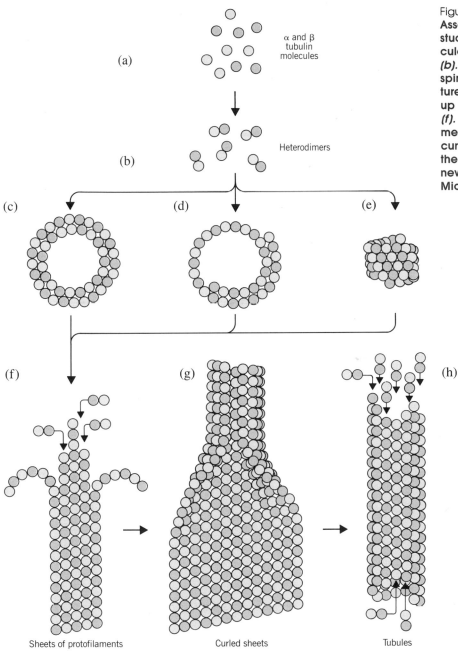

(a)

α and β tubulin molecules

(b)

Heterodimers

(c)

(d)

(e)

(f)

(g)

(h)

Sheets of protofilaments

Curled sheets

Tubules

Figure 23–12

Assembly of microtubules. In in vitro studies, alpha and beta tubulin molecules *(a)* **combine to form** *heterodimers* *(b).* **The dimers associate to form rings, spirals, and other intermediate structures** *(c, d, e)* **which eventually open up to form strands or** *protofilaments* *(f).* **Side-by-side assembly of protofilaments creates sheet-like structures that curl to form a tube** *(g).* **Elongation of the tube occurs by direct addition of new dimers** *(h).* **(Modified from P. Dustin, Microtubules.** *Sci. Am. 243,* **66, 1980.)**

microtubule is accompanied by the loss of dimers from the other end.

Assembly of tubulin into dimers requires that these polypeptides bind GTP. GTP-activated dimers can then combine with other dimers or with the growing microtubule. Attachment of a dimer to the microtubule is accompanied by the hydrolysis of the GTP but the resulting GDP and phosphate remain bound to the tubule. The tubulin dimer also

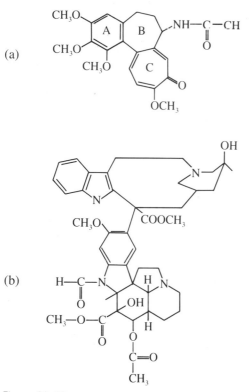

(a)

(b)

Figure 23–13
Microtubule assembly inhibitors *colchicine (a)* and *vincristine (b).* Vinblastine, another inhibitor, is similar to vincristine but has a methyl group in place of the formyl group.

has sites that can bind the drugs *colchicine, vincristine,* and *vinblastine* (Fig. 23–13) and these substances inhibit microtubule assembly. Tubules that are already present at the time of addition of these inhibitors disassemble. Calcium has long been recognized as an important ion in the microtubule assembly and disassembly process. Calcium may act either directly or in association with the regulatory protein *calmodulin.*

Recently, a number of proteins have been identified that associate with the surface of microtubules (Fig. 23–14). These proteins, called **microtubule-associated proteins,** or **MAPs,** facilitate microtubule assembly; that is, microtubules are formed considerably faster and at lower tubulin concentrations in the presence of MAPs. MAPs also protect microtubules from disassembly by colchicine and low temperatures.

Distribution and Functions of Microtubules

As noted earlier in the chapter, microtubules are an integral part of the cytoskeleton, where their interaction with cytoplasmic filaments and the microtrabecular lattice gives shape and form to the cell. Microtubules are also involved in endocytosis and exocytosis, in the movements of cilia and flagella, in the movements of the chromosomes during mitosis and meiosis, and in amoeboid movement.

Centrioles

Centrioles and **basal bodies** belong to a group of cell structures referred to as **microtubule-organizing centers.** These centers are involved in the elaboration of microtubules. While the basal bodies are located at the bases of cilia and flagella, centrioles are usually found near the cell nucleus and occur in pairs; structurally, both organelles are identical. The typical centriole is comprised of nine sets of *triplets,* each triplet consisting of one complete microtubule and two C-shaped ones. The triplets are arranged parallel to one another and create a cylindrical body having a diameter of 150 to 250 nm. Although rigorous proof is still lacking, it is generally believed that centrioles are involved in the production of the microtubules that form the spindle of a dividing cell. However, not all cells that form a spindle during nuclear division have centrioles; for example, cone-bearing and flowering plants do not. In cells that do have paired centrioles, the centrioles separate at the onset of nuclear division and move to diametrically opposite positions around the nucleus. Subsequently, as the chromatin condenses to form chromosomes and the nuclear envelope disappears (Chapter 20), fibers of the spindle make their appearance, extending from an area adjacent to one centriole through the cell to the other centriole (Fig. 23–15). As division proceeds, a new centriole appears near each original one; growth of the new centriole is always *perpendicular* to the long axis of the original centriole. By the time division is complete, each daughter cell has two complete centrioles.

Centrioles also play a role in the formation of microtubules of flagella and cilia. Here the centrioles are more often referred to as *basal bodies* or *kinetosomes* (also *blepharoplasts, basal granules,* or *basal corpuscles*).

Structure of the Centriole. While not all cells contain centrioles, in those that do the structure of the centriole is the same. Most algal cells (but not red algae), moss, some

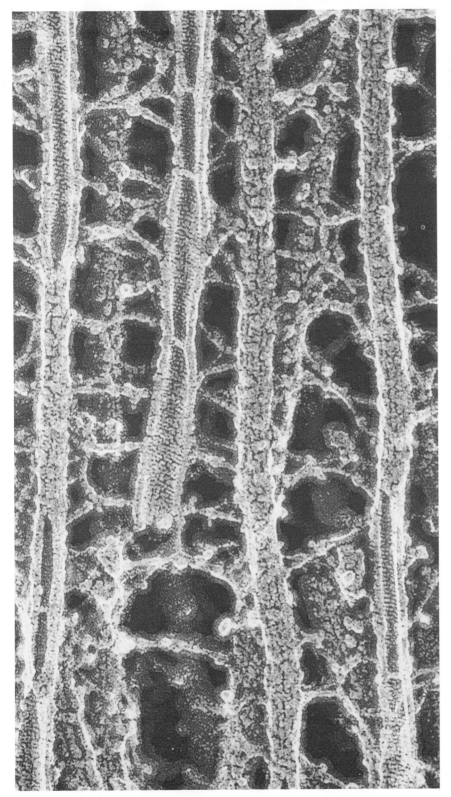

Figure 23–14
Replica of fractured and etched microtubules. Portions of three of the microtubules seen in this view were fractured open, revealing their inner walls. The reticulum surrounding the microtubules is believed to consist of unpolymerized tubulin and microtubule-associated proteins (MAPs). (Courtesy of Drs. J. Heuser, and M. W. Kirschner. Copyright © 1980 by The Rockefeller University Press, *J. Cell Biol. 86,* 215.)

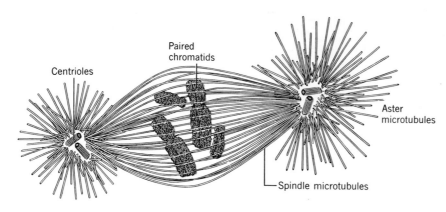

Centrioles

Paired chromatids

Aster microtubules

Spindle microtubules

Figure 23–15
Diagram of centriole, aster microtubules, spindle microtubules, and chromosomes during nuclear division.

fern cells, and most animal cells have centrioles, but cone-bearing and flowering plants, red algae, and some nonflagellated or nonciliated protozoans (like amoebae) do not. Some species of amoebae have a flagellated stage as well as an amoeboid stage; a centriole develops during the flagellated stage but disappears during the amoeboid stage.

The most notable structural characteristic of a centriole is its nine sets of microtubules called **triplets.** Each triplet contains three microtubules, which in cross section appear to be arranged like the vanes on a pinwheel (Fig. 23–16). Although there is no surrounding membrane, the nine triplets appear to be embedded in an electron-dense material.

The nine triplets are identical. The innermost (or *a*) microtubule of each triplet is a complete, round tubule, but the middle (*b*) and outer (*c*) microtubules are C-shaped and share the wall of the neighboring tubule. Also, the outermost (*c*) tubules may not run the full length of the centriole as do the *a* and *b* tubules. The triplets, although generally parallel to each other, may be closer together at the proximal end of the centriole (that end when observed "end-on" which has the triplets tilted inward in a clockwise direction, as shown in Fig. 23–16). The triplets may also spiral somewhat about the axis of the centriole or individually twist. Less twisting of the microtubules is observed in the basal body form of the centriole. Strands of material extend inward from each *a* tubule and join together at the central *hub*. These strands, when seen in cross section, give the centriole the appearance of a cartwheel (Fig. 23–16*b*).

The idea prevalent several years ago that centrioles divide or that new centrioles are formed from preexisting ones (i.e., one centriole acts as the template for the assembly of another) is no longer accepted. The development of cen-

trioles can be inferred from electron photomicrographs, and it is clear that centrioles arise de novo, without association with preexisting centrioles. However, **procentriole granules** have been described. New centrioles develop *near* and at right angles to preexisting ones, the two centrioles separated by a 50 to 100 nm space. Development of a new centriole begins near the proximal end of the original centriole.

In ciliated cells, development of new basal bodies (i.e., centrioles) is detected first by the presence of a small, amorphous, electron-dense body. About this body develop a number of microtubule-like *procentrioles*. In cases where a single centriole develops, the procentriole forms within the amorphous body. During development, the electron-dense body disappears, as if it were being consumed in the production of the developing procentriole.

Basal body development has been studied in the ciliates *Paramecium* and *Tetrahymena* and in tracheal epithelia of *Xenopus* and chicks. The stages of development are virtually the same in all of these. Development of the basal body begins with the formation of a single microtubule in the amorphous mass. Microtubules are added one at a time until there is an equally spaced ring of nine (Fig. 23–17). There is some evidence that a "connector" exists between the microtubules, which could act to set the distance between them. Each of the nine microtubules in the ring are *a* tubules. The *b* tubules develop next, and finally the *c* tubules. Before the tubules reach the doublet stage, the cylinder is rarely longer than 70 nm, but after this stage, the tubules elongate. At the same time, the hub and cartwheel are added in the center (Fig. 23–17*b*). The *a*–*c* links are not formed until the end of development.

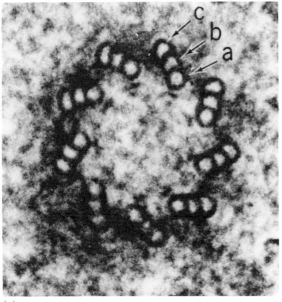

(a)

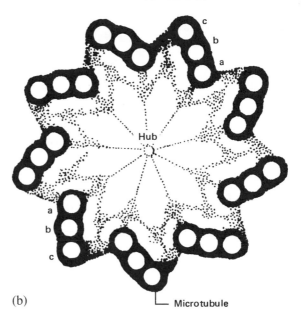

(b)

Microtubule

Figure 23–16

(a) Cross section of a centriole as revealed by electron microscopy. The spokes of the "pinwheel" turn inward in a clockwise manner, indicating that this view is from the proximal end of the centriole. (Courtesy of D. Fawcett.

Copyright © 1966 by W. B. Saunders and Co., *Atlas of Fine Structure*, p. 185.) *(b)* Diagram of a centriole showing the nine triplets that form the "pinwheel" and the electron-dense granular connectives.

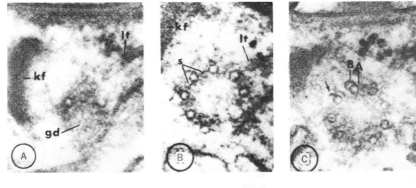

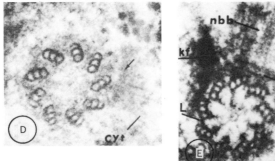

Figure 23–17

Development of a basal body in *Paramecium. (a)* The first microtubules form; *(b)* a developing ring of seven microtubules; *(c)* ring containing nine *a* microtubules, with some *b* tubules forming; *(d)* complete set of nine triplets; *(e)* mature basal body with new basal body *(nbb)* seen in longitudinal section above. (Courtesy of R. Dippell. *Proc. Nat. Acad. Sci. 61*, 461, 1968.)

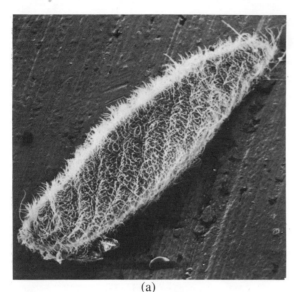

(a)

Figure 23–18

(a) Scanning electron photomicrograph of cilia on the dorsal sur-face of *Paramecium.* The cilia beat rhythmically, creating a wave-like appearance (a *metachronal* wave). (Courtesy of E. Vivier. Copyright © 1974 by Elsevier Publishing Co., *Paramecium,* p. 14.) *(b)* Cilia at the lumenal surface of epithelial cells lining the ovi-duct. (Courtesy of J. Rhodin, *Histology—A Text and Atlas* (Oxford University Press. Copyright © 1974, New York.) In Paramecium, the beating cilia propel the cell, whereas the cilia of the oviduct epi-thelium serve to push the egg released from the ovary toward the uterus.

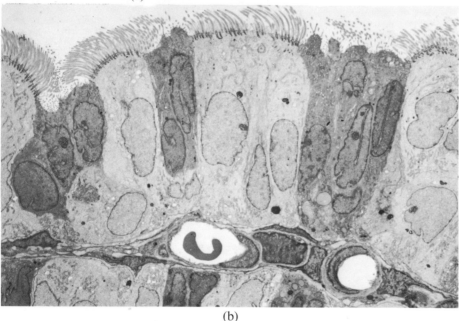

(b)

Basal bodies act as organizing centers for the develop-ment of the microtubules of cilia and flagella. New basal bodies form adjacent to centrioles. While still not associated with a flagellum or cilium, the basal body is more properly called a centriole, but after it migrates to a position just underneath the plasma membrane and acts as a center for flagellum or cilium development, it is called a basal body.

The synthetic functions of centrioles and basal bodies are not clear. It has been suggested that these bodies may contain DNA and carry out transcription. However, the few reports of the occurrence of DNA in centrioles have been severely criticized. Despite a number of studies showing that specific structural changes occur in the centrioles of cells treated with RNAase, there is still considerable skepticism about the presence of RNA in centrioles.

Cilia and Flagella

Cilia and **flagella** are organelles that project from the surface of certain cells and beat back and forth or create a corkscrew action (Fig. 23–18). In many instances, ciliary or flagellar movements propel cells through their environment. In other cases, the cell remains stationary and the surrounding medium is moved past the cell by the beating of its cilia (as in the layer of epithelial cells that lines the trachea or the collar cells lining the internal chambers of sponges).

Cilia are generally shorter than flagella (i.e., 5–10 μm vs. 150 μm or longer), and are present in far larger numbers per cell. Flagella usually occur alone or in small groups; occasionally they are present in large numbers, as in a few protozoa and the sperm cells of more advanced plants. The distinction between cilia and flagella is somewhat arbitrary, because other than differences in their lengths, the structure and action of cilia and flagella of eucaryotic cells are identical. (Bacterial flagella differ in structure and action; see below.)

A eucaryotic cilium or flagellum is composed of three major parts: a central *axoneme* or shaft, the surrounding plasma membrane, and some cytoplasm (Fig. 23–19). The axonemal elements of nearly all cilia and flagella (as well as the tails of sperm cells) contain the same ''9 + 2'' arrangement of microtubules. In the center of the axoneme are two singlet microtubules that run the length of the cilium (Fig. 23–20). Projections from the central microtubules occurring periodically along their length form what appears to be an enclosing **sheath.** Each of the central microtubules is composed of 13 *protofilaments*.

Nine doublet microtubules surround the central sheath. One microtubule of each doublet (i.e., the **A subfiber**) is composed of 13 protofilaments. The adjoining **B subfiber** is ''incomplete,'' consisting of 11 protofilaments (Fig. 23–20). **Radial spokes** occurring at regular intervals along the axoneme extend from each *A* subfiber inward to the central sheath. Adjacent doublets are joined by **interdoublet** or **nexin links;** these links occur irregularly along the length of the axoneme. Each *A* subfiber has sets of two ''arms'' composed of enzymes called *dynein*. The *outer* dynein arm

points away from the center of the axoneme, while the *inner* is directed somewhat into the axoneme.

Each beat of a cilium or flagellum involves the same pattern of microtubule movement. The beat may be divided into two phases, the *power* or *effective* stroke and the *recovery* stroke (Fig. 23–21). The power stroke occurs in a single plane, but recovery may not occur in the same plane as the power stroke.

The *sliding microtubule* model of ciliary movement is accepted by most investigators. In this model, the doublet microtubules retain a constant length and slide past one another in such a manner as to produce localized bending of the cilium. This activity is powered by ATP hydrolysis. The localized bending takes the form of a wave that begins at the base of the organelle and proceeds toward the tip. The localized bending is produced through the cyclic formation and breakage of links between the dynein arms of one doublet and the neighboring doublet that accompanies the hydrolysis of ATP. The protein filaments that make up each doublet are rows of *tubulin* molecules that apparently contain the sites to which the dynein will bind. The fact that the sliding of microtubules past one another results in bending of the cilium may be explained by the behavior of the radial spokes that connect the outer nine doublets to the central sheath. In straight regions of the axoneme, the radial spokes are aligned perpendicular to the doublets from which they arise, whereas in the bent regions they are tilted and stretched (Fig. 23–22). Firm attachment of the radial spokes at both ends provides the *resistance* necessary to translate the sliding of the doublets into a bending action. Indeed, if the radial spokes and nexin links of sperm tails are destroyed by exposure to trypsin, addition of ATP results in the axonemes becoming *longer* and *thinner,* since microtubule sliding is no longer resisted. In effect, sliding is *uncoupled* from bending by elimination of the connections between the doublets and the central sheath.

The biochemical reactions taking place in conjunction with ciliary movement are generally considered to be similar to the reactions occurring during muscle contraction (Chapter 24). Analogies are evident between the dynein-tubulin system and the actin-myosin system. However, whereas Ca^{++} appears to activate the actin-myosin system, these ions have the opposite effect on the dynein-tubulin system. The regulation of the Ca^{++} level in a cilium or flagellum probably involves the plasma membrane surrounding the axoneme. Under normal circumstances (i.e.,

(a)

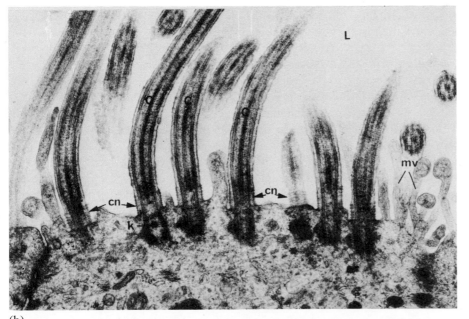

(b)

Figure 23–19

Cilia. The cilia have been fractured *(a)* and sectioned *(b)* in a plane that
runs parallel to the long axis of the organelles. *C*, cilium; *mv*, microvilli; *L*,
lumen, *k*, kinetochore or basal body; *cn*, ciliary necklace. (Courtesy of Dr. E.
Boisvieux-Ulrich. Copyright © 1977 by *Biol. Cellulaire 30*, 245.)

Figure 23–20

Diagram of a cilium or flagellum seen in cross-section (see text for explanation). (Courtesy of F. D. Warner and P. Satir. Modified from *J. Cell Biol. 63*, 40. Copyright © 1974 by The Rockefeller Press.)

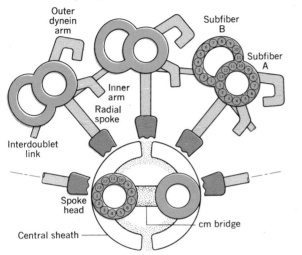

Figure 23–21

Two typical patterns of flagellar and ciliary motion. (a) Successive waves move toward the tip of the flagellum propelling the cell (e.g., sperm) in the opposite direction. (b) The power stroke of a cilium is similar to the action of an oar in a rowboat; the recovery stroke (color) may not take place in the same plane as the power stroke. The large arrows show the direction of movement of the surrounding liquid.

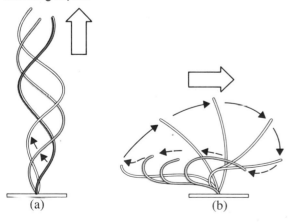

(a)   (b)

during periods of continuous beating), the internal Ca^{++} level is low (about 0.1 μM) while Mg^{++} (necessary to stimulate the ATPase of the membrane) remains in the millimole range. When the membrane is depolarized, the Ca^{++} level inside the cilium increases and beating ceases. ATP is clearly the source of energy for movement and is produced by cellular respiration. In many cells, mitochondria are located adjacent to the basal body of the cilium or flagellum, and ATP diffuses toward the tip of the organelle. In sperm, a large mitochondrion is an integral part of the tail (Fig. 23–23) and is wrapped in a spiral about the middle piece of the axoneme.

Bacterial Flagella. Flagella of bacteria are different from cilia and flagella of eucaryotic cells. The bacterial flagellum consists of a **spiral filament** about 13.5 nm in diameter and 10 to 15 μm long composed of the protein *flagellin*. It is not covered by the plasma membrane. The filament is attached to a "hook," which in turn is connected to a **rod** penetrating the bacterium wall and membrane (Fig. 23–24). A number of rings connect the rod with the membrane and wall layers.

Bacterial flagella work by rotation of the rod and hook, which causes the filament to spin. When the filament spins in a clockwise direction, the cell is propelled smoothly, but when the spin is counterclockwise, an irregular tumbling motion of the cell is observed. The source of energy for the rotation (believed to be generated at the cell membrane) is an electrochemical gradient established by an electron transport system that acts across the plasma membrane. ATP is not the source of energy.

The Mitotic Spindle

Most studies of chromosome movement during mitosis have focused on the role played by the microtubules that make up the mitotic spindle fibers. The spindle fibers cause three distinct chromosome movements during mitosis (Chapter 20): (1) orientation of sister chromatids, (2) alignment of the centromeres on the metaphase plate, and (3) separation of centromeres and movement of sister chromatids (segregation) to opposite poles of the spindle. The microtubules that occur in the spindle include (1) the **centromere microtubules,** which terminate in a centromere; (2) the **polar microtubules,** which terminate at the poles; and (3) the **free microtubules,** which do not terminate in either a pole

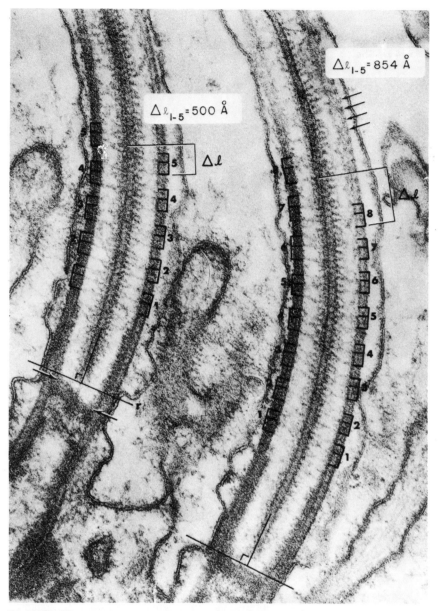

Figure 23–22

Median longitudinal section through the basal end of beating cilia. Successive radial spokes occur in groups of three (numbered 1, 2, 3, 4, etc.). The spokes in each group are directly opposite one another in a straight cilium and are perpendicular to the doublet microtubules. In a bending cilium, the radial spokes are tilted and stretched. In the cilium to the right, displacement of the spokes results in group 4 being opposite group 5. $\Delta\ell$ is a measure of the amount of displacement taking place as diametrically opposite doublets slide tipward on the inner concave side and toward the base on the outer convex side. (Courtesy of R. Warner and P. Satir, *J. Cell Biol. 63*, 25. Copyright © 1974 by The Rockefeller Press.)

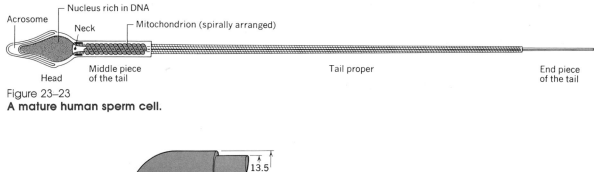

Figure 23–23
A mature human sperm cell.

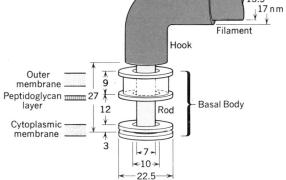

Figure 23–24
Diagram of the basal region of a bacterial flagellum. (Courtesy of J. Adler. Copyright © 1971 by American Society of Microbiology, *J. Bact. 105*, 395.)

or a centromere. All three types can be dissociated into tubulin subunits by colchicine or cold temperatures.

Over the years, several models have been proposed to account for the movements of the chromosomes during anaphase. For example, it has been suggested that the chromosomes are *pushed* apart by spindle fibers developing between centromeres, that they are *pulled* apart by spindle fibers extending between the centromeres and the poles of the spindle, and that chromosomes *migrate* along spindle fibers. The various models are discussed in some detail in Chapter 20. Although individual spindle fibers do not stretch or contract per se, they do change in length through either addition or removal of subunits. S. Inoue has shown that free microtubules alternately grow and decrease in length. His in vitro studies indicate that the microtubules assemble at one end and disassemble at the other. The centromere microtubules are required for movement, for if they are dissociated by colchicine or cold temperatures, chromosome movement stops. During movement, the centromere microtubule becomes shorter.

Other Cell Movements

The references at the end of this chapter may be consulted for additional examples of the diverse forms of cell and cytoplasmic movement. Regardless of whether the movements are "internal" (such as cyclosis in plant cells) or result in a major change in shape or position of the cell, the present evidence indicates that microfilaments and/or microtubules are fundamental to these activities. In most cases, an interaction between proteins—such as in the actin-myosin or dynein-tubulin systems—with the simultaneous involvement of an ATPase is the underlying biochemical phenomenon.

Summary

The cytoplasmic *ground substance* is composed of two phases, the **microtrabeculae** and the **intertrabecular space.** The **cytoskeleton** is made up of **microtubules** and **cytoplasmic filaments. Microtrabeculae, cytoplasmic filaments,** and **microtubules** function in concert to bring about movement or contribute to the cell's structural framework. Cytoplasmic filaments have been shown to play a role in *phagocytosis, pinocytosis, exocytosis, cytokinesis,* and *cytoplasmic streaming.*

Microtubules are cylindrical structures whose walls are composed of heterodimers of α and β *tubulin.* These globular proteins appear to be arranged in 13 chains or protofilaments that run parallel to each other and to the hollow axis of the tube. Microtubules are involved in the structure and function of **centromeres, spindle fibers, flagella,** and **cilia.**

References and Suggested Reading

Articles and Reviews

Allen, R. D., Cell motility. *J. Cell Biol. 91,* 148s (1981).

Dentler, W. L., Microtubule-membrane interactions in cilia and flagella. *Inter. Rev. Cytol. 72,* 1 (1981).

Dustin, P., Microtubules. *Sci. Am. 243*(6), 66 (Aug. 1980).

Gibbons, I. R., Cilia and flagella of eukaryotes. *J. Cell Biol. 91,* 107s (1981).

Goldman, R., Berg, G., Bushnell, A., Change, C-M., Dickerman, L., Hopkins, N., Miller, M., Pollack, R., and Wang, E., Fibrillar systems in cell motility, in *Locomotion of Tissue Cells, Ciba Foundation Symposium,* Vol. 14, Associated Scientific Publishers, New York, 1973, p. 83.

Haimo, L. T., and Rosenbaum, J. L., Cilia, flagella, and microtubules. *J. Cell Biol. 91,* 125s (1981).

Inoue, S., Fussler, E., Salmon, E., and Ellis, G., Functional organization of mitotic microtubules. *Biophys. J. 15,* 725 (1975).

Margolis, R. L., and Wilson, L., Microtubule treadmills—possible molecular machinery. *Nature 293,* 705 (1981).

Nagai, R., and Rebhun, L., Cytoplasmic microfilaments in streaming *Nitella* cells. *J. Ultrastruc. Res. 14,* 571 (1966).

Porter, K. R. and Tucker, J. B., The ground substance of the living cell. *Sci. Am. 244*(3), 56 (Mar. 1981).

Roberts, K., Cytoplasmic microtubules and their functions. *Prog. Biophys. Mol. Biol. 28,* 273 (1974).

Satir, P., How cilia move. *Sci. Am. 231*(4), 44 (April 1974).

Books, Monographs, and Symposia

Borgers, M., and DeBrabander, M., *Microtubules and Microtubule Inhibitors,* North-Holland, New York, 1975.

Goldman, R., Pollard, T., and Rosenbaum, J., *Cell Motility.* Book A: *Motility, Muscle and Non-muscle Cells.* Book B: *Actin, Myosin and Associated Proteins.* Book C: *Microtubules and Related Proteins,* Cold Spring Harbor Laboratory, Cold Spring Harbor, N. Y., 1976.

Reinert, J., and Ursprung, H., *Origin and Continuity of Cell Organelles.* Springer-Verlag, Berlin, 1971.

Schweiger, H. G. (Ed.) *International Cell Biology 1980–81,* Springer-Verlag, Berlin, 1981.

Part 6
SPECIAL CELL FUNCTIONS

Chapter 24
CELLULAR DIFFERENTIATION AND SPECIALIZATION

Differentiation is but one aspect of the more general field of developmental biology. Developmental biologists are concerned with the changes that organisms and their cells and molecules undergo in making the transition from cells unspecialized in structure and/or function to forms having a permanent and specific structure and function. Four component processes characterize these changes: **determination, differentiation, growth** and **morphogenesis.**

New cells are usually not "committed" to a specific function. For example, following fertilization of an egg cell or **ovum** by a sperm cell, the resulting **zygote** divides mitotically many times to produce a ball of cells called a **blastula.** Each cell or group of cells in the blastula subsequently gives rise to specific tissues and organs in the developing embryo. Usually, the entire blastula gives rise to a single embryo. However, if one cell separates from the others at an early blastula stage, that cell may also develop into a complete embryo. In humans, it is this phenomenon that results in identical twins, identical triplets, and so on. At later stages of embryo development, cells that normally become epidermal tissue (i.e., **ectoderm**) can be surgically transplanted to another part of the embryo and there develop into **mesodermal** or **endodermal** tissue. At some early stages of organismal development, all cells of a given species have the potential to develop into any of the variety of different tissue and cell types of that species. This potential, called **totipotency,** is based on the genetic comple-

ment of the cells and ultimately on the specific sequence of DNA nucleotides in each gene. The generalization can also be made that at some point in development, cells become *committed* to a specific course of differentiation. The process that establishes the fate of a cell is called **determination.** During determination, some alternative modes of gene expression become permanently "turned off" while others are sequentially expressed, further and further narrowing the course of differentiation of the cell.

During **differentiation** cells take on new and specific properties. These can be *structural* (such as the formation of actin and myosin filaments in muscle cells) or *biochemical* (as in the appearance of enzymes of a new metabolic pathway). Differentiation may also take the form of *loss* of preexisting structures or biochemical processes. For example, in the differentiation of mammalian red blood cells, the nucleus and other cellular organelles are lost, together with the biochemical processes that these structures provided.

The determination and differentiation of the cells of a developing organism are accompanied by *growth*—that is, an increase in the size and number of cells comprising the organism. In humans, for example, embryonic and fetal development proceeds from the hundred or so cells comprising the small blastula to a fetus weighing several pounds and containing hundreds of millions of cells. Growth, in turn, is accompanied by **morphogenesis**—the generation of form and shape in the developing organism. In this process, differentiating and growing cells give rise to the characteristic organizational pattern of the organism. Small masses of cells take on the form and shape of specific and identifiable structures, such as bones, appendages, the brain, and other organs.

Above all, development and differentiation rest with the DNA of the cell nucleus. Before a cell can develop into a hair cell of a mammal, a feather cell of a bird, or a scale cell of a reptile, there must be a genome whose transcription and translation into enzymes and other proteins allow the cell to differentiate in a specific direction. Moreover, given the appropriate genetic complement, conditions must allow these genes to be expressed. Gene expression is regulated at three levels. The first level involves molecular or metabolic interactions, such as mass action, feedback control, and allosteric enzyme function. These regulatory mechanisms were discussed in Chapter 11 and will not be pursued here. The second level of control is effected through the interaction between the cell nucleus, the cytoplasm, and the cytoplasmic organelles. The third level of control involves the interactions between the cell as a whole and its environment.

Intrinsic (Nucleocytoplasmic) Interactions

Transplantation of Amphibian Cell Nuclei

In 1892 August Weissman proposed that cellular differentiation and specialization might be the consequence of the progressive loss of certain genes from a cell's genome during successive mitotic divisions. This hypothesis retained widespread acceptance for more than 50 years until a series of experiments carried out with amphibian egg cells indicated otherwise. The egg cells of amphibians (e.g., frogs and toads) are large enough to be manipulated by microsurgery. In the 1950s R. W. Briggs and T. H. King enucleated unfertilized egg cells of *Rana pipiens* and replaced the haploid nucleus with a diploid nucleus taken from a cell at the blastula stage of development that would have developed into specific embryonic tissues. The cells produced by the nuclear transfer developed into normal

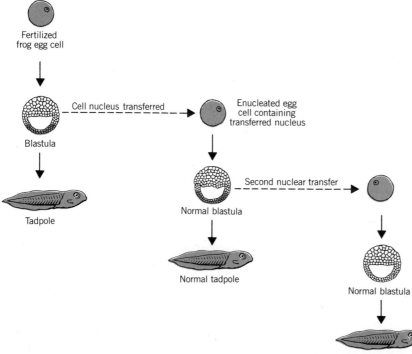

Figure 24–1

Effects of serial transplantation of nuclei from blastula stage cells to enucleated egg cells. The transplanted nuclei retain the capacity to express all genes, and as a result, normal embryos and tadpoles develop from these eggs.

frog embryos and normal tadpoles. This demonstrates that determination had not yet taken place in the nucleus from the blastula stage cell. Moreover, the cell still contained all of the genes necessary to produce the different types of cells of the complete embryo. The totipotency of the blas-

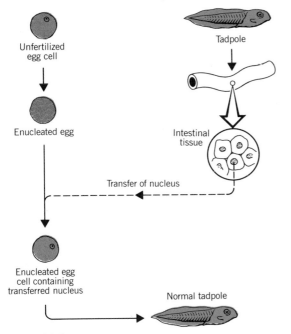

Figure 24–2
Effects of transplantation of nuclei from intestinal epithelial cells to enucleated egg cells. Normal embryos and tadpoles develop, indicating that even the differentiated intestinal cell's nucleus contains a full set of genes. Moreover, the capacity to express these genes was not permanently lost during differentiation into intestinal tissue.

tula cell nucleus can be perpetuated even over a series of experimental transfers (Fig. 24–1); that is, if a nucleus is withdrawn from a blastula stage cell following the first transfer and transferred to a second anucleate egg, the embryonic development of that egg proceeds normally; if a nucleus is removed from a cell at this blastula stage and transferred to a third anucleate egg, normal embryonic development again occurs, and so on.

While the observations of Briggs and King demonstrated that nuclear totipotency exists in cells at the blastula stage of embryonic growth, a series of experiments carried out a decade later by J. B. Gurdon showed that totipotency is retained through much later stages of development. Gurdon removed nuclei from intestinal cells of *Xenopus laevis* tadpoles and implanted them in enucleated unfertilized egg cells. Some of the resulting cells developed into normal tadpoles and normal, fertile adults (Fig. 24–2). Thus, even the nuclei of differentiated intestinal cells contain the full complement of genes essential to the production of a complete organism. The experiments with amphibian cells indicate that there is neither a loss of genes during the many mitoses that lead from the undifferentiated fertilized egg cell to the highly differentiated cells of later developmental stages *nor an irreversible loss of the capacity to express these genes*.

Experiments with Acetabularia

Classic experiments with the unicellular alga *Acetabularia* demonstrated that the cytoplasm has a direct effect on gene expression. During the life cycle of *Acetabularia* (Fig. 24–3), newly formed zygotes attach to the sea bottom by

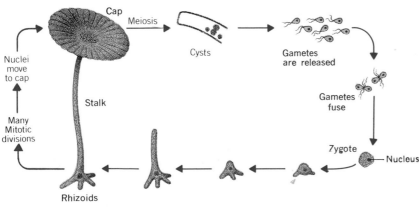

Figure 24–3
The life cycle of *Acetabularia*.

developing rootlike structures called **rhizoids.** A slender **stalk** several centimeters long grows upward from the rhizoid and develops an umbrellalike **cap.** During growth of the cell, which usually takes several weeks, the nucleus remains in the rhizoid. When growth is complete, the nucleus divides mitotically many times, and the resulting nuclei migrate up the stalk into the cap, where meiosis produces haploid nuclei that become encysted. Eventually, the cysts are released and break open, and the gametes swim out. Fusion of two gametes to form a new zygote initiates a new cycle.

Acetabularia is capable of regenerating lost parts. For example, if the cap is cut off, the cell grows a new cap. If the new cap is removed, another cap is grown, and this cycle can be repeated many times with the same individual. If the cap is cut off and the cell nucleus then removed by microsurgery a new cap is regenerated, but regeneration will not occur if the experiment is repeated with the same anucleate organism. Regeneration of the second cap in the absence of a nucleus indicates that nuclear information had previously been transferred to the cytoplasm and persisted there, at least through the time required to form the new cap. The information in the cytoplasm for regenerating the cap takes the form of messenger RNA.

Different shape caps are formed by different species of *Acetabularia*. *A. mediterranea* has a complete cap, and *A. crenulata* has a fingerlike cap. The interactions of the cytoplasm and nucleus are strikingly illustrated by grafting experiments using these two species (Fig. 24–4). If the foot of *A. crenulata* (which contains the nucleus) is joined to an anucleate piece of the stalk of *A. mediterranea*, the two fuse and a cap develops at the top of the stalk. The cap assumes characteristics of both species, presumably resulting from the combined influences of the cytoplasmic mRNA that was present in the *A. mediterranea* stalk and the nucleus and cytoplasmic mRNA that were in the *A. crenulata* foot. If the cap produced by the grafted cell parts is removed and a second cap allowed to form, the new cap has the *A.*

Figure 24–4
Results of grafting experiments using nucleated and anucleate segments of
***Acetabularia mediterranea* (top) and *Acetabularia crenulata* (bottom).**
(See the text for description of the experiments.)

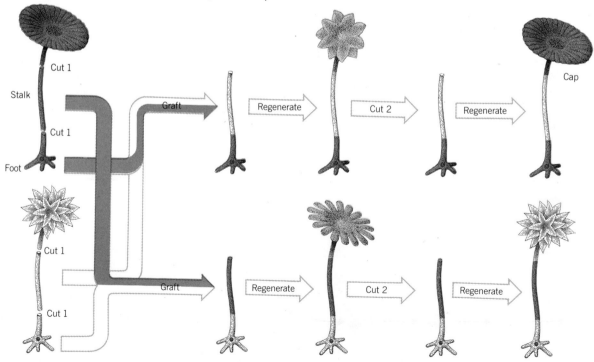

crenulata shape. Thus, the cap information (mRNA) in the original cytoplasm of the stalk from *A. mediterranea* was ''lost'' after one regeneration, whereas the cap information (mRNA) of the *A. crenulata* nucleus is expressed.

Extrinsic (Environmental) Effects on Differentiation of Cells

Neighboring cells and the surrounding environment have a direct effect upon a cell's differentiation. The fates of the late blastula–early gastrula stage cells of frog embryos (as well as embryos of a number of other species) have been determined and fate maps have been made of the types of tissues formed from each group of cells (Fig. 24–5). Even by the late blastula stage, cell determination has not yet taken place. For example, if the cells that normally form an eye lens are surgically exchanged with cells that form the gut, embryonic development is still normal. The transplanted presumptive lens cells are influenced by their new position in the embryo and develop into gut tissue. Likewise, the transplanted presumptive gut cells develop into lens tissue. Apparently influenced in some manner by their new surrounding, the transplanted cells follow an altered course of development.

The direct effects of environmental factors on simple plants like *Fucus,* a brown marine alga, can be vividly demonstrated. The fertilized *Fucus* egg cell has no apparent topological polarity. Fifteen hours after fertilization a *rhizoid* develops and this is followed by nuclear division and formation of a wall that separates the rhizoid from the *apical* cell. This first division establishes a polarity for the embryo (Fig. 24–6). Environmental factors can be shown to influence *where* on the fertilized egg the rhizoid will

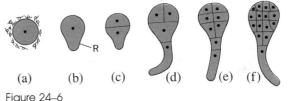

Figure 24–6
Early stages of development of the *Fucus* embryo.

develop. In a cluster of fertilized *Fucus* eggs, the rhizoid develops on that side of each egg that is closest to the center of the group. When cells are placed in a temperature gradient, the rhizoids develop on the warm side of each egg. In a pH gradient, they develop on the more acid side, and in white light, they form on the shaded side.

As we have seen, the expression of a cell's genes is influenced by its cytoplasm, the adjacent cells, and the environment. The cellular differentiation that results from these influences produces a number of uniquely specialized cells. Specialized cell structure and organization is an overt reflection of specialized cell functions. Specialized cellular functions are in turn founded on the differentiation of a specialized structure. Among specialized cell types, muscle cells, nerve cells, and blood cells have been more extensively studied than any other form. Some of their specializations are considered in the following sections.

Red Blood Cells

The differentiation of mammalian red blood cells in bone marrow (a process called **erythropoiesis**) is one of the most striking examples of specialization occurring in nature. In adults, erythropoiesis begins with a **pluripotent stem cell**

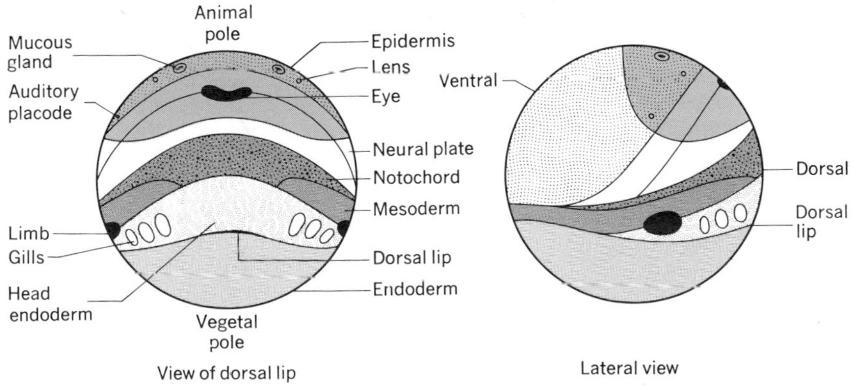

Figure 24–5
Fate map **of the different cell areas of the frog blastula. The cells in each area of the blastula are associated with the development of specific types of tissues and organs.**

Mucous gland
Auditory placode
Limb
Gills
Head endoderm

Animal pole
Epidermis
Lens
Eye
Neural plate
Notochord
Mesoderm
Dorsal lip
Endoderm

Vegetal pole

View of dorsal lip

Ventral
Dorsal
Dorsal lip

Lateral view

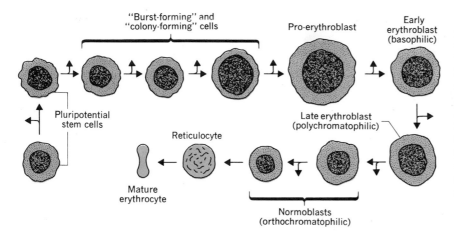

"Burst-forming" and "colony-forming" cells

Pro-erythroblast

Early erythroblast (basophilic)

Pluripotential stem cells

Late erythroblast (polychromatophilic)

Reticulocyte

Mature erythrocyte

Normoblasts (orthochromatophilic)

Figure 24–7

Differentiation and maturation scheme of erythrocytes of mammals. Differentiation begins with pluripotent stem cells in the bone marrow and ends seven to 10 days later with the release of reticulocytes or mature erythrocytes into the bloodstream. Each double-headed arrow represents a mitotic division into two daughter cells, only one of which is shown. Normoblasts become reticulocytes and then erythrocytes *without* division (the cell nucleus and other organelles are lost at the normoblast stage).

that gives rise in seven to 10 days to mature, hemoglobin-filled erythrocytes. As noted earlier in the chapter, differentiation and specialization of cells may be accompanied not only by the acquisition and development of specialized structures but also by the *loss* of internal structures or physiological properties. The latter is the case during erythrocyte differentiation, for mature red blood cells lack nuclei, mitochondria, endoplasmic reticulum, Golgi bodies, ribosomes, and most other typical cell organelles.

The mature erythrocyte is a simple cell, delimited at its periphery by a plasma membrane and containing internally a highly concentrated, para-crystalline array of hemoglobin molecules for oxygen transport. The apparent simplicity of the maturing erythrocyte is the principal reason for its selection over most other kinds of cells as the preferred object for studying plasma membrane structure and protein (i.e., hemoglobin) structure and biosynthesis. These subjects, together with the contributions made by intensive study of the erythrocyte, are treated in some depth in Chapters 4, 15, and 22 and will not be dealt with here.

Erythropoiesis

The development and differentiation of the mammalian red blood cell is depicted in Figure 24–7. Development takes place in the extrasinusoidal stroma of the bone marrow and begins with pluripotent stem cells capable of proliferating leucocytes (white blood cells), as well as erythrocytes. When primitive stem cells undergo division, one of the daughter cells remains undifferentiated and pluripotent, so that depletion of marrow stem cells does not normally take

place. The erythropoietic activity of the bone marrow is under hormonal influence, increasing or decreasing according to the level of circulating **erythropoietin** (produced in the cortical region of the kidneys and secreted into the bloodstream). The erythrocyte progenitors show an increasing sensitivity to erythropoietin through the proerythroblast stage and a parallel, ever-decreasing proliferative potential (Fig. 24–8). As a result, by the proerythroblast stage, the cells are irreversibly committed to the maturation sequence leading to erythrocytes.

Figure 24–8

Relationship between the time course of differentiation and maturation of erythrocytes in the bone marrow and the proliferative potential of progenitor cells, sensitivity to erythropoietin influence, accumulation of mRNA for globin chains, and the buildup of cellular hemoglobin.

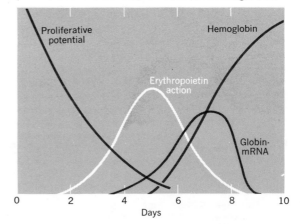

Proliferative potential

Hemoglobin

Erythropoietin action

Globin-mRNA

Days

The mRNAs for the various globin chains of hemoglobin appear and increase in quantity in the proerythroblast and erythroblast stages, and this is followed by the synthesis and accumulation of hemoglobin. By the late erythroblast and normoblast stages, the synthesis of hemoglobin accounts for more than 95% of all protein synthesis occurring in the cells. Hemoglobin synthesis is concluded in the reticulocyte stage and is accompanied by the progressive dissolution of internal cell structure (including the loss of the nucleus) and residual nucleic acids. Late reticulocytes leave the bone marrow and enter the circulating blood, where in the ensuing hours they lose their granulation and are transformed into the biconcave disks typical of mature erythrocytes. Complete differentiation and maturation from stem cell to the erythrocyte takes from seven to 10 days.

Because of its highly differentiated state, the mature erythrocyte is incapable of further proliferation. In humans, the average life span is 120 days. Since there are 5 billion red blood cells in each cm^3 of blood, a few simple calculations* quickly reveal that in a typical adult the differentiation and maturation of 3 million erythrocytes is completed *each second!* Obviously, an appreciable proportion of the body's energy and resources is continuously consumed to support erythropoietic activity. This is in stark contrast with other highly differentiated cells such as those of muscle and nerve, whose proliferation ceases shortly after birth.

Genetic and Molecular Basis of Erythrocyte Differentiation

Because of the intensity with which the red blood cell has been studied by biochemists, cell biologists, molecular geneticists, physiologists, and others, the differentiation and specialization of this cell is better understood in molecular terms than any other; accordingly, erythrocyte differentiation may serve as a model for the genetic and molecular events that underlie cell differentiation generally.

*Since about 8% of the body weight is blood and blood contains 5 billion red cells per cm^3, a person weighing 78 kg (172 lb) contains 6.24 kg or about 6.3 liters of blood. Therefore, altogether there would be 3.15×10^{13} circulating erythrocytes. Since all of these are turned over in an interval of 120 days or 1.04×10^7 seconds, this means that 3.03×10^6 red blood cells reach full maturity and are released into the bloodstream each second.

As discussed in Chapter 4, development in humans is accompanied by the sequential appearance of six different hemoglobins comprised of various combinations of seven different globin chains. The structural genes coding for these chains represent only a miniscule fraction of the total genetic complement of the pluripotent progenitor cells. In humans, there are two pairs of alleles (i.e., four genes) for alpha globin chains, one pair for beta chains, two pairs for gamma chains, one pair for delta chains, one pair for epsilon chains, and two pairs for zeta chains. These genes account for less than 0.0002% of the total DNA content of the nucleus, and yet globin chain synthesis accounts for more than 95% of all protein synthesis taking place in the cell by the later stages of maturation. It is clear that during differentiation, the *progeny of the stem cells are committed to the highly selective expression of only a small number of genes.*

The globin chain genes are distributed between two pairs of homologous chromosomes—alpha and zeta chain genes on one chromosome pair and beta, gamma, delta, and epsilon chain genes on the other pair. The zeta, epsilon, and gamma chain genes are expressed only during embryonic and fetal development, whereas the beta and delta chain genes are expressed from mid-fetal stages through adulthood. Alpha chain genes are expressed throughout life beginning at an early embryonic stage. Therefore, superimposed on the highly selective expression of globin chain genes in maturing erythrocytes is the temporal expression of certain members of this family of genes during organismal development. Expression of these genes is also coordinated between the two different chromosomes bearing the members of this gene family. A most interesting finding of recent years is that the order in which the globin chain genes are arranged on their respective chromosomes parallels the order in which the various genes are expressed during embryonic and fetal development. Also present in the chromosome regions containing these genes are nucleotide sequences encoding polypeptide chains that are very similar in primary structure to the globin chains, but that are never expressed. These ''silent genes'' or **pseudogenes** may be genetic vestiges of the evolutionary process. Certainly, the inclination to draw a parallel between phylogeny and ontogeny at the organismal and cellular levels is unavoidable.

Cellular DNA includes not only structural genes but also sequences vital to DNA organization within the chromosomes and the coordination of gene expression during differentiation. About 60% of all the nucleotide sequences in the DNA of stem cells belongs to the "unique sequence" class in which only a few copies of each sequence are present. The structural genes for the globin chains belong to this class. Highly repetitive DNA accounts for about 10% and consists of short sequences repeated in tandem many thousands of times. These regions are not transcribed and are believed to be located in condensed chromatin (see Chapter 20). The balance of the DNA—about 30%—is moderately repetitive (repeated up to several hundred times) and includes the genes for histones and the rRNAs (see Chapter 22).

Some moderately repetitive DNA sequences are located adjacent to the structural genes for globin and other proteins and may be involved in the coordination of gene activity. For example, by binding to the moderately repetitive sequences, regulatory substances could "turn on" the transcription of *physiologically related* genes. In the case of the red blood cell, such a gene set would be represented by (1) the globin chain genes, (2) the genes for the heme synthetic enzymes, and (3) the genes for the cell surface antigens (which serve as the basis for blood typing). Models accounting for gene regulation at this level were considered in Chapter 11.

As noted in Chapter 20, the genes of eucaryotic cells are associated with histones and other basic proteins to form complexes called chromatin composed of repeating *nucleosome* units. Euchromatin has the more open structure and is transcriptionally active, while the condensed heterochromatin is not transcribed. Experiments with developing erythrocytes indicate that the globin chain genes are included in nucleosomes of condensed chromatin during the early stages of differentiation and that these are converted to the open form just before globin mRNA synthesis begins. The transcription of globin DNA sequences by RNA polymerase that occurs when these genes are "turned on" can be traced to a modification of the associated histones and an interaction with nonhistone proteins. A selective expression of globin mRNA genes occurs in erythroblasts, with the result that globin mRNA is produced in amounts 100 times greater than expected on the basis of the proportion of total template DNA represented by the globin genes. Globin chain synthesis parallels the appearance of globin mRNAs in erythroblasts.

Although globin chain synthesis accounts for the over-whelming majority of all protein synthetic activity in the maturing erythrocyte, a number of other structural genes are expressed (those for the enzymes of glycolysis, enzymes for initiation, elongation, and termination of globin chain assembly, enzymes of the metabolic pathway for heme synthesis, enzymes and structural proteins of the plasma membrane, and so on). Although all protein synthesis comes to a halt during reticulocyte maturation, a red cell retains a limited metabolic capacity after release from the bone marrow. For example, glycolytic activity provides the ATP needed to maintain the sodium and potassium pumps of the erythrocyte membrane (see Chapter 15) and for other energy-requiring processes. What limited metabolism is retained by the mature cell serves to sustain it during its 120-day and 700-mile journey through the circulatory system.

The progressively selective expression of genes in the maturing red blood cell is accompanied by the loss of internal structure so that in the mature state the only remaining organelle is the plasma membrane itself. Internally, the cell consists primarily of a highly concentrated (30% by weight) crystal-like arrangement of hemoglobin molecules suited to the cell's principal function—oxygen transport.

Morphological and Physiological Specialization of Red Blood Cells

Little cell growth occurs during the periods between successive mitotic divisions of erythropoiesis. As a result, the mature erythrocyte is among the smallest cells of the body. In humans, the average red blood cell has a volume of about 100 μm^3 and is disk-shaped (Fig. 24–9). The factors that maintain this rather unusual cell shape have fascinated scientists for more than two centuries. Two points of view dominate this controversial area. One hypothesis is that the biconcave shape results from the chemical nature, arrangement, and *interaction* of the proteins and lipids in the plasma membrane. In spite of positive hydrodynamic pressure inside the cell—a pressure that would be expected to impose spherical shape to a body encapsulated by a flexible membrane—tensional and compressional forces in the membrane are translated into biconcavities on opposing membrane surfaces. Such an explanation has been substantiated using artificial membranes as models.

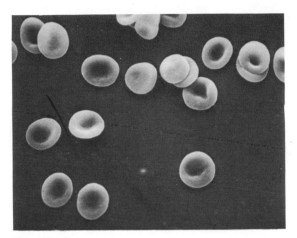

Figure 24–9
Scanning electron photomicrograph of erythrocytes showing their biconcave disk shape. (Courtesy of Dr. M. A. Lichtman.)

A second and broadly supported view is that the internal para-crystalline organization of hemoglobin molecules imposes an overall shape on the cell. Such a view is supported by the finding that erythrocytes of individuals with *abnormal* hemoglobins (hemoglobins containing one or more amino acid substitutions) usually lack the normal biconcavity. The most notorious example is the "sickle" or crescent shape of erythrocytes in individuals with *sickle-cell anemia*. In this genetically determined disease, a single substitution occurs in the beta globin chains of the hemoglobin, and under conditions of oxygen deprivation or shortage, the relative positions of neighboring hemoglobin molecules within the cell change. These changes alter the normal shape of the cell.

Oxygen enters and leaves the erythrocyte by diffusion; the biconcave shape facilitates oxygen flux by increasing the surface area-to-volume ratio of the cell. Oxygen inside the cell forms a reversible combination with hemoglobin (see Chapter 4), binding to hemoglobin as the blood circulates through the capillary networks of the lungs (where the net oxygen flux is directed into the erythrocyte) and being released as the blood circulates through oxygen-deficient tissues. The disk shape of the erythrocyte induces the formation of **rouleaux** or long *stacks* of cells; when arranged in such a regimented manner, far greater numbers of cells can pass through the narrow capillaries than when the cells are freely and independently suspended. Experi-

mental elimination of the biconcave shape using hypotonic solutions eliminates the ability of erythrocytes to form rouleaux.

Lymphocyte Differentiation

A dramatic example of the degree of specialization that can accompany cellular differentiation is provided by a family of white blood cells (or **leucocytes**) called **lymphocytes.** Lymphocytes are one of several different kinds of leucocytes derived from stem cells in the bone marrow, spleen, thymus, and other lymphoid tissues. These cells play a protective role in the body by synthesizing and secreting **antibodies** or **immunoglobulins** into the bloodstream in response to the sudden appearance of bacteria, viruses, or other antigen-containing particles during infection or injury. By combining with these antigens, the antibodies trigger a process that eventually leads to the destruction of the antigen-containing material (see Chapter 4 for a discussion of immunoglobulin structure and action). One class of lymphocytes, called *B-lymphocytes,* is derived from stem cells in the spleen. These stem cells give rise to more than a million different lines of B-lymphocytes, each line developing an independent capacity for producing antibodies against a specific antigenic determinant. The antibodies secreted by lymphocytes are the products of the transcription and translation of two types of genes called *V* and *C* genes. Despite the presence of many hundreds of different *V* and *C* genes in each lymphocyte, only one pair of *V* and *C* genes is expressed by each line of lymphocytes. The specific *V* and *C* genes expressed by an antibody-producing lymphocyte are not only expressed to the exclusion of other *V* and *C* genes but also to the exclusion of most other structural genes in the cell.

Hybridomas and Monoclonal Antibodies.   Because of the highly specific nature of the antibodies produced by differentiated B-lymphocytes, cultures of the many lines of these cells have long been contemplated as a source of antibodies for clinical application. While the highly differentiated B-lymphocytes cannot be cultured, hybrid cells formed by fusing malignant *myeloma* cells with B-lymphocytes do proliferate in culture and synthesize their respective antibodies. These **hybridomas** can be separately cloned to yield large cultures of cells producing specific antibodies (called **monoclonal antibodies,** see also Chapter 21).

Muscle Cells

Vertebrates possess three types of muscle tissue: (1) **smooth** muscle, the contractile elements of most of the digestive system and most visceral organs, (2) **cardiac** muscle, found only in the heart, and (3) **striated** (or **skeletal**) muscle, responsible for most of the gross movements of the body. Each of these tissues is composed of cells called **muscle fibers,** which contain filaments of *actin, myosin,* and other proteins responsible for the contractile nature of muscle. The filaments are arranged differently in each of the tissue types. The arrangement is most highly organized in striated muscle, and it is this type of muscle tissue that has been most extensively studied.

The organization of striated muscle is shown diagrammatically in Figure 24–10. During development, the muscle fiber is formed by the end-to-end fusion of many cells into a continuous tubelike structure. This explains the fact that striated muscle fibers are multinucleate and contain many more mitochondria than most other cells. The plasma membrane of the muscle fiber is called the **sarcolemma;** in addition to its exceptional length, the sarcolemma is characterized by numerous porelike invaginations that extend into the sarcoplasm (cytoplasm) at right angles to the long axis of the cell. These transverse or **T-system** extensions of the sarcolemma (Fig. 24–11) make contact with most of the internal **myofibrils**—the contractile units of the cells. Each myofibril contains a large number of *thick* (150 Å diameter) and *thin* (60 Å diameter) filaments called **myofilaments** that run parallel to each other and to the long axis of the cell. Seen in cross-section (Fig. 24–12), the filaments are arranged in a repeating geometric pattern. The thick filaments are equidistant from each other with each surrounded by six thin filaments in an hexagonal array (Fig. 24–12*b*). The thick and thin filaments slide past one another during contraction. When examined microscopically in longitudinal section, each myofibril reveals areas of different density (Fig. 24–13). The alternating light and dark areas are called the *A* (i.e., **anisotropic**) and *I* (i.e., **isotropic**) bands. At the center of each *I*-band there is a dark line called a *Z-line* and those portions of a myofibril that extend from one *Z*-line to the next are termed **sarcomeres.** One end of each actin filament is anchored in the *Z*-line while the other end projects toward the center of the sarcomere. The thick myosin filaments are sandwiched between the actin filaments but are not attached to the *Z*-lines. Within each sarcomere, the *A* bands extend from one end of a

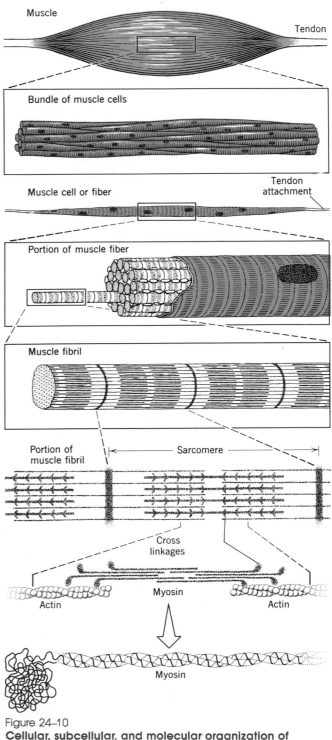

Figure 24–10

Cellular, subcellular, and molecular organization of striated muscle tissue.

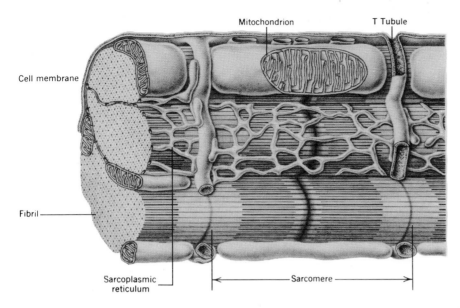

Figure 24–11
The sarcoplasmic reticulum and T-system of striated muscle cells.

Labels: Mitochondrion, T Tubule, Cell membrane, Fibril, Sarcoplasmic reticulum, Sarcomere

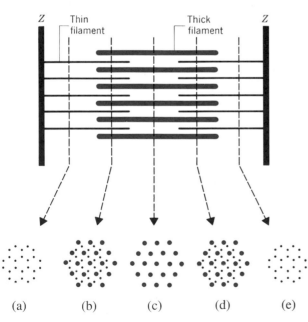

(a)   (b)   (c)   (d)   (e)

Labels: Z, Thin filament, Thick filament, Z

Figure 24–12
Cross-sections through various positions of a sarcomere. The thin (actin) filaments are arranged parallel to the thick (myosin) filaments and form an hexagonal array in regions of overlap (i.e., regions *b* and *d*).

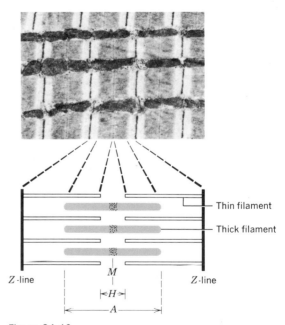

Labels: Thin filament, Thick filament, Z-line, M, H, A, Z-line

Figure 24–13
Relationship between the light and dark zones seen in electron photomicrographs of striated muscle tissue and the arrangements of thick and thin filaments. (Photomicrograph courtesy of R. Chao.)

stack of thick filaments to the other. As seen in Figures 24–12 and 24–13, there is a region within each *A*-band that is devoid of actin filaments; this region appears somewhat less dense than the remainder of the *A*-band and is called an *H-zone*. Finally, at the center of the *H*-zone is the somewhat darker *M*-line. As contraction occurs, the thick

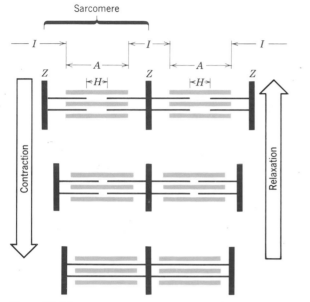

Figure 24–14
Changes in the relative positions of the thick and thin filaments in neighboring sarcomeres during contraction and relaxation.

and thin filaments slide past each other, so that the *H*-zone disappears and the *I*-band becomes much narrower (Fig. 24–14). The *Z*-lines are drawn closer together, and the myofibril as a whole becomes thicker.

The myosin molecules that comprise the thick filaments are fibrous proteins containing a "head" and "tail" portion (Fig. 24–15). The tail is formed by two alpha helical polypeptide chains twisted around each other to form a right-handed superhelix (Chapter 4). The chains extend into and form a portion of the head along with two other short polypeptide chains. Myosin can be cleaved into three pieces using proteolytic enzymes. Treatment with trypsin releases part of the tail (i.e., **light meromyosin** [LMM]) from the molecule. Light meromyosin has no ATPase activity and cannot combine with actin. The remaining portion of the molecule, called **heavy meromyosin** (HMM), contains ATPase activity and binds actin. Treatment of HMM with the enzyme *papain* severs the head (the portion containing the ATPase activity) from the tail.

In thick filaments, the myosin molecules are arranged with their tails parallel to each other and their heads projecting away from the long axis of the filament at intervals (Fig. 24–16). No heads are present in the center of the filament, this region coinciding with the **H-zone.** The heads of the myosin molecules form cross-bridges with adjacent actin filaments.

Thin filaments are composed primarily of actin but also present are small amounts of *tropomyosin, troponin,* α *actinin,* and β *actinin* (see Fig. 24–17). In the ionic envi-

Figure 24–15
The myosin molecule. The enzyme trypsin cleaves the molecule into two pieces called *light* meromyosin and *heavy* meromyosin. The enzyme papain severs the head region (i.e., the region containing ATPase activity) from the tail. (Redrawn with permission. Copyright © 1979 by Worth Publishing, Inc.; A. L. Lehninger, *Biochemistry,* p. 512.)

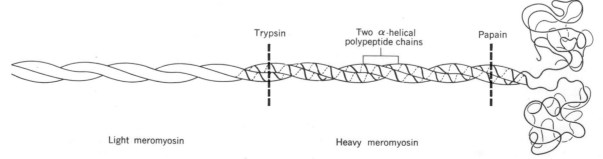

Trypsin   Two α-helical polypeptide chains   Papain

Light meromyosin   Heavy meromyosin

to be red blood cells, as 95% of all protein synthesis is directed to hemoglobin. The buildup of hemoglobin is then followed by the dissolution or loss of cell structures such as the nucleus and residual nucleic cells. The cells are irreversibly differentiated into red blood cells.

Muscle cells develop a complex mechanism for contraction. **Myofibrils** containing *actin* and *myosin* are geometrically differentiated parallel to the long axis of the cell and are spatially arranged about one another with a number of bonds formed between their constituent filaments. When stimulation is applied or ATP and Mg^{++} are added, the bonds between the actin and myosin filaments progressively break and reform as the filaments slide past one another and the fibril (and cell) shortens.

Nerve cell differentiation is also an excellent example of the degree of specialization that may accompany cell development. Neurons develop *processes* many thousands of times the diameter of the **cell body.** Impulses are *conducted* between remote parts of the body along these processes and are *transmitted* from one cell to the next as a result of specialized properties of the nerve cell membrane and the axoplasm.

References and Suggested Reading

Articles and Reviews

Baker, P. F., The nerve axon. *Sci. Am. 214*(3), 74 (March 1966).

Buchthal, F., and Schmalbruch, H., Motor unit of mammalian muscle. *Physiol. Rev. 60*, 90 (1980).

Eisenberg, E., and Greene, L. E., The relation of muscle biochemistry to muscle physiology. *Annu. Rev. Physiol. 42*, 293 (1980).

Huxley, H. E., Muscle contraction and cell motility. *Nature 243*, 445 (1973).

King, T. J., and Briggs, R., Changes in the nuclei of differentiating gastrula cells as demonstrated by nuclear transplantation. *Proc. Natl. Acad. Sci. 41*, 321 (1955).

Nienhuis, A. W., and Benz, E. J., Regulation of hemoglobin synthesis during the development of the red cell. *N. Engl. J. Med. 297*, 1318 (1977).

Patterson, P. H., Potter, D. D., and Furshpan, E. J., The chemical differentiation of nerve cells. *Sci. Am. 239*(1), 50 (July 1978).

Rogart, R., Sodium channels in nerve and muscle membrane. *Annu. Rev. Physiol. 43*, 711 (1981).

Schwartz, J. H., The transport of substances in nerve cells. *Sci. Am. 242*(4), 152 (April 1980).

Wallick, E. T., Lane, L. K., and Schwartz, A., Biochemical mechanism of the sodium pump. *Annu. Rev. Physiol. 41*, 397 (1979).

Weber, A., and Murray, J. M., Molecular control mechanism in muscle contraction. *Physiol. Rev. 53*, 612 (1973).

Werz, G., Determination and realization of morphogenesis in *Acetabularia. Brookhaven Symp. Biol.*, Upton, N.Y. (1965).

Books, Monographs, and Symposia

Aidley, D. J., *The Physiology of Excitable Cells,* Cambridge University Press, Cambridge, 1971.

Bessis, M., Weed, R. I., and Leblond, P. F. (Eds.), *Red Cell Shape,* Springer-Verlag, New York, 1973.

Browder, L. W., *Developmental Biology,* Saunders, Philadelphia, 1980.

Goldman, R., Pollard, T., and Rosenbaum, J., *Cell Motility* (Books A and B), Cold Spring Harbor Conferences on Cell Proliferation, Vol. 3, Cold Spring Harbor Laboratory, Cold Spring Harbor, New York, 1976.

Hodgkins, A. L., *The Conduction of the Nervous Impulse,* Charles C. Thomas, Springfield, Ill., 1964.

Stein, R. B., *Nerve and Muscle—Membranes, Cells and Tissues,* Plenum Press, New York, 1980.

GLOSSARY

Acetyl coenzyme A An intermediate in energy-transfer reactions of metabolism; reactant in initial step of Krebs cycle reactions.

Actin Protein found in thin filaments of striated muscle cells and microfilaments of many nonmuscle cells.

Actinomycin D An antibiotic that inhibits the elongation of RNA molecules.

Activation energy Energy required to initiate a chemical reaction.

Active site Region of the enzyme's surface that binds and alters the substrate molecule.

Active transport The movement of molecules through a membrane that requires cellular energy.

Adenosine triphosphatase ATPase; an enzyme that hydrolyzes ATP to form ADP and inorganic phosphate.

Adenosine triphosphate ATP; one of the nucleoside triphosphates; a high-energy intermediate in many energy-transfer reactions.

Adenylcyclase (adenylate cyclase) An enzyme that catalyzes production of cyclic AMP from ATP.

Aerobes Cells that live in and utilize oxygen.

AIA Anti-immunoglobulin antibodies. Antibodies produced in response to the presence of *foreign* antibodies in an organism.

Allele One of the alternative forms of a gene.

Allosteric effectors Small molecules, usually metabolites, that bind to allosteric proteins at a site other than the active site and cause a change in protein shape and activity.

Allosteric enzymes (allosteric proteins) Enzymes whose activity is modulated by the binding of allosteric effectors at sites other than the active site.

Alpha-actinin A protein found in the Z-line of muscle fibers.

Aminoacyl adenylate In protein synthesis, an intermediate in the formation of a covalent bond between an amino acid and its tRNA.

Amphipathic Molecules that have both hydrophilic and hydrophobic regions.

Anabolism Energy-requiring synthesis of more complex molecules from simpler ones.

Anaerobes Cells that can live in the absence of oxygen.

Anaphase Stage of mitosis or meiosis in which the chromosomes move toward the poles of the spindle.

Anaplerotic Reactions that replenish intermediates depleted by other metabolic pathways.

Aneuploidy Chromosome number that is not an exact multiple of the haploid number.

Ångstrom (Å) A unit of length usually used to describe molecular dimensions; equal to 10^{-8} cm.

Antibody Protein formed by lymphocytes in response to the presence of a foreign substance (antigen).

Anticodon   The three-base group of a tRNA molecule that recognizes and pairs with the three-base codon of mRNA.

Antigen   A substance, often a protein, that upon injection into a vertebrate stimulates the production of antibodies.

Apoenzyme   The protein component of an enzyme; apoenzyme + coenzyme = holoenzyme.

Aromatic amino acids   Amino acids whose side chains include a derivative of a phenyl group. In proteins, the aromatic amino acids are phenylalanine, tyrosine, and tryptophan.

Aster   The region at the poles of a dividing cell, composed of microtubules and a pair of centrioles.

ATP   See *adenosine triphosphate*.

ATPase   See *adenosine triphosphatase*.

Attenuator region   A region of DNA within an operon at which most RNA polymerase molecules stop transcription. Receipt of a specific antitermination factor will cause transcription to proceed.

Autophagic vacuoles   Membrane-lined vacuoles containing morphologically recognizable cytoplasmic components. They include autolysosomes (which are secondary lysosomes) and autophagosomes (which are vesicles sequestering cytoplasmic organelles).

Autophagy   The process of sequestration of intracellular components in vacuoles.

Autoradiography   Determination of the location of radioactive tracers introduced into cells by exposure of photographic emulsions placed in contact with the cells.

Autosomes   Chromosomes other than the sex chromosomes.

Autotrophic cells   Cells than can synthesize macromolecules from simple nutrient molecules, such as carbon dioxide, ammonia, and water.

Axoneme   The microtubular organization of a cilium or flagellum.

B-DNA   Double helical DNA in which the helices are right-handed (see also *Z-DNA*).

Bacteriophage   A virus for which the host cell is a bacterium.

Basal body   An organelle located at the base of cilia and believed to be involved in the organization of ciliary microtubules.

Beta-galactosidase   An enzyme that catalyzes the hydrolysis of lactose into glucose and galactose.

Bioenergetics   Thermodynamics as applied to living cells or whole organisms.

Bivalent   A synapsed pair of homologous chromosomes.

Calmodulin   A protein that binds four calcium ions and regulates various metabolic processes in eucaryotic cells.

Calorie   A unit of energy; the amount of heat required to raise the temperature of 1.0 g of water from 14.5° to 15.5° C.

Capsid   A shell-like structure, composed of aggregated protein subunits, arranged in a regular pattern about the nucleic acid component of some viruses.

Capsomeres   An individual protein subunit of a capsid.

Carcinogen   An agent that causes cancer.

Carotenoids   Accessory photosynthetic pigments.

Catabolite Activator Protein (CAP)   A dimeric protein that controls glucose-sensitive operons.

Catabolite repression   In bacteria, the decreased synthesis of specific enzymes concerned with degradation of glucose or similar catabolites. Caused by low levels of cyclic AMP in such cells.

Catalyst   An agent that increases the rate of a chemical reaction without altering its equilibrium point.

C_4-cycle (C_4 photosynthesis)   A photosynthetic CO_2-reducing pathway in some green plants; also known as the Hatch-Slack pathway.

Cell culture   A population of cells grown in vitro.

Cell cycle   The cyclic sequence of events in dividing cells including the G_1, S, G_2, and M periods.

Cell division   Formation of two daughter cells from a parent cell by enclosure of the two nuclei in separate cell compartments.

Cell plate   New cell wall material laid down in the center of a dividing plant cell.

Cell fusion   Formation of a hybrid cell (or *heterokaryon*) from the nuclei and cytoplasm from different cells. Induced by treatment of a mixed cell culture with *Sendai* virus.

Cell wall   A rigid or semirigid structure enclosing the protoplast of most plant and procaryotic cells.

Central dogma   The relationship between DNA, RNA, and protein: DNA serves as a template for both its own replication and the synthesis of RNA; RNA, in turn, serves in protein synthesis.

Centriole   Microtubule-containing structures formed near the nucleus of many cells, at the spindle poles in most dividing cells, or forming the basal portion of a cilium or flagellum.

Centromere The primary constriction of a chromosome to which the spindle fibers attach.

Chemiosmotic Theory A hypothesis that accounts for the coupling of electron transfer across a membrane and ATP formation by postulating the formation of H^+ gradients across the membrane.

Chiasma Site of DNA exchange between two chromatids of a bivalent chromosome.

Chlorophyll Light-capturing pigment, located in chloroplast thylakoids of plant cells or in procaryotic cells.

Chloroplasts Membranous organelles of plant cells containing chlorophyll and the sites of photosynthesis.

Chromatids Identical members of a replicated chromosome, joined to each other at the centromere.

Chromatin The nuclear material visible by light microscopy of stained cells and seen as dense masses in transmission electron photomicrographs.

Chromosome The gene-containing structure in the nucleus or nucleoid of a cell.

Cilia (sing. **cilium**) Locomotor organelles located at the cell surface, composed of a precise arrangement of microtubules.

Cisterna (pl. **cisternae**) A flattened, membrane-bordered channel.

Clone A group of identical cells descended from one cell by mitosis.

Codon A sequence of three nucleotides of messenger RNA that codes for an amino acid or chain termination.

Coenzyme A small organic molecule weakly bound at or near the active site of the enzyme and required for enzyme activity.

Colchicine A drug that binds to tubulin and causes the breakdown of microtubules; colchicine also prevents formation of spindle fibers and is used to prevent anaphase separation of chromosomes during mitosis.

Complement A series of blood serum proteins that can lyse cells to which antibodies have been bound.

Complementary base-pairing Hydrogen bond formation between a particular purine and a particular pyrimidine in nucleic acids; for example, guanine and cytosine and adenine and thymine.

Concanavalin A A lectin used to study the surface properties of cells.

Conjugate base A base derived from an acid through the removal of one or more hydrogen ions.

Connexon See *gap junction*.

Constitutive enzyme An enzyme maintained at a constant level because the structural gene(s) for this enzyme is continuously expressed.

Contact inhibition A halt in cell division that occurs when freely growing cells from a multicellular organism physically contact one another.

Copolymer A polymeric molecule containing more than one kind of monomer unit.

Corepressor A metabolite that combines with repressor protein and blocks transcription of messenger RNA.

Coupled reactions Two chemical reactions that have a common intermediate through which energy can be transferred from one reaction to the other.

Coupling factor F_1 factor; the headpiece of the mitochondrial inner membrane sphere that has ATPase activity.

Covalent bond Bond between atoms formed by sharing electrons.

Cristae Foldings of the mitochondrial inner membrane and the site of enzymes of oxidative phosphorylation and electron transport.

Crossing over Exchange of homologous chromosome segments leading to genetic recombination.

Cyclic adenosine monophosphate Cyclic AMP; adenosine monophosphate in which the phosphate group is bonded between the 3' and 5' carbon atoms to form a cyclic molecule; this nucleotide is active in regulating certain reactions in cells.

Cycloheximide An inhibitor of protein biosynthesis.

Cytochromes Electron-transport intermediates containing heme (or related prosthetic groups) in which the iron undergoes valency changes during electron transfer.

Cytoplasm The protoplasmic contents of the cell, exclusive of the nucleus.

Cytosol The fluid portion of the cytoplasm in which the organelles are suspended.

Cytoskeleton An intracellular framework comprised of filaments and microtubules.

Dalton Unit of molecular weight approximately equal to the weight of a hydrogen atom or $\frac{1}{16}$ the weight of an oxygen atom.

Dark reactions Photosynthetic reactions that can proceed in the absence of light. Carbon dioxide is reduced ("fixed") to form sugars and other carbohydrates, using ATP and NADPH produced in the light reactions.

Deletion A loss of part of a chromosome or DNA molecule from the genome of a cell.

Denaturation Change in the native configuration of a macromolecule resulting, for example, from heat treatment, extreme pH changes, or chemical treatment. Denaturation is usually accompanied by the loss of the molecule's biological activity.

Density-dependent inhibition A halt to cell division in a layer of cells when their edges touch one another.

Deoxyribonucleic acid (DNA) The genetic material of all cells and many viruses.

Desmosome One of several different kinds of plasma membrane specializations that occurs where adjacent cells in a tissue contact one another.

Dictyosome A stack of cisternae that forms part of the Golgi apparatus. In plant cells, the term is often used for the entire Golgi body.

Diffusion The net overall movement of molecules in the direction of a lesser concentration.

Dimer Structure resulting from association of two identical subunits.

Diploid A cell or individual or species having two sets of homologous chromosomes in the nucleus of somatic cells.

Disulfide bond (disulfide bridge) A covalent bond between two sulfur atoms in separate amino acids of a protein.

DNA-RNA hybrid A double helix in which one polynucleotide is DNA and the other is RNA.

Dynein Protein component of microtubules.

Effector A regulatory metabolite that activates or inhibits an enzyme by binding to its allosteric site.

Electron carriers Intermediates such as flavoproteins and cytochromes that reversibly gain or lose electrons.

Electron transport The movement of electrons from substrates to oxygen via the oxidative respiratory chain intermediates.

Electrophoresis A method of separating macromolecules or particles according to their charge, size, and shape as they migrate through a gel or other medium in an electrical field.

Endergonic reaction A chemical reaction with a positive standard free energy change; an energy-consuming reaction.

Endocytosis Intake by a cell of solutes or particles by enclosing them in an infolding of the plasma membrane.

Endonuclease An enzyme that makes internal cuts in a polynucleotide.

Endoplasmic reticulum ER; folded membrane system distributed within the cytoplasm of eucaryotic cells; the ER frequently has attached ribosomes (rough ER).

Endothermic process A process that absorbs heat.

End-product repression A control mechanism in which the production of an enzyme required in a metabolic pathway is inhibited by the final product of that metabolic pathway.

Entropy The randomness or disorder of a system.

Episome A genetic element that can exist either free or as part of the normal cellular chromosome. Examples of episomes are the F factors of *E. coli*.

Eucaryotic cells Cells in which the nucleus is separated from the cytoplasm by a pair of membranes.

Euchromtain Noncondensed, active regions of chromosomes in the interphase nucleus of a cell.

Exergonic reaction A reaction with a negative standard free energy change; an energy-releasing reaction.

Exocytosis A mode of transport of substances out of the cell by enclosure in a vesicle, fusion of the vesicle with the plasma membrane, and subsequent expulsion of the vesicle's contents.

Exons Polypeptide encoding portion(s) of structural genes that are separated from each other by intervening sequences (see also *introns*).

Exonuclease An enzyme that cleaves polynucleotide chains at their ends.

Exothermic process A process in which heat is evolved.

Facilitated diffusion Assisted transport of molecules through the membrane along a concentration gradient.

Feedback (end-product) inhibition Inhibition of the activity of the first enzyme in a metabolic pathway by the end product of that pathway.

Fermentation Oxidation of carbohydrate in non-oxygen-requiring pathways such as glycolysis.

Ferritin An iron-rich protein found in the liver, spleen, and bone marrow.

Feulgen reaction Staining reaction for DNA used in microscopy.

First law of thermodynamics   Energy cannot be created or destroyed; statement of the principle of the conservation of energy.

Flagellum (pl. **flagella**)   A cell organelle ultrastructurally similar to a cilium but usually longer and present in smaller numbers per cell.

Flavin adenine dinucleotide (FAD)   An electron carrier molecule that acts in energy-transfer reactions as a coenzyme.

Fluid mosaic membrane   Model of the plasma membrane in which proteins are distributed in a phospholipid bilayer.

Fluorescent antibody technique   Detection of selected antigens in cells by staining with a specific antibody conjugated with a fluorescent dye.

Formylmethionyl-tRNA   fmet-tRNA; the initial aminoacyltransfer RNA complex that reacts with the small ribosome subunit at the beginning of polypeptide chain synthesis.

Free energy   A component of the total energy of a system that can do work under conditions of constant temperature and pressure.

Freeze-fracture   Procedure for preparing materials for electron microscopy by rapid freezing and fracturing of the tissue; the exposed fracture faces are used to create a replica that is observed and photographed in the electron microscope; the fracture faces may or may not be further sublimed before the replica is made.

Furrowing   A cell division mechanism that involves a pinching-in, or cleavage, to form two daughter cells from the parent cell.

Galactosidase, β (beta)   See *beta-galactosidase*.

Gametes   Haploid cells (e.g., ova and sperm) which unite at fertilization to produce the diploid zygote.

Gap junction   Portions of the plasma membranes of adjacent cells that are bridged by narrow, tubular channels that permits cell-to-cell communication. Also called a nexus or connexon.

Gene amplification   Differential replication of specific genes.

Generation time   The interval of time in which growing cells double their numbers or mass.

Genome   The genes associated with a haploid set of chromosomes.

Gluconeogenesis   Synthesis of carbohydrates from non-carbohydrate precursors such as fats or proteins.

Glycogen   A polysaccharide formed by polymerization of glucose units through $1 \rightarrow 4$ glycosidic linkages between adjacent residues. The most common form in which carbohydrate is stored in animal cells.

Glycolipids   Lipids that contain polar, hydrophilic carbohydrate groups.

Glycolysis   A pathway for glucose catabolism not requiring oxygen.

Glycoprotein   A protein containing one or more chains of sugar residues.

Glyoxylate cycle   An anaplerotic metabolic pathway replenishing 2-carbon metabolites; the pathway is associated with glyoxysomes.

Group-transfer reactions   Reactions (excluding oxidations or reductions) in which molecules exchange functional groups.

Growth curve   The change in the number or mass of cells in a growing culture as a function of time.

Growth factor   A specific substance that must be present in the growth medium to permit cells to grow and divide.

Hairpin loops   A folded region of single stranded DNA or RNA formed by the pairing of two contiguous complementary stretches of bases.

Haploid   Cell or individual having but one copy of each homologous pair of chromosomes.

Haptens   Small nonantigenic molecules that are capable of stimulating specific antibody synthesis when chemically coupled to a larger molecule.

Heavy isotope   Forms of atoms containing greater than the common number of neutrons and thus heavier and more dense than the commonly observed isotope (e.g., ^{15}N, ^{13}C).

HeLa Cells   An established line of human cervical carcinoma (cancer) cells derived from Helen Lane.

Helix   A spiral structure having a repeating pattern described by two simultaneous operations—rotation and translation. It is the natural conformation of many regular biological polymers.

Helix-destabilizing proteins   Proteins that bind to and stabilize the unwound DNA strands during replication.

Heme   An iron-containing porphyrin that serves as the prosthetic group of the hemoglobins and enzymes such as catalase and cytochromes.

Hemoglobin   Protein carrier of oxygen found in red blood cells; composed of two pairs of identical polypeptide chains and an iron-containing heme group.

Hemolysis   Rupture of a red blood cell and release of its hemoglobin.

Heterochromatin   Highly condensed chromatin regions of the interphase cell nucleus.

Heteroduplex   Double-stranded DNA molecule in which the two strands are not entirely complementary; produced by mutation, by recombination, or by annealing DNA single strands in vitro.

Heterokaryon   See *cell fusion*.

Heterotrophic cells   Cells requiring complex nutrient molecules (e.g., glucose, amino acids) for energy and for the synthesis of their own macromolecules.

High-energy bond   A bond that yields a large amount of free energy upon hydrolysis.

High-energy phosphate compound   A phosphorylated compound having a highly negative standard free energy of hydrolysis.

Histone   A protein component of the chromosome having a high content of basic amino acids. Eight histones comprise the core of the nucleosome.

Holoenzyme   The complete form of an enzyme (coenzyme + apoenzyme = holoenzyme).

Homologous   Having the same or similar gene content.

Homologous chromosomes   Chromosomes that pair during meiosis, have the same morphology, and contain genes governing the same characteristics.

Hormone   A chemical substance released into the bloodstream by one tissue or organ that modulates biochemical functions in the cells of another tissue or organ.

Hydrogen bond   A weak electrostatic force between one electronegative atom and a hydrogen atom covalently linked to a second electronegative atom.

Hydrolysis   The cleavage of a molecule into two or more molecules by the addition of a water molecule.

Hydrophilic   Molecules or parts of molecules that readily associate with water; usually containing polar groups that form hydrogen bonds in water.

Hydrophobic   Used to describe molecules or groups in molecules that are poorly soluble in water because of their nonpolar nature.

Hydrophobic bond   The associations formed by hydrophobic groups in aqueous solution.

Hyperchromic shift   An increase in UVL absorption by a solution of DNA caused by disruption of hydrogen bonds of a DNA duplex to yield the single-stranded structure.

Immunoglobulin   See *antibody*.

Inducible enzymes   Enzymes whose structural genes can be activated by the presence of specific inducer molecules in the cell.

Initiation factors   Proteins required for the initiation of protein synthesis in procaryotic and eucaryotic cells.

Inner membrane sphere   Spherical particles attached to the matrical surface of the mitochondrial inner membrane.

Integral protein   Protein molecule embedded in the lipid bilayer of the plasma membrane.

Interphase   The state of the eucaryotic nucleus when it is not engaged in mitosis or meiosis; consists of G_1, S, and G_2 periods of the cell cycle.

Intervening sequences   See *introns*.

Introns   DNA sequences within a structural gene that are transcribed but do not give rise to amino acid sequences in the protein. Intron transcripts are excised during RNA processing.

In vitro (Latin: ''in glass'')   Experiments carried out with isolated cells, tissues, or cell-free extracts rather than in situ, in place within the organism.

In vivo (Latin: ''in life'')   Experiments carried out using the intact organism.

IS element (insertion sequence)   A short sequence of nucleotide pairs in DNA that facilitates transposition of itself and/or other genetic elements in the genome.

Isomers   Alternative molecular forms of a chemical compound.

Isoosmotic   Having the same salt concentration.

Isopycnic density gradient centrifugation   A centrifugal method used to separate particles in density gradients on the basis of differences in particle density.

Isotonic   Producing equivalent osmotic pressures.

Isotopes   Alternative nuclear forms of an atom, all having the same atomic number (proton number) but different atomic weights (i.e., the number of neutrons varies).

Isozymes   Alternative molecular forms of an enzyme.

Joule (J)   A unit of work or energy; 1 joule = 1 newton-meter = 0.239 calories = 10^7 ergs.

Karyotype   A photograph or diagram of a complete complement of chromosomes from a cell or individual.

Krebs cycle   Most common pathway for oxidative metabolism of pyruvic acid, which is an end-product of glucose fermentation; also known as the citric acid cycle or the tricarboxylic acid cycle.

Label (radioactive)   A radioactive atom, introduced into a molecule to facilitate observation of the molecule's metabolic transformations.

Lectins   Cell-agglutinating proteins. Most lectins are isolated from plant seeds.

Ligase   Enzyme that joins together the parts of single strands of DNA between the 5′ end of one strand and the 3′ end of another.

Light reactions   Photosynthetic reactions in which light energy is converted to chemical energy in the form of ATP and NADPH.

Lipid   Class of organic compounds that are poorly soluble or insoluble in water but soluble in nonaqueous (organic) solvents such as ether and acetone.

Lipid bilayer   An early model for the structure of cell membranes based upon the hydrophobic interactions between phospholipids. The polar head groups face outwardly, while the hydrophobic tails are clustered in the interior.

Lymphocyte   A class of leucocytes that protects the body against infection by the production and secretion of antibodies.

Lysis   The bursting of a cell by the rupture of the plasma membrane.

Lysogenic bacterium   A bacterium that contains a prophage.

Lysogenic viruses   Viruses that can become prophages.

Lysosomes   Intracellular organelles that contain a large variety of hydrolytic enzymes; these fuse with ingested food vacuoles and break down their contents.

Lysozymes   Enzymes that degrade the polysaccharides found in the cell walls of certain bacteria.

Lytic infection   Viral infection leading to the lysis of the host cell.

Lytic viruses   Viruses whose proliferation within the host cell leads to the cell's lysis.

Macromolecules   Molecules having molecular weights in the range of a few thousand to hundreds of millions of molecular weight units (daltons).

Macrophage   A large, phagocytic white blood cell.

MAPs   Proteins associated with the surfaces of microtubules.

Matrix   The essentially unstructured substance of a cell or organelle consisting of a suspension of molecules and particles in an aqueous medium.

Meiosis   The reduction division of the cell nucleus in sexual organisms; produces daughter nuclei having half the number of chromosomes as the original nucleus.

Melting   The separation of the two strands of duplex DNA to form single strands by disruption of hydrogen bonds between the duplex strands.

Mesosome   An extensively infolded portion of the procaryotic plasma membrane.

Messenger RNA (mRNA)   The complementary copy of DNA that is made during transcription and that codes for protein during translation.

Metabolic pathway   A set of consecutive cellular enzymatic reactions that converts one molecule to another.

Metabolism   The type and diversity of various chemical reactions occurring in a living cell; the sum total of these reactions.

Metaphase   The stage of mitosis or meiosis when chromosomes are aligned along the equatorial plane of the spindle.

Microbody   A membrane-bounded cytoplasmic organelle with varied enzyme content and functions; e.g., peroxisomes and glyoxysomes.

Microfilaments   Long, intracellular fibers that contain polymerized actin and that are thought to function in maintenance of cell structure and movement.

Micron (micrometer, μm)   A unit of length used for describing cellular dimensions; it is equal to 10^{-4}cm or 10^4 Å.

Microsomes   A membrane-rich fraction of a tissue homogenate produced during high-speed centrifugation of the mitochondrial supernatant.

Microtrabecular lattice   A gossamer network of filaments radiating through the cytosol and anchored to the plasma membrane and organelles.

Microtubule   An unbranched cylindrical assembly of protofilaments involved in cell movement; microtubules are present in spindle fibers, cilia, and centrioles.

Microvilli   Fingerlike projections of the plasma membranes of cells.

Mitosis   The division of the nucleus that produces two daughter nuclei exactly like the original parental nucleus; somatic nuclear division.

Mole A gram molecular weight of a chemical substance.

Molecular weight The sum of the atomic weights of all of the atoms that comprise a molecule.

Monolayer A single layer of cells, molecules, or other particles.

Monomer The basic subunit from which, by repetition of a single reaction, polymers are made. For example, amino acids (monomers) yield polypeptides (polymers).

Multienzyme system A group of enzymes that catalyze sequential steps of a metabolic pathway and are in physical proximity to one another.

Mutagens Physical or chemical agents, such as radiation, heat, or alkylating or deaminating agents, that raise the natural frequency of mutation.

Mutation A change in the gene structure of a chromosome.

Myofibril Parallel units of a muscle cell composed of bundles of myofilaments.

Myofilament Individual thick (myosin) and thin (actin) filaments of the myofibril.

Myosin Protein molecules, each composed of two coiled subunits (M.W., 220,000), that can aggregate to form a thick filament, globular at each end.

NAD, NADP Nicotinamide adenine dinucleotide and nicotinamide adenine dinucleotide phosphate; carriers of electrons in many enzymatic oxidation-reduction reactions.

Neutral fats Glycerides; fatty acid esters of glycerol; a major storage form of fats.

Nexus See *gap junction*.

Nucleases Enzymes that cleave the phosphodiester bonds of nucleic acid molecules.

Nuclear envelope The double membrane surrounding the eucaryotic cell nucleus.

Nucleic acid Polymer of nucleotides in an unbranched chain; DNA and RNA.

Nucleoid A region of DNA in procaryotic cells that is not separated from the cytoplasm by a membrane.

Nucleolar organizing region (NOR) The specific part of the nucleolar organizing chromosome containing rRNA genes.

Nucleolus Structure found in nucleus of eucaryotic cells. Involved in rRNA synthesis and ribosome precursor formation.

Nucleoplasm The unstructured matrix portion of the nucleus in which the chromosomes and nucleoli are suspended.

Nucleoside Molecule containing a purine or pyrimidine linked to a pentose sugar.

Nucleosomes Spherical particles seen along decondensed chromatin; comprised of eight histones around which there are nearly two turns of DNA.

Nucleotide A phosphorylated nucleoside.

Operator A specific nucleotide sequence in the operon that binds the repressor and exerts control over transcription of adjacent structural gene(s).

Operon A cluster of associated genes and recognition sites on the chromosome that participates in regulating transcription and includes regulatory gene, promoter site, operator site, and structural gene(s).

Organelle A subcellular component; a discrete structural differentiation of the cell containing particular enzymes and performing particular functions for the whole cell, e.g., mitochondria, ribosomes, etc.

Oxidant An oxidizing agent; a substance that accepts electrons, or hydrogens, from a reducing agent or reductant.

Oxidation The loss of electrons or hydrogen from an atom, ion, or compound.

Oxidation-reduction reactions Reaction which involves electron transfer from a reductant to an oxidant. The reductant is said to be *oxidized* and the oxidant *reduced*, as a result of the transfer.

Oxidative phosphorylation In mitochondria, the enzymatic phosphorylation of ADP to form ATP that is coupled to electron transport along the respiratory chain to oxygen.

Palindrome A stretch of DNA in which the base sequences read the same from the 3′ or 5′ end.

Peptide bond A covalent bond between two amino acids in which the alpha-amino group of one amino acid is bonded to the alpha-carboxyl group of the other.

Permease A carrier protein in the plasma membrane that is involved in transport of specific substrate molecules across that membrane.

Peripheral protein Protein molecule loosely attached to the lipid bilayer of the plasma membrane.

Peroxisomes Intracellular organelles that contain a fine granular matrix and often crystal-like cores. They contain enzymes involved in hydrogen peroxide metabolism, including catalase.

pH A measure of hydrogen ion concentration in aqueous solutions.

Phagocytosis A form of endocytosis in which large amounts of particulate material, even whole cells, are enclosed in endocytic vesicles.

Phase-contrast microscope An instrument that translates differences in the phases of transmitted or reflected light into gradations of contrast.

Phosphodiester linkage A covalent linkage involving esterification to phosphoric acid.

Phospholipids Lipids that contain charged, hydrophilic phosphate groups; primary components of cell membranes.

Photophosphorylation Process of formation of ATP from ADP and inorganic phosphate in the light reactions of photosynthesis; occurs by a cyclic or noncyclic pathway involving photosystems I and II.

Photorespiration Uptake of oxygen and release of carbon dioxide by photosynthetic cells or whole plants in the light.

Photosynthesis The enzymatic conversion of light energy into chemical energy by forming carbohydrates and oxygen from CO_2 and H_2O in green plant cells.

Photosynthetic phosphorylation The enzymatic formation of ATP from ADP in green plants coupled to light-dependent transport of electrons from excited chlorophyll.

Photosystem I (PS I) A photochemical reaction system in photosynthesis; coupled with photosystem II.

Photosystem II (PS II) A photochemical reaction system in photosynthesis; coupled to photosystem I.

Phycobilin An accessory photosynthetic pigment present in red and blue-green algae.

Pinocytosis Endocytosis of small molecules together with aqueous medium into small endocytic vesicles.

Plaque Round, clear areas in a confluent sheet of cells; results from the killing or lysis of clusters of cells by several cycles of virus growth.

Plasmalemma See *plasma membrane*.

Plasma membrane The membrane that surrounds the cell and encloses the cytoplasm. This membrane is semipermeable and largely composed of lipid and protein.

Plasmids Cytoplasmic, autonomously replicating, small, circular chromosomal elements in bacteria.

Plasmodesmata Cytoplasmic channels through the cell walls connecting the protoplasts of adjacent plant cells.

Plastid Eucaryotic organelle that stores pigments or carbohydrates.

Polyacrylamide gel electrophoresis A method of molecular separation that relies on the differential migration of molecules, usually proteins or polynucleotides, through a polyacrylamide matrix upon application of an electrical potential.

Polymer An association of monomer units into a large molecule.

Polymerase An enzyme catalyzing the synthesis of DNA or RNA from nucleoside triphosphate precursors.

Polynucleotide A linear sequence of nucleotides in which the sugar of one nucleotide is linked through a phosphate group to the sugar on the adjacent nucleotide.

Polynucleotide ligase Enzyme that covalently links free 3′ and 5′ ends of polynucleotide chains.

Polypeptide A long, unbranched polymer of amino acids.

Polyploid Cell or individual having one or more extra complements of chromosomes.

Polyribosome Complex of ribosomes joined together by a messenger RNA molecule (number of ribosomes depending on size of mRNA); polyribosomes are engaged in polypeptide synthesis.

Polysomes See *polyribosome*.

Polytene chromosome Giant chromosome arising from successive rounds of chromatid duplication.

Pore An opening in a membrane or other structure; often referring to the nuclear pore complex of the nuclear envelope.

Primary constriction Location of centromere on chromosome.

Primary protein structure The number of polypeptide chains in a protein, their sequences of amino acids, and the location of interchain and intrachain disulfide bridges.

Primer A structure that serves as a growing point for polymerization.

Procaryote Simple unicellular organism, such as a bacterium or cyanobacterium (blue-green algae), having no nuclear membrane.

Procentriole An immature centriole.

Procaryotes Organisms such as bacteria, blue-green algae, and mycoplasmas in which the nucleus is not separated from the cytoplasm by membranes.

Promoter A specific nucleotide sequence in the operon to which RNA polymerase binds.

Prophase The first stage of mitosis or meiosis, after DNA replication and before chromosomes align on the equatorial plane of the spindle.

Proplastid An immature plastid.

Prosthetic groups Coenzymes that are more or less permanently bound to their enzymes.

Protamines A class of proteins rich in the basic amino acid arginine. They are found complexed to the DNA of sperm in many invertebrates and fish.

Protist Unicellular eucaryotic organisms such as protozoa and algae.

Protoplasm The living material of the cell.

Protoplast The living structure of the cell, made of protoplasm, contained within but including the plasma membrane.

Provirus The state of a virus in which it is integrated into the genome of a host cell and is transmitted from one cell generation to another.

Puff A region of a chromosome that is undergoing active transcription, usually observed in giant polytene chromosomes.

Pulse-chase experiment A radioactively labeled compound is added to living cells or a cell extract (pulse), and a short time later, an excess of unlabeled compound is added. Samples are then taken at periods after the pulse to follow the course of the label as the compound is metabolized (chase).

Purine Parent compound of the nitrogen-containing bases adenine and guanine.

Puromycin Antibiotic that inhibits polypeptide synthesis by competing with aminoacyl tRNAs for the *A* binding site.

Pyrimidine Parent compound of the nitrogen-containing bases cytosine, thymine, and uracil.

Quantum The energy of a photon.

Quaternary structure The manner in which the separate polypeptide chains of a protein are held together and oriented in space with respect to one another.

Radioactive isotope Isotope with an unstable nucleus that emits ionizing radiation; important as labels or tracers in biology.

Reannealing Renaturation; specifically, the restoration of duplex DNA regions through complementary base pairing of single-stranded DNA molecules.

Recombination The appearance of characteristics in the offspring that were not found together in either of the parents.

Recombinant DNA technology Methods used to transfer genes from one organism to another, usually employing a vector such as a bacterial plasmid or a viral nucleic acid.

Redox couple Compounds that occur in both the oxidized and reduced forms and that are participants in oxidation-reduction reactions (e.g., NAD^+-NADH).

Reductant A reducing agent that donates electrons or hydrogens in oxidation-reduction reactions.

Reduction The gain of electrons or hydrogens by an atom or molecule.

Regulatory genes Genes whose primary function is to control the rate of synthesis of other genes.

Release factor Specific macromolecule involved in the reading of the ''stop'' signal during protein synthesis.

Renaturation The return of a protein or nucleic acid from a denatured and nonfunctioning state to its ''native,'' functioning configuration.

Repetitive DNA Repeated sequences of DNA nucleotides in the chromosomes of eucaryotic cells.

Replicating fork Y-shaped region of chromosome that acts as growing point in DNA replication.

Replicating forms (RF) The structure of a nucleic acid during its replication; most frequently used to refer to double-helical intermediates in the replication of single-stranded DNA and RNA viruses.

Replicon A unit of DNA replication; contains one origin and two termination sites.

Repressor A protein product of the regulator gene of the operon that binds to the operator site and prevents transcription of structural genes.

Residual bodies Secondary lysosomes containing undigested residues, membrane fragments, and whorls.

Respiration The oxidative breakdown and release of energy from molecules by reaction with oxygen in aerobic cells.

Restriction enzymes Enzymes that cleave double-stranded DNA at specific sequences that exhibit twofold symmetry about a point.

Reticulocyte Immature red blood cell still capable of limited hemoglobin synthesis.

Reverse transcriptase   An enzyme encoded by the RNA of certain viruses that is able to make complementary single-stranded DNA chains from RNA templates.

RIA   Radioimmunoassay procedure; a highly sensitive method employing radioactive isotopes to assay antibodies.

Ribonuclease   An enzyme that cleaves the phosphodiester bonds of RNA.

Ribonucleic acid (RNA)   Nucleic acids that function in transcription and translation. The genetic material of certain viruses.

Ribosomal DNA (rDNA)   The genes at the nucleolar organizing region that code for ribosomal RNA.

Ribosomal proteins   A group of proteins that combine with rRNA and give the ribosome its three-dimensional structure.

Ribosomal RNA (rRNA)   Ribonucleic acids that are part of the ribosome structure and that function in protein synthesis.

Ribosomes   Small cellular particles made up of rRNA and protein. Ribosomes are the site of protein synthesis; in eucaryotic cells, they are often attached to the endoplasmic reticulum.

RNA   See *ribonucleic acid*.

RNA polymerase   Enzyme that catalyzes the formation of RNA from ribonucleoside triphosphates, using DNA as a template.

Rough ER (RER)   Portion of the endoplasmic reticulum bearing ribosomes.

S period   Interval of the cell cycle in which DNA replication occurs.

Sarcolemma   The plasma membrane of a muscle cell or fiber.

Sarcomere   The contractile unit of muscle fiber, extending from one *Z*-line to an adjacent *Z*-line.

Sarcoplasm   The cytoplasm of a muscle cell or fiber.

Sarcoplasmic reticulum   Endoplasmic reticulum of a muscle cell or fiber.

Scanning electron microscope (SEM)   Electron-microscopic technique that permits observation of a specimen's surface structure.

Secondary constriction   Any pinched-in site along a chromosome other than the primary constriction at the centromere.

Secondary structure   Structure of a polypeptide chain describing the location, extent, and types of helices (as well as nonhelical regions).

Second law of thermodynamics   The principle that all physical and chemical changes proceed in a direction such that the entropy of the universe increases.

Sedimentation coefficient   A quantitative measure of the rate of sedimentation of a given substance through water at 20°C in a unit centrifugal field; expressed in Svedberg units *(S)*.

Semiconservative replication   The usual mode of duplex DNA synthesis resulting in daughter duplex molecules that contain one parental DNA strand and one newly formed strand.

Sex chromosome   Any chromosome involved in sex determination, such as the *X* and *Y* chromosomes.

Smooth ER (SER)   Portion of the endoplasmic reticulum devoid of ribosomes.

Spindle   Aggregation of microtubules during nuclear division that functions in the alignment and movement of chromosomes at anaphase.

Spindle fiber   A microtubule in mitotically or meiotically dividing cells that extends from one pole to an attachment in the centromere region of a chromosome or that extends from pole to pole.

Spontaneous process   A process accompanied by a decrease in free energy.

Standard electrode potential E'; the oxidation-reduction potential of a substance relative to a hydrogen electrode; expressed in volts.

Standard free-energy change $\Delta G°$; a thermodynamic constant representing the difference between the free energy of the reactants and the free energy of the products of a reaction.

Standard state   Most stable form of a pure substance at 1.0 atmosphere pressure and 25°C (298 K). For reactions occurring in solution, the standard state of a solute is a 1.0 M solution.

Steady state   A nonequilibrium state of an open system through which matter is flowing and in which all components remain in constant concentration.

Stem cell   A cell from which other cells arise by differentiation.

Stereoisomers   Molecules that have the same structural formula but different spatial arrangement of dissimilar groups bonded to a common atom. Stereoisomers have differences in their crystal structures and differ in the direction in which they rotate polarized light; they also differ in their ability to be used in an enzyme-catalyzed reaction.

Steric (stereochemical) Relating to the arrangement in space of the atoms in molecules.

Steroids Compounds that are derivatives of a tetracyclic structure composed of a cyclopentane ring fused to a substituted phenanthrene nucleus.

Streptomycin An antibiotic isolated from *Streptomyces griseus* (a soil bacterium) that binds specifically to bacterial 30 *S* ribosomal subunits, thereby blocking protein biosynthesis.

Stroma Unstructured matrix of the chloroplast in which the grana and stroma thylakoids are suspended.

Substrate Specific compound acted upon in the active site of an enzyme.

Supercoiled DNA Twisted forms taken by covalently closed, circular, double-stranded DNA molecules.

Suppressor gene A gene that can reverse the phenotypic effect of a variety of other genes.

Svedberg unit The unit of sedimentation equal to 10^{-13} seconds. The number of *S* units of a molecule or particle in a given centrifugal field is related to the weight, shape, and density of the molecule or particle.

Synapsis Specific pairing of homologous chromosomes, typically during zygotene of prophase I in meiosis.

Synaptinemal complex A complex structural component linking a pair of synapsed homologous chromosomes.

Telophase Last stage of mitosis in which each daughter nucleus reestablishes its interphase structure.

Tertiary structure The three-dimensional folding of a macromolecule into a complex structural form, brought about by interactions among side chains of amino acids.

Thermodynamics The science that deals with exchanges of energy.

Thylakoid A membranous sac present in chloroplasts that may be disk-shaped (in the grana) or elongated; it is the site of the light-requiring reactions of photosynthesis.

T_m The midpoint melting temperature; temperature at the midpoint of the transition of duplex DNA molecules to single strands during melting.

Totipotent Capable of giving rise to all of the various cell types that comprise the tissues of an organism.

Transcription Process by which the base sequence of DNA is converted to a complementary RNA molecule.

Transduction The transfer of bacterial genes from one bacterium to another.

Transfer RNA (tRNA) The RNA molecule that carries an amino acid to a specific codon in messenger RNA during translation.

Transferases Enzymes that catalyze the exchange of functional groups between molecules.

Transformation The genetic modification induced by the incorporation into one cell of DNA from another cell (or a virus).

Translation Process by which amino acids are assembled into a polypeptide on the ribosome, under the direction of the base sequence transcribed from DNA into messenger RNA.

Translocation A structural rearrangement involving parts of an entire nonhomologous chromosome.

Tritium ^3H; a radioactive isotope of hydrogen; extremely useful in tracer studies.

Tropomyosin A muscle protein that associates with actin to form long, thin fibers; plays a role in muscle contraction.

Troponin Protein associated with actin filaments in muscle cells.

Trypsin A proteolytic enzyme secreted by the pancreas; cleaves peptide chains at peptide bonds that are adjacent to the basic amino acids arginine and lysine.

T-system Invaginations of the sarcolemma in muscle fibers of striated muscle, producing a system of transverse tubular infoldings.

Tubulin Globular protein subunits (M.W., 55,000 and 57,000) whose regular helical packing forms microtubules.

Ultracentrifuge Centrifuge capable of producing rotor speeds up to 80,000 rpm and able to rapidly sediment tiny particles and macromolecules.

Ultraviolet light Electromagnetic radiation having a wavelength shorter than that of visible light (390–200 nm). Causes DNA base-pair mutations and chromosome breaks.

Uncoupling agent A substance (e.g., 2,4-dinitrophenol) that can uncouple phosphorylation of ADP from electron transport.

Unit membrane Membrane showing a three-layer or dark-light-dark pattern of electron density in the electron microscope; a former incorrect model of membrane structure proposing that phospholipid bilayer is coated on its outer and inner surfaces by proteins.

Vacuole A membrane-enclosed sac in the cell cytoplasm filled with molecules and particles in a watery medium; common in plant cells.

Van der Waals force A weak attractive force between atoms; particularly important in hydrophobic bonding of amino acids in proteins.

Vesicle A small, spherical, membrane-bordered intracellular chamber.

Viroids Pathogenic agents that consist only of short RNA molecules.

Viruses Infectious, disease-causing particles that require a host cell for replication and that contain either DNA or RNA as their genetic material.

Wobble Ability of third base in tRNA anticodon (5′ end) to hydrogen bond with any two or three bases at 3′ end of codon. Thus, a single tRNA species can recognize several different codons.

X-ray crystallography The use of X-ray scattering by crystals to determine the three-dimensional structure of molecules, especially proteins and nucleic acids.

Z-DNA Double helical DNA in which the helices are left-handed (see also *B-DNA*).

Zygote The product of fusion of two gametes during fertilization; the cell from which a new diploid individual develops in each sexual generation.

Zymogen A digestive enzyme precursor lacking catalytic activity which can be converted to the active form.

INDEX

COMMON ABBREVIATIONS

A	Adenine or adenosine	FMN, FMNH$_2$	Flavin mononucleotide and its reduced form
Å	Angstrom		
ACTH	Adrenocorticotrophic hormone	FP	Flavoprotein
AMP, ADP, ATP	Adenosine 5'-mono-, di-, and triphosphate	g	Gram, gravity
		G	Gauss, guanine, or guanosine
cAMP	Cyclic AMP	G	Giga, 10^9
ala	Alanine	$\Delta G^{\circ\prime}$	Standard free-energy change at pH 7
arg	Arginine		
asn	Asparagine	Gal	D-Galactose
asp	Aspartic acid	GDH	Glutamate dehydrogenase
ATPase	Adenosine triphosphatase	GLC	Gas-liquid chromatography
C	Cytosine	glc	Glucosamine
c	Centi, 10^{-2}	GlcNAc	N-Acetyl-D-glucosamine
c	Velocity of light (vacuum), 2.997×10^{10} cm sec^{-1}	gln	Glutamine
		glu	Glutamic acid
^{14}C	Carbon 14	gly	Glycine
cal	Calorie	GMP, GDP, GTP	Guanosine mono-, di-, and triphosphate
cm	Centimeter		
CMP, CDP, CTP	Cytidine mono-, di-, and triphosphate	G3P	Glyceraldehyde-3-phosphate (3-phosphoglyceraldehyde)
CoA, CoA-SH, acyl-CoA, acyl-S-CoA	Coenzyme A and its acyl derivatives	G6P	Glucose-6-phosphate
		hr	Hour
CoQ	Coenzyme Q; ubiquinone	h	Planck's constant, 6.624×10^{-34} joule-sec or 1.584×10^{-34} cal-sec
CPM	Counts per minute		
CsCl	Cesium Chloride		
cys	Cysteine	h	Hecto, 10^2
cyt.	Cytochrome	H+	Hydrogen ion
d	deci, 10^{-1}	^{3}H	Tritium
DNA	Deoxyribonucleic acid	Hb	Hemoglobin
DNAase	Deoxyribonuclease	his	Histidine
DPN+, DPNH	Same as NAD+, NADH	ile	Isoleucine
E_0	Standard electrode potential	IMP, IDP, ITP	Inosine mono-, di-, and triphosphate
e−	Electron		
EDTA	Ethylenediaminetetraacetic acid	J	Joule (1.0 J $= 0.239$ cal $= 10^7$ erg)
ER	Endoplasmic reticulum	K	Kelvin (degrees)
ESR	Electron spin resonance	k	Kilo, 10^3
ETP	Electron transfer particle (from mitochondrial membrane)	kcal	Kilocalorie
		l	Liter
$\mathscr{F}$	Faraday	λ	Lambda phage or wavelength of radiation
f	femto, 10^{-15}		
FAD, FADH$_2$	Flavin adenine dinucleotide, oxidized and reduced forms	LDH	Lactate dehydrogenase
		leu	Leucine
FDP	Fructose-1, 6-diphosphate	LH	Luteinizing hormone
fmet	N-formylmethionine	ln	Logarithm to the base e